*S*pringer *M*onographs in *M*athematics

For further volumes:
www.springer.com/series/3733

Pavel Pudlák

Logical Foundations of Mathematics and Computational Complexity

A Gentle Introduction

Springer

Pavel Pudlák
ASCR
Prague, Czech Republic

ISSN 1439-7382 Springer Monographs in Mathematics
ISBN 978-3-319-34268-9 ISBN 978-3-319-00119-7 (eBook)
DOI 10.1007/978-3-319-00119-7
Springer Cham Heidelberg New York Dordrecht London

Mathematics Subject Classification: 03D15, 03E30, 03E35, 03F03, 03F20, 03F30, 03F40, 68Q15

Printed on acid-free paper

Springer is part of Springer Science+Business Media (www.springer.com)

Dedicated to my parents
Anna Pudláková and Ján Pudlák

Preface

As the title states, this book is about logic, foundations and complexity. My aim is to present these topics in a readable form, accessible to a wide spectrum of readers. The message that I want to convey is that complexity, either in the form of computational complexity or in the form of proof complexity, is as important for foundations as the more traditional concepts of computability and provability are. Rather than presenting my own philosophical doctrine in the foundations, my goal is to isolate the most important problems and invite the reader to think about them.

The foundations of mathematics has always attracted mathematicians and philosophers. There were periods of time when many mathematicians were involved in the discussion of foundations. The most important such period was at the beginning of the 20th century. At that time the set-theoretical foundations were laid down, but set theory itself ran into problems—paradoxes were found showing that the intuitive use of set theory sometimes leads to contradictions. This problem was solved by accepting a particular axiomatic system for set theory, and things settled down. Later the interest in the foundations was stirred by several events. In the 1930s, it was Gödel's Incompleteness Theorem that showed that Hilbert's program to prove the consistency of the foundations was not possible. The second major event was Cohen's proof of the independence of the Continuum Hypothesis in the 1960s. This was an open problem concerning a basic question about the cardinality of the real numbers, posed by Cantor already in the 1870s. Also in the late 1960s a new field emerged that seemed to be somehow connected with foundations. This was the computational complexity theory.

Achievements in foundations can be viewed as solutions of important problems, but in fact they present us with much deeper open problems. Do the axioms of set theory describe the real universe of sets? Can we trust the axiomatic system for set theory to be free of contradiction? When the consistency of a theory is only provable in a stronger theory, according to the Incompleteness Theorem, what are we going to do with the consistency problem? How are we going to decide the Continuum Hypothesis, when it is independent of the axioms of set theory? In computational complexity there are a number of open problems. They may just be very difficult solvable problems, but their nature, which is similar to logical problems, and their resilience with which they resist any attempts to solve them, rather suggest that there are more fundamental reasons why they are still open.

These examples show that, in spite of all the progress that has been achieved, there are problems in the foundations that are still widely open. Many mathematicians and philosophers are aware of this fact and are thinking about the problems. But not only them; also physicists have realized that they must know something about the foundations of mathematics if they want to find the unified foundations of physics. One can observe a renewed interest in the foundations in the past decade notwithstanding the fact that there has been no breakthrough result obtained recently.

However, a mathematician with a deeper interest in this subject does not have much choice of suitable sources: on the one hand, there are many popular books that present the subject in a very superficial manner, and often incorrectly; on the other hand, there are monographs about various parts of logic, set theory and computational complexity theory that can only be read with considerable effort. Furthermore, these monographs always cover much more than is needed for understanding the basic questions about the foundations, and someone not acquainted with the field does not know what to read and what to skip.

This book is intended to fill this gap by presenting a survey of results related to the foundations of mathematics and complexity theory in a readable form and with a sufficient amount of detail. It focuses on explaining the essence of concepts and the ideas of proofs, rather than presenting precise formal statements and full proofs. Each section starts with concepts and results that can easily be explained, and gradually proceeds to more difficult ones. The idea is that the readers should not be lost before they get to the heart of the matter. But since mathematicians are always curious how the things are actually done, some formal definitions and sketches of proofs are provided in the notes to the sections.

The prospective readers of this book are mathematicians with an interest in the foundations, philosophers with a good background in mathematics and, perhaps, also philosophically minded physicists. Most of the book should be accessible to graduate students of mathematics. Logicians may find much of the material familiar, but they can profit from the chapters about computational and proof complexities, unless they also work in these fields.

I should also say what the reader should not expect from the book. Although the style of the presentation is often light (such as in the quotations from science fiction stories), the book is not popular science—its primary aim is not to entertain, but to educate the reader. So the readers will need to stop from time to time and ponder what they have read, or even to skip a part and return to it later. But the book is also not a typical dry monograph consisting of definitions, theorems and proofs. Concerning the history of mathematics, the facts that I occasionally mention are only meant to make the text more readable and are not intended to give a complete picture of the development of the field.

The book consists of seven chapters. The first two chapters are an introduction to the foundations of mathematics and mathematical logic. The material is explained very informally and more detailed presentation is deferred to later chapters. For example, set theory is introduced by means of several informal principles that are presented more precisely as the axioms of Zermelo-Fraenkel Set Theory in Chap. 3. Similarly, the Incompleteness Theorem is only stated and the proof and the consequences are discussed in Chap. 4.

Chapter 3 is devoted to set theory, which is the most important part of the foundations of mathematics. The two main themes in this chapter are: (1) higher infinities as a source of powerful axioms, and (2) alternative axioms, such as the Axiom of Determinacy.

Proofs of impossibility, the topic of Chap. 4, are proofs that certain tasks are impossible, contrary to the original intuition. Nowadays we tend to equate impossibility with unprovability and non-computability, which is a rather narrow view. Therefore, it is worth recalling that the first important impossibility results were obtained in different contexts: geometry and algebra. The most important result presented in this chapter is the Incompleteness Theorem of Kurt Gödel. I believe that the essence of the proof of this theorem can be explained with very little formalism and this is what try to I do in this chapter. Due to the diversity of results and connections with concrete mathematics, this is probably the most interesting chapter.

Proofs of impossibility are, clearly, important in foundations. One field in which the most basic problems are about proving impossibility is computational complexity theory, the topic of Chap. 5. But there are more connections between computational complexity and the foundations. I think that one cannot study the foundations of mathematics without understanding computational complexity.

In fact, there is a field of research that studies connections between computational complexity and logic. It is called '*Proof Complexity*' and it is presented in Chap. 6. Although we do have indications that complexity should play a relevant role in the foundations, we do not have any results proving this connection. In the last section of this chapter I present some ideas of mine about the possible nature of these connections. I state several conjectures which, if true, would give an explicit link between these two areas.

Every book about the foundations of mathematics should mention the basic philosophical approaches to the foundations of mathematics. I also do it in Chap. 7, but as I am not a philosopher, the main part of the chapter rather concentrates on mathematical results and problems that are at the border of mathematics and philosophy. Since I feel that the field lacks innovative approaches, I present one at the end of the chapter. It is based on the idea that natural numbers that can be represented in the physical universe are different from those studied in mathematics.

I tried to be as neutral as possible, but one cannot avoid using a certain philosophical standpoint when explaining the foundations. At the beginning of the book I assume the point of view of a realist, because it is easier to explain logic to a beginner from this viewpoint. My actual philosophy is the one of a moderate formalist, which is certainly apparent from my comments throughout the book. The only special feature of my philosophy is the stress on the importance of the complexity issues.

Even a thick volume like this cannot cover everything that is relevant to the foundations of mathematics. The main omission that I am aware of concerns intuitionistic type theories. These theories play a central role in the current research into the intuitionistic foundations of mathematics. The reasons for this omission is my lack of expertise in this field and the fact that the book is already fairly long as it is.

Prague, Czech Republic
January 2013

Pavel Pudlák

Acknowledgements

I would like to thank all who helped me by reading parts of the manuscript, pointing out errors, suggesting improvements or answering questions related to the text: Paul Beame, Arnold Beckmann, Lev Beklemishev, Samuel Buss, Lorenzo Carlucci, Stephano Cavagnetto, Dmitri Gavinsky, Stefan Hetzl, Edward Hirsch, Radek Honzík, Pavel Hrubeš, Peter Koellner, Leszek Kolodziejczyk, Jan Krajíček, Sebastian Müller, Jan Nekovář, Adam Nohejl, Jeffrey Paris, Ján Pich, Michael Rathjen, Zenon Sadowski, Neil Thapen, Iddo Tzameret, Eva Vachková.

I am also grateful to the anonymous reviewers of the manuscript, whose critical remarks were very useful and helped me to correct several errors.

My thanks are further due to Julie Cismosky, Sean Miller and Neil Thapen for correcting the English, and to Petr Pudlák for helping me with computer related issues.

The photographs were kindly provided by: Fachbereich Mathematik, Universität Hamburg (Cantor, Dedekind); Kurt Gödel Society, Vienna (Gödel); Archives of the Mathematisches Forschungsinstitut Oberwolfach (D. Hilbert); King's College, Cambridge University (Turing); Princeton University Library (A. Church); Bertrand Russell Archives, McMaster University Library (Russell); Universitätsarchiv Zürich (Zermelo).

I appreciate the help of Lynn Brandon, Lauren Stoney and Catherine Waite from Springer-Verlag London during the preparation of the manuscript for publication.

Through all the work I was supported by the Institute of Mathematics of the Academy of Sciences of the Czech Republic and received additional support from several grants of the Grant Agency of the Academy of Sciences.

Last, but not least, I want to thank my wife Věra for her understanding and encouragement over the years of writing this book.

Contents

Chapter 1
Mathematician's World

The real universe arched sickeningly away beneath them. Various pretend ones flitted silently by, like mountain goats. Primal light exploded, splattering space-time as with gobbets of junket. Time blossomed, matter shrank away. The highest prime number coalesced quietly in a corner and hid itself away for ever.

Douglas Adams, *The Hitchhiker's Guide to the Galaxy*

FOR an ordinary person, it is a strange, imaginary world. At the entrance we meet very familiar creatures, such as the natural numbers $0, 1, 2, \dots$, but further on there will appear many strange aliens, like the imaginary unit i, the first uncountable cardinal number $\aleph_1$ and things even stranger than these. In some sense it is like the artificial worlds of science fiction, or like a detective story made up of mysteries with logical solutions, but still in many respects it is very different. The main difference is, perhaps, not in the artificial nature of the things that we encounter in mathematics, which apparently have very little to do with our everyday life, but in the strict rules that they obey. In a good detective story the detective eventually solves a mysterious crime by applying logical deduction. The author usually pretends that you could also have deduced who the murderer was already at the beginning of the story, knowing only the basic data presented on the first few pages. But in fact this is not true; on the contrary, the author chooses the most unlikely person. In mathematics you really can solve problems using only deduction and, in fact, no initial data are needed, except for the statement of the problem; the only things you need are patience and determination.

If you read a good novel or regularly watch a TV series, you enter into the world of the heroes of the story and often forget, at least for a while, that it is not real. In science fiction stories you can even experience a completely different world than ours here on Earth. Science fiction gives writers the opportunity to construct new worlds, even worlds that are in contradiction with firmly established laws of physics. There is nothing wrong with this if it has its own logic. Similarly, mathematicians invent worlds which are sometimes completely alien to ordinary people. In their minds they create mental pictures of the concepts about which they are thinking, as if they could really see numbers, sets, functions, infinitely dimensional spaces and a lot more, and move in this environment arranging these objects until they construct

P. Pudlák, *Logical Foundations of Mathematics and Computational Complexity*, Springer Monographs in Mathematics, DOI 10.1007/978-3-319-00119-7_1,

the one they were looking for. Active mathematicians actually spend a big portion of their lives in this world. The more time they spend there, the more real this world seems to them. Like many teenagers who spend a lot of time in the virtual realities of computer games, mathematicians live part of their lives in what I would call *real virtuality*. Whereas virtual reality is pretend reality, what mathematicians do is the opposite: their worlds seem virtual, but are in some sense very real.

So is the mathematical world real or not? Most mathematicians would defend the true existence of at least some mathematical objects; in fact, most people would agree that the numbers $0, 1, 2, \ldots$ in some abstract sense do exist. As I will explain later, this is not just an important philosophical question, it is a question which is very important for the foundations of mathematics independent of our philosophical view, or our lack of interest in philosophy. But before we discuss such problems we have to know what kind of "things" mathematicians deal with.

1.1 Mathematical Structures

In biology we study animals, plants, bacteria, etc., in astronomy stars, planets, etc. So we can define biology as the science studying living organisms, astronomy as the science of the universe, and so on. But how can we describe mathematics? The answer to this question used to depend on what the main topic in contemporary mathematics was. For ancient Greeks, mathematics was essentially geometry and thus mathematics was the science of space. In the 18th century, when mathematics was tightly connected with physics, an answer to the question '*What is the subject of study of mathematics?*' would most likely be that it is *quantities and the relations between them*. A '*quantity*' was a real number that possibly depended on other numbers. For example, when describing a motion of a physical object, quantities could be position, speed, and momentum, all depending on time. The views on what the subject of mathematics is changed gradually. In roughly the 19th century mathematicians realized that there could be other objects of study on top of the traditional ones. The discovery of non-Euclidean geometries was an important step towards realizing that one does not have to study only objects which occur naturally in real life. An especially dramatic shift happened in algebra, where mathematicians realized that the usual number-theoretic structures are merely special instances from classes of structures sharing properties with the standard ones. Later on, new mathematical fields appeared where the objects studied had little to do with numbers or geometry. A systematic treatment of all mathematical objects became possible only after calculus had been given rigorous foundations and when there was a sufficiently general tool at hand: the concept of set.

I will describe the current standard approach to the question of what mathematical objects are. It is based on the concept of a *mathematical structure*, which gradually developed in the first half of the 20th century and was finally adopted as a key concept by the Bourbakists. *Nicolas Bourbaki* was a pseudonym under which, in 1939, a group of young French mathematicians started publishing an encyclopedic series of monographs covering the main fields in mathematics. Naturally, an attempt to give a unified treatment to the whole of mathematics needed a general concept such as the concept of mathematical structure.

This is certainly not the only possible view of contemporary mathematics. If I were not interested in foundations and wanted rather to explain the source of ideas which led to the most profound results, I would choose a different vantage point. Quite often it is difficult to formalize general ideas by a single mathematical concept. In fact, the main progress in modern mathematics has in most cases been achieved by realizing that the same idea was present in several fields and thus results and proof techniques could be transferred from one field to another. A prominent example is algebraic geometry, a field which applies geometric ideas to various non-geometric objects, including some discrete structures. Mathematics has always been a never ending struggle to express general ideas in a comprehensible, general and rigorous way and thus it cannot be explained completely by a single concept such as the mathematical structure. Nevertheless, Bourbaki's structuralist approach is the best that we have.

The ancient mathematicians considered only a few structures: the natural numbers, the plane and three dimensional space. Gradually new structures appeared in mathematics, although it was not an easy process to accept them. For instance, the complex numbers turned out to be very useful, but for a long time they were treated as a strange auxiliary means to solve problems about real numbers. We still use the terminology of *real* and *imaginary* numbers, but now we treat these words as purely technical terms and do not attribute more existence to real numbers than to complex ones. In mathematical analysis people realized that functions can be added, multiplied, etc. just as numbers can be, though they are not numbers. An important turning point was when mathematicians realized that they did not have to study only one of the few standard structures, instead they could choose any structure from a large variety. It was as if the objects of study were not given to them, but they could design them according to their own will and need, just following certain rules. (Whether one views it as the possibility to choose, or the possibility to create, depends on one's philosophical standpoint.)

Let us turn to the definition of a structure. Roughly speaking, a mathematical structure is a toy or a gadget that you can play with. You push or turn knobs and something happens. It is also like a painting where a single brush stroke makes no sense, but together the strokes give some meaning. You can also think of a structure as a game. In a game you have certain *objects*, and *rules* that determine what you can do with the objects.

A nice example is Rubik's cube, the well-known toy: the objects are the 26 small cubes and the rules are fixed by the ingenious mechanism of Rubik's cube that allows you to move only certain groups of small cubes together, namely those that form a face of the cube. Though it was important to design the mechanical construction of the cube, so that it worked well and could be mass produced, the essence of it is not the mechanism. The only thing that is important is that you have 26 pieces and particular rules how to move them. You can do "mathematical research" on Rubik's cube by studying what configurations are possible, which are symmetric, how many steps you need to transform a particular configuration into another one and so on. This is, in fact, what mathematicians actually do with structures.

There are many different structures; some are, in some sense, unique, while some are just members of large classes of similar structures. Let us consider the most familiar structure which is the natural numbers $0, 1, 2, 3, \ldots$. The structuralist point of view is that a single number, say 4, does not have any meaning. It has a meaning only as a part of the structure, namely, that *there are four numbers less than it*. Notice that we need the relation '*less than*', without it we could not distinguish 4 in this way. Furthermore we can add and multiply numbers (this is the 'playing with a toy' alluded to above). Thus we arrive at the following description of the natural numbers as a mathematical structure: they consists of

1. the *set* of nonnegative integers $\{0, 1, 2, 3, \ldots\}$, called the *universe*, or the *base set*, or the *underlying set* of the structure;
2. the *operations* of addition $+$ and multiplication $\cdot$;
3. the *relation* of being less than or equal $\leq$.

Notice the stressed words *set*, *operations*, and *relation*. This is, in fact, the form of all basic structures: they consist of a set on which there are some operations and relations defined. We do not restrict the number of operations and relations, except that their number must be finite. In particular, a structure can have only relations or only operations. For example, we may consider the natural numbers only with the ordering relation, or, on the contrary, we may add more operations. The natural numbers with no operations and $\leq$ as the only relation form a much simpler structure, but they are important when we are interested in a particular class of structures, namely, ordered sets.

In our example above the operations are *binary*, which means that they produce an element from 2 elements. Obviously, one can consider operations with this parameter 2, called the *arity*, replaced by any natural number.[1] In particular, operations of arity 0 are called *constants* and operations of arity 1 are called *functions*. Operations with arity greater than 2 are rare. The arity of a relation can be any number greater or equal than 1. A unary relation, that is a relation of arity 1, is usually called a *predicate*, or a *property*. An example of a ternary relation is the relation '*x is between y and z*' used in the formalization of plane geometry.

It probably required a considerable psychological effort for mathematicians to realize that the underlying set, the universe of a structure, does not determine the relations and operations. For example, originally people thought of the natural numbers as something intrinsically associated with the natural ordering and the two basic operations. The realization that we are completely free to choose operations and relations (and that the resulting structures can be interesting and useful) led to a dramatic development of mathematics in the 19th century, especially in algebra. A similar revolution occurred in physics one century later. In the 20th century theoretical physicists discovered that mathematics offers not only the classical structures of

[1] 'Arity' is not an English word, but it is common in mathematical jargon. The word is derived from the suffix *-ary*.

mathematical analysis, but many more, and they can be very useful in physics. This started with Einstein's use of the tensor calculus on manifolds in general relativity theory and Heisenberg's use of matrices in quantum mechanics.

Now, what happened with quantities? Modern mathematics has replaced this informal term by the concept of *function*. When describing some real phenomenon by two numbers x and y, where the number y is uniquely determined by the number x, we say that *y is a function of x*. This is formally written as

$$y = f(x).$$

We call x a *variable* and y the *value*, and f is a symbol by which we denote the function. The basic functions have names, such as '*square of*', '*sine*', '*exponential*', ..., and they are often expressed using special notation,

$$x^2, \quad \sin x, \quad e^x, \quad \ldots .$$

More generally, y may depend on several variables. Thus, in particular, operations are also functions. We use the word '*operation*' in situations when the function of several variables possesses some "nice" properties. This is the case of the operations of addition and multiplication on the natural numbers: they are commutative and associative (which means that the sums and products do not depend on the order in which they are computed).

If f is, say, a function defined on the real numbers, then it can be studied as the structure consisting of

1. the universe $\mathbb{R}$, which is the set of real numbers, and
2. the function f, as an operation.

It may seem that I am too fastidious about details when mentioning the universe. Isn't the structure already determined by the function? When we are stating theorems about the structure, it must be clear what the elements we are talking about are. We use the universe to determine the range of elements. It is a sort of a universe in which things concerning the structure take place.

I assume that the reader already knows most of these elementary concepts, but it is good to recall the terminology before discussing more difficult ones.

Ordered Sets

Let us now consider an example of a *class* of structures. The structures in the class are called *ordered sets*. This is probably the most ancient kind of structure. As soon as people started to organize their things they made lists by ordering the items that they considered. In fact this structure is imposed on essentially all data people use. We use language which is a sequence of words; written records are also sequences. So things are communicated in some order, whether we want to stress it or not. It is also interesting to note that the word '*ordering*' comes from '*order*' which also

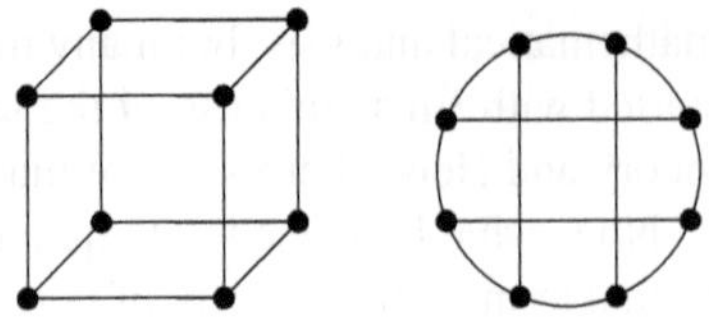

Fig. 1.1 Two drawings of the graph of the cube

means that things are properly organized, the opposite of disorder. And this is in fact the main purpose of mathematical structures, namely, to organize things, to introduce some order into our observations and data, so that we are able to manipulate them efficiently, physically and mentally.

The most obvious example of an ordered set is the set of natural numbers with the ordering relation $\leq$ that I mentioned above. Other familiar examples are the structure of the integers with ordering and the structure of the real numbers with ordering. These three structures are essentially different, not only because they have different universes, but because they have different structures, now using the word in the usual meaning. It does not matter how we represent the natural numbers, the integers and the real numbers, there will always be something different. The natural numbers are distinguished from the integers and the reals by the fact that they contain a smallest element. In the integers there are pairs of numbers such that there is no element between them, for example, there is no element strictly between 0 and 1. This is not true for the structure of reals: for every two elements, there exists an element between them (their mean is such an element).

Graphs

The word '*graph*' is used in two meanings. The traditional one is the diagram of a function, such as the dependence of the price of some commodity on time. It has a different meaning in the modern branch of mathematics that studies discrete structures, *the theory of graphs.* In this theory a *graph* consists of *points* and *arcs* that connect some points. This looks like a geometric concept, and it did originate in geometry, but it has more to do with topology than geometry. Consider for instance a cube. A cube determines in a natural way a graph, where we take the vertices of the cube as points and the edges of the cube as arcs, see Fig. 1.1. In fact, the standard terminology uses '*vertices*' and '*edges*' for all graphs. The reason why graph theory is so different from the classical fields of mathematics is that we completely abstract from the nature of vertices and edges and we only consider facts that depend on information about which vertices are connected and which are not. So if our cube is made of rubber and we twist it, the graph will be the same.

As another example of a graph, let us consider the graph of the flight connections of an airline. You can think of it as cities on a world map connected by arcs. On most such maps the arcs have little to do with the actual routes that an aircraft takes when flying between the two cities. An actual route must follow particular corridors, which is irrelevant for a passenger who only wonders whether there is a direct flight from city X to city Y.

Groups

If mathematicians voted for the most important class of structures, they would probably elect *groups*. The name is just a historical accident, so do not try to guess the meaning from normal use of the word. This concept is slightly more difficult, but worthwhile to learn. A group is a structure with one binary operation which in some sense behaves nicely. What this means precisely can be defined by postulating some simple laws that the operation must meet, which I will state shortly (page 10). Here I will only explain the concept in plain words.

The best way to imagine a group is to think of the elements of the group as *reversible actions* and the group operation as the composition of actions. As usual in mathematics, taking no action also counts as an action, called the *unit element*. Note that there is an important conceptual shift here: the actions themselves are elements, not the objects on which they act. Rubik's cube and similar toys are excellent examples. For Rubik's cube group, an action is, for example, turning the front face clockwise 90°, or turning the top by 180°. These are just some elementary actions. An action, however, may be more complex. For instance, we can compose the first one with the second one and this is also an action. We will get a different action, if we start with the second one and then apply the first one. The trick to solving this puzzle is to have several complex actions which do some particular things, such as turning two neighboring corners in opposite directions while keeping the rest the same. To transform a particular position into the original position is also an action. The goal is to compose this action from the elementary ones.

As you can imagine, the group of Rubik's cube is not a very simple one, it has $2^{12} \cdot 11! \cdot 3^8 \cdot 8!$ elements. There are groups which have infinitely many elements, but whose structure is simpler. Namely, one of the basic groups is the group of integers where the group operation is addition. To visualize it as a group of actions think of it as adding money to and withdrawing money from an account, say, starting with balance 0. Adding money is represented by positive integers, withdrawing by negative ones. This structure is the additive part of the structure of the integers that we considered earlier.

Groups are also essential in the study of *symmetries*. Consider a simple symmetric object, say an equilateral triangle A, B, C. We call a rigid action which transforms the triangle to itself a symmetry. There is a trivial symmetry corresponding to "no action", which we have, in fact, for any geometrical object. A nontrivial symmetry is the rotation where A goes to B, B goes to C and C goes to A. We can describe it by the list $A \to B$, $B \to C$, $C \to A$, or by saying that we rotate counterclockwise by 120°. We have one more rotation for 240°. Then we have another type of symmetry—we can flip the triangle along its axes of symmetry. For instance, flipping along the axis going through A can be described as interchanging B with C while A does not move. Another natural way of representing the same group is by permutations of three elements. The six permutations

$$(A, B, C), \quad (C, A, B), \quad (B, C, A), \quad (B, A, C), \quad (C, B, A), \quad (A, C, B)$$

are the elements of the group. They correspond to the identity, the transformation that does not move anything, and the symmetries denoted by a, b, c, d, e in Fig. 1.2.

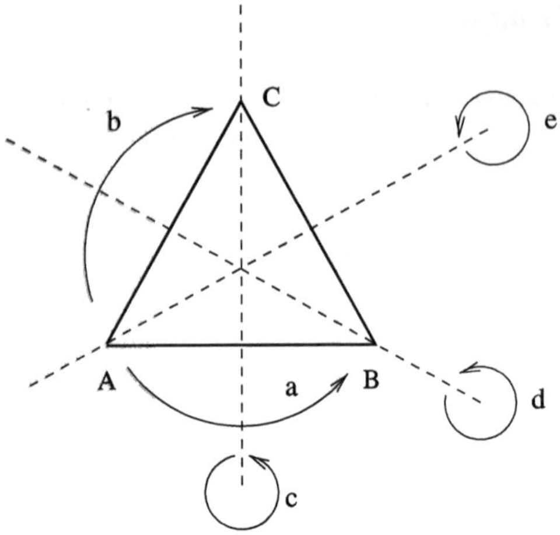

Fig. 1.2 The symmetries of an equilateral triangle

Fig. 1.3 The multiplication table of the group of symmetries of an equilateral triangle. The unit element of the group is denoted by 1

	1	a	b	c	d	e
1	1	a	b	c	d	e
a	a	b	1	e	c	d
b	b	1	a	d	e	c
c	c	d	e	1	a	b
d	d	e	c	b	1	a
e	e	c	d	a	b	1

The identity is the unit element of the group and is denoted by 1. The group operation is the composition of two permutations. For example, (C, A, B) is the transformation $A \mapsto B$, $B \mapsto C$, $C \mapsto A$ and (B, C, A) is the transformation $A \mapsto C$, $B \mapsto A$, $C \mapsto B$. Hence their composition is the identity (A, B, C).

These two representations use specific properties of the group. A general way by which we can represent any binary operation is the multiplication table. The multiplication table of the group of symmetries of an equilateral triangle is in Fig. 1.3.

Finally, we consider a way of representing groups that plays an important role in the study of finite groups—representations by matrices. In this way problems about finite groups can be translated into problems about matrices. Matrices form a very rich structure with a lot of interesting concepts and important theorems. The study of such representations is so useful that it forms a separate field called the *group representation theory.* Here is one such representation of the group of symmetries of an equilateral triangle.

$$\begin{pmatrix} 1 & 0 \\ 0 & 1 \end{pmatrix}, \begin{pmatrix} 0 & -1 \\ 1 & -1 \end{pmatrix}, \begin{pmatrix} -1 & 1 \\ -1 & 0 \end{pmatrix}, \begin{pmatrix} -1 & 1 \\ 0 & 1 \end{pmatrix}, \begin{pmatrix} 1 & 0 \\ 1 & -1 \end{pmatrix},$$
$$\begin{pmatrix} 0 & -1 \\ -1 & 0 \end{pmatrix}.$$

(For the definition of the matrix product see page 396.)

We have seen four representations of the same group. Each of them determines the structure of the group, but the group as an abstract object cannot be identified with any of them.

Why do we need various structures, why do we not just use numbers? The examples of graphs and groups show that there are practical situations which cannot be described only by numbers. We can think of structures as *models* of real and potentially realizable situations. Another possible view is that structures give us ways to classify objects. One useful way of classifying collections is to count the number of elements. We count our pieces of luggage to check that we have them all, which clearly does not ensure that we have all *our* luggage. But this test usually works. Numbers are not the only kind of structure used for such a classification. In particular groups are very good for this purpose. They are used in crystallography, to name a practical application. The symmetry group can be used to distinguish various objects, but it does not necessarily determine them completely.

In mathematics such a use of groups is almost ubiquitous. Returning to our example, we can distinguish the triangle from other geometrical objects by its group of symmetries. It is rather awkward here, as the triangle is much simpler than its symmetry group, but for larger objects it makes sense. In this case we would rather use the triangle to define the group.

One of the most beautiful pieces of mathematics, which I will consider in some detail in Chap. 4, is also based on this concept. This is the famous result that algebraic equations of degree 5 are not solvable using radicals. This means that there is no explicit formula using basic arithmetic operations and roots, expressing a solution to the equation in terms of the parameters. Here we have a natural scale given by the degree of the equation. But this gives us no clue why equations should be solvable up to degree 4, and unsolvable from degree 5 on. It was a great insight of Galois that one should assign groups to equations. The kind of groups that can be associated with equations of degree 5 and higher do not occur for equations of lower degree, and this gives the distinction between the solvable and the unsolvable.

Let me finally mention a result which belongs among the major achievements of twentieth century mathematics. The result is interesting also because it is a theorem with the longest proof ever written by mathematicians. It is called *the classification of simple finite groups.* The word "simple" is a little misleading; it is a technical term which specifies groups that are in some sense basic building blocks for constructing all finite groups. Naturally, having a description of them is very important, if we want to understand finite groups. The whole result is contained in a series of papers produced by a number of first rate mathematicians. The total number of pages amounts to several thousands. Some simple groups had to be described explicitly, the smallest one with $2^4 3^2 5 \cdot 11 = 7\,920$ elements, the largest one having $2^{46} 3^{20} 5^9 7^6 11^2 13^3 17 \cdot 19 \cdot 23 \cdot 29 \cdot 31 \cdot 41 \cdot 47 \cdot 59 \cdot 71$ (approximately $8 \cdot 10^{53}$) elements, called *the Monster*. The enormous length of the proof and the huge size of the groups that it describes are certainly remarkable, but what is also interesting is a strange kind of irregularity. We are used to the fact that in mathematics things tend either to be very regular, or to look very random; if there is regularity with some exceptions then the exceptions are small. Here, in contrast, we have 26 exceptions that share very few common properties.

Types of Structures

In order to give a more precise meaning to the concept of a structure, we have to use more technical means of mathematics, some notation, and a few symbols. Formally, a structure is given by a list that consists of several sets. The first set is the universe, the set of objects of the structure. The remaining sets are relations, functions and operations on the universe. Let us denote by $\mathbb{N}$ the set of all natural numbers. Taking $\mathbb{N}$ as the universe, we can define various structures. The universe by no means determines the structure, however, there are some structures with this universe that we like more than the others. On the set $\mathbb{N}$ we usually take the following one $(\mathbb{N}; +, \cdot, \leq)$. To stress the special role of the universe, it is separated from the other sets by a semicolon. In this structure the binary relation $\leq$ is superfluous because we can define it from the operation $+$ (namely, $x \leq y$ if and only if there exists a z such that $x + z = y$), but we may have other reasons for keeping it. This structure has two binary operations and one binary relation—this information is what we call the *type of the structure.* Let $\mathbb{R}$ denote the set of all real numbers. We can define a structure of the same type as the natural numbers by taking $(\mathbb{R}; +, \cdot, \leq)$.

A different example is a directed graph. It is determined by a set of vertices and a general binary relation. Hence we can say that directed graphs are structures of the type consisting of one binary relation.

Structures with one binary operation also have a special name; they are called *magmas*, or *groupoids*.[2] Groups can be defined as those structures with one binary operation that satisfy the following axioms:

1. there exists a *unit element* (an element, usually denoted by 1, that satisfies $x \cdot 1 = 1 \cdot x = x$);
2. the operation is *associative* ($(x \cdot y) \cdot z = x \cdot (y \cdot z)$ for all elements x, y, z);
3. every element has its *inverse* (the inverse of x is usually denoted by x^{-1} and satisfies $x \cdot x^{-1} = x^{-1} \cdot x = 1$).

Groupoids and groups belong to a large class of structures, called *algebraic structures*, or *universal algebras*, which are structures that only have functions and operations, but no relations.

All the structures that we have considered so far are *first-order structures*. There are structures that use more complex objects; such structures are called *second order, third order*, etc. In second order structures we have sets of subsets of the universe and relations between such subsets. This can be explained as follows. In a second order structure we have two universes, one consists of the elements that we want to study, the other consists of *sets of elements*, which we call *second order elements*. In a second order structure we also have relations and functions defined on second order elements. In order to imagine second order elements, think of subsets of the universe as properties of elements and sets of these subsets as properties of properties.

---

[2]Not to be confused with groupoids in category theory.

Example Let us take the color *navy blue* as an example of a property of real objects. Then we can take *dark colors* as an example of a property of properties that contains navy blue as an element.

If we attempt to define second order structures in full generality things become quite complicated. We can consider not only relations between subsets, but also between subsets and elements. Furthermore we should allow talking about properties of relations. But that is still not enough, since functions are also first-order objects, so we should allow relations between functions and so on. It is rather complicated, but it is only a technicality. The essence is that we have certain levels: the zero level are elements, the first level are relations and operations. In a second order structure we can define relations and operations on all objects of the first two levels.

The simplest example of a class of second order structures is the class of topological spaces. Topological space consists of a set of *points* A (this is the universe), and a set of subsets of A, called *open sets* that must satisfy some laws. For instance, the real numbers as topological space (called *the real line*) have the universe $\mathbb{R}$ and the open sets are subsets of reals which are unions of open intervals. (An open interval is the set of numbers between r and s *not* including the endpoints r, s.) The empty set is defined to be open too. Intuitively an open set is a set which does not contain a point on its border.

Let us proceed to the third order. This essentially means that we allow subsets of all subsets of the universe. There is no reason to stop at the third order, but already there it is hard to find nice examples. Let us take the first-order structure of reals $(\mathbb{R}; +, \cdot, \leq)$. Extend it to a second order structure by adding the set of all continuous functions of one variable, denoted by $\mathcal{F}$. Then we would like to consider the limits of the continuous functions, so we add a topology on $\mathcal{F}$ by taking the set of all open subsets of the functions, denoted by $\mathcal{X}$. This results in a third order structure $(\mathbb{R}; +, \cdot, \leq, \mathcal{F}, \mathcal{X})$. In this structure $+, \cdot, \leq$ are first-order concepts, $\mathcal{F}$ is second order and $\mathcal{X}$ is third order.

The types of structures are associated with certain set-theoretical constructions. The first one is the *Cartesian product* of sets. The Cartesian product of two sets X and Y is denoted by $X \times Y$ and it is the set of all pairs of elements (x, y) where x is an element of X and y is an element of Y. The reason for using $\times$ is that the size of the Cartesian product is the product of the sizes of the two sets; otherwise this set operation shares very little with the corresponding operation on numbers. Clearly, we can iterate this operation to get the product of a finite number of sets. The name '*Cartesian*' is in honor of the French mathematician and philosopher René Descartes (1596–1650), to whom we attribute the invention of coordinates and analytic geometry (although some analytic methods in geometry had already been used in ancient Greece). In modern terms it means that one dimensional space can be identified with the set of real numbers, and higher dimensional spaces are simply the products of copies of one dimensional space. His invention was probably the first step in the process of formalization of mathematical objects by mathematical structures. Mathematicians very often use pictures to visualize structures that they are thinking about. In the case of the Cartesian product $X \times Y$ the picture is the familiar one with X drawn as the coordinate x, Y the coordinate y and the product being the points on the plane. The Cartesian product corresponds to relations, since we can define relations on a set A as subsets of the products of A with itself. Thus a subset of A is a unary relation, a subset of $A \times A$ is a binary relation, etc.

The second set operation is related to exponentiation and thus it is denoted by Y^X. It is the set of all functions f defined on X and having values in Y. Instead of saying that f is an element of Y^X, we prefer to express it by $f : X \to Y$. The Cartesian product enables us also to define functions with several variables, which we call operations. Thus, for example, a binary operation f on a set A is an element of $A^{A\times A}$, or using the other notation $f : A \times A \to A$ (which is read as '*f maps $A \times A$ into A*'). For higher order structures, we need yet another set operation. Let us denote by $\mathcal{P}(A)$ the *power set* of the set A, the set of all subsets of A. Thus, for example, relations between second order elements are subsets of $\mathcal{P}(A) \times \mathcal{P}(A)$.

This notation can be used to define types of structures, but for this book we do not need a formal definition. Moreover, there are types of structures that do not quite fit into this schema. In classical parts of mathematics real numbers play a key role, thus many structures are somehow connected with them. Consider, for instance, a *real vector space*. It is a set of vectors A and a binary operation on A, usually denoted by $+$, satisfying certain axioms (namely $(A; +)$ is a commutative group). Furthermore, for every real number r, we can multiply any vector a of A by r and thus obtain another vector of A. This does not fit into the above schema, as the real numbers are not in $(A; +)$, they are *external*. In order to define this structure we have to take the union of the two structures—the real numbers and the group of vectors. The resulting object can be denoted by $(\mathbb{R}, A; +_{\mathbb{R}}, \cdot_{\mathbb{R}}, +_A, \cdot_{\mathbb{R},A})$. I have distinguished the two additions and two multiplications by subscripts, (to be more precise, we should write specifications such as $\cdot_{\mathbb{R},A} : \mathbb{R} \times A \to A$ which is multiplication of a vector by a real number, etc.). So we have to generalize the concept of a structure further and allow more than one universe. Also notice that in this particular example the roles of the two universes $\mathbb{R}$ and A are different: while A may vary arbitrarily, $\mathbb{R}$ is fixed for all real vector spaces.

For understanding the foundations of mathematics we do not have to study the whole ramified variety of structures. The most interesting phenomena can be observed in simple first-order structures.

Structures of Structures

In order to understand structures, it is important to realize that only *the form* is important, not the content. This means that the nature of the elements is irrelevant. The word '*structure*' denoting this concept is chosen appropriately, as we would like to identify two objects that have *the same structure*. Thus to get the whole point we only need to define what '*the same structure*' means. Intuitively it means that we can move one structure so that it completely coincides with the other. To move the structure means to move the points of the universe, the rest, the relations and operations will move along because it is attached to the points. In mathematics structures do not live in space, so the transformations from one structure into another one are not continuous transitions (unless we incidentally study topology). Thus we only need to specify the beginning and the end of the movement. This is done by the concept of *mapping*. (A mapping and a function are the same things; we use different names only because of the different context.) Such a mapping should be *one-to-one*, which means that no two points are mapped onto one, and it should be *onto*, which means that every point of the universe of the second structure is an image of a point of the first structure. The mapping translates in a natural way relations and operations from one structure into the other. If the resulting image is identical with the second structure, we say that the structures are *isomorphic*. Isomorphism is the mathematical concept of having the same form. We often do not distinguish structures that are isomorphic and often say that '*two structures are the same, up to isomorphism*'.

To understand the above definition, think of the problem of comparing two pictures on a film in order to check if they are the same. First you have match them correctly. This means that you need some special points, in this case two are enough, which determine the correct position. If you put the pictures so that the points coincide, then it suffices to check if every line, every spot, etc. coincides.

The study of mappings of one structure into another is not restricted to isomorphisms. Given a class of structures one defines a more general concept, called *homomorphisms* or just *morphisms*, by using more general mappings. In particular, a homomorphism does not have to be a one-to-one mapping, hence it can map several elements on one. In this way some information about the structure on which it is defined may be lost in its image. Homomorphisms enable us to formalize the intuitive concept of similarity. The ability to recognize similarities is one of the most important features of human and animal thinking. Thus it is not surprising that in modern algebra many important results can be stated purely in terms of morphisms. A class of structures and morphisms is in some sense also a structure; it is called a *category*. We can study a class of structures by studying its category.

The Four Color Theorem

I will conclude this section with a couple of mathematical results that will be used as examples in the following chapters.

In 1852 an English mathematician, Francis Guthrie, conjectured that every map can be colored by four colors so that no two neighboring countries have the same color. This is, perhaps, the most famous problem in combinatorics, or at least it had been so until it was solved by Kenneth Appel and Wolfgang Haken in 1975 [5, 6]. The original statement talks about the topology of the plane, but it can be stated as a problem about certain graphs. Given a map, represent countries as vertices, say choose a point inside every country. Then connect by an arc every two vertices that come from neighboring countries. Then, instead of coloring countries, we will color vertices. The restriction is that two vertices connected by an arc must have different colors. This simple transformation shows why graphs are so useful. We can transform a rather complicated statement to a simple combinatorial one.

This reduction alone does not suffice to translate the problem to graph theory. Not every graph corresponds to a map and it is very easy to construct a graph that is not colorable by four colors (take five vertices and connect every pair of vertices). Thus we need a characterization of graphs that come from maps; these graphs are called *planar*, as they come from maps in the plane. Such a purely combinatorial characterization was found by Kazimierz Kuratowski (1896–1980), a Polish set theorist and topologist; thus the problem has been reduced to finite combinatorics.

Whether or not every map can be colored by four colors has no bearing on the foundations of mathematics. What has is the way the problem was solved. Appel and Haken did not write down a proof of the conjecture, they only tested by computer that a proof exists. Following some earlier results they reduced the problem to a finite number of cases that were possible then to check by computer. Each particular case can be checked "by hand", but the total number of cases is too large for a human, even with the more recent improvements that have reduced the number of cases. This raised a discussion as to whether such proofs are legitimate. Certainly, such a proof conveys less to a mathematician than a usual proof. Typically, a proof is based on a small number of ideas that one can memorize so that it is possible to reconstruct the formal proof when needed. The experience of mathematicians with long proofs is that they are very likely wrong if such a set of basic ideas cannot be extracted from them. I agree with that, as this concerns proofs that are written by people and such proofs are never completely formal. Once the things are done formally, computers are much more precise than people. By now the validity of the theorem has been verified by running the programs on different machines and by using alternative proofs written by different people. What remains a mystery is why we do not have a 'normal' proof, a proof sufficiently short to be understood by people. As we will see later, there are theorems that do not have short proofs. But our mathematical tools are still very limited and thus we are not able to prove for such concrete theorems almost anything about the lengths of their proofs.

Note that there is a generalization of this problem to all orientable surfaces. Interestingly enough, the generalization had been solved for surfaces of all genera, except for the plane, without using a computer and before the original problem was solved.

The Four Color Theorem was not the first case in which an infinite problem was reduced to a finite number of cases. The famous Goldbach Conjecture, probably the oldest unsolved problem in mathematics, says that every even natural number greater than 2 can be expressed as the sum of two prime numbers. A weaker conjecture states that every odd number greater than 7 is a sum of three odd primes. In the 1930s, the Russian number theorist Ivan M. Vinogradov proved the weaker conjecture for all odd numbers starting from some large number N_0 [299]. Thus, theoretically, it sufficed to check all odd numbers less than N_0 in order to complete the proof. Unfortunately the number N_0 was so large (the original estimate was $e^{e^{e^{42}}} \approx 10^{10^{10^{17}}}$) that there was no chance to check the remaining cases by computation. This is still the case, in spite of the bound on N_0 being substantially reduced and in spite of the possibility to use contemporary powerful computers.[3]

More recently another famous problem has been solved using a computer in a similar way as in the Four Color Theorem. It is the Kepler Conjecture that the densest arrangement of equal balls is, in fact, the one that people have always been using. In 1998 S.P. Ferguson and T.C. Hales announced a proof of the conjecture [112]. It is based on a reduction proposed by L. Fejes Tóth in the 1950s. Since the computations used computer arithmetic, some doubts about the completeness of the proof still persist.

One may expect that computer aided proofs would be quite widespread by now, but it is not so. It turns out that there is a very narrow window where computers may help mathematicians. If ever a proof can be reduced to a finite number of cases, then, usually, either the problem can be solved completely by a mathematician, or the number of cases is so huge that it cannot be checked even by a computer.

Ramsey's Theorem

Frank P. Ramsey (1903–1930), a British mathematician and philosopher, proved a lemma that he needed in order to solve a certain problem in logic (the decidability of a certain part of first order logic) [235]. The lemma was later rediscovered by Paul Erdős and Gyorgy Szekerés working in a totally different field [69]; since then it became one of the main parts of combinatorics. Today this lemma is called *Ramsey's Theorem* and plays an equally important role also in logic and set theory. Therefore this theorem is very useful when we want to illustrate the connections between various fields of mathematics.

The essence of the theorem can easily be explained to anybody. Suppose that you have a symmetric binary relation on a finite set. Such a relation is also called an 'undirected graph', or just a 'graph'. Traditionally, for this theorem, one takes a random group of people and the relation of knowing each other as an example of a graph. The question that this theorem addresses is to what degree the relation can be chaotic, or put positively, must there be at least some order in any such relation? There are many ways to define the degree of order, but the extreme cases are clear: if every pair is connected by the relation, then clearly it has the maximal order; by the same token, if no two are connected it also has the maximal order. Ramsey's theorem, roughly speaking, says that total chaos is impossible. More precisely, we can always find a small subset of vertices where either all elements are connected in the graph, or all elements are not connected. For example, if there are at least 6 people in the group, there must be at least 3 that all know each other or all do not know each other (see Fig. 1.4). Similarly, if the group has at least 18 people, then

[3] The very recent result of T. Tao [289] that every odd integer greater than 1 can be represented as a sum of 5 or fewer primes uses the fact that the Goldbach conjecture has been verified by computation for all numbers up to $4 \cdot 10^{14}$.

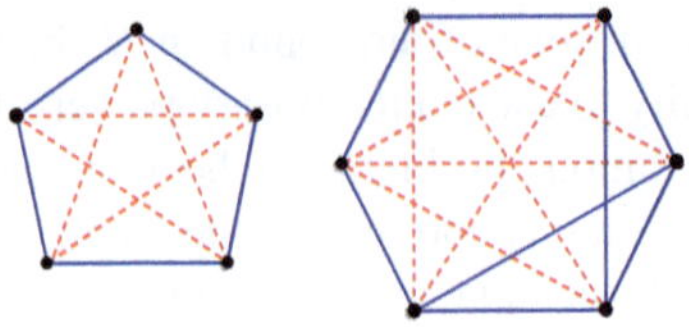

Fig. 1.4 Examples of a colorings of pairs of elements of a 5-element set and a 6-element set. The coloring of the 5-element set shows that $R(3) > 5$ because there is no 3-element monochromatic set. Since $R(3) = 6$, there must exist 3 points connected by lines of the same color in any coloring of a 6-element set. In the example there are two such triples, both form blue (solid line) triangles

there must be at least 4 that all know each other or all do not know each other. For 5, it suffices to have 46 people in the group.

In general, for every number n, we can find a number r, such that a graph on r vertices contains a subset of size n where either all elements are connected or no pair is. The *Ramsey number* $R(n)$ is defined as the least r such that every graph on r vertices contains a subset of size n where either all elements are connected or no pair is. The theorem says that this is a correct definition, such a number exists for every n.

The above examples can be stated as $R(3) \leq 6$, $R(4) \leq 18$ and $R(5) \leq 46$. In fact we know that $R(3) = 6$ and $R(4) = 18$, but we do not know the exact value of $R(5)$. We only know that $43 \leq R(5) \leq 46$. This is remarkable since to determine the value of $R(5)$ is a finite problem, one has "only" to check all the graphs on 43, 44 and 45 vertices. Testing a single graph is not so difficult (though it is quite a time consuming task—there are more than one million subsets of size 5 of a set of size 45), the problem is that there are too many graphs to be tested.

The classical infinite version of the theorem states that for every graph on the natural numbers, there is an *infinite* subset of the natural numbers such that either all elements in the subset are connected, or no pair is. A remarkable fact is that the finite version of the theorem can be derived from the infinite one. The advantage of such a proof of the finite version is that we do not have to bother with counting. The disadvantage, the price for the simplification, is that we do not get any bounds on the Ramsey numbers.

Notes

1. *General structures.* A general structure is defined by an *echelon construction.* The construction starts with base sets (universes) $A_1, \ldots, A_n$. Then we can apply operations of the Cartesian product $\times$, the power set operation $\mathcal{P}$ and the operation of taking the set of all functions from one set into another set B^A. This means that we successively produce sets such that every new set is obtained from $A_1, \ldots, A_n$ and the already produced ones by applying one of the three operations. We identify products of several sets if the order of the sets in them is the same; for instance, we do not distinguish between $(B_1 \times B_2) \times B_3$ and $B_1 \times (B_2 \times B_3)$. Thus we can omit parentheses in the products. A structure is a sequence of the form $(A_1, \ldots, A_n; B_1, \ldots, B_m)$ where $B_1, \ldots, B_m$ are subsets of the sets obtained by the echelon construction or mappings between them.

For example, our third order structure considered above $(\mathbb{R}; +, \cdot, \leq, \mathcal{F}, \mathcal{X})$ is produced from the sequence $\mathbb{R}, \mathbb{R} \times \mathbb{R}, \mathbb{R}^{\mathbb{R}}, \mathcal{P}(\mathbb{R}^{\mathbb{R}})$, where the operations $+$ and $\cdot$ are mappings from $\mathbb{R} \times \mathbb{R}$ to $\mathbb{R}$, the relation $\leq$ is a subset of $\mathbb{R} \times \mathbb{R}$, the set $\mathcal{F}$ is a subset of $\mathbb{R}^{\mathbb{R}}$ (the set of all real functions) and $\mathcal{X}$ is a subset of $\mathcal{P}(\mathbb{R}^{\mathbb{R}})$ (the set of all subsets of real functions).

In a precise definition of a structure we have to associate a *type* to each of the sets. In particular, in first-order structures this means determining if it is a relations or an operation and then its arity. first-order structures are those where neither of the operations $\mathcal{P}(X)$, X^Y is used. In second order structures these operations can be applied, but not iterated, in third order structures they may be iterated once etc.

It is possible to simplify the matter by considering operations and functions as a special kind of relations (for example, a binary operation is a ternary relation). However, quite often, it is an advantage to have operations as a primitive concept.

2. *Higher type functionals.* General structures can use all three operations: Cartesian product, power set operation, and the operation of taking all functions from a given structure to another one. We can get, however, very interesting objects by considering only the last one. This means to concentrate on functions and not to use relations and sets. We start with elements as the basic type of objects; the set of elements is the universe of the structure. The next type consists of functions. A function is a mapping from the universe to itself. Then we can define functionals, which are mappings that map functions to elements. We can use also mappings that assign functions to elements and mappings that assign functions to functions and so on. We will simply call all such objects *functionals* and distinguish them by their *types*. As the types do not have linear structure, we cannot use numbers for denoting types, we need to introduce special notation. The type of elements will be denoted by o ('o' for 'objects'). Given types τ and σ, the type of functionals that map objects of type τ to objects of type σ will be denoted by $\tau \to \sigma$. Thus functions are functionals of type $o \to o$, the lowest level functionals are $(o \to o) \to o$, etc. Note that functionals of type $o \to (o \to o)$ can be identified with binary operations, that is, functions of two variables.

   Now we will consider some important classes of structures.

3. *Ordered sets.* An ordered set is a structure with one universe and one binary relation on it denoted usually by $\leq$ (ambiguously, because the relations in different structures are different). By an ordered set we usually mean a *partially* ordered set which means that there may be incomparable elements. The axioms of partially ordered sets are

   a. $x \leq x$—reflexivity,
   b. $x \leq y$ and $y \leq z$ implies $x \leq z$—transitivity,
   c. $x \leq y$ and $y \leq x$ implies $x = y$—antisymmetry.

   The ordered sets where every two elements are comparable are called *linear orderings*; they satisfy also

   d. $x \leq y$ or $y \leq x$.

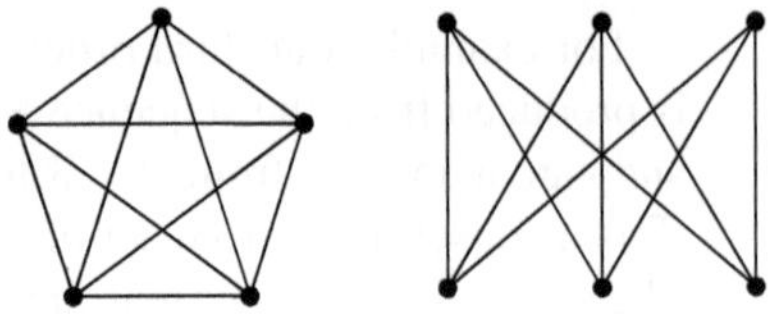

Fig. 1.5 The graphs K_5 and $K_{3,3}$

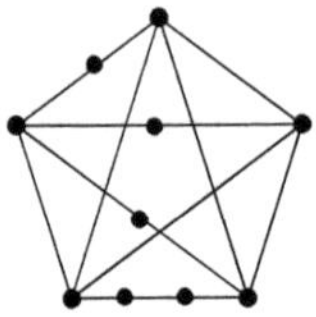

Fig. 1.6 A subdivision of K_5

4. *Graphs.* A general graph is a binary relation on the set of vertices. It is called a *directed graph* because we may have a directed edge (u, v) without having the opposite (v, u). Edges of the form (u, u) are called loops. For instance, partially ordered sets are a subclass of graphs. Graphs in the narrow sense are symmetric, which means (u, v) is an edge if and only if (v, u) is, and loops are prohibited. We denote by (u, v) an ordered pair. For symmetric graphs, we can take unordered pairs which are two-element sets. They are denoted by $\{u, v\}$. (Sometimes a more general concept is considered where there can be more than one arc between two vertices.)

   Kuratowski's characterization of *planar graphs* is based on forbidden subgraphs. He found a set of graphs such that planar graphs are exactly those that do not contain a graph from the set. The set of forbidden graphs consists of the two graphs in Fig. 1.5 *and all their subdivisions.* A *subdivision* of a graph is obtained by refining edges into paths; pictorially, we put several dots on an edge (see Fig. 1.6).
5. *Groups.* A group is usually considered as a structure with one binary operation, one unary operation (a function) and a constant. These are called *multiplication, the inverse element function* and *the unit element.* Thus we write $(G; \cdot, {}^{-1}, 1)$. The inverse element and the unit is definable from multiplication, but having these two additional primitives enables us to write axioms as equations:

   a. $1 \cdot x = x \cdot 1 = 1$,
   b. $x \cdot x^{-1} = x^{-1} \cdot x = 1$,
   c. $x \cdot (y \cdot x) = (x \cdot y) \cdot x$.

   Note that we do not postulate commutativity. You can check that the symmetry group of a triangle is not commutative. The groups where the commutative law $x \cdot y = y \cdot x$ holds are called *commutative* or *Abelian* groups. For commutative groups, one often uses additive notation, thus instead of calling the binary operation '*multiplication*' we call it '*addition*'.

   We will now define some concepts needed to explain the meaning of simple groups.

A *group homomorphism* is a mapping f of a group G_1 into a group G_2 which preserves the operation. This simply means

$$f(x \cdot y) = f(x) \cdot f(y),$$

for every two elements of G_1. (As customary, we use the same dot for both groups, though these are different operations in general.) This condition implies that f preserves 1 and inverses. The *image* of the group G_1 under f, denoted by $Im(f)$, is the set of all elements of G_2 to which some element of G_1 is mapped. This set is, as you can easily check, closed under the operations of multiplication and inverse and contains the unit element. So it is a *subgroup* of G_2. The *kernel* of f, denoted by $Ker(f)$, is the set of elements of G_1 which are mapped onto 1. It is also a subgroup. $Ker(f)$ is the trivial one element subgroup of G_1 if and only if f is a one-to-one mapping, in which case G_1 is isomorphic to $Im(g)$. On the other hand, if the image is the trivial one element subgroup of G_2, then G_1 is equal to $Ker(f)$.

The two groups $Ker(f)$ and $Im(f)$ do not give full information on G_1 in general, but the structure of G_1 can be very well understood if we know them.[4] Take an element g of G_1 which is not mapped to 1. Then the set of all elements which have the same image, namely $f(g)$, is the set of elements of the form $g \cdot h$, where h runs through the elements of $Ker(f)$. Thus G_1 can be decomposed into *cosets* which have a structure similar to $Ker(f)$, every coset corresponding to an element of $Im(f)$. Also, if we know that g_1 and g_2 are from cosets determined by h_1 and h_2, that is, $f(g_1) = h_1$ and $f(g_2) = h_2$, then the element $g_1 \cdot g_2$ is from the coset determined by $h_1 \cdot h_2$.

As an example consider Rubik's cube. We have Rubik's group, let us denote it by G_1, which consists of the transformations of the whole Rubik cube. Note that we consider only transformations that can be physically realized without breaking the cube into pieces (there would be 12 times more transformations, if we allowed decomposing and reassembling the cube). Further we can consider the same transformations, but look only at the small cubes at the edges, which means that we identify the transformations which act in the same way on edges. Let us denote this group by G_2. Then we have a mapping, in fact a group homomorphism $f : G_1 \to G_2$, given by '*forgetting the vertices of Rubik's cube*'. In this case $Im(f) = G_2$, the group of transformations on edges. $Ker(f)$ is the group of the transformations which are mapped on the identity element of G_2 which are transformations which move only the small cubes on vertices while preserving the edges. This decomposition is actually used by Rubik cube solvers.

Now comes the crucial definition. A group G is called *simple*, if for every homomorphism f from G to another group, either f is one-to-one, or f maps G to the trivial one element group. By the remarks above, this is equivalent

[4]In order to get full information about G_1, we need the groups $Ker(f)$ and $Im(f)$ and, furthermore, a homomorphism from $Im(f)$ into the group of automorphisms of $Ker(f)$. It would take us too far afield to explain this relation.

to the condition that either G is equal to $Ker(f)$ and $Im(f)$ is a one element group or G is isomorphic to $Im(f)$ and $Ker(f)$ is a one element group. In other words, we cannot decompose a simple group into smaller groups using a group homomorphism, which makes the study of simple groups more difficult.

6. *Rings and fields.* A *ring* is a structure with two binary operations and one constant on one universe. The operations are usually denoted by $+$ and $\cdot$ (the $\cdot$ is almost always omitted when writing terms), the constant is denoted by 0. The axioms of the rings express that for a given ring $(A; +, \cdot, 0)$ the structure $(A; +, 0)$ is a commutative group, $\cdot$ satisfies the associative law and the two operations are connected by the distributive laws

$$x \cdot (y + z) = x \cdot y + x \cdot z \quad \text{and} \quad (x + y) \cdot z = x \cdot z + y \cdot z.$$

A ring $(A; +, \cdot, 0)$ is a *field*, if the nonzero elements with the operations $\cdot$ form a group, which means that there is a multiplicative unit element and nonzero elements have multiplicative inverses. The most familiar fields are rational numbers, the real numbers, and the complex numbers. Integers form only a ring.

7. *Universal algebra.* This is the field of mathematics that studies algebraic structures in general, without assuming any special properties of them. The aim is to generalize theorems that are known for various special classes of algebras such as groups, semigroups, rings, fields, lattices, Boolean algebras, etc. Except for fields, these classes can be defined by equations, as we have done for groups and implicitly for rings (the problem with fields is that 0 has to be treated separately). So it is natural to study the equations valid in classes of such structures, the equational theories. Furthermore, there is a natural concept of homomorphisms for universal algebras, namely, as in groups, the mappings which preserve operations.

8. *Topological spaces.* A topological space is a structure of the form $(A; \mathcal{X})$ with $\mathcal{X} \subseteq \mathcal{P}(A)$ where $\mathcal{X}$ contains $\emptyset$ and A and it is closed under arbitrary intersections and finite unions. The sets in $\mathcal{X}$ are called *open sets.* Note that it is a second order structure. Moreover, the condition that the open sets are closed under intersection is even of a higher order (namely the third order) since it talks about arbitrary subsets of sets of subsets of A.

9. *Special structures.* There are several structures that play a special role in mathematics. The reasons for their exceptional status are in that they appear in many problems, or they are in some sense universal, or it is simply the tradition. Examples of such structures are: the ring of integers, the field of real numbers, the ordering of rational numbers, the topology of real numbers, etc. The classes of structures were often defined by choosing some general properties of these special structures.

10. *Real vector spaces.* A real vector space is a structure of the form $(\mathbb{R}, A; +_{\mathbb{R}}, \cdot_{\mathbb{R}}, +_A, \cdot_{\mathbb{R},A})$. It has two types of objects: the real numbers $\mathbb{R}$, called scalars, and vectors A. On the real numbers there are the two basic arithmetical operations $+_{\mathbb{R}}, \cdot_{\mathbb{R}}$; further, there is a binary operation $+_A$ on vectors, called addition, and an operation of multiplying a vector by a scalar $\cdot_{\mathbb{R},A}$. In a real vector space

the real numbers are determined uniquely, so one has to postulate only axioms which determine the structure of vectors and bind it with real numbers. The axioms of the real vector spaces say that $(A; +_A)$ is a commutative group and for every $r, s \in \mathbb{R}$ and $a, b \in A$,

a. $1 \cdot a = a$,
b. $(r + s) \cdot a = r \cdot a + s \cdot a$,
c. $r \cdot (a + b) = r \cdot a + r \cdot b$,
d. $(r \cdot s) \cdot a = r \cdot (s \cdot a)$.

Here 1 stands for the real number 1, (in order to be able to write these axioms as equations, we should include 1 into the definition of the structure as a special constant). I have omitted the subscripts since it is clear from the context which operation is meant. Thus real vector spaces are described by equations, but, clearly, we cannot derive all of their properties from these equations since they say nothing about real numbers. They determine this concept assuming that we know what the real numbers are.

11. *Finite automata.* This is one of the basic concepts of the theory of computation. A finite automaton is a structure of the form $(A, B, Q; q_0, \delta, \sigma)$ where $q_0 \in Q$ is a constant, and $\delta : A \times Q \to Q$, $\sigma : A \times Q \to B$ are operations. The interpretation is that A is the input alphabet, B is the output alphabet, Q is the set of the states of the automaton and q_0 is the initial state. Such an automaton works in discrete steps. In every step it receives a letter a from the input alphabet. It reacts by changing its state from its current state q to the state $\delta(a, q)$ and it writes the output $\sigma(a, q)$. This concept is not only similar to algebras, but it actually is amenable to algebraic methods.
12. *Boolean functions.* A Boolean function is mapping of the form $f : \{0, 1\}^m \to \{0, 1\}^n$, in words, a function that maps strings of zeros and ones to strings of zeros and ones. This is the main structure studied in theoretical computer science. It is the prototype of finite functions, functions with finite domain and finite range.
13. *Boolean algebras.* A Boolean algebra is an algebra with three operations meet $\wedge$, join $\vee$ and complement $'$ and two constants 0 and 1. It satisfies the laws of propositional logic, which can be expressed by the following axioms.

    a. commutative and associative laws for $\wedge$ and $\vee$;
    b. both distributive laws for $\wedge$ and $\vee$: $x \wedge (y \vee z) = (x \wedge y) \vee (x \wedge z)$ and $x \vee (y \wedge z) = (x \vee y) \wedge (x \vee z)$;
    c. $x \wedge x' = 1$, $x \vee x' = 0$, $x \wedge 1 = x$, $x \vee 0 = x$;
    d. De Morgan's laws: $(x \wedge y)' = x' \vee y'$, $(x \vee y)' = x' \wedge y'$.

    A Boolean algebra has an ordering which we will denote by $x \leq y$ and which is defined by $x = x \wedge y$; 1 is the top element and 0 is the bottom element.

    One can show that this theory is in a certain sense the algebraic theory of the two element set $\{0, 1\}$. Namely, one can define every Boolean function $f : \{0, 1\}^n \to \{0, 1\}$ using the basic operations of the two element Boolean algebra, and all equalities between terms formed from Boolean functions are provable in this theory. However, this theory has also larger finite and infinite models.

14. *Manifolds.* Sometimes it is rather difficult to present a concept in question as a structure, sometimes we even need to generalize further the concept of a structure. A *real manifold* is a topological space where we have locally real coordinates. Intuitively, it is a patchwork assembled from pieces of an n-dimensional Euclidean space. This is formalized by the concept of an *atlas* of *charts*. The charts are homeomorphisms (= mappings preserving the topology) from open subsets of the manifold into a d dimensional real space. A topological space with an atlas is a nice structure, but *it is not a manifold*; it is only one of the infinitely many ways to determine it. To resolve this purely formal problem, one defines a manifold as a topological space with *all admissible charts* for an atlas, charts that are compatible with the charts in the atlas. The atlas itself is not a part of the structure.

    Is it necessary to use such awkward definitions? The point is that there is no problem with an intuitive definition, if we work with typical objects. Once we need to consider extremal cases or when we need to generalize theorems as much as possible, we are in trouble without a rigorous definition.
15. *Categories.* When studying structures we are interested only in their form, but when we prove their existence we need to construct them. This amounts to choose particular elements for the universe and defining which are in the particular relation or what a particular operation does with them. Thus we define the rational numbers as pairs of natural numbers, the real numbers as certain sets of rational numbers, etc. Then, of course, we can forget what the actual elements of the universe are. We are interested only in the shape, but we have to use material to realize it. Can we not avoid the ad hoc part of choosing material and, instead, get the shape directly?

    The theory of categories is at least a partial remedy. In this theory, instead of individual structures, we study *a category*, which is the overarching structure of a *class* of structures. Thus a category is a large structure whose elements are some structures in the usual sense.

    In a category we have two kinds of basic elements: *objects* and *morphisms* between objects. For every object A, there is an identity morphism i_A from A to A. Given a morphism f from A to B and a morphism g from B to C we can form their composition fg which goes from A to C, otherwise the composition is not defined. The axioms are

    a. $i_A f = f$ and $g i_A = g$, whenever defined,
    b. $f(gh) = f(gh)$, whenever defined.

    (Note that it is not an algebraic structure in the strict sense since the composition is not defined for every pair of morphisms.)

    A typical category is the class of all groups with homomorphisms. In general, the intended meaning of morphisms is some kind of mappings, but there are categories where morphisms are not represented by mappings. For example, any partially ordered set is a category, if we interpret $x \leq y$ as a unique morphism from x to y. A morphism f from A to B is defined as being an *isomorphism*, if there exists a morphism g from B to A such that $fg = i_A$ and $gf = i_B$.

Here is an example of an important concept that can be defined purely using the language of categories. We define that an object C is the product of the objects A and B, if there are morphisms f, g from C to A, B respectively such that for any other object C' and morphisms f', g' from C' to A, B, there exists a *unique* morphism h from C' to C such that $hf = f'$ and $hg = g'$. In the theory of categories, such definitions are often presented in the form of *commutative diagrams.* The 'commutativity' means that if we compose morphism along two paths of arrows that start and end in the same objects, then the resulting two morphisms are equal. Below is a commutative diagram for the definition of the product.

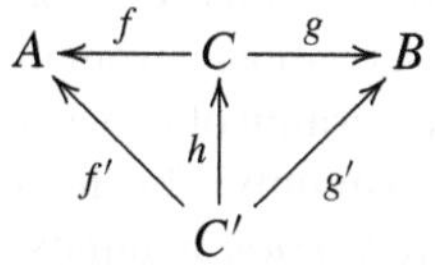

The product does not have to always exist, but when it does then it is unique in the sense that any two such objects are isomorphic. In the category of all sets the product is just the Cartesian product of the two sets. Thus we are able to define it without mentioning pairs! Also in the categories of algebras defined by equations the categorical product is the naturally defined product. It is instructive to realize that the product of two elements in a partially ordered set interpreted as a category is their greatest lower bound.

Categories behave like a special kind of structure except that their universe can be too big to be considered as a set. For instance, all groups do not form a set but a proper class (I will explain this concept later). There is a natural concept of morphisms between categories, they are called *functors*. Functors preserve identity morphisms and the operation of composition of morphisms. The operations used in the construction of structures (product, power-set, the set of all functions from one set to another) can be extended to functors.

For instance, the power-set operation $\mathcal{P}$ can be extended to a functor from the category of sets into itself. As $\mathcal{P}$ is defined for sets, the objects of the category, we only need to define $\mathcal{P}(f)$ for morphisms, which are mappings between sets. Suppose $f : A \to B$, then $\mathcal{P}(f) : \mathcal{P}(A) \to \mathcal{P}(B)$ is defined by putting $\mathcal{P}(f)(X)$ equal to the image of X under the mapping f.

16. *Proof of Ramsey's theorem.* Since the role of edges and non-edges in the theorem is symmetric, one uses colorings of unordered pairs instead of graphs. Assume a coloring of pairs of the set $\{1, \dots, r\}$ by two colors is given. We want to construct a *monochromatic* subset, which is a subset in which every pair has the same color. We start with the pairs of the form $(1, x)$ with $1 < x \le r$ and consider their colors. One of the colors has to occur at least $(r-1)/2$-times. We take the subset X_1 of $\{1, \dots, r\}$ of those elements $x > 1$ for which the pair $(1, x)$ has the prevailing color. (If both colors occur the same number of times, then it does not matter which color we choose.) Let v_1 be the least element of X_1. In the next step we consider all pairs (v_1, x) with $v_1 < x$ and $x \in X_1$. There must be a color that occurs at least $((r-1)/2 - 1)/2$-times. Note that

this color may be different from the color that we selected in the previous step. Take the subset X_2 of those elements $x > v_1$ of X_1 such that the pair (v_1, x) has the prevailing color. We can continue this process until the sets X_i shrink to an empty set. Thus we have chosen elements $1, v_1, \ldots, v_n$ for some n that is approximately the binary logarithm of r. Now look at the colors of the pairs of elements of the set $\{1, v_1, \ldots, v_n\}$. As noted above, both colors can occur, but the coloring has a very special property: the color of every pair (x, y), with $x < y$, depends only on the smaller element x. In the Ramsey theory jargon this property is called '*combed to the right*'. For $x = 1, v_1, \ldots, v_{n-1}$, associate this color with x. Then, again, take a color that occurs at least $(n-1)/2$ times and select the elements associated to this color (and add v_n if you wish to have one more element). Thus you get a monochromatic subset. The size of the subset is approximately $1/2$ of the logarithm of r, hence, if r goes to infinity, then the size of such a subset also goes to infinity. This finishes the proof of the finite version.

The proof of the *infinite version* is almost the same. The only changes are that we take the coloring of pairs of the infinite set $\{1, 2, 3, \ldots\}$, and we do not talk about the prevailing color but a color that occurs infinitely many times. It may happen that both colors considered at some stage occur infinitely many times. In such a case we can choose any of them.

The proof of the finite Ramsey theorem is quite easy, so it is not a matter of economy to deduce it from the infinite version. The reason for doing it is to show how one can get a finite statement from an infinite one. Later we will see nontrivial applications of this proof. The proof is based on the following result. A *tree* is a connected graph without cycles. A *rooted tree* is a tree with a distinguished vertex, called *the root*. (Examples of a rooted trees are in Fig. 4.4 on page 325.) We consider infinite trees. A tree is called *finitely branching*, if the degree of every vertex, that is the number of edges incident with the vertex, is finite. An *infinite branch* is an infinite chain starting in the root (a sequence of pairwise distinct vertices starting in the root in which every two consecutive vertices are connected).

König's Lemma *Every infinite finitely branching tree has an infinite branch.*

Proof Start the construction of an infinite branch with the root of the tree. Since the degree of the root is finite, the subtrees that are connected to it cannot be all finite. So choose as the next vertex the root of an infinite subtree connected to the root. Apply the same argument recursively to subtrees. Thus we obtain an infinite branch.

To derive the finite Ramsey theorem from the infinite one, assume that the finite Ramsey theorem fails. This means that for some n the following holds. For every r, there is a coloring of pairs of elements of $\{1, \ldots, r\}$ by two colors such that no subset of size n is monochromatic. Having these colorings we would like to construct a coloring of the infinite set $\{1, 2, 3, \ldots\}$ that has no monochromatic subsets of size n, hence no infinite monochromatic subsets. Thus we have colorings of pairs on arbitrary large segments that satisfy some property and we would like to construct a coloring of all pairs. König's lemma is a perfect tool for it. The vertices of the tree will be the colorings for which there is no monochromatic subset of size n. This will include the trivial empty coloring of the empty set of pairs of elements of the one element set $\{1\}$, which will be the root. Two colorings will be connected by an edge in the tree if, for some r, one is on $\{1, \ldots, r\}$, the other is on $\{1, \ldots, r+1\}$ and the second one is an extension of the first one to the larger set. It is quite straightforward to prove that this is a tree. Assuming the finite Ramsey theorem fails, it is an infinite tree. It is finitely branching since there is only a finite number of colorings on a finite set. By König's lemma, there is an infinite branch. This is a sequence of gradually extending colorings. We take their union as the coloring of all pairs. Clearly, if there were a monochromatic subset of size n, it would already be in one of the colorings. □

The above theorems are in fact only special cases of a more general result proved by Ramsey. The general result concern not only pairs, but also k element subsets for every fixed finite k. Furthermore, but less important, one can consider an arbitrary fixed finite number of colors. Here is the general form of the infinite theorem.

The Ramsey Theorem *For every k, m positive integers, and for every coloring of k-element subsets of the natural numbers by m colors, there exists an infinite monochromatic subset X (a subset in which all k-element subsets have the same color).*

The proof of the theorem goes by induction on k. One reduces the case of $k+1$ to the case k by a 'combing' argument similar to the one that we have used above.

1.2 Everything Is a Set

No one shall be able to drive us from the paradise that Cantor created for us.

David Hilbert, *On the infinite*

The concept of a set was introduced in mathematics by Georg Cantor (1845–1918). Similar ideas had been considered before him, in particular in philosophy and logic. Cantor was not a logician and he arrived at the concept working on mathematical problems. Thus he was the first one to realize that sets are not only a good methodological tool but they are also useful for obtaining mathematical results. In spite of his success in proving results using set theory, mathematicians of his time, except for a few, ignored Cantor's results. His first paper on this subject was published in 1878, but it took several decades before set theory was accepted by the mathematical community.

Cantor's first major result in mathematics concerned functions on the real numbers. He proved a theorem about trigonometric series, which are series of sine and cosine functions. In physics this theory is used to decompose sounds into pure sounds. He proved that such a decomposition is unique, if the series of functions converged at every point. Then he realized that he could weaken the assumption so that the series converged everywhere except for finitely many points. But that was still not the best possible result. He found out that it is also possible to allow certain infinite sets. He went on to describe more and more complex sets. For that he needed to make the concept of a subset of a real line precise. Furthermore, he needed to apply a certain operation (derivation of a set) transfinite number of times, which led to his discovery of transfinite ordinal numbers. (I will explain these concepts in Chap. 3). He realized that set theory was a new, completely unexplored field of research and thus he devoted the rest of his scientific career to this subject.

Georg Cantor
Courtesy of Universität Hamburg

I will not follow Cantor's development of the theory. Instead I will introduce this concept assuming the reader knows almost nothing about it. The concept of a set seems familiar: a set is an arbitrary collection of arbitrary elements. However in order to understand the way this concept is used in mathematics, we have to describe it more precisely. I will state several basic principles that determine the concept of the set. I will state them in plain language so that they are easily understandable. When stated formally they are postulates of set theory, but to obtain an axiomatization that is possible to use in mathematical practice, one needs more postulates. I will present the remaining ones in Chap. 3, which will be devoted to set theory.

The first basic principle is the following one:

The Principle of Extensionality *A set is uniquely determined by the elements that it contains. Thus two sets are considered equal if they have the same elements.*

This means that it does not matter how we specify the elements that belong to a set, what matters is only which elements are selected. We can determine the elements by some property, say described by a formula, or by an algorithm that decides if an element belongs to the set or not, or just list the elements of the set, etc. A particular definition is only a way of specifying the set; the set is simply the collection of elements that satisfy the definition.

This principle is not so easy to accept, unless you already have some experience with set theory since in natural language we tend to identify the sets with their definitions. Suppose I say *the red things in this room.*[5] What I mean is *the set* of the red things in this room. The set consists of a lamp, a pen, a control light on my computer and some books on the shelves. Would you say that the first definition and the list of the objects define the same thing? The principle of extensionality says that it is so: there is an abstract thing, a set, which is defined by both the condition of being red and the list. The reason why in a natural language we do not interpret extensionally definitions in the manner mentioned above is that the

[5] I assume that we agree on what red things are.

same word construction is used in different occasions by different people at different times, etc., thus the actual set of the things it defines varies depending on context.

Let us consider a different example: *the set of people over 60 in the sample of patients that we have treated.* If you write a report on your medical research, you want to give as much information as possible, so you will certainly include the definition of the set. On the other hand, if you process your data on a computer it is a different task. The information on the patients will be most likely stored with their birthdays, thus you can write a program to determine who is over 60. But you can also simply list the patients over 60; the computer does not care and the output will be the same.

At this point it is worthwhile to digress to history. The principle of extensionality is, clearly, the most distinctive feature of the set theoretical approach started by Cantor. It is interesting to compare it with other ideas which appeared or became popular at about the same time. The philosophical doctrine of *logical positivism* is a modern version of positivism developed by the Vienna circle in the 1920s and 30s. According to this doctrine the only meaningful statements are those that talk about observable events. The concept of the *black box* is much more recent, but it can be used to explain the essence of logical positivism. By a black box, we mean a device which we can observe only from outside and we cannot open it in order to see how it functions. Positivism tells us that if we cannot open the box, any theories about what is going on inside are meaningless. We can only make theories about *how it behaves.* A mathematical description of the behavior of a black box is a function. Such a function f tells us that, given an input x to the box, we will get $f(x)$, the value of f on x, as the output. Thus the black box is described by a structure consisting of the set of possible inputs, the set of possible outputs and the function f. A positivist would interpret the concept of function in the same way as contemporary mathematicians: the function is just the set of pairs of input x and the corresponding output $f(x)$. Particular definitions, or algorithms are only auxiliary means of determining the function. Extensional interpretation of functions was probably well established in mathematics before logical positivism appeared, but it cannot be a mere coincidence that similar ideas appeared in different fields of science in a relatively short period of time.

How is this related to sets? Think of a set A as a black box. For a given element x the set A tells us whether or not x belongs to it. If we have another set B which behaves in the same way, then B is equal to A. This is exactly the principle of extensionality.

Extensionality, in a broader sense, is a fundamental principle of all mathematics. It does not concern only set theory because what we call '*abstraction*' can often be explained by extensionality. When we count we only use properties of numbers and we forget about the concrete collections that correspond to the numbers. A number, such as 5, is the same 5 whether it is represented by five apples or five oranges. This concerns every mathematical structure—we abstract from the nature of elements, we only use the shape, the structure that the elements form. In this extensional understanding of structures, a relation is merely a set of pairs, it is *not* the definition that determines which pairs are related. Similarly a function is also a set of pairs, it is *not* a mechanism that produces $f(x)$ from x. It is worthwhile to restate the principle for relations and functions.

The Principle of Extensionality for Relations and Functions *A relation is uniquely determined by the pairs of elements that are related, a function f is uniquely determined by pairs x, an argument, and $f(x)$, the value. Thus two relations are considered equal if they relate the same pairs of elements and two functions are equal if they give the same values for the same arguments.*

We can state such a version of extensionality for every mathematical structure. In set theory the Principle of Extensionality is never postulated for relations, functions or other structures because in set theory all structures are sets, hence the principle for them follows from the principle for sets.

The second basic principle is:

The Principle of Comprehension *For any* reasonable *property, there exists a set which contains exactly those elements that satisfy the property.*

There is a modifier '*reasonable*' in this statement that makes it rather vague. I will explain shortly (in the next section) why we have to use it. For now, let us ignore it. In any case, the meaning is very general: for instance, a property can be determined by an algorithmic procedure, or simply by a list of elements, etc.

This principle seems intuitively completely clear. If we can distinguish some elements, we can name the '*collection*' of these elements. This means that we have a '*name*', or a '*concept*', so we can treat it as an entity. One of the ideas behind sets is to simplify our language by sticking to a single word '*set*' instead of '*a collection*', '*a concept*', '*a class*', etc. (though sometimes we will distinguish between '*classes*' and '*sets*').

Now we can make our first deduction and prove that there exists at least one set, namely an empty set. For that, we simply need a property which is never satisfied (such properties are abundant) and apply the comprehension principle. The extensionality principle, on the other hand, tells us, that the empty set is unique.

To get a good picture of how sets are used we have to stress one more fact, which, perhaps, should not be called a principle, but which is still very important.

The Principle of Being an Element *A set can be an element of another set.*

More formally this means that we do not distinguish elements and sets, thus we have only one *type* of object. Why is it important? If we could not form sets of sets, set theory would be just a kind of descriptive language. We need to produce many different elements to be able to combine them into various structures. If we could not use sets, we would have to postulate the existence of elements somehow. Furthermore, in modern mathematics there are powerful methods that are based on constructions that use the possibility of forming a set from other sets as elements. We build new structures by using parts of, or just whole structures as the elements of the new structures.

Let us note that it follows from the above principles that every set *is* an element of another set. Namely, a set x is an element of a set with the unique element x.

We denote the set with elements $a_1, a_2, \ldots, a_n$ by $\{a_1, a_2, \ldots, a_n\}$. Hence the set with the unique element x is $\{x\}$. (Note that these two sets are in general different: while $\{x\}$ has only one element, x can be empty or have more than one element![6]) In particular this enables us to construct a new set from the empty set. So we have two sets. Then we can form a two element set. This is another set, it has more elements than either of the two. It is not difficult to realize that we can go on and create infinitely many different sets. With infinitely many elements we can construct infinite structures.

It is remarkable that we are creating everything from the empty set. Does this fact have a deeper meaning, or it is just a technical ad hoc trick? In theoretical physics matter dissolves more and more into empty space. Particles are just some probability amplitudes or vibrating strings which themselves have no volume...

Well, so far we have only infinitely many elements and we have to work more to get, say, the natural numbers. To construct a structure we need not only a set, but also relations and operations. Recall that the extensionality principle applies to relations and operations as well. Once we accept the extensionality principle for relations the next step is to realize, that relations are just *sets of pairs*. Similarly we can identify binary operations with certain sets of triples and so on. Thus, for example, the relation $\leq$ on the natural numbers is just a set of pairs (a, b) where a is less than or equal to b and $+$ is the set of triples (a, b, c) such that $a + b = c$.

So extensionality easily gives us an explanation of what are relations and operations. What still remains to be defined are pairs, triples, etc. We could assume that these are primitive concepts determined by axioms, such as the concept of a set, but there is a better solution. It turns out that pairs, triples, etc., can easily be constructed from sets. I will describe it in more detail in the next section; for now, let us only observe that an *unordered* pair of elements a and b can be simply identified with the *set* $\{a, b\}$ having the two elements a and b.

Note that it is just a matter of convenience that we reduce the concept of a function to the concept of a set. We could do it otherwise too. Historically people were first interested in the concept of a function and only much later the concept of a set emerged in mathematics. In Newton's time people thought of functions as physical quantities and thus attributed to them properties which are common in physical phenomena such as continuity and the existence of derivatives. When the theory developed, further examples of functions with some bad properties were found. The most striking among them is a continuous function with no derivative in any point. This means that the curve that the function defines does not have a tangent in any point, so to say, any point is a like a sharp edge. The question arose then: what is an arbitrary function? This question is closely related to the question about arbitrary sets, as sets can be defined as the points where a function is zero; on the other hand, a function is, as we understand it today, a set of pairs.

---

[6]It is consistent to assume the existence of *some* sets x which are equal to $\{x\}$, but usually they are prohibited by other axioms, as they are rather unnatural.

The Natural Numbers

Richard Dedekind
Courtesy of Universität Hamburg

The numbers which count the number of elements in a set are called *cardinal numbers*, or simply *cardinals*. Later on I will also talk on infinite cardinal numbers, but here I will only consider the finite ones. These are the numbers that we denote by $0, 1, 2, 3, \ldots$ and call the *natural numbers*. The nature of numbers was studied by many philosophers and mathematicians. The first rigorous foundations of the natural numbers was given by Richard Dedekind (1831–1916) in the book *What are and what should the numbers be?* published in 1888 [59]. In 1889 Giuseppe Peano (1858–1932) published a book *The principles of arithmetic presented by a new method* [216] where he presented Dedekind's formalization in a more precise form. According to Peano, the natural numbers are defined as a structure with a universe N, a function S and a constant 0 satisfying:

1. for every x, $S(x) \neq 0$,
2. if $x \neq y$, then $S(x) \neq S(y)$,
3. for every set $X \subseteq N$, if $0 \in X$ and $x \in X$ implies $S(x) \in X$, then $X = N$.

The function $S(x)$ is the *successor function*, which is the unary operation of adding one: $x + 1$. The notation with $+$ looks as if we implicitly used $+$ to define it, therefore logicians prefer to use a special symbol for it.

The third axiom is the basic principle of the natural numbers: *mathematical induction*. This principle is usually stated as the following rule:

Mathematical Induction *For a given property of the natural numbers $\varphi(x)$, if φ holds for* 0 *and $\varphi(x)$ always implies $\varphi(x+1)$, then all numbers satisfy φ.*

Giuseppe Peano [7]

Note that the only difference between 3. and the statement of Mathematical Induction is that sets are replaced by the informal concept of properties.

This obvious and seemingly trivial principle is used in many proofs, simple and difficult ones as well. In fact, it is a universal principle—since this axiom determines the natural numbers, *all* results in number theory and finite combinatorics can be rewritten so that they only use this principle.[8]

The system based on the three axioms above is called *Dedekind-Peano Arithmetic*. The structures satisfying these

[7] This media file is in the public domain in the United States.

[8] More precisely, we must also use definitions of arithmetical operations and axioms about sets.

axioms are uniquely determined up to isomorphism. But note that it is not an axiomatization in logic since the third axiom speaks about sets. In other words, the natural numbers defined in this way are a second order structure. Thus when using these axioms, we must also use set theory.

It is possible to approximate this system by axiomatizations in logic, but then we can never achieve uniqueness up to isomorphism. The most natural axiomatization based only on logic is traditionally called *Peano Arithmetic* (see page 116).

The German logician Gottlob Frege (1848–1925) studied the question from a more philosophical point of view. His idea is very natural (later also used by Russell and others): a number n is the property shared by all sets with this number of elements. Using set-theoretical terminology, *a number is just the set of all sets of the same cardinality*. This presupposes knowing what it means to be of *the same cardinality*. But this causes no problems; two sets have the same cardinality, if we can find a one-to-one assignment of elements of the first set to the second set so that every element is matched with an element from the other set. In set theory a one-to-one assignment is a function and this in turn is just a set of pairs. So this can be expressed purely in terms of sets. Note that instead of saying that two sets have the same cardinality, we also say that they are *equinumerous*.

When formalizing the natural numbers in set theory we need to represent numbers by sets. Frege's definition of the natural numbers is not suitable for the formal system currently accepted as the standard. In Zermelo-Fraenkel Set Theory the class of all sets of a given cardinality greater than 0 is never a set. We can solve this problem by choosing one representative from each class of equinumerous sets. Another possibility is to use the Dedekind-Peano definition and just say that the natural numbers are one of the structures satisfying the three axioms above. Since in Zermelo-Fraenkel Set Theory we have to state an axiom of infinity anyway, we could just state the axiom saying that such a structure exists. But it is better to use more esthetically pleasing construction. Such a construction is due to the Hungarian mathematician John von Neumann (1903–1957). He defined the number n to be the set of numbers $0, 1, \ldots, n-1$; thus n is identified with the set of numbers smaller than n. Note that this works very well: as there are no natural numbers smaller than 0, 0 *is* the empty set; 1 contains only 0; 2 has two elements 0 and 1 etc. We get the next number by adding it to itself as an element. In set theoretic notation the numbers $0, 1, 2, \ldots, 5$ are: 0 (zero) is $\emptyset$ (the empty set; the two objects are the same), 1 is $\{0\}$, 2 is $\{0, 1\}$, 3 is $\{0, 1, 2\}$, 4 is $\{0, 1, 2, 3\}$, 5 is $\{0, 1, 2, 3, 4\}$. Since zero and the empty set are the same, set-theorists prefer to use 0 instead of $\emptyset$. If we substitute for the numerals their definitions, we can express all numbers only using the symbol 0 for the empty set, braces and commas. Thus $0, \ldots, 5$ become:

$$
\begin{aligned}
&0,\\
&\{0\},\\
&\{0, \{0\}\},\\
&\{0, \{0\}, \{0, \{0\}\}\},\\
&\{0, \{0\}, \{0, \{0\}\}, \{0, \{0\}, \{0, \{0\}\}\}\},\\
&\{0, \{0\}, \{0, \{0\}\}, \{0, \{0\}, \{0, \{0\}\}\}, \{0, \{0\}, \{0, \{0\}\}, \{0, \{0\}, \{0, \{0\}\}\}\}\}.
\end{aligned}
$$

This notation is, of course, not good for practical purposes.

So far we have only described a few small numbers, but we need a general definition. First we note that the successor is defined very easily by

$$S(x) := x \cup \{x\},$$

which is the key idea of this definition of the natural numbers. We want to say that every number n can be obtained from 0 by applying the successor function a finite number of times. We cannot do it directly because defining a '*finite number of times*' is equivalent to defining the natural numbers. So such a definition would be circular. Thus instead we use Dedekind's idea and say that

n must be contained in all sets which contain 0 *and which are closed under the successor function.*

For a set a to be closed under S means that whenever it contains m it also contains $S(m)$. Note that the condition stated above is a property of elements n, hence, by the comprehension principle, there exists a set consisting of such elements. This set is the smallest set that contains 0 and is closed under the successor function. So it is natural to think of it as the set of numbers that can be obtained from 0 by applying the successor function. We define the set of natural numbers $\mathbb{N}$ to be this set.

To prove that the principle of mathematical induction holds for $\mathbb{N}$ defined in this way, we argue as follows. Suppose φ is a formula such that φ holds for 0 and $\varphi(x)$ implies $\varphi(S(x))$. Let N' be the set of numbers satisfying φ. By the assumptions of induction, N' contains 0 and it is closed under S. Hence, by definition, $\mathbb{N} \subseteq N'$, which means that all n satisfy φ.

Once we have defined the set of all natural numbers, the universe of the structure, and the successor function, it remains to define the ordering relation and the operations. The ordering is defined very simply: $a \leq b$ if and only if a is a subset of b. Here we see the advantage of having sets as elements: the structure of the elements enables us to define some relations very easily.

We define addition by saying that $n + m$ is the number whose cardinality is equal to the union of two disjoint sets A and B, where A is equinumerous to n and B is equinumerous to m. Unfortunately the sets representing the two numbers are not disjoint (unless one of them is 0), but it is a trivial task to construct such A and B. As regards the multiplication we are luckier, we can define $n \cdot m$ as the number equinumerous to the Cartesian product $n \times m$ of the sets n and m.

These constructions use special properties of the two operations. There is a much more general way of defining arithmetical function called *definition by recursion*. Let us consider a recursive definition of addition. Addition is determined by the following equations

$$\begin{aligned} x + 0 &= x, \\ x + S(y) &= S(x + y). \end{aligned} \tag{1.1}$$

Here we define what it means to add 0 and then we define how to add a number bigger than 0 using the successor function and the addition for a smaller number.

So these equations uniquely determine the operation. Having addition, we can give a recursive definition of multiplication:

$$x \cdot 0 = 0,$$
$$x \cdot S(y) = x \cdot y + x.$$

We can go on and define x^y and other functions. (See also the general form of recursion on page 142.)

The Real Numbers

I will skip the constructions of the integers and the rational numbers from the natural numbers because they are easy and well-known. A more interesting problem is to construct the real numbers.

In the 18th century, mathematicians were aware of the fact that calculus, the theory of real functions, limits, integrals, infinite series, etc., needs some axioms of continuity.

Example Let f be a continuous function defined on the closed interval $[0, 1]$. If $f(0) < 0$ and $f(1) > 0$, then there exists a real number a, $0 < a < 1$ such that $f(a) = 0$.

No one doubts that principles such the one above are true. But in order to develop the theory, we either have to state them as axioms, or we need a definition of the real numbers from which they follow. When using set theory as the foundations, we do not want to add axioms that are not about sets. We would like to derive everything from the basic axioms about sets, so only the second option is of concern to us. To this end we must define a mathematical structure representing real numbers, in a similar way as we defined a mathematical structure representing the natural numbers.

The classical approach is based on *Cauchy sequences*, named after the French mathematician Augustin-Luis Cauchy (1789–1857).The starting point is, as in all constructions of the real numbers, the rational numbers. An infinite sequence of rational numbers $r_0, r_1, r_2, \ldots$ is called a *Cauchy sequence* if the elements of the sequence get closer and closer as n increases. This is a rather subtle concept that needs a more precise explanation. It does not suffice that the distance between consecutive elements decreases. What we need is that if n is large, then the distance between r_n and *all* r_m, for $m > n$, is small. Once we know that all Cauchy sequences converge, then all the properties of real numbers follow. So the idea is to ensure that a Cauchy sequence converges by choosing an object representing its limit. Clearly, we have to choose the same object for all Cauchy sequences that should converge to the same limit. Thus the whole construction boils down to the definition of what it means that two Cauchy sequences converge to the same real number, which must be stated without mentioning the real number itself. The formal definitions of these concepts are in Notes.

For formalization in set theory, Cauchy sequences are as good as any other formalization. However, the option preferred by set-theorists is based on Dedekind's cuts. It is also a more acceptable answer to the philosophical question '*what are the real numbers?*' Dedekind used the fact that we only need to add irrational numbers to the rationals. Then an irrational number is defined as a '*hole*' in the line of rational numbers. Set theory enables us to easily define what a hole means: it is a partition of the rational numbers into two parts, one below the hole, the other above. Arithmetic operations with holes are done by shifting these partitions appropriately.

Example $\sqrt{2}$ is thus identified with the pair of sets (X, Y), where X is the set of all rational numbers less than $\sqrt{2}$ and Y is the set of all rational numbers bigger than $\sqrt{2}$. However, it would be a circular definition if we used this to define $\sqrt{2}$. Therefore we must say that Y is the set of all positive rational numbers y such that $2 < y^2$ and X is the complement of Y.

If we only want to show that the real numbers can be formalized in set theory, we can ignore tradition and philosophy and use some simple straightforward construction, such as decimal representation. In this representation a real number is an infinite sequence of numbers $0, \dots, 9$ with a period and a sign. In order to get uniqueness, we disallow sequences that end with a tail of 9s.

Interestingly, the formalization of structures in set theory is a similar task as the formalization of structures for computers. Programming languages seldom use sets, they rather use lists and arrays, in which elements are given in some order, but this is not essential. The only essential difference between representing objects in set theory and in computers is that in computers we do not have infinite structures.

Notes

1. *Urelements.*[9] It is possible to develop set theory using true elements which are not sets. Such elements are called *urelements*. Another possibility is to use sets which have themselves as the only one element, sets that satisfy $x = \{x\}$. Thus we can mimic urelements while preserving extensionality for all objects, which we cannot do in the first case. The standard approach is, however, to use neither of the two kinds of urelements since we do not need them for practical purposes and the theory is simpler without them.
2. *Pairs and sequences.* The pair (a, b) is defined, following Kuratowski, by $\{\{a\}, \{a, b\}\}$. If $a = b$, then (a, b) contains one element that contains one element a. If $a \neq b$, then it contains two elements; one element is a one element set containing a, this determines a; the other is a two element set containing both elements, this determines b as the element that is not in the one element set.

[9] *Ur-*, originally a German prefix now also used in English, means *primitive, original*.

To represent a finite sequence with n elements $a_1, \ldots, a_n$ we take the set of pairs $\{(1, a_1), \ldots, (n, a_n)\}$. This is, in fact, a function defined on the set $\{1, \ldots, n\}$. Other indexed structures (matrices, infinite sequences, etc.) are done in similar way.

3. *Recursive definitions in set theory.* It is not difficult to prove in set theory that functions defined by recursion exist. We can reduce it to induction, which we already have. For example, we prove by induction that, for all n, there exists a unique partial operation defined on the interval $[0, n]$ satisfying the equations (1.1). Then the operation of addition defined on all natural numbers is the union of these partial operations.
4. *Cauchy sequences.* A sequence $r_0, r_1, r_2, \ldots$ is Cauchy, if for every $\varepsilon > 0$, there exists n such that for all $k, m > n$, the inequality $|r_k - r_m| < \varepsilon$ is satisfied.

   Two Cauchy sequences $r_0, r_1, r_2, \ldots$ and $s_0, s_1, s_2, \ldots$ converge to the same real number if for every $\varepsilon > 0$, there exists n such that for all $m > n$, $|r_m - s_m| < \varepsilon$. Note that we are able to define it without knowing the number to which they converge. Thus we can use the above condition to define an equivalence relation on Cauchy sequences. Then we define the real numbers as equivalence classes.

   The advantage of this construction is that it works for all metric spaces. Thus one can prove that every metric space can be extended to a complete metric space.
5. *Dedekind's real numbers.* Let $\mathbb{Q}$ denote the set of rational numbers. Dedekind's definition can be simplified by considering only one set of rational numbers for every real number. Thus we define a real number to be a nonempty proper subset of $\mathbb{Q}$ which is closed downwards (with any rational number it contains all smaller ones) and which does not have the largest element. For two real numbers r, s, we say that r is less than or equal to s, if r is a subset of s; $r + s$ is defined as the set of rational numbers which are less than or equal to $a + b$ for some $a \in r$ and $b \in s$; multiplication is defined in a similar way. The rational numbers $\mathbb{Q}$ are not a subset of $\mathbb{R}$ constructed in this way, but they are embedded in $\mathbb{R}$ by the assignment $a \mapsto (-\infty, a)$.
6. *Other structures.* As regards a small finite structure there is no problem to construct it now. We take, say, an initial segment of the natural numbers as the universe and to define a subset, relation or function, we simply list the elements. In the case of infinite structures, we have to find a particular construction in each case. This may depend on the axioms of set theory that we use! We can talk freely about classes of structures satisfying some properties, but to prove that there exists at least one such structure we need a construction.

So far we are only using naive set theory, which is inconsistent, if taken strictly logically. We will have to restrict the general principles to get consistency and then add new axioms to retain the necessary strength. For instance, the existence of the power set $\mathcal{P}(X)$ for every set X is a consequence of the Principle of Comprehension, but it will be postulated as an axiom later. In order to prove that $\mathbb{N}$ is a set, we will also need an axiom—the Axiom of Infinity.

1.3 Antinomies of Set Theory

The decadent mood of the end of the 19th century influenced also the views on the future of science and technology. People thought that all important inventions had been discovered and there were no substantial discoveries going to happen in physics. Mathematics has always been different because it has had famous open problems. They will never be exhausted, as new problems arise at least as fast as old problems are solved. The foundations of mathematics are, however, a different thing. In foundations there is a clear convergence to more complete and more precise systems. From this point of view the state of the affairs in mathematics was similar to physics. During the 19th century all concepts of mathematics were reduced to natural numbers. This process, called arithmetization, started with Descartes's introduction of analytic geometry, continued with the formal definitions of convergence, derivations and integrals, and ended with the introduction of sets. Set theory was able to reduce even the remaining natural numbers to the abstract concept of a set.

For mathematicians this was a positive thing. Except for a few, they are interested in doing research on real mathematical problems. The problems on foundations are seldom clear cut and often it is more philosophy than science. Having firm foundations meant that they could discard those pseudoproblems forever. But even before set theory became generally accepted, it received a serious blow. This was because a contradiction was derived from basic principles.

Before considering the contradictions, I will briefly digress to explain why a contradiction is fatal for any theory. A contradiction is a pair of statements such that one is the negation of the other. When we derive such statements we can derive also their conjunction (also called a contradiction) which is logically false. It follows from the rules of logic that any statement is a consequence of a false statement. In Latin this rule is referred to as '*ex falso sequitur quodlibet*'.[10] This is also used in natural language. When we want to stress that something is blatantly false, we say that if that is true then something ridiculous is also true. However, the natural human interpretation of implication is that the parts of the implication, the *antecedent* and the *consequent* share some content. Therefore it is not easy to accept that a single statement can imply everything. The best way to see that a contradiction implies everything is to use a proof by contradiction. In such proofs we assume that the statement that we want to prove is false and derive a contradiction. Then we argue that therefore it is not possible that the statement is false, hence it is true. Now, if we are able to derive a contradiction without any assumptions (except for the basic principles), then it is formally derivable from any assumption. Thus any assumption can be rejected, hence everything can be proved.

Once we know that we can prove everything, there is no point in actually proving anything. Such a system gives us no information and certainly does not describe any real phenomenon, as in the real world a statement cannot be true and false at the same time.

---

[10] From falsehood, it follows anything you like.

Here we are, of course, concerned with mathematical truth. In our life it is quite different. We get a lot of contradictory information. One source of contradiction is unreliable information, another one is the use of generalization based on partial data. We are always ready to reject such statements and recompute our model of the world.

Contradictions in set theory are often called *paradoxes* or *antinomies* because they contradict our intuition.[11] The simplest and the most important one is *Russell's Paradox* discovered by the philosopher and logician Bertrand Russell (1872–1970) in 1901. He showed that already one particular instance of the Comprehension Principle is contradictory. Namely, he applied this principle to the property '*of not containing itself as an element*'. The principle asserts that there is a set, let us call it R, whose elements are just the sets with this property. For example, the empty set belongs to R since it does not contain any set as an element. On the other hand, the set of all sets (suppose we proved that it existed) contains any set, hence also itself, thus it does not belong to R. We obtain a contradiction if we consider the question, whether R is an element of R. For suppose R is an element of R, then R does not satisfy the defining property of R, hence it does not belong to R. This contradiction shows that R cannot belong to itself. But if it does not, then it does satisfy the defining property of R so it must belong to itself. Thus we get a contradiction in any case.

Russell was probably not the first to discover this paradox. Logicians in Hilbert's circle knew this paradox and attributed it to Ernst Zermelo (1871–1951). Zermelo did not publish the paradox, but according to his recollections, he thought about it around 1900. He used it to prove that the largest cardinality does not exist. But Cantor had been aware of the problems with certain sets already before Zermelo. He said that they "*cannot be conceived as determinate, well-defined, finished sets*". He also called them "*absolutely infinite sets*".[12] However, there is an essential difference between the approaches of Russell on the one hand, and Cantor and Zermelo on the other. While Cantor and Zermelo studied sets as mathematical entities, Russell's focus was on the principles of logic. Cantor and Zermelo viewed the paradoxes as proofs that "very large sets" do not exist. In contrast, Russell presented his paradox as a proof that the principle of comprehension is not a universally valid logical principle. If we want to have a consistent system, we must restrict the class of properties to which it is applied. Presenting the paradox in this way had a decisive impact on the further development of set theory.

In fact, Russell arrived at his paradox analyzing an earlier paradox found by Cantor. Cantor proved that for every set, the set of all subsets of it is strictly larger. The problem then is with the set of all sets. This set exists by the Comprehension Principle, where one uses as the defining property any property which is generally

---

[11]Strictly speaking, we should distinguish between paradoxes—apparent contradictions, and antinomies—actual contradictions, but when using informal reasoning it is difficult to make this distinction. Therefore, these words are used interchangeably.

[12]Letters to Hilbert, September 26 and October 2, 1897. See [65], page 42.

true (say, the property of being equal to itself). This set is, by definition, the largest set, so it contradicts to Cantor's theorem.

Apparently most mathematicians were not very impressed by the antinomies. They felt that what they were doing was sufficiently well tested and they used mathematical objects that were in some sense more real than sets. In any case, the historical experience suggested that even if a part of the present mathematics would have to be abandoned because of its contradictory character, it would be only a small part. Still, it was rather disturbing that the contradiction was derived from what seemed an intuitively obvious principle.

At this point several other paradoxes were discussed. One of them, whose roots go back to the ancient Greeks, is the well-known *liar's* or *Epimenides* paradox.' The story says that Epimenides was a Cretan who said: "*All Cretans are liars.*" Was he a liar?[13] A modern version of this paradox is *the paradox of the barber: "There is a man in a village who shaves all men in the village who do not shave themselves, and only those. Does he shave himself?*"

Another, known as *Berry's paradox*, goes as follows. We know that any nonempty subset of the natural numbers has the first element. (This is just an equivalent form of the induction principle.) Also it is clear that there are only finitely many English sentences with at most 100 letters, hence there are natural numbers which cannot be defined by such sentences. Thus we can define a number n to be *the first number that cannot be defined by an English sentence with at most one hundred letters.* This is a contradiction, as we have just defined n by such a sentence!

At first it may seem that the problem with the paradox may have something to do with infinity. After all, we have no idea how large the largest number definable in this way is. But in fact we can easily give an upper bound on the numbers that have to be considered. With 26 letters used in English and one more character for the space between words (or at the end of the sentence) we can estimate the number of English sentences with at most 100 letters by 27^{100}. Thus the alleged number should be amongst the numbers $0, 1, 2, \ldots, 27^{100}$ since at least one of these numbers cannot be defined using 100 letters.

The number 27^{100} is, unfortunately, too big even for a computer. Furthermore, English is rather complex, so it would be difficult even to generate all syntactically admissible sentences. But you can design, or at least imagine, your own special purpose language with a simple and precise syntax and such that one can state the paradox using a sufficiently small number instead of 100.

Another version of this paradox is based on the assumption that our universe is finite. Under this assumption we do not have to give an explicit estimate of the length of the definition.

These two paradoxes belong to a class of paradoxes, called *semantic paradoxes*, that are based on natural language and use words such as '*true*' and '*defined*', which are not precisely defined. However, it is possible to formalize these concepts when we have a formal language. Then, the paradoxes are resolved by strictly distinguish-

[13] This is the traditional version of the paradox which assumes that a liar is *always* lying.

ing between the object language and the language that we use to define these concepts. These paradoxes have never been perceived as a real threat. After all, such paradoxes have been known for thousands of years and they never interfered with mathematics.

Paradoxes in Mathematics

There are paradoxical results in mathematics that do not present inconsistencies. They are exact theorems, except that they are counterintuitive. A classical example of a paradoxical object is the function, constructed by Bolzano and Weierstrass which is continuous but not differentiable at any point, which I already mentioned above. Another classical example is a curve constructed by Peano in 1890 that completely fills a square.

Paradoxical results are present in many fields of mathematics, the more the field is connected with our a priori intuition the more likely we can find some. For instance, human understanding of the geometry of three dimensions, which is to a large extent inborn, is quite good. When thinking about higher dimensions we try to use our three dimensional intuition, but it often fails badly. It is an easy exercise to construct two circles C_1 and C_2 in four dimensions such that the distances between all pairs of points one on C_1 and the other on C_2 are the same. We can do it using analytical geometry, but we are not able to visualize it because in three dimensions it is impossible.

Since a lot of our everyday decisions are based on estimating probabilities of various events, one would expect that our intuition about probability is fairly good. But there are examples of the failure of our intuition also in this field. Perhaps the most popular is the well-known *Birthday Paradox* of Richard von Mises. It seems very counterintuitive that with a probability greater than $1/2$ among 23 randomly chosen people there are two with the same birthday. One would expect to need essentially more to get this probability, but the above fact can easily be shown by a simple calculation.

A more recent and more tricky one is the following nice puzzle, so nice that it made it into the pages of the New York Times as *the Hat Problem*. There are n people each having a blue or red hat. Each person can see the color of everybody else's hat, but not his or her own. According to the rules of the game they play, at some point they are asked to guess the colors of their own hats. They have to answer at once and independently of each other, but anybody can abstain. If everybody abstains or one of them guesses incorrectly they loose. If at least one does not abstain and everybody who answers gives the correct answer, they win. The players can agree on a suitable strategy beforehand, but once the game starts they are not allowed to communicate. It is clear that every strategy may fail since either everybody abstains, which is a failure, or at least one player always answers, but the player surely may give the wrong answer, as the players do not know their colors. The question is, what is the best strategy when one wants to get the highest chance of winning. We are, of course, assuming that the colors of the hats are completely random. Intuition tells us that it is not possible to get the probability of winning greater than $1/2$. The argument is that one player makes a wrong guess with a probability of $1/2$; if more players answer at the same time, the probability that one of them answers wrongly is even larger. Thus it seems that the best they can do is to choose one player who will be the only one who answers and who makes a random guess. In reality, there are strategies for which the probability of winning goes to 1 with n going to infinity.

There are paradoxes that do not lead to contradictions also in set theory. In Chap. 3 I will talk on consequences of the Axiom of Choice that look paradoxical. I will consider the famous Banach-Tarski paradox that says that it is possible to decompose a sphere into finitely many pieces from which we can reconstruct *two* spheres. Though the Axiom of Choice does not introduce a contradiction, such paradoxical consequences undermine its position among the basic mathematical principles.

Notes

1. *Russell's paradox.* Let us analyze Russell's paradox more formally. The contradiction will be more apparent, if we use a bit of logical notation. The paradoxical set R is defined by the condition

$$x \in R \quad \text{if and only if} \quad x \notin x. \tag{1.2}$$

To find out if R is an element of R, we substitute R for x. Then we get

$$R \in R \quad \text{if and only if} \quad R \notin R.$$

This is a false proposition as it says that a proposition is equivalent to its negation.

Note that we have not used any non-logical assumption other than (1.2). Therefore we can conclude that whatever the relation $\in$ is, the sentence (1.2) cannot be true for all x. Thus we can interpret the argument as a general theorem on structures with a binary relation. Let us restate it in this way.

Proposition 1 *Let $(A; B)$ be a structure with one binary relation B. Then there is no element r of A such that the proposition*

$$B(x,r) \quad \textit{if and only if} \quad \neg B(x,x), \tag{1.3}$$

holds for every x in A.

If we interpret A as the men of the village and $B(x, y)$ as '*y shaves x*', then it shows that there cannot exist a man claimed in the Paradox of a barber. It is much harder to imagine sets as the structure where the universe is all sets and $\in$ is a binary relation. We feel that $\in$ is something special. But as far as logical deductions are concerned, it is only a binary relation.

We can view this result as based on *loops* in the graph defined by the relation R. It has been observed that one can also use longer cycles. For cycles of length 2, we get the following version of the proposition above.

Proposition 2 *There is no element r of A such that the proposition*

$$B(x,r) \quad \text{if and only if} \quad \text{there exists no } y \text{ such that } B(x,y) \text{ and } B(y,x), \tag{1.4}$$

holds for every x in A.

This case appeared in one of the popular puzzles of Raymond Smullyan. These generalizations are not only used in recreational mathematics, but also for serious results. W.V.O. Quine observed that one can use cycles of arbitrary lengths and made the bold conjecture that it suffices only to prohibit these cycles in order to obtain a consistent set of axioms of set theory. I will say more about it in Chap. 3. S. Žák used cycles of increasing lengths to prove hierarchy theorems for nondeterministic complexity classes (not knowing Quine's observation; the hierarchy theorems for deterministic computations are explained in Chap. 5, page 377). A version with an infinite cycle was found by D. Mirimanoff already in 1917.

2. *Cantor's Paradox.* Before analyzing Cantor's Paradox, let us prove his theorem on the cardinality of powersets.

Theorem 1 *For every set X, there is no one-to-one mapping from $\mathcal{P}(X)$ to X.*

Proof Suppose f is such a mapping. Define $Y = \{f(Z);\ Z \subseteq X \wedge f(Z) \notin Z\}$. Then $f(Y) \in Y$, by the definition, if and only if $f(Y) \notin Y$. Thus the assumption of the existence of such a mapping leads to a contradiction. □

Clearly, there exists a one-to-one mapping from X to $\mathcal{P}(X)$ (namely $x \mapsto \{x\}$). Thus we conclude that $\mathcal{P}(X)$ has larger cardinality than X. The paradox is now obtained by taking the set of all sets V. We have $\mathcal{P}(V) \subseteq V$, which means that $\mathcal{P}(V)$ is mapped to V by the identity mapping. For $X = V$ and f equal to the identity mapping, the set Y in the proof of Cantor's theorem is exactly Russell's set R. Thus Russell's Paradox is the pure logical essence of Cantor's Paradox.

The type of argument used in Russell's paradox and Cantor's theorem is called *diagonal*, or *self-referential*. It can actually be used to obtain nontrivial mathematical results. In (1.3) we can avoid the contradiction, if r is not in A. So another interpretation of the argument of the paradox is that any r satisfying formula (1.3), for all x in A, must lie outside of A. This argument is used also in the most important theorem in the foundations of mathematics, Gödel's incompleteness theorem. I will return to it and discuss it in more detail.

Another paradox discovered by Cantor concerns cardinal arithmetic (see Chap. 3 for the basic concepts). Suppose the that set X of all "alephs" (more precisely, sets $\aleph_\alpha$, where α are ordinals) exists. Then it is a well-ordered set, hence its cardinality must be some $\aleph_\beta$. But also $\aleph_\beta$ must be larger than all $\aleph_\alpha$ in X. This is a contradiction.

3. *Hilbert's Paradox.* Consider two operations: the union of a set X (denoted by $\bigcup X$) and the set of all mappings on a set X (denoted by X^X). Start with a non-empty set and let U be the set obtained by applying these operations in all possible ways. Then U^U must be a subset of U, by definition. But U^U has a larger cardinality than U (as one can show by the same argument as in the theorem above). Hence we cannot consistently assume the existence of such a set.

   Hilbert considered this paradox to be more serious than others because the set U is apparently constructed from below only using basic set-theoretical operations. But this is only what it looks like if we do not use a precise definition. As soon as one tries to define U precisely, one sees that it is not possible to avoid referring to "large entities". For example, the standard way to define such a U is to use transfinite recursion. To this end, however, one has to use *all* ordinals. Ordinals do not form a set, hence also U will not be a set. In set theories with classes, ordinals form a proper class (a class which is not a set) and so will also be U.
4. *Burali-Forti's Antinomy.* For the sake of completeness we mention also Burali-Forti's antinomy. It was published by Burali-Forti, but had been known to Cantor before. After all, it is not so much different from Cantor's paradox. By the theory of ordinal numbers, which we will consider later, an initial segment of ordinal numbers has an ordinal number which is bigger than any element of the segment. Thus the existence of the set of all ordinal numbers leads to a contradiction.
5. *The Hat Problem.* The reason why the intuitive argument is wrong is the following. While it is surely true that when a player answers, he gives the right answer with probability $1/2$ and the wrong one with $1/2$, this is only the conditional probability with the condition being that the player answers. If we take into account that a player sometimes abstains, we have the probability ε of the correct answer, ε of the wrong answer and $1-2\varepsilon$ that he abstains, for some $0 \leq \varepsilon \leq 1/2$. If ε is small, then the player gives wrong answer with small probability. Now the trick is that with more players it may be possible to arrange it so that the bad cases overlap, so the probability of failure remains small, but the good cases are distinct, so the probabilities add up. Indeed, in the optimal solution that exists for n of the form $2^k - 1$,

   a. either all players answer incorrectly, and this happens with probability $\frac{1}{n+1}$,
   b. or exactly one player answers correctly while others abstain; each of the players does so with probability $\frac{1}{n+1}$, hence they win with probability $\frac{n}{n+1}$

   The solution is based on Hamming codes, which is a hint for the reader who wants to solve it.
6. *Paradoxes in computational complexity theory.*

   a. Consider computations of Boolean functions $f : \{0,1\}^n \to \{0,1\}^n$ by Boolean circuits. This model of computation will be introduced later. For now, think of it as a piece of hardware consisting of electronic gates that works as follows. If you fix the input values on input wires, the circuit computes for a while and when the values on all gates stabilize, you get the output value of the function f that the circuit computes. Suppose, for some function f, the minimal size of a Boolean circuit computing f is S. Now suppose that you want to compute f in parallel on two independent inputs. This means that you want to compute the function F that from $2n$ bits x, y produces $2n$ bits $f(x)$, $f(y)$. Intuition tells us that the minimal size of a circuit computing F should be $2S$. The following is an intuitive reason that it cannot be less. Given a circuit for F we can think of it as two overlapping circuits, one computing $f(x)$ the other $f(y)$. The overlap consists of the gates that depend on both inputs x and y. But if a gate may have an arbitrary value depending on y, then it should be useless for computing $f(x)$ and symmetrically with x replaced by y. Thus the gates from the overlap should be useless, hence the best we can do should be to take two disjoint circuits. Yet, one can show that, for some functions f, one needs only a tiny fraction more than S to compute F (namely $(1+\varepsilon_n)S$, where $\varepsilon_n \to 0$ as $n \to \infty$). (See [295].)

b. Consider the following three player communication game. Player 1 gets a bit string x of length n, $x = (x_0, \ldots, x_{n-1})$ and a number i; Player 2 gets the same string x and a number j; Player 3 gets i and j. Their information is private, so, for example, Player 1 does not know j. Then Player 1 and Player 2 send independently of each other messages to Player 3. They have agreed beforehand on what messages they will send in all possible situations and they have done so in such a way that Player 3 is always able to say correctly what is the value of x_k for $k \equiv i + j \bmod n$. The question is what is the total length of the messages they would have to send to Player 3 in the worst case. Clearly, a possible protocol on which they may agree is that they would send all the bits of x to Player 3, which is n bits. Intuitively this seems the best possible thing they can do. The argument is as follows. For Player 1, the information about i is totally irrelevant, as for a given i the $k \equiv i + j \bmod n$ may be completely arbitrary. Hence the only relevant information Player 1 can send concerns x. Similarly for Player 2. So they will send some information about x, independently on the indices i, j. But then they have to send at least n bits, as the information on x cannot be compressed. Yet, the minimal number of bits that the players have to exchange is bounded by a function $f(n)$ such that $f(n)/n \to 0$ as $n \to \infty$. (See [230].)

1.4 The Axiomatic Method

At the age of eleven, I began Euclid, with my brother as my tutor. This was one of the great events in my life, as dazzling as first love. I had not imagined that there was anything so delicious in the world.

Bertrand Russell, *The Autobiography of Bertrand Russell*[14]

The oldest mathematical texts contain examples of mathematical problems with solutions. They served as guides of how to solve equations, how to construct geometric figures etc. The first proofs of mathematical theorems appeared in ancient

[14] [254], Vol. 1, page 36.

Greece probably in the 6th century BCE. They are attributed to Thales and members of the Pythagorean School (for example, the proofs of Thales' Theorem and the Pythagorean Theorem). Convincing evidence that mathematical proofs had been used in the ancient Greece in the 5th century BCE is the discovery of the incommensurability of the side and diagonal of a square. (This is essentially the fact that $\sqrt{2}$ is not a rational number, see page 257.) This is a kind of statement that requires a proof; you cannot claim that it is *impossible* to write $\sqrt{2}$ as a fraction of two whole numbers, unless you can prove it.

This was not only the time when first proofs appeared, but also the time when western philosophy emerged. According to tradition, it was Pythagoras who coined the term *philosopher*. The emergence of philosophy meant that science ceased to be considered to be a tool serving to efficiently accomplish practical tasks, but rather an environment for intellectual activity, disregarding any possible applications. Once people started to ask, not only '*how?*', but also '*why?*', they could not have been satisfied with mere statements of mathematical facts. They needed *proofs*.

Aristotle (384–322), the greatest philosopher of Antiquity, studied logic and the scientific method in general. He determined a set of logical rules, which he called *syllogisms* and described logical deductions as successive applications of these rules starting from some basic assumptions. By this, he described what we now call the *axiomatic method*.

Aristotle distinguished between two types of basic assumptions: *postulates* and *axioms*. Postulates are those that are common to all sciences, whereas axioms are special for a particular field. In the modern terminology of mathematical logic we do not use the word '*postulate*'; however, we do distinguish between *logical axioms* and *mathematical axioms*.

A prime example of an application of the axiomatic method are *Elements* written by Euclid of Alexandria around 330 BCE. Euclid starts by explaining the basic concepts such as '*A point is what does not have a part.*' Part of these statements are not definitions in the modern mathematical sense; they relate the abstract mathematical concepts to reality. We would rather call them *intended interpretation*. Then he presents two lists of statements. The first one can be interpreted as geometrical axioms, the second as logical and arithmetical axioms. The results are presented as theorems, constructions and algorithms.

We know about some gaps in proofs and that the postulates in *Elements* are not sufficient to derive all theorems. Yet, it is an impressive work, whose style is surprisingly close to present-day mathematical monographs. Many mathematicians used *Elements* as a prototype for their treatment of geometry. In fact, this book is among the most influential ones of Western civilization. Finally, a modern axiomatization of geometry was given by the great German mathematician David Hilbert (1862–1943) in his *Foundations of Geometry*[15] in 1899.

The axiomatic method is a way to reduce assumptions used in a theory to a few basic principles. But this does not only concern assumptions; at the same time,

[15]Grundlagen der Geometrie, [124].

we are also reducing concepts to simpler ones. Thus the reduction goes on in two parallel lines: on the one line we are reducing the assumptions, on the other we are reducing the concepts.

axioms	theorems
primitive concepts	defined concepts

Reducing the assumptions means that we show that they are derivable from others; reducing concepts means that they are definable from others. Eventually no further reduction is possible and then we talk about *axioms* and *primitive concepts*.[16] The primitive concepts are those which are not defined. The main reason is that they cannot be further reduced, but we usually also assume that they are clear and do not need further explanation. Similarly axioms are statements that we are not able to reduce to more primitive ones.

In principle, we could develop theory only using primitive concepts, but it would be very cumbersome. Definitions enable us to use short terms to express more complicated concepts and thus we can express ideas more efficiently.

An ideal mathematical text starts with axioms, followed by definitions, theorems and proofs of theorems. Definitions do not have to be all at the beginning. Furthermore, proofs may use auxiliary theorems, which are called lemmas. Proofs may also use auxiliary concepts that are not used in the axioms and the statements of the theorems. Although we use a special word 'lemma' for auxiliary theorems, we do not have words distinguishing auxiliary terms and their definitions from the genuine concepts and their definitions. However, mathematical articles and monographs do not only consist of definitions and theorems. Reading a completely formal mathematical text would be difficult and readers need to know the motivation for the theorems, how the results relate to those in other articles etc. It also helps to give informal descriptions of difficult proofs.

Example In elementary plane geometry the primitive concepts are points, lines and the incidence relation between points and lines. Thus we have two kind of objects, *points* and *lines*, and the relation 'a point *lies* on a line'. The basic axioms of plane geometry are:

1. *for every two different points, there is a unique line incident with them*;
2. *every line has at least two points*;
3. *any two different lines have at most one point in common*;
4. *there are three points which do not lie on one line.*

Using these basic concepts one can define other objects, such as triangles, quadrilaterals, etc., but also relations such as two lines being parallel ($\Leftrightarrow$ no point lies on both lines). These axioms are only a part of the list that Euclid needed, but already

[16] Sometimes it is useful to keep some redundancy; sometimes we are not able to prove that further reduction is impossible, but it is.

using these axioms one can prove many theorems. Also the concepts available in this system are rather simple and we have to add more primitive ones and more axioms to get interesting theorems. In particular we need the relation of congruence in order to be able to say that two line segments have the same length.

The main reason for using the axiomatic method is that we want to understand the subject that we study, we want to know what is essential—we need a *theory*. By a theory we usually understand a collection of statements which explain certain phenomena. It is very difficult to define what it means *to explain*. There are, however, some attributes that are quite clear: simplicity and universality. Thus a good theory must be based on a small number of general statements. The simplest theories may consist of a single postulate. The law of free fall asserts that the speed of falling objects is proportional to the square of the elapsed time. The universal nature of this theory is in its applicability to any object. A more general theory is Newton's theory of gravity. It explains much more than just the attraction of bodies to the Earth. It can also be given by a single equation asserting that the attraction of bodies is proportional to the product of their masses and to the square of the distance. Maxwell's theory unifies electrostatic and magnetic forces using a few differential equations. The ultimate goal of theoretical physics is a unification of all physical theories, dubbed the *Theory of Everything*; presented more modestly, it should be one theory for all forces in nature.

This is just to name a few examples from physics. Theories are present in all scientific disciplines. They are not always called theories; sometimes they are called *models* (when there are alternative theories), sometimes they are called *laws*. Formally, they are all just axiomatic systems.

Ancient Greeks not only discovered that one can axiomatize mathematics, but also the striking fact that one needs only a small number of very basic principles to do that. This also concerns some other fields of science. If nature were evil, we would need to get more experimental data every time we wanted to get more knowledge. That would mean accepting more and more axioms, which eventually would make the axiomatic method almost useless. But on the contrary, especially in physics, we are witnessing a reduction to fewer and fewer basic principles, one needs fewer and fewer absolute constants, etc. Already the present physical theories are able to reduce all chemistry to a few physical laws. In principle, it is possible to compute the chemical properties of all atoms and molecules only using quantum electrodynamics. We can go on and reduce molecular biology to chemistry etc. These are, of course, only theoretical reductions. In practice, the computational problems involved are so difficult that it is unlikely that one will ever be able to do without experiments.

In the foundations of mathematics the axiomatic method plays an extremely important role. Russell's paradox taught us a lesson: set theory cannot be based only on intuitive principles. In particular, it is necessary to restrict the use of the Principle of Comprehension. In this situation, it is reasonable to present the modification as precisely as possible. Although stating axioms of set theory explicitly does not guarantee the consistency of the resulting theory, it gives us at least something that we can test. In 1908 Zermelo published a set of axioms for set theory [318]. This was the beginning of axiomatic set theory. With a small, but important, addition by Abraham A. Fraenkel (1891–1965), this axiomatic system has been commonly used as *the* foundations of mathematics to the present day. The theory is called *Zermelo-Fraenkel Set Theory.* (I will describe it in detail in Chap. 3.) A large body of results have been proved in this theory and the theory has passed these "tests"—no contradiction has been found.

Axiomatic approach is praised by logicians, but not everybody shares their view. There has always been resistance to formalization of mathematics. The strongest opposition was declared by the intuitionistic movement at the beginning of the 20th century. Even today some mathematicians would prefer to treat mathematical concepts informally instead of axiomatizing them. Their main argument is that concepts such as the natural numbers are clear to everybody and are more fundamental than logic. I will discuss these views in more detail in Chap. 7.

Axiomatic Theories

The purpose of axioms is to describe some part of the real world and, accepting the structuralist view, the world of mathematics consists of structures. Thus axioms can be viewed as describing some structures. In mathematical logic we call a set of axioms *a theory*, and structures that satisfy the axioms *models* of the theory. For example, integers with the constant 0, the operations $-x$ and $x + y$ are a model of the theory of groups.

There are two basic situations where we use axioms:

1. to describe a class of structures,
2. to describe a single structure.

We have already considered various classes of structures defined by axioms. In particular ordered sets are defined by the axioms 1., 2. and 3. on page 17. The subclass of linear orderings is defined by adding axiom 4. Another important class of structures is groups. This class is defined by the axioms on page 18. The subclass of commutative groups is defined by adding the axiom of commutativity $x \cdot y = y \cdot x$.

The most important examples of single structures that we would like to axiomatize are the basic algebraic structures: the natural numbers, the integers, the rational numbers and the real numbers. To axiomatize such structures is usually a much harder task. Whether it is possible at all depends on the language and logic that we use. These problems are among the most important problems in the foundations of mathematics; I will deal with them in the following chapters.

Many important classes of structures were obtained by generalizing some standard structures. The prime example is the concept of a field. This class of structures was defined by taking the basic properties of the structure of the real numbers as the axioms, (see page 20 for the axioms). Another example comes from geometry. Consider plane geometry with the basic objects being points and lines and the basic

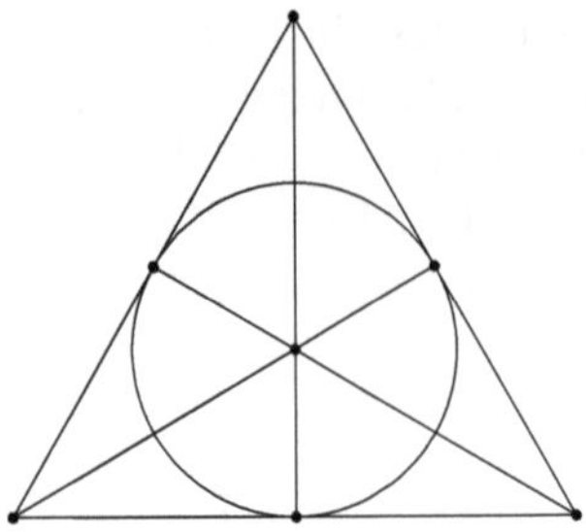

Fig. 1.7 Fano plane. Points are the seven points on the intersections of the lines; lines are the triples of points lying on the six lines and on the circle. The lines and the circle are drawn only to indicate which triples of points form lines

relation of incidence. Formally, it is a structure with two universes, P for points and L for lines and the incidence relation $R \subseteq P \times L$, where $R(x, y)$ means that the point x is on the line y. The relation satisfies the axioms on page 45. However these axioms do not describe the Euclidean plane uniquely. There are infinitely many essentially different structures satisfying these axioms, in fact, there are even finite structures. An example of such a structure is the *Fano plane*, see Fig. 1.7. Although discrete planes look like "pathological objects", they are, in fact, useful. Finite fields are even more important.

An axiomatic theory, in the strict sense, should only use logic; it should not use any higher order, or external concepts, such as sets and numbers, unless they are also axiomatized by the theory. However, this term is often used with a broader meaning. A theory whose axioms are stated in logic is called *elementary;* otherwise it is called *nonelementary.* This distinction is irrelevant from the practical point of view since the most interesting theorems usually refer to nonelementary concepts, such as subgroups in the theory of groups, ideals in the theory of rings, etc., whose definitions require the concept of set, but it is very important from the point of view of logic. An example of an elementary theory is the elementary group theory defined on page 18.

A nonelementary theory may use some standard structure as a sort of a primitive concept. Thus it is assumed that all true statements on that structure are given. An example is the concept of a real vector space. The standard structure in a real vector space is the structure of the real numbers with addition and multiplication; so this is the external part. The axioms (see page 20) talk only about the internal part and its relation to the real numbers. Also all physical theories refer to the integers, the real and the complex numbers as basic concepts and leave the job of defining them to mathematicians. Most theories studied in mathematics are nonelementary because they use the concept of set.

In principle, one can transform every nonelementary theory into an elementary one simply by adding axioms describing the nonelementary concepts. But in practice this is often impossible because there is no way to present the axioms in an efficient way.

The latter comment concerns a fundamental question: how should the axioms of a theory be presented? The ideal situation is when we can write down a *finite list of axioms.* When this is not possible, we may still be able to present the axioms in a very reasonable way, namely by an *axiom schema.* An axiom schema is a formula containing a metavariable for formulas. We obtain an instance of the schema, a concrete axiom, by substituting a formula for the metavariable. The two most important theories axiomatized by schemata are Peano Arithmetic and Zermelo-Fraenkel Set Theory, which I will describe in the following chapters. One can relax the condition on the set of axioms to the mere requirement that there is an algorithm for deciding whether or not a given formula is an axiom or not. But this is as far as we can go; if the set of axioms is algorithmically undecidable, we cannot consider it to be a formal system. In a formal system, we should be able to decide whether or not a given text is a proof; if we are not able to decide if a sentence is an axiom, then this is impossible.

In this book, I will only consider theories that are axiomatized by an algorithmically decidable set of axioms. To stress the latter fact, I will sometimes use the term '*formal theory*' or '*formal system*'. The latter one has a little broader meaning—the system does not have to be based on logic. I will also use '*axiomatic system*', '*axiomatization*', etc. with the same meaning as '*formal theory*'. The reader not familiar with the concept of decidability can simply imagine a formal theory as a theory axiomatized by a finite set of axioms and schemas since for a large class of theories, axiomatizability by a decidable set of axioms is equivalent to axiomatizability by a schema (according to a result of R.L. Vaught [298]).

The assumption that the set of axioms must be algorithmically decidable has profound consequences. It implies that certain structures cannot be axiomatized. This concerns, in particular, the structure of the natural numbers, as well as all structures that contain the natural numbers. This fact is the essence of the Gödel Incompleteness Theorem, which I will explain in Chap. 2 and then in more detail in Chap. 4. An important consequence is that nonelementary theories that use the natural numbers as primitive concepts cannot be fully formalized. In particular, none of the currently used physical theories can be fully formalized.

Properties of Theories

1. The most important property of an axiom system is its *consistency*. This means that the system is free of contradiction. In an inconsistent system one can derive any sentence, hence such a system is useless, as we noted in the section on antinomies in set theory. Actually, axiomatization of set theory was historically the first case where the question of consistency became important. Before people axiomatized concrete structures. Assuming that a particular structure exists, we get the consistency of any set of sentences that are satisfied in the structure. In particular, we believe that the natural numbers exist, therefore the axioms about them are consistent. For sets there is no such "canonical" structure. The only place where they occur is our natural language, which is imprecise and inconsistent in many ways. There is nothing to which we could reduce the consistency of set theory.

   Upon closer inspection, we realize that the situation is not much better even if we have a canonical structure for the theory. For example, we may firmly believe that the natural numbers are a real object and as such they must be consistent. But how can we test that a sentence that talks about all numbers is true in the structure? We cannot test all infinitely many numbers. So our argument that the axioms about natural numbers are consistent is based on the *belief* that the axioms are satisfied in this structure. What we, however, can do completely formally is to reduce the consistency of one theory to another one. Thus, for example, we can reduce the consistency of an axiomatic system for the natural numbers to the consistency of an axiomatic system for set theory.

Consistency is the key concept in the foundations, so we will learn more about it later; it will occupy us essentially for the rest of this book.

2. The second most important property of axiomatic systems is the *completeness.* A system is complete, if we can derive all sentences that are true in the structure that we are axiomatizing. In the case the system should describe a class of structures, we require that any sentence which is true in all structures of a given class is derivable in the system. For some simple structures, it is possible to find a complete axiomatization, for more complex ones, it is impossible. Note that completeness depends on the language that we consider. Thus, for example, it is possible to give a complete axiomatization of elementary geometry of the plane in the style of Euclid and Hilbert. However, if we want to study deeper problems, say differential geometry, the task becomes impossible. Another example is the natural numbers with addition as the only operation, which we denote by $(\mathbb{N}; +)$. This structure is axiomatizable, whereas if we also include multiplication, that is, if take the structure $(\mathbb{N}; +, \cdot)$, it is not.

   In the case of classes of structures defined by axioms we get completeness automatically. For example, groups are precisely those structures (with one binary operation, one unary operation and a constant) that satisfy the three axioms on page 18. Thus the three equations form a complete set of axioms. This looks terrific, as if we could just let a computer generate all the theorems about groups from these axioms. Unfortunately there is again the problem of the language that one considers. If we only use the elementary language of group theory $\{1, \cdot, x^{-1}\}$ we get only trivial theorems. In order to express interesting concepts, for example, to define a simple group, we need either to use a higher order language, or work in set theory. In both cases a complete axiomatization is elusive.

   A more technical remark concerns *relative completeness.* I touched on this subject already above when talking on real vector spaces. The set axioms of real vector spaces is complete relative to the structure of the real numbers $(\mathbb{R}; +, \cdot)$, which means that we can derive all true sentences about real vector spaces using the axioms and sentences true in $(\mathbb{R}; +, \cdot)$. Incidentally, there is a complete axiomatization of $(\mathbb{R}; +, \cdot)$, which implies that we can also completely axiomatize real vector spaces. But again, interesting problems concern *sets* of vectors.

3. We say that a collection of axioms is *independent*, if no axiom can be derived from the others. Put otherwise, axioms are dependent, if they can be further reduced to a smaller set. So it is important to know, if a given set is independent.

The famous case of the fifth postulate of Euclid concerns this property. The original statement of this axiom was that *two lines a, b intersecting a line c so that at one side of c the sum of inner angles is less than* 180° *("two right angles") must intersect at that side of c*. This is equivalent, using the other axioms, to: *for a line a and a point B not on the line, there is a unique line b through B which does not intersect a*. A lot of people tried to derive this axiom from the others. It took a long time for people to accept the possibility that it cannot be done. A positive outcome of this were new structures, the *non-Euclidean geometries*. We will come back to this topic later and I will explain how is possible to show independence. For now, let us just say that one needs to construct a structure which satisfies all axioms except the one that we want to show to be independent.

Independence is not as important as consistency and completeness. If we want to axiomatize a structure or a class of structures, we are satisfied with any consistent and complete set of axioms. We are interested in the dependence of axioms only because we want to fully understand the concept and, possibly, find its generalizations.

Notes

1. *First-order logic.* In this chapter I have been using the term '*logic*' for what is more precisely called '*first-order logic*'. The name stems from the fact that the logic uses *first-order language*, the language for first-order structures. I will explain this connection and the key role of first-order logic among other logics in the next chapter.
2. *The axioms of Euclidean geometry on a plane.* Above I have stated only the most basic axioms, the axioms about the incidence relation between points and lines. To develop elementary geometry one needs axioms about two more relations:

   a. "*point A is between points B and C*";
   b. "*segment AB has the same length as segment CD*"; we say that *AB* is *congruent* to *CD*.

   There are two groups of axioms one for each of the two relations. These are a few cleverly chosen statements that rather surprisingly suffice to derive all that one needs. What they say can be informally described as follows.

   a. The axioms about the relation 'between' say that on every line, once we fix a direction by taking two points, we can define a linear ordering that is dense and does not have the largest or the smallest elements.
   b. The axioms about the congruence relation say, roughly speaking, that we can drag a segment on a line and to any line and that all distances in congruent triangles are preserved.

   Once we have congruence on segments, we can define congruence on angles.

   A large part of elementary geometry can be developed using these axioms and only using logic. In particular, although we do not have the circle as a primitive concept, we can emulate it by a point C that determines the center and a segment CD whose length determines the diameter. Essentially the same can be done with the ellipse and other quadrics. However when using more complex objects one has to resort to set theory. For example, one cannot express in logic concepts such as *polygon* and *connected*, and cannot define curves that are not determined by algebraic equations. (Below I will show that connectedness of graphs is not expressible in logic.) Hence, in order to get a completely formal system in which we can develop more advanced parts of geometry, we have to accept some axioms of set theory on top of the Euclidean axioms.

This set of axioms is not complete. To make it complete one has to add axioms about the topology; it suffices to do it on lines. This is usually done by talking about sets of points. Adding axioms on sets results in a system that cannot be completed, due to Gödel theorems, but if we restrict ourselves to the primitive concepts of these axioms and only use order logic, one can get a complete theory. The idea is to replace the axiom on sets by an infinite schema that states it for every formula. For example, one can take the following set of axioms for every two formulas ϕ and ψ.

> Let a line be given and an ordering on the line be fixed. Suppose that on the line every point that satisfies ϕ is before every point B that satisfies ψ, then there exists a point A on the line that is between the points that satisfy ϕ and ψ (A may satisfy one of the two formulas).

Note that this is very much related to the axiomatization of the structure $(\mathbb{R}; +, \cdot)$.

3. *Gaps in Elements.* One kind of important missing axioms are instances of the continuity principle. In particular, the axioms telling when a circle and a line intersect and when two circles intersect. This axiom is needed already in the first theorem that proves the existence of an equilateral triangle with a given side AB. Euclid relied on the intuitively clear fact that if we want to connect a point inside of a circle with a point outside using a line, we have to intersect the circle. This is correct, but it does not follow from his axioms.
4. *Connected graphs is a nonelementary class.* We will use the class of connected graphs to illustrate some limitations of the axiomatic method.

   A graph is connected if every two different vertices are connected by a path. This is a clear and natural definition, but there is a problem: we need the concept of a path. We can define connected graphs equivalently by saying that *a graph is* not *connected, if there is a partition of the vertices into two nonempty disjoint sets such that there are no edges between the two blocks.* In this definition we need the concept of a set. Without using such concepts we cannot define connected graphs.

   Suppose that connected graphs can be defined by a first-order sentence Φ. Consider an infinite sequence of symbols $v, u_1, u_2, u_3, \ldots$, which will be interpreted as vertices of a graph. Furthermore, consider the following infinite set of axioms:

   a. $u_i \neq u_j$, (for all $i \neq j$);
   b. $u_i \neq v$, (for all i);

c. *u_1 is connected only to u_2;*
d. *u_{i+1} is connected only to u_i and u_{i+2},* (for all i).

A graph satisfying these axioms is not connected since it contains an infinite path of vertices that are not connected with any vertex outside the path, and there is a vertex v outside. Therefore the axioms must be incompatible with Φ, which means that one should be able to derive a contradiction from the union of the axioms and Φ. However, we will show that it is not possible to derive a contradiction from these axioms and Φ. Suppose there is such a proof P of contradiction. Every proof is *finite*, hence there is the largest i such that u_i is mentioned in the proof. Now, take the graph which is just a path of $i+1$ vertices; interpret the first i vertices as $u_1, \dots, u_i$ and the last one as v. This is a connected graph, and it satisfies all axioms used in the proof P. Thus using the axioms which occur in P we can derive only sentences which are valid in this graph, in particular, we cannot derive a contradiction.

The conclusion is that connectedness is a property that we cannot study without using some higher order concepts such as sets.

5. *Theories in physics.* A typical theory in physics is based on differential equations. Consider *Maxwell's equations* for the electromagnetic field. For the sake of simplicity, we restrict ourselves to the case of vacuum.

$$\operatorname{curl}\mathbf{H} = \varepsilon_0\frac{\partial \mathbf{E}}{\partial t}, \qquad \operatorname{curl}\mathbf{E} = -\mu_0\frac{\partial \mathbf{H}}{\partial t},$$
$$\operatorname{div}\mathbf{E} = 0, \qquad \operatorname{div}\ \mathbf{H} = 0.$$

Here $\mathbf{H}$ (the magnetic field) and $\mathbf{E}$ (the electric field) are vector fields depending on time, curl and div are well-known vector differential operators, whose explicit form is not important for us here. Thus each of the first two equations written in coordinates splits into three equations with partial derivatives with respect to space and time coordinates. A solution of this system of differential equations can be presented as a structure $(\mathbb{R}^3, \mathbb{R}, \mathbf{H}, \mathbf{E})$ where $\mathbf{H} : \mathbb{R}^3 \times \mathbb{R} \to \mathbb{R}^3$ and $\mathbf{E} : \mathbb{R}^3 \times \mathbb{R} \to \mathbb{R}^3$ are everywhere differentiable and satisfy Maxwell's equations. Hence the theory of electromagnetic field in vacuum is the theory of such structures. Clearly, Maxwell's equations are not an axiomatic system for which we would only need logic to make deductions. To get conclusions from the equations we have to use *mathematics*. If we want obtain some global statements, for example, to show the possibility of electromagnetic waves, we have to construct or prove the existence of particular solutions to the equations. Finding a solution of a set of differential equations is a nontrivial task that often requires a lot of ingenuity, in fact it is the subject of study of separate fields of mathematics. In practical applications we are given some partial information about the solution, such as boundary conditions or an initial state, and we want to compute the solution at every point.

Einstein's Relativity Theory is different. This theory cannot be completely reduced to a set of equations. The key concepts concern observers and what they see. Thus the problem of axiomatizing Relativity has attracted a lot of researchers. Most axiomatic systems proposed so far only formalize *Special Relativity.* Axiomatizations help us understand what are basic principles and what are their consequences. Then one can clearly see that specific mathematical concepts, such as the Lorenz transformation and the Minkowski norm, follow from the assumption that the speed of light is the same for all inertial observers and a few other basic principles.

General Relativity is a much more difficult theory. In this theory space-time is described by *Einstein's Field Equations*, which are nonlinear partial differential equations. One can use some axiomatizations of Special Relativity and extend it by adding Einstein's Field Equations to obtain an axiomatization of General Relativity.[17] It would be more interesting to have a theory in which Einstein's Field Equations would logically follow from basic principles.

1.5 The Necessity of Using Abstract Concepts

Building a good theory is the main goal in any field of science. Having a theory we can give explanations of a variety of phenomena and make predictions. Making predictions means that we are able to compute what happens more precisely and more efficiently. A characteristic feature of theories is that they use more abstract concepts than those that we can observe immediately. Philosophers argue whether or not one should use concepts that do not correspond to things that we can observe. The *Occam's Razor*, also called *the law of parsimony*, tells us that we should avoid any use of concepts that are not inevitable for describing the situations that we study. Logical positivism was based on a similar axiom, the aim being to avoid meaningless 'metaphysical' considerations. In mathematics essentially all concepts are abstract, so these problems may seem irrelevant, but it is not true. What should be called '*abstract*' and what should not is difficult to decide and mathematicians do not care anyway. What is however undeniable is that there is a hierarchy of mathematical concepts. The words '*more abstract*' and '*higher order*' correspond to our feelings about the concepts higher in the hierarchy. Furthermore, mathematics, being the most precise of all fields of science, gives us the possibility to study the role of abstract concepts systematically. More than that, we can even prove that abstract concepts help in several ways. In fact, the field of logical foundations is all about it.

A Tough Nut for Computers

It's high time now to be less abstract and give some concrete examples. I will start with a very elementary example, which is a well-known problem from recre-

[17] This is not quite precise. One has to first generalize the theory and only then it is possible to add Einstein's Field Equations. The generalizations without the Field Equations are also interesting theories and can describe nontrivial phenomena.

ational mathematics, called *the Mutilated Chess-Board Problem*. Consider an ordinary chess-board with two opposite corners cut out, say `a1` and `h8`. The problem is whether or not it is possible to tile such a board with dominos. This is a finite problem, so, in principle, one can solve it by trying systematically all possibilities. But there are so many possibilities to partially tile the board until we get stuck that one cannot enumerate them all, even with the help of the fastest computer. The answer to this puzzle is no, it is not possible to tile completely the mutilated chess-board. Hence if we merely use the method of trial and error, we have to try all partial tilings in order to be sure that there is none. However, there is a really simple *proof* that it is not possible. Every domino covers one black and one white square. Hence a complete tiling is only possible if the number of black and white squares is the same, which is not the case.[18] The solution is simple enough so that most people find it very quickly. But now imagine that we were not so familiar with the chess-board, for example, suppose the question was put for the *go*-board whose squares are of the same color. Then it would be harder to guess this solution. The concept of coloring squares is not part of the problem. In order to solve the problem, we have to recall this concept and connect it with the problem. As I said, we have the advantage of knowing the concept of coloring, but in mathematics, quite often, one has to invent a completely new concept to solve a problem. Note that coloring is higher in the hierarchy of concepts than the concept of tiling. Tiling is defined by local conditions, while coloring concerns the whole board. But there is an even more abstract concept that is used in the solution. It is the concept of a natural number. That it is really more abstract than the concept of tiling can be demonstrated practically. You can teach very small children to tile; a little older ones will even understand that some areas can be tiled and some cannot, but understanding the concept of a number in such a way that it can be used to show the impossibility of some configuration, requires substantially more intelligence.

I do not know if anybody has tried this on children, but this problem is well-known in the area of computer science called *artificial intelligence*. The ultimate goal of artificial intelligence is to develop intelligent computer systems. But computers are still not able to do a lot of things that are simple for humans. In 1964 John McCarthy proposed the Mutilated Chess-Board Problem as an example of a statement that can easily be presented to computer theorem provers, but which will be a "*tough nut*" for them (see [195]). Let me stress that he did not mean that computers would not be able to do even such simple proofs. What he meant was that for the Resolution Calculus, the proof system most commonly used in automated theorem proving, the problem was hard because proofs in this proof system are very much like trying all possibilities. In other words, the proof system can handle efficiently only problems that can be solved by sort of local manipulations. Whether or not the Mutilated Chess-Board Problem is hard for the ordinary 8×8 chess-board is still open, but we do have evidence that it is probably hard. What we know for sure is

[18] For this argument one can also use the simpler version in which only one square is cut out. The reason to cut out two is simply to make the puzzle a little harder. With only one square away the idea of counting the parity of the number of squares comes to one's mind immediately.

that the generalized problem for boards $n \times n$, with n an even number of the order of thousands, any proof in this proof system is so large that it cannot be practically performed. For showing that the argument using coloring cannot be done in the proof system, this is enough because if one could use this argument, the proof would be still relatively short even for large boards.

Transcendental Numbers

Let us consider something more serious. Most examples of using abstract concepts for solving problems that are stated in elementary terms come from number theory. One of the popular problems in number theory is proving that a number is not a solution of an algebraic equation with integer coefficients. Numbers that are solutions of such equations are called *algebraic*; those that are not are called *transcendental*. For example, $\sqrt{2}$ is a solution of the equation

$$x^2 - 2 = 0,$$

so $\sqrt{2}$ is algebraic. On the other hand π is not a solution of any such equation, hence it is a transcendental number. Proving that a particular number is transcendental is usually hard. The first proof that a number is transcendental was given as late as in the 19th century. Later, when Cantor discovered set theory, he showed that the existence of such numbers can be proved very easily using set theoretical concepts. He proved that the cardinality of the set of all real numbers is not countable, whereas the cardinality of the set of algebraic numbers is countable. Therefore, there are transcendental numbers.

Notice the similarity with the previous problem. Again the main idea is counting. Such counting proofs are often very simple, but we have to pay for it: such proofs do not give us explicit examples of the objects claimed to exist. We will encounter proofs that prove the existence without giving explicit examples again later.

Diophantine Equations

There are many problems about natural numbers that can be stated only using the basic arithmetical operations. Problems of this type were studied by Diophantus of Alexandria, who lived in the 3rd century. The problems he solved can be presented in modern terms as follows. Given an equation with integer coefficients, find a solution that is also an integer (or several integers, if the equation contains more variables). A classical problem, solved already in antiquity, is to give all such solutions to the Pythagorean equation

$$x^2 + y^2 = z^2.$$

There are infinitely many such triples (3, 4, 5 is the smallest one) and they have a simple characterization. The proof is completely elementary. One may be tempted to conjecture that it should be always so: once a problem uses only elementary operations on numbers, the solution must be elementary in the same way. But there are other equations for which no elementary proofs exist. For example, the following simple equation has only a finite number of solutions (according to Thue's theorem), but no elementary proof of this fact is known

$$x^3 - 2y^3 = 11.$$

There is an abundance of such examples in number theory. Another classical problem is, for a given number d, to characterize prime numbers p for which the equation

$$x^2 + dy^2 = p$$

has a solution with x, y integers. Pierre de Fermat (1601–1665) found solutions of this problem for $d = 1, 2, 3$. For example,

$$x^2 + y^2 = p,$$

which is the above equation for $d = 1$, has a solution if and only if p has residue 1 when divided by 4 (thus $p = 5, 13, 17, 29 \ldots$). A number of great mathematicians contributed to this problem by finding solutions for more numbers d. A complete solution of this problem is known, but the proof uses a substantial part of modern number theory. For example, the set of primes for which $x^2 + 14y^2 = p$ has a solution has a fairly simple characterization, but no elementary proof of the correctness of this characterization is known.

One of the most famous mathematical problems is *Fermat's Last Theorem.* It is the following theorem:

Fermat's Last Theorem *For $n \geq 3$, the equation*

$$x^n + y^n = z^n$$

does not have a solution with x, y and z positive integers.

When a name is associated with a theorem, usually this means that the theorem was proved by the person of this name. This not the case here. The theorem was proved by Andrew Wiles in 1994. The history of this problem is well-known. Fermat wrote this statement in his copy of Diophantus's book and added that he had a wonderful proof of the statement, but the margin was too narrow for it. We have enough evidence that the note was either just intended to tease other mathematicians, what he often did, or he simply had a wrong proof. Our evidence that Fermat did not have a proof is based on the results obtained when trying to prove the theorem. For some numbers n, the problem is elementary; for example, the case $n = 4$ is treated in elementary books on number theory. For other values of n, only nonelementary proofs are known. The complete proof of Wiles is a masterpiece of mathematics. What is the most interesting is that it uses a major part of the deep theories developed in number theory before and more theory is introduced along with proving the theorem. It should be noted that contemporary number theory uses tools from a number of other fields—algebra, geometry, topology, analytic function theory, etc.

Thus Wiles's theorem is a great example of using abstract concepts. Almost none of the powerful results used today were available to Fermat; some fields, such as topology and algebraic geometry did not exist back then. Therefore, we believe he could not prove it.

This is true, but, frankly, we are not completely sure. More precisely, in all other fields the evidence would be accepted as a "clear proof", but in mathematics we have higher standards. When we claim something, we must have a *mathematical proof.* We have to admit that we do not have a proof that Fermat's Last theorem does not have an elementary proof. Logic has not developed suitable tools for this purpose yet. Fermat's problem is one of the most difficult ones and Wiles's proof is very deep (in the sense of using abstract mathematics). It seems that a proof that Fermat's Last theorem does not have an elementary proof could be of a comparable complexity. There are, however, some other cases, in which we can prove that abstract means, in a certain sense, are necessary; I will show some examples in Chap. 4. Thus I do believe that eventually logicians will be able to handle theorems of this type too. Then, perhaps, we will also have a proof that Fermat's Last Theorem was rather "*Fermat's Last Joke*".

The Reasons why Abstract Concepts Are Needed

When trying to find reasons for the use of abstract concepts the following three come to mind naturally.

1. An elementary solution exists but it is too long; it can happen that it is so long that it cannot be written down. Using more abstract concepts we get a shorter proof that is feasible.
2. There is no solution that only uses elementary concepts.
3. Any solution needs axioms that surpass the concepts present in the problem.

There is a subtle difference between 2. and 3. A solution that uses nonelementary concepts, meaning concepts that are not present in the statement of the problem, may still only use elementary axioms. The second reason seems to be the most frequently occurring one, but in fact it can never occur. There are plenty of nonelementary proofs that do not use nonelementary axioms. But according to an important result in logic, we can, so to say, always apply Occam's Razor to such proofs. Roughly speaking, this means that we can eliminate all statements that are not mere modifications of the theorem that we want to prove, or modifications of an axiom that we want to use in the proof. This is may look surprising because it implies that lemmas can also be eliminated from proofs. A lemma is an auxiliary theorem that can be formally unrelated to the theorem that we want to prove. In particular, it may refer to concepts that do not occur in the theorem. Unless a proof is completely straightforward, it contains lemmas. According to that result in logic, we can eliminate them and thus the only reason for using them is our concern for brevity. However elimination of lemmas may result in a tremendous increase of size. So it may be impossible to write down such a proof. This is why we have to use lemmas in practice.

I will talk about this result later (in Chap. 6); for the time being let us just remember that all cases that look like the second case are in fact the first case (or there is an elementary and short proof that we do not know of).

The third case also needs a caveat. We cannot claim that a proposition ϕ is only provable using axioms that contain some concepts not present in ϕ. This is surely not true because we can simply take ϕ as an axiom and then ϕ follows trivially. The meaning of 3. is that there is no *natural* axiom system that uses the same concepts and that is strong enough for ϕ. The question then is what '*natural*' means. Best is to take concrete theories used in practice and find the weakest one in which ϕ is provable. Such a scale can be constructed as follows. Take some basic axioms of set theory, but *without* the axiom of infinity. This will be the bottom element. The next will be the set theory, in which we add the axiom of infinity. Then we gradually add stronger and stronger set-theoretical axioms. One can show that, for example, there are finite statements (meaning statements about finite sets) that are not provable in this 'finite set theory'. One can show statements that have a similar relation to higher levels. We will see such examples in Chap. 4.

Of course, this does not solve the classification problem completely. Instead of saying '*natural axioms*', I said '*basic axioms*', so one can still object that, although we use some standard axiom systems, there is no unambiguous justification of these systems. However, the scale becomes much less ambiguous if we go higher in the hierarchy. There the axioms correspond to infinite cardinal numbers, so, in particular, they are linearly ordered.

I will discuss these fundamental questions in more detail later, after I explain the necessary concepts.

Logical Classification of Concepts

In logic, there are tools to measure the complexity of a concept. The mere length of the definition does not necessarily mean that the concept is abstract, but it does give an approximation of this property. Using a particular measure of the complexity of the definitions, we get a corresponding hierarchy of concepts. We can think of the concepts higher in a hierarchy as being more abstract. A definition is formally a logical formula, hence we need measures of *the complexity of formulas*.

The most basic classification is according to the *order* of the language used. I have already mentioned order of structures. To each order we have a corresponding language. Thus first-order concepts are defined using formulas that only talk about elements, second order concepts are those that we can define only using subsets or functions, third order concepts require subsets of subsets, etc.

Another basic classification is according to the number of quantifiers. We can count the total number of quantifiers, but a more important hierarchy is based on the number of *alternations* between the universal quantifier and the existential one. I will explain it in the next chapter.

We also need to measure the strength of the assumptions. As noted above, to prove some theorems we may need strong assumptions in spite of the fact that the statement of the theorem uses only elementary concepts. Often it is possible to get a hierarchy of theories from a hierarchy of formulas.

Let us consider Peano Arithmetic as an example (see page 116). In this theory the elements are numbers and the only relations and functions are the ordering and the operations of addition and multiplication. So the formulas express facts about numbers using only ordering, addition and multiplication. This language is usually called *the language of arithmetic* and, correspondingly, the formulas are *arithmetical formulas*. The theory has a few very basic axioms and an infinite set of axioms of induction, for every formula in the language one axiom expressing induction for this formula. In spite of the fact that we have the principle of induction for every formula, Peano Arithmetic is incomplete; there are true arithmetical sentences that are not provable in the theory. The reason is not that the principle of induction does not suffice to characterize the natural numbers, but that in Peano Arithmetic this principle is stated *only* for arithmetical formulas. Some true sentences need higher order induction. Therefore a theory called *Second Order Arithmetic* has been defined, in which one can also talk about subsets of numbers and in which induction is stated for formulas in the enriched language (see page 295). This theory is stronger than Peano Arithmetic, but it is still incomplete. So we can define *Third Order Arithmetic*, in which one can talk about sets of subsets of numbers, and so on.

As an example of a hierarchy of theories based on quantifier complexity, take, for every number n, Peano Arithmetic with the induction axioms only for formulas with at most n quantifier alternations. This gives a sequence of theories below Peano Arithmetic the strength of which increases with n.

Notes

1. *Resolution calculus.* The calculi used in logic do not seem very suitable for automated theorem proving. Therefore a very compact system was developed in the 1960s, with the main contribution by J.A. Robinson. It is called *Resolution.* This proof system is used for proving first-order sentences (sentences with quantifiers). The propositional part (also called Resolution) has been widely studied since it is the simplest proof system for propositional logic. Rather than defining it formally, I will explain its application to the Mutilated Chess-Board Problem. Our atomic propositions will be statements saying that a domino is placed on particular two adjacent squares. The proposition that a domino is put on squares, say, `b2`, `c2` will be denoted by `[b2,c2]`. A disjunction of atomic and negated atomic propositions is called a *clause.* We define a set of clauses that express that the board is completely tiled. The set consists of two parts. The first one expresses that every square is covered. It contains a clause for every square saying that the square is covered. Say, for the square `b2` the clause is

   `[a2,b2]` *or* `[b1,b2]` *or* `[b2,c2]` *or* `[b2,b3]`

   because these are all four possible ways in which it can be covered. The second set consists of clauses saying that no two dominos overlap, more precisely, no square is covered twice. We have several clauses for every square, depending on its position on the mutilated board. For example, for b2 the clauses are

not [a2,b2] *or not* [b1,b2],
not [b1,b2] *or not* [b2,c2],
not [b2,c2] *or not* [b2,b3],
not [a2,b2] *or not* [b2,b3],
not [a2,b2] *or not* [b2,c2],
not [b1,b2] *or not* [b2,b3].

Our aim is to disprove the assumption that there is a tiling, therefore we want to derive a contradiction from the clauses. A contradiction is a pair of clauses one being a proposition and the other its negation (for example, [a2,b2], *not* [a2,b2]). The advantage of Resolution is that it has only one simple rule. This rule allows us to combine two clauses in the following way. The condition for applying the rule is that one clause contains a proposition (not negated) and the other contains the negation of this proposition. Then we can remove the two complementary propositions and unite the clauses into one. For example, from

[a2,b2] *or* [b1,b2] *or* [b2,c2] *or* [b2,b3],
not [b2,b3] *or not* [b3,c3]

we can derive

[a2,b2] *or* [b1,b2] *or* [b2,c2] *or not* [b3,c3]

Furthermore, we always replace multiple occurrences of a proposition or a negation of a proposition by a single one. (Hence the result of an application of the rule may be smaller than both clauses.)

It was this proof system, Resolution, for which the bound on the lengths of proofs was proved. A theorem of M. Alekhnovich says that the length of the proofs increases exponentially with the size of the board (see [3]). Above I said that every proof in Resolution is like considering possible cases. To see this relation we have to look at the proof from the bottom up. So when we are at some clause in the proof we can go up to the two predecessors, the two clauses from which it was derived (provided that it is not an initial clause). Recall that one clause contains a proposition and the other contains the negation of the proposition. Thus this splitting corresponds to considering two cases; in one the proposition is true and in the other one it is false.

2. *Analytic number theory.* In analytic number theory analytic functions are applied to study problems about the natural numbers. A classical problem in number theory is the distribution of prime numbers. A particular case of this problem asks to estimate the function $\pi(x)$, defined as the number of prime numbers less than or equal to x. The well-known *Prime Number Theorem* (proved by Jacques Hadamard and Charles de la Vallée-Poussin in 1896) says that $\pi(x)$ is asymptotically $x/\ln x$. It was first proved using analytic means, but in 1949 P. Erdős and A. Selberg found an elementary proof [68, 263]. (Here I am using the word 'elementary' to say that the proof did not use analytic functions; it

does not mean that the proof was easy.) Elementary proofs have been found for several other theorems that had been proved by analytic means. However, analytic functions seem to be a very strong means that is not possible to replace by an elementary approach in general.

Probably the most famous problem in mathematics, *the Riemann Hypothesis*, is a problem from analytic number theory, stated by Bernhard Riemann in 1859. The ζ-function is defined by

$$\zeta(x) = \sum_{n=1}^{\infty} \frac{1}{n^x},$$

for every complex x such that its real part $\mathrm{Re}(x) > 1$. The function can be analytically extended to the whole complex plane without $x = 1$, where ζ has a pole. The ζ-function has zeros $\zeta(-2) = \zeta(-4) = \zeta(-6) = \cdots = 0$, called *trivial zeros*, and infinitely many zeros in the strip $0 < \mathrm{Re}(x) < 1$, called *non-trivial zeros*. The Riemann Hypothesis is the following statement:

The Riemann Hypothesis *All nontrivial zeros of ζ are on the line* $\mathrm{Re}(x) = \frac{1}{2}$.

So even to state the hypothesis we have to refer to some results on analytic functions. However the consequences of the hypothesis that we are interested in are elementary statements. The Riemann Hypothesis implies very good estimates on the distribution of primes. The ζ-function was defined by Leonhard Euler (1707–1783), who also discovered the formula

$$\sum_{n=1}^{\infty} \frac{1}{n^x} = \prod_{p} \frac{1}{1 - \frac{1}{p^x}},$$

where p ranges over prime numbers. This shows a relation of ζ function to primes. Some estimates can be derived using this formula; for more precise estimates, the following formula is used

$$-\frac{\zeta'(x)}{\zeta(x)} = \sum_{n=1}^{\infty} \frac{\Lambda(n)}{n^x},$$

where $\Lambda(n) = \ln p$, if n is a power of a prime p, and $\Lambda(n) = 0$ otherwise. In fact, the Riemann Hypothesis is equivalent to a bound on the number of primes less than a given number n:

Equivalent statement 1 *There exists a constant C such that, for every $n \geq 3$,*

$$\left| \int_3^n \frac{dx}{\ln x} - \pi(n) \right| \leq C\sqrt{n} \ln n.$$

Thus, though originally stated using analytic functions, there is an elementary statement about the natural numbers that is equivalent to the Riemann Hypothesis.

Another equivalent statement can be stated using the *Möbius function* μ. This function classifies natural numbers into three classes:

a. $\mu(n) = 0$, if n is divisible by a square of a prime;
b. $\mu(n) = -1$, if n is a product of an *odd* number of distinct primes;
c. $\mu(n) = 1$, if n is a product of an *even* number of distinct primes.

If n is not divisible by a square of a prime, we say that it is *square-free*. The number of square-free numbers in the initial segment $[1,n]$ is asymptotically $\frac{6}{\pi^2}n \approx 0.608n$. The crucial quantity is the difference between the numbers less than or equal to n that are products of an even number of primes and those that are products of an odd number of primes. This can be conveniently expressed by $\sum_{i=1}^{n} \mu(i)$.

Equivalent statement 2 *For every $\varepsilon > 0$, there exists n_0 such that for all $n > n_0$,* $|\sum_{i=1}^{n} \mu(i)| \leq n^{1/2+\varepsilon}$.

Note that for a random sequence r_i of ± 1 s of length n, the standard deviation of $\sum_{i=1}^{n} r_i$ is $\sqrt{n}$. It has been observed that for an infinite series of random ± 1 s, the condition of the equivalent statement is satisfied with probability 1 for every $\varepsilon > 0$. Thus the Riemann Hypothesis says that the Möbius function on square-free numbers shares some properties with random sequences of ± 1 s, although it does have some regularities that do not occur in truly random sequences; for example, $\mu(2x)$ is always either $-\mu(x)$ or 0.

The Riemann Hypothesis problem has fascinated generations of mathematicians, therefore it is unlikely that it has an elementary proof. The evidence so far is that it should be true, but if it is not so, it is at least a good project—better bounds on how close the nontrivial zeros are to the $\mathrm{Re}(x) = \frac{1}{2}$ line give better estimates on the distribution of the primes. Already the proofs of the Prime Number Theorem of Hadamard and de la Vallée-Poussin were based on proving that no zero has $\mathrm{Re}(x) = 1$.

However strong analytical number theory seems to be, it is not clear that from the point of view of our current logical classification it transcends "finite means". Complex functions are higher order objects, but this does not automatically mean that we cannot simulate them by finite objects. Often, it is possible to replace infinite objects by their names, which are finite objects. Very little research has been done on showing formally that analytic means are stronger.

3. *Experimental testing of the Riemann Hypothesis.* Computer experiments with the Riemann Hypothesis are also interesting from the point of view of foundations. There are several ways how to test the Riemann Hypothesis. Nontrivial zeros of ζ are usually enumerated in the order of their absolute values. Approximations of zeros can be computed quite efficiently and, moreover, the computation also determines if the zero is on the line $\mathrm{Re}(x) = 1/2$ or not. The first 15 zeros were calculated in 1903. At the time of writing these lines almost 60,000,000,000 have been known, all being on the line $\mathrm{Re}(x) = 1/2$. (Check the Internet for the current record.) Let me stress that in this way one can only refute the conjecture. No matter for how many zeros it is verified, it still may be false.

The results of testing the equivalent statement 2. are a good warning to all who may be tempted to make deductions about the truth of a hypothesis from such experiments. For $n > 200$, the value of $|\sum_{i=1}^{n} \mu(i)|$ keeps being less than $\frac{1}{2}\sqrt{n}$ for a very long time, often being very close to it. Thus one is led to the conjecture that it is always so. But then suddenly, for $n = 7,725,038,629$, it exceeds $\frac{1}{2}\sqrt{n}$. This computation, however, did not disprove the *Mertens Hypothesis*' that for all n, $|\sum_{i=1}^{n} \mu(i)| \leq \sqrt{n}$. The Mertens Hypothesis (conjectured by Stieltjes in 1885) is slightly weaker than the inequality stated above because $1/2$ is replaced by 1, but it is still stronger than the Riemann Hypothesis. In 1985 A.M. Odlyzko and H.J.J. te Riele disproved the Mertens Hypothesis [210], but not by computations. Later, upper bounds on the first counterexample n were computed; the present best bound is $n < \mathrm{e}^{1.59 \cdot 10^{40}}$. Thus the conjecture is false, in spite of all empirical evidence that we have so far. It may well happen that we will never find any concrete number that violates the Mertens Hypothesis.

Let me stress, however, that much more evidence for believing that the Riemann Hypothesis is true comes from theoretical results. In particular, deep results about finite fields have been proved that can be interpreted as a version of the hypothesis.

Main Points of the Chapter

- A structure is given by a set of elements (the universe) and relations and operations (defined on the universe). There are infinitely many types of structures.
- Mathematicians study some standard structures, such as the natural numbers and the real numbers, and various classes of structures, such as groups. The basic structures were introduced a long time ago; others were defined more recently.
- Two basic principles for sets are extensionality and comprehension.
- Infinitely many sets can be constructed starting with a single set, the empty set.
- Relations and operations can be defined as sets of pairs, triples, etc. Therefore, we only need sets to formalize mathematical structures.
- Russell's paradox destroys the hope of having a consistent set theory based solely on our intuition. We have to use axiomatic set theory.
- In mathematics, theories are defined by axioms. To derive theorems in an elementary theory we only need logic. However, in order to be able to state and prove interesting results, logic alone does not suffice; we have to use set theory.
- There are several reasons for using the axiomatic method: 1. it is precise, 2. it is fair because we state the assumptions explicitly, 3. it is useful because we can test whether the theory can be applied to a particular phenomenon, 4. it helps us to explain the studied concepts because a short list of basic axioms explains the essence better than a long complicated description.
- There are mathematical problems that can be stated in a completely elementary way, but cannot be solved without applying very abstract concepts.
- Mathematical logic has the means to measure the degree of abstractness of concepts and to prove that such concepts are indispensable. Until now, however, we have succeeded in proving the necessity of using abstract concepts only in a few, rather simple instances.

Chapter 2
Language, Logic and Computations

Realizing that the aliens would not understand us, we were continuously sending out the Pythagorean Theorem and other simple geometrical propositions. But our call into space remained without an answer.

Stanisław Lem, *Magellan's Cloud*

It seems difficult to define mathematics. A possible definition, or rather an explanation of mathematics, could be that mathematics is an extension of our language that enables us to perform rigorous deductions. The problem with this explanation is that it does not take into account the theorems and proofs that are the main products of this field. Nevertheless, the role of mathematics as a means of expressing our ideas precisely is unquestionable. It is also possible to observe the growing number of mathematical terms in common language. Of course, many of these terms were present before the advent of mathematics and they just obtained a more precise meaning in the course of the development of mathematics. The first thing that comes to ones mind are numerals. The etymology of numerals 11 and 12 in many languages witnesses that the decimal system had been accepted at their early stage. (The most likely etymology, say, of the English word *eleven* is '*one left over*'.) There are many geometrical concepts commonly used, for example, *triangle, square, line, curve, cylinder*. There are also many more modern concepts, like *function, minimum, set, probability*. This is not the privilege of mathematics; scientific terms from all fields penetrate common language. Mathematics is special in that it is applied in other sciences, but not conversely, other sciences can provide problems and motivation for mathematics, but cannot be used there.[1] Because of its universal nature, mathematics has been proposed as a communication means with extraterrestrial civilizations.

A language is not merely a collection of words. The words must have some *meaning*. A very primitive language, perhaps a language of some species of animals, may consist of words and their meaning, without any complicated constructions. Human languages allow combinations of words to talk about arbitrarily (arbitrarily at least in principle) complicated things. This requires some rules, rules that say how we

[1] There is one exception: computer science is used in experimental mathematics.

P. Pudlák, *Logical Foundations of Mathematics and Computational Complexity*, Springer Monographs in Mathematics, DOI 10.1007/978-3-319-00119-7_2,

can combine words into sentences—*the syntax*, and rules by which we deduce the meaning of sentences—*the semantics*. Furthermore, there is a special, and very basic, type of meaning of a sentence, its *truth*. This brings us to logic, the art of putting pieces of truth together in order to get new evidence.

'*Mathematics as the foundations of logical reasoning*' sounds good, but we are looking for the foundations *of mathematics* in this book. So let's try to invert it: *logic as the foundations of mathematics*. There is no doubt that logic is essential for mathematics. But do we know precisely what logic is? Or is the logic that we are using the right one? We definitely must learn more about logic to be able to answer such questions. Anticipating what we should learn in this chapter, let me say that it turns out that the logic that we use in mathematics is a very clear and unambiguous concept. There is no problem with logic as far as the foundations are concerned. It does not mean that it is a simple thing, on the contrary, there are computationally unsolvable problems in logic and a lot of problems for mathematical research. A different question is if logic suffices for the foundations. At the turn of the 20th century, several philosophers and mathematicians tried to find the foundations of mathematics based only on logic. This stream in the philosophy of mathematics is called *logicism*. The incompleteness theorems of Gödel proved that this goal (or at least its original version) cannot be achieved. We have to introduce specific axioms about sets in order to be able to develop mathematics. We will see how much we have to add to logic in Chap. 3 that deals with set theory.

Having a formal language and formal rules for logic one can do certain operations mechanically. '*Mechanically*' is an obsolete term; the 20th century's version should be '*electronically*', but we know what it means: one can find an algorithm, or put differently, write a program for that task. Such a task is, in particular, *proof checking*. To check the correctness of a formal proof does not require any intellectual ingenuity, it can be done by a machine. This fact has practical applications, but we are not interested in them. For us, it documents that the concept of a proof is so clear that one can check mechanically whether or not a given text is a correct proof. There are, however, tasks that are not algorithmically solvable, in particular the problem of finding a proof of a given sentence cannot be mechanized. Thus we have arrived at the concept of a computation, another concept that is very relevant for foundations. And again, the same questions: what is it, can it be precisely defined, etc? We'll see.

2.1 The Language of Mathematics

Now we will leave aside the use of mathematics as a language, and talk about the language used in mathematics. Let us look at a typical mathematical text. Nowadays it is written most often in English, but this is the least important property, as it can be translated into any language. The first most striking feature of a mathematical text is its very restricted vocabulary. Potentially it may contain any words and often words that are used in everyday life for things and properties that have nothing to do with mathematics, (for example, *ring, field, group, good, dense*) which, however, have a completely different meaning in mathematics. The reason for that is that sometimes people use analogies from real life to explain mathematical ideas. If mathematicians wrote papers for computers, whose life is restricted to solving given problems, the vocabulary would be even smaller. The grammar of mathematical texts is also only rudimentary. For instance, mathematics uses verbs only in a very restricted way. Surely, the verb *to be*, is used, but it only connects a noun with an adjective, or it is used to express existence. There are several other verbs that you can find in mathematical texts, for example, *suppose, assume, follow, imply*. Those can be avoided using a different sentence construction; we can replace them by connectives (namely *if... then*). In real life time is very important and thus a lot of the grammatical structures concern time. Most verbs express changes in time. Verbs come with various tenses expressing the relation to the present. Mathematics, on the other hand, expresses facts independent of time. Even if the problem concerns time we consider it as an entity; for example, instead of motion mathematicians consider its trajectory. Time, in the form we perceive it, is not present even in physics. It is just the fourth dimension. Mathematicians are like historians: they do not have to think what will happen, they do not have to perform experiments, they have the complete history of the phenomenon at hand. Like historians only need the past tense, mathematicians only need the present.[2]

So what remains of the natural language in mathematics? Here is what is used:

- nouns (*number, point, set, ...*);
- adjectives (*straight, continuous, large, ...*);
- prepositions and verbs expressing relations (a point *on* a line, a function *vanishes* at zero, ...);
- connectives (*and, or, not, ...*);
- two quantifiers '*there exists*' and '*for all*'.

Nouns are necessary because we should be able to speak about things. Things in mathematics are elements of structures, structures and sets. Typically we study a particular mathematical structure and then the things that we consider are elements of the structure. But also when speaking about structures and sets, we can always assume that they are elements of a larger structure, say, the structure of all sets. Hence nouns are elements.

Adjectives express properties. A property is something that concerns one element. Then we have relations, which, in the simplest case, concern pairs of elements and are called *binary relations*. Interestingly, there is no single word class that corresponds to relations. In most cases relations are expressed by prepositions or verbs, but they can also be expressed by certain combinations of words. From the point of view of logic, these are linguistic nuances and we put all relations into one class. Moreover, logicians also include properties in the same class and call them '*unary relations*'. Another word that is sometimes used for properties is '*predicates*'. This

[2]There are languages in which grammatical structure for expressing time is only rudimentary (for example, Chinese), other languages may lack other structures. The language of mathematics is certainly the poorest of all.

is, however, confusing because grammatically 'predicate' refers to a particular part of a sentence which, from the point of view of logic, may describe a binary relation like the following sentence:

Point P lies on line ℓ.

According to grammar '*lies on line ℓ*' is a predicate, according to logic the sentence expresses that a certain binary relation holds between a point and a line. To see the reason for thinking about the sentence as expressing a relation, notice that one can say the same as follows:

Line ℓ goes through point P.

The last two items on the list above concern logic. There are various ways of expressing the same connectives and the same quantifiers. Furthermore, the interpretation of connectives in a natural language is often a little different from the one of formal logical. The connective *or* is usually interpreted as the *exclusive* or, hence in law one often uses '*and/or*' to express the non-exclusive or. In logic *or* is the non-exclusive, which means that '*A or B*' is true also when both *A* and *B* are true; but the *exclusive or*, a different connective, is sometimes used too. In some languages and in slang, double negation is used just to stress simple negation, while in logic two negations cancel each other. All the connectives used in a natural language reduce to the following four *not, and, or, if... then.* The two quantifiers can also be expressed in many ways. For the universal quantifier, there are essentially the same expressions in all languages, which are words such as *all, every, any, each.* These words differ slightly in their use and meaning, but in logic, again, they are considered to be synonyms. For the existential quantifier, some languages have special constructions such as '*there is*', '*il y a*', '*es gibt*', but the most common way is simply to use the verb *to be*.

Natural languages also use *modalities*, which are words expressing how much we believe in what we are saying. Such words include *surely, necessarily, maybe, probably*. Modalities can be considered as unary connectives, similarly as we treat negation as a unary connective. Their meaning is not precise, according to the mathematical standards, therefore they are not used in mathematics.[3] But it does not mean that mathematics is completely deprived of the possibility of expressing such things. Consider the modalities *probably, likely, unlikely*, for example. Mathematics has developed a whole theory in order to be able to express such statements, the probability theory. Mathematicians, of course, do not use these vague words, instead they quantify numerically the probability of an event. Sometimes this is accepted also in our everyday life. For instance, in the USA they often forecast that the probability of rain is so many percent, instead of just saying that it is likely, unlikely, as they do elsewhere.

Similar is the role of the various quantifiers that are present in natural languages such as *a few, some, many, almost all.* In fact, it is one of the main goals of mathematics to replace these imprecise expressions by precise statements. This is achieved

[3]Modal logics are not used to state and prove mathematical results, but they have interesting applications, see provability logic on page 297.

by *counting*, which means using numbers to say precisely, or to estimate the quantity in question. Thus what remains from the language are the precise quantifiers *for all* and *there exists*.

It is important to realize that the possibility of replacing modalities and imprecise quantifiers by precise statements is the main reason why the former ones are not used in mathematics. Once in a while suggestions are made to enrich mathematics by modalities, other quantifiers, etc. These are futile proposals; this will never catch up because it is against the spirit of mathematics!

The aforementioned list is the result of a superficial analysis based on grammatical categories, thus one key item is missing on the list: *variables*. In the traditional classification there is no word class corresponding to variables, although it is one of the most important concepts in mathematics. Still, variables are present in the everyday use of language. Suppose for instance that you want to say that there are two elements without referring to the numeral 2. You can say: '*There is an element and there is another one different from the previous one*.' With three elements we would use something like: '*...and yet another one...*'. With four elements it will become rather messy. Here we have considered only the task of expressing the existence of a certain number of elements, but we have to handle the same problem in other situations where we need to talk about several elements. In a natural language we have very limited means for that; we have words like '*another, yet another, this, that*' which are good for two elements, and may be used for three, but for more they are not practical. What one uses then are descriptions like '*the tall man, the blond girl, the one who came first*'. However, the most efficient way is to use names. The most ancient mathematical texts used the awkward way of distinguishing elements of natural language. The simple trick of assigning letters as temporary names to the investigated entities, used already in antiquity, must have had dramatic consequences for the development of mathematics. It enabled mathematicians to treat much more complex problems than before.

Nowadays we call the temporary names *variables* and the permanent ones *constants*. Variables are most often x, y, z; examples of constants are π and e.[4] The number of variables that may be needed in an average length paper is surprisingly large. Therefore, people use various alphabets, indices, primes, and other marks.[5]

There is one more class, *functions*, which I will explain later.

You see that linguists study languages from a different perspective; grammatical categories do not render the substance of the language of mathematics. The correct classification for the language of mathematics is the following:

- constants and variables;
- functions;
- relations;

---

[4]The term *variable* is often used with another meaning, namely as a function; for example, a *random variable* is a function defined on a probability space.

[5]In one of my papers I used all lower case Latin letters and on top of that several Greek letters and some upper case Latin letters. This is not unusual.

- connectives;
- quantifiers.

This is the classification of mathematical logic, the field of science that studies the language of mathematics.

Why Is the Language of Mathematics so Restricted?

I spoke about *what* the language of mathematics looks like. Now the difficult question is '*Why?*'. Why do we use this particular fragment of our natural language? Why don't we use more? Does it have to be a fragment of a natural language? These are very difficult questions, yet we can give at least some partial answers.

Consider, for example, the modalities '*maybe, probably, surely, necessarily*'. These words sound alien to mathematicians, since in mathematics statements are either true or false, there is no third possibility, *tertium non datur*. Thus the question can be rephrased as: '*why only two truth values?*'. The answer is simple: *because two suffice and one is not enough.*

It is obvious that we have to use at least two truth values. The reason for using only two is not so much economy; the true reason is that we want to analyze the concepts as much as possible in order to understand their essence. Whenever it is possible to decompose a phenomenon into simpler components we do so and it always helps us. Thus physicists decompose molecules into atoms, in order to find laws about molecules, then they decompose atoms into elementary particles, in order to understand atoms better, etc. Concerning truth values, we cannot say that three truth values can be 'decomposed' into two, but we can describe all arguments of non-classical logics using just the classical one, the logic with two truth values.

Before going on I have to make a terminological digression. When new trends in the foundation of mathematics first appeared, such as intuitionistic mathematics, a term was needed to denote the standard approach. Thus we use phrases '*classical mathematics*' and '*classical logic*'. This is not very fortunate, since meanwhile a lot of modern mathematical fields appeared and one would hesitate to call such mathematics classical. In logic it is, however, very common to use this word to distinguish the standard approach from various alternatives, especially when talking about alternative logical calculi. Therefore I will stick to this word, (but remember, 'classical mathematics' includes almost all modern fields).

There are various ways we can translate uncertainty, fuzziness etc. into classical logic. For instance, for uncertain statements we can consider all possible cases where the uncertainty is replaced by either truth or falsehood. In the case of *fuzzy sets*, sets where elements are contained in the set with various values between 0 and 1, we can think of them simply as functions in the classical set theory.

The two values are not a dogma that is imposed on each new generation of mathematicians, it is simply the most convenient way of thinking. Another reason for classical two valued logic is its uniqueness among minimal possible logics. The connectives are not simply *some* connectives that can be used in two valued logic. By taking combinations, the connectives of classical logic generate all possible Boolean functions.[6] It would be much harder to agree on a non-classical logic, as there are many and there is no natural choice among them. The usual argument for someone who proposes a non-classical logic is that their logic contains the classical one and on top of that has other means of expressing statements and making deductions. But when they are describing the "new" logic, they use classical logic. So using classical logic the new system can always be described as another mathematical structure.

I used the truth values only as an example of a general principle. What mathematicians actually do is more general: they look for the simplest universal system that suffices for a given purpose. In the problem considered here, it is the language of mathematics. I argued that two truth values are the right choice for truth values. There are more thing to decide, for example, quantifiers. We need them in order to be able to talk about unspecified entities, to be able to make generalizations and eventually to be able to talk about infinity. One quantifier suffices, say, the quantifier '*for all*'. The existential quantifier '*there exists*' is dual to '*for all*', and it is definable from it. Hence, once we take one of them we actually accept also the other one. But we do not need more! If we want to quantify in another way, we can use mathematical structures such as probability spaces, numbers, etc.

Such a reductionistic trend in science can be documented by many examples. The axiomatic method is a good example: a theory, for example, a physical theory, is considered better if based on smaller number and simpler equations. We think it is better, not because we save space and time for its presentation, but because it *explains better* the studied phenomenon, because it is *more likely to be true*, or simply because it *looks nicer*. We prefer to explain a certain phenomenon by referring to a single principle, rather then to several.

Is Logic Simply a Part of Natural Language?

In order to explain the language of logic, I presented it as a fragment of a natural language, but I do not claim that natural language is primary and logic is derived from it. A language is necessary for logic, but, in a sense, logic is absolute. There are many natural languages and there are huge differences between languages that are not related, such as the difference between English (or other Indo-European languages) and Chinese (dialects). But there is no difference between the logics used by the people using different languages—their logic is the same.

When I say that language is necessary for logic I mean it in a very broad sense. The language does not have to be a human language. There are several species of very intelligent animals that only use a few sounds for communication between themselves. So their 'vocabulary' is very small and does not contain essentially any logic. But they are able to perform deductions. The language that they are using

---

[6]We use conjunction, disjunction, implication and negation, but it suffices to only take one of the three binary plus negation.

for this purpose is an internal language of each individual, the language of *concepts* that they are able to form in their brains. I am sure animals think very much like we do, namely, they imagine possible situations, in particular situations that may occur after they perform a particular action, and choose the one that is the most favorable for them. Intelligence is the ability to see similarities between situations and thus be able to better estimate the consequences. Logic is a means of organizing this process.

The Language of Mathematical Logic

Mathematical logic is the field that deals with the language used in mathematics, the proofs and the truth of mathematical statements. Therefore, we will call a formal language for mathematics a *logical language*. Later I will also talk on deduction, so we will also have some rules to derive true sentences. Such systems are called *logical calculi*.

People often associate mathematical logic with symbols (and often it is the reason why they do not like it). The role of symbols in logic is emphasized more than needed. At early stages the name *symbolic logic* was used to distinguish mathematical logic from its sister branch in philosophy. But symbols are not any more important in logic than in any other branch of mathematics. Symbols enable us to express things more compactly, more precisely and sometimes they are useful for calculations. But in spite of this and in spite of the tendency of mathematicians towards brevity and precision, logic symbols are almost never used in mathematical writings with the exceptions of mathematical logic itself and set theory.

The same can be said about the syntactical rules, the rules describing how formulas are constructed. One of the first things that you encounter in a textbook on logic is a lengthy and boring description of the syntax. But unless you are going to study a particular formal system, you do not need it. The syntax of a formal logical language is only a simplified syntax of a natural language. Nevertheless, there is a good reason for presenting it so formally in textbooks. What is needed is to show that a formal language can be presented precisely and thus studied as a mathematical concept. A natural language is a rather vague and complicated thing, moreover, it is changing in time; it may be anything except a precise mathematical concept. Mathematicians, who use language as a tool not as a subject of their study, do not mind that the language is not precisely defined. But if you want to state and prove a theorem concerning language, you need something precise. Therefore, you need to do the boring job of explicitly writing down all the syntactical rules.

However, it depends on the person studying logic; if somebody is willing to accept that a formal concept of a language can be constructed, they do not need to consider an example of such a formalization. Once the primitives are described, it is not necessary to describe how a sentence is formed from these primitives as it is essentially the same as in natural languages. The actual syntax may vary for different logical languages, but this is not important. There are a few things that one has to secure; the main thing is to structure sentences in such a way that they allow only unique reading, which is usually done using parentheses. In natural languages we use pauses in speech instead of parentheses, and punctuation in writing; also there are some words for separation. But it does not work so perfectly, there are sentences that can be interpreted ambiguously, (which is a problem well-known among lawyers).

Actually, a large part of the contemporary population has a very intimate experience with formal languages. Those are the people who know a programming language. For such computer minded readers, I do not have to describe a logical language at all. Logical languages have been prototypes for programming languages, the difference being only what the texts in the languages expresses: statements in logic, algorithms in a programming language.

In order to complete the picture of the logical analysis of the language of mathematics, we need some more detail and, after all, I also have to mention at least the basic symbols used in mathematical logic.

The description in the last section was simplified, as we did not consider an important type of concepts, which are functions. Functions are present in natural languages too, for example, '*the mother of*' is the function that assigns to a person his or her mother. Though functions can be considered to be only a special kind of relations, it is more practical to use them as primitives. Functions are the main subject of studies in many branches of mathematics. The corresponding syntactical concept in logic, the names of functions, are called *function symbols*.

Having function symbols we get a new type of syntactical concept the *terms*. A term is an expression formed from function symbols, variables, and constants. What children use already at elementary school as *algebraic expressions* are terms that use function symbols for the basic arithmetical operations. Terms are, of course, used in other structures too. The interpretation of a term is a function if the term contains variables, or an element of a structure if it only contains constants. For example, $1+1$ is one of the names for the number 2, while $2x+1$ denotes a linear function. For binary functions, one often uses *infix* notation and calls them operations; this concerns mainly $+$ and $\times$ and other group operations. Again, for a programmer the concept of a term is very familiar, as this part of logic completely penetrated programming languages. No wonder, functions are here to be computed.

Now, having terms and a relation symbol, we can form an *atomic* formula, which is a formula that contains no logical symbols, namely, no connectives and no quantifiers. For example, take terms $x+2$ and $x \cdot y$ and the binary relation symbol $<$ and form an atomic formula $x+2 < x \cdot y$.

In algebra we often need only one relation, the relation of equality. Equality is often treated as a logical primitive, but again this is only a matter of convenience. It is a relation that occurs very frequently, so it is more practical to assume that it is present in logic. Furthermore, it is interesting to study just equations (expressions of the form $s = t$, with s, t terms). Already this fragment of logic is sufficiently complex.

The next step is to combine atomic formulas using connectives and quantifiers. Taking the above atomic formula and, say, $x = z^2$, we can form, for example, the conjunction of the two $x+2 < x \cdot y \wedge x = z^2$.

The role of quantifiers is very important, so let's consider some examples of their use.

All people are good.

This is a sentence from life, so the variable is not denoted by a letter. The variable there is hidden in *people*, which at the same time specifies the range of the variable. This sentence has the same structure as:

Every number is prime.

This sentence is false, but this is a positive fact: we are able to decide its truth value. Now take:

A number is prime.

Here we cannot decide, if it is true or false. The variable *number* is not quantified, it is *free*. There are two more possibilities to make a sentence with a definite truth value. One is to use the existential quantifier:

There is a prime number.

The other is to substitute a concrete element for the variable, say:

5 *is prime.*

Bertrand Russell explained the meaning of formulas with free variables as being *propositional functions*. As when we write $\sin x$, we denote a function that can have any value between -1 and 1, so the sentence *x is prime* can have either of the two values *true* or *false*. In fact, one of the names for the logical calculus with predicates, variables and quantifiers used to be the *functional calculus*. This denotation is not used anymore; we call it *first-order logic*.

Theoretically we could avoid formulas with free variables, but it is quite convenient to use them inside of proofs. For example, a proof in number theory may start with:

Let x be a number. Suppose x is prime...

But a theorem must be true under all circumstances, hence all variables must be *bound* by quantifiers. Therefore, in logic we reserve the word '*sentence*' for formulas in which all variables are bound and hence have a definite truth value.

The quantifiers are usually denoted by $\forall$ (for All) and $\exists$ (there Exists). Quantifiers bound variables exactly in the same way as, say, summation and integration operators. For example, consider a binary function $\sin(x + y)$. When we integrate along x, say, $\int_0^1 \sin(x + y)\,dx$, we get a function of only one variable y. We say that x is bound in $\int_0^1 \sin(x + y)\,dx$. A similar thing happens if we take a formula $\varphi(x, y)$ with two free variables and apply, for instance, the universal quantifier to x. The resulting formula is $\forall x\ \varphi(x, y)$. While the truth of $\varphi(x, y)$ depended on x and y, the truth of $\forall x\ \varphi(x, y)$ only depends on y.

Single quantifiers or several quantifiers of the same kind are easy to understand. The complexity arises when we alternate the two kinds. One alternation, such as

$\forall x \exists y\ \varphi(x, y)$, can be understood fairly easily. In this way we can express things like the infinitude of the prime numbers:

For every *prime*, there exists *a larger prime*.

To grasp the meaning of $\forall x \exists y \forall z\ \psi(x, y, z)$ is not so easy. In natural speech it is never used. When it is needed it is somehow circumvented. Consider, for example, the sentence:

In every *town* there is *the tallest building*.

There are only two quantifiers in this sentence, but the third one is hidden in '*the tallest*'. The latter fact is expressed using the comparative relation *taller* as follows:

The building taller than every *other building*.

Plugging this definition into the last sentence we get a sentence with three alternating quantifiers, which shows a typical way of avoiding several alternations of quantifiers in speech. Three alternating quantifiers occur in the definition of the limit of a function. This is one of the first definitions that students encounter when starting with the calculus. The inability to grasp the meaning of such a definition is used to sort out those who do not have a talent for mathematics. But the concept of the limit can be explained, not quite precisely, without the three quantifiers as follows. y_0 is the limit of the function f at point x_0, if

whenever x is very close *to x_0, then $f(x)$ is* very close *to y_0*.

The quantifiers are hidden in the imprecise expression '*very close*'. (It is possible to develop a theory in which this apparently vague expression has a precise meaning, see Chap. 3, *Robinson's Nonstandard Analysis and Vopěnka's Alternative Set Theory*.)

To imagine the meaning of four alternations of quantifiers seems as difficult as to imagine four-dimensional space. But there is a way out: using the concept of a game we can imagine even long alternations of quantifiers. I will explain it in the next chapter.

Notes

1. *The language of propositional logic*. This language uses

   a. an infinite set of propositional variables,
   b. connectives, and
   c. parenthesis.

   The standard connectives are in Table 2.1.

   All connectives can be defined using one of the first three and negation. Other connectives are also used sometimes; e.g., XOR (exclusive or) and propositional constants: true ($\top$, 1) and false ($\bot$, 0).

   In intuitionistic logic only equivalence can be defined from the other connectives.

Table 2.1 Standard connectives

Symbols	English word	Name
$\wedge$, &	and	conjunction
$\vee$	or	disjunction
$\rightarrow, \supset$	if ... then	implication
$\neg, \sim$	not	negation
$\equiv, \Leftrightarrow$	if and only if	equivalence

2. *The language of first-order logic.* This language has two parts: logical and non-logical. The *logical symbols* are the symbols of propositional logic except for propositional variables. Further, it uses (first-order) variables and quantifiers $\forall$ (universal) and $\exists$ (existential). In classical logic it suffices to use one quantifier.

   The *nonlogical symbols* are constant symbols (which may be treated as 0-arity function symbols), function symbols and relation symbols of various arities. A typical theory uses a finite set of non-logical symbols. I will present an example after I define context-free languages. Equality ($=$) is usually treated as a logical symbol, but it can also be used as a special binary relation.
3. *Context-free languages.* The theory of formal languages was founded by Noam Chomsky in the 1950s. I will start with an important concept from this theory, *context-free languages*. It is worth noting that the concept was discovered when studying natural languages. Nowadays it is an important concept in computer science. In logic it is not used, but the way logicians define a formal language of a logical calculus is equivalent to a use of a context-free language. I will use a definition of logical formulas based on a context-free language as it shows a connection between natural languages and formal ones.

   In the mathematical language theory a *language* is a very general concept, it is any set L of finite strings of symbols from a finite set A. The set A is called the *alphabet* of the language L; the strings are called *words*.

   To define a context-free language we need another finite set N, disjoint with A, whose elements are called *nonterminal symbols*, with one distinguished nonterminal symbol s. Furthermore we need a finite set of pairs consisting of a nonterminal symbol and a word in the alphabet $A \cup N$. Such a pair (a, w), $a \in N$, w a word, is usually written as $a \Rightarrow w$, indicating that a can be replaced by w, and it is called a *rewrite rule*. The set of such rules is called a context-free *grammar* of the language. (Using general rewrite rules one defines general grammars, but we do not need them here.) The language determined by the grammar is the set of all words that can be obtained starting with the initial symbol s and applying rewritings.

   The terminology that I am using here is common in computer science, but does not apply to natural languages. When considering natural languages we should call A the *vocabulary* and the strings of symbols *sentences*. The nonterminal symbols are *grammatical categories* such as *subject, predicate, object, adverbial clause*. The initial symbol s represents the class of all sentence in the language. Let us consider a couple of rules that are probably valid for any human language.

$$
\begin{aligned}
&< \text{sentence} > \Rightarrow < \text{subject} >< \text{predicate} >\\
&< \text{predicate} > \Rightarrow < \text{verb} >< \text{adverbial clause} >
\end{aligned}
$$

The meaning of the angled brackets $< \cdots >$ is that the compound expression denotes a single symbol. Thus <sentence>, <subject>, etc. are elements of the set of nonterminal symbols. If we only were interested in the structure of sentences, we would use only the grammatical categories such as *noun, verb, adverb*, etc. as terminal symbols. If we want to get actual sentences, we need also rules that transform nonterminal symbols into terminal ones, such as

$$
\begin{aligned}
&< \text{verb} > \Rightarrow abandon\\
&< \text{verb} > \Rightarrow abase\\
&< \text{verb} > \Rightarrow abash\\
&\quad \ldots
\end{aligned}
$$

The context-freeness means that we are describing grammatically correct sentences, with no regard to their meaningfulness. So sentences such as '*A yellow poem lies in the air.*' are considered to be in the language. In fact, probably no natural language is context-free because there are always some dependencies between the forms of words due to declension, conjugation, etc., that cannot be completely described by context-free rules.

Programming languages are usually defined as context-free languages, but there are often additional exceptions that spoil this property. The famous programming language ALGOL, created soon after the birth of the formal language theory, was based on a context-free grammar.

4. *The language of first-order logic, cont'd.* We will consider a language with two connectives $\wedge, \neg$ ('and' and 'not', which suffice to define all other connectives), universal quantifier $\forall$ ('for all', the other quantifier, $\exists x \ldots$' is definable by $\neg\forall x \neg \ldots$), equality $=$ and an infinite set of variables (say $x, x', x'', x''', \ldots$). In our simple example, there is one binary relation symbol R, one binary function symbol F and a constant c.

   A context-free grammar for this language has three nonterminal symbols: <formula>, <term> and <variable>, with <formula> being the initial symbol. The rules are

$$
\begin{aligned}
&< \text{formula} > \Rightarrow \forall < \text{variable} > (< \text{formula} >)\\
&< \text{formula} > \Rightarrow (< \text{formula} >) \wedge (< \text{formula} >)\\
&< \text{formula} > \Rightarrow \neg(< \text{formula} >)\\
&< \text{formula} > \Rightarrow R(< \text{term} >, < \text{term} >)\\
&< \text{term} > \Rightarrow F(< \text{term} >, < \text{term} >)\\
&< \text{term} > \Rightarrow < \text{variable} >\\
&< \text{term} > \Rightarrow c\\
&< \text{variable} > \Rightarrow < \text{variable} >'\\
&< \text{variable} > \Rightarrow x
\end{aligned}
$$

This system has superfluous parentheses around atomic formulas, which can be avoided by having a nonterminal symbol for atomic formulas and a few more rules.

5. *Higher-order languages.* In a second-order language we have variables for relations and for functions. Usually we distinguish first-order symbols from second order symbols by using lower case letters for first-order symbols and upper case letters for the second-order symbols. This is not enough for all the bookkeeping that one would need if all done quite formally. To this end we have to declare for each second-order symbol whether it is a relation or a function and then of what number of variables. Note that the relations and functions of the first-order language can be viewed as first-order constants in higher order languages.

   As examples, I will write formally the axiom of induction and an axiom of topological spaces.

   a. Suppose we are describing the natural numbers, thus all elements are natural numbers. Then we do not have to mention the set $\mathbb{N}$ and the axiom of induction reads:
   $$\forall X\big(\big(X(0) \wedge \forall x\big(X(x) \to X(x+1)\big)\big) \to \forall x\; X(x)\big).$$
   b. To write down the second axiom of the two axioms of topological spaces, I will use the same symbol $\mathcal{O}$ for the predicate expressing that a set is open. Thus the predicate $\mathcal{O}$ is a third order constant. I am using capital calligraphic letters for third order objects. Then the axiom reads:
   $$\forall \mathcal{X} \forall Y\big(\big(\forall X\big(\mathcal{X}(X) \to \mathcal{O}(X)\big) \wedge \forall x\big(Y(x) \equiv \exists X\big(\mathcal{X}(X) \wedge X(x)\big)\big)\big) \to \mathcal{O}(Y)\big).$$

6. *Propositional logic.* Propositional logic is the part of logic that uses neither quantifiers nor equality. Then the structure of atomic formulas does not matter; the only thing that matters is which atomic formulas are the same. Thus we can use any symbols for atomic formulas, preferably we use *propositional variables*. There is a good reason for referring to them in this way, as their meaning is simply *true* or *false*. We can view propositional logic as the study of a two element structure. The two elements represent *true* and *false* and they are usually denoted by 1 and 0. On this structure we study operations (but not relations). The operations are called *Boolean functions*. Boolean functions corresponding to negation, conjunction, disjunction, etc. are defined by the familiar truth tables.

   The idea of studying propositional logic as the theory of a two element set is due to George Boole (1815–1864) and therefore we talk about Boolean functions, Boolean algebras, etc. Gottfried Wilhelm Leibniz (1646–1716) was very close to this discovery. He noticed that, when interpreting true as 1 and false as 0, conjunction is the product. He thought that disjunction should be the sum, but that did not work. That was before mathematicians realized that it is not necessary to stick to familiar structures and that one can invent new interesting ones.

   A set of operations is called *a complete set of connectives*, if every operation on $\{0, 1\}$ can be expressed using operations from the set. For example, $\{\neg, \wedge\}$ and $\{\neg, \vee\}$ are complete. All these are simple facts, but they are important for realizing that at least propositional logic is uniquely determined: *it is the theory of the simplest nontrivial set.*

7. *Normal forms.* Many problems on formulas become simple if we can use a normal form, that is, if we can transform a general complicated formula into a formula having a nice structure. Everybody knows that (due to the distributive law) it is possible to write any polynomial as a sum of products of variables and constants (called *monomials*). A similar fact holds for propositional logic, where we have two distributive laws $(x \wedge (y \vee z) \equiv (x \wedge y) \vee (x \wedge z)$ and $x \vee (y \wedge z) \equiv (x \vee y) \wedge (x \vee z))$, and De Morgan's laws $\neg(x \wedge y) \equiv (\neg x \vee \neg y)$ and $\neg(x \vee y) \equiv (\neg x \wedge \neg y)$. This enables us to write every proposition as a disjunction of conjunctions of propositional variables or negated propositional variables. This is called the *Disjunctive Normal Form* or simply DNF. There is, of course, the dual version, the *Conjunctive Normal Form*, or CNF. Note that the interpretation of a DNF is that we list all cases when the formula is true. This is not a very efficient way of expressing a given Boolean function, in fact, the reduction to a DNF or CNF often results in an exponential blow-up in the size. Thus DNFs and CNFs simplify problems in propositional logic only theoretically.

   In first-order logic we also have a nice normal form. First we move all quantifiers to the beginning of the sentence. This is enabled by rules such as $\neg\forall x\ \phi \equiv \exists x \neg\phi$ and $\phi \wedge \forall x\ \psi \equiv \forall x\ (\phi \wedge \psi)$, where x does not occur in ϕ. Then the inner part of the formula contains no quantifiers, thus we can transform it into a DNF (or a CNF). The resulting formulas are called *prenex normal forms*. This is a useful and efficient reduction, but the prenex normal forms are often more difficult to understand than the original sentences where the quantifier occur in places to which they actually refer.

   Having all quantifiers in a prefix enables us to define the *quantifier complexity* of sentences. Rather than counting the number of quantifiers we count the *number of alternations* of quantifiers. Furthermore, it is important to know what is the first quantifier in the prefix. If the number of alternations is small we denote the class simply by listing the quantifiers.

   Example $\exists x \forall y \forall z \exists u\ \phi$, with ϕ quantifier-free, is a $\exists\forall\exists$ formula.

   If we have more alternations, we write only the first quantifier indexed by the number of alternations. So the above formula is a $\exists_3$ formula. Sometimes people use Σ and Π instead of $\exists$ and $\forall$, but that may lead to confusion with other hierarchies.

8. *Equational theories.* I spoke about natural languages as a motivation for the language of first-order logic. This concerns propositional connectives and quantifiers. Function symbols, terms and equations come from mathematics. Function symbols describe elementary operations, terms describe computations and equality is a basic binary relation. In first-order logic we call equations atomic formulas. It may seem that they are too simple to be of any interest, but the contrary is true. Using equations one can express quite a lot, in fact we can, in some sense, simulate the whole first-order logic. To be quite precise we should note that a connective and a quantifier is implicit in equational theories. When we talk about a set of equations we mean, in fact, the conjunction of the equations.

When we say that an equation with variables is true, we mean that it is true for any possible value of the variables, which means that we implicitly assume that the variables are universally quantified. (For example, we state the commutative law as $x + y = y + x$, meaning that $\forall x \forall y (x + y = y + x)$.)

An important example of an equational theory is Boolean algebras. This is the equational theory of the two element set $\{0, 1\}$. We can take all operations that can be defined on this set (that is, all Boolean functions), or only a finite complete set of them. Thus propositional logic can be viewed as the equational theory of a two element set.

9. *Communication with extraterrestrials.* Mathematics would certainly be useful, but it is naive to expect that use of mathematical language would automatically solve the problem of communication. This problem was studied by the Dutch mathematician Hans Freudenthal. In his book *Lincos: Design of a Language for Cosmic Intercourse*, he presented a language for communication with extraterrestrials. His idea is roughly as follows. In order to be able to communicate with intelligent beings, we need a common language, but we cannot agree on a common language because it is impossible to exchange messages. Therefore we have to *design* a suitable language such that we can *teach* the other party this language. He proposed to teach by examples, starting with concepts from number theory, logic and set theory. When they learn the language, they will understand any messages that we send them.

   When designing messages for aliens the first thing one should do is to realize what we want to achieve. If the message should only convey that we are intelligent creatures, we do not have to send the Pythagorean Theorem, as the mere fact that we are able to send electromagnetic signals proves that our knowledge exceeds such trivial theorems. In such a case we only need to send signals that can be distinguished from those naturally occurring in space. A more difficult task is to persuade a potential recipient about our achievements in science (other than understanding electromagnetic waves), in particular, about our successes in mathematics. An especially interesting problem, but rather theoretical one, is how to persuade someone about having advanced computing technology. Problems of this kind have been studied in computational complexity theory in the case of the two parties exchanging information in both directions. In the situation when the recipient is hundreds of light years away, one has to assume only unidirectional communication and thus the problem is different.

   Naturally, it is more promising to look for incoming signals, but a number of signals have also been sent out.

2.2 Truth and Models

The Definition of Truth and Satisfaction

To define the concept of truth in general is a difficult philosophical problem. In mathematical logic, however, there is a precise definition of this concept. Truth is a relation between sentences and reality. I have described "mathematical reality" in the first chapter; it is the realm of mathematical structures. In the previous section I explained the formal language of mathematics as studied in mathematical logic. So the definition of truth is the definition of a certain relation between these two things. More precisely, it is a definition of the relation that '*a sentence ϕ is true in a structure M*'. It is more common to say that '*a sentence ϕ is satisfied in a structure M*' and reserve the word '*truth*' for a special situation that I will describe shortly. Thus we rather talk about the definition of *satisfaction*.

The definition of satisfaction is based on defining the meaning of the parts of the sentence. When we decompose a sentence, which is a formula in which there are no free variables, we get parts that do have free variables.

Example Consider the sentences

'*For all x, $x \leq 1$, or $x^2 > x$*'

which is true in the natural numbers. This sentence contains a subformula

'*$x \leq 1$, or $x^2 > x$*'

in which the variable x is free. Therefore we have to define satisfaction also for formulas and particular values of their free variables. In our example, we first define the satisfaction of the subformula for $x = 0, 1, 2, 3, \ldots$ and then the satisfaction of the sentence.

Given a formula $\phi(x_1, \ldots, x_n)$ a structure M and elements $a_1, \ldots, a_n$, the definition of satisfaction of ϕ by the elements $a_1, \ldots, a_n$ in M proceeds inductively, starting with atomic subformulas. We define the satisfaction of atomic formulas according to the relations and functions in M. The satisfaction of compound formulas is defined by interpreting connectives and quantifiers in the natural way. The formal definition is in Notes; here I will only consider an example.

Example The formula above has two atomic subformulas $x \leq 1$, and $x^2 > x$. The first one is satisfied by 0 and 1, otherwise it is not satisfied. The second one is satisfied for $2, 3, \ldots$. The subformula '*$x \leq 1$, or $x^2 > x$*' is satisfied for every natural number because: for 0 and 1 the first term is true, for $2, 3, \ldots$ the second term is true, and a disjunction of two formulas is satisfied if at least one of them is. The sentence '*For all x, $x \leq 1$, or $x^2 > x$.*' is true in the natural numbers because the subformula '*$x \leq 1$, or $x^2 > x$*' is true for every natural number.

At first glance, this looks like a circular definition because we are defining satisfaction assuming that we already know what it is. We are defining '*for all x*' by saying that it holds for all x. Certainly, if somebody did not understand the sentence, the definition would not help them to understand it. The Polish logician Alfred Tarski (1901–1983), who invented this definition, made this point by saying:

The sentence 'It's snowing.' *is true if it's snowing.*

So what is the matter? In order to understand this definition, one has to realize two things. Firstly, we are not in the position of philosophers who want to find the *meaning* of the concept of truth. We are defining truth and satisfaction as a *mathematical* concept. Forget about meaning and look at it as a mathematical definition.[7] There are sentences on one side and structures with their elements on the other. Since the sentences are formalized as certain strings, we are defining a relation between finite strings representing formulas and strings of elements of a structure. Thus this is a perfectly legitimate mathematical definition that can be formalized in set theory. Also it is a general definition that works for every structure and tells us in which structure a sentence is true and in which it is false.

It is instructive to consider the special case of finite structures. In this case, one can even write a program to determine, if a given sentence is true in a given structure. Programming languages often contain at least part of the propositional logic, so the task is simpler if we use such a language. The quantifiers will be tested by searching all elements of the structure, using constructs such as `do ... while ...`. Note that when writing such a program we are doing essentially the same as what we did above. In particular, we are programming how to test that a formula is satisfied for all x by letting the computer check it for all x. Because we are considering finite structures, there is no doubt that it makes sense—the computer will be able to tell us whether or not the formula is satisfied. Now, imagine an ideal computer that is able to do infinite computations. Then the definition of satisfaction for general structures can be viewed as a program for such a computer.

Secondly, we have to realize is that there are two levels of discourse. The lower level is the formal language for which we are defining the concept of satisfaction; it is called *the object language*. The upper level is the language that we use to make this definition; it is called *the metalanguage*. We have already observed that not distinguishing between the two levels leads to paradoxes, which would result in contradictions in formal systems. On the other hand, having this distinction, the definition makes sense: although we are using the same logical operators, such as '*or*' and '*there exists*', they appear in different places. In particular, we are defining '*or*' in the object language using '*or*' in the metalanguage. We suppose that we understand '*or*' in the metalanguage, so we can use it to define it in the object language.

The psychological factor that makes this definition difficult to accept is that we are defining something that is completely clear to us. Thus it seems that there is nothing to define. Therefore, we should view it as a *formalization* rather than a definition.

Let us now fix some terminology. Instead of saying that a sentence ϕ is satisfied in a structure M, one often says that *M is a model of ϕ*. This is further extended to theories. We say that M is a model of a theory T if all axioms of T are satisfied in M. We also often say '*models*' instead of '*structures*'. *Model theory,* an important

[7] And read the quotation from Isaac Asimov's *Imaginary* at the beginning of the next chapter (page 157).

field of logic, studies mathematical structures, that is, models, from the point of view of logic. The concepts of truth and satisfaction are the basic notions in this field.

When stating a theorem we often assume that the particular structure is clear from the context. For example, if we state that an arithmetical sentence is true, we mean that it is satisfied in the natural numbers. Surely, there are many other structures in which we can interpret the sentence.

This leads us back to philosophical questions. Although we do have a formal definition of truth, the meaning of this concept is a matter of philosophical views. Saying that a sentence is true presupposes an objective reality where the sentence should be satisfied. But what is mathematical reality? Specifically, are mathematical structures real? If not, how can we then talk about mathematical truth? Another question is, assuming we believe in mathematical reality, how do we acquire mathematical knowledge and how do we learn what is true? We cannot empirically test sentences that talk about an infinite number of elements. We can only decide the validity of a sentence in small finite structures. For large infinite structures, as well as for large finite ones, we use proofs instead of empirical tests. But proofs need axioms; logic alone does not suffice. So we need to justify axioms. How do we justify axioms if we cannot test their validity? And so on...

I leave these questions without an answer for the time being. What I am going to present further in the book should give us more ideas on which we can base our opinion. Finally, in Chap. 7, I will address these questions directly.

Logically Valid Sentences

When talking about truth we always imagine something absolute. What we have considered so far, was only relative truth: a sentence being true relative to a particular structure. So here is an important concept. There are sentences that are true in all structures. One may get the impression that such sentences are very simple and uninteresting. Also one of their names, *tautologies*, has such a connotation. But in fact, these are the sentences that we are mostly interested in, the substance of logic. In logic we are interested in absolute truth, not sentences that are true only in a particular situation, those are the subject matter of other sciences. We call the sentences true in all structures[8] *logically valid*. The term '*tautology*' is mostly used for logically valid formulas of propositional calculus.[9]

Logically valid sentences express all that logic can say about truth. If we want to know, if a sentence φ follows from another sentence ψ, we can just check, if the implication '*if* ψ, *then* φ' is logically valid. In the same way we can reduce the question whether a sentence is a consequence of a set of axioms to logical validity

[8] More precisely, true in all structures of an appropriate type.

[9] Some authors distinguish between logically valid sentences and tautologies in first-order logic and call tautologies only the sentences whose validity can be established by means of propositional logic.

of certain sentences. For example, if we want to know whether the sentence $x \cdot y = y \cdot x$ is a consequence of the axioms of groups, we just need to know whether the sentence $x \cdot y = y \cdot x$ is true in all structures that satisfy the axioms of groups, which simply means we need to know if it is true in every group. (This particular sentence expresses the commutative law and is not true in every group.)

All this looks very simple, but the conclusion that we get is extremely important: *we can define logical validity.* This is a consequence of the fact that we can define satisfaction. By saying '*define*' I mean it in the strongest sense, namely, *logical validity is a mathematical concept.* There is no other basic concept of gnoseology, the science of knowledge, that can be so unambiguously defined! At this point logic departs from philosophy and becomes a part of mathematics.

Proving Consistency and Independence by Constructing Models

One of the most important problems studied in logic is the consistency of an axiomatic system. This problem is also relevant in other theoretical fields, but the closer the field is to practice the less important it is. This is because physical reality is considered the best test of consistency. In mathematics too we can test the consistency of sentences on small finite structures, small enough to be handled by computers. Thus, for example, we can show the consistency of certain axioms of geometry using the Fano plane. But the majority of the interesting theories concern infinite structures. Therefore, we have to substitute physical reality by the world of mathematical structures.

Having a definition of satisfaction, we can formally prove that testing an axiomatic theory on structures suffices to prove its consistency. Indeed, one can prove that a sentence ϕ is either true or false in a given structure M, *but not both.* Further, one can show that if axioms are satisfied in M, then so are all their logical consequences. Therefore, if we want to prove that an axiomatic theory T is consistent, it suffices to find a structure in which all axioms of T are true, in other words, to find a model of T.

A theory without models is certainly strange. One often calls a concept void if there is no example of it. What is a void theory? It turns out that such a theory is inconsistent. So having a model and being consistent are equivalent things. This is a nontrivial fact; it is called the Completeness Theorem. I will talk about it in the next section because it concerns both semantics, the topic of this section, and the syntax of proofs, which will be the topic of the next section.

Let us see how this reduction is used in practice. The simplest situation is when the axioms are satisfied in one of the structures that we already know. If this is not the case we try to adapt or combine the existing structures to get a model that we need. In other words, we construct the model from available models. Thus, for example, we prove the consistency of the complex numbers using the *Gaussian plane*, which is the ordinary plane with axis x used for the real numbers and the axis y for the imaginary numbers. A point with coordinates (a, b) corresponds to the complex number $a + ib$.

The problems arise when we need models of strong theories, in particular, set theories. Except for some very weak set theories, their models cannot be constructed from the classical standard structures such as the natural numbers and the real numbers. We will see that in these cases we have to accept the consistency as a hypothesis.

If we analyze consistency proofs more closely, we find out that we are using certain assumptions even in such simple cases as the case of the complex numbers. When proving that the theory of complex numbers is consistent our assumption is that the theory of real numbers is consistent. We may say that the latter assumption is obvious, but, strictly speaking, we are only reducing the consistency of the complex numbers to the consistency of the real numbers. Essentially in all proofs of consistency we are reducing the consistency of a theory T to the consistency of another theory S. To stress this fact we say that *T is consistent relative to S*. The consistency of S can be justified by our belief that S axiomatizes a structure that is real. Then we also believe that T is consistent and do not talk about relative consistency.

To prove that a sentence is independent of a system of axioms one can also use models. This is due to the simple relation between consistency and independence:

Proposition 3 *A sentence ϕ is not provable from a consistent set of axioms A, if and only if the set A supplemented with $\neg\phi$ is consistent.*

Consequently, it is possible to prove that ϕ is not provable from A by constructing a model in which the sentences of A are true and sentence ϕ is false. This simple proposition is the basic tool in many proofs of independence. Its power lies in the fact that it replaces a negative task—to show that no proof gives ϕ, by a positive one—to construct a model.

A nice example is Euclid's fifth postulate. Recall that this axiom is equivalent to the statement that for a line and a point not on the line there is a unique line through the point that is not incident with the given line. In Euclidean geometry, the line through the point is parallel to the given line. For centuries, people believed that this axiom is superfluous because it can be derived from the others. Only at the beginning of the 19th century did some mathematicians realize that it may not be the case. Indeed, this axiom does not follow from the others. The solution of the problem is attributed to János Bolyai, Carl Friedrich Gauss and Nikolai Ivanovich Lobachevsky. Lobachevsky and Bolyai published their works in 1829 and 1831; the only evidence about Gauss' work that we have is from his letters, but it is convincing enough. Lobachevsky and Bolyai developed what is nowadays called *hyperbolic geometry* or *Lobachevsky-Bolyai geometry*. Lobachevsky studied fairly non-trivial problems such as the volumes of polyhedra in three-dimensional hyperbolic geometry.

In logical terms, they studied the theory in which the fifth postulate is replaced by its negation. In order to prove that the fifth postulate is independent, it sufficed to show that the new theory was consistent. Lobachevsky and Bolyai considered the fact that the theory leads to meaningful results as sufficient evidence that the new theory is consistent, but they did not have a proof. It was still possible that when the theory was developed further it would run into a contradiction. A genuine proof of independence only appeared later, in 1868, when Eugenio Beltrami constructed a model of this theory. In Beltrami's model all axioms of Euclidean geometry were true, except the fifth postulate, which showed that the fifth postulate was independent. There is no doubt that the insight of Bolyai, Gauss and Lobachevsky was the major step in the solution of the problem, but the problem was solved by Beltrami.[10]

In popular expositions of the problem of Euclid's fifth postulate you can still read that "Gauss, Lobachevsky and Bolyai proved the existence of non-Euclidean geometries". Let us ponder what a proof of the existence of a concept means. In contemporary mathematics it means precisely this: to prove the existence of a structure that is a model of the concept in Zermelo-Fraenkel Set Theory. So the fact that one can develop a meaningful theory about the concept does not count as a proof; it may only be accepted as a piece of evidence supporting the conjecture. There is, however, one important exception: set theory itself. As we will learn in the next section, one cannot prove the consistency of Zermelo-Fraenkel Set Theory in itself, hence also one cannot construct a model of Zermelo-Fraenkel Set Theory in itself. So for the existence of this concept, we only have arguments based on a having "well-behaved theory" etc., arguments of the kind that we dismissed in the case of non-Euclidean geometries.

Models Are not Uniquely Determined by Theories

Assuming that structures are the main subject of our study and logic only serves to describe them, we would like logic to be able to determine each structure as much as possible. Clearly, logic cannot determine a particular structure uniquely because for a given structure there are infinitely many isomorphic ones. That is all right, we do not want to distinguish isomorphic copies of the same structure. So our concern is if one can determine a structure *up to an isomorphism*. In general, this is not possible. More precisely, one can determine only finite structures by the sentences that are true in them. For an infinite structure, there are structures that are essentially different, but which satisfy exactly the same sentences. The best way to demonstrate it is to consider the sizes of structures. A classical result of Leopold Löwenheim (1878–1957) [186] and Thoralf Skolem (1887–1963) [270] says that for an infinite structure there are structures of any infinite cardinality satisfying the same sentences. (For a more precise statement see Notes.)

Example Consider the natural numbers and the real numbers. The first structure is countable and the second is uncountable. We think of the set of natural numbers as the canonical example of a countable infinite set (in fact, '*countable*' comes from the possibility of enumerating the set by numbers). Yet, there are structures that are uncountable and they satisfy the same sentences. Similarly, the real numbers are a

[10]Gauss did important work in the study of the concept of curvature. We may thus speculate that he could have realized that curved surfaces are models of non-Euclidean geometry, but we do not have any historical evidence of that.

prototype of a higher type infinity, the continuum, but there are countable structures that are logically indistinguishable from them.

There is an even more striking example. Consider an axiomatic system for set theory, say Zermelo-Fraenkel Set Theory. Assuming it is consistent, it has a model, but then, according to Löwenheim-Skolem's theorem, it also has a countable model. It has a countable model, in spite of the fact that in this theory there are many much larger cardinalities! This looks really weird, but in fact, it is also a good example for explaining how it is possible. The crucial thing is to realize that a structure is a world that is different from ours. People living there see things from a different perspective, from a much narrower one. It is like popular explanations of higher dimensions. People living in a two-dimensional world could not escape from a circle. Watching them in a three-dimensional world, we see that it is possible to use the extra dimension to jump over the border. But only we can use the three dimensions, the rules of the game do not allow the people from the two dimensions to do so.

Turning back to the countable model of set theory, let us call it M. Take the natural numbers of M and the reals of M. As the whole model M is countable, both the natural numbers and the real numbers of M are countable sets. This seemingly contradicts the theorem of set theory saying that the two sets have different cardinalities. To resolve this apparent contradiction we have to recall the definition of what it means that two sets have the same cardinality. The definition says that they have the same cardinality, if there exists a one-to-one mapping f from one of the sets onto the other one. So let such an f be the mapping of the natural numbers of M onto the real numbers of M. To conclude that the two sets have the same cardinality from the point of view of M we would need to prove that such an f is in M, but there is no reason why it should be. So the contradiction is only apparent. Exactly like the action of jumping from the circle is not allowed in the two-dimensional world, the mapping f is not allowed in the model M.

The reason for this discrepancy is that logic is in some sense finite (the technical term for this property is *compact*). In particular, each proof is finite and therefore it cannot use more than a finite number of axioms. Hence everything that we can prove about a structure only depends on local properties. The cardinality, however, is a global property.

A Nonstandard Model of Arithmetic

Not only are there models of different cardinalities that satisfy the same sentences, but also in one given cardinality there may be different ones. This may be viewed as a drawback of logic, but also as an advantage: it gives us interesting structures. In particular such interesting structures are the *nonstandard models of arithmetic*. By arithmetic I mean a theory that partially describes the structure $(\mathbb{N}; +, \cdot, \leq)$, or more precisely, a theory T one of whose models is $(\mathbb{N}; +, \cdot, \leq)$. We call such theories *arithmetical*. An important arithmetical theory is *Peano Arithmetic* (see page 116). This is a theory based on a small set of simple axioms and an infinite set of axioms stating the principle of the mathematical induction for every formula in the language of $(\mathbb{N}; +, \cdot, \leq)$. In model theory the word '*theory*' is used for any consistent set of sentences, even if there is no effective procedure to determine if a sentence is an axiom of this theory. Such a theory is *True Arithmetic*, which is simply the set of all sentences true in $(\mathbb{N}; +, \cdot, \leq)$. We will see in Chap. 3 that nonstandard models of True Arithmetic are very useful.

To define nonstandard models, we rather define its opposite, the standard model. The standard model is simply $(\mathbb{N}; +, \cdot, \leq)$ and all models isomorphic to it. Hence, M is nonstandard, if it is not isomorphic to $(\mathbb{N}; +, \cdot, \leq)$. Thus each uncountable model of arithmetic is nonstandard, but there are also countable ones. Nonstandard models are very complex structures and they cannot be obtained by an explicit construction (except for some very weak subtheories of Peano Arithmetic). To get at least an idea of what they look like, one should look at the ordering of elements of such a model M. The model starts with a copy of the standard model. These are the elements that can be denoted by numerals; they are called *standard numbers* or *finite numbers*. Since M is not standard, there must be other elements. They are all after standards numbers and they are called *nonstandard* or *infinite*. Clearly, there is no largest nonstandard number, but also there is no smallest nonstandard number. In fact, for a nonstandard number α, the number $\alpha - 1$, and the integer parts of $\alpha/2$, $\sqrt{\alpha}$, etc. are also nonstandard. Note that the fact that there is no least nonstandard number does not contradict the least number principle because the set of nonstandard numbers is not definable in M.

Notes

1. *The definition of satisfaction.* Assume that a finite list of nonlogical symbol is given. To simplify the definition I will only use relation symbols R. Further, I will assume that the logical symbols are only $\wedge$ (conjunction), $\neg$ (negation), and $\exists$ (existential quantifier). Let $\mathcal{L}$ be such a language. Given a relation symbol R from $\mathcal{L}$ and a structure M for $\mathcal{L}$, I will denote by R^M the relation of M that is the intended interpretation of the relation R in M.

   For a formula ϕ with free variables $x_1, \ldots, x_n$ and elements $a_1, \ldots, a_n$ from M, we want to define the relation '*ϕ is satisfied by elements $a_1, \ldots, a_n$ in M*'. To simplify the definition I will further use the standard notation $M \models \phi[a_1, \ldots, a_n]$ to denote this relation of satisfaction. The definition goes by induction on the complexity of the formula ϕ:

   a. for a k-ary relation symbol $R(x_1, \ldots, x_k)$, $M \models R[a_1, \ldots, a_k]$ if the elements $a_1, \ldots, a_k$ are in the relation R^M;
   b. for formulas ϕ and ψ, $M \models (\phi \wedge \psi)[a_1, \ldots, a_n]$ if $M \models \phi[a_1, \ldots, a_n]$ and $M \models \psi[a_1, \ldots, a_n]$;
   c. for a formula ϕ, $M \models \neg\phi[a_1, \ldots, a_n]$ if it is not true that $M \models \phi[a_1, \ldots, a_n]$;
   d. for a formula ϕ, and a variable x_1, $M \models \exists x_1 \phi[a_2, \ldots, a_n]$ if there exists an element a_1 in M such that $M \models \phi[a_1, a_2, \ldots, a_n]$.

Since this is a very important definition, I will describe in more detail how it is formalized in set theory. Let M be a fixed model. Let $\mathcal{F}$ be all first-order formulas in language $\mathcal{L}$ and let A be all finite sequences of elements of M. Formulas are formalized by finite strings of symbols from a finite alphabet. Let $\mathcal{F}_i$ denote the subset of $\mathcal{F}$ consisting of all formulas of logical complexity i. Thus $\mathcal{F}_0$ are atomic formulas, $\mathcal{F}_{i+1}$ are all formulas from $\mathcal{F}_i$, plus those that are obtained from them using a connective or a quantifier. We have $\mathcal{F}_0 \subseteq \mathcal{F}_1 \subseteq \mathcal{F}_2 \cdots$, and $\mathcal{F}$ is the union of the sets $\mathcal{F}_i$, $i = 0, 1, \ldots$. Then we define the relation $\models$ between $\mathcal{F}_i$ and A by recursion on i. Condition 1. defines the base case. Conditions 2.–4. define how to extend the relation $\models$ from $\mathcal{F}_i$ to $\mathcal{F}_{i+1}$. Thus we obtain a sequence of relations, where the first relation is defined explicitly and each succeeding one is defined explicitly from the previous one. Finally, the relation $\models$ is the union of all these partial relations.

Note that the definition of truth of sentences is a special case of this definition: a sentence ϕ is true in M, if it is satisfied by the empty string in M.

The definition of satisfaction for higher order structures and languages is analogous.

2. *The Löwenheim–Skolem Theorem.*

Theorem 2 *If a theory T has an infinite model, then it has models of arbitrary infinite cardinalities.*

The theorem is a consequence of the proof of the completeness theorem, which we will present in the next section. Here I will only sketch a proof of a weaker theorem:

If T has an infinite model, then it has a countable model.

Furthermore I will only consider the special case where the theory has a single axiom of the form $\forall x \exists y\, \phi(x, y)$, where $\phi(x, y)$ is quantifier-free. Assume the language of the theory does not contain function symbols. Let M be an infinite model of this theory. Let us define a function f on the universe of M by choosing, for every x in M, an element $f(x)$ such that $\phi(x, f(x))$ holds in the model. (f is called the *Skolem function* for $\forall x \exists y\, \phi(x, y)$.) Pick an element a of the model and take M' to be the submodel of M with universe $\{f^n(a);\ n = 0, 1, 2, \ldots\}$, where f^n denotes n-times iterated f. Since ϕ is quantifier free, $\phi(f^n(a), f^{n+1}(a))$ is true also in the submodel M', thus M' satisfies the axiom. M' is either countable infinite or finite. If it is finite, repeat this process with another element, etc.

If the language does contain function symbols, we have to take the universe of M' to be all elements generated from a by the functions of M and f.

The above proof can be explained as follows (the same idea is used to prove the theorem in general). First we replaced the theory by a universal theory, namely, the theory axiomatized by $\forall x\, \phi(x, f(x))$. Then we have taken a substructure of M generated from a single element. A substructure satisfies all universal sentences that the structure does, hence it is also a model of the original theory. For more detail, see also *Compactness* on page 115.

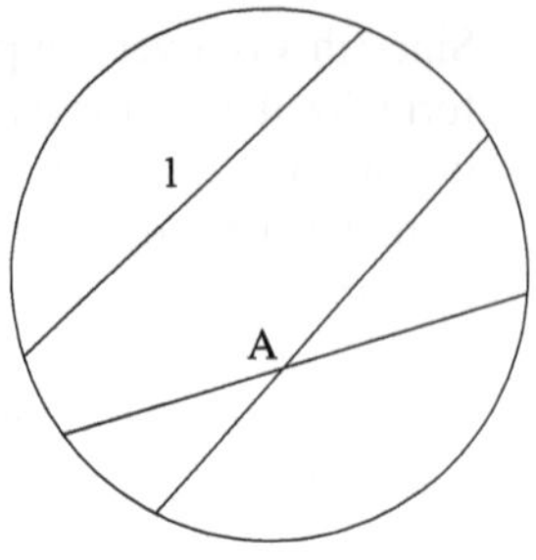

Fig. 2.1 The Beltrami-Klein model of the hyperbolic plane

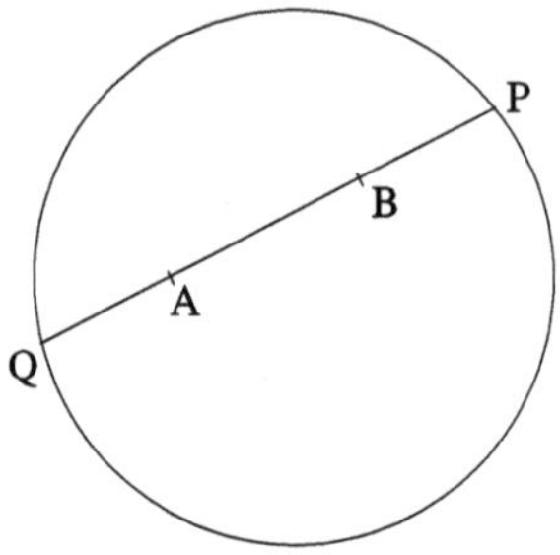

Fig. 2.2 The hyperbolic length of the segment AB is $\frac{1}{2}\log\frac{|AP|\cdot|BQ|}{|AQ|\cdot|BP|}$, where $|\cdots|$ denote the Euclidean lengths of the segments

3. *Models of planar hyperbolic geometry.* In hyperbolic geometry the fifth postulate is replaced by

   For every line l and every point A not on l, there are at least two different lines going though A that are not incident with l.

   The first model of non-Euclidean plane geometry found by Beltrami was on surfaces of negative curvature. Later he realized that one can project it on a disc in a Euclidean plane. This model was popularized by Felix Klein, thus it is often called the Beltrami-Klein model. The points of this model are points inside of the circle (the points on the circle are excluded). The lines of the model consist of the inner points of segments whose endpoints are on the circle, see Fig. 2.1. The verification of the axioms of incidence, including the axiom that replaces the fifth postulate, and the axioms about the relation '*between*' is very easy because both relations are the same as in the Euclidean plane from which is the model constructed.

   The tricky part is to define the relation of congruence between segments and to prove that the corresponding axioms are true. To this end, one defines the hyperbolic length of a segment, see Fig. 2.2. Then two segments are congruent (have the same length) in the new sense if the real numbers assigned to them are the same.

   A different model was discovered by Poincaré. The points of the model are again the inner points of a circle. Lines of the model are arcs (sections of circles) that start and end on the circle and that are perpendicular to the circle at both sides. Moreover, every diameter is a line too. The advantage of this model is that

Fig. 2.3 Poincaré's model

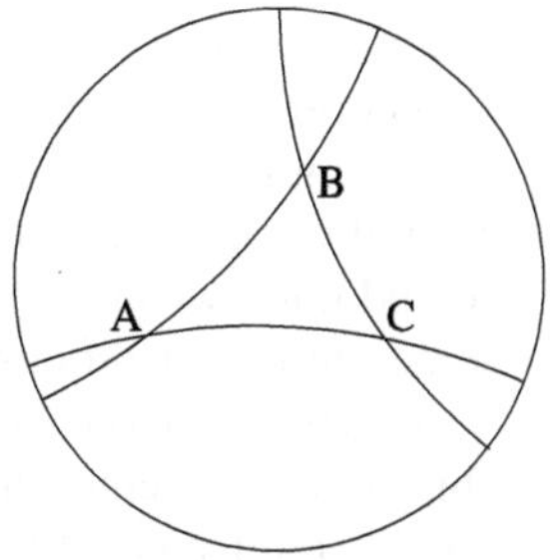

the angles in the model are the true angles. Thus you can clearly see that the sum of the angles of the triangle ABC in Fig. 2.3 is strictly less than 180°.

Note that both models are constructed from the Euclidean plane. Thus this proves *relative consistency*: if the axioms of the Euclidean plane are consistent, then so are the axioms of the hyperbolic plane.

4. *Nonstandard models of* $(\mathbb{N}; +, \leq)$ *and* $(\mathbb{R}; +, \cdot, \leq)$. In contrast to the arithmetic of the natural numbers with addition and multiplication, these two structure determine fairly weak theories. Therefore, it is possible to construct explicitly nonstandard models of them.

   Here is a nonstandard model M of the addition of integers, the theory of $(\mathbb{N}; +, \leq)$. The universe consists of pairs (q, n), with n an integer and q a non-negative rational number; however we exclude pairs $(0, n)$ for $n < 0$. The ordering of M is the lexicographic ordering determined by the orderings of integers and rational numbers. The addition is defined componentwise, $(q, n) +_M (r, m) = (q + r, n + m)$.

   We know that Peano Arithmetic certainly does not have such a simple model. One result that proves this is a theorem by S. Tennenbaum. This theorem gives a lower bound on the complexity of countable nonstandard models of Peano Arithmetic: they cannot be constructed using computable relations and operations.

   The theory of $(\mathbb{R}; +, \cdot, \leq)$ is called the *Theory of real closed fields* because it is axiomatized by the axioms of ordered fields and the axiom schema that says that every polynomial of an odd degree has a root. To get a countable model of this theory, one can simply take the submodel of this structure determined by algebraic numbers (numbers that are solutions of algebraic equations with integer coefficients).

   Note, however, that all this is due to the restricted language. In the structure $(\mathbb{R}; +, \cdot, \leq)$ it is impossible to define integers. If we enrich the structure by predicates or functions that allow us to define integers, axiomatization by a decidable set of axioms becomes impossible, as well as the construction of computable nonstandard models.

5. *Free algebras.* Let a theory T be given. We would like to have a most general model of T, so that we can study what is provable in T and what is not. In general there is no canonical way to assign any such model to T, but in some cases it is possible. The problem is that while in a theory T there may be undecidable sentences, in a model, for every ϕ, we must have either ϕ or $\neg\phi$; but there is

no rule that would tell us which of the two is better. The most important class of theories that do have such canonical models are equational theories. These are theories that are given by a set of elementary positive statements—equations. Then if we are to decide which of the two independent sentences $s = t$ or $\neg s = t$ should hold, we always take the second. Thus we obtain an algebra that has in some sense the least possible number of dependences among its elements.

This is only a rough description. To make it work one has to start with a countable set of elements, called *generators*. The resulting algebra is called the *free algebra* (with countably many generators). Then an equation $t = s$ with variables $x_1, \ldots, x_n$ is derivable in the equational theory if and only if it is satisfied by n distinct generators of the free algebra.

6. *Logics without semantics and logics without syntax.* I have presented the standard approach that assumes that there is a world of structures (or models) and logic is a means of talking about them. Hence a logical system that does not have semantics is considered meaningless. Nevertheless, other philosophical directions in the foundations of mathematics refute this basic dogma. They say that the only real entities are proofs. Then having semantics is not considered important at all. There are logical systems that have a nice syntax, which means that they have certain plausible properties, but for which finding semantics was a problem (see, for example, Church's λ-*calculus*, page 146).

   On the other hand, there are logics that miss an important part of syntax. Any system called logic must have formulas to express statements, but some systems do not have the concept of a proof, or if they have a concept of a proof, there is no completeness result saying that all logically valid sentences are provable. This concerns all higher order logics starting with second-order logic. There are also logics that use infinitely long formulas and infinitely long proofs. This can be accepted as syntax, but it is not a kind of syntax that we can use for practical purposes.

2.3 Proofs

The definition of truth determines the set of logically valid sentences, but it does not give us a way to determine which are these sentences. Applying the definition would mean testing the truth on each structure. There is no effective procedure for that even for a single structure. This does not mean that the definition is completely useless for that purpose. For a particular sentence, we might be able to prove, using mathematical considerations, that it is true in all structures. But why should we look for a proof of the fact that a sentence φ is true in all structures, if we can prove φ itself?

This suggests that there is another way to define when a sentences is true—by a proof. The concept of a proof is an old one. It has been used mainly in two fields: mathematics and law. The meaning of the concept is a collection of pieces of evidence that is able to persuade any person about the truth of a sentence in question. The pieces of evidence alone do not suffice, they have to be arranged in such a way that they fit together. In criminal investigations and lawsuits the evidence is the facts related to the case and testimonies of reliable witnesses. In mathematics the evidence is axioms, previously established theorems, and computations. The pieces of evidence are just some sentences, thus the essence of the concept of a proof hinges on how they are connected.

In lawsuits one should prove the accusation *'beyond a reasonable doubt'*. Clearly, it is unreasonable to ask for proof without any doubt, as nobody would be convicted. The problem is in that any presented evidence and testimony can be questioned as nothing in the world is one hundred percent sure. To support the questioned evidence new proofs may be required and so on resulting in a never-ending process. But even if all the facts and testimonies are accepted as unquestionable, there are still a lot of hidden assumptions that one has to use. People who try to simulate human reasoning on computers know that even in reasoning about simple situations we unconsciously use hundreds of assumptions.

Gottlob Frege[11]

In mathematics the first proofs also were very incomplete. But already in Euclid's *Elements* the proofs are essentially the same as in contemporary mathematics (except that they sometimes use assumptions not stated explicitly as axioms). The concept of a proof, however, was not defined back then. Aristotle studied *syllogisms*, which are some rules for propositional logic, but he did not have a complete system. In the 17th century the German mathematician and philosopher Gottfried Wilhelm Leibniz (1646–1716) proposed the idea of designing a calculus for logic in which one could present logical reasoning and check its correctness in the same way as we calculate in algebra. It was, however, only two centuries later, when his vision was realized. The first complete system for first-order logic was constructed by Frege in his *Begriffsschrift* in 1879 [77].[12] Peano presented another system in his *Principles of Arithmetic* in 1889 [216]. A.N. Whitehead and Russell published their fundamental work *Principia Mathematica* on the foundations of mathematics in three volumes in 1910, 1912 and 1913 [313–315]. Their formal system for proofs in first-order logic became the prototype for many systems that appeared later.

We say that the concept of a proof is now *formal*, meaning that it can be treated as a rigorous mathematical concept. Note that formalization of the concept of a proof was only needed for understanding the foundations of mathematics. It is remarkable that every mathematician has a perfect sense for what constitutes a proof and what does not. So to say, mathematicians learn the rules of the game by playing it.

Claiming that we have a formal definition of a proof we should give such a definition quite formally. I will do it in the sequel; here I only give a short description how it can be done. Firstly, we take some simple sentences which are obviously tautologies as *logical axioms*. A typical one is the *law of excluded middle*, which is φ *or not* φ; we postulate it for every formula φ. Then we need *logical rules* that enable

[11] This media file is in the public domain in the United States.

[12] The title is translated as *Concept Script*.

us to derive tautologies that are not axioms. The rule *modus ponens* used already by Aristotle is a case in point. This rule says:

Suppose 'φ' and 'if φ, then ψ' *has been established. Then we can derive 'ψ'.*

A proof is constructed by successively deriving sentences. This means that at each step we either simply write down an axiom or derive a sentence from previously obtained ones using a logical rule. Note that at each step we have several options what to do next. The sentence that we want to prove gives us only hints, but there is no general strategy. The daily experience of mathematicians is that sometimes it is extremely hard to find a proof of a given sentence.

Once we list all symbols used in the language, all axioms and all rules, the concept becomes completely formal. The proof is a sequence of symbols satisfying certain syntactical rules. It is simpler to describe the syntax of a proof than, say, the syntax of an English sentence, even not counting the complexity of the English vocabulary. All logicians agree that the concept of a proof can be made completely formal. However, there are some mathematicians who doubt that all actual proofs produced by mathematicians can be converted into such formal proofs. The purpose of a real mathematical proof is not to present a sequence of symbols that can be mechanically checked, but to convey the idea that leads to a proof and persuade the reader that the idea can be realized. Thus for instance, if the proof splits into considering several similar cases, a proof for only one is given and the others are left to the reader to prove, or it is only stated how the proof for the first case can be modified to work for the other cases, etc. Furthermore the steps in such a proof are not elementary steps as in a formal proof, they are more like small jumps. Depending on how big these jumps are, the proof requires more or less effort and ingenuity on the part of the reader. Careful checking of a ten-page article may take several days, but there are some that need many more. If we have problems with short proofs, what would happen if we tried to convert a really long one into a formal proof?

Can All Mathematical Proofs Be Turned into Formal Proofs?

I contend that, if needed, we would be able to formalize any proof with only a moderate amount of work. I think that our experience with computers demonstrates it sufficiently clearly. That experience shows that any algorithm can be formalized. Before the advent of computers nobody cared to write down algorithms formally. Nowadays, when we need to communicate a lot of algorithms to computers, there are algorithms of extreme complexity written formally. This was certainly assisted by the use of high level programming languages. While people write programs routinely, writing a mathematical proof in a formal system so that a computer can check it is only done by researchers in the field of proof checking and a few mathematicians. Proof checkers and automated theorem provers had been written almost as soon as computers became available. Already in the late 1960s, Hao Wang (1921–1995) wrote a program that proved all the approximately 350 theorems of *Principia Mathematica*, (see [303]). Since then the power of computers has increased dramatically and many proof checkers and programs for automated theorem proving have been written. The interest of working mathematicians in these programs is growing and the demand for such programs in industry is significant. It is just a matter of time when applications of formal logic become an important part of the computer industry. Then writing formal proofs will be as routine as writing programs.

The book *Proofs and Refutations, The Logic of Mathematical Discovery* [178] by Imre Lakatos is an interesting treatise on mathematical proofs. The leading idea is that mathematical theories can turn out to be wrong in the same way as it happens with physical theories. A physical theory, however nice it is, has to be refuted, if there is an experiment which is in contradiction with the prediction of the theory. In physics it is not a theoretical possibility, but rather a typical process—theories are proposed and refuted only to be replaced by more accurate ones. This goes on and on. Lakatos's claim is that the same happens in mathematics. As a case example, he considers proofs of the well-known Euler formula which asserts that for each polyhedron the number of vertices plus the number of faces equals to the number of edges plus 2, which is expressed as

$$v + f = e + 2.$$

He shows how proofs of this theorem were found and then refuted by counterexamples, then fixed again and so on. He hints that this is a never-ending process. I do not think that by claiming that proofs are never quite correct he wanted to harm mathematics. Apparently, he was inspired by Karl Popper's *Conjectures and Refutations* whose main thesis is that a scientific theory must be refutable at least in principle. Ideologies claiming their infallibility are good only for manipulating people. This is a great idea of a great philosopher, but should it be applied to mathematics too? There is a fundamental difference between mathematical results and physical theories. When a physical theory fails, it means that the given equations are not applicable to the reality they should have represented. It is not that the theory as a mathematical work is wrong. Mathematical results, though inspired by our physical experience, are derived without referring to it; they may have one or more useful applications, but some do not have any, and this is not considered as a failure.

Lakatos talks on refutations of proofs, but it is not the concept of a proof which is not rigorous, the problem is in the definitions of the concepts used in the proofs. The troubles are caused mainly by the definition of polyhedra. There are several classes of three-dimensional shapes that we might be willing to call polyhedra. If we take the most restrictive definition, a simple proof works well. For more general forms, the proof fails and we may need a different proof or the theorem is not true at all. Proofs are refuted because some tacit assumptions turn out to be wrong. It is not logic that fails.

Some mathematicians may still doubt that it is possible to spell out all the assumptions in deeper mathematical proofs. This was indeed the case until the 19th century. By the end of that century all mathematical concepts had been precisely defined and that became a standard for the future. The concepts introduced into mathematics later were always constructed rigorously. Furthermore also the concept of a proof itself became a part of mathematics. I admit that some proofs would need a considerable amount of work to be turned into formal proofs, but I am sure it would be less work (and, perhaps, more interesting) than writing an average size computer program. According to current estimates, based on the experience with writing formal proofs for computer checkers, the size of a formal proof is on the average about three to four times bigger than a plain "mathematical" proof. This may seem too much for being accepted by mathematicians. But a four-times longer text does not mean that one needs four times more *time* to write it. Every mathematician knows that writing up a mathematical result is always the easier part of the job, usually much easier than discovering the result. There is another argument that suggests that mathematicians might be willing to devote a little more time to writing their proofs if they could gain something. It is the experience with typesetting mathematical papers and, in particular, mathematical texts. Nowadays mathematics (including books like this) is typeset in the system TEX (or various versions of it) designed by the American mathematician and computer scientist Donald Knuth. If you compare the difficulty of writing a formula in TEX with writing it by hand you probably get a similar factor (about 3:1) for the ratio between formal and informal proofs. Yet most mathematicians prefer typing their papers in TEX themselves.

In 1892, Peano started an ambitious project called *Formulario Mathematico*, [217]. The aim of the project was to write a collection of all known mathematical theorems in a formal logical language. He expected that the advantage of presenting mathematics in this way would be brevity, precision and uniformity. The succinctness of logical formulas would enable him to cover a large amount of material, which otherwise would be impossible. The project was finished in 1908, but it was rather a failure—nobody was using it, except for Peano. Although he described the main theorems also in words, it was very difficult to read it because most of the text were just formulas containing a lot of new symbols. Nevertheless, some symbols introduced by Peano have been accepted and have been used since then; for instance the set-theoretical symbols $\in, \subset, \cup, \cap$.

Peano's idea was that mathematics would be taught using *Formulario*. He gave such courses of calculus, but the students were desperate. His great mistake was that he did not realize that mathematical texts have a twofold role. The first one is to verify the truth of the stated theorems. Therefore we need precise language, precise definitions and precise proofs. It is true that the highest level of precision is only attainable by formal systems, but they do not have to be based solely on symbols. The second role, the one that Peano neglected, is to communicate ideas to other people. It is possible to memorize a formal proof, but it is useless. Mathematicians need to understand the proof. They need to make a "mental picture" of the proof and put it into context of related results etc. Again, there is a parallel with computer programs. A computer only needs a formally precise program, but studying an uncommented complicated program is a programmer's nightmare.

One reason some mathematicians do not believe that proofs can be written formally is that introductory textbooks most often use formal systems for first-order logic that are simple to describe, but difficult to use. These systems were developed with the aim to have as simple definitions of the formal systems as possible. There are, however, other systems that are aimed at practical use. Proofs in these systems are indeed very close to the way mathematicians argue in their proofs, therefore they are called *natural deduction* systems. I will describe such a system in Notes; below I will only give an example of a proof in this system.

Example Let us compare a mathematical proof and a fully formalized proof in a natural deduction system. We want to prove the tautology:

If it is not true that A is true and B false, then either A is false or B is true.

Here is how a mathematician would prove it. (Mathematicians would do it only for didactic reasons because the tautology is "obviously true".) The numbers in parentheses are added for the sake of comparing the mathematical proof with the formal proof below.

Suppose that it is not true that A is true and B is false (1.). *Consider two cases.*
Case 1, *A is true* (2.). *Arguing by contradiction, suppose that B is false* (3.). *Then A is true and B is false, which contradicts to our initial assumption. Hence B must be true* (5.). *This implies that either A is false or B is true* (6.).
Case 2, *A is false* (7.). *Then we have again that either A is false or B is true* (8.).
Since A is either true or false (9.) *and in both cases either A is false or B is true, the latter fact is always true* (10.).
Since we have derived that either A is false or B is true from the assumption that it is not true that A is true and B is false, we have proved the implication (11.).

In logic the tautology is expressed by

$$\neg(A \wedge \neg B) \rightarrow (\neg A \vee B).$$

Here is a formal proof in the natural deduction system presented on page 113.
1. $\neg(A \wedge \neg B)$ [assumption]
2. A [assumption]
3. $\neg B$ [assumption]
4. $A \wedge \neg B$ [from 2. and 3. by rule 1]
5. B [from 1. and 4. by rule 8]
6. $\neg A \vee B$ [from 5. by rule 3]
7. $\neg A$ [assumption]
8. $\neg A \vee B$ [from 7. by rule 3]
9. $A \vee \neg A$ [axiom]
10. $\neg A \vee B$ [from 9., 6. and 8. by rule 4]
11. $\neg(A \wedge \neg B) \rightarrow (\neg A \vee B)$ [from 1. and 10. by rule 5]

The main problem of proof checkers (and automated theorem provers as well) is their lack of the basic knowledge that all mathematicians have. For most results, should they be represented formally, one needs to introduce a lot of concepts and prove a lot of elementary theorems. This depends on the fields of mathematics. In set theory it is not such a problem, since what we start with are axioms of set theory. That is why Wang was able to prove all theorems of *Principia Mathematica* [303]. In other fields one has to go a very long way from the axioms of set theory. Thus some proof checkers are only built for a single mathematical field. To overcome this problem we need to create a large "mathematical library" of basic concepts and theorems, so that then authors can use some concepts and refer to some theorems instead of proving everything from scratch.

The second big problem is the lack of a sufficiently strong versatile automated theorem prover. When writing a proof, a mathematician leaves a lot of simple logical deductions to the reader (sometimes they are not that simple, sometimes they are gaps that cannot be fixed). It would be very boring, also for the reader, if every simple detail were spelled out. So this has to be done by computer and eventually will, but it will take time.

Now that I have spoken so much about the exactness of proofs the reader may have become worried that I want to hide something more fundamental, without which the above question does not have a proper meaning, namely:

Is the Concept of Logically Derivable Sentences Uniquely Determined?

One can propose a large number of different logical calculi, thus the question is whether they differ also in the theorems that they are able to prove. This is a really fundamental question, it is the question about what logic is. Recall that in the previous section we gave a semantical definition of true sentences. Our point of view is that structures are primary and logic is a means to describe them. Therefore, our aim should be to show that true sentences are exactly those that are provable, which will automatically solve the problem of the uniqueness. Showing that the semantical definition of logically valid sentences is the same as the syntactical one entails two things:

1. *one can only prove logically valid sentences, and*
2. *every logically valid sentence can be proved.*

The first property is called the *soundness* of the logical calculus and it is fairly easy to prove it. One shows that the logical axioms are true and that logical rules preserve truth, and then we apply the principle of mathematical induction to prove that all derivable sentences are true. Checking axioms and rules is easy, as they have a very simple structure. Let us look, for instance, at *modus ponens* (see page 94). We want to show the following: if φ is true and φ *implies* ψ is true, then also ψ is true. But that is obvious, because if φ *implies* ψ is true and we know that φ is true, then also ψ must be true.

It is as if I heard you saying: '*Wait a moment, what kind of proof is this! You are proving the soundness of modus ponens by using modus ponens!*'. Indeed, this is what we do. But notice that this is exactly the same situation as it was in the definition of satisfaction. We are just translating the logical axioms and rules from the metalanguage into a formal system which is the object of our study.

Also here it helps to imagine that the formalization is intended for a computer. Suppose that we have written a program for automated theorem proving and now we want to check that it is correct. Namely, we want to prove that it can only prove logically valid sentences. Then we would argue in exactly the same way as sketched above. We would say that it can only start with logical axioms, which are obviously logically valid, and derive sentences by rules that are obviously logically sound.

Note that we do use a mathematical argument in this proof, although a very easy one—we are using mathematical induction to prove that all derived sentences are logically valid.

Those who doubted the usefulness of logic before starting to read this book may view this as a confirmation of their opinion. Indeed, it looks like all logicians do is just rewriting obvious things in an obscure formalism, but we will shortly see that this is not the case.

Let us now turn to the second property. It concerns the question about the *completeness* of logical calculi: are our logical calculi complete in the sense that one can derive all true sentences from them? Here we meet Kurt Gödel (1906–1978) for the first time with his first fundamental result:

The Completeness Theorem *The calculi for first-order logic are complete.*

In his doctoral dissertation from 1929 (published in a journal in 1930 [95]), Gödel proved the theorem for the calculus presented in *Principia Mathematica.* But the choice of the particular calculus was not important because all proposed calculi are equivalent.

Kurt Gödel
Courtesy of Kurt Gödel Society, Vienna

The soundness and completeness means that the semantical definition of logical validity coincides with the syntactical definition. This has one profound consequence. Since proofs are finite entities that we can explicitly construct, logical validity is a concept that is fully accessible to us.

Maybe completeness does not come as a surprise to you and you expect a similar kind of a "silly argument" that I used to prove soundness, but completeness is far less obvious and requires a nontrivial proof. The reason is that the two properties, soundness and completeness, are not treated in the same way. The soundness is clearly a necessary condition that we require from any logical calculus, and that we had better put it in the definition of a logical calculus. Of course, we would like to have both properties, but we would not sacrifice soundness for the sake of completeness. This is not a purely academic question, as in general completeness does not hold always. It holds for the most common logic, first-order logic, but there is no way to extend it to stronger logics which contain set theoretical notions. I will give an informal sketch of the proof of the Completeness Theorem in the sequel.

At this point it is difficult to fully appreciate the fact that it is possible to completely characterize the true sentences by the syntactical concept of a proof (expressed briefly, that we have soundness and completeness). It is a great achievement that the question of what logic is has been solved once and for all. We will see in the next section that, in spite of similarities between the concepts of proofs and computations, we are not able to prove that the commonly accepted definition of computations is the right one.

The First Incompleteness Theorem

We know what logic is, we know what proofs are, and we can prove every logically valid sentence. So why are we not yet content? The problem lies in the distinction between logical validity and mathematical truth. First-order logic, which works so perfectly, tells us what is true in *all* first-order structures. Though this is an important part of mathematical truth, there are other things in mathematics. The most interesting results in mathematics concern single structures. For example, in number theory, we study the structure of natural numbers. When proving theorems about numbers we, of course, use logic, but we also need specific properties of the structure of numbers that are not of a pure logical nature. On top of logic, we also need axioms about numbers. We need axioms in geometry, we need axioms about the real numbers, the real valued functions, and so on. By accepting set theoretical foundations we reduce the problem of finding axioms for particular structures to finding axioms for sets, but as we have seen, it is not a simple task to find axioms for sets. Thus we may give up set theory and try to find the axioms at least for the theory of numbers. When we restrict the project of finding a complete axiomatization only to the natural numbers, it does not look too ambitious. After all, natural numbers are the most familiar entities that we find in mathematics and they are so concrete. Furthermore, we know the basic principle that governs the natural numbers: the principle of mathematical induction. Still, it turns out that it is impossible to find such an axiomatization. Let me stress that the problem is not caused by an extreme complexity or difficulty, it is simply impossible.

David Hilbert
Courtesy of Mathematisches Forschungsinstitut Oberwolfach

This, certainly, needs more explanation, but let me make a brief digression into history. The aim of *Hilbert's Program* was to find axioms for all the basic mathematical structures and proofs of the consistency of these axiomatic systems. The program was presented in Hilbert's address at the Third International Congress of Mathematicians, held in Heidelberg in 1904. Rudiments of his program are also the content of the second problem of the famous twenty three open problems presented on the previous International Congress of Mathematicians, held in Paris in 1900. The problem asked for proving the consistency of an axiomatic system for the real numbers.[13] A prototype of axiomatizations that Hilbert asked for was his axiomatization of geometry. Proofs of consistency were also quite common in geometry, where the consistency was shown by constructing a model for a given set of axioms. If successful, the program would guarantee that whatever meaning mathematical results have, we would

[13]His goal was not to find axioms for the structure of real numbers with only operations $+$ and $\times$. He wanted to axiomatize a richer structure in which at least a basic mathematical analysis could be done.

at least know that proving theorems is not a completely futile activity. In a consistent theory the provable theorems are separated from refutable ones, so by proving a theorem we get a piece of information. Having a complete list of axioms for a given mathematical structure would, furthermore, give us the confidence that at least potentially we can prove every true theorem. This would not solve the problems of foundations completely, as there still would remain questions such as what mathematics has to do with reality, where do the structures come from, etc. But, in some sense, it would leave the problems to philosophers—the remaining problems would not be mathematical problems.

Hilbert was very enthusiastic about this project and several mathematicians, in particular the Swiss logician and philosopher Paul Bernays (1888–1977) joined him. They founded a new field of logic, *proof theory*,[14] the main aim of which was to achieve the goals of Hilbert's Program. But after initial success came a blow. In 1930 at a conference in Köningsberg Gödel presented his incompleteness theorem, which we call the *First Incompleteness Theorem.* This theorem showed that it is impossible to achieve the goal of Hilbert's Program because every theory that formalizes a sufficiently large part of mathematics is necessarily incomplete. The Second Incompleteness Theorem had even more profound impact on Hilbert's Program, but let me first explain the simpler of the two theorems.

In order to state the First Incompleteness Theorem more formally, we need to understand the concepts involved. Recall that a formal theory is a theory axiomatized by a decidable set of axioms. The decidability means that using an algorithm we can decide whether or not a given sentence belongs to the set of axioms. In practical theories, the decision procedure is very simple because a typical theory has a finite list of axioms and, possibly, a finite list of axiom schemata.

The second thing that needs specification is what part of mathematics should the theory axiomatize. In his paper Gödel used the Theory of Types of Whitehead's and Russell's *Principia Mathematica* (as the title of his paper suggests "*On formally undecidable sentences of Principia Mathematica and related systems I*") [96]. Focusing on *Principia Mathematica* was a natural choice, as this was the most important of the formal theories that aimed at formalizing the whole of mathematics. Gödel showed that not only this theory is incomplete, but so is any extension.

Requiring the entire Theory of Types to be in the theories about which the Incompleteness Theorem speaks is a fairly strong assumption. Weakening this assumption is an important problem because one may hope to be able to completely formalize at least some parts of mathematics, and it was known that in some cases it was possible (for example, elementary geometry). Gödel was aware of the fact that his theorem can be proved for other theories and he explicitly mentioned Zermelo-Fraenkel Set Theory, von Neumann's theory and a version of Peano Arithmetic. Analyzing Gödel's proof, it is not difficult to see that the proof only needs some *finite* mathematics to be a part of the theory. The theory should be able to prove some basic facts about finite structures—finite sets, or finite strings, or natural numbers. Of the

[14] *Die Beweistheorie* in German.

three mentioned concepts, the last one is part of the most traditional mathematics. Therefore, the modern presentation of the theorem assumes that the theory formalizes the basic concepts of arithmetic and certain basic theorems about arithmetic are provable in it. One should, however, keep in mind that using arithmetic in the statement is only an elegant way of presenting the theorem and equivalent theorems can be stated using finite sets, or finite strings.

Another important condition is the consistency of the theory. If a theory is inconsistent, then it is complete in a really bad way—everything is provable. When talking about complete theories we usually have in mind those that are also consistent. For such theories, exactly one of the sentences ϕ or $\neg\phi$ is provable for every ϕ, whereas in inconsistent theories both are provable. Although we are only interested in consistent theories, when stating a general theorem we have to take care of the undesirable case of inconsistent theories. Gödel used a stronger assumption, called ω-consistency, which was later weakened to mere consistency of the theory by J.B. Rosser.

Now we know all we need to state the theorem.

The First Incompleteness Theorem *Any consistent formal theory T that is able to formalize a certain part of arithmetic is incomplete. More precisely, there is an arithmetical sentence ϕ such that neither ϕ nor its negation $\neg\phi$ is provable in T.*

The remarkable fact is that the unprovable sentences concern finite structures and numbers—the concepts that are at the heart of mathematics, the entities that most mathematicians view as real (real at least in the sense that there should be no ambiguity what is true and what is false).

Higher Order Logics and Theories

What we have considered so far is first-order logic. 'First-order' because we have variables and quantifiers only for the lowest order objects, relations and functions are fixed. What happens if we introduce variables for relations and allow quantifying them? As far as the language and its interpretation are concerned, there is no problem; we can expand the language in this way. Thus we obtain second-order languages—the languages for second-order structures. Second-order languages enable us to define interesting things; for example, we may define that 'two objects are the same if every property that one has, the other also has'. The next step is to define the satisfaction of second-order formulas in second-order structures. This also causes no problems; the definition of satisfaction for first-order formulas extends naturally to the second order. Having the definition of satisfaction, we can define logically valid second-order sentences—the sentences satisfied in all second order structures.

In order to define second-order logic, it still necessary to find axioms and rules of this logic. But here we run into an essential problem. The incompleteness phenomenon concerns also this logic: *there is no formal system by which the set of logically valid sentences could be defined.* In other words, we may propose some system of logical axioms and rules, but if such a system is sound, then it is incomplete. Thus what is called '*second-order logic*' is only the set of logically valid second-order sentences; there is no second-order calculus.

This paradigm can be used to define logics of any order. But all these logics suffer the same problem as second-order logic because they contain it as a subsystem. One can also take a system that contains all these finite order logics, which looks like the ultimate system, the most general logic of all. For such a logic, Whitehead and Russell introduced the Theory of Types. But their system is incomplete, and, due to Gödel, we know that it is incomplete for an essential reason.

Therefore, we should not classify these higher order logics as logics at all. The possibility to characterize logically valid sentences using a formal system is an essential property of logics. The most natural place to draw the line between logics and set theories is between first-order logic and second-order logic. Note that already in the third order theory of arithmetic one can express some important sentences and concepts of set theory such as the Axiom of Choice, the Continuum Hypothesis, the determinacy of infinite games, etc.; in weaker forms they can also be defined in the second-order theory of arithmetic.

Higher order logics were intended to be the base logics of theories about higher order structures. Since higher order logics are not axiomatizable, first-order logic is also used as the base logic for higher order structures. For example, there is a very natural (but incomplete) axiom system for the second-order structure of natural numbers, which is the standard structure of natural numbers extended by subsets. This theory, called *Second Order Arithmetic*, has two sorts of objects, numbers and sets, an both are treated as first order elements (see page 295). This is the same as in set theories, where sets of any type are treated as elements and the membership relation $\in$ is a binary relation defined on these elements.

The realization that there is only one logic, namely, first-order logic, was an important step in the development of mathematical logic. This idea is due to Skolem [271].

The Second Incompleteness Theorem

The next natural question is: what is the meaning of the independent sentence constructed in the proof of the First Incompleteness Theorem? Gödel found the answer soon after he proved the First Incompleteness Theorem (in the same year 1930). He proved a strengthening of the First Incompleteness Theorem, which is called the *Second Incompleteness Theorem*, that specifies the nature of the incompleteness. According to this theorem the unprovable sentence expresses the formal consistency of the theory.

The Second Incompleteness Theorem *If T is a consistent formal theory which is able to formalize a certain part of arithmetic, then T does not prove its own consistency.*

John von Neumann, who learned Gödel's First Incompleteness Theorem in Köningsberg, discovered the Second Incompleteness Theorem independently too. The letter in which he announced his result to Gödel arrived just three days after Gödel sent his paper [96] with both theorems to the publisher.[15]

The Second Incompleteness Theorem is not a strengthening of the First. There are consistent theories which prove sentences expressing their *inconsistency*. For such a theory, the Second Incompleteness Theorem does not show that T is incomplete.

It is clear that the first theorem destroys the hopes of achieving the first goal of Hilbert's Program, the axiomatizations. A little more refined argument is needed to show that the second incompleteness theorem kills also the second goal: proving the consistency of axiomatic systems used for the foundations of mathematics. To prove the consistency of such systems seems a more modest goal. We know we cannot completely axiomatize structures such as the natural numbers and the sets, but we may be satisfied with a nice and sufficiently strong axiomatic system. The fact that we are not able to completely axiomatize these structures, however, shows that we are not quite sure what are the theories that describe true arithmetic, true set theory etc. The more arbitrary the axiom systems look, the more pressing the question about consistency is. The last thing that we would want is an inconsistent system, a system in which everything is provable, which gives us no information whatsoever. In particular, in set theory we want to be sure that there will be no new generation of paradoxes that will force us to abandon the currently used axiomatic system.

Let us return to Hilbert at the beginning of the 20th century to see how the second incompleteness theorem works. Based on the original works of Frege, Russell, Whitehead, and others, Hilbert's proof theory established that a proof is a finite mathematical structure described by elementary combinatorics. So the mathematics needed for formalizing the syntax of logic is very simple. The consistency of a theory is simply another syntactical concept. Thus Hilbert was naturally led to the belief that the consistency problem can be handled with the same sort of mathematics. Why would one need to use infinite sets to prove something about finite sets? Hilbert had been thinking about the problem for several years, thus it was clear to him that proving consistencies would not be a completely trivial thing, and he knew very well that he had to use some assumptions, some means. He never specified precisely his idea of *finitary means* that he proposed for proving the consistencies, but the name tells us enough: one should only use numbers, finite structures, etc., infinite sets were disallowed. He hoped that with such restricted means one could eventually prove the consistency of arbitrary consistent theories, even theories such as Zermelo-Fraenkel set theory, in which there are extremely large infinite sets. Suppose we can present the finitary assumptions that he had in mind as a formal system,

[15]November 20 and November 17, 1930 respectively, see [58], p. 87. Gödel's priority has never been seriously challenged. The only fact worth mentioning is that in 1905, Poincaré conjectured (without stating it formally) that one cannot prove the consistency of mathematical induction without using mathematical induction itself [220]. Gödel's Second Incompleteness Theorem applied to Peano Arithmetic implies a possible formal version of Poincaré's conjecture, but it is even stronger.

a theory T based on first-order logic. Then Gödel's second theorem tells us that T is not capable of proving its own consistency. So T already fails to prove the consistency of a theory that only talks about finite objects, let alone the consistency of a theory that talks about infinite ones.

We may modify the approach by saying: well, we cannot only use finitary means, so let's look for *any* theory that would prove the consistency of all consistent theories. But Gödel's theorem also prevents this; in fact, Gödel's theorem says exactly that this is not possible. Every theory T fails to prove the consistency of some theory, namely of itself. Note that if a theory S contains theory T, then the consistency of S immediately implies the consistency of T. Thus T not only does not prove the consistency of itself, but it also does not prove the consistency of any theory that contains T.

Another consequence of the incompleteness theorem is that we can always expand our theory by adding to it the statement of its formal consistency as a new axiom. The expanded theory again does not prove its consistency, it only proves the consistency of the original one. (But, perhaps, this may be a way to produce new useful axioms? I will elaborate on this in Chap. 7.)

Gödel's theorems were the end of Hilbert's Program, but they did not destroy proof theory. This field is still flourishing and results related to the incompleteness theorems are an important part of it.

After this brief acquaintance with the incompleteness theorems, I defer the proofs to Chap. 4, where I will also explain the theorems in more detail. It is impossible to fully understand the incompleteness theorems without knowing at least the basic ideas of proofs.

Misconceptions About the Incompleteness Theorems

The incompleteness theorems are very important results and many people outside of logic know them, but they are often wrongly interpreted. The most frequently occurring misinterpretation concerning the first incompleteness theorem is that '*there are unprovable sentences*', meaning that there are mathematical theorems that can never be proved. The mistake is in that Gödel's theorem does not claim an *absolute* impossibility of learning truth; it only says that *relative* to a theory something is impossible to prove. Take an arbitrary true sentence ϕ. Then, trivially, there is a theory in which it is provable, namely the theory whose axiom is ϕ. Or, if you do not like this, add ϕ to the currently used axioms of set theory. Thus the theorem does not exclude that we can gradually expand our axiom systems and the resulting theories can potentially prove any given sentence. Nowadays most mathematicians accept Zermelo-Fraenkel set theory as the foundations of mathematics, so one may suggest interpreting '*provable*' as '*provable in Zermelo-Fraenkel set theory*'. The problem is, however, that it is more natural to accept this set theory as an open system that we can gradually expand by new axioms. A natural way to expand Zermelo-Fraenkel set theory is to add large cardinal axioms, which are a kind of higher infinity axioms (I will consider this topic in the following chapters). What the incompleteness theorem does say is that we cannot expand theories in a systematic way. Certainly, humankind will only be able to prove finitely many theorems, thus there will be some that we will never be proved. But we cannot tell which are those we will never prove.

A related misconception is that '*we cannot prove the consistency of the foundations of mathematics*'. Suppose we only needed finite structures, so we could use Peano Arithmetic or Finite Set Theory (defined on page 116). For proving the consistency of these theories, it suffices to use essentially any set theory that postulates the existence of infinite sets, much less than the full strength of Zermelo-Fraenkel Set Theory. This also explains '*the paradox that we know that formal arithmetic is consistent, but we cannot prove it*'. Our belief in the consistency of formal arithmetic (represented by Peano Arithmetic) is based on the fact that we need very simple set-theoretical assumptions to construct its model. The correct statement is that *we cannot prove the consistency of some formal foundations of mathematics in the same, or a weaker system.* This is, in fact, the Second Incompleteness Theorem restated in different words.

Here I should remind the reader that finding the foundations of mathematics in which their consistency would be provable was *not* the goal of Hilbert's Program. Having a formal system T in which its consistency would be provable would not help us to justify its consistency. If one believes that theorems of T are true, then one implicitly assumes that T is consistent. Therefore, if we use some T as the foundations, it does not matter whether the consistency of T is provable in T.

The main problem in understanding the incompleteness theorem stems from the fact that the theorem, along with asserting that there is an unprovable sentence, also gives us a concrete true sentence that is missing in the theory (the consistency of the theory). Therefore, it seems that while theories are incomplete, this phenomenon does not affect our knowledge of the true facts. One is lead to the conclusion that we, people, have the ability to overcome the limitations posed on formal systems by the incompleteness theorem. A careful analysis of such arguments immediately reveals the fallacy. Essentially, one has only to spell out the assumptions that are used. These arguments are always based on the assumption that the theory in question is sound, which means that it only proves true sentences. So the fact that we can consistently extend the theory by adding the Gödel sentence to it is already contained in the assumption. This, certainly, needs a longer discussion, which I defer to Chap. 7.

Shortly after the incompleteness theorems were published, there were attempts to avoid the incompleteness phenomenon by replacing first order theories by something else, but very soon it turned out that this cannot work. The reason is that the incompleteness of the theories to which Gödel's theorem is applicable is caused by the complexity of the sets of sentences provable in these theories. Every theory in which it is possible to prove certain elementary propositions about numbers must have this complexity and, therefore, cannot be complete. Consequently, it is not possible to avoid the incompleteness by changing logic and formalism.

On the Proof of the Completeness Theorem

Proving completeness is not as trivial as proving soundness, but once the concepts involved are sufficiently familiar, the proof is not difficult. To prove the theorem Gödel considered the contrapositive implication: if a sentence does not have a proof, then it is not true in all structures. Thus we only need, for a given unprovable sentence, to construct a structure that does not satisfy the sentence: it satisfies its negation. Consider such a hypothetical structure M that satisfies the negation of the sentence. What else should the structure satisfy? Certainly, all sentences that logically follow from the negation of the sentence. This still leaves a lot of sentences that we do not know whether they should or should not hold in M because the sentence does not decide everything. Here comes the first trick. If we cannot decide a sentence φ we can (expressing it in trendy terms) break the symmetry by choosing φ or *not* φ arbitrarily. In both cases the enlarged set of sentences will be consistent. The enlarged set will imply more sentences, but there will still be some undecided. So we repeat this enlarging process on and on. Doing it in a systematic way will ensure that after infinitely many steps we decide all sentences and still have a consistent theory of what should hold in M.

Now we have all sentences that should be true in the hypothetical model M, but we still do not have the model. The crucial idea of this proof is that we can use logical symbols as the elements of the structure. I have warned about the distinction between syntax and semantics so you may be worried that this would break this law, but what Gödel did is permissible. The logical calculus is a mathematical entity and lives in the same world as structures. We are now looking at this world from the outside and consider relations between the elements of the logical calculus and structures. So both proofs and models are the subject of our study, they are at the same level, they are in the world of structures. Proofs, as sequences of symbols, are simply another type of structures. We know that the substance of the elements of structures is irrelevant, so the elements of the structure M may coincidentally be symbols used in the calculus. To construct M, we assume that we have sufficiently many constant symbols and take the universe of M to be these symbols. More precisely, we have to ensure that for every existential statement, we have a constant which testifies this statement. Then it remains to define relations and operations. This turns out to be easy, as we already have all the sentences that should be true in M. To decide, for instance, whether c should be in relation R with d, we just look at the theory and see whether the corresponding sentence $R(c, d)$ is there.

See Notes for more detail on the proof of the Completeness Theorem.

Constructive Mathematics—Proofs Instead of Structures

As it is natural to assume that there is the real world that we perceive by our senses and about which we reason, it is also natural to think about mathematics in this way. Therefore, I have started with describing structures and went on by describing logic as a means of studying structures. But mathematical reality is elusive. When we study the natural numbers, we do not perform experiments, instead we use axioms, for example, the induction axiom. We can try a few examples to see if a conjecture is reasonable, but the only way we can convince ourselves that it is a theorem is to prove it. A proof may use facts that we know about other structures, or facts about sets, but if we analyze carefully all the assumptions that we use, we find out that we are proving it from basic axioms such as axioms for sets and induction. Thus, while pretending that we are working with structures, we in fact are just doing proofs. Hence, shouldn't we consider a proof as the basic entity in mathematics, all the rest being derived from it?

There is a stream in mathematical logic that pursues such a direction; it is called *constructive mathematics*. The main sources from which it originated are the *intuitionism* represented by the Dutch mathematician L.E.J. Brouwer (1881–1966) and the *constructivism* of the Russian mathematician Andrey A. Markov Jr. (1903–1979). Intuitionism started around 1900 as a reaction to two events. The first was the emergence of nonconstructive proofs, which are proofs that show the existence of a mathematical object without explicitly constructing it. The second was the so called foundation crisis caused by the discovery of Russell's paradox. Intuitionists claimed that paradoxes were just samples of symptoms of the ill foundations of mathematics.

The most prominent figure among those whose views were connected with intuitionism at its early stages was the French mathematician Henry Poincaré (1854–1912). Very soon Brouwer became the most important proponent of intuitionism and it was he who coined the name of this stream.

Intuitionists proposed to found mathematics not on formal systems, but on mathematicians' intuition. Their argument does have some appeal: whatever problems occurred in the foundations, they had little impact on everyday mathematics. Most concepts, originally based solely on intuition, survived all the foundation crises without problems. While formal foundations always have had some trouble, intuition apparently hasn't. This argument is acceptable, but if one rejects the classical set theoretical foundations of higher order structures, such as the real numbers, one has to propose something else. Therefore, Brouwer started an ambitious program of revising the whole of mathematics in an intuitionistic manner. What he proposed was too restrictive for most mathematicians and they did not accept it.

One of the main differences between classical mathematics and constructive mathematics is logic. Describing logic went against the aims of intuitionism, as they reject formal systems in favor of intuitive reasoning, but it turned out that the logic they used is a very natural system. It is now called *intuitionistic logic*. Intuitionistic logic uses the same connectives and quantifiers but some of them are interpreted differently. In particular, the disjunction $\phi \vee \psi$ is interpreted as follows. One can assert $\phi \vee \psi$ to be true, only if one can assert that ϕ is true or that ψ is true. In classical logic we can assert $\phi \vee \psi$ without knowing which of the two propositions is true; in intuitionistic logic it is not allowed. The main example is the *law of excluded middle*, which is the classical tautology $\phi \vee \neg\phi$. This is true, in classical logic, no matter whether or not we know which of the propositions ϕ and $\neg\phi$ is true. In intuitionistic logic we must know which of the two is true. Thus one of the basic classical laws is rejected in intuitionistic logic.

Example In intuitionistic mathematics a real number is identified with a sequence of rational numbers that approximate it with arbitrary precision. If we are given two real numbers r_1 and r_2 by such sequences, we do not know how far we must go to establish that $r_1 < r_2$ or $r_1 = r_2$ or $r_1 > r_2$. (In fact, to establish that $r_1 = r_2$ no approximations suffice; we need a proof.) Therefore the trichotomy law does not hold for real numbers in intuitionistic mathematics.

Here are examples of some pairs that start as if they were the same.

$$\pi = 3.14159265\ldots$$
$$355/113 = 3.14159292\ldots$$

$$\sqrt{5} + \sqrt{6} + \sqrt{18} = 8.928198\ldots$$
$$\sqrt{4} + \sqrt{48} = 8.928203\ldots$$

These examples concern two important problems studied in mathematics and computer science: 1. the problem of rational approximations of real numbers, 2. the problem about the complexity of deciding inequalities between sums of square roots of integers.

Intuitionists distinguish two relations that are the same in classical mathematics: *non-equality* $r_1 \neq r_2$ and *apartness* $r_1 \# r_2$. The first means that we only know that r_1 cannot be equal to r_2, the second means that we have an estimate on how much the two real numbers differ. Thus knowing that π is irrational, we can conclude that $\pi \neq 355/113$; to establish that $\pi \# 355/113$ we need a concrete estimate such as $355/113 - \pi > 0.0000002$.

At first glance intuitionistic interpretation of disjunction looks as if it completely eliminates this connective. Why should we assert $\phi \vee \psi$ when we know that ϕ is true? But it is not so. When combining disjunction with other connectives we can get intuitionistic tautologies that use disjunction essentially. Consider the proposition $\alpha \vee \beta \to \phi \vee \psi$. The intuitionistic interpretation of this proposition is: if one can decide which of α and β is true, then one can decide which of ϕ and ψ is true. Hence, for instance, $\phi \vee \psi \to \phi \vee \psi$ is an intuitionistic tautology.

Similarly the existential quantifier is reinterpreted in intuitionistic logic. In order to be able to assert that there exists an element x with a property ϕ, one has to provide an example of such an element. This is quite a natural requirement when one interprets disjunction in the way above, since the existential quantifier is like an infinite disjunction. Disallowing the law of excluded middle is a drastic restriction, as it results in disallowing proofs by cases and proofs by contradiction. But intuitionistic interpretation of the existential quantifier is not so alien to a working mathematician. Mathematicians have always preferred constructive proofs of existential statements. 'Constructive' means that an element with a given property is somehow constructed explicitly. Nonconstructive proofs, proofs in which the existence of an element is deduced indirectly, appeared only at the end of the 19th century. At that time some mathematicians rejected such proofs. I will mention some specific nonconstructive results in Chap. 5 (see page 391).

Intuitionistic logic prohibits certain proofs, so one may hope that foundations based on this logic would be safer. However, Russell's Paradox can be derived in this logic too, therefore changing only logic is not sufficient. While intuitionistic logic is a very natural and stable concept, the other changes proposed by intuitionism are not so well justified. In fact research is still going on with the aim of producing constructive foundations that could compete with classical ones. Nevertheless, there are some interesting and very abstract structures that are based on intuitionism.

The Russian constructivist school came later, in the late 1940s. It was based on more formal concepts. It shares with intuitionism the intention to prohibit nonconstructive reasoning and to this end it uses the same logic. But unlike intuitionism, it gives a precise meaning to the disjunction and the existential quantifier, a meaning based on the concept of algorithm. For example, a sentence

For every x, $P(x)$ or not $P(x)$.

is true if and only if there is an algorithm to decide the property P. Or one can say that

For every x, there exists y such that $Q(x, y)$.

only if it is possible to construct y from x such that $Q(x, y)$, which precisely means that there is an algorithm that does it. The mathematics of constructivism is based on the strict requirement that only constructive structures should be studied. Again, 'constructive' means algorithmically computable. As a result all real valued functions must be computable in the following sense. Given a good approximation of x by a rational number r, we should be able to compute a good approximation of the function value $f(x)$ by a rational number q. A consequence is that all such functions are continuous. This is an appealing feature, as it returns to the original intuition of the function and very well corresponds to what is going on in reality. In the real world all physical processes are more or less continuous. But such little positive pieces are completely overwhelmed by a host of negative results such as the result saying that there is a piecewise linear function that is not integrable.

Thus most mathematicians do not find the constructivists' approach to be very useful. Although constructive mathematics has contributed a lot to the study of computations, it seems that more has been done in the classical fields of mathematics; this concerns especially computational complexity. On the positive side, constructive systems are very useful for designing programming languages and in automated theorem proving.

Intuitionistic logic can also be used to obtain new theorems in classical mathematics. Remarkable examples of such applications can be found in the area of *proof mining*. This is a research program whose goal is to extract more information from existing proofs. It was initiated in the 1950s by Georg Kreisel who asked the famous question: "*What more do we know if we have proved a theorem by restricted means, than if we merely know that it is true?*" Many logicians have contributed to this program (for example, J.-Y. Girard analyzed Van der Waerden's Theorem [93], H. Luckhardt analyzed Roth's Theorem [187]). U. Kohlenbach studied the logical structure of certain proofs in functional analysis and succeeded in strengthening and generalizing a number of theorems in this field [158]. The key tool that he used was intuitionistic theories. In spite of using intuitionistic systems, the results he obtained are standard mathematical theorems that have nothing to do with intuitionism.

There are various branches of constructive mathematics, but most logicians working in constructive mathematics share the view that the principal objects in mathematics are proofs. For Brouwer, mathematics was a mental activity independent of any objective reality and the products of this activity were proofs. Some constructivists go as far as to reject semantics at all. Contemporary constructive mathematics focuses on connections between algorithms and proofs. The connections discovered so far confirm the key role of proofs in mathematics. This is in contrast with the traditional view that the main objects are mathematical structures and proofs only serve to gain more knowledge about them.

Notes

1. *Algebra of logic and the logic of relations.* George Boole (1815–1864) partially realized Leibnitz's idea in his book *An Investigation of the Laws of Thought.* He axiomatized the theory of classes with the basic set-theoretical operations, intersection, union and complement. From the point of view of the systems currently used in mathematical logic, we can view his system either as propositional logic, or first-order logic restricted to the use of predicates, i.e., unary relations, or as an axiomatization of Boolean algebras. Logic, however, cannot be based only on unary relations. Therefore, Ernst Schröder (1841–1902), Charles S. Pierce (1839–1914) and others developed *algebras of relations*. This line of research culminated in the work of Tarski and his students, who introduced the concept of *cylindric algebras* [119, 120]. Cylindric algebras formalize first-order logic in a purely algebraic fashion, like Boolean algebras do in the case of propositional logic.
2. *Axioms and rules for propositional logic.* The first thing we should realize is that there is a simple algorithm to test whether a propositional formula is a tautology. By definition, a formula is a tautology if and only if it is true for all truth assignments to propositional variables. Since every variable can only have two values there is only a finite number of truth assignments and one can systematically check all of them. Why then do we need any axiomatization at all? There are several reasons. One reason is purely aesthetic: we have to axiomatize first-order logic, so it would be awkward if the propositional part of it was presented in a different way. Another one is that the way people think is closer to an axiomatic system than to truth value checking. Last but not least, for propositions with many variables, the exhaustive algorithm would run too long, while the proposition may still have a short proof. One of the main open problems in proof complexity is whether or not there exists a proof system for propositional logic in which every tautology has a proof of at most polynomial length.

   Here are the axiom schemata of a system of axioms for propositional logic (from [110]).

$P \to (Q \to P)$
$(P \to (Q \to R)) \to ((P \to Q) \to (P \to R))$
$(P \vee Q) \to (Q \vee P)$
$(P \wedge Q) \to (Q \wedge P)$
$P \to (P \vee Q)$
$(P \wedge Q) \to P$
$(P \to (Q \to (P \wedge Q))$
$((P \to R) \wedge (Q \to R)) \to ((P \vee Q) \to R)$
$(P \to (Q \wedge \neg Q)) \to \neg P$
$(P \wedge \neg P) \to Q$
$P \vee \neg P$

The system has a single rule, modus ponens:

From P and $P \to Q$ derive Q.

In the above axiom schemata and in the rule the letters P, Q, R are variables representing arbitrary propositions. In the propositional calculus we can substitute for them arbitrary propositional formulas to get an instance of an axiom or the rule. Thus the formulas above represent infinitely many axioms.

Examples 1. Let us interpret P as the propositional variable p and Q as $p \to p$. Then $p \to ((p \to p) \to p)$ is an axiom by the first axiom schema.
2. Interpret P as p, Q as $p \to p$ and R as p. Then

$$\big(p \to ((p \to p) \to p)\big) \to \big((p \to (p \to p)\big) \to (p \to p)\big)$$

is an axiom by the second axiom schema. From this, we can derive $p \to p$ by applying two instances of the first schema and twice modus ponens.

A proof in this system is a sequence of propositional formulas in which every formula is either an instance of an axiom schema or follows from two formulas occurring before by the rule. Any formula in a proof is a tautology.

There are many ways in which the propositional calculus can be axiomatized. The above one belongs to a group in which the aim is to minimize the number of rules, namely, there is only one rule and many axiom schemata. In such systems modus ponens is the most common rule used. The other extreme is to take a minimal number of axiom schemata with many rules. Such systems are called *sequent calculi* and are extremely useful for analyzing proofs. Note that due to the fact that in the axiom schemata above the main connective is the implication, one can easily present them as rules (except for the last one).

Axiomatizations also depend on the set of connectives that we want to use. We can take a small set of connectives that forms a complete set of connectives (for example, $\neg$ and $\to$) and thus we get a simpler set of axioms.

There are several other types of systems that are not based on axioms and rules in particular natural deduction systems, see below.

3. *Axioms and rules for first-order logic.* To get an axiomatization for first-order logic, we can take the set of axiom schemata from the previous paragraph, *modus ponens* (now applied to first-order formulas) and add only two axioms and two schemata specific for quantifiers. The axiom schemata are:

   $$(\forall x \phi(x)) \to \phi(t)$$
   $$\phi(t) \to \exists x \phi(x)$$

   where x stands for an arbitrary first-order variable and t stands for an arbitrary term whose variables are not quantified in ϕ and it is substituted for all occurrences of the variable x. The rules are:

   From $\phi \to \psi(x)$ *derive* $\phi \to \forall x \psi(x)$.
   From $\psi(x) \to \phi$ *derive* $(\exists x \psi(x)) \to \phi$.

   In both rules there is a restriction that the variable x must not occur in ϕ. (One can weaken it to: every occurrence of x in ϕ is in the scope of a quantifier that binds x).

   Formal systems of this form are called *Hilbert style systems.*
4. *Natural deduction systems.* The main feature in which they differ from others is that a proof may contain a subproof. A subproof uses some additional hypotheses that are valid only in the subproof. An example is a subproof where an implication $\alpha \to \beta$ is proved. The subproof starts with the hypothesis α and ends with β. The subproofs may contain other subproofs and so on. One can use formulas from outer subproofs in inner subproofs, but not conversely. This is very much like in programming languages where one can use procedures with local variables. Most rules come in pairs associated with connectives or quantifiers. One member of the pair is used to obtain a more complex formula using the connective or the quantifier, the other does the opposite.

   Here is a list of the axioms and rules of a natural deduction proof system.
   Axiom schema: $A \vee \neg A$.
   Rules:

   a. *From* A *and* B *derive* $A \wedge B$.
   b. *From* $A \wedge B$ *derive* A *and* B.
   c. *From* A *derive* $A \vee B$ *and* $B \vee A$.
   d. *If* C *was derived from hypothesis* A *in a subproof and* C *was also derived from hypothesis* B *in a subproof, then from* $A \vee B$ *derive* C.
   e. *If* B *was derived from hypothesis* A *in a subproof, then derive* $A \to B$.
   f. *From* A *and* $A \to B$ *derive* B (*modus ponens*).
   g. *If* B *and* $\neg B$ *were derived from hypothesis* A *in a subproof, then derive* $\neg A$.
   h. *If* B *and* $\neg B$ *were derived from hypothesis* $\neg A$ *in a subproof, then derive* A (*proof by contradiction*).
   i. *From* $A(y)$ *derive* $\forall x A(x)$ (*y must not occur in any hypothesis*).
   j. *From* $\forall x A(x)$ *derive* $A(t)$ (*for any term t whose variables are not quantified in* A).
   k. *From* $A(t)$ *derive* $\exists x A(x)$ (*for any term t whose variables are not quantified in* A).

1. *If B was derived from hypothesis $A(y)$ in a subproof, then from $\exists x A(x)$ derive B (y must not occur in other hypotheses).*

The first natural deduction system was introduced by a Polish logician S. Jaśkowski in 1934.

5. *The proof of the Completeness Theorem—more detail.* Recall that given an unprovable sentence ϕ we want to construct a model M in which ϕ is false. Note that ϕ being unprovable is equivalent to $\neg\phi$ being consistent. So, for a consistent sentence we need to construct a model. I will consider a more general task: for a given consistent set T of sentences to construct a model in which they are satisfied. Thus, in particular, I will show that every consistent formal theory has a model.

   The idea is to gradually enlarge the set T to a set T' in such a way that we can use terms of the language as the elements of a model. For the sake of simplicity, let us assume that the language of T contains one binary relation R and, possibly but not necessarily, function symbols and constants; furthermore, we will assume that equality is not in the logical calculus.

   I will start by stating the properties that T' should satisfy:

   a. T' contains T;
   b. T' is consistent;
   c. T' is complete;
   d. for every sentence $\exists x\varphi(x)$ in T', there exists a closed term (a term without variables) t such that $\varphi(t)$ is in T'.

   Given T' with the above properties, define a model M as follows. The universe consists of all closed terms of T'. The interpretation of R in M, denoted by R^M, is defined by

$$R^M(s,t) \quad \text{if and only if} \quad R(s,t) \text{ is in } T',$$

   where s and t are terms without free variables. Similarly, the interpretation of an n-ary function symbol F, denoted by F^M, is defined by

$$F^M(t_1,\ldots,t_n)=s \quad \text{if and only if} \quad F(t_1,\ldots,t_n)=s \text{ is in } T'.$$

   The main lemma is the following:

   Lemma 1 *A sentence is true in M if and only if it is in T'.*

   This lemma is easily proved by induction on the logical complexity of the sentence (the number of connectives and quantifiers). The base case, which is the case of atomic formulas, follows immediately from the definition of M.

   Consider $\neg\varphi$, and suppose that the lemma is true for all sentences of smaller complexity. In particular, the lemma must hold for φ. By the consistency and the completeness of T', $\neg\varphi$ is in T' if and only if φ is not in T'. Thus we get the lemma for $\neg\varphi$.

   Now suppose we want to prove it for $\exists x\varphi(x)$ assuming it holds for all sentences of smaller complexity. If $\exists x\varphi(x)$ is in T', then for some term t, $\varphi(t)$ is in T'. Since we assume the lemma for formulas of smaller complexity, $\varphi(t)$ is true in M, hence also $\exists x\varphi(x)$. Proving the converse implication, assume that $\exists x\varphi(x)$ is true in M. Then, for some closed term t, $\varphi(t)$ is true in M. By the induction assumption, this implies that $\varphi(t)$ is in T'. Hence (by the completeness and the consistency of T') also $\exists x\varphi(x)$ is in T'. I leave the other cases to the reader.

To finish the proof of the completeness theorem, it is sufficient to construct T'. Indeed, if we have such a T', then the model M from the lemma is a model of T by the first condition. The construction of T' is done by an infinite process in which we at each step either add a sentence φ if it is still independent, or we add a new constant symbol c and a sentence $\varphi(c)$ if $\exists x\varphi(x)$ is already in the set so far constructed. The technical details how to do it are inessential.

6. *Compactness.* The most important property of first-order logic, called *compactness*, is a consequence of having finitary syntactical means. It is the following fact:

Compactness of First-Order Logic *A set of sentences Φ is consistent if and only if every* finite *subset of Φ is consistent.*

The proof of this proposition is easy. Φ is inconsistent means that there exists a proof of contradiction using Φ as assumptions. But a proof is a *finite* sequence of formulas, hence it only uses a finite number of sentences from Φ. Note that there are also (more difficult) proofs that do not use the Completeness Theorem.

As an easy application, I will show that an infinite model M can always be expanded to a larger model M' (which means that the universe of M' is a proper superset of the universe of M) so that M' satisfies the same sentences as M. To this end, let Φ be the set of all sentences true in M. Let us introduce a symbol, a constant, for every element of M; let c_a, $a \in M$ be these constants. Let c be another constant symbol. Now we are using a language with infinitely many symbols, but all the facts that we need hold true also in this case. Let Ψ be Φ augmented with axioms $c \neq c_a$, for all $a \in M$. I claim that Ψ is consistent. Suppose not, then we have a finite set Ψ' of sentences from Ψ that are inconsistent. Those finitely many sentences can only use a finite number of constants c_a. But then M is also a model of Ψ' because we can interpret the new constant c as an element not mentioned in Ψ'. Hence Ψ' is consistent. By the Completeness Theorem it has a model M'. The universe of the model M' does not have to contain the universe of M. However, as we have a constant for each element of M and these constants are interpreted in M' we have a natural embedding of M into M': element a of M is mapped on the interpretation of c_a in M'. Since M' satisfies all new axioms $c \neq c_a$, the interpretation of c is different from the interpretations of constants c_a. Hence no element of M is mapped on the interpretation of c.

The constructed model M' satisfies the following stronger property: if ϕ is a formula with k free variables, and $a_1, a_2, \ldots, a_k$ are elements of M, then $M \models \phi[a_1, a_2, \ldots, a_k]$ if and only if $M \models \phi[a_1, a_2, \ldots, a_k]$. Such a model M' is called an *elementary extension* of M. If we start with the standard model of arithmetic $(\mathbb{N}; +, \cdot, \leq)$, we obtain a nonstandard model in which exactly the same arithmetical sentences are true as in the standard one.

One can perform this construction with more than one c, in fact, with an arbitrarily large set of constants. If one adds also the axioms $c \neq c'$ for every pair of new distinct constants, then a model of arbitrary large cardinality is obtained. This is the *upward* version of the Löwenheim–Skolem Theorem.

7. *Peano Arithmetic and Finite Set Theory.* One of the most studied theories in mathematical logic is *Peano Arithmetic* (abbreviated as *PA*). It is a theory whose language has one constant 0, one unary operation S and two binary operations $+, \cdot$.[16] The operation $S(x)$ is interpreted as $x + 1$, the successor of x. It is axiomatized by the following axioms:[17]

$$
\begin{aligned}
&S(x) \neq 0\\
&S(x) = S(y) \to x = y\\
&x \neq 0 \to \exists y(x = S(y))\\
&x + 0 = x\\
&x + S(y) = S(x + y)\\
&x \cdot 0 = 0\\
&x \cdot S(y) = x \cdot y + x
\end{aligned}
$$

and for every formula $\phi(x)$, the following is an axiom

$$\big(\phi(0) \wedge \forall x\big(\phi(x) \to \phi\big(S(x)\big)\big)\big) \to \forall x\, \phi(x).$$

This is called the *induction axiom schema* for arithmetical formulas.

The first seven axioms of Peano Arithmetic is a theory called *Robinson Arithmetic*.

Robinson Arithmetic is usually presented with the binary relation $\leq$ and an axiom, which is, in fact, a definition of this relation:

$$x \leq y \equiv \exists z(z + x = y).$$

This theory is very weak (it does not prove even such basic facts as the commutativity and associativity of the operations and the transitivity of $\leq$), but is important in logic.

In algebra one uses the constant 1 instead of the successor function from which the successor function is definable. The successor function is only used in logic and some programming languages. The reason for using it explicitly in axiomatizing Peano Arithmetic is that it is the most primitive function in this system. Note that the axioms for $+$ and $\cdot$ are, in fact, the natural recursive definitions of these operations. Thus it is more natural to have the successor function explicitly. Note, however, that the axioms for addition and multiplication are not definitions in the sense of first-order logic. We can use these formulas as definitions only in higher order logics, or in set theory.

Peano Arithmetic is not quite suitable as a theory on which we could base the foundations of mathematics because the only objects in it are numbers. Even elementary mathematics needs set theory; one should be able to talk at least

---

[16] In the sequel I will also often denote multiplication by $\times$.

[17] I am using the standard convention that the universal quantifiers in front of formulas are omitted.

about finite sets. Although we cannot talk directly about finite sets in Peano Arithmetic, it is possible to code sets by numbers. One of the possible encodings is the following. For a number n, define a set D_n by $D_0 = \emptyset$ and for $n > 0$ let

$$D_n = \{D_{k_1}, \ldots, D_{k_m}\}$$

where $n = 2^{k_1} + \cdots + 2^{k_m}$, $k_1 < \cdots < k_m$. This mapping is a bijection between the natural numbers and *the hereditarily finite sets*. These are sets that are finite, its elements are finite, the elements of its elements are finite, etc. Having this coding we can speak freely about finite structures in Peano Arithmetic. Consequently, we can talk about formulas and proofs, about Turing machines, etc.

Let us now consider the structure consisting of hereditarily finite sets with the usual membership relation $\in$. In the same way, as the natural numbers are the canonical structure for number theory, this is a canonical structure for theories of finite sets. There are several equivalent axiomatic systems whose model is this structure. One such system is the usual Zermelo-Fraenkel set theory with the axiom of infinity replaced by its negation, an axiom saying that all sets are finite. But in fact we do not have to take all the axioms of Zermelo-Fraenkel; for instance, the Axiom of Choice can be derived from others in this theory. The theory is denoted by ZF_{fin} and I will call it *Finite Set Theory* (for lack of a better name). The problem with this name is that it suggests that the theory completely describes the structure of hereditarily finite sets, which is impossible by the incompleteness theorem. But this theory is such a natural system that the name is justified. In particular, if we translate its axioms using the above encoding, we get arithmetical statements provable in Peano Arithmetic. Hence, everything that we can prove in Finite Set Theory can also be proved in Peano Arithmetic. And vice versa, if we define the natural numbers in Finite Set Theory in the usual way, then we can prove all the theorems of Peano Arithmetic. Thus Peano Arithmetic and Finite Set Theory are the same theories up to suitable translations.

This shows that Peano Arithmetic and Finite Set Theory are very natural theories.

8. *Proving relative consistency by interpretation.* We say that a theory T is interpretable in theory S if it is possible to represent the relations and function symbols of T by the formulas of S so that the translations of the axioms of T are provable in S.

   I will define it more precisely in the special case in which the language of T contains only one binary relation R. Let ρ be a formula in the language of S with two free variables. Then for every formula ϕ in the language of T, we can translate ϕ into a formula of the language of S by replacing all atomic subformulas of the form $R(x, y)$ by formulas $\rho(x, y)$. We say that ρ defines an interpretation of T in S, if the translations of the axioms of T are theorems of S.

   Notice that the last condition implies that the translations of all theorems of T are theorems of S. In particular, if T is inconsistent, then so is S. Thus we get an easy, but very important fact:

Theorem 3 *If S is consistent and T is interpretable in S, then T is also consistent.*

Proving that T is interpretable in S is a very efficient way of proving the relative consistency of T with respect to S, especially when T is axiomatized by a finite set of axioms. Then the proof of the relative consistency is given by the proofs of the translations of the axioms. Thus we have a completely finite object that shows the relative consistency.

Usually we need a slightly more general concept of interpretation, in which the domain of the interpretation of T may be a proper subdomain of S. The above relation between Finite Set Theory and Peano Arithmetic are such mutual interpretations. Many relative consistency proofs in set theory are also by interpretation.

9. *A consistent theory that proves its own inconsistency.* According to the Second Incompleteness Theorem a consistent theory T (and sufficiently strong to express such things) does not prove its consistency. This is equivalent to the following statement: the theory T augmented by the statement that T is inconsistent is consistent. I will use $+$ to denote the operation of adding an axiom to a theory and denote by Con_T the formal statement that T is consistent. Thus the Second Incompleteness Theorem can be equivalently stated as follows:

   If T is consistent, then $T + \neg Con_T$ is consistent.

   Now, let T be consistent. Then also $T + \neg Con_T$ is consistent and this theory proves that T is inconsistent. Since $T + \neg Con_T$ is an extension of T, if T is inconsistent, $T + \neg Con_T$ is inconsistent too. This can be formalized in $T + \neg Con_T$, hence this consistent theory proves its own inconsistency.

   What are models of this theory? By the Completeness Theorem we know that this theory has a model. The natural numbers of the model cannot be the standard model, since in the standard model there is no inconsistency of T. Thus the numbers of this model form a nonstandard model and the number representing the proof of contradiction in T is a nonstandard number.

10. *Reducing the consistency of Peano Arithmetic to the well-ordering of ε_0.* In 1936, not long after Gödel published his seminal results on incompleteness, results that essentially destroyed Hilbert's Program, Gentzen obtained a fundamental result in proof theory. His result is in the direction that Hilbert was aiming at: it gives a proof of the consistency of Peano Arithmetic using an argument that seems acceptable from the point of view of finitism [91, 92]. Gentzen proved the consistency of Peano Arithmetic using transfinite induction over ε_0. The ordinal ε_0 is a more complex structure than the natural numbers, but it can be presented as efficiently as the natural numbers, including the operations of addition, multiplication and exponentiation (see Chap. 3).

    It is disputable whether Gentzen's result is a realization of Hilbert's Program in the special case of Peano Arithmetic, but in any case it is an important result. As far as the consistency problem is concerned, it gives us some justification to believe that PA is consistent, as the consistency is based on an assumption that is not directly linked with PA. Another consequence is that we get an independent sentence that is different from the one given by the second incompleteness theorem and which is more "mathematical".

    I will say more about this subject in Chap. 6.

11. *Automated proof checking and theorem proving.* I will explain later why it is not possible to automate the problem of finding a proof of a theorem. Since mathematicians are able to prove difficult theorems nevertheless, and everybody (including animals) can do more or less difficult logical reasoning, it should be possible to do something also on computers. People had wondered to what extent our thinking can be simulated by computers even before the first computers were constructed. Nobody knows how far we are from the day that one will be able to say that a computer thinks, but it is, perhaps, the biggest challenge of humankind.

   Automated theorem proving started in the 1950s with systems designed by M. Davis, P.C. Gilmore, H. Wang and others. It is interesting to note that some of them were quite successful though they had been written before key concepts were introduced in the area (namely the *Davis-Putnam procedure* of Martin Davis and Hilary Putnam, in 1960 [56], and the *resolution*[18] of John Alan Robinson, in 1965 [248]). One of the successes of automated theorem proving is a proof of *Robbins' Conjecture*. This conjecture, which is now a theorem, states that the following equations suffice to axiomatize Boolean algebras:

$$x \vee y = y \vee x$$
$$(x \vee y) \vee z = x \vee (y \vee z)$$
$$\left((x \vee y)' \vee \left(x \vee y'\right)'\right)' = x.$$

   This means that the equational theory defined by these equations, augmented by definitions $0 := (x \vee x')'$, $1 := x \vee x'$ and $x \wedge y := (x' \vee y')'$, is equivalent to the standard axiomatizations, such as the one given on page 21. The conjecture was stated in 1933 and was proved by W. McCune in 1996 [197] using his computer program *EQP* (related to his more well-known *OTTER*). It is not surprising that computers proved to be superior to people in this particular area. After all, manipulating with equations, which in most cases have no meaning, is a rather mechanical work. Yet it is a remarkable fact, since some very good mathematicians tried to solve the problem.

   The proof of Robbins' conjecture should not be confused with *computer assisted proofs*. In computer assisted proofs, such as the proof of the famous Four Color Conjecture, the problem is reduced to a search of a large number of special structures. Except for this part, the proof is completely designed by a mathematician. In the case of Robbins' Conjecture, a general program for proving equations was applied to the conjecture. So computer assisted proofs require formalization of only some special concepts, while theorem provers need formalization of all possible proofs in the field considered. The advantage of Robbins' Conjecture was that it only sufficed to formalize equational proofs.

   One of the first languages for formalizing proofs so that they can be used in computers was N.G. de Bruijn's *AUTOMATH*,[19] whose development started in

[18] See page 60.

[19] http://www.cs.ru.nl/~freek/aut/

1967. In 1994 an initiative *QED* (*Quod Erat Demonstandum*, an abbreviation often used to mark the end of a proof) was launched to unify forces on building a proof checker that would be accepted as the international standard. In the *QED Manifesto* many arguments for such an enterprise had been collected. Unfortunately the project died out eventually without producing significant outputs, leaving various groups to compete, as before. (You may remember the similar, but much more successful initiative called *ALGOL* in the early history of programming languages.) A remarkable event was the verification of the proof of the Four Color Theorem by B. Werner and G. Gonthier in 2004 [106]. It was done using the proof assistant *Coq*, initially developed by T. Coquand and G. Huet in 1991. The Prime Number Theorem was verified by J. Avigad and his students using the system *Isabelle* in 2004. One of the most successful on-going projects of formalizing and verifying proofs is *MIZAR*.[20] Its library of formalized and verified theorems has more than 49 000 items at the time of the writing of these lines.

12. *Intuitionistic logic.* The axioms and rules of classical logic are often chosen so that one only needs to remove something to get intuitionistic logic. This is also the case with the system on page 111; to get an axiomatization of intuitionistic logic we only need to remove the axiom of excluded middle $P \vee \neg P$. The axioms and rules for quantifiers are the same. Thus intuitionistic logic is weaker than classical logic. On the other hand one can interpret classical logic in intuitionistic logic. For classical propositional logic, it is very simple: write two negations before the sentence. So a propositional sentence ϕ is a classical tautology, if and only if $\neg\neg\phi$ is an intuitionistic tautology.

    Intuitionistic logic is more complex than classical logic. Unlike in classical logic, it is not possible in intuitionistic logic to express one of the basic connectives $\neg, \vee, \wedge, \rightarrow$ using the others. The most attractive feature of intuitionistic logic is that one can extract algorithms from certain proofs.

    The presence of double negations that cannot be reduced to a single one shows a relation to modal logics. Both ϕ and $\neg\neg\phi$ express that ϕ is true, but in $\neg\neg\phi$ with lesser emphasis. Fortunately, three negations reduce to one. The connection to modal logics is also seen in semantics.

13. *Modal logics.* Modalities are not used in mathematics; in mathematics we only use precise statements. Still it is interesting to study modal logics. Applications of these logics are in the study of human reasoning and in artificial intelligence, but also in some branches of mathematics. In mathematics we use modal logics as mathematical entities, say, formal systems, or structures, but we reason about them using classical logic without modalities. Attempts have been made to base set theory on a modal logic, in order to avoid Russell's paradox while keeping an unrestricted schema of comprehension, but such proposals will never be accepted by mathematicians (as I explained in Sect. 2.1).

    The basic modalities are *necessarily*, denoted by $\Box$, and *maybe*, denoted by $\Diamond$. Hence, for a proposition ϕ, $\Box\phi$ is read '*ϕ is necessarily true*' or simply

[20] http://mizar.org/project/

'*necessarily* ϕ' and $\Diamond\phi$ is pronounced '*maybe* ϕ'. Using negation we can define one from the other: $\Box\phi$ is equivalent to $\neg\Diamond\neg\phi$ and $\Diamond\phi$ is equivalent to $\neg\Box\neg\phi$. Unlike in classical or intuitionistic logic, there is no canonical unique modal logic. In the spectrum of systems that have been studied there are some that seem to be more natural than others. I will describe one of these here, the system called S_4, and mention another one, *Provability Logic* in Chap. 4 (page 297). I will only talk about propositional logic.

System S_4 is an extension of classical propositional logic. Hence we accept all the axioms and rules of classical logic for formulas in the language expanded by $\Box$. The other modality $\Diamond$, will be treated as an abbreviation of $\neg\Box\neg$. For example, we accept the law of excluded middle for modal formulas. If we apply this law for the formula $\Box\phi$, we get the following tautology $\Box\phi \vee \neg\Box\phi$. Using $\Diamond$, this is equivalent to $\Box\phi \vee \Diamond\neg\phi$. The meaning is: *either* ϕ *is true necessarily, or maybe* ϕ *is false.* To get something interesting, it is important to add some axioms and rules specific for the modality. In S_4 the additional axioms are:

(1) $\Box(\phi \to \psi) \to (\Box\phi \to \Box\psi)$,
(2) $\Box\phi \to \Box\Box\phi$,
(3) $\Box\phi \to \phi$,

and an additional rule, called *generalization*, is

from ϕ, *derive* $\Box\phi$.

Note that there are formulas ϕ such that $\phi \to \Box\phi$ are not tautologies (for example, if ϕ is a propositional variable). This seems counterintuitive because of the rule of generalization. The reason is that in this logic we cannot derive an implication $\phi \to \psi$ by deriving ψ from the assumption ϕ.

14. *Kripke's "possible-world" semantics of modal and intuitionistic logics.* A crucial moment in the history of modal logics was the discovery of *possible world semantics*. The idea is very old, perhaps, this interpretation of modalities was known already to Aristotle, but it was only in the late 1950s when the American logician and philosopher S.A. Kripke (as a teenager) found a mathematically sound approach to possible world semantics, which is now called *Kripke semantics* [170]. The basic idea is that *true* should mean *true in a particular world*, while *necessarily true* should mean *true in all worlds*. In particular, this explains the generalization rule: if we derive a proposition without any additional assumptions, then it must always be true. This idea has to be developed further, since in general, formulas may have many occurrences of the modality and the occurrences can be nested, etc. Therefore, we not only need possible worlds, but also a relation between the worlds. It is a binary relation with some properties; we say that W_2 is an *alternative world* to W_1, if the relation holds between W_1 and W_2. Then one defines inductively the truth value of modal formulas. For propositional variables, we assume that they get a definite value true or false in each world; they can have different values in different worlds. If we already know the truth values of some formulas and we combine them by Boolean connectives, then the truth value of the compound formula is defined

in the usual way. The key step is when we know the truth value of ϕ for each world and we need the truth value of $\Box\psi$ in a world W_1. Then we define:

> *$\Box\psi$ is true in a world W_1, if ϕ is true in every world W_2 which is alternative to W_1; otherwise it is false.*

In order to avoid confusion with the usual concept of satisfiability, one often says that *W forces sentence ϕ* and writes $W \Vdash \phi$ when the sentence is true in the world W in the sense of possible-world semantics.

Example Suppose the relation on worlds is a linear ordering; we can think of it as one world in different times. Then *$\Box\phi$ is forced in a world W_1* means that ϕ is forced in W_1 and in all future worlds.

Formally, in Kripke semantics one model is replaced with a set of models and a binary relation on this set. In the case of propositional logic that we consider here, single models are simply truth assignments to variables. The relation is called the *frame* of the model. The most interesting thing is that the properties of the frame determine the modal logic. In particular, reflexive and transitive relations determine S_4. What it precisely means is this.

Theorem 4 *A formula is derivable in S_4 if and only if it is forced in all models with reflexive and transitive frames.*

Notice that the link between the axioms and properties of frames is quite direct. The axiom schema (2) expresses transitivity and (3) expresses reflexivity. Other frames determine other systems of modal logics.

Example Suppose that ϕ is forced in a world W_1, but it is not forced in an alternative world. Then in W_1 $\Box\phi$ is false. This shows that $\phi \to \Box\phi$ is not a tautology.

Intuitionistic propositional logic can be interpreted in S_4 as follows. For an intuitionistic formula ϕ, we define its translation ϕ' into S_4, essentially, by putting boxes everywhere. More precisely, we define inductively: a propositional variable x is translated as $\Box x$; the translations of $\neg\phi$, $\phi \vee \psi$, $\phi \wedge \psi$ and $\phi \to \psi$ are respectively $\Box\neg\phi'$, $\Box(\phi' \vee \psi')$, $\Box(\phi' \wedge \psi')$ and $\Box(\phi' \to \psi')$, where ϕ', ψ' are the translation of ϕ, ψ.[21] Then a sentence ϕ is an intuitionistic tautology if and only if its translation ϕ' is an S_4 tautology.

We can check that the law of excluded middle is not valid in intuitionistic logic. Translating the formula $x \vee \neg x$, x a propositional variable, into S_4 we get $\Box x \vee \Box\neg x$. If this were true, then we would have either x true in all alternative worlds, or x false in all alternative worlds. But a model can easily be constructed such that in a world alternative to a fixed world W_1 x is true and in another it is not.

[21] Boxes in front of disjunctions and conjunctions can be omitted.

Notice that this interpretation in turn gives possible world semantics for intuitionistic propositional logic. It is, however, easy to define Kripke models for intuitionistic logic directly. For example, the defining clauses for the negation and the universal quantifier are:

$M \Vdash \neg\phi$ if for no N alternative to M, $N \Vdash \phi$;
$M \Vdash \forall x\ \phi(x)$ if for all N alternative to M and all $a \in N$, $N \Vdash \phi(a)$.

In Kripke models of first-order intuitionistic sentences the worlds are structures with relations and functions for the corresponding symbols of the sentences. It is further assumed that if W_2 is an alternative world for W_1, then the structure W_1 is a substructure of W_2. The equality relation has to be interpreted as an equivalence relation because, in general, equality is not preserved in alternative worlds.

2.4 Programs and Computations

The history of computations is as old as mathematics. As a matter of fact, early mathematics was just the art of computation. When proving more and more general results, mathematicians gradually started to consider problems that were not computable practically, but computable only *in principle*, which means computable having unlimited time for computations. Eventually they arrived at problems that are not computable even in principle (but the proofs of such non-computability were found much later). The algorithmic aspect has been, however, always eminent. You cannot apply a theory to practical problems, if there is no way to compute the results. We believe that our interaction with the real world is of a computable nature, thus any theory should produce instructions for computations, otherwise we cannot test it. Of course, there are many theoretical results that do not lead to algorithms, for example, results showing the non-existence of certain structures, the impossibility of some computations, etc. (some of these will be treated in Chap. 4). There we do not expect any output of the form of an algorithm, they only serve as warnings: *do not try!* Others give us insights, explanations, etc., but these insights should eventually help us to solve very concrete problems.

In the second half of the 20th century the theory of computing, or rather its applied parts, split from mathematics. Computing has become a new type of industry. Like physics, computer science studies very concrete things: computers and computations with them. Yet there are many purely theoretical problems concerning computing, which rather belong to mathematics. The main problem is what can be computed with limited resources (resources being time, the size of computers and their memory). The reason why this is a fundamental question which is not relevant only for computer science is that our interaction with the world is subject to the same type of limitations. Our brains are some sort of computer too and we too have very limited time for using them as individuals but also as humankind. Because of that these questions are very relevant for the foundations of mathematics.

The main property of algorithms is that for each particular instance of the problem for which they are designed, shortly *for each input*, they should only use a *finite* amount of time and space. This is clearly a necessary condition, otherwise we cannot realize computations physically. For practical computations, this is surely not enough. There are problems that need so much time that they are practically not computable. This leads to the study of the amount of resources, namely time and space, needed for computations, which is the subject of the *computational complexity theory*. Before I consider such subtle questions about computations, I have to explain the general concept whose only limitations are the finiteness of resources used.

In the theory of computation there is the same duality that we observed in logic. There is a description of computation called *algorithm* or *program* and the actual run of the algorithm or program, which are mechanical or electronic operations performed by a person or a computer. There is a slight difference in the use of the words 'algorithm' and 'program'. An algorithm is usually a higher level description which is not quite formal; it is assumed that the details of what to do precisely will be filled in by people using it, or programmers who transform the algorithm into an actual program, the *implementation* of the algorithm. A program is, on the other hand, a precise definition (in a programming language) of what the computer should do. When we want to compute something, we have to divide the task into a sequence of elementary simple tasks. People not only know a lot of problems that can easily be computed, but also are able to fill in gaps, thus the "elementary" tasks need not to be so simple as those used by computers. Therefore, people only need an algorithm, while a computer needs a program.

In natural languages there is a special class of sentences that corresponds to programs, the imperative. Cooking recipes are also some kind of program, so are various instructions for use, etc. The real programs work with symbols and numbers; on the lowest level they always use only bits, which means two values. In mathematics children start with algorithms for addition and multiplication. Then they learn some algebra. An algebraic expression is an algorithm. In order to compute the value of an algebraic expression for particular numbers, we perform the operations using the algorithms for addition and multiplication as subroutines. But only some computations can be defined by algebraic expressions. One of the most ancient algorithms is attributed to Euclid; it is an algorithm for finding the greatest common divisor of two integers. For this problem, there is no algebraic expression as for many others. Incidentally, Euclid's algorithm is very efficient and still used in practice. The word '*algorithm*' is also old; it stems from the name of Al-Khwarizmi, an important mathematician of the 9th century Baghdad school.

What a computer computation is everybody knows, but let me quickly describe it anyway, in order to introduce a few basic concepts. A computation starts with input data. During the computation new internal data are created. I will call both just *data*; these are strings of symbols, numbers or bits. A program contains commands and tests. Commands specify simple operations that should be performed with data. Tests specify simple properties of the data. The computation proceeds according to the result of the tests. A part of the program can be skipped, or the computation returns to a command before (this is called *a loop*). The existence of loops is an important feature. A loop may occur because there is an instruction to return to a previous part of a program (the *go to* control statement) or there can be a *repeat* statement.

Consider a "practical program" for driving a nail into wood: *Keep* (repeat statement) *hitting the nail* (command) *until it does not stick out* (test). It consists of a simple loop where we hit the nail and check, if it still sticks out. The reason for the loop is that we cannot specify *a priori* how many times we have to hit the nail. This kind of construction enables programmers to describe something that they do not know how much time it will actually need. In such cases the programmer must have some reason to believe that the computation will eventually stop; for instance, the programmer may have an upper bound on how many times the computation will go over the loop in the worst case. (In our "nailing" program 100 seems to be a safe upper bound.)

The concept of an algorithm had been defined in mathematics more than a decade before electronic computers were built. It was not only an advance in electronics that led to the construction of the first computer, but also an advance in mathematics. Several definitions of what is computable were proposed of which I will mention the three most important ones, two mathematical definitions (Turing machines and recursive functions) and one more practical definition, in fact a whole class of definitions (programming languages).

Turing Machines

Alan Turing
Courtesy of King's College, Cambridge University

The first approach to defining computations that I want to mention here, though a mathematical one, is friendlier for non-mathematicians. This concept is due to the English mathematician Alan M. Turing (1912–1954) and it bears his name: *the Turing machine*. A similar concept was defined independently by the American logician (born in Poland) Emil L. Post (1897–1954). Both definitions appeared around 1936 [221, 293]. The description of a Turing machine looks as if it were a technical device, very much like a primitive computer. A Turing machine consists of a control device connected to an infinite tape. The tape has distinct squares that can hold a symbol from an alphabet. The machine can read the symbol in the square that is currently scanned by the head and possibly rewrite it to another symbol. The control device is a finite machine which has a finite number of states (in mathematical terms, it is a finite automaton). We do not care how the machine is constructed, we only require that the current state and the currently read symbol determine what the machine does, including the next state to which it switches. A Turing machine operates in discrete time intervals, like a real computer, and in each step it reads, possibly rewrites a square on the tape, moves the tape one square to the right or to the left and switches the state of the control device.

The amazing thing is that, however primitive and cumbersome it is, it can perform any algorithm! Of course, the input data must be suitably presented; for example, when computing with natural numbers, we have to use the binary or a similar representation. Once I had the opportunity to admire the collection of physical models of Turing machines of a German professor who was fascinated by this fact. As they said, every new assistant researcher in his department got as the first assignment the task of constructing one. The exposition rather than giving insight into the concept of Turing machines, nicely showed the development of electronic components. The first machines were built from mechanical telephone components, while the last one used modern electronics. Those were the old days, nobody would do it that way nowadays. The best way to visualize such a concept nowadays is on a computer screen. This is not only an entertaining experiment, but also shows that the concept of a Turing machine is not a technical one, it is as mathematical as other mathematical definitions of computation. Consider the following argument: if you can represent a Turing machine in a computer, then you can represent it as a mathematical structure as well.

A Turing machine is a very cumbersome device. Even if you only want to design it to do such simple tasks like the addition of two numbers, you need to do a lot of simple but tedious work. So why has this concept been chosen and why is it still being used in theoretical computer science? The reason is that the simpler definition we are able to find for a given concept, the more the definition will reveal about the essence of the concept that is defined.

I will now give an informal justification of the concept of the Turing machine. I will argue that computations on any physical device can be simulated by a Turing machine.

Firstly, such a device should be able to store the input data, auxiliary information obtained during the computation and it should be able to present the output data. These can be represented using various data structures, but the simple data structure are strings in a finite alphabet. This suggests the use of a *tape*. As we cannot *a priori* bound the size of memory needed during the computation and the size of the output, we need an infinite tape. It makes little difference if the tape is infinite in both directions or only in one.

Secondly, the changes of the data recorded on the tape have to be simple. So why not to change just one symbol at a time? The place where the change is done must be determined by a simple rule. This requirement alone does not lead directly to the concept of the head of the machine, as we can easily think of various natural rules how to determine where the change should be done. However, if we think more mechanistically, it does not seem natural to make a change at one place and then in a single step to jump to another distant place. Thus the most natural thing is to have a pointer that defines the place in the data structure that is to be processed and allow the pointer to move only one step per unit time.

Finally, the whole process of local changes in data has to be controlled somehow. Saying that we have a finite list of rules that the machine follows would correspond to the intuitive concept of an algorithm as used in mathematics. However, to get a picture of a physical device that performs computations independently of us, it is better to talk about a mechanical device that does the elementary steps.

Turing's original justification in his seminal paper "*On Computable Numbers*", from 1936, was different. The main difference is that he analyzed what algorithms can be done by a *human*, rather than talking about physical devices. He had two reasons for that. Firstly, the idea that the human brain is just a very complex physical device was not so widely accepted at that time. Secondly, the available computing machines were able to perform only very simple algorithms. Interestingly, he did talk about "computers", but the meaning of this word back then was "a person who computes". When he reached the conclusion that a computation of a "computer" can be simulated by his concept, he finished by saying: "*If this is so, we can construct a machine to write down the successive formulae, and hence to compute the required number.*" Hence, humans are no better than machines.

Programming Languages

The construction of first computers in the 1940s changed the situation dramatically. Before we needed to communicate algorithms only to people and computations were performed by people (with some help of calculators at the later stage). Thus it sufficed to describe an algorithm in a natural language, using, of course, some mathematical terms. The precise definitions, like the Turing machine, were needed for pure research, for instance, to show that there are problems that are not solvable using an algorithm. Computers, first of all, needed precise definitions of algorithms. They are not able to fill in gaps and use hints instead of a full description, nor are they able to understand our language full of vague words. But that was only part of the problem. What was needed was a communication means between a person and a computer, a language that *both* can understand and that is sufficiently efficient in order to be used practically.

The first programming languages were more tailored for computers. Computer time was expensive, compared to the price of the time of researchers experimenting with them. Also computers were worshiped by many as being capable of doing miracles. Gradually the roles interchanged and now, in most cases, the time of a programmer is more expensive. Using the machine code for programming was a short episode. Soon a lot of effort was exerted to make the task of writing a program as simple and as efficient as possible.

A programming language borrows words from a natural language (almost exclusively from English) in the same way as logic does. These are words such as *for, do, repeat, stop, go-to*, etc. Their number is small, but it is not because we want to use as few words as possible. Furthermore, one uses mathematical symbols, mainly for arithmetic, as some numerical computations occur in most programs. The languages also offer *libraries* of functions that are often used, so that programmers do not have to program the functions again and again that often occur in programs. For instance, JAVA includes a function for finding the greatest common divisor (whose implementation is very likely based on Euclid's algorithm), but it also contains much more difficult ones; for example, you can ask for a random prime number in an interval. Not only libraries of functions simplify programming, but also the types of constructions that are possible in a given language are very important. One of the most important recent inventions is the concept of an *object*. It documents very nicely the convergence to human languages. Formally, it allows the programmer to impose a certain hierarchical structure on the program. Practically it means that the programmer can use features of human thinking that are not based on logical or mathematical deduction but on the vague concept of similarity. The point is that similar *real* objects can be treated in the same way. Thus in object oriented programming we group similar function and we allow the same procedures to be applied to them. Such features, however important they may be for practical programming, are irrelevant from the point of view of theory. In order to be able to define all functions in a programming language, one needs a very tiny fraction of it; we can do with very elementary commands and operations, the more complex ones can be programmed from the simpler ones.

What is the most amazing fact on programming languages is how many people "speak" these languages. Using a programming language is not the same as knowing it; it is an ability similar to a good command of a natural language; it is the ability of *thinking* in terms of the programming language. But, unlike natural languages, programming languages are completely formal systems, which shows that people can use a completely formal language, if there is a strong incentive to do so. Obviously, these languages can only express algorithms, but that is already quite a rich domain.

We should also note that there are big differences between architectures of different languages. So we should consider each single programming language as a different model of computable functions. Another thing to note is that a program not only defines a computable function, but also determines to a large extent how the function is computed. However, this is a common property of all definitions of computable functions.

Noncomputable Functions

Our practical experience with computation suggests that there are easy functions and difficult ones. The easy ones are computable practically. For the difficult ones, the problem seems to be that we do not have enough time and memory to compute them, but, perhaps, they are computable in principle? In fact there are functions that are very difficult, but in principle computable, but there are also functions, that cannot be computed even with arbitrary resources available. The main example of such a function is the *halting problem*. The task is to compute whether a given program applied to given data will ever terminate. This is, actually, an important question because a computer that is not producing any output is considered to be stuck, no matter how ingenious the computations it may be performing inside. Given a program and data we can perform an experiment by running the program on the data. If it stops, we know the answer, but how long should we wait? In practice we wait at most a couple of minutes and if nothing changes on the screen we know something has gone wrong. It can be proved, however, that there is no way to estimate how long it is necessary to wait. Thus sometimes we have to interrupt the experiment without getting a definite answer. So this is not a way to find out whether a program stops on given input data. The result about the halting problem says even more: not only can we not solve the halting problem by trying to run the program for a limited amount of time, but there is no way at all to decide it in finite time. A curious fact is that even an apparently much more restricted version of the halting problem is algorithmically unsolvable: to decide if a program *running on its own code* will eventually stop.

Some programmers may be surprised by the fact that finite problems such as the halting problem are not algorithmically solvable because these problems almost never occur in their daily practice. The halting problem is not computable because there is no bound on for how long a time the program may run. In practical problems there are usually some implicit bounds on how big the number tested should be, for how long, etc. The halting problem is a decision problem, thus it can be represented as a function with two possible outputs YES and NO (if we want to have mathematical objects, we can use 1 and 0 instead). The above observation about bounding the running time suggests the following noncomputable function f, which I will call *the halting bound*. For a natural number n, we define $f(n)$ to be the maximum of all running times of programs of length at most n applied to their own code.[22] We do not consider the programs that do not stop when applied to their own code. If f were computable, we could solve the halting problem as follows. Given a program P of length n, we would first compute the bound $f(n)$ and then we would run the program on its code for $f(n)$ steps. If P stops, then we are done, we know it stops. If, however, if does not stop within the time limit, we know nonetheless that it will never stop. So again we can give a correct answer. What is interesting in this example is that we actually do not need the precise value of $f(n)$. The same argument works for any function g such that $g(x) \geq f(x)$. Hence the reason why the halting bound is not computable is not that the values of it have a special property, but it is the growth of the function. The halting bound function increases so rapidly that there is no program that can compute such a function.

The halting bound is closely related to the well-known *busy beaver function*. This function, denoted by $\Sigma(n)$, is defined as the maximal number of steps that a Turing machine with n states and an empty tape can make and halt. Notice that we take the maximum over all n state machines that *halt*, which is a condition that we cannot algorithmically test. Further, we assume that the machine uses two-letter alphabet on the tape. This function is also noncomputable.

Some of the nicest examples of algorithmically unsolvable problems come from number theory. Given an equation, it cannot be decided by an algorithm if the equation has a solution in the domain of natural numbers. Recall that equations with integer coefficients are called Diophantine equations, and we are interested only in integral solutions of these equations. I gave examples of such equations in Chap. 1, page 56. Let me mention one more important example, the Pell equation

$$x^2 - dy^2 = 1,$$

[22]For the sake of simplicity I restrict myself to the more specific version of the halting problem.

where x, y are unknowns and d is a given natural number. This equation has been studied and we know, for example, that it has a solution for every d that is not a square. There are several other classes of equations that are relatively well understood (in particular some classes of quadratic equations), but there is no general theory of all Diophantine equations. In fact already the work of Diophantus was criticized for giving many ingenious proofs, but no general theory. The algorithmic unsolvability of Diophantine equations gives an explanation why he could not do it. This result does not exclude that in the future we will have a nice general theory, but if this ever happens, it will not be the kind of theory that gives us formulas or algorithms for solutions. I will show explicit examples of algorithmically undecidable Diophantine equations in Chap. 4.

In logic there are several problems that are not computable. The most basic one concerns first-order logic; for a given sentence ϕ it is undecidable by an algorithm whether or not ϕ is logically valid (= provable). By the same argument as for the halting problem, it implies that we also cannot bound the lengths of the shortest proofs of valid sentences. This gives an intuitive explanation, why provability in first-order logic is hard. I will say more about it shortly.

Universal Machines

Of the many other important results in computation theory I will mention only one. It says that there are in some sense *universal machines* or *universal programs*. This means that there is a program U such that for any program P and input data x, if we use x and the program P as input data for U, the universal program U will produce the same output as P produces on x, assuming P stops and produces an output. We say that, given an input (x, P), the machine U *simulates* the computation of P on the input x. This looks like a nice statement for mathematical recreation or for impressing philosophers, but, in fact, it has very important practical consequences. Suppose you use real Turing machines for computations, then the result says that you need only one. It is much easier to write additional data on the tape than to construct a new machine for each new task. This is also what computers do: a computer uses one universal program, called the operation system. When you run a concrete program, you give a file with the program to the computer and the operation system simulates your program. The same phenomenon can be observed on a higher level: a compiler (a program that enables you to run your programs on a computer), say, for the programming language C, can be written in C. Such a compiler is a universal C program.

It is important to stress that a universal program U behaves in the same way as the given simulated program P also when P does not stop, which means U also does not stop. In order to be able to define a universal program, we have to consider all programs, not only those that always stop. Hence, U will not stop on some inputs. We cannot require U to stop on every input because then U would solve the halting problem.

In order to prove that, say, a universal Turing machines exist, one defines such a machine explicitly. This is the best way to prove it and the constructed machine is not terribly complicated, but it is rather boring to write down the description of such a machine and it is even more tedious to check that it does what it should. To write down a description of such a machine with no explanations is a rather cheap trick to impress readers. So instead, I will use a simple explanation why such a universal machine should exist, an explanation based on the assumption that Turing machines can compute any computable function. Suppose your task is, for a given Turing machine, input data and a number n, to find out what the machine will do with these data within n steps of computation. This is, clearly, something that you can solve algorithmically: you simply plug in the input into the machine, switch it on and see what happens. Therefore, this should be a computable function. Let me spell it out:

> *The function that for a given description of a Turing machine M, an input string x and a number n, gives the result of the n steps of the computation of the Turing machine M on the input x is a computable function.*

Hence this function is also computable on a Turing machine. So take a Turing machine that computes this function and modify it so that instead of using a fixed number n, it successively tries $n = 1, 2, \dots$ until the simulated machine stops; if the simulated machine does not stop, we let it run forever. The resulting machine is a universal Turing machine.

We can turn this argument around and say that having a universal function is a natural requirement for any class of functions that we would like to call computable because the action of executing a given program on given data should be computable.[23]

Considered from a broader point of view, universality is also an attribute of intelligence. Knowing a solution of a practical problem, we can instruct another person how to do it (provided we speak the same language). Dogs are quite intelligent, but they are able to interpret only single commands. Still universality is present in the brains of even less intelligent animals at least to some extent. Of course, we cannot communicate programs to them, but we can *train* them to perform quite complex tasks. What we train them to do is by far much simpler than what the animals need for their life in nature. Universality of the brain is the ability to learn to do things that the body is capable of doing. Of course, universality is not sufficient for a system to be considered intelligent; we do not consider our computers to be intelligent. But, perhaps, there is a kind of higher universality that is the essence of intelligence.

Universality sounds sublime, but it is not a luxury. Special purpose chips are used less and less, as mass production prefers universal ones. Most computers are used exclusively for word processing, yet it is not profitable to produce machines only for this purpose. Universality of the brain was the main selective advantage of *homo sapiens*. One may wonder how a brain shaped for a primitive hunter can play chess, do mathematics, program computers, etc. The explanation is that nature has discovered that a universal brain is much better if the conditions change unpredictably.

[23]This only concerns partial functions, functions that may be not always defined, see Notes.

Animals with universal brains can easily adopt a particular useful behavior without the long process of natural selection that is otherwise needed to "hardwire" it into the brain.

The Undecidability of First-Order Logic

Alonzo Church
Courtesy Princeton University Library[24]

Once a formalization of proofs in first-order logic was achieved, the next natural question was how difficult is to decide whether or not a given sentence is logically valid. In particular, is there an algorithm for deciding if a sentence is logically valid? This problem became famous as Hilbert's *Entscheidungsproblem*, which means *decision problem.* The importance of this problem stems from the fact that if the answer were positive, then it would be possible to automatize mathematics, at least in principle. The undecidability of first-order logic was proved by the American logician Alonzo Church (1903–1995) and independently by Turing; their papers [43, 293] appeared in 1936 and 1937, respectively.

Let us see how the decidability of first-order logic would help solve mathematical problems. Suppose that we are working in a theory that can be axiomatized by a finite list of axioms $\alpha_1, \alpha_2, \ldots, \alpha_n$ and we want to find out whether or not a sentence ϕ is a theorem. Then ϕ is a theorem in the theory if and only if the sentence

$$\alpha_1 \wedge \alpha_2 \wedge \cdots \wedge \alpha_n \rightarrow \phi,$$

(which expresses that the conjunction of axioms implies ϕ) is a logically valid sentence. Hence having an algorithm for the predicate calculus, we would also be able to solve the decision problem whether or not a sentence is a theorem in this theory.

Zermelo-Fraenkel Set Theory cannot be axiomatized by a finite set of axioms, but this does not matter. We will see that there is a finitely axiomatizable theory (Gödel-Bernays Set Theory) that proves exactly the same sentences about sets. Since Zermelo-Fraenkel Set Theory suffices for essentially all mathematics, a positive solution to the *Entscheidungsproblem* would enable us, in principle, to solve all mathematical problems mechanically.

This shows the theoretical importance of the problem, but for practice, it is not so essential. Note that although we do not have an algorithm for the *Entscheidungsproblem*, we can still do something: we can systematically generate all possible proofs and create an infinite list of theorems such that a sentence is a theorem if and only if it appears on the list. But, if are interested in a particular sentence which is not a theorem, we will wait forever and never learn the answer. So this does not

[24] Alonzo Church Papers. Manuscripts Division. Department of Rare Books and Special Collections. Princeton University Library.

give us a decision procedure. What is, however, more important is that generating theorems in this way is extremely inefficient. As researchers in artificial intelligence know, one can obtain only very simple theorems in this way, thus such a method of generating theorems is practically useless. But the same could have been the case if it had turned out that first-order logic was decidable—a decision algorithm for first order logic would have been of no practical use either. The bottom line is that, from the practical point of view, it is not important whether there is an algorithm for the decision problem, but whether there is an *efficient* algorithm. (All of Chap. 5 will be devoted to the concept of efficiency of computations.)

The undecidability concerns pure first-order logic. If we add some axioms, the set of provable sentences may be decidable. For example, if the axioms are inconsistent, all sentences are provable, thus the decision problem is trivial. There are, however, also nontrivial examples. Recall that there are complete axiomatizations of the natural numbers with $+$ as the only operation, and of the real numbers with the operations $+$, $\times$ and the relation $<$. Both theories are also decidable.

It is not just a fluke that completeness and decidability appears at the same time in these examples. Any complete theory is decidable. This is an easy fact based on the procedure that generates all proofs in a theory. Recall that if a sentence ϕ is provable, then eventually the procedure will find the proof and thus certify that ϕ is a theorem. But if T is complete, we know that if ϕ is unprovable, then $\neg\phi$ is provable. So we always find in finite time either a proof of ϕ (when ϕ is a theorem) or a proof of $\neg\phi$ (when ϕ is not a theorem).

Notice that this only gives us an algorithm without any estimate on its running time. Considering how inefficient is to generate all proofs, one may suspect that the decision algorithms for decidable theories could also be very inefficient. Indeed, this is the case; for the two problems above, the complexity is doubly exponential and exponential, respectively, and for some other decidable theories the lower bounds are even higher.

There are also undecidable theories. first-order logic, as a theory with no axioms, is an example. More interesting examples are Peano Arithmetic and Zermelo-Fraenkel Set Theory. In fact, any theory in which one can prove some basic propositions about numbers is undecidable. Also, the proofs of the undecidability of the predicate calculus are based on proving that some finitely axiomatized theory is undecidable. There will be more about it in Chap. 4.

Recursive Functions

Here is a purely mathematical definition of computability. The concept of a *recursive function*[25] is due to the American logician Stephen C. Kleene (1909–1994) [153]. This approach is very mathematical, as we only consider computations with natural numbers and define the class of functions that can be computed, the re-

[25] a.k.a. *general recursive function*

cursive functions. By functions we mean functions of one or more variables, hence, what we usually call arithmetical *operations* are also functions. Arithmetical functions are not only important examples of possibly computable objects, but also they have a sufficiently rich structure. Thus it is possible to encode other computations only using operations on natural numbers. In general we have to choose a suitable domain for data. In the case of Turing machines the domain consists of sequences of symbols, while the domain of recursive functions is the set of all natural numbers.

In logic it is very common to define a class inductively by saying that some initial elements belong to the class and giving some operators that produce new elements in the class. The class consists of those elements that can be produced in this way. (Think of, say, proofs as defined by axioms and derivation rules.) Kleene, as a logician, used this approach. He took some basic functions and considered several operators producing new functions from given ones. The basic functions are quite simple, such as the constant function 0, the successor function $x + 1$, etc., so they clearly should be considered computable. We can add other functions, for example, addition and multiplication, that we surely consider computable, but it is not necessary as the operators enable us to produce them from the basic ones. The operators are also simple. In particular, we take the operator of *composition* (also called *substitution*) of functions. Having two functions f and g of one variable, we may first apply f to the input number and then apply g to the value produced by f. The resulting function is the composition of f and g. We may also compose binary functions. Thus we can get, for example, the function $x + yz$ from addition and multiplication.

If we started with the basic arithmetic operations (addition, multiplication and constants) and only used composition we would only obtain the functions that can be expressed by algebraic terms (polynomials). Therefore, we need more operators. Another basic operator is the operator of *recursion*. The name '*recursive functions*' comes from this operator, but it is misleading, as this operator is still too weak to produce all computable functions. A special case of it is the operator of *iteration*, by which we compose a function with itself a given number of times. The most powerful operator is the *minimization* operator. It allows us to search for the smallest number satisfying a condition.

We can imagine recursive functions as follows. We expand our "algebraic" language by taking more operators on top of the composition. Having a sufficiently powerful set of basic functions and operators enables us to define all computable functions by an expression in this language.

A formal definition of recursive functions is in Notes.

The Church-Turing Thesis

Having definitions of computable functions the next natural question is how good these definitions are. What seems clear is that each of the definitions only describes functions that can be (at least in principle) computed. But do these concepts (Turing machines, programs, recursive functions, etc.) cover all computable functions? This is the same kind of a property that we studied in logical calculi and called completeness, but here we have a problem: we do not have a class of functions that we would consider as the computable functions and that would be defined independently of a concrete computational model. We do not have a purely semantical definition of computable functions. We do have a fairly clear idea about computable functions and all the computation models are in good agreement with it, (the idea is that a computation should use a finite number of elementary operations), but to make this idea precise we have to opt for one of the computation models. Before we choose one, it is, surely, worthwhile to compare them. In particular, is the class of numeric functions computable by Turing machines the same as the class of recursive functions? I have already said that all algorithms can be done on Turing machines, so it should not come as a surprise that the two classes coincide. In fact, if you think about these concepts, after a while you will realize that they are not so different as they appear at first glance. If you try to implement algorithms on a Turing machine you will soon realize that a programming language for it would be handy. That is exactly like asking for a higher level language instead of a machine code for your computer. So isn't a Turing machine rather a primitive programming language? In some sense it surely is. It is a programming language with a single data structure which is a linear array and a single pointer whose position can be incremented or decremented only by one. If you analyze it more, you may find an even closer correspondence, for example, the program lines correspond to the state of the control in the Turing machine, etc.

What about recursive functions? Also this definition can be interpreted as a kind of a programming language. It is a programming language for computations with natural numbers. It has some simple functions as primitive concepts, as most programming languages do. Then it has certain operators that we can interpret as possible constructions that can be used in a program. One of these is *composition*, that is used to form terms; this is also available in most programming languages. Another is *recursion*, again this construction is very common in programming languages. It helps when writing very short programs, but professional programmers try to avoid it whenever possible, as it does not give them good control of the computation. *Minimization* is not present in programming languages as a basic construct (it gives even less control of what is going on during the computation), but it can be programmed easily. On the other hand, *iteration* corresponds to a simple loop in a program and this is the most frequent construction.

Such mutual interpretations were found not only for the aforementioned three concepts, but for all that had been proposed. This is not a proof, but sufficiently good evidence that the concepts have been chosen correctly. The claim that these concepts characterize computable functions is called the *Church-Turing Thesis* (although neither Church nor Turing stated the thesis precisely in the way it is presented nowadays).

Can the Church-Turing Thesis be proved or disproved? Firstly, it cannot be proved or disproved as a *mathematical* statement because it is not a mathematical statement. It relates mathematical concepts to part of our practical experience for which we do not have a rigorous definition. Theoretically, it is possible that somebody might come up with a programming trick, an operator on functions, etc. that we would like to call computable but that would not be covered by the current definitions of computations. In such a case we would have a reason to abandon the thesis, but strictly speaking this would not be disproving. Considering the years of experience with programming, such a possibility is almost excluded.

The only way to make this thesis a little more formal statement is to interpret it as a postulate in *physics*. It perfectly makes sense to take the current physical theories and look whether the phenomena there have a computable nature and how they can be used for computations. The conjecture that all *physically* computable functions are computable on Turing machines, or their equivalents, is called the *Physical Church-Turing Thesis*. This question has been studied and lead to the new concept of quantum computing. Quantum Turing machines are a new important concept and it seems that they can solve some problems faster than classical Turing machines (as we will see in Chap. 5), but when computational complexity is ignored, the two models are equivalent. This is not a proof, but a strong evidence that the Physical Church-Turing Thesis cannot be disproved using quantum theory.

Naturally, also general relativity was used in the attempts to refute the Physical Church-Turing Thesis. There the situation is less clear. There are enthusiasts who believe that some likely occurring phenomena, such as black holes, can be used to solve problems that are not computable on Turing machines, others are more sceptical. In any case, this is only a theoretical discussion; nobody believes that such schemes can ever be used to help people compute. This is in contrast with quantum computing, where several teams of experimental physicists are working on constructing quantum computers. Nevertheless, the research into relativistic computations is extremely important because it concerns the fundamental question of what can physically be computed. (For more about it, see Notes.)

It has also been suggested that a noncomputable function could be encoded in fundamental physical constants. For example, the decimal digits of some constant could define a noncomputable set. If there also were a way to measure the value of the constant with arbitrary high precision, we would obtain a refutation of the Physical Church-Turing Thesis. In the most optimistic scenario we would also know that the digits of the constant encode a particular noncomputable set and we could use it to decide Church-Turing undecidable problems.

It should be noted that the Physical Church-Turing Thesis is meaningful only if the universe is infinite, which we still do not know. If space and time were finite, then there could only be finite computations. In such a case the fundamental role of computability would be replaced by the role of computational complexity. But I believe that computational complexity is equally important for physics even if the universe is infinite.

The Syntax and the Semantics of Computations

The distinction between syntax and semantics is essentially the same as in logic. Syntax is (descriptions of) Turing machines and programs in programming languages. In the case of recursive functions it is a little bit more subtle. There the syntax is a definition of a function in terms of basic functions and operators. Using a suitable notation for operators we can write these definitions as algebraic terms.

Semantics is given simply by the corresponding functions. We assume that a program starts on input data, computes, and then outputs the result of the computation and stops. The function is the assignment of the output data to the input data. Recall that in mathematics we treat function in the purely extensional way: we think of a function as a set of pairs *input-output*. When defining recursive functions, the functions are part of the definition, so it is even simpler (no wonder, it is a mathematical approach to computability). A Turing machine determines a function on strings in the given alphabet. We assume that, except for a finite segment, the tape contains blank squares. When the machine starts, the input to the machine is the word on the non-blank squares; similarly, when the machine stops, the output is what is written on the non-blank segment of the tape. Thus the machine always works with finite strings.

In logic we have not only formulas and models, but also *proofs*. A similar concept in the theory of computations are *computations*. When we are only interested in the question whether or not a function is computable, we do not have to define computations, instead we can use the definition of a computable function. But if we want to study practical computations, it is important to have a definition of a computation. In particular, time and memory requirements are of great concern for difficult problems, and those can only be studied using the concept of a computation. Mathematically, a computation is a sequence of elementary steps. What is a single elementary step depends on the model. For Turing machines, it is scanning the current square, printing a new symbol on it, moving the head to an adjacent square and changing the state of the control.

Furthermore, one needs the concept of computation for a single purpose program. Suppose we need to solve a concrete problem, but the program that we write probably will never be used again. Then the semantics based on functions does not make any sense, as we only need one output. The distinction between computable and noncomputable functions can only be made if we have *infinitely many* inputs. Computability only concerns the distinction between finite and infinite, namely, computable means that the process always ends after finitely many steps.

The Matrix Model of Computations

I am going to introduce yet another model of computations. My aim is to capture the combinatorial character of a computation. This model is useful for analyzing and proving theorems about computations, in particular, it can also be used for studying finite function, which can be done using the classical models too, but in a more cumbersome way. Furthermore, the main parameters of the model, the dimensions of the matrix, correspond to the main complexity measures, time and space, which also suggests a relation to physics.

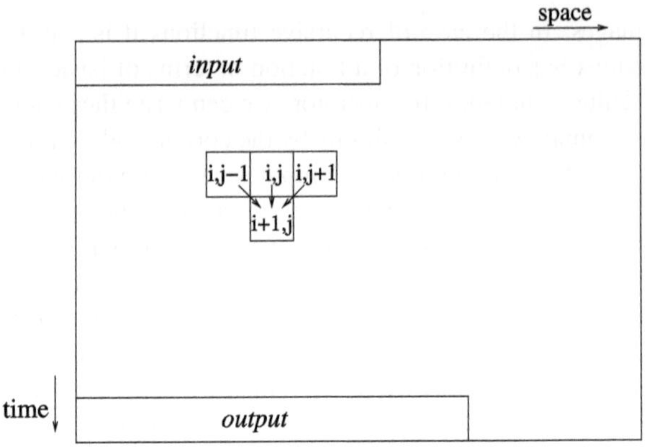

Fig. 2.4 The matrix model

The main twist is to focus on *computations* instead of devices that do computations. I define a computation to be a matrix M with entries from a finite alphabet satisfying the following condition. There is a rule R that uniquely determines the entry $m_{i+1,j}$ (the entry in row $i+1$ and column j) from the entries that are in the row above (the row i) that are adjacent to it, see Fig. 2.4. Thus the entries $m_{i,j-1}, m_{i,j}, m_{i,j+1}$ determine the entry $m_{i+1,j}$, provided j is not the index of the first or the last column; if it is, then it depends only on two entries above (on $m_{i,j}, m_{i,j+1}$ if j is the first column, that is, $j=1$, and on $m_{i,j-1}, m_{i,j}$ if j is the last column). Let me stress that the rule is uniform for all entries; it does not depend on the position in the matrix, except that it is sensitive to the sides of the matrix.

The computational interpretation of this model is as follows. We put input data on the first row. Then fill in the matrix row by row and read the output data in the last row. So the idea that we use to solve a particular problem, the algorithm, is hidden in the rule R. The advantage of viewing a computation in this way is that we eliminate the dynamical aspect of the computation by representing it by a fixed structure. A column j describes the content of the memory location j during the computation, the entries in a row i encode the content of memory location in time i. Thus the physical interpretation of the two dimensions of the matrix is time and space. The matrix is the trajectory of a computation. Furthermore, the rule that determines the system is local, as it should be in physics. Filling in the matrix can be viewed as a discrete version of a classical problem of solving an ordinary differential equation with a boundary condition; the boundary condition is the input data. An example of computation is given in Fig. 2.5.

In order to implement all algorithms in this model, we must allow arbitrarily large alphabets (because in a finite alphabet we have only a finite number of possible rules R). The letters in the alphabet will be used not only to code the bits in the memory, but also the currently implemented instruction from the program. We can avoid this problem by relaxing the definition a little. If we allow rules that determine the entry not only by at most three entries above, but at most k entries in the row above (adjacent to the entry right above), then we can do with a two letter alphabet (0 and 1). Various other generalizations are possible, such as using more dimensions

2	3	0	1	0	3	1	2
2	0	3	0	1	1	3	2
0	2	0	3	1	1	2	3
0	0	2	1	3	1	2	3
0	0	1	2	1	3	2	3
0	0	1	1	2	2	3	3

Fig. 2.5 The matrix of a computation that sorts the sequence 23010312 (the top row) into 00112233 (the bottom row). The computation is done by swapping every two consecutive numbers when the first is bigger than the second (this is the rule R). Thus each entry of the matrix, except for the first row, is uniquely determined by the two or three neighbors above it. This example is contrived, since there are no ambiguities as to which elements should be swapped. (Such an ambiguity would appear, for example, in 321, in which we can swap either 3 and 2, or 2 and 1.) In order to be able to sort any sequence we would have to use a rule that would decide which of the possible conflicting replacements should be done. Furthermore, this is not a very efficient way to sort. Fast sorting algorithms are based on a special choice of which pairs are compared and swapped; then one has to consider also pairs of non-consecutive elements

for space. But such improvements bring only insignificant gain in the efficiency of computations.

We should note that often more space is needed than is allocated for input data. In fact, we often want to keep the input data intact during the whole computation and use other memory locations for the data created by the computation. Thus we allow the use matrices with more columns than is the length of the input string. We put the input data at the beginning of the first row and fill in the rest uniformly by a symbol not present in the data (the 'blank' symbol). Then the number of columns is the size of memory used in the computation—the space complexity.

I will use this model for explaining computational complexity in Chap. 5.

Beyond Computability—Complexity Hierarchies

Naturally we wonder what is beyond the range of computable functions. Because things become more difficult, I will restrict myself to decision problems and I will only speak about sets of natural numbers. Formally, a decision problem is simply a set; the problem is, for a given element, to decide if it belongs to the set. We say that the decision problem is *algorithmically solvable*, or that the set is *recursive*, or *decidable* if there is an algorithm to decide this problem. Equivalently, it means that the characteristic function of the set is computable.

For example, let P be a program and let the problem be to decide for a given input, if the program will terminate on the input. Thus we are interested in the set of inputs on which P terminates. We know that this set is not decidable for some programs P (in particular for the universal program). Nevertheless, this set is not very far from computable ones. It can be defined using a single existential quantifier. It is the set of inputs x such that *there exists* a terminating computation on x, which we can represent as a string of symbols y. Notice, that the only noncomputable thing

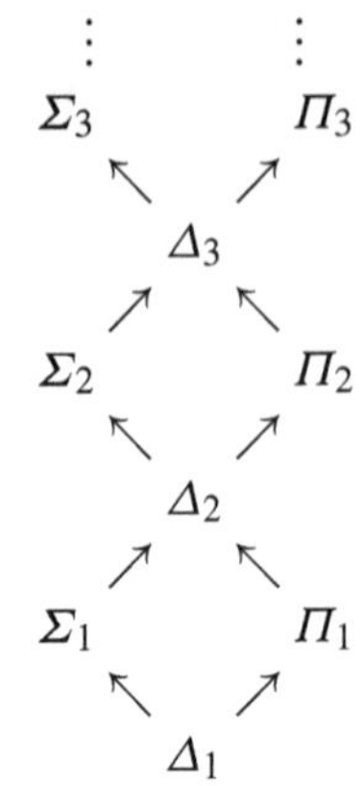

Fig. 2.6 The diagram of the Arithmetical Hierarchy. Δ_1 is the class of recursive sets, $\Delta_n = \Sigma_n \cap \Pi_n$, for $n = 1, 2, \ldots$

is how to get the string y; once we have it, we can verify that it has the required property. Another such set is the set of arithmetical equations that have a solution. Again, having a solution we can check by simple computation that it satisfies the equation, the problem is only how to find it. A very important example is the set of logically valid sentences of first-order logic. This set can be defined, by the completeness theorem, as the set of provable sentences. The latter of the two definitions is based on one existential quantifier: the set all sentences ϕ such that *there exists* a proof of ϕ.

Formally the relations with one existential quantifier can be defined by a formula of the form

$$\exists y\, \phi(x, y)$$

with ϕ denoting a computable binary relation. They are called Σ_1 or *recursively enumerable* sets. The Σ_1 stands for a single existential quantifier. The name '*recursively enumerable*' stems from the property of such sets that their elements can be enumerated by a recursive function. We can similarly define the class of sets definable using a single universal quantifier. This class is denoted by Π_1. As the negation of the existential quantifier is the universal quantifier, Π_1 is the set of the complements of sets in Σ_1. The classes Σ_1 and Π_1 are incomparable; there are sets that belong to Σ_1 but not to Π_1 and *vice versa*. Then we can use two quantifiers. Taking two quantifiers of the same kind does not produce new sets, but if we take two different ones, we get more complex sets. Thus we get the class Σ_2 by using formulas of the form $\exists y \forall z\, \phi(x, y, z)$ and class Π_2 by using formulas of the form $\forall y \exists z\, \phi(x, y, z)$. In general the class Σ_n (respectively Π_n) is the class of sets definable by formulas with n alternating quantifiers starting with $\exists$ (respectively $\forall$). A diagram showing relations between the classes is in Fig. 2.6.

One can show that at each step we get more sets, thus this hierarchy is increasing (but the pairs Σ_n and Π_n are incomparable). Hence using more *alternations* of quantifiers we can define more. This has an important consequence for logic. It proves that in general we cannot reduce the number of alternations of quantifiers in a formula.

It seems that we have exhausted logical means of defining sets, but that is not quite true. It is true that using only first-order formulas we cannot define more sets of natural numbers in the structure $(\mathbb{N}; 0, 1, +, \cdot)$. The sets of numbers definable by arithmetical formulas are called *arithmetical*, they are sets that belong to classes of the *arithmetical hierarchy* $\Sigma_1, \Pi_1, \Sigma_2, \Pi_2, \dots$.[26] But using higher order languages we can define more sets; we can define sets that are outside the arithmetical hierarchy. In higher order languages we can talk not only about numbers, but also about sets of numbers, sets of sets of numbers, etc. Let us look at the simplest example. Consider a formula of the form: *there exists a set* Y *such that* $\phi(x, Y)$ *holds true*. Formally, it is written as

$$\exists Y \ \phi(x, Y).$$

We use the capital letter Y to distinguish sets from numbers. We would like to use this formula to define the set of all x that satisfy it. It is, of course, important to specify what relations we should allow for ϕ because otherwise we could trivially define any set of numbers. When defining sets in the Arithmetical Hierarchy, we used formulas ϕ that were recursive relations. But now the formula speaks not only about numbers, but also about sets. Therefore, we allow ϕ to be any formula constructed from recursive relations and relations of the form $z \in Y$ with arbitrary first order quantifiers, by which I mean quantifiers binding numbers. (The first-order quantifiers are much weaker than a single second-order quantifier, so in fact, it would suffice only to use one universal quantifier for numbers.) Though Y may stand for arbitrary sets, a formula such as $\exists Y \phi(x, Y)$ cannot define arbitrary sets. Notice that Y is quantified in the formula, it is not a parameter, therefore it is as if we only used generic properties of all sets, not a particular set. Hence again, we only get some subset of all sets of numbers. However, we get more than we have in the Arithmetical Hierarchy.

This is only the first step after the Arithmetical Hierarchy. Then we can use more alternations of the second-order quantifiers—the quantifiers for sets, then third order quantifiers—quantifiers for sets of sets, etc. At each step of this construction we get more and more definable sets. After exhausting all levels of higher order languages we can invent a new way of defining sets again and go on as far as our imagination will allow. We will always get more and more sets. The process can never stop because the number of formulas will always be only countable, while the number of all subsets of natural numbers is uncountable.

Thus we can draw a crucial conclusion: even if we are only studying numbers, we may need higher order concepts, namely sets, sets of sets, etc.

Notes

1. *Total and partial recursive functions.* To make it simpler I have concealed an important distinction, the distinctions between partial and total functions. A *partial*

[26] These symbols are often used with the superscript 0: $\Sigma^0_1, \Pi^0_1, \Sigma^0_2, \Pi^0_2, \dots$.

function is like a function, but it is not defined for all arguments. We say that a function is *total* to stress that it is not partial. When we have a program that does not stop on all inputs it defines only a partial function. Formally, we can easily extend a partial function f to a total one, but this does not solve the problem. The problem is that it is not algorithmically decidable whether or not a program stops on a given input (the halting problem). So while we can easily redefine the function, we cannot rewrite a program P to P' so that P' prints the symbol ∞ when P does not stop. It is also not possible to change the syntax of the programming language so that only programs that always terminate can be written without essentially reducing the class of functions that can be programmed. Furthermore, partial functions are interesting things *per se*, therefore we should not prohibit them. In particular, a universal program defines a partial function that cannot be completed to a total computable function.

When talking about a total function, we say *a computable function* or *a recursive function* meaning the same (if we talk about numeric function); if it is a partial function we call it a *partial recursive function*.

It is essential to use partial functions when defining universal functions because of the following:

Theorem 5 *The class of total recursive functions does not have a universal function.*

This can be proved very easily. Suppose $U(x, y)$ is a universal function for total recursive functions of one variable. We fist transform U to a univariate function u by defining $u((x, y)) = U(x, y)$, where (x, y) denotes a pairing function, (say $(x, y) = ((x + y)^2 + 3x + y)/2$). Thus for every unary recursive function f, there exists a number n such that for all m, $f(m) = u((n, m))$. We claim that u cannot be recursive. If it were, then v, defined by $v(x) = u((x, x)) + 1$, would also be recursive. But then, for some n, $v(n) = u((n, n))$, which is a contradiction because $v(n) = u((n, n)) + 1$.

Note that although the theorem explicitly talks about recursive functions, it concerns total computable functions in any model. Thus

- there exists no Turing machine U that is universal for Turing machines that halt on every input and such that U also halts on every input,
- there exists no programming language in which all total computable functions can be defined and only such functions,

and so on. The proof of the theorem is an application of *Cantor's diagonal method*, which we will encounter again in the sequel.

2. *Definition of recursive functions.* The set of recursive functions is the minimal set of functions defined on the natural numbers that satisfies the following four conditions:

   a. it contains the following functions: the successor function $S(x) = x + 1$, the identity function $I(x) = x$, the projection function $I(x, y) = y$ and the constant zero (function) 0;

b. if $f(x_1, \ldots, x_n)$ and $g(y_1, \ldots, y_m)$ are recursive functions (the sets of variables of these functions do not have to be disjoint), $1 \le i \le n$, then

$$f\big(x_1, \ldots, x_{i-1}, g(y_1, \ldots, y_m), x_{i+1}, \ldots, x_n\big)$$

is also recursive;

c. if $f(y)$ and $g(x, y, z)$ are recursive, then the function $h(x, y)$ that satisfies (is defined by)

$$h(0, y) = f(y),$$
$$h\big(S(x), y\big) = g\big(x, y, h(x, y)\big)$$

for every x, y, z, is also recursive;

d. if $f(x, y)$ is recursive and satisfies the condition that *for every x there exists y such that* $f(x, y) = 0$, and $h(n)$ is defined to be the least m such that $f(n, m) = 0$, then h is also recursive.

Thus the set of recursive functions is obtained from initial functions of (1) by applying operators of composition (2), recursion (3) and minimization (4). Note that the condition in (4) cannot be effectively tested, we can only *prove* it using some set-theoretical assumptions. To get an effective definition we have to consider the larger class of *partial* recursive functions; then we do not have to test the condition in (4). Notice that minimization corresponds to loops in the program, such as `do ... while ...`.

The name '*recursive*' probably originated by mistake. Gödel introduced a class of numeric functions only using 1., 2. and 3. hoping that it would contain all computable functions. He was shortly corrected by Herbrand, who pointed out that 4. is also needed. Among these three the most important was 3., the operator of recursion—therefore recursive functions. The class defined by 1.–3. is a proper subclass of all recursive functions, called *primitive recursive functions* nowadays.

3. *Definition of Turing machines.* A Turing machine is determined by the tape alphabet and the control device. The control device is simply a finite automaton. Thus, formally, a Turing machine is a structure (Σ, Q, τ), where

$$\tau : \Sigma \times Q \to \Sigma \times Q \times \{L, R\}.$$

The set Σ is the *tape alphabet*, Q is the set of *states* of the finite automaton. Given a symbol $a \in \Sigma$ and a state $q \in Q$ the value of the *transition function* $\tau(a, q) = (b, r, d)$ determines the symbol b printed by the machine, the new state r and the direction of the movement of the head d.

A *configuration* of the machine is determined by (1) the content of the tape, which is represented by an infinite sequence indexed by integers, (2) the state of the finite automaton, which is $q \in Q$, and (3) the position of the head, which is an integer.

To define a computation of the machine, one has to define one step *transition* from one configuration to the next one. This is easy, provided that we have in mind the interpretation of the concepts. Then a computation is a sequence of configurations where the consecutive configurations are obtained by single transitions.

In order to use such a machine as an algorithm, we have to make several provisos. Firstly, we want only to process finite sequences of symbols (these are called *words*). Thus we define an initial configuration to contain such a word on a specified part of the tape with the ends of the word marked and the rest of the tape being uniformly filled with the same symbol. The most convenient way to do this is to reserve a special symbol of Σ as the blank symbol and require that the input and output words do not use it. Then we have to fix an initial state of the automaton, a final state of the automaton and the initial position of the head, say the index 0. The output is the maximal length word of non-blank symbols containing the 0 position that appears at the moment when the automaton reaches the final state. Thus the machine determines a partial mapping on the set of all words of non-blank symbols.

4. *The matrix model of computation—continued.* Let Σ be the alphabet of symbols used in the matrix. To define this model formally, we have to solve the problem with the boundaries. The standard solution is to add columns on both sides filled in with an extra symbol, say, #. Then we can say that each entry inside is determined by the three entries above. Hence the model is given by an alphabet Σ, a rule $\rho : \Sigma' \times \Sigma \times \Sigma' \to \Sigma$, where $\Sigma' = \Sigma \cup \{\#\}$, and the two dimensions of the matrix.

A computation of a Turing machine can be represented by such a matrix, except that we may need infinitely many columns and rows. In the first approximation we take simply the tapes with their content in the order of the computation of the machine. This is not quite correct as the content of the next tape depends also on the state of the control and the position of the head. This can be fixed by taking a larger alphabet in which we can encode also the state of the control. Using the notation for a Turing machine above, we take as the larger alphabet the disjoint union of Σ and $\Sigma \times Q$. Then an entry $a \in \Sigma$ in the matrix means that a is on the tape and the head is elsewhere, and $(a, q) \in \Sigma \times Q$ means that a is on the tape, the head is currently on the same square and the control is in state q. Thus rows encode configurations of the machine. We can make the matrix finite, if we consider only computations on inputs of lengths up to a fixed number n and assume that all computations terminate. Then the machine uses only a finite portion of the tape and we can bound the number of steps.

Other combinatorial models are related to the matrix model, in particular arrays of automata and Boolean circuits. A *Boolean circuit* consists of gates and connections between them, called wires. It has input gates, inner gates and output gates. The role of input and output gates is clear; the inner gates process information that reaches them and send it further by wires. '*Boolean*' means that information comes in bits, denoted usually by 0 and 1. In practical circuits feed-back is very important, but in theoretical studies we mostly use circuits without any feed-back, just to make the model simpler. A matrix model with the alphabet $\{0, 1\}$ is a Boolean circuit (the first row being the input gates, the last row the output gates, etc.). The difference between the two models is not big. As I have noted, we can restrict the alphabet to $\{0, 1\}$ by allowing dependence on a little more distant places in the matrix. The restriction to a rectangular grid is also inessential. When it is needed to send a bit to a more distant place, we can do it in several steps.

5. *Relativistic computations.* Relativistic computations were proposed independently by I. Németi, D. Malament, I. Pitowski and M.L. Hogarth in the late 1980s and early 1990s. The aim of the proposed schemas was to show that it is consistent with General Relativity that there is a physical process that computes something that is not computable on Turing machines. I will outline the basic idea using the schema proposed by Németi [206] that uses a highly probable assumption that there are large rotating black holes. Note that all schemas assume that the universe is infinite in size and time, or at least that the size of the universe grows beyond any limit.

   Consider two entities P (programmer) and C (computer), such that P wants to learn, using C, something that is not classically computable. Namely, P wants to know if some efficiently computable predicate $R(n)$ is satisfied by all numbers. For example, $R(n)$ can express that the nth nontrivial zero of the ζ-function is on the line $Re(x) = \frac{1}{2}$. Then $R(x)$ is satisfied by all numbers if and only if the Riemann Hypothesis is true. To this end we need a black hole (with suitable parameters) and let P fall into it (on a suitable trajectory). Meanwhile C will use a Turing machine to check all numbers for the property R. One cannot construct an infinite tape for the Turing machine, but since the universe is infinite, it is possible to gradually extend the tape beyond any bounds, which is all that C needs.

   From the point of view of C, P will never reach the event horizon of the black hole. So if C finds a number that does not satisfy R, it can send a signal to P while P is still before the event horizon.[27]

   From the point of view of P, nothing special happens at the event horizon. In particular, he will cross the event horizon in finite time according to his watch. But he can calculate time t_1 when he will be crossing the event horizon. So at time t_1 he will know that C tested all numbers (and ceased to exist). If everything is set up properly, at some time $t_2 > t_1$ P will know the answer: if a signal has arrived, then there is a number that violates R, otherwise R holds true for all numbers.

   This shows that such computations are consistent with general relativity. Though some effort has been made to include quantum theory into consideration, it is not clear that they are consistent with other theories that describe basic physical phenomena. Note that we still do not have a consistent theory that would unite relativity and quantum theory, so the concept of consistency with known physical phenomena is not precisely defined.

6. *Second-order logic is not axiomatizable.* This means that there is no formal system in which a second-order sentence would be provable if and only if it is sat-

[27] More precisely, we should talk about the event horizon as seen by C because C is in a finite distance from the black hole.

isfied by every second-order structure. This implies that higher order logics also cannot be axiomatized.

One can prove this fact by considering the complexity of the set of logically valid second-order sentences. Given a formal system Π, the set of sentences provable Π is recursively enumerable. Recall that recursively enumerable is the same as being Σ_1 in the arithmetical hierarchy. The set of sentences ϕ provable in Π can be defined by the condition informally stated as follows:

There exists a Π-proof of ϕ.

So there is only one existential quantifier, which explains Σ_1. But the set of second-order logically valid sentences is much more complex—it is not contained in any of the classes of the arithmetical hierarchy. Therefore it cannot be axiomatized.

The reason this set has such a large complexity is that in second-order logic one can define the standard model of arithmetic. Consequently, all true arithmetical sentences can be derived in second-order logic. In more detail, let Φ be the conjunction of the three second-order axioms of Dedekind-Peano Arithmetic (page 30) and the first seven axioms of Peano Arithmetic (page 116). Then for every first-order arithmetical sentence ψ, ψ is true if and only if the second-order sentence $\Phi \to \psi$ is logically valid. Thus for any arithmetical set X, the decision problem for X can be reduced to the decision problem of the set of logically valid second-order sentences.

2.5 The Lambda Calculus

The λ-calculus is a very interesting formal system whose main features are universality and flexibility. As such, it can be used for various purposes. It also has many versions which split into two main branches: the *type-free λ-calculus* and the *typed λ-calculus*. The λ-calculus can be used as a logical calculus, a formalism for defining algorithms and as a formal system for the foundations of mathematics. In the type-free λ-calculus all of the elements are of the same type. The typed λ-calculus is based on a hierarchy of functionals (see page 17), in a similar fashion as types are used in Russell's Theory of Types, which we will consider shortly, and thus it can be viewed as a generalization of Russell's theory. Because of important connections with proof theory, it is often viewed as a branch of it. The typed λ-calculus also has applications in computer science, especially in functional programming and automated theorem proving.

Church introduced the λ-calculus in the early 1930s [42]. A little later, he introduced the typed version [44]. Related calculi, called *combinatory logic* were invented independently by M.I. Schönfinkel and H.B. Curry a little before, but the connection between the two approaches was discovered later. As the title of his seminal paper '*A set of postulates for the foundation of logic*' suggests, Church intended to use his formal system for the foundations of mathematics in a similar way as Frege,

Whitehead, Russell and other logicists did. Thus the λ-calculus was conceived as a kind of general logic. His first attempts were plagued with inconsistencies, but soon after that he found the sound version and proved its consistency.

We will start with the type-free version. The basic idea of the λ-calculus is that we can think of everything as being a function, an argument and a value at the same time. Computer programs and computer data are a good example. Since both are just some files, you can run a program on data, but you can also run a program on a file that is a program as well, which is sometimes very useful. You can take a file which is not a program and ask your computer to execute a file like a program, in which case the computer also does something, though it is hardly of any use. The paradigm that objects are functions, arguments and values at the same time, is similar to the one used in set theory, where an object is a set and an element at the same time. In a sense, it is the opposite of the structural approach to mathematics that we considered in Chap. 1 (while, in contrast to it, the typed λ-calculus is fully in line with it).

The λ-calculus has one binary operation as the only nonlogical symbol, the interpretation of which is application of a function to an argument. Usually no symbol is used for the binary operation of application, instead we simply use juxtaposition of the arguments. So ab means that a function a is applied to an argument b (outside of the λ-calculus the most common notation for this operation is $a(b)$, but juxtaposition is also frequently used). This notation stresses the identification of objects and functions. We think of functions in the λ-calculus as being unary, as we apply them only to one argument, but they can also be treated as functions of arbitrarily many arguments. For instance, if you want to apply f to x and y, then first apply f to x. Then, since the result is again a function, you can apply it to y. Formally, this is written as $(fx)y$, or simply fxy, using the convention that parenthesis grouped to the left are omitted.

The essence of the λ-calculus is the principle that *every term defines a function that is an object of the theory.* This is not such an obvious principle as it may look at the first glance and, in fact, in some context it may even be contradictory (recall the related Comprehension Principle). To formalize this principle, Church introduced an operator, denoted by λ, from which the name of the calculus stems. This operator can be applied to any variable x and any term t, not necessarily containing a variable x, resulting in the expression

$$\lambda x.t.$$

In this expression, λ binds the variable x, like a quantifier. The dot between x and t is a convenient way of separating the operator from the term. (Dots were used instead of parenthesis very often at the beginning of the 20th century.) The meaning of the expression is:

the function which to x assigns the value obtained by evaluating the term on x.

Example $\lambda x.x$ is the identity function; $\lambda x.y$ is the function constantly equal to y. Notice the difference between the expressions x and $\lambda x.x$—the first one is a variable and can represent any object, the second one is a constant symbol and it represents a unique object, the identity function.

The λ-notation is very useful in general, but, unfortunately, most mathematicians do not know about it, although in mathematical practice we often need to distinguish a function from its value.

Example One can distinguish a *number* which is a square x^2, from the quadratic *function* $\lambda x.x^2$. In the expression $\int_0^1 x^2\,dx$ we use dx to say that we integrate the quadratic function, but we do not use $x^2\,dx$ for this purpose elsewhere. If the λ-notation was used, we would have a more systematic notation. The integral above would be expressed by $\int_0^1 \lambda x.x^2$.

Combinatory Algebras

For main-stream mathematicians, it may be easier to grasp the main ideas of the λ-calculus using a class of first-order structures that are models of the λ-calculus. The *Extensional Combinatory Algebras* is a class of structures defined by the following axioms:

The Axiom of the Existence of at Least Two Elements *There are two different elements.*

This axiom in combination with the others ensures that there are infinitely many elements.

The Axiom of Extensionality *If $fx = gx$ for every x, then $f = g$.*

In words, if function f gives the same value as function g for every argument x, then $f = g$. This is essentially the same as the Axiom of Extensionality for sets.

The Axiom Schema of Combinatory Completeness *For every term t and every sequence of variables $x_1, \dots, x_n$ that includes all variables of t, the following is an axiom: There exists an f such that for every $x_1, \dots, x_n$, $fx_1 \dots x_n = t$.*

In words, every function that can be defined by a term is present in the universe. This schema formalizes the basic principle of the λ-calculus. Using the λ-notation it can be stated as the following schema.

β-conversion *For every term t,*

$$(\lambda x_1 \dots \lambda x_n.t)x_1 \dots x_n = t.$$

In first-order logic we can view the expression $\lambda x_1 \dots \lambda x_n.t$ as a new constant introduced to represent the function f from the schema above.

Curry discovered that it suffices to take only three instances of the Axiom Schema of Combinatory Completeness. For these special cases, he introduced constant symbols denoting the particular functions. Then all other functions resulting from the schema can be written as terms using these basic functions, called *combinators*. The three combinators are denoted by **I**, **K**, **S** and the three instances of the Axiom Schema of Combinatory Completeness are:[28]

1. $\mathbf{I}x = x$
2. $(\mathbf{K}x)y = x$
3. $((\mathbf{S}x)y)z = (xz)(yz)$

The simplicity of this axiom system is, indeed, striking. Moreover, the first axiom of the three axioms above is redundant, as one can express **I** by **SKK** (an easy exercise).

In order to get some feeling for this calculus, it is worthwhile to interpret at least the two simplest combinators **I** and **K**. **I** is clearly the identity function. **K** is the function which on argument x produces the function that is constantly equal to x, (namely, if we apply $\mathbf{K}x$ to any y we get x). In the λ-notation the combinators **I**, **K** and **S** are defined by

$$\mathbf{I} = \lambda x.x, \qquad \mathbf{K} = \lambda x \lambda y.x, \qquad \mathbf{S} = \lambda x \lambda y \lambda z.(xz)(yz).$$

Nevertheless, to construct a model of an extensional combinatory algebra is not easy. The consistency of the λ-calculus was proved using combinatory means. The proof is based on a concept of a normal form of a term, which plays a central role in the theory. Since we can interpret the theory of extensional combinatory algebras in the λ-calculus, this implies, by the completeness theorem, that there is a model of extensional combinatory algebras. However, an explicit construction of such a model was found only much later by Dana Scott (see Notes).

The following is an important theorem in the type-free λ-calculus.

The Fixed Point Theorem *For every f, there exists g such that*

$$fg = g.$$

Proof Let $h = \lambda x.f(xx)$, let $g = hh$. Then

$$fg = f(hh) = \big(\lambda x.f(xx)\big)h = hh = g. \qquad \square$$

This looks like a meaningless manipulation with terms, but it is not. I will not explain this proof because its essence is self-reference which can be more easily explained in the context of logic, which I will do in Chap. 4.

---

[28] For the sake of clarity I do not use the convention of omitting parentheses here.

The λ-Calculus as Logic

To explain how the λ-calculus can be used as logic, it is better to use the typed λ-calculus. The objects of this calculus are also functions, but in contrast to the type-free version, we can apply a function to an argument only if the types match. Starting form some basic types, new types are formed by taking the set of all functions mapping objects of a type τ to objects of a type σ. This type is denoted by $\tau \to \sigma$. If a is of type $\tau \to \sigma$ and b is of type τ, we can apply a to b to obtain an element ab of type σ. In order to formalize functions of two (and more) variables, one can either introduce the product of types $\tau_1 \times \tau_2$ and use $\tau_1 \times \tau_2 \to \sigma$, or use the type $\tau_1 \to (\tau_2 \to \sigma)$.

In the standard formalization of first-order logic we distinguish formulas from terms. The reason is that formulas express the truth or the falsehood, while terms denote objects. If t is a closed numerical term, then it has a value in the domain of the numbers; if $t(x)$ is a numerical term with one free variable x, then $t(x)$ denotes a function of one variable, etc. If ϕ is a sentence, then the value of ϕ is either truth or falsehood; if $\phi(x)$ is a formula with one free variable x, then it represents a *propositional function* of one variable or a *set*, etc. Church observed that it is only the range of values that distinguishes these two concepts, and since one can consider various ranges, there is no need to distinguish the Boolean one from others.

So let us assume that we have the Boolean type $B = \{\mathbf{0}, \mathbf{1}\}$ and let us use only terms. Since we have eliminated sentences, we have to say what we will use instead of them and what will replace proofs. Sentences are easy: these are just the terms that are of type B. In order to define proofs, we have to view a proof not as a text that presents evidence that a sentence is true, but as a *process* by which we obtain this evidence. Consider a proof by contradiction of a sentence ϕ. It starts with writing $\neg\phi$, then we derive new formulas using logical rules and eventually we derive contradiction. This process is very much like evaluating a numerical expression: we start with the expression, apply rules to transform it, and eventually obtain the value. Again, there is no reason to treat these two kinds of process as being distinct. So in the λ-calculus we prove a sentence represented by a term t by evaluating it, which means that we apply rewriting rules until we obtain the Boolean value $\mathbf{1}$.

In order to embed first-order logic into the λ-calculus, we must define translations of connectives and quantifiers. Since connectives are, by definition, Boolean functions, their translations are straightforward. For example, negation is represented by a constant of type $B \to B$. Since quantification always ranges only over objects of a given type, we need a constant Π_τ for every type τ. This constant is of type $(\tau \to B) \to B$. Recall that $\tau \to B$ are propositional functions of one variable, functions defined by a formula $\phi(x)$ with one free variable x. Hence Π_τ maps such functions to the Boolean constants. The sentence $\forall x\, \phi(x)$ is true if and only if the propositional function is constantly equal to $\mathbf{1}$. So Π_τ is the function that assigns $\mathbf{1}$ to all functions constantly equal to $\mathbf{1}$, and $\mathbf{0}$ to all other functions.

Let a be a term representing a formula $\phi(x)$, where x is of type τ. Then

$$\forall x\, \phi(x) \quad \text{is represented by } \Pi_\tau \lambda x.a.$$

The use of the letter Π is quite natural: in the standard notation we would express the same by $\Pi_{x\in\tau}a(x)$.

Describing axioms and rules would take us to far afield and it is not interesting anyway because we only need to translate the usual axioms into the new formalism. Let us just observe that axioms are rules that rewrite an expression directly to **1**.

Formulas as Types

In the typed λ-calculus we need different combinators for different types. Thus we have one combinator $\mathbf{I}_\tau$ for every type τ, one combinator $\mathbf{K}_{\tau,\sigma}$ for every pair of types τ, σ and one combinator $\mathbf{S}_{\tau,\sigma,\rho}$ for every triple of types τ, σ, ρ. Let us look the types of these combinators.[29]

$$\begin{aligned} \mathbf{I}_\tau &: \tau \to \tau \\ \mathbf{K}_{\tau,\sigma} &: \tau \to (\sigma \to \tau) \\ \mathbf{S}_{\tau,\sigma,\rho} &: \big(\tau \to (\sigma \to \rho)\big) \to \big((\tau \to \sigma) \to (\tau \to \rho)\big) \end{aligned}$$

The striking fact is that the types look like propositional tautologies; indeed, if we interpret the type variables as propositional variables they are tautologies. Moreover, these three tautologies are often used as axioms of propositional calculus. And this is still not all, look at this:

If $a : \tau$ *and* $b : \tau \to \sigma$, *then* $ba : \sigma$.

This rule corresponds to the propositional rule of *modus ponens.* Thus suddenly, an axiomatization of a fragment of propositional logic pops up. The fragment is the restriction of intuitionistic logic to formulas built of implications, the *implicational fragment of intuitionistic logic.*

The connection with logic is very close. Not only these three combinators have types that are intuitionistic tautologies, but *all terms* have such types. *Vice versa*, for every tautology ϕ of the implicational fragment of intuitionistic logic, there exits a term whose type represents ϕ. Hence we can view terms as proofs of the tautologies that are represented by their types. This correspondence is represented by the following diagram.

type ↔ *formula*
term ↔ *proof*

Viewing it semantically, as a complex structure, we have types that are empty, which correspond to non-tautologies, and types that are *inhabited*, which correspond to tautologies. To prove a tautology means to show that the corresponding type is inhabited.

---

[29]The column is used to express membership in a type.

Example In Curry's formalism variables and terms are used without specifying types. Then there are terms that do not have a type, hence cannot be interpreted as proofs. Show, as an exercise, that one can assign types to the combinators in the following term so that it has the specified type:

$$\big(\mathbf{S}(\mathbf{KS})\big)\mathbf{K} : (\sigma \to \rho) \to \big((\tau \to \sigma) \to (\tau \to \rho)\big).$$

The corresponding tautology is a version of the transitivity of implication. The term $(\mathbf{S}(\mathbf{KS}))\mathbf{K}$ is a proof of this tautology.

To extend this correspondence to full intuitionistic propositional logic, one has to use more operations on types, products $\tau \times \sigma$ for conjunctions, sums $\tau \oplus \sigma$ for disjunctions, and one has to add the empty type $\bot$ for *false*. (The negation of ϕ is represented by $\phi \to \bot$.)

This correspondence is called the *Curry-Howard isomorphism*. It has been extended in many directions. Concerning logic, the correspondence has been extended to first-order intuitionistic logic, classical propositional logic and some other logics. Concerning proof theory, connections have been found between certain transformations of proofs in logic on the one hand, and transformations of terms in the λ-calculus on the other. In this way the computational nature of the λ-calculus has been translated to proof theory and one can view certain proof-theoretical operations as computations.

Notes

1. *Composition of functions in the λ-calculus.* The binary operation of the λ-calculus, the application of a function to an argument, should not be confused with the composition of functions. The latter is definable by means of the combinator $\mathbf{B}$ defined by $((\mathbf{B}x)y)z = x(yz)$ (or by the λ-term $\lambda x \lambda y \lambda z.x(yz)$). The composition of function y with function x is the function $(\mathbf{B}x)y$. One can show that $\mathbf{B}$ can be expressed as $(\mathbf{S}(\mathbf{KS}))\mathbf{K}$.
2. *Combinatory logic.* The classical presentation of the λ-calculus does not include the Axiom of Extensionality. The system defined by the axioms 1.–3. on page 149 with the axiom $\mathbf{K} \neq \mathbf{S}$ is Curry's *Combinatory Logic*. It is slightly weaker than the λ-calculus, but it can be made equivalent to it by adding five more equalities. By adding four other equalities it can be made equivalent to the λ-calculus with the Axiom of Extensionality. Furthermore, there exists a presentation of the Combinatory Logic with a single combinator.
3. *The Church-Rosser Theorem.* The great appeal of the λ-calculus stems from the possibility of manipulations with terms, which can be used to define computations. In fact, the λ calculus was one of the first systems by means of which computable functions were defined. The two basic operations are *λ-abstraction* and *β-conversion*. The first of the two is the operation considered above: given a term we construct another term that represents the function defined by the given

term. β-conversion is in a sense the opposite operation; it is the transformation of a term $\lambda x.\tau(x)a$ into the term $\tau(a)$. This follows from the definition: applying the function defined by a term $\tau(x)$ to an argument a we must obtain the value of the term on the argument a. At first glance it looks stupid to introduce a λ-term in order to eliminate it in the next moment, but it is not as simple as it may appear, and quite interesting things may happen. The point is that τ may also contain occurrences of the λ-operator. Thus if we have a complex term, there may be several different places where we can apply β-conversion, hence we can reduce it in different ways. Furthermore, β-conversion may reduce the size of the term, but it also may increase it. (If σ is a long term and x occurs in τ more than once, the term $\tau(\sigma)$ is clearly longer than $\lambda x.\tau(x)\sigma$.) Consequently, the process of successive applying β-conversion may run indefinitely when started on some terms. The key result of the theory, the *Church-Rosser Theorem*, says that if the process converges, then it converges to a unique term, whatever strategy we choose. By converging we mean that we arrive at a term to which β-conversion is not applicable anymore. This uniquely determined term is called *the normal form*.

There are several interesting applications of this basic result. In particular, one can use it to prove the *consistency of the λ-calculus*. One can prove that two terms that have normal form are equal in the λ-calculus if and only if the normal forms are equal. Since it is easy to construct two different normal forms we get immediately the conclusion that the λ-calculus is consistent (meaning that one cannot prove that all elements are equal).

4. *The λ-calculus and computations.* In order to define computations, terms of the λ-calculus will be interpreted in two ways: as programs and as data. The computation of program τ on input data σ is the β-conversion process on the term $\tau\sigma$. The output is the resulting normal form, provided it exists, otherwise the computation diverges. To compute functions of more variables we apply program τ to a string of data $\sigma_1, \dots, \sigma_n$ as follows: $(\dots((\tau\sigma_1)\sigma_2)\dots\sigma_n)$ (according to the standard convention, the parentheses can be omitted).

   In order to compare it with other computational models, we need to encode some standard structures by terms that are in the normal form. One can define natural numbers $0, 1, 2, 3, \dots$ by the terms

$$\lambda x\lambda y.y, \quad \lambda x\lambda y.xy, \quad \lambda x\lambda y.x(xy), \quad \lambda x\lambda y.x\big(x(xy)\big), \quad \dots .$$

   So, for example, 0 is the function that on every argument gives the identity function as the value. It is a remarkable fact that every partial recursive function can be defined by a term. For example, addition can be defined by the term $\lambda x\lambda y\lambda z\lambda u.((xz)(yz))u$ and multiplication by $\lambda x\lambda y\lambda z.x(yz)$ (which is the combinator **B**).

5. *Propositional calculi from the λ-calculus.* It is interesting to see what the logical calculi resulting from the Curry-Howard isomorphism are. If we only use constant terms, as in the example of transitivity of implication, the calculus is essentially a Hilbert-style calculus (also called Frege system). If we allow variables and λ-abstraction, the proof system is like the natural deduction calculus.

A variable plays the role of an assumption and when λ-abstraction is applied to it, the assumption is discharged.

β-reduction has a different role. It corresponds to normalization of proofs, a transformation related to cut-elimination (which we will consider in Chap. 6).

6. *A model of the λ-calculus.* The first natural model of the λ-calculus was constructed by D. Scott in 1971 [261]. Scott found a general construction that can be applied to any nontrivial complete lattice. We will only consider the simplest case, where one starts with the two element lattice. We define a sequence of lattices $D_0, D_1, D_2, \ldots$ as follows. D_0 is the two element lattice. Having D_n, define D_{n+1} as the set of all order preserving mappings $f : D_n \to D_n$ with the ordering given by the condition that $f \leq g$ if $f(x) \leq g(x)$ for all $x \in D_n$. Furthermore, define order-preserving mappings $\phi_n : D_n \to D_{n+1}$ and $\psi_n : D_{n+1} \to D_n$ as follows. $\phi_0(x)$ is the constant function equal to x, $\psi_0(x)$ is $x(\bot)$, where $\bot$ is the bottom element of D_0. Having ϕ_n and ψ_n, define ϕ_{n+1} and ψ_{n+1} as follows. For $x \in D_{n+1}$, i.e., $x : D_n \to D_n$, put
$$\phi_{n+1}(x) = \psi_n \circ x \circ \psi_n,$$
and for $y \in D_{n+2}$, (i.e., $y : D_{n+1} \to D_{n+1}$) put
$$\psi_{n+1}(y) = \psi_n \circ y \circ \phi_n,$$
where $\circ$ denotes composition of functions. Note that ϕ_n is an embedding, ψ_n is a mapping onto D_n, and $\psi_n \circ \phi_n$ is the identity. Put
$$D_\infty = \{(x_0, x_1, \ldots);\ \forall n\ \psi_n(x_{n+1}) = x_n\}.$$
D_∞ is the universe of the algebra. It is a limit of D_n's in a well defined sense. In particular, every D_n is naturally embedded in D_∞ by the assignment
$$\iota_n(x) = \left(\psi_n^n(x), \ldots, \psi_n^2(x), \psi_n(x), x, \phi_n(x), \phi_n^2(x), \ldots\right).$$
Furthermore, D_∞ with the natural ordering is a complete lattice. We denote the supremum of a set $X \subseteq D_\infty$ by $\bigvee X$; to compute it, take suprema on all coordinates. Now we are ready to define the operation of application. For $a, b \in D_\infty$, where $a = (a_0, a_1, \ldots), b = (b_0, b_1, \ldots)$,
$$ab = \bigvee_n \iota_n\left(a_{n+1}(b_n)\right).$$
The trick is that we can think of an element of D_∞ as a sequence of elements and as a sequence of functions at the same time. (It is tempting to define the nth coordinate of ab simply as $a_{n+1}(b_n)$, but this does not work.)

To prove the combinatory completeness, one needs to have a natural property that is satisfied by all functions that are definable by terms of the algebra. To this end Scott uses a natural topology defined on D_∞, and the corresponding concept of continuous functions. Continuous functions are characterized as the functions that preserve suprema of directed sets. (A set is directed if for every two elements there is an element which is larger than both.) For finite lattices, continuous functions are the order preserving functions. So in the simplified case here, where all D_n's are finite, we only needed the property of order preserving.

It is possible to prove that

(i) The operation $a, b \mapsto ab$ is continuous in D_∞.
(ii) For every continuous function $f : D_\infty \mapsto D_\infty$, there exists $a \in D_\infty$ such that for all $x \in D_\infty$, $f(x) = ax$.

In order to get the combinatory completeness, we need a little more in (ii). The function f may depend on other parameters, and then a is a function of these parameters. We need a, as a function depending on these parameters, to be continuous. Then the two statements imply combinatory completeness.

Main Points of the Chapter

- The language of mathematics developed by converging to a minimal and universal language.
- The primitives of a logical language are: constants, variables, relation symbols, connectives and quantifiers. The syntax uses the same principles as the syntax of natural languages.
- The truth of a sentence in a structure can be precisely defined.
- Consequently, the logical validity of a sentence can be defined by the condition that the sentence is true in all structures.
- The concept of a proof can be made quite precise by postulating syntactical rules that must be satisfied by every proof.
- The logical calculus is complete: every sentence that is true in all structures can be proved.
- People are willing to use a completely formal language, if there is a strong incentive to do so.
- We have several precise mathematical concepts that define a class of computable functions; for example, Turing machines. We believe that this class comprises all computable functions (the Church-Turing Thesis), but we cannot prove it, as it is not a mathematical statement.
- There are very concrete noncomputable functions. The problem is not that our devices are too inefficient or that we do not have enough time to compute these functions, it is because they are not computable even in principle.
- There are universal Turing machines that can simulate every Turing machine. Computers are constructed so that they are universal in this sense.

Chapter 3
Set Theory

Meaning! Listen to the mathematician talk. Great space, man, what has mathematics to do with meaning? Mathematics is a tool and as long as it can be manipulated to give proper answers and to make correct predictions, actual meaning has no significance.

Isaac Asimov, *The Imaginary*

I BRIEFLY considered sets in Chap. 1. It should be clear to the reader by now that this topic is extremely important for the foundation of mathematics and thus deserves more attention. Since the time of Cantor, set theory has developed into a reputable mathematical field, so it is possible to cover only a small part of the results in the limited space I have here. I will restrict myself only to the main ideas and results in this field.

Following the history of set theory in Chap. 1 we paused at the moment when Russell published his paradox. Cantor did not perceive Russell's paradox as a disaster; he knew about paradoxes in set theory before Russell and, in fact, Russell extracted his paradox from Cantor's. Cantor thought of sets as reality, therefore he did not see the problem in the concept of sets but in our approach to them. He concluded that some sets, in particular very large sets, are 'paradoxical' and we have to treat them somehow differently. But merely to ignore the paradoxes because they refer to strange sets which we do not need was not a solution; set theory needed to be rehabilitated. It was necessary to explicitly say what had to be given up and what should be preserved. The axiomatic approach was at hand at that time and it was the most explicit way of saying which assumptions should be used. But in contrast to geometry, where the object of study was the space in which we live, the universe of sets seemed elusive. There are no physical objects, situations or phenomena that we could associate with sets of higher cardinalities.

Frege, a contemporary of Cantor, was more interested in the logical foundations than in the mathematics of sets. He further developed his ideas of *Begriffsschrift* in *Grundgesetze der Arithmetik I, II*[1] published in 1893 and 1903 [78, 79] where he defined a logical system that, in effect, allows the use of the unrestricted Principle of

[1]In English: *Basic Laws of Arithmetic.*

P. Pudlák, *Logical Foundations of Mathematics and Computational Complexity*, Springer Monographs in Mathematics, DOI 10.1007/978-3-319-00119-7_3,

Comprehension. The way the theory is presented is more complicated, but, essentially, one can interpret in it the naive set theory with full comprehension schema, which is, as we already know, inconsistent. Russell discovered his famous paradox just when the second volume of *Grundgesetze* was to appear and notified Frege. Frege wrote an appendix to the volume in which he explained the contradiction and proposed a modification of the system that, as he believed, would avoid it. Unfortunately the modified system was also found inconsistent, but Frege did not try to fix it second time. After years of struggling to have his work published, he apparently did not have enough energy and optimism left to revise his system.

The logical system that Frege introduced in *Grundgesetze* was considered inspiring and important for the further advancement of the logical foundations of mathematics, but the general opinion was that it was a dead-end—an inconsistent system that in principle cannot be fixed. In the 1980s, researchers started looking at *Grundgesetze* with a more positive attitude. It turned out that one only needs to state one of the laws in a weaker form and then Frege's approach to foundations can be salvaged.

Those who started later than Frege had the advantage that they knew what they should avoid. They also were luckier because no new generation of antinomies appeared. The origin of the axiomatic foundations of set theory is due to two logicians, Bertrand Russell and Ernst Zermelo. Russell, who read Euclid's *Elements* at the age of eleven, naturally aimed at an axiomatic system. Zermelo, who himself contributed a key result to Cantorian set theory, rather tried to give firm foundations to set theory in the way it was imagined by Cantor.

I will start with the two approaches and briefly mention some alternative ones. The main theme of this chapter is: what are the additional assumptions that one has to use on top of logic in order to be able to develop mathematics? There are some elementary parts of mathematics on which we can test our hypothesis. An example is geometry which describes something very real—physical space. If more general concepts are considered, in particular those that involve infinity, there is very little we can do to test our theories by experiments. Therefore, mathematicians proposing axiomatic systems for set theory use various intuitive arguments to justify them. It may be a 'higher order' principle, it may be just an ad hoc way to satisfy the needs of mathematicians. It is important to realize that there are different and incompatible ways in which one can axiomatize set theory and mathematics developed in these systems may also be different. The only test for the theories is that mathematics 'works' in them. Eventually mathematics is used for solving problems in physics and other real world problems. If one system fails in these applications, while another works, then the former one is rightly rejected. But if they do not differ on the practical type, then it may depend on trends, historical coincidence, etc. which will be accepted by a majority of mathematicians.

By Gödel's theorem we know that we will never be able to collect all necessary axioms needed in mathematics, so the process of extending our axiomatic system is a never-ending one. In set theory this is connected with large infinite cardinal numbers. The study of these cardinals is like exploring the limits of the mathematical universe. This will be another important topic treated in this chapter.

The last two sections will be about other possible directions in the development of set theory. I will speak about the slightly controversial Axiom of Choice and axioms that may be used instead of it. Finally I will briefly mention two different axiomatizations of set theory that were proposed as alternatives to the standard axiom system.

3.1 The Axioms of Set Theory

The Theory of Types

Bertrand Russell
Courtesy of McMaster University Library

Russell was among the first who tried to resolve set theoretical paradoxes. He realized that it does not suffice to ask '*what are sets?*'—one should also ask '*what causes paradoxes?*'. His thoughts led him to the conclusion that most paradoxes are caused by what he called *the principle of the vicious circle*. The "vicious circle" appears whenever one wants to define a set x that may contain itself. In order to form a set x, we have to determine for every element whether or not it belongs to x. In this process we need to determine it also for x, namely, we have to answer the question whether x is an element of x, and here is the problem: we have not finished the construction of x, so we cannot answer this question. Hence, concluded Russell, we have to prohibit such definitions. To this end, Russell proposed a theory in which there are different types of sets, hence the theory is called the *Theory of Types*. The first version of this theory appeared in an appendix to his *Principles of Mathematics* published in 1903 [251]. A fully developed theory was presented in Whitehead and Russell's *Principia Mathematica* [313–315], a milestone in the foundations of mathematics. Here, I will deal with a simplified version of Russell's Theory of Types, which will be called *Simple Type Theory*. The reader should be warned that there are different systems in the literature that use the same name.

The basic idea is to stratify sets into types: the first type are elements, the second type are sets, the third consists of sets of sets, the fourth of sets of sets of sets, etc. An element is not a set, a set can only contain elements, a set of sets can only contain sets, etc. Thus we have various *types* of objects. We can assign numerals to the types, thus elements are of type 0, sets of type 1, sets of sets of type 2, etc. An object of type i may only contain objects of the type $i-1$. Then the comprehension axiom schema is modified as follows:

The Typed Comprehension Axiom Schema *For every type i and every formula ϕ which mentions only sets of type i, there exists a set x of type $i+1$ which contains exactly those sets of type i that satisfy ϕ.*

The stratification of sets into types is not unnatural. In natural language we tend to distinguish objects from properties. Objects correspond to type 0, properties to 1.

Next we can talk about properties of properties, type 2. We can go on, but higher types practically do not occur in normal speech, even type 2 is not very common. The doctrine of the Theory of Types is that different types should not be mixed and each definition leads to a higher type. Such a hierarchical structure occurs also elsewhere in nature (or, maybe, we impose such a structure on various phenomena to understand them better). The smallest particles of matter, according to the present knowledge, are quarks. Combining quarks we get elementary particles, from elementary particles we get atoms, from atoms molecules.

How does the Theory of Types avoid Russell's paradox? Let x be the set of all sets which do not contain itself. Then x contains all sets, by which we mean all objects of type 1, since a set can only contain elements (objects of type 0), so it cannot contain itself. Does x contain x? Of course not, as x is of type 2, hence it can only contain objects of type 1. This shows that we cannot reproduce the paradox on the lowest type; the same argument works on any type. But remember, this is not a proof of consistency.

When introducing types we impose restrictions on how sets can be defined. As a result we have to sacrifice certain principles available in naive set theory. In particular, since we have elements that are not sets, we have to abandon the extensionality, as we cannot distinguish elements by what they contain. However, this only concerns type 0. This is not counterintuitive at all, as in the real world we have objects which are not sets, and the sets are rather our constructions. Another problem is that we have to postulate an axiom of infinity. We have to assume the existence of infinitely many elements of type 0, otherwise we would have only finite sets on all types. Intuitively, it seems that an axiom of infinity should not create problems: we encounter infinity in our physical world, so why not to assume it in set theory. What is disturbing is that we have to add it explicitly, and such an ad hoc action may hide some inconsistency in itself. In fact, some philosophers claimed that the source of inconsistencies was a too liberal usage of infinity. After a century of experience with axioms of infinity we are not so much worried anymore; yet it is good to keep it in mind.

The Theory of Types also reflects the idea that mathematical objects are mathematical structures. Recall that a structure has a universe, which is a set of elements, then some relations, for instance unary relations which are subsets of the universe, then it may contain higher order relations, for instance sets of sets of elements, and so on. So the universe of objects in the Theory of Types is like a huge universal structure, except that in structures we only allow objects up to a finite order, while in the Theory of Types we have all orders. It is not surprising then that the Theory of Types is able to serve as a foundation for most of mathematics, as has been shown in *Principia Mathematica.*

Impredicative Definitions, Semantical Antinomies and the Hierarchy of Languages

A possible explanation of the vicious-circle principle is by the concept of *predicative and impredicative definitions.* Roughly speaking, a set, or a class X, is defined by a predicative definition if it is defined by a condition ϕ that only speaks about objects different from X. Formally, this means that the range of quantifiers in ϕ must not include X as an element. Otherwise the definition is impredicative.

Example Let x be the intersection of all sets y that satisfy a given condition ϕ (in symbols, $x = \bigcap\{y; \phi(y)\}$). This set can be defined as the set of all elements z that satisfy the following condition:

z is an element of every set y that satisfies ϕ,

(in symbols, $z \in x \equiv \forall y(\phi(y) \rightarrow z \in y)$). The set x is not mentioned explicitly in the condition, but it may happen that x satisfies ϕ. Thus the condition *implicitly* refers also to x. Hence this definition of x is impredicative. This example also shows that there are quite natural and innocent looking definitions which are impredicative.

Although the Theory of Types avoids Russell's paradox, it does use impredicative definitions—there is no restriction on the quantifiers used in the Typed Comprehension Axiom Schema. If we restricted the schema to predicative definitions, the resulting theory would be very weak. For example, we would not be able to define the natural numbers as the least set containing 0 and closed under the successor function S because such a set is defined as the intersection of all sets containing 0 and closed under the successor function S.

Impredicative definitions were one reason for Russell to introduce a more complicated version of the theory; another reason was the aim to avoid semantical antinomies. Although semantical antinomies seem to be less dangerous, we cannot ignore them. It does not make sense to make a system proof against a particular type of antinomy while allowing a contradiction of another type to sneak in.

Recall Berry's paradox: *the first number that cannot be defined by an English sentence with at most one hundred letters*. One can easily dismiss this paradox by pointing out that English is a tremendously complicated object, especially when we want to talk about meaning (semantics), and the meaning of many words and sentences is vague, so we cannot say that they really *define* anything. Nevertheless, we do not have to use a natural language in this paradox. We can take a formal logical language for which the semantics is clear and which is strong enough to express all the necessary concepts. So what happens if we state the paradox in such a precise way?

Recall what we have learned in Chap. 2:

1. for a given language L it is possible to precisely define what it means 'definable in L',
2. but it is not possible to write this definition in the language L itself.

Thus the paradox can be solved by distinguishing between the language and the meta-language; in other words, by using a hierarchy of languages.

Russell decided to incorporate the hierarchy of languages in his Theory of Types. For a mathematician, it may seem a useless complication, but for a philosopher, it perfectly makes sense. Why should one use a formal system with logical derivations defined precisely in order to talk about sets, while leaving considerations about the hierarchy of languages on an intuitive level? The system that he introduced is called *the Ramified Class Calculus* [252]; it is an extension and a refinement of the simple Theory of Types described above. In order to make his system immune against semantical paradoxes, Russell proposed viewing the comprehension principle not as a property of the set universe, but as an act by which we define a new set. So the set is not shown to exist in the universe that we study, but merely defined by us. This implies that the set is on a higher level of the hierarchy of languages no matter what level of types we are considering. Formally, this means introducing another stratification into infinitely many *orders*. Hence every set has two indices: one for its type and another for its order. The order does not influence the relation of membership $\in$, so again a set of type $i+1$ contains sets of type i, the orders do not matter. The order is important only in the comprehension axiom schema. The variables of the language of ramified class calculus also have the two indices; for example, ${}^3x^2$ is a variable for sets of type 2 and order 3. When applying the comprehension to a formula ϕ we take the maximum of the orders occurring in ϕ and the set which it defines will have the next order.[2] Thus each type 'ramifies' into infinitely many orders.

Let's look at what happens with Berry's paradox. When we define the least number satisfying a condition, we have to talk about numbers only up to an order j. The number that we define will be of order $j+1$ so it will be different from those from which we took the minimum. More precisely, its value as a number will be the minimum, but it will be a different object. This looks like a very good safeguard against antinomies and it certainly is, but the price is too high. In particular the least number principle is the minimum that we would like to have in the arithmetic of natural numbers. However, the form of this principle that we obtain in the Ramified Class Calculus is very weak, so weak that it is practically useless.

The Ramified Class Calculus is too weak to be used as the foundations of mathematics, unless we add more axioms. In an attempt to save his system Russell introduced the *Axiom of Reducibility* which says that for every set, there is a set of the same type and of order 1 (the least order) containing the same elements. Unfortunately, this axiom cancels the effect of orders—one can derive the impredicative comprehension axiom schema for sets of order 1 and any type. Hence the Ramified Class Calculus with the Axiom of Reducibility contains the unramified Theory of Types and, therefore, it is not better in avoiding paradoxes.

The bottom line of this story is that it is hopeless to look for one hundred percent safe foundations. In order to propose something that mathematicians might accept, one has to risk that the system will be found inconsistent.

The Theory of Types is not used by mathematicians, but various systems derived from it play an important role in proof theory, the λ-calculus, intuitionistic foundations and functional programming. What might have been seen as a failure turned out to be inspiration for many important results.

---

[2] More precisely, we have only to worry about the variables that are quantified, as they talk about the 'totality of all sets of a given order', while parameters do not matter, as they talk about particular elements.

Zermelo Set Theory

Ernst Zermelo
Courtesy of
Universitätsarchive
Zürich[3]

Zermelo's aim was to preserve the results in set theory proved by Cantor and to solve the problem with paradoxes using Cantor's suggestion that they are caused by very large sets. Therefore, we should start with small sets and gradually produce larger and larger ones, but not create a huge one all at once. In order to disallow such big jumps, we have to restrict the comprehension axiom. Let us look at when it should be safe to use the comprehension axiom according to the theory that contradictions are caused by large sets. Suppose we already have produced some sets, so to say, we have them well in hand. Then for any property ϕ we should be able to single out those which satisfy ϕ. The paradox appeared because we wanted to decide property ϕ for all sets, *including those which had not yet been constructed.*

This seems very plausible, except that we have to say more precisely what it means 'to have some sets well in hand'. There may be different answers leading to different systems. What Zermelo proposed was very simple and elegant: it means that all sets from which we want to construct the new set must be contained already in a previously constructed set. Thus knowing that the previously constructed set was 'small', we also know that the new one is small. Therefore, he proposed restricting the comprehension axioms as follows.

The Restricted Comprehension Axiom Schema *For every property of sets ϕ, if x is a set, then there exists a set y which contains exactly those elements of x which satisfy ϕ.*

The difference between the unrestricted and this axiom schema is that we can only define subsets of a given set. Russell's paradox cannot be reproduced using such a restricted comprehension principle, but it is less obvious than in the Theory of Types, since in Zermelo Set Theory it is possible to have kinds of sets that do not exist in the Theory of Types.

Because such a restricted comprehension schema does not increase the size of sets, we cannot prove much from this principle alone. We do need axioms that gradually increase sets. First of all we have to postulate the existence of at least one set.

The Axiom of the Existence of Sets *There exists a set.*

Sometimes this is assumed to be an axiom of logic; furthermore, it follows from the Axiom of Infinity. Zermelo postulated the existence of the empty set.

Now we need some way of getting larger sets. Let us first consider axioms which enable us to construct larger sets from those that we already have.

---

[3]Item No. AB.1.1165.

The Axiom of Pairing *For every pair of sets x and y, there exists a set z whose elements are precisely x and y.* (The set z is denoted by $\{x, y\}$.)

The Axiom of the Union *For every set x, there exists a set y that is the union of the elements of x. Formally, $z \in y$ if and only if there exists $u \in x$ such that $z \in u$.*

These two axioms enable us to produce the union of two sets. Other boolean operations can be defined using the Comprehension Axiom Schema. The Axiom of the Union, in this general form, is not essential for making Zermelo Set Theory strong, but it will play an important role in Zermelo-Fraenkel Set Theory.

The Axiom of the Power-Set *If x is a set, then there exists a set which contains all subsets of x as elements and no other elements.* (The new set is called *the power-set of x* and it is denoted by $\mathcal{P}(x)$.)

This is a very powerful axiom; it produces very large sets. If x is a finite set with n elements, the power-set $\mathcal{P}(x)$ has 2^n elements. In the case of an infinite set x, Cantor proved that $\mathcal{P}(x)$ has larger cardinality than x, so to say, it is 'more infinite' than x. Thus, by repeating the power-set operation, we get infinitely many types of infinity in Zermelo's universe of sets. But in spite of the strength of the power set axiom, it still produces finite sets from finite sets. To get infinite sets, we have to postulate the existence of at least one such set. Hence we start the process of constructing sets from an infinite set. So, again, we have to postulate:

The Axiom of Infinity *There exists an infinite set.*

Zermelo Set Theory, as presented in his seminal paper [318] in 1908, contained furthermore the Axiom of Extensionality and the Axiom of Choice, an important invention of Zermelo. I will not state the Axiom of Choice now, as it is not important from the point of view of the consistency and strength of the axiomatic systems. I only mention a property which distinguishes it from other axioms: *All the axioms added on top of the restricted comprehension schema are special instances of the schema.* The Axiom of Choice is the only axiom of Zermelo Set Theory that is not of this form. Thus the whole theory can be presented as the Axiom of Extensionality, some cases of the comprehension schema and the Axiom of Choice.

Today we present Zermelo Set Theory as a theory based on first-order logic, but Zermelo did not mean it so. He was a platonist, like Cantor, and believed that there is an actual universe of sets and that the axioms that he proposed were true in this universe. In particular, he did not think it was necessary to say what the properties ϕ in the comprehension schema are. In a truly axiomatic system we must describe explicitly what we mean by the properties. It was Skolem who in the early twenties proposed to present Zermelo's system as a theory based on first order logic and properties expressible in a formal logical language. Zermelo was strongly against such an interpretation.

Zermelo-Fraenkel Set Theory

For most results in mathematics, Zermelo Set Theory would amply suffice. In some cases, however, we do need more. Thus what is now accepted by the majority of mathematicians as the set theoretical foundations is an extension of Zermelo Set Theory called *Zermelo-Fraenkel Set Theory*. The extension was proposed independently by Abraham Fraenkel (1891–1965) [74] and Skolem [271] in the early 1920s. The extended system appeared in Zermelo's paper [319].

The new idea is just another explication of Cantor's suggestion that the paradoxes are caused by large sets. Zermelo allowed turning collections contained in a set into a set. The idea was that the collection is small as it is contained in a set. Skolem and Fraenkel took a different interpretation of being small: small means to have a small number of elements. The point is that for some collection Y which is not known to be a set, we can define a mapping assigning elements from some set x onto Y. Then, clearly, if x has a small number of elements, then also Y has a small number of elements. Thus, according to Cantor, it should be safe to assume that Y is actually a set.

At first glance it is not clear how strong this principle is, or even it is not obvious that it allows us to prove the existence of more sets. It turns out that it is a very strong principle, provided that we add another natural axiom, the axiom of the existence of the union of sets contained in a set. For now, I will confine myself to stating the new principle as an axiom schema.

The Replacement Axiom Schema *Let F be a function defined on a set y. Then the image of y, which consists of all $F(x)$ for $x \in y$, is a set too.*

To state the axiom formally one has to specify the meaning of 'function'. This is done in the same way as in the case of the Comprehension Axiom. Namely, one has to state it as a schema for every possible definition of a function by a formula of the language of set theory. I do not know why exactly the word 'replacement' is used here, but the following description of the principle is a likely explanation: *If we replace elements of a set by some other elements, the resulting entity is still a set.*

I will use the standard abbreviation *ZFC* for this theory. The letters 'Z' and 'F' are, clearly, the initials of Zermelo and Fraenkel. The letter 'C' stands for 'choice'; it is intended to stress that the axiom of choice is among the axioms.

Set Theories with Classes

Ordinary mathematicians do not like talking about logical formulas; they prefer real mathematical objects. The axiom schemas of Comprehension and Replacement, when stated properly, need the concept of a formula, but there is a way to avoid it. It requires introducing a new type of objects, called *classes*. The first such system, introduced by von Neumann [207], was based on the concept of a function, instead of sets and classes. Bernays proposed a version based on sets and classes [23], which was later simplified by Gödel [97]. The resulting system is called *Gödel-Bernays Set Theory*, or *von Neumann-Bernays Set Theory*.

In this system we have two kinds of objects: sets and classes. For the sake of convenience, it is better to call a class either of the two and say that sets are a special kind of class. In fact, we *define* which classes are sets: *sets are those classes which are elements of a class.* (Thus what we called *The Principle of Being an Element* for sets in Chap. 1, becomes the definition of a set.) A class which is not a set is called a *proper class*. We think of proper classes as being too big to be used as elements of a set. The idea of proper classes corresponds to Cantor's view that "large sets" may be paradoxical. Those "large sets" are now called proper classes.

The comprehension axiom schema is modified as follows:

The Axiom Schema of Predicative Comprehension for Classes *For every property of* sets ϕ, *there exists a* class *which contains exactly those sets which satisfy property* ϕ.

In particular, we can prove that there exists the *universal class* V, the class of all sets, by taking some property that is trivially satisfied by all sets. In the formal presentation of this axiom schema the property ϕ should be a formula in which only sets are quantified. If we treated sets and classes as separate entities, this would mean that we only allow *predicative definitions of classes.*

The reason why this formulation of the comprehension principle avoids Russell's argument is the same as in the Theory of Types. Indeed, let ϕ be the property that the set is not an element of itself. This defines a class and Russell's argument yields that the class is not a set, it is a proper class. One can prove that the class of all sets exists and it is a proper class too. The relation between sets and classes is similar to the relation between sets of two consecutive types in the Theory of Types, except that we treat sets as a special kind of class, while the types are disjoint.

The advantage of the Schema of Comprehension for Classes is that it allows us to identify properties of sets with classes and thus to avoid talking about formulas. Let us look what happens with the Restricted Comprehension Axiom Schema. We simply replace 'property ϕ' by 'a class Y'. (Note that it is customary to denote sets by lower case letters and classes, which are not a priori sets, by upper case letters.) As a result we get a single axiom instead of a schema which reads:

> *For every class Y and every set x, there exists a set z containing exactly those elements of x which are elements of Y.*

But this is only an awkward way of saying a very simple statement:

The Axiom of Subsets *The intersection of a class and a set is a set.*

In the same manner the replacement axiom schema is substituted by a single axiom.

Clearly, we do not get such a simplification for free; we have to postulate some axioms about classes. What we need is the above Schema of Predicative Comprehension for Classes. Too bad, we have again 'properties', 'formulas', etc.! Fortunately, no—we can now replace the schema by a small number of concrete axioms, a few instances of the schema. These axioms postulate the existence of the class that encodes the membership relation $\in$, the intersection of two classes, the complement of a class and a few more slightly more complicated constructions of classes. Thus eventually the whole system can be presented using a finite number of relatively simple axioms and no reference to formulas, etc. is needed (see page 174).

Such a system is certainly more appealing to working mathematicians. From the point of view of logicians, however, the difference between Zermelo-Fraenkel and Gödel-Bernays set theories is not essential. In particular one can prove that what is provable about sets in one system is also provable in the other one (a consequence of the fact that the principle of comprehension for classes is stated only for predicative definitions). So classes do not make the system stronger.

Zermelo's Universe

Let us now try to imagine the universe of sets as given by Zermelo's axioms. All axiomatic systems for set theory, including Zermelo's system, are incomplete in an essential way, hence the universe is not determined uniquely.[4] On the one hand, there are sets that we know for sure that they must exist, namely those for which we can prove it. On the other, we can disprove the existence of sets satisfying certain definitions. Such a definition of a set is Russell's formula from his paradox. We also know that the universal set, the set of all sets, does not exist. If it existed, we could use it as the parameter which restricts the comprehension schema and then the schema would not be restricted, as every set is an element of the universal set. Thus, using Russell's paradox, or rather the argument in it, we can derive a contradiction from the assumption that the universal set exists, hence it does not exist. Finally there are sets which may but need not exist. This means that we can neither prove nor disprove their existence. What I am going to describe will be in some, rather weak, sense a minimal universe.

By the consideration above, the universe will not be symmetric; it will grow from small sets into larger, infinite, with no largest set. Having at least one set, no matter whether it is infinite or not, we can apply the comprehension schema with a property which is never satisfied and we get the existence of the empty set. Now consider the following operation defining a new set from a given set. Let x be given, take the power-set of x and then take the union of it with x. Using set-theoretical notation it is the operation which produces the set $x \cup \mathcal{P}(x)$ from x. In words, we extend x by adding to it all its subsets. Starting with the empty set we will apply the operation

[4]Even if it were complete, it would not guarantee that there was a unique model.

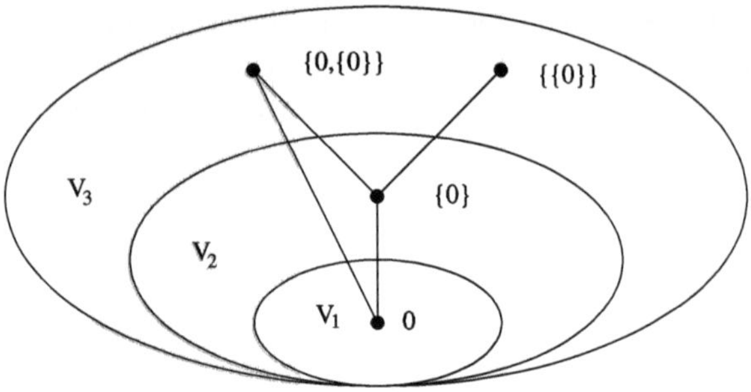

Fig. 3.1 The three bottom levels of the cumulative hierarchy. *Points* represent sets, *lines* the membership relation

repeatedly. Let us denote by V_i the ith element in the sequence thus obtained. So V_0 is the empty set $\emptyset$, then V_1 is $\emptyset \cup \mathcal{P}(\emptyset)$, which is just $\{\emptyset\}$ the one element set containing the empty set (as you can easily verify). Here are two more sets from the sequence, see Fig. 3.1:

$$V_2 = V_1 \cup \mathcal{P}(V_1) = \{\emptyset\} \cup \mathcal{P}(\{\emptyset\}) = \{\emptyset, \{\emptyset\}\}$$
$$V_3 = V_2 \cup \mathcal{P}(V_2) = \{\emptyset, \{\emptyset\}, \{\{\emptyset\}\}, \{\emptyset, \{\emptyset\}\}\}.$$

It looks rather complicated, but, in fact, V_i is determined by a simple condition: it contains all sets that can be written using at most $i-1$ nestings of braces. These sets form the *cumulative hierarchy*. 'Hierarchy' because they contain more and more complex sets, and 'cumulative' because the sets preceding V_i are contained in it as subsets. I am using the more complex operation instead of just the power-set operations in order to make the cumulative character of the hierarchy quite apparent. Since we start with the empty set, the power-set operation alone would suffice.

If we do not assume the Axiom of Infinity and instead we only postulate the existence of at least one set, then every set in the minimal universe is contained in some V_i. Hence the universe is the infinite union $V_0 \cup V_1 \cup V_2 \cup \cdots$. Let us call this union V_ω. Here the letter ω stands for the first infinite ordinal number. I will introduce infinite ordinals shortly; for the time being, we can treat it as a convenient symbol.

Now assume that we do have an infinite set, say a set a. We can construct the cumulative hierarchy in the same way starting with a. The resulting universe will, unfortunately, depend very much on the initial set a. The most natural thing then is to take V_ω as the infinite set. This is in accord with the rule of a thumb used to make set theories stronger: once we can somehow visualize the whole universe given by our axioms, we can add the axiom that such a collection is also a set.

In order to get a nice picture, we have to introduce more notation. We will denote by

$$V_{\omega+1} = V_\omega \cup \mathcal{P}(V_\omega)$$
$$V_{\omega+2} = V_{\omega+1} \cup \mathcal{P}(V_{\omega+1})$$
$$\text{etc.}$$

So in this way we have extended the cumulative hierarchy of sets beyond the finite levels. The resulting universe is the union of all levels.

Zermelo-Fraenkel's Universe

There is no reason not to extend the cumulative hierarchy beyond the levels $V_\omega, V_{\omega+1}, V_{\omega+2}, \ldots$ by defining the next level to be $V_{\omega+\omega} = \bigcup_n V_{\omega+n}$. Once we have $V_{\omega+\omega}$, the power-set axiom yields $V_{\omega+\omega+1}, V_{\omega+\omega+2}, \ldots$. The axioms of Zermelo Set Theory do not enable us to deduce that $V_{\omega+\omega}$ is a set. The operation that we need is the union of an arbitrary number of sets. More precisely, we need the universe of sets to be closed under taking such an infinite union. In set theory, if possible, we prefer to break operations into more elementary ones. This union can be composed of two operations. First we take the *set* of the sets that we want to unite, say X. In our case, we can schematically denote it by $X = \{V_\omega, V_{\omega+1}, V_{\omega+2}, \ldots\}$. Then we take the *union of the set* X, which means that we unite the elements of X. In our case it is $\bigcup X = \bigcup_n V_{\omega+n}$.

Thus we need two axioms. One should guarantee the existence of the set X, the other the existence of the union. For the first one, we use the Replacement Axiom Schema. The assignment $n \mapsto V_{\omega+n}$ is definable and every n is in the *set* V_ω. Thus we can deduce that X is a set. The second is the Axiom of the Union.

Equipped with these tools and having $V_{\omega+\omega}, V_{\omega+\omega+1}, V_{\omega+\omega+2}, \ldots$, we can go on and take the union of this sequence. Let us denote it by $V_{\omega+\omega+\omega}$. Then we can apply the power-set operation and so on.

In order to understand this process, we need to study the concept of a transfinite ordinal number. Furthermore, we should know that each new step in the cumulative hierarchy introduces a higher cardinality and along with it also larger ordinal numbers. I will explain transfinite ordinals and the hierarchy of infinite cardinal numbers shortly. For now, let us use these concepts intuitively.

Having a countably infinite set, we can show that the cumulative hierarchy contains V_α for every countable ordinal α. It is clear that we obtain a lot of sets and lot of cardinal numbers in this way. However complicated the structure of sets looks like already at these stages, this is only a beginning. The first uncountable set appears in $V_{\omega+1}$; we denote its cardinal number by $\aleph_1$. From this set, we get levels V_β for all ordinals of cardinality $\aleph_1$. But we are still extending the hierarchy only using cardinalities that appear in V_α for α countable, while producing higher and higher cardinalities. We produce a huge amount of new cardinalities before we use up ordinals whose cardinalities appear in levels V_α, for α countable. Then we will use the ordinals whose cardinalities appear in V_β, for β of cardinality $\aleph_1$, and so on.

Watching the beginning of the process of constructing sets in the cumulative hierarchy, we see that there is a very strong positive feedback in it that causes a tremendous explosion of the universe of sets. So the apparently minor extension of Zermelo's system of axioms by the Replacement Axiom Schema has a very dramatic effect.

In mathematics, cardinalities larger than $\aleph_1$ are not used very often. Nevertheless, human curiosity does not have any limits. The question whether there are even larger cardinals than those that we obtain in this process has attracted attention since the very beginning of set theory.

Cleaning Up the Universe

According to one theory an explosion of a supernova near our solar system swept away the cosmic dust from our neighborhood. Therefore, our lucky civilization can watch stars on bright nights. This certainly has greatly helped the development of all science.

But let's get back to set theory. The cleaning process that we are going to consider has little to do with that theory. It is rather related to the well-known *Occam razor* which suggests getting rid of all unnecessary concepts. Following Cantorian tradition, it is unpopular to prohibit something in set theory. If a set can exist, then in "*Cantor's Absolute*" the ideal world of sets, it does exist. Hence, by forbidding some sets, we get narrow-minded, and decide to study only a part of reality. Still there are sets which most set theorists give up voluntarily. Consider, for example, a set x which has a unique element which is itself; so $x = \{x\}$. Let y be another set with the same property. By extensionality they are different because they contain different elements $x \neq y$. If we take the elements of their elements, it is the same and so on. Structurally they are the same, but still they are different. The axioms considered so far do not exclude such sets, but such sets will never appear in the cumulative hierarchy of sets $\{V_\alpha\}_{\alpha \in ON}$, where ON denotes the class of all ordinal numbers. On the other hand, those which are in the hierarchy are nice, as they are in some sense constructed from the canonical set $\emptyset$. Therefore, we prefer to have:

The Axiom of Foundation *There are no sets outside the cumulative hierarchy.*

This axiom may give an impression that we have restricted the universe of sets as much as possible, which is not true. We will shortly meet another hierarchy, called the constructive hierarchy, which restricts the universe much more—so much that set theorists do not consider reasonable to accept the axiom that would only allow sets from this hierarchy.

All this is only a brief description of what set universes of these theories can look like. In order to get a more precise idea about it, one has to study models of set theories. I will have more to say about it in Sect. 4.5.

Notes

1. *The contradiction in Frege's system and how to fix it.* In modern terms the logical system that Frege used can be explained as follows. There are two kinds of entities: elements and classes connected by a membership relation $\in$. As usual, I will use lower case letters for elements and upper case letters for classes.

   The first basic principle is Comprehension.

   For every formula $\phi(x)$,

$$\exists Y \forall x \left(x \in Y \equiv \phi(x)\right).$$

The second principle, Frege's *Basic Law V*, states the existence of a function F from classes to elements such that

$$\forall X \forall Y \big(F(X) = F(Y) \equiv \forall z(z \in X \equiv z \in Y)\big).$$

Assuming extensionality for classes, F is a one-to-one mapping from classes to elements and we can *define* a binary relation ε on elements by

$$x \;\varepsilon\; y \equiv x \in F^{-1}(y).$$

Given an arbitrary formula $\phi(x)$, we get a class Y such that $\forall x\ (x \in Y \equiv \phi(x))$ from the Comprehension Principle for classes. Let $y = F(Y)$. Then

$$x \;\varepsilon\; y \;\Leftrightarrow\; x \in F^{-1}(y) \;\Leftrightarrow\; x \in Y \;\Leftrightarrow\; \phi(x).$$

This shows the Comprehension Principle for elements and the relation ε, hence Russell's paradox can be derived. (In this proof we have used extensionality for classes, but one can prove the same without it.)

Frege's *Basic Law V* is a special instance of a more general principle: for any equivalence relation $\mathcal{E}$, one can represent the classes of $\mathcal{E}$ by elements. This can be formalized by introducing functions $F_{\mathcal{E}}$ such that

$$\forall X \forall Y \big(F_{\mathcal{E}}(X) = F_{\mathcal{E}}(Y) \equiv X \;\mathcal{E}\; Y\big).$$

In the *Basic Law V*, the equivalence relation is extensionality. Frege considered another instance of this general principle where the equivalence $\mathcal{E}$ is equinumerousity. (Equinumerousity is defined using binary relations on elements.) This axiom, called *Hume's Principle* by contemporary authors, says that there exists a function $X \mapsto \#X$ that gives the cardinality of a set X as an element. In the 1980s, researchers studying Frege's work discovered that in order to develop arithmetic in Frege's system, *Hume's Principle* suffices and one does not need to use the *Basic Law V* (see [30]). The theory with this restriction is equivalent to Second Order Arithmetic (see page 295 for the definition), which is believed to be consistent and in which most of main-stream mathematics can be developed.

2. *The axioms of the Simple Type Theory.* I will only describe a simplified version of the original system. I will present it as an axiomatic theory based on first order logic. This is not the way Russell did it. He did not separate set theory from logic. For him, the entire system was logic; it was logic in a broader sense than we view logic today.

   The Simple Type Theory has infinitely many types of variables, one type for every natural number. The type of a variable is given by its superscript. Then the language contains one binary relation $\in$ of membership. Only elements of consecutive types can be in this relation. The elements of different types are distinct. Therefore, atomic formulas are required to be of the form $x^i = y^i$, or $x^i \in y^{i+1}$.

The Axiom Schema of Extensionality *For every type i,*

$$\forall x^{i+1}, y^{i+1}\big(\forall z^i\big(z^i \in x^{i+1} \equiv z^i \in y^{i+1}\big) \rightarrow x^{i+1} = y^{i+1}\big).$$

The Axiom Schema of Comprehension *For every type i and every formula* $\phi(x^i)$ *not containing variable* y^{i+1},

$$\exists y^{i+1} \forall x^i \left(x^i \in y^{i+1} \equiv \phi(x^i)\right).$$

Here I am assuming that the formula ϕ may contain free variables other than x and I am using the convention that free variables are implicitly universally quantified. Doing it more formally, one can list all free variables $x^i, p_1, \ldots, p_n$ of ϕ and write the axiom as follows:

$$\forall p_1 \ldots \forall p_n \exists y^{i+1} \forall x^i \left(x^i \in y^{i+1} \equiv \phi(x^i, p_1, \ldots, p_n)\right),$$

where $p_1, \ldots, p_n$ are variables of arbitrary type and y^{i+1} does not occur among them. We call $p_1, \ldots, p_n$ *parameters*.

The Axiom of Infinity *There are infinitely many elements of type* 0.

The most natural way to state this axiom is to postulate the existence of the successor function: there exists a function S defined on elements of type 0 (an object of type 1) that satisfies the first two axioms of Peano, see page 30.

Russell, however, considered such an approach unacceptable. He wanted to derive the axiom from logical axioms, but since this was not possible, he wanted at least to state it in such a way that it would not postulate the existence of any special structure. So Whitehead and Russell observed that one can *define* a function S on type 2 objects such that S satisfies the two axioms if and only if there are infinitely many objects of type 0. Hence, in order to postulate the existence of infinitely many objects of type 0, one does not have to state the existence of a special set, one only needs to postulate a property of a definable structure.

The idea of this definition is to think about a number n as a set of all sets with precisely n elements (of type 0). Thus 0 is the set that contains only the empty set; 1 is the set of all one-element sets etc. The next idea is that one can define $S(n)$ by saying that it is the set of all sets that have "one more element" than a set in n. More precisely, x is in $S(n)$ if and only if there exists a $y \in n$ and an element $a \notin y$ such that $x = y \cup \{a\}$. Then, one defines the natural numbers as the least set $\mathbb{N}$ of type 3 that contains 0 and is closed under S. So to state the axiom of infinity, one only needs to say that *this S and this $\mathbb{N}$ satisfy the two axioms of Peano.*

This is certainly only a historical curiosity. There are many ways to state the axiom of infinity and one may argue about which is more "logical" and "natural".

In the original system Russell and Whitehead considered not only sets, but also relations of arbitrary large numbers of arguments. In particular, above we had only sets of type 1, while they had sets, binary relations, ternary relations, etc. of type 1 and similarly for higher types. As far as the strength of systems is concerned, we do not loose anything by restricting ourselves to sets. Relations can be coded by sets of pairs, triples, etc., as we have seen in the previous chapter. But note that using the standard definition of a pair, which is $\{\{a\}, \{a, b\}\}$, we code binary relations on type 0 by sets of type 3, which is rather unnatural.

Clearly, if one wants to *use* the system, one takes the system with relations, while if one wants to *study* the system, the simpler one is better.

3. *The axioms of Zermelo-Fraenkel Set Theory.* The language of the theory has one binary relation $\in$, the membership. Unlike the Theory of Types it has only one sort of object. I use the same convention about free variables (parameters) in the axiom schemas as above.

The Axiom of the Existence of Sets $\exists x(x = x)$. (Usually this is already assumed in logic.)

The Axiom of Extensionality $\forall z(z \in x \equiv z \in y) \to x = y$.

The Axiom Schema of Comprehension (also called, *of Subsets* or *of Separation*) *For every formula* $\phi(x)$ *that does not contain the variable* z,

$$\forall y \exists z \forall x \big(x \in z \equiv (x \in y \wedge \phi(x))\big).$$

The Axiom of Pairing $\forall z \forall u \exists y \forall x(x \in y \equiv (x = z \vee x = u))$.

The Axiom of the Union $\forall y \exists z \forall x(x \in z \equiv \exists u(x \in u \vee u \in y))$.

The Axiom of the Power-Set $\forall y \exists z \forall x(x \in z \equiv \forall u(u \in x \to u \in y))$.

The Axiom of Infinity $\exists y(\emptyset \in y \wedge \forall x(x \in y \to x \cup \{x\} \in y))$.

The Axiom Schema of Replacement (also called, *of Collection*) *For every formula* $\phi(u, v)$ *that does not contain the variables* w, z,

$$\forall u \forall v \forall w\big((\phi(u, v) \wedge \phi(u, w) \to v = w\big) \\ \to \forall y \exists z \forall v\big(v \in z \equiv \exists u\big(u \in y \wedge \phi(u, v)\big)\big)\big).$$

The Axiom of Foundation (also called *of Regularity*)

$$\forall x\big(x \neq \emptyset \to \exists y(y \in x \wedge y \cap x = \emptyset)\big).$$

The Axiom of Choice

$$\forall x\big(\forall y(y \in x \to y \neq \emptyset) \wedge \forall y, z(y \in x \wedge z \in x \wedge y \neq z \to y \cap z = \emptyset) \\ \to \exists u \forall y\big(y \in x \to |y \cap u| = 1\big)\big).$$

I denote by $|x|$ the cardinality of the set x. I stated the condition on cardinality in this way for the sake of brevity, but it can easily be expressed using only the basic notions: $\exists v(v \in y \wedge v \in u \wedge \forall w((w \in y \wedge w \in u) \to v = w))$. The Axiom of Foundation is stated differently than on page 170, but the two statements are equivalent (using the remaining axioms).

The axioms of the Schema of Comprehension are easily derivable from the axioms of the Schema of Replacement, thus the first schema can be omitted. Even such a reduced set of axioms is still not independent, but one can prove that it cannot be reduced to a finite set—Zermelo-Fraenkel Set Theory cannot be axiomatized by a finite number of axioms.

One can replace the Axiom of Infinity by the following instance of the comprehension schema:

$$\exists u \forall v \big(v \in u \equiv \big(\exists y \big(\emptyset \in y \wedge \forall x \big(x \in y \rightarrow x \cup \{x\} \in y\big)\big) \\ \rightarrow \forall y \big(\big(\emptyset \in y \wedge \forall x \big(x \in y \rightarrow x \cup \{x\} \in y\big)\big) \rightarrow v \in y\big)\big)\big).$$

The Axiom Schema of Replacement can also be stated as instances of the unrestricted comprehension schema:

$$\forall y \exists z \forall x \big(x \in z \equiv \big(\forall u, v, w \big(\big(\phi(u, v) \wedge \phi(u, w) \rightarrow v = w\big) \\ \wedge \exists u \big(u \in y \wedge \phi(u, x)\big)\big)\big)\big).$$

4. *Gödel-Bernays Set Theory (von Neumann-Bernays Set Theory).* The objects of the theory are called *classes*; they will be denoted by upper case letters. Sets are defined as those classes that are elements of classes. Sets will be denoted by lower case letters. Thus $\forall x\, \phi(x)$ means $\forall X (\exists Y (X \in Y) \rightarrow \phi(X))$, and similarly for $\exists$.

   The axioms are the same as for *ZFC* with the following changes. The Extensionality is stated for all classes. The Comprehension is replaced by the following schema and an axiom.

The Axiom Schema of Predicative Comprehension for Classes *For every formula $\phi(x)$ that does not contain the variable Y and where all quantifiers bind only sets,*

$$\exists Y \forall x \big(x \in Y \equiv \phi(x)\big).$$

The Axiom of Subsets $\forall X \forall y \exists Z (X \cap y \in Z)$ (*The intersection of a class and a set is a set.*)

Furthermore, the axiom schema of Replacement is substituted by:

The Axiom of Replacement *If F is a (class) function and x is a set, then the image of x by F is also a set.*

Note that in the axiom schema of Predicative Comprehension only a special type of formula is allowed. The reason is not that a system with such an unrestricted schema would be inconsistent. Such a stronger system has been considered (it is called *Kelly-Morse Set Theory*). The reasons are different.

Firstly, the way the schema is stated satisfies the property of *predicativity*. Given some classes (the class parameters of the formula ϕ) we assert that another class should exist. So in order to obtain this "new" class, we refer only to some given classes, we do not need to know *all* classes.

Secondly, it is possible to replace such a schema by a finite number of its instances (which is not possible in the case of Zermelo-Fraenkel and Kelly-Morse set theories). Thus *Gödel-Bernays Set Theory can be presented by a finite number of axioms.* The following eight axioms can be used instead of the Axiom schema of Predicative Comprehension for Classes:

The Axioms for Class Formation

$$\exists X \forall x (x \in X \equiv \exists y \exists z (x = (y, z) \wedge y \in z))$$
$$\forall X \forall Y \exists Z \forall x (x \in Z \equiv x \in X \wedge x \in Y)$$
$$\forall X \exists Y \forall x (x \in Y \equiv x \notin X)$$
$$\forall X \exists Y \forall x (x \in Y \equiv \exists y ((y, x) \in X))$$
$$\forall X \exists Y \forall x (x \in Y \equiv \exists y \exists z (x = (y, z) \wedge z \in X))$$
$$\forall X \exists Y \forall x (x \in Y \equiv \exists y \exists z (x = (y, z) \wedge (z, y) \in X))$$
$$\forall X \exists Y \forall x (x \in Y \equiv \exists y \exists z \exists u (x = (z, (u, y)) \wedge (y, (z, u)) \in X))$$
$$\forall X \exists Y \forall x (x \in Y \equiv \exists y \exists z \exists u (x = (y, (u, z)) \wedge (y, (z, u)) \in X))$$

Recall that we define the ordered pair (x, y) as the set $\{\{x\}, \{x, y\}\}$. The first axiom says that there is a class that encodes the binary relation $y \in z$; the second axiom says that for every two classes their intersection is a class; the third axiom speaks about the existence of the complement; the fourth axiom postulates the existence of the projection of a binary relation; the rest are "structural" axioms. Clearly, the axioms correspond to syntactical operations on formulas: the atomic formula, logical connectives ($\wedge$ and $\neg$, which is a complete basis), the existential quantifier, a dummy variable and permutations of variables.

We say that this theory is *conservative* over *ZFC*. This means that a sentence that mentions only sets is provable in *ZFC* if and only if it is provable in the Gödel-Bernays Set Theory. Nevertheless, the richer language enables us to state some new axioms that do not have equivalents in *ZFC*. An example is the *Axiom of Global Choice*. It postulates that there exists a class function that chooses an element from every nonempty set.

Classes are not used in traditional parts of mathematics, but they are useful in some modern fields (for example, the theory of categories).

5. *Finite Set Theory.* This theory is closely related to Peano Arithmetic, as I mentioned on page 116. It is obtained from Zermelo-Fraenkel Set Theory by replacing the axiom of infinity by its negation:

The Axiom of Finiteness $\neg\exists y (\emptyset \in y \wedge \forall x (x \in y \rightarrow x \cup \{x\} \in y)).$

It is not terribly important how we state the Axiom of Finiteness, but one may prefer a more elegant way than this. The above statement only says that there do not exist infinite sets of a particular form and we have to prove that there are no infinite sets at all. Therefore, a more explicit way of forbidding infinite sets may be more appealing. Furthermore, we may omit the Axiom Schema of Comprehension, the Power-Set Axiom and the Axiom of Choice because they are derivable from the remaining ones.

The 'universe' (the correct term used in logic is '*the standard model*') is V_ω, the set of all hereditarily finite sets (finite sets whose elements are finite sets, elements of elements are finite sets, etc.).

3.2 The Arithmetic of Infinity

Infinity is one of the most intriguing concepts. Of course, philosophers have thought about it a lot and mathematicians have always assumed that there are infinitely many natural numbers and that lines 'stretch to infinity', etc., but as a mathematical concept it seemed elusive for a long time. Cantor was the first one to present a mathematical theory of infinity. His most important result is that there is not only one infinity, but there are higher and higher types of infinity. Another thing for which he became famous is not a result, but a problem that he tried to solve unsuccessfully. It is the *Continuum Hypothesis* which we will consider shortly. Roughly speaking it is the question whether there is a type of infinity which is between the type of natural numbers and the type of real numbers. This shows that the realm of infinities is fairly complex and deserves our attention if we want to understand the foundations of set theory.

Recall that Cantor used sets in an intuitive way; he did not state and use axioms. Therefore, we have to consider his results with some reservations. His results may be false in some axiomatic systems; for instance, an axiomatic system may postulate precisely which infinities occur. In this section, therefore, we assume Zermelo-Fraenkel Set Theory (with the Axiom of Choice), which is the axiomatic system that corresponds to Cantor's idea of the *Absolute* in the best way.

What Is Infinity?

Our everyday experience keeps reminding us of our limitations, our actions take place in finite time and finite space. We cannot truly perceive infinity, like we cannot imagine the fourth dimension, but as an arbitrary large number of dimensions can be perfectly treated in mathematics, also infinity can be precisely defined once we accept some set-theoretical principles. Mathematicians accepted the infinitude of the natural numbers already in ancient Greece. In *Elements*, for example, Euclid proved that the number of prime numbers is infinite, although he avoided the use of the word 'infinite': "*Prime numbers are more than any assigned multitude of prime numbers.*"[5]

In fact infinity has often been considered the border where problems become nontrivial and thus interesting. It was used without being defined, which is not very surprising, since the same was true about other mathematical concepts. However, infinity was perceived more like a metaphysical problem than an object of mathematical research. Back then, it seemed paradoxical that infinite sets have some properties that finite sets cannot have. Thus the nature of writings about infinity was rather philosophical.

Galileo Galilei (1594–1642) in his *Discourses and Mathematical Demonstrations Relating to Two New Sciences* [88], discussed the concept of infinity in arithmetic and geometry. He also suggested how to resolve some paradoxes.

[5] Book 10, Proposition 20 [117].

An essay that deals exclusively with infinity is *Paradoxes of Infinity* [28] of the Czech[6] mathematician and philosopher Bernard Bolzano (1781–1848). Bolzano noticed that infinite sets have the special property that they have the same number of elements as some proper subset of them. This is what distinguishes them from finite sets. Somewhat later Dedekind accepted this property as the definition of infinite sets. The popular explanation of this "paradox" is as follows. There is an infinite hotel whose rooms are numbered $1, 2, 3, \dots$ which is fully occupied. When a new guest comes, it is still possible to accommodate him. He will get the room number 1 and each guest already staying in the hotel will be moved to the room with the next higher number. We can say that this is a "paradox of infinity" as it contradicts our experience with finite objects.

The property used in Dedekind's definition of infinite sets, stated more formally, is that one can move elements of the set inside of the set so that at least one position remains free. In terms of the arithmetic of infinite sets (which we will consider shortly) it can be stated as the equation

$$1 + x = x,$$

where x stands for the size (cardinality) of an infinite set. Note that the choice of the property is not completely arbitrary. The minimal requirement is, of course, that at least one infinite set has such a property. As regards Dedekind's definition, clearly, the set of natural numbers has this property. A less trivial requirement is that all infinite sets satisfy this property. Informally we can argue as follows. Since the cardinality of the natural numbers is the least infinite cardinality, we should be able to find a copy of the natural numbers in every infinite set. So we can write the set in the form $x + y$, where x is a copy of the natural numbers and y is the rest. Then we have $1 + x + y = x + y$ because we already know that $1 + x = x$.

Now I hope that you have spotted something odd. We want to *define* infinite sets, so how can we prove that the property is satisfied by all infinite sets? Yes, if the statement above is interpreted literally, it is wrong. To be more precise, I should have said that we need to show that there is no ambiguity in the definition, which means that all natural alternative definitions are equivalent to this property. This can really be proved.

Philosophers, since the time of Aristotle, distinguish two kinds of infinity: *potential* and *actual*. A *potential infinity* is a process that creates larger and larger *finite* sets, sets whose sizes grow beyond any bound. If we view the natural numbers in this way, it means that at any moment there is only a finite segment of them existing, but we can extend this segment arbitrarily. If the universe started at certain point of time and if it is never going to end, then the infinity of time is similar—at any point in the history only finite time will elapse from the beginning, but still time will go to infinity.

In *Discourses*, Galileo considered the following paradox. The set of square numbers, $1, 4, 9, 16, \dots$ seems clearly to be smaller than the set of all positive integers $1, 2, 3, 4, \dots$. Yet there is a one-to-one correspondence between the two sets:

[6] Born and lived in the Kingdom of Bohemia.

$n \leftrightarrow n^2$. His explanation (in the words of Sagredo) is that we should only allow potential infinity. Then we only look at finite intervals $[1, n]$. As n goes to infinity, the density of square numbers in these intervals goes to zero, which demonstrates that the set of squares is smaller than the set of all positive integers. The answer of modern mathematicians to the question whether the set of squares is smaller than the set of all positive integers is '*It depends*'. If you use cardinality as your measure, then they have the same size. If you use density, then the set of squares is smaller. One reason why the view that infinity can only be potential prevailed for such a long time was Euclid's authority. The *Common Notion No. 8* of his *Elements* reads: "*The whole is greater than the part.*" This explicitly prohibits actually infinite sets.

An *actual infinity* is an infinite set that has already been entirely created. Viewing mathematical objects as structures requires accepting actual infinity. The natural numbers viewed as a structure is an object that is finished; it is an actual infinite structure. Probably, one thing that significantly contributed to the shift from accepting only potential infinity to accepting actual infinity was the discovery that the real numbers can be defined from the natural numbers. Recall that it is easy to define the rational numbers from the natural numbers. Then the real numbers can be defined using certain sequences of rational numbers or using certain subsets of rational numbers. Each of these definitions requires infinite series or infinite sets of rational numbers. For example, if we define a real number by its decimal expansion (which means as the limit of decimal fractions) we need the entire sequence to specify a real number. This shift culminated in the work of Cantor whose theory not only used actual infinity, but also introduced an infinite hierarchy of infinities.

A revival of the philosophy of potential infinity occurred at the beginning of the 20th century as a reaction to the paradoxes in set theory. The proponents of this view blamed the paradoxes on using actual infinity. Such an approach, however, requires finding a different way of defining real numbers. Essentially, one has to abandon the reduction of the real numbers to the natural numbers and introduce the real numbers as a primitive notion. Most mathematicians prefer the comfort of using actual infinity, as well as its power.

Counting Infinite Sets

The *cardinality* of a set is, as usual, the number of elements of the set, except that now we are considering infinite sets. From now on, I will speak about *infinite cardinalities* instead of the less precise *types of infinity*, as it is common in set theory. It is convenient to represent every infinite cardinality by a fixed set, in the same way we did for finite numbers; these representatives will be called *cardinals*. Furthermore, we need symbols to denote infinite cardinals. For this purpose, the first letter of the Hebrew alphabet has been chosen, the letter $\aleph$ called *aleph*. Cardinals will be distinguished by indices at aleph; for example, $\aleph_\alpha$.

The first technical problem that we have to solve is how to measure the cardinality of a set. For finite sets we can use counting which is assigning numbers to the elements. This immediately suggests extending this procedure to infinity by saying that a set is *countably infinite*, or simply *countable* if one can assign natural numbers to the elements of the set in such a way that no element of the set is omitted in the process. This kind of set is also called '*denumerable*', but I will stick to the term '*countable*'. Let us assume that there are *uncountable* sets. Such sets can also be enumerated in a way, but we need more numbers. This is done using *transfinite ordinals*. (We prefer to say '*transfinite*' instead of '*infinite*' when speaking about ordinal numbers.) Before explaining this important, but more complicated concept, let us ask a more basic question: when are two sets of the same size? This can be, in fact, defined without much theory. Recall that we say that two sets have the *same cardinality*, or are *equinumerous*, if it is possible to assign elements of one set to the elements of the other set in a one-to-one way without omitting any element from either of the sets. This definition also suggests a way to compare the cardinalities. A set X has cardinality *at most as large as* the cardinality of Y if we can assign elements of Y to elements of X in a one-to-one way without necessarily exhausting the set Y.

Based on our experience with finite sets, these definitions seem obvious and trivial. They are easy, no doubt, but the point is to make the right decision about which of the concepts that we use for finite sets should be used for infinite ones. We will see soon that there are big differences in the arithmetic of finite and infinite numbers, hence we must not expect that everything generalizes in a straightforward way.

It is not difficult to show that countable sets have to have the smallest infinite cardinality. This cardinal is denoted by $\aleph_0$. One can prove that there is the next largest cardinality and this will be denoted by $\aleph_1$. You have surely guessed that there should be $\aleph_2, \aleph_3, \ldots$, but let me stress that it is not obvious that the smallest infinite cardinals can be enumerated in increasing order in such a way. As noted above, Cantor showed that for every cardinal number, there is a larger one, but one has to prove that among the cardinals larger than $\aleph_0$ there is the smallest one (and so on for $\aleph_1, \aleph_2, \ldots$).

Once we accept the definition that two sets have the same cardinality if there exists a one-to-one mapping from one set onto the other, we are led to a canonical definition of finite sets and this in turn gives a canonical definition of infinite sets. Thus a set is finite, if it has n elements, for n a natural number. This means precisely that a set is finite if it is empty or it can be one-to-one mapped onto the set $\{1, 2, \ldots, n\}$ for some natural number $n > 0$. A set is defined to be infinite, if it is not finite. To show that Dedekind's definition is equivalent to this one, take an arbitrary set X and pick different elements from X one after the other. Thus we construct a sequence $x_1, x_2, \ldots$. Now either we exhaust X after a finite number of steps, then X is finite (the assignment $x_i \mapsto i$ gives the required one-to-one mapping), or we get an infinite sequence of elements. Having such an infinite sequence $x_1, x_2, \ldots$ we can show that X satisfies Dedekind's definition: define a mapping by $x_n \mapsto x_{n+1}$ for the elements of the sequence and $x \mapsto x$ for the remaining elements. It is a one-to-one mapping and no element is mapped on x_1. In particular the smallest cardinality $\aleph_0$ is the cardinality of the set of natural numbers.

...					
10	...				
6	11	...			
3	7	12	...		
1	4	8	13	...	
0	2	5	9	14	...

Fig. 3.2 The figure shows a way to enumerate all pairs of natural numbers. The fact that it can be done was known already to Galileo Galilei. The simple formula for the number of a pair (m, n) in this enumeration $((m+n)^2 + 3m + n)/2$ is attributed to Cantor

Some Simple Arithmetic

Before we proceed to larger cardinals, let us play with those that we already have for a while. We need to learn at least some simple properties of infinite cardinals in order to get an idea about them, although the concrete facts from this section will not be used later.

We would like to add and multiply infinite cardinals. In order to define these operations, we have to find set theoretical interpretations of these operations. Again, it is quite easy. Let X and Y be two sets and suppose they are disjoint (which means that they do not share a common element). Then the sum of the cardinalities of the two sets should be the cardinality of the union of them. The product of the cardinalities should be the cardinality of the Cartesian product of the two sets $X \times Y$ (which is the set of all pairs of elements one from X the other from Y).

Let us consider some simple cases. Let n be a finite cardinal (a natural number). Then $n + \aleph_0 = \aleph_0$. This is obvious, if we add a finite set to a countable set it is still countable: we can enumerate it by first enumerating the finite set and then the infinite one. It is not much harder to prove that $\aleph_0 + \aleph_0 = \aleph_0$. Namely, two countable sets can be enumerated by alternatively taking elements from the first one and from the second one. This can be expressed by $2 \cdot \aleph_0 = \aleph_0$. This implies $3 \cdot \aleph_0 = \aleph_0$, $4 \cdot \aleph_0 = \aleph_0$, etc., but in fact we have even $\aleph_0 \cdot \aleph_0 = \aleph_0$. A possible way to visualize this relation is in Fig. 3.2.

These relations may look familiar to you. You have surely seen equations such as $\infty + \infty = \infty$ and $\infty \cdot \infty = \infty$ when you studied the calculus. The essence is the same, but in the calculus ∞ is merely an auxiliary symbol, it does not represent the cardinality of a set.

There is nothing special about $\aleph_0$ as far as addition and multiplication are concerned. We have the same relations $\aleph_n + \aleph_n = \aleph_n \cdot \aleph_n = \aleph_n$ for any n. In general both operations reduce to the operation of taking the maximum:

$$\kappa + \lambda = \kappa \cdot \lambda = \max(\kappa, \lambda),$$

for every two cardinals such that at least one is infinite. Consequently the arithmetic of infinite cardinals is trivial when only plus and times is used. If you need to evaluate an arithmetical expression, just look for the largest term. However, this only concerns addition and multiplication. If we include more operations, then there are many difficult problems about the arithmetic of infinite cardinals.

The arithmetical operations do not help us to get larger cardinals, unless we take infinite sums or products. Let us consider

$$\aleph_0 + \aleph_1 + \aleph_2 + \cdots.$$

The meaning of the infinite sum is the same as of the finite one: take the cardinality of the disjoint union of sets of these cardinalities. It is not difficult to see that this number must be bigger than all cardinals $\aleph_n$ (for n a natural number), and is the smallest one with this property; so it is the next cardinal after all cardinals $\aleph_n$.

We need a name for it and you may propose to call it $\aleph_{\aleph_0}$. Though $\aleph_0$ and ω are usually represented by the same set, we prefer to call it $\aleph_\omega$ for a good reason. Consider the next largest cardinal, which is $\aleph_{\omega+1}$. If you wrote $\aleph_{\aleph_0+1}$, then this could mean simply $\aleph_{\aleph_0}$, since $\aleph_0 + 1 = \aleph_0$. So this would not be a good notation. But it is not just a matter of notation; you should realize that the indices do not express cardinalities, they express the position in an *order*. Therefore, we use ordinals for indices and you see that we cannot understand infinite cardinal numbers without understanding *ordinal numbers*. This is a special property of infinite cardinals; for finite cardinals, we do not have to develop the theory of ordinal numbers, since the two concepts coincide.

How to Get a Larger Cardinal Number—Cantor's Diagonal Argument[7]

Isn't it nonsense to have so many infinite cardinalities? In the real world we meet only finite structures. Maybe, space and time is infinite, but even this is not clear. Shouldn't we allow at most one kind of infinity?

Mathematicians, already before set theory had been developed, thought that there were different types of infinity. Their intuition was based on experience with continuous structures. They felt that there is a big difference between discrete structures such as the natural numbers and continuous ones such as the real numbers. On the real line we can do things which are impossible on integers. (For instance, we can move smoothly from one number to another, or we can sum an infinite number of positive reals and still get a finite real number.) This intuition turned out to be correct, but it is not as obvious as it may seem. Similar intuition suggested that spaces of higher dimension should have more points; say, there should be more points in a plane than on a line (more means that the cardinality should be bigger), but this is wrong; a line, a plane, a 3-dimensional space, etc. all have the same number of points. This follows from the general law $\aleph_\alpha \cdot \aleph_\alpha = \aleph_\alpha$.

The reason for accepting more infinities is the following theorem [37].

[7]The history of the invention of the diagonal argument is not quite clear. Some historians attribute the idea to Paul du Bois-Reymond.

Cantor's Theorem *For every set x, the cardinality of the power-set $\mathcal{P}(x)$ (the set of all subsets of x) is bigger than the cardinality of x.*[8]

A nice feature of the theorem is that it holds for any set, finite or infinite. For finite sets, we know that for a set of cardinality n the power-set has cardinality 2^n. Thus for finite sets the theorem can be proved simply by showing that $n < 2^n$ for every n. In fact, we interpret the theorem in this way for every set: for an infinite cardinal $\aleph_\alpha$ we *define* $2^{\aleph_\alpha}$ to be the cardinality of the power-set of a set of cardinality $\aleph_\alpha$. This also explains the use of the term '*power-set*' for the set of all subsets. Hence Cantor's theorem is the inequality $\kappa < 2^\kappa$ for every cardinal number κ. Since the Power-Set Axioms ensures the existence of the power-set for any set, we can get larger and larger cardinalities. Consequently, *there is no largest cardinal number.*

I will explain the proof in a small example which proves the theorem for $n = 4$. Suppose your task is to designate a name for a new product. The main condition that you need to satisfy is that the name is different from those already being used. Thus you get a list of the names of existing products and you want to designate a name which is different. Suppose the list looks like this.

$$\begin{matrix} A & B & B & A \\ B & B & B & A \\ B & A & B & A \\ B & A & A & A \end{matrix}$$

There are two more restrictions that need to be satisfied: we need a word using only As and Bs and it should have length 4. We will use the special property of the list that the number of words equals the number of letters in the words. Take the word that appears on the diagonal. It is ABBA; coincidentally it occurs on the list. Let us switch the first letter to B. The resulting BBBA is on the list, but if we keep B in the first position from now on, the word will be different from the first word on the list. If we switch the second letter, we ensure that the word will be different from the second word on the list, etc. So, we simply flip all letters and get a word different from all words on the list. In our example the word obtained by this algorithm is BAAB.

This construction is not very practical, as it can only be used when the number of words is at most the number of letters in the words, but it is a very general one and thus it can also be used in the case of infinite words. Furthermore, there are many other applications of this '*diagonal method*'. The connection with sets is the following one. Words of length 4, in our example, can be thought of as representing subsets of a four-element set x. The letter A in the ith position of a word w means that the ith element of the set x belongs to the set represented by word w, while B means that it does not belong to it. Hence in general the construction shows that, given n subsets of an n-element set, we can always construct another subset, which implies that the number of subsets is bigger than n. The same argument can be used for infinite sets, hence subsets of a set of cardinality κ cannot be enumerated by κ, thus $2^\kappa > \kappa$.

---

[8]See also Theorem 2 on page 41.

Before going on, let us pause and ponder the meaning of such results. Once we know that there is no canonical way how to choose axioms of set theory and we are only trying to collect some reasonable and hopefully consistent ones, such theorems on cardinalities seem of dubious importance. Why don't we simply first accept axioms that decide the cardinalities following our intuition and then add other axioms so that the system is still consistent? There are various reasons for not doing it in this way. In particular, Cantor's proof is simple and clear, thus any system which somehow prohibits such an argument cannot be very natural. The most important reason, however, is that big cardinalities give big power to the theory, while axioms prohibiting large cardinals do not make the theory substantially stronger. As long as they do not lead to a contradiction, it is good to have such strong means. We will see that in the section on large cardinals.

The Continuum Hypothesis

Cantor denoted the cardinality of the real numbers by $\mathfrak{c}$, where c stands for the "continuum". He knew that $\mathfrak{c} = 2^{\aleph_0}$. Having proved that $\mathfrak{c} > \aleph_0$ he wondered whether $\mathfrak{c}$ is the next infinity after $\aleph_0$, that is, whether $\mathfrak{c} = \aleph_1$. As he was not able to construct a subset of reals that was neither countable nor equinumerous to $\mathfrak{c}$, he conjectured that $\mathfrak{c}$ is, indeed, the next cardinal number. This became known as the famous *Continuum Hypothesis*. Hilbert included this conjecture as the first problem, "*Cantor's Problem on the cardinality of Continuum*", in his list of open problems presented at the International Congress of Mathematicians in 1900 [125]. Set theory was not accepted very eagerly by most mathematicians at that time, so it is rather surprising to find a problem on set theory on the list of the most important problems of 1900. It was not only because logic and foundations were areas in which Hilbert was interested, but also because the question is really fundamental. Stated not quite precisely, but understandable for everyone, the question is: *How many real numbers are there?*

Using notation for infinite arithmetic the hypothesis can be stated simply as $2^{\aleph_0} = \aleph_1$. Thus it can be regarded as a hypothesis about arithmetic of infinities. Soon it turned out that there were many similar statements that nobody was able to prove or disprove. In particular for every $\aleph_\alpha$, the only available information on $2^{\aleph_\alpha}$ was Cantor's theorem: $\aleph_\alpha < 2^{\aleph_\alpha}$.

Again it was Gödel who made the first big step towards solving the problem. In 1938 he proved that the Continuum Hypothesis is consistent with the Zermelo-Fraenkel axioms (with the Axiom of Choice) [97]. In fact he proved it for the Generalized Continuum Hypothesis which states $2^{\aleph_\alpha} = \aleph_{\alpha+1}$ for every α. This meant that either the Continuum Hypothesis was provable, or independent. One more generation of logicians had to grow up before the problem was solved. In 1963 the American mathematician Paul J. Cohen (1934–2007) showed that the Continuum Hypothesis is independent from the Zermelo-Fraenkel axioms, provided that they are consistent [46]. His work has had a tremendous influence on set theory because he not only solved a deep problem, but he solved it by introducing a method which afterwards was used to prove independence results in set theory for many other statements. (I will give a brief exposition of his proof in Chap. 4.)

Is this a happy ending of a famous problem? For some people, it is, for others, it is not. Gödel, a prominent figure in this story, was definitely not satisfied because he wanted to know, whether or not the Continuum Hypothesis is *true*. He considered Zermelo-Fraenkel Set Theory as a very imperfect means to learn something about the true world of sets. Since we can never obtain a complete axiom system, we have to gradually augment Zermelo-Fraenkel Set Theory in order to decide the truth of statements such as the Continuum Hypothesis. But how can we tell which axiom is right and which is not? Gödel's theorem implies that there is no rule or algorithm which would enable us to gradually extend the theory into a complete one. Thus we have to rely on 'intuition', a 'naturalness' of axioms, etc. (whatever these words mean).

It may seem very easy to reject such a point of view in set theory, as based on very vague concepts, some experience does support such a 'realism'. It happens sometimes that we cannot solve a problem using available theories, but when we sort of step back and look at it from a higher perspective, we are able to solve it. What realists assume is that this is not a rare event, but a general phenomenon. Namely, there should always be a more general and stronger principle we can use to enrich our system.

On the other hand, there are mathematicians and philosophers that do not share this view. They believe that there is no true world of sets—there are only imaginary worlds given by the axioms we choose, but none of them is more real than others. For them, the independence of the Continuum Hypothesis on the axioms of Zermelo-Fraenkel Set Theory and our inability to decide whether or not it is true by stronger axioms is evidence that supports their belief.

Ordinal Numbers

Ordinal numbers are one of the most important concepts in the foundations of mathematics. In set theory they are used to number cardinal numbers, construct models of set theory, in proof theory constructive ordinals are used to measure the assumptions that we use in mathematics—the strength of theories, and there are many more uses of them. The least nonconstructive ordinal (denoted by ω_1^{CK}) is a rather mysterious object—it is "the first one that we are not able to describe".

The distinction between cardinal and ordinal numbers is important, since it is present in natural language. We have *one, two, three, etc.* and we also have *first, second, third, etc.* We have noticed that the two structures are isomorphic in the finite case. This includes the basic arithmetical operations. For cardinal numbers, the meaning of $+$ is the cardinality of the disjoint union. For ordinal numbers, the meaning of $+$ is a little different. Suppose you have two disjoint ordered sets. Then you can subordinate all members of the first set to the members of the second one. What you get is an order on the union of the two sets. For example, suppose your team is fifth in the second league and the last team in the first league is twentieth. Then your team is 25th in the absolute ranking. Though defined differently the operation gives the same numbers.

For infinite ordinals the correspondence between ordinals and cardinals breaks down. We have already considered the ordinal ω, the first infinite ordinal. While adding one to the first infinite cardinal does not change it, $\omega + 1$ is a new ordinal. Let us look at the process by which we have obtained it. There are two basic steps. First we have obtained ω. This was produced by a *limit* process; it was a result of starting with zero and adding one infinitely many times. We imagined that we could view the whole history of the process as far as infinity and then we applied again the operation of adding one.

The two operations can be used to define general ordinals. Let us state it explicitly.

1. There is the smallest ordinal 0.
2. For every ordinal, there is a successor of this ordinal.
3. For any set of ordinals there is its *supremum*, which means the least ordinal larger than all ordinals in the set.

In the third condition we speak about a set of ordinals and a supremum, but it is more natural to think of it as follows. For every process of producing larger and larger ordinals, one can find an ordinal that is beyond this process. I will demonstrate it in some examples.

Let us suppose that we already have ω. Now we can again apply the process of adding one, according to condition 2. Thus we get ordinals $\omega,\ \omega+1,\ \omega+2,\ \omega+3,\dots$. According to condition 3, there is a limit of this process. We will denote it by $\omega+\omega$ or $\omega\cdot 2$. Take $\omega\cdot 2$ and apply the process of adding one. We get $\omega\cdot 2,\ \omega\cdot 2+1,\ \omega\cdot 2+2,\ \omega\cdot 2+3,\dots$. The limit is $\omega\cdot 2+\omega=\omega\cdot 3$, etc.

We see that the limit of the process of adding one results in adding ω. Let us take this as a single step and let us apply it to 0. Thus we get $0,\ \omega,\ \omega\cdot 2,\ \omega\cdot 3,\ \omega\cdot 4,\dots$. It is natural to denote the limit of this process by $\omega\cdot\omega$ or by ω^2. Let us apply the operation of adding ω to ω^2. We get

$$\omega^2,\ \omega^2+\omega,\ \omega^2+\omega\cdot 2,\ \omega^2+\omega\cdot 3,\ \omega^2+\omega\cdot 4,\dots.$$

The limit of this process is $\omega^2+\omega^2$ which we can also denote by $\omega^2\cdot 2$. Doing the same once again yields $\omega^2\cdot 3$. Thus we have described a process of adding ω^2. We can step up to the next level again and take the limit of this process. This will give us ω^3.

Now I could just wave my hands and say that in the same way we could get $\omega^4,\ \omega^5,\dots$ and eventually ω^ω. It seems obvious that it should be possible, but doing it properly would require some work. Recall that even getting ω^3 was not quite easy. But we have barely got off the ground by producing ω^ω. We would need to define exponentiation of ordinals in general in order to get to interesting ordinals.

Countable ordinals can be represented as subsets of the real axis. I have drawn some in Fig. 3.3.

The picture, presented in Fig. 3.4, is different; it shows schematically ordinals less than ω^ω. First we have finite ordinals, these are in the first row. Then we have

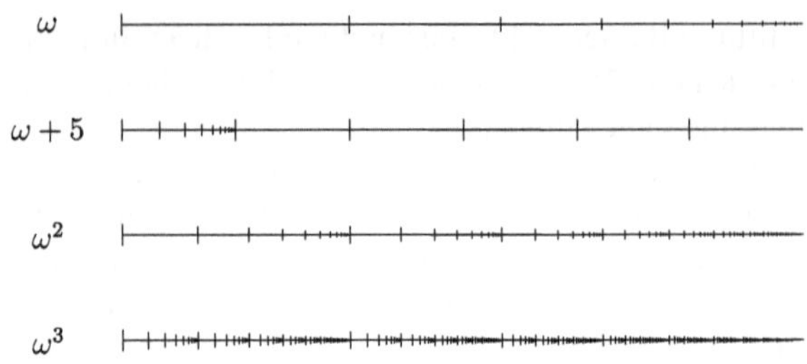

Fig. 3.3 Ordinals ω, $\omega+5$, ω^2 and ω^3

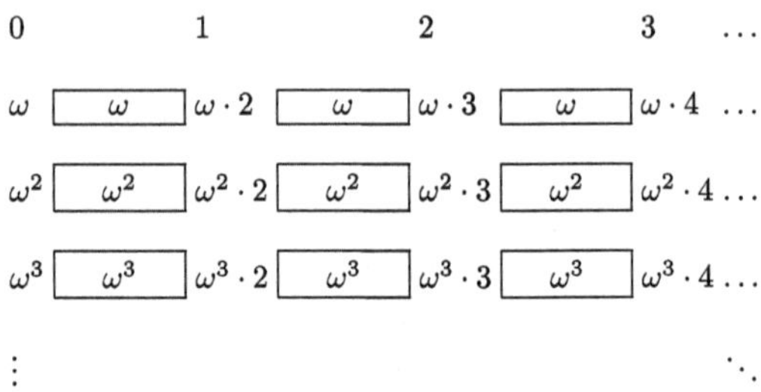

Fig. 3.4 Ordinals less than ω^ω

finite multiples of ω. Between each of them we have ordinals ordered as ω. For example, between ω and $\omega \cdot 2$ we have $\omega+1, \omega+2, \omega+3, \ldots$. This is shown schematically by the box $\boxed{\ \omega\ }$. In third row we have finite multiples of ω^2. Between each of them we have ordinals ordered as ω^2. Between multiples of ω^3 we have ordinals ordered as ω^3, etc. Though the boxes are of the same size in the picture we should think of the rows as getting denser and denser as we go down, since in every row all the previous structure is repeated between the boxes.

It was necessary to go through a rather technical construction of these small ordinals in order to observe an important phenomenon. It is the possibility of changing the point of view and looking at things from a higher perspective. We start with a simple process (adding one). Then we step up and view the whole infinite process as a single step and thus we can define a new process whose steps are the infinite runs of the simple one. We can repeat it and get higher and higher processes. Then we can look at the hierarchy of these processes and think of going from a lower order to the next highest order process as a single step of a new process. It may seem that one can go on in this way forever, but it is only an illusion. Defining a higher order process is simple only on low levels. When we go far enough, it requires coming up with a new principle.

The ordinals that we have so far considered are *constructive*. It means, roughly speaking, that they can be explicitly described. The theory of constructive ordinals is very interesting. Although it may seem that constructive ordinals are just some linear orders and thus have a rather simple structure, the opposite is true. I will explain their use in Chaps. 6 and 7.

Constructive ordinals form only a small part of all countable ordinals. A fundamental theorem, proved by Zermelo, says that there are ordinals of all possible cardinalities. Consequently, the ordinals play an important role in the universe of sets. The standard approach is to represent a cardinal κ by the least ordinal of cardinality κ. In this representation cardinal numbers are some special ordinal numbers. Consequently, cardinals are well-ordered. Furthermore the cardinal numbers can be indexed by ordinal numbers and for every ordinal, there is a cardinal with that index. Using the aleph notation we can state it as follows:

Proposition 4 *For every infinite cardinal number κ, there exists an ordinal number α such that $\kappa = \aleph_\alpha$ and, vice versa, for every ordinal α, there is an infinite cardinal number $\aleph_\alpha$.*

In set theory an ordinal number can be used to encode the process of constructing sets. We can think of ordinals as forming the spine of the set universe with other sets as limbs and flesh around it.

Interlude—Archimedes and Ordinals

Mathematics is not only what is written. It is not possible to express all mathematical thinking with words and formulas. The same idea may be realized by several definitions of different concrete concepts, but it is hard to define one common generalization. In the same way proofs and algorithms are based on tricks and strategies which one can learn only by reading or talking a lot about mathematics. One of the main goals of mathematics, and the same is true for all theoretical sciences, is to develop concepts and theories so that more and more originally different facts are covered by general results.

Here I want to show that an abstract idea that we use to construct some infinite ordinals was in some sense used already by Archimedes, the greatest ancient mathematician (287–212 BCE). In his letter addressed to the King of Syracuse (the letter known as *The Sand Reckoner*) he showed that contrary to the prevailing belief, the number of grains of sand is finite. Ancient Greeks were very good in mathematics, so surely many of them accepted the finiteness of the number of grains as a true fact, but there was another misconception, a more subtle one, that though being finite the number of grains of sand is so big that such a number cannot be named. To disprove it was the main goal of Archimedes. In order to refute it once and for all, Archimedes decided to name a number bigger than the number of grains of sand that would fill the whole universe. He, of course, assumed that the universe was finite.

In more detail, Archimedes accepted the heliocentric system of Aristarchus of Samos (approximately 310–230 BCE) and assumed that the universe was a sphere around the Sun. To estimate the radius of this sphere, he assumed that the ratio of the distance of the Earth from the Sun to the radius of the Earth is equal to the ratio of the distance of the Earth from the Sun to the radius of the universe. For us, this is a ridiculously small estimate, but for the argument it makes little difference, all the more so that he developed names for much larger numbers than he needed. Most likely, his numbers suffice even for the size of the visible universe.

As a matter of fact, the ancient Greeks did not have a system of numerals sufficient for this purpose and Archimedes had to develop a new system in order to be able to express such a number. This is in parallel with infinite ordinals: to define a large ordinal α in a constructive way, one has to develop a system of notation for all ordinals less than α. The largest number for which the Ancient Greeks had a name was a *myriad* which is 10,000. Archimedes started by observing that using the current system one could name numbers up to a myriad myriads, which is $100{,}000{,}000 = 10^8$ ('myriad myriads' is a legal language construction). The actual number that he started with is not important, but having the initial number large certainly helps us get bigger numbers. He called the numbers less than a myriad myriads *first numbers*. He called *units of the second kind* the numbers that are myriad myriads of first numbers. Then *second numbers* were composed (as sums) of units of the second kind and first numbers. A myriad myriads of second numbers were *units of the third kind* and *third numbers* were composed from units of the third kind and second numbers. And so on.

To see how one can use Archimedes' system to name numbers, let us compare it with the system used nowadays. We use a *thousand* (10^3) as the initial number instead of a myriad myriads (10^8). So our first numbers are numbers less than a thousand. Our units of the second kind are numbers

$$1{,}000,\ 2{,}000,\ 3{,}000,\ \ldots,\ 999{,}000.$$

Our second numbers are numbers from 1,000 to 999,999. Our units of the third kind are

$$1{,}000{,}000,\ 2{,}000{,}000,\ 3{,}000{,}000,\ \ldots,\ 999{,}000{,}000.$$

Take as an example

$$22,\ 354,\ 781.$$

It consists of 22 units of the third kind, 354 units of the second kind and a first number 781. Thus the name of this number would be something like:

twenty two third units, three hundred fifty four second units and seven hundred eighty one

Instead we call it:

twenty two millions, three hundred fifty four thousand and seven hundred eighty one

Compared to Archimedes', our system looks rather stupid. The drawback is not the fact that we use a smaller initial number, it is the fact that we have only a limited number of special names for the kinds of numbers. We use *thousand, million, milliard, billion, billiard*, etc. in the European system, and *thousand, million, billion, trillion*, etc. in the American system. Theoretically we could have a kind of number for every number less than one thousand, and maybe more, but with increasing numbers it becomes a linguistic problem. Thus only a few of these names are used. In Archimedes' system the kinds of numbers are directly numbered by numbers, so there is no problem to number the kinds of numbers up to a myriad myriads. In this way, Archimedes was able to name numbers up to $(10^8)^{10^8} = 100000000^{100000000} = 10^{8\cdot 10^8}$.[9]

Archimedes did not stop at this point, although these numbers sufficed for his purpose. He suggested how one could go on using a new name '*period*'. He called the numbers up to $10^{8\cdot 10^8}$ the *first period.* Then he repeated the process described above with the first numbers replaced by the first period in order to obtain the *second period.* These were the numbers up to $10^{8\cdot 10^8}.10^{8\cdot 10^8} = (10^{8\cdot 10^8})^2$. Similarly, the *third period* were the numbers up to $10^{8\cdot 10^8}.10^{8\cdot 10^8}.10^{8\cdot 10^8} = (10^{8\cdot 10^8})^3$, and so on. The number of periods introduced in this way was myriad myriads. In this way, he introduced names for all the numbers up to

$$\left(10^{8\cdot 10^8}\right)^{10^8} = \left(10^8\right)^{(10^8)^2},$$

which is $10^{8\cdot 10^{16}}$.

The reason for such a lengthy description was not to calculate the largest number that Archimedes considered. This is not important, as he was surely able to construct larger numbers if he had any use for them. What is important is the structure of the construction. Let us see how it corresponds to infinite ordinals. The point is to replace the initial (finite) number a myriad myriads by "infinity", namely by the first infinite ordinal ω. With ω as the initial number, first numbers are finite ordinals all natural numbers $0, 1, 2, \ldots$. Units of the second kind are $\omega, \omega\cdot 2, \omega\cdot 3, \ldots$. Second numbers are ordinals from ω up to (but not including) $\omega\cdot\omega$. Third numbers are ordinals from ω^2 up to ω^3, etc. Thus we obtain ordinals up to ω^ω.

The next step was to define periods. Continuing the analogy, the first period are ordinals up to ω^ω. The second period are ordinals up to $\omega^\omega\cdot\omega^\omega = (\omega^\omega)^2$, the third period are ordinals up to $\omega^\omega\cdot\omega^\omega\cdot\omega^\omega = (\omega^\omega)^3$, and so on repeating this process ω-times. In this way we obtain the ordinal

$$\omega^{\omega^2}.$$

We did formally everything the same with the ordinal ω what Archimedes did with 10^8. It is also possible to do the opposite. Given names for numbers in an initial segment of natural numbers and a construction of an ordinal one can develop a system of names for large numbers.

Another possible interpretation of Archimedes's construction is in terms of computation. We can think of what he did as computation of n^{n^2} with $n = 10^8$. There are several ways of computing this number; the one that corresponds to Archimedes's construction is shown in Fig. 3.5. The outermost loop corresponds to *periods*; y is the number of *first numbers* in the zth *period*; the next loop corresponds to *kinds*, etc.

What may seem a little puzzling is that the program uses only the successor functions, whereas I used other operations to define ω^{ω^2}. The reason is that I did not bother to define arithmetical operations on ordinals using the successor function

[9]I am using the standard convention that x^{y^z} means x raised to the power y^z; it should not be confused with $(x^y)^z$ which is equal to x^{yz}.

```
n:=100000000
y:=n
for a=1 to 2 do                         % repeating exponentiation
          z:=1
          for b=1 to n do                   % raising y to the power n
                    u:=0
                    for c:=1 to y do          % multiplying z by y
                              v:=u
                              for d:=1 to z do    % adding z to u
                                        v:=v+1
                              u:=v
                    z:=u
          y:=z
print(y)
```

Fig. 3.5 A program for computing $(10^8)^{(10^8)^2}$

and fixed points. If this is done, one gets a complete correspondence between the program and the ordinal.

Thus you see that there is also a link from ordinals to computations, more precisely, from notations for ordinals to programs.

Notes

1. *The definition of finite sets.* Instead of defining infinite sets we can define finite sets because having such a definition we will define infinite sets simply as those that are not finite. I will use the following two postulates:

   a. The empty set is finite.
   b. If x is finite, then by adding one element to x we obtain again a finite set (formally, x finite implies $x \cup \{y\}$ is finite for every y).

   It is clear that every reasonable definition of finiteness must imply these two postulates. It is also obvious that if we have two definitions such that there are more sets that satisfy the first one than those that satisfy the second one, then the second definition is better. Hence, if we prove that there is a definition which postulates the smallest class of finite sets and still satisfies a. and b., then we are done: this will be the canonical definition of finiteness.

   As it is more convenient to talk about classes instead of formulas, I will use Gödel-Bernays Set Theory. However, essentially the same result can be proved also for Zermelo-Fraenkel Set Theory. Our goal is to prove that there exists the smallest class X with respect to the ordering by inclusion whose elements satisfy a. and b. Taking the natural definition of X, which is the intersection of all classes that satisfy a. and b. does not work. Such a definition uses quantification of classes, which is not permissible in Gödel-Bernays set theory, so we have to define X indirectly. The indirect definition is very simple: X is the class of all sets y such that, for some $n \in \mathbb{N}$, y has n elements. The formal definition of 'having n elements' is as usual: there exists a one-to-one mapping from y onto n (which is $\{0, 1, \ldots, n-1\}$). It is a trivial fact that such an X satisfies a. and b.; the minimality of X is proved by induction. Thus the crucial property that one needs is that induction holds for $\mathbb{N}$.

Now it is not difficult to prove that Dedekind infinite sets are exactly the sets that are not in the X as defined above. However, the proof requires the Axiom of Choice. Without this axiom, the definition of finiteness is a more complicated matter.

2. *What is "the true" cardinality of the continuum.* There is no test that can confirm or refute a conjecture about infinite sets. The only facts on which we can rely are proofs from a set of axioms, and proofs of independence from a set of axioms. Since the Continuum Hypothesis is independent from the axioms of Zermelo-Fraenkel Set Theory, we can only use either intuitive arguments in favor of a particular solution of this problem, or add new axioms that imply a solution and argue that these axioms are "natural". Several proposals of what the cardinality of $2^{\aleph_0}$ should be have been made. Surprisingly there are few advocates of the validity of the Continuum Hypothesis, whereas there are several arguments supporting the conjecture that it is false. One of these arguments is due to Gödel [98]; another one is based on the axiom *Martin's Maximum*, a strengthening of Martin's Axiom (see page 363). Cohen's point of view was more radical. He argued that the Power Set Axiom, which is needed to obtain the continuum, is a very powerful axiom and thus it is unreasonable to expect that the continuum can be easily reached from below. Thus $2^{\aleph_0}$ should be very large (namely *weakly inaccessible*, which means that it is $\aleph_\alpha$ with α a limit ordinal and it is not the limit of any set X of smaller cardinals, $|X| < 2^{\aleph_0}$). I will mention some arguments in Chap. 7.
3. *Infinite ordinal and cardinal numbers.* An ordered set is said to be *well-ordered* if every nonempty subset has the least element. In this definition we use the most important property of the ordering of natural numbers. An ordinal number is classically defined as an isomorphism type of a well-ordering. As is the case with cardinal numbers, it is more convenient to take a representative for each type of well-ordering and define it to be an ordinal number. The most natural way of choosing such representatives is to extend the definition of natural numbers from page 31 to infinite ordinals. Thus an ordinal α is the set of all ordinals less than α. In particular

$$\begin{aligned}
&\omega = \{0, 1, 2, \ldots\},\\
&\omega + 1 = \{0, 1, 2, \ldots, \omega\},\\
&\omega + 2 = \{0, 1, 2, \ldots, \omega, \omega + 1\},\\
&\ldots
\end{aligned}$$

Formally, von Neumann's ordinals are defined as follows. Say that a set x is *transitive*, if for all y and z, $z \in y \in x$ implies $z \in x$.

Definition 1 An ordinal is a transitive set α such that every element of α is also transitive.

Note that this implies that elements of elements of α are transitive and so on. The apparently mysterious fact that such a simple definition implies that every ordinal is a well-ordered set (where $\in$ is the 'less than' relation), has a simple explanation: it is a direct consequence of the Axiom of Foundation. The latter axiom says that the relation $\in$ is *well-founded*, which is a natural generalization of 'well-ordered' to relations that are not necessarily orderings. In particular, if x is a set on which $\in$ defines a linear ordering (and this is the case of ordinals), then this ordering is well-ordered.[10]

Since every ordinal α is the set of all ordinal less than α and α is well-ordered, the class of all ordinals is also well-ordered. Let us apply this property to the set of ordinals of a given cardinality. Thus we get that there exists the smallest ordinal of this cardinality. Then it is quite natural to define that an ordinal number α is a cardinal number, if no smaller ordinal has the same cardinality as α. For example, $\aleph_0$ is ω and $\aleph_1$ is the first uncountable ordinal. We have to prove that every set X can be "measured by such cardinals", which means that it is equinumerous to such a cardinal. The idea of the proof is as follows. Pick successively distinct elements $x_0, x_1, \dots$ from X. But instead of stopping after defining the sequence for all natural numbers, if there are still elements left, we continue and construct a *transfinite sequence*, which is a sequence indexed by ordinal numbers. In this way we construct a bijection between the elements of X and an initial segment of ordinal numbers. As every initial segment of ordinal numbers is an ordinal, we thus obtain an ordinal α. Then we take the least ordinal that has the same cardinality as α, which is the cardinal that is equinumerous to X. To do this argument formally is not quite easy and it is necessary to use the Axiom of Choice. The reason for using the Axiom of Choice is that there is no way to determine uniquely the next element when defining the sequence; we have to *choose* an arbitrary one.

It is important to realize that cardinal numbers, as a subclass of ordinal numbers, are also well-ordered. Thus, for instance, there is the smallest cardinal number larger than $\aleph_0, \aleph_1, \aleph_2, \dots$, which we denote by $\aleph_\omega$. In particular, every two cardinal numbers are comparable and for every ordinal α, there exists a cardinal $\aleph_\alpha$ (which is typically a much bigger ordinal).

We have been assuming the Axiom of Choice in this note; without it the structure of cardinalities is much more complicated.

4. *Transfinite induction* is, essentially, a generalization of mathematical induction. The latter can be viewed as the special case of transfinite induction for ω. It can be stated as the following theorem.

Theorem 6 *Suppose that for a property of ordinals Φ the following is true. For every ordinal α, whenever Φ holds for all ordinals less than α, then it holds for α. Then the property Φ holds for all ordinals.*

Property Φ holds for 0 because there are no ordinals less than 0. We cannot only use an assumption that $\Phi(0)$ and $\Phi(\alpha) \to \Phi(\alpha + 1)$, as this does not work

---

[10]In theories without the Axiom of Foundation we have to add explicitly that the elements of α are well-ordered by the membership relation $\in$.

for limit ordinals. Transfinite induction is just the disguised least number principle, so the theorem is an immediate consequence of the definition of ordinals.

5. *Constructive ordinals and ordinal notations.* In order to define *concrete* ordinals, we need to introduce some structure on the set of all countable ordinals. This is done mainly using continuous functions and operations. The topology on ordinals is determined in a natural way by the ordering. It simply means that limit ordinals, those that are not successors, are the limits of the set of ordinals below. Thus when defining continuous operations we only need to define them on non-limit ordinals. The operations $+, \cdot$ and exponentiation are defined by the same clauses as for natural numbers. For example, addition is defined by the equations $\alpha + 0 = \alpha$ and $\alpha + (\beta + 1) = (\alpha + \beta) + 1$, where $+1$ is the successor function, and by the requirement that it is a continuous operation. Note that addition and multiplication are not commutative, in particular we have *absorption* $1 + \alpha = \alpha$ for every infinite α.

An important set of ordinals is the set of ordinals that can be generated by the three basic operations from 0 and ω. Let us study this set of ordinals. Observing that $\omega^{\alpha} + \omega^{\beta} = \omega^{\beta}$, if $\alpha < \beta$, we get a unique representation in the form

$$\omega^{\alpha_1} + \cdots + \omega^{\alpha_n}, \tag{3.1}$$

where $\alpha_1 \geq \cdots \geq \alpha_n$ for every nonzero ordinal, called the *Cantor normal form.* For ordinals that are generated by the basic arithmetical operations, we can recursively represent the exponents in Cantor normal form and eventually obtain an expression built only using symbols for 0, ω, $+$ and exponentiation. Furthermore, we only need exponentiation with the base ω. To simplify notation, we write 1 instead of ω^0, ω instead of ω^1, $n \cdot \alpha$ instead of α added n-times and n instead of $n \cdot 1$, for n a positive integer. For example,

$$\omega^{\omega^2+1} + \omega^3 + \omega \cdot 2 + 5.$$

There is a canonical sequence of such ordinals which eventually gets larger than any element of the set, which is the sequence $\omega, \omega^{\omega}, \omega^{\omega^{\omega}}, \omega^{\omega^{\omega^{\omega}}}, \ldots$. The limit of this sequence, which is the limit of the set of all ordinals constructed in this way, is denoted by ε_0. The ordinal ε_0 is considered to be a constructive ordinal because we can explicitly describe all ordinals below. We have a unique representation for every ordinal below ε_0 and moreover we have an algorithm that for every pair of distinct terms, determines which of the terms denotes the larger ordinal. These two properties define constructive ordinals.

Though, as an ordinal number, ε_0 is well beyond the set of natural numbers ω, it is by no means less explicit than ω. In particular, if we, for some reason, only trust in natural numbers, we can represent the ordinals below ε_0 very easily by natural numbers. The following is an elegant way of coding ordinals below ε_0 by natural numbers, which not only assigns a unique natural number to every ordinal below ε_0, but also every natural number is a code of such an ordinal. Let $p_0, p_1, p_2, \ldots$ denote the enumeration of primes in the increasing order (thus $p_0 = 2$, $p_1 = 3$, $p_2 = 5$, etc). For ordinals $\alpha < \varepsilon_0$, we define their number codes $N(\alpha)$ by recursion. We put $N(0) = 0$ and for an ordinal α in the Cantor normal form (3.1) we put

$$N(\alpha) = p_{N(\alpha_1)} \cdot p_{N(\alpha_2)} \cdot \dots \cdot p_{N(\alpha_n)}.$$

Let us compute the code of $\omega^\omega + 2 \cdot \omega$. First we have to rewrite it using 0, ω and $+$ as follows:

$$\omega^{\omega^{\omega^0}} + \omega^{\omega^0} + \omega^{\omega^0}.$$

Then we can compute

$$N(\omega^\omega + 2 \cdot \omega) = p_{p_{p_0}} \cdot p_{p_0} \cdot p_{p_0} = p_{p_2} \cdot p_2 \cdot p_2 = p_5 \cdot 5 \cdot 5 = 13 \cdot 5 \cdot 5 = 325.$$

This is, of course, only a technical means given by the need to represent ordinals as numbers. Once we have represented ordinals below ε_0 by terms, it was clear that we could do it. Representing ordinals by terms is the standard way of proving that an ordinal is constructive. The system of such terms is called an *ordinal notation*. Thus ε_0 (as the set of ordinals less than ε_0) has an ordinal notation based on constant symbols 0 and ω and the operations of addition and exponentiation. As ε_0 cannot be represented by such a term, we have to use a new name when we want to talk about it.

Going back to our interpretation of Archimedes's number system in terms of infinite ordinals, we can think of it as a system based on $0, 1, \omega$, addition, multiplication and ω new names. If he had not used the new names, he would have only got ordinals below ω^ω. Thus, in order to continue, he needed a name for ω^ω. Instead of giving a name to this ordinal, Archimedes introduced a new name '*numbers of the second period*' for the whole segment of ordinals from ω^ω to $\omega^{\omega \cdot 2}$. Using a name for the operation $\alpha \mapsto \omega^\alpha$ instead of his ω names, he would get all ordinals below ε_0.

Next we define the *Feferman-Schütte ordinal* Γ_0, another of the milestones in constructive ordinals. Suppose we have reached ε_0 and we want to continue. First we do the same as we did with ω: we start with ε_0 and close it off under the arithmetical operations. Then we repeat this construction again and again. Think of closing off under the three arithmetical operation as a single step in a new process. Let us denote ε_1 the first that we get after ε_0, let ε_2 be the next and so on. Precisely it means that we define ε_α to be the αth number ξ such that $\omega^\xi = \xi$. These numbers are called ε-numbers. Thus we have defined a continuous function $\xi \mapsto \varepsilon_\xi$. We can find names for a lot of new ordinals using this function, but since every continuous function has a fixed point, there is a number α such that $\varepsilon_\alpha = \alpha$ (namely, it is the limit of $\varepsilon_0, \varepsilon_{\varepsilon_0}, \varepsilon_{\varepsilon_{\varepsilon_0}}, \dots$). We do not have a name for this epsilon number, as its index is the same as the number. We have to introduce a new name (which is not Γ_0, be patient). Clearly we can do the same as we did with exponentiation—we can enumerate the fixed points of the ε-function. Then we want to do the same again and again. Clearly, we have to introduce notation to keep this process under control.

Let us call ε-numbers ϕ_1-numbers, and denote by $\phi_1(\alpha) = \varepsilon_\alpha$. Let us call ϕ_2-numbers the fixed points of the ϕ_1 function (which is the ε-function) and denote by $\phi_2(\alpha)$ the αth ϕ_2-number. In general, $\phi_\alpha(\xi)$ is defined to be the function which enumerates the ordinals that are fixed points of all functions $\phi_\beta(\xi)$ for $\beta < \alpha$. This sequence of functions, presented as a function of two variables, $\phi(\alpha, \beta) = \phi_\alpha(\beta)$, is called the *Veblen function.*

The Veblen function is continuous, hence there is a fixed point also of this process of generating new ordinals. This is the Γ_0. One can show that it is the first ordinal γ such that $\phi(\gamma, 0) = \gamma$, or the first ordinal that cannot be expressed in terms of 0, ω, the basic arithmetic functions and the Veblen function. To show that Γ_0 is a constructive ordinal one defines a normal form for the ordinals below Γ_0 and from that it is not difficult to define an algorithm for comparing two ordinals represented in the normal form.

A good exercise is to try to define an ordinal essentially larger than Γ_0. It is not easy, it requires a new idea!

The first nonconstructive ordinal is denoted ω_1^{CK}. The meaning of this notation is that it is like ω_1 from the point of view of constructive processes: though countable, it cannot be enumerated by a computable function. The *CK* stands for Church and Kleene. Saying that it is the first ordinal that cannot be described is not precise and sounds paradoxical. It is the first ordinal that cannot be presented as a recursive ordering of the natural numbers. It is determined by a short definition, which, however, has a highly nonconstructive nature—it mentions all recursive orderings.

6. *Kripke-Platek Set Theory.* This theory is not used as the foundations of mathematics, but it is interesting for other reasons. Kripke-Platek Set Theory is a subsystem of Zermelo-Fraenkel Set Theory obtained by omitting the Axioms of Power-Set and Choice, and restricting the Axiom Schemas of Comprehension and Replacement to *bounded set formulas*. In a bounded set formula the range of each quantifier is restricted to a set. Thus every quantification has the form

for every x such that $x \in y \dots$

or

there exists x such that $x \in y \dots$

Formally, this is written as $\forall x(x \in y \rightarrow \cdots)$, respectively $\exists x(x \in y \wedge \cdots)$ (x and y are distinct variables and dots represent the subformula that is the scope of the quantification,) which is usually abbreviated to $\forall x \in y(\cdots)$, respectively $\exists x \in y(\cdots)$.

So the axioms of Kripke-Platek Set Theory are:

The Axiom of Extensionality

The Axiom Schema of Comprehension for bounded set formulas

The Axiom of Pairing

The Axiom of Union

The Axiom of Infinity (sometimes this is not included)

The Axiom Schema of Replacement for Bounded Set Formulas

The Axiom Schema of Foundation *Because the Axiom Schema of Comprehension is limited, we have to state this principle as a schema for all formulas* $\phi(z)$:

$$\exists z\, \phi(z) \to \exists y\big(\phi(y) \wedge \forall x \in y(\neg\phi(x))\big).$$

An important class of models of *ZFC* and its subtheories is the class of *transitive models*. A transitive model is a model in which the membership relation is the actual $\in$ relation and such that the universe of the model is a transitive set. The existence of transitive models of *ZFC* can be shown in a mild extension of *ZFC* (*ZFC* with the axiom postulating the existence of an inaccessible cardinal; see the next section, page 198, for the definition of this cardinal). In such models a number of important concepts are *absolute*, which means that a set x has a given property in the model if and only if it has it actually. In particular, ordinal numbers of a transitive model are the usual ordinal numbers.

Transitive models of Kripke-Platek Set Theory play an especially important role. I will mention only one remarkable fact about them. There exists the least transitive model of Kripke-Platek Set Theory and the ordinals of this model are exactly all constructive ordinals. In other words, an ordinal is constructive if and only if it is in the minimal model of Kripke-Platek Set Theory.

This is a striking result because the axioms do not mention any computations, yet the theory characterizes constructive ordinals, a concept inherently tied in with computations. But looking more closely at the theory one can see that it has a very constructive nature, which explains at least why all ordinals in the minimal model are constructive. Let us see why the particular restriction of the axiom system used in Kripke-Platek Set Theory makes the theory look distinctly constructive. In the full *ZFC* we also construct sets starting from small ones and gradually producing larger ones. When applying Comprehension or Replacement it seems that we determine a new set uniquely by a formula $\phi(x)$, but there is a catch. If ϕ contains unbounded quantifiers, we do not know if $\phi(x)$ holds for a particular set x because ϕ mentions *all* sets, which means it mentions also those sets that are yet to be constructed. This problem is avoided by taking only bounded set formulas in the schemas. The second difference between *ZFC* and Kripke-Platek Set Theory is in omitting the Axiom of Power-Set. Since the Continuum Hypothesis is an independent sentence of *ZFC*, we know that we cannot exhaust all subsets of $\mathbb{N}$ by a reasonably constructive process. Thus the power-set operation has a very non-constructive nature because it posits the existence of a set whose elements are not constructed yet. The last difference, the omission of the Axiom of Choice, is not very important because for an explicitly defined set we can also explicitly define a choice function.

For more detail, see Barwise [14].

3.3 What Is the Largest Number?

We have arrived at one of the most interesting problems in the foundations of mathematics. At first it may look like a puzzle from recreational mathematics, but in fact it has bearings on fundamental philosophical questions about mathematics.

Already in ancient times people wondered how large numbers could be. Mathematicians in antiquity had essentially the same opinion concerning this question: there is no largest natural number because operations, such as the simplest operation of adding one, produce from any number a larger one. So the question is not interesting if we only talk about finite numbers. But we can also ask: How large can *infinite cardinal numbers* be? Still it is not clear that we are asking a meaningful question. Cantor's theorem says that for every cardinal number, there is a larger one, hence also there is no largest infinite cardinal number. However, the structure of infinite cardinal numbers is different from the structure of the natural numbers. Since there are infinite cardinal numbers of various kinds, it makes sense to ask which kinds of infinity are possible. This is the question that I am going to discuss now.

Let us explain it in a concrete example. Using the Axiom of Infinity we get that $\aleph_0$, the smallest infinite cardinal, exists. Using Cantor's theorem repeatedly we get that also $\aleph_1, \aleph_2, \aleph_3, \ldots$ exist. I have also mentioned the limit of this set which is the cardinal denoted by $\aleph_\omega$, but does such a cardinal really exist? It is possible to show that the existence of such a cardinal cannot be proved using only Zermelo's axioms. It can, however, be proved using the Replacement Axiom Schema. Thus, while unprovable in the weaker Zermelo Set Theory, it is a theorem of Zermelo-Fraenkel Set Theory that such a cardinal number exists.

In general the pattern is the same. We have some description of a property of a cardinal number and we ask whether we can prove or disprove that such a cardinal exists. The most interesting case is when we can neither prove it, nor disprove it. This means that the existence of such a cardinal number is a new principle, a new kind of infinity. In fact, this problem appears already on the lowest level, namely, when we add the Axiom of Infinity. The process of adding the Axiom of Infinity to the basic axioms that only ensure the existence of finite sets is quite analogous to adding assumptions about the existence of large cardinals. So we can view large-cardinal axioms as *higher axioms of infinity*.

A popular way of presenting the large-cardinal theory is by a game where players are trying to say the largest number. The hypothetical large-cardinal game has the following rules. Players alternate in turns, where every turn consists of a proposal of a large cardinal and, if it is not the first turn, a proof that it is larger than the last one played. Moreover, instead of playing a new cardinal, the player can prove that one of the considered cardinals is not consistent, in which case it is replaced by the previous largest one. The game goes on forever, so there is no winner; the only goal is to have the record.[11] But notice that it is worse than in sports: your record can not only be beaten, it can be completely erased because of inconsistency. Yet, it is more than just a game: we are trying to reach absolute truth, though we never know where the border is between the truth and absurdity.

Talking about truth in the context of large cardinals requires to accept the philosophical view of mathematical realism, called platonism. So let us imagine that there is a universe of sets that exists independently of our theories and the axiomatic

[11] The explanation of the game by Kenneth Kunen is a little different: "… *to try to completely demolish your ego by transcending your number via some completely new principle*", [172] page 396.

systems for set theory that we are studying describe this reality. Gödel's Incompleteness Theorem implies that axiomatization of set theory is an open-ended process; we will never be able to fully describe the universe of sets. Incompleteness may concern various properties of the universe among which the questions about its size seem to be the most important. Essentially, there are two such questions:

1. How "wide" is the universe of sets?
2. How "long" is the universe of sets?

These are certainly very vague questions, but we can make them more concrete by asking specific problems. Examples of the first kind are problems such as the Continuum Hypothesis. Roughly speaking, these are problems of how many sets of limited size there are.

The second kind of problems is of what kind of infinities there are, which is the subject of the present section. In general, we do not know why some sentences are independent, but for this kind we do have a natural explanation. The universe of sets is so long, that is, the types of infinities are so abundant, that we are not able to describe it by our finite means. We can only approximate its length from below by describing some types of infinity. Describing more types of infinity requires a lot of ingenuity because it cannot be a mechanical process.

Let us now assume a pragmatic point of view. The main reason for axiomatizing set theory was the need to avoid paradoxes. Introducing new, apparently very strong, axioms goes in the opposite direction—the theories become more likely inconsistent. Do we gain anything by taking higher risk?

Large-cardinal axioms do have important consequences. We cannot ignore large cardinals even if we reject the platonistic view of the universe of sets. We know positively that they imply true arithmetical sentences that we are not able derive without using them. This is because large-cardinal axioms imply the consistency of theories, the consistencies that we are not able to prove by other means. In principle these axioms may also be used to decide other problems of low logical complexity, but we have only a few such examples. However, already the fact that large-cardinal axioms decide consistencies makes them of prime interest for the foundations of mathematics.

In this chapter I will describe some large cardinals and give examples of their applications. Large cardinals will be discussed again in Chap. 7 in connection with the philosophy of mathematics.

The Inaccessible Cardinal

To get some feeling for large cardinals, we will first study how far we can go in Zermelo-Fraenkel Set Theory. While in Zermelo Set Theory we could only prove the existence countably many infinite cardinals $\aleph_0, \aleph_1, \aleph_2, \ldots$, in Zermelo-Fraenkel theory infinite cardinals have a very rich structure. The additional Axiom Schema of Replacement increases the power of the theory tremendously.

To start, we observe that we have alephs for every countable ordinal, for example, $\aleph_{\omega^2}$, $\aleph_{\omega^\omega}$, $\aleph_{\varepsilon_0}$ (where ε_0 is $\omega^{\omega^{\omega^{\cdot^{\cdot^{\cdot}}}}}$). This is only a slow beginning. Then we can take the first uncountable ordinal ω_1 and get $\aleph_{\omega_1}$. There are various operations which produce larger cardinals from given ones. One that gives a very large increase is the operation of using a given cardinal as the index of an aleph: for a cardinal $\aleph_\alpha$ we take the smallest ordinal of this cardinality, which is denoted by ω_α, and produce $\aleph_{\omega_\alpha}$.[12] Starting with the smallest infinite cardinal $\aleph_0$ and the corresponding smallest infinite ordinal ω we get by iterating this construction

$$\aleph_\omega, \aleph_{\omega_\omega}, \aleph_{\omega_{\omega_\omega}}, \ldots .$$

These seemingly large cardinals, hardly ever needed for concrete mathematical constructions, still do not exhaust the possibilities that we have in Zermelo-Fraenkel Set Theory. In fact, we are still rather close to the bottom. The reason is that we have iterated this construction only countably many times. The power of the replacement axiom schema lies in the possibility to iterate any construction transfinitely many times using ordinals whose existence we have already proved. Notice that this gives a strong positive feedback effect: the larger cardinals we get, the longer we can iterate the construction; thus we get even larger cardinals, so we can iterate even more and so on.

This self-enhancing principle seems so powerful that one may conjecture that it is all that we can ever do. One may even suspect that a theory which allows it cannot be consistent. So far no contradiction has been found and many believe that Zermelo-Fraenkel set theory describes the true world of sets. Assuming that it does, there is also a good reason to accept the possibility that the above principle is still not sufficient to produce all cardinals. If there are cardinals that cannot be reached in this way, they are certainly extraordinary entities and deserve a definition and a name.

The Inaccessible Cardinal *An inaccessible cardinal is an uncountable cardinal κ which cannot be reached by any limit process of length less than κ and for every smaller cardinal λ, its power 2^λ is smaller than κ.*

The condition about the limit process, stated formally, means that κ is not the supremum of any sequence of ordinals $\{\mu_\alpha\}_{\alpha<\lambda}$ indexed by ordinals less than λ, where λ is some ordinal less than κ.

By defining inaccessible cardinal we have reached our goal to determine how far we can go in Zermelo-Fraenkel Set Theory. The existence of such a cardinal is not provable in the theory. The reason for that is that in Zermelo-Fraenkel Set Theory we have two means to produce larger cardinals: taking a limit of a sequence of cardinals and taking the power of a cardinal. The first operation is enabled by the Replacement Axioms, the second by the Power-Set Axiom. We cannot obtain an inaccessible cardinal κ as a limit because the sequence itself would have to have

[12] Recall that in fact the ordinal ω_α and the cardinal $\aleph_\alpha$ are represented by the same set.

length κ, neither can we get it in the second way because that is explicitly prohibited by the definition.

A terminological explanation is needed at this point. The definition of an inaccessible cardinal uses a property which can be satisfied by many cardinals. What is important is how large *the smallest one* is. As we cannot prove the existence of any inaccessible cardinals in Zermelo-Fraenkel Set Theory, even the smallest one is beyond the reach of this axiomatic system. Other properties will be used to define larger cardinals. We will compare the concepts by comparing always the smallest one from every class.

The theory of large cardinals started as early as in 1908 when Felix Hausdorff defined inaccessible cardinals. Soon after Paul Mahlo defined a family of cardinals which are named after him. Mahlo cardinals are strictly larger than the least inaccessible cardinal. Roughly speaking a *Mahlo cardinal* is a cardinal number κ such that inaccessible cardinals are very frequent (in a precisely defined sense) below κ. This was a new idea how to get essentially larger cardinals, but in some sense it was still based on a process of going up from below. Now I turn to examples in which a cardinal is defined by a property. Such properties can be found in various fields of mathematics, the one that I am going to consider next comes from the measure theory.

The Measurable Cardinal and the Measure Problem

This is another example of a large cardinal. The name may be misleading, as one would expect that 'measurable' means 'not too large because it can be measured', which is not true. In fact, the smallest measurable cardinal is extremely large compared to the smallest inaccessible and Mahlo cardinals. Again there are a lot of Mahlo cardinals below the smallest measurable cardinal, but we have more reasons to claim that it is much larger. In particular, several types of large cardinals in between have been defined, each of them being much larger than the previous ones. The name comes from *the measure problem.* The problem is to define measure of all subsets of the real numbers. The measure of a set of reals is a nonnegative real number which expresses the size of the set; for example, the size of the interval [0, 1] should be 1 and the size of a single point (one element set) should be 0. This does not seem very interesting at first glance, as one would rather like to define the size of areas in a plane and volumes of bodies in a three-dimensional space. But, as it is the case with many problems in mathematics, the general problem can be reduced to the special one concerning subsets of the space of dimension 1.

Our practical experience tells us that it should not be a problem to *define* the measure, though it will often be difficult to *compute* the measure for particular sets. For three-dimensional bodies, we can use the famous trick discovered by Archimedes and measure the amount of water spilled over from a full vessel when an object is submerged in it. This story reminds us that the concept of the measure of a body is also a physical concept. Therefore, it seems that we only need to render it properly by a mathematical definition. This was the view of mathematicians until complicated sets appeared in their investigations, which happened roughly in the second half of the 19th century.

In the early 1900s Henry Lebesgue gave a very general definition of measure. He essentially solved the problem, as his definition is the best one in a certain well-defined sense. It works perfectly for everything that we use in analysis. Not only that, it defines the measure of every set that is explicitly defined. It has only one defect: it does not define the measure of *every* set. This was shown quite soon afterwards (in 1905) by Giuseppe Vitali. He proved that there are sets for which the Lebesgue measure is not defined. Vitali's result was actually more general, he proved that no measure satisfying natural axioms can be defined for all sets of reals. Thus the measure problem was solved positively for practical purposes by Lebesgue and negatively, from the point of view of pure theory, by Vitali.

One of the axioms used by Vitali was that the measure was *translationally invariant.* This means that the measure of a set does not change if it is shifted or rotated. After Vitali's 1905 result there was still an open problem left: can there be a measure which is not translationally invariant, but defined on all sets of reals? Translational invariance is certainly an important property and it is not clear if there is anything useful that we can do with a measure which does not have this property, but it happens so often in mathematics that strange structures turn out to be useful. (Recall the story about non-Euclidean geometries which was also about omitting one important axiom, the fifth postulate.) Perhaps such a strange measure could eventually be useful, especially if we could get it as an extension of the Lebesgue measure to all sets.

We are primarily interested in measures on the real numbers, but since we now do not require translational invariance, we can consider measures defined on subsets of any set. In order to avoid confusion, we need to introduce some notation.

Since the standard definition of a measure does not require μ to be defined for all subsets, we will call a measure μ on a set X *total* if $\mu(Y)$ is defined for every subset Y of X. We are interested in the *total-measure problem*, which is the question whether there exists a set with a total measure.

In order to answer this question, we must say what precisely a total measure is.

Definition 2 We will say that *μ is a total measure on a set X* if it is a function that assigns a real number from the interval $[0, 1]$ to every subset $Y \subseteq X$ and such that

1. $\mu(Y) = 0$ if Y is finite,
2. $\mu(X) = 1$, and
3. for every countable family of pairwise disjoint sets Y_n, $n \in \mathbb{N}$,

$$\mu\Big(\bigcup_n Y_n\Big) = \sum_n \mu(Y_n).$$

The first two conditions ensure that a total measure is not trivial. The third condition is called *countable additivity*.

If we do not require other conditions to be satisfied, such as translational invariance, then the only relevant property of X is its cardinality. So *the total-measure problem* is:

Problem 1 For which cardinalities of X, does there exist a total measure?

If we consider different cardinalities, it is also natural to consider different types of additivity. For a cardinal number κ, we say that μ is κ-additive, if condition 3. holds true for any family of cardinality less than κ.

One important class of measures are 0/1-*measures*, the measures that only use 0 and 1. It is like using words *small* and *large* to describe the size, instead of using real numbers. Observe that κ-additivity in this case reduces to the condition: *if the measure of the union of less than* κ *pairwise disjoint sets is* 1, *then exactly one of these sets has measure* 1.

Let us start with total 0/1-measures. On a countable set there are total 0/1-measures provided that we relax the additivity condition to finite additivity. For higher cardinalities, the measure problem for total 0/1-measures is "solved" not by a theorem and a proof, but by a definition.

Definition 3 A cardinal $\kappa > \omega$ is *measurable*, if there exists a κ-additive total 0/1-measure on a set of cardinality κ.

Why is this accepted as a solution? Can't we solve every problem in set theory by such a silly trick? The reason is that this definition is a typical definition of a large cardinal. We do not have a precise definition of large cardinals, but there are some basic properties that must always be satisfied. First, the cardinal must be so large that its existence does not follow from the axioms of Zermelo-Fraenkel Set Theory. Second, it should fit into the hierarchy of large cardinals. This means that the assumption of the existence of the cardinal must be comparable with other large-cardinal axioms. If this is the case, then we see that the largeness is an essential property of this definition. If we know that the cardinal is less than some previously studied large cardinals, then we have some evidence supporting its consistency.

To sum up, it is possible that sets with total 0/1-measures exist, but if they do, they must have very large cardinalities, so large that its existence is not provable from the basic axioms.

The general total-measure problem remained open until the mid-1960s when it was solved by Robert Solovay. The way it was solved is a nice example of an application of large cardinals. Solovay showed that if a measurable cardinal exists, then we can consistently assume that the Lebesgue measure can be extended to a total measure on the real numbers [279]. He also showed that the assumption about the measurable cardinal is necessary. Formally, his result is stated as follows.

Theorem 7 [279] *The following two theories are equiconsistent.*

1. *Zermelo-Fraenkel Set Theory with the existence of a measurable cardinal,*
2. *Zermelo-Fraenkel Set Theory with the existence of a total measure extending the Lebesgue measure on the real numbers.*

Recall that the axioms of Zermelo-Fraenkel Set Theory include the Axiom of Choice. We will see shortly that if one is willing to sacrifice the strong form of this axiom, then there is an even better solution to the total-measure problem (Theorem 10, page 224).

New Types of Infinity

Each higher cardinal is a new type of infinity. From the point of view of large cardinals these new infinities are not so different from previous ones if we only use the bottom up process of generating them. For example, we know that for every cardinal $\aleph_\alpha$, there is its successor $\aleph_{\alpha+1}$ whose existence is automatically ensured. On the other hand, the existence of the smallest one, the $\aleph_0$, has to be postulated. To postulate it, we have used the property schematically expressed by the equation $1+x=x$, which formally means that a set is equinumerous with its proper subset. In this way we were able to define essentially new type of sets, the infinite sets. It naturally comes to mind that other properties may give rise to higher order infinities. Paraphrasing Bolzano, there may be even more paradoxical infinities than the usual ones. In the previous subsection we discussed a large cardinal defined by a property coming from measure theory. Here we will consider another example. It is interesting because the defining property is satisfied by $\aleph_0$ and is based on Ramsey's Theorem, which is already familiar to us.

The cardinal is called *weakly compact*. It is a cardinal κ which is larger than $\aleph_0$ and Ramsey's Theorem is true when we replace the word 'infinite' by 'cardinality κ'. More precisely:

Definition 4 A cardinal κ is weakly compact if it is uncountable and if for every set X of cardinality κ the following condition is satisfied. If pairs of elements of a set X of cardinality κ are colored by two colors, then it is possible to find a subset Y of X of cardinality κ, such that all pairs of elements of Y are colored by the same color.

It is interesting that the Ramsey property in the definition holds for $\aleph_0$ (this is the usual Infinite Ramsey Theorem), but to find the next cardinal with this property we have to go beyond the cardinals whose existence is provable in Zermelo-Fraenkel Set Theory. Thus a weakly compact cardinal is inaccessible. One can prove that the least weakly compact cardinal is bigger than the least Mahlo cardinal, but it is smaller than the least measurable cardinal.

Extensions of Zermelo-Fraenkel Set theory are not always based on postulating the existence of a large cardinal. Recall that adding Fraenkel's Replacement Axiom Schema to Zermelo's original set of axioms also had a tremendous effect on the cardinal numbers. Using only Zermelo's axioms we cannot prove even the existence of $\aleph_\omega$. Thus the Replacement Axiom Schema can be considered a sort of large cardinal assumption over Zermelo Set Theory. If we wished to state it as a single axiom, we would have to use the slightly stronger assumption of the existence of an inaccessible cardinal. (Such a theory, though strong enough, would not be very practical, as it does not imply the Replacement Axiom Schema that is used in many specific mathematical constructions.) Similar situations have occurred on a higher level. For example, there is an axiom schema called *Vopěnka's Principle*, proposed by Petr Vopěnka. This schema, like the Replacement Schema, cannot be stated as a single axiom nor does it speak explicitly about cardinalities. In order to compare it with large cardinal assumptions, a related concept of the *Vopěnka cardinal* was defined.

The Vopěnka cardinal is one of the largest that have ever been studied (and not found inconsistent). The history of this cardinal is quite amusing. When Vopěnka discovered his new principle, he thought he could disprove it. Since he did not like large cardinals, it occurred to him that he could tease the large cardinal community by proposing it as a large cardinal hypothesis and then, later, showing that it was not consistent. But when the time came to show the inconsistency, he discovered a gap in what he thought was a proof. So in spite of his negative attitude to the large-cardinal theory, his name remains attached to one of the large cardinals.

Cardinals and Braids

The strongest large cardinal axioms are defined by means of elementary embeddings $j : V_\alpha \to V_\alpha$. Such an embedding is a one-to-one mapping that preserves all true sentences—if a sentence is true about some elements it is also true about the images. There is a trivial elementary embedding, the identity mapping. The postulate that for some ordinal number α there exists an elementary embedding, different from the identity, is one of the strongest axioms ever proposed. As such it is also one of the axioms that are more likely to be inconsistent than the other large cardinal axioms. In order to justify such axioms, one has to study their consequences very systematically to make sure that there is no easy way to get a contradiction. This should not be viewed as mere empirical tests of the axioms. The aim is also to see that it conforms to our expectations about it, and to see that the consequences are "natural". One could also say that we want to develop "intuition" about it.

A natural way to study elementary embeddings is to consider them as elements of an algebraic structure in which the operation is the composition of mappings. This operation is associative, hence elementary embeddings form a familiar type of a structure—a *monoid*. Such an operation appears naturally in many situations and is well understood. What is peculiar to elementary embeddings is another natural binary operation, called *application*. (I will define it in Notes, for now we only need its properties.) This operation is neither associative nor commutative, instead it satisfies the following equation, called the *left self-distributive law*,

$$x * (y * z) = (x * y) * (x * z).$$

It is like the familiar distributive law, but it is applied to a single operation, instead of two. An algebra with a binary operation satisfying this equation is called a left self-distributive system. Such algebras are much less understood than monoids.

The algebraic approach to elementary embeddings of sets V_α was especially pursued by Richard Laver [179]. He discovered remarkable properties of these algebras which in particular enabled him to solve the *word problem* for self-distributive systems in 1989. This means that he found an algorithm to decide which equalities are derivable from the left self-distributive law, or equivalently which equalities are true in every self-distributive system. An example of a derived equality is

$$(x * y) * \big(x * (z * u)\big) = x * \big((y * z) * (y * u)\big),$$

(as you can easily check). The striking fact was that for proving that his algorithm always stops and gives the correct answer, he needed the assumption that there exists a non-identical elementary embedding $j : V_\alpha \to V_\alpha$. Thus to prove a theorem about finite strings of symbols and algorithms, he needed one of the strongest large cardinal axioms (the axiom called I3).

For logicians this was a truly remarkable result, but at the same time a type of result that was expected. Logicians knew that large cardinal axioms have low complexity consequences, consequences about finite sets. Thus it was only a matter of time when such a concrete result was found. Rather unfortunately for logicians, this situation did not last long. In 1992 Patrick Dehornoy succeeded in eliminating the large cardinal assumption [60]. It is a pity that we do not have an example of an algorithm that requires such a strong axiom, but for the problem itself it is good because we know "for sure" that the algorithm exists and we do not have to worry about the consistency of an extremely strong axiom.

In fact Dehornoy's proof had more positive consequences. The spin-off of the proof were new concepts and new results in the classical field of braids. A *braid* is a concept related to a *knot*. These formal mathematical concepts correspond very well to the natural meanings of these words. The main difference between knots and braids is that instead of one string in the case of knots, braids use several strings and the problem is how they are mutually interlaced. More precisely, a braid consists of two bars to which n strings are attached. The basic question is, similarly as for knots, for two braids to determine if they are equivalent, which means that we can transform one into the other without cutting the strings or disconnecting them from bars. Two equivalent braids are shown in Fig. 3.6.

Again, the natural way of studying braids is to consider an associated algebraic structure. This structure is the *braid group* whose elements are transposition of consecutive strings. It is natural to think of such transpositions as operations that form a group. This group is very complex, but it is an object that has been studied for quite a long time. It turned out that some problems about left self-distributive systems can be reduced to braid groups and then solved using means available in this theory. As I said, this connection also enriched this classical field.

This story shows that new axioms of set theory may be not needed for proving new results, but sometimes may help very much to discover them.

In any case this is not the whole story. Laver and other mathematicians also studied some finite left self-distributive systems and proved theorems about their structure using the same large cardinal assumption. Whether or not these theorems can be proved without this assumption is still open. (For more detail, see page 215.)

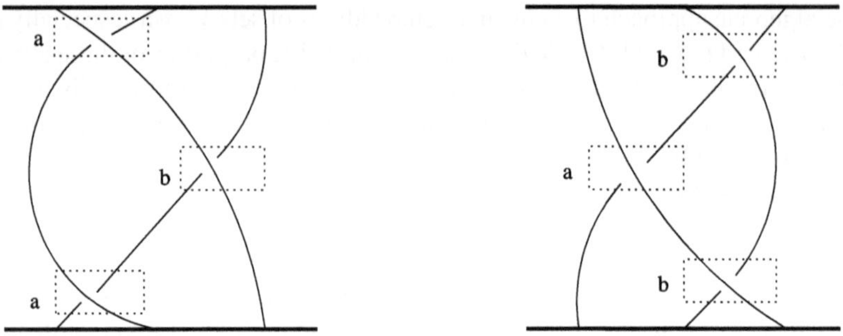

Fig. 3.6 Two equivalent braids. This equivalence is algebraically expressed by $aba = bab$. This equation holds for every pair of transpositions that share a common string, but it is not true in general. Two transpositions a and b that do not share a string satisfy $ab = ba$. The braid group is determined by these two sets of equalities

The Remarkable Linearity of Large Cardinals

A lot of work done in large-cardinal set theory concerns the comparability of large cardinals. This means that, for a given pair of large-cardinal definitions, the aim is to decide if the least cardinal satisfying the first definition is smaller, or larger, or equal to the least cardinal satisfying the second definition. In many cases the answer is known. In other cases it has been established at least that the consistency of one axiom implies the consistency of the other one. There are rather few exceptions, where the axioms are not known to be comparable. (Figure 3.7 shows some large cardinals that are all comparable with one exception.) Past experience suggests that eventually those will also be decided. Is there a reason why every two large cardinal axioms should be comparable?

If our philosophy is platonism, the answer is easy. According to this view, we are not merely studying axiomatic systems, but we are exploring the true universe of sets. The axioms are just useful facts that we have found out about it. So large cardinals are real objects resting in this universe. As all cardinals are linearly ordered, our study of the universe of sets will gradually reveal how the large cardinals corresponding to our axioms compare to each other. Sometimes it may be hard, but there must be a definite answer in each case.

There is a weaker property that large cardinals satisfy, now, without exceptions. It is the *confluence property:* for any pair of large cardinals, there is another large cardinal that is larger than any of the two. Again, assuming the universe of sets is real this property is a trivial fact.

Can one give an explanation of these empirical facts without referring to the reality of sets? We can simplify this problem by considering not the large cardinal themselves, but only their *consistency strength.* This means that we now only compare the logical strength of the axioms postulating the existence of large cardinals. For the linearity of the consistency strength, the evidence is even stronger. A possible explanation, not based on realism in set theory, is that each time one defines

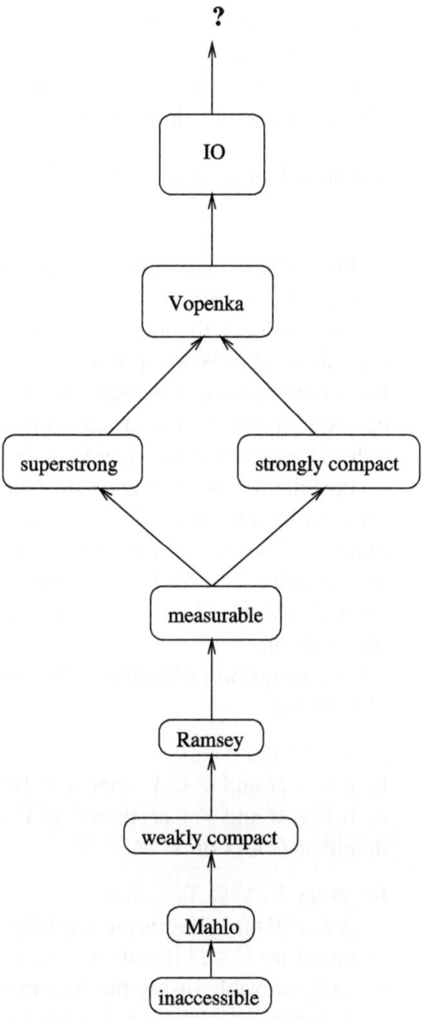

Fig. 3.7 Some large cardinals. I have chosen those that are mentioned in the text, plus a pair of cardinals whose relation is unknown

a new large cardinal, one postulates an assumption so much stronger than previous ones that it encompasses all of them.

Notes

1. *Grothendieck universes.* A typical category is a proper class because we want to include all the structures of a given type in it. For example, the category of sets consists of all sets, as objects, and all mappings between pairs of sets, as morphisms. Alexander Grothendieck developed certain important theories using the concept of category, for which he needed to construct categories from categories. Namely, the objects of a new category are original categories and the

morphisms are functors. To this end, he introduced the concept of universes. Leaving the definition aside, we will only consider what is needed to add to ZFC in order to formalize universes. This is an axiom first considered by Tarski (though stated in a different form):

Tarski's Axiom *For every cardinal κ, there exists an inaccessible cardinal λ, $\lambda > \kappa$.*

In set-theoretical jargon this is also stated as: *There exists a proper class of inaccessible cardinals.*

Grothendieck theories play an important role in number theory. In particular they are used in Wiles' proof of the Fermat's Last Theorem. All experts believe that Grothendieck universes are just a convenient environment to work in and can be eliminated. But, strictly speaking, Wiles' proof as it stands is a proof in a theory essentially stronger than *ZFC*. See discussion in [196].

Postulating the existence of a proper class of inaccessible cardinals may seem to be an extremely strong axiom, but in fact it isn't. All consequences that talk about sets of bounded cardinality, in particular arithmetical sentences, are derivable from *ZFC* augmented with the axiom postulating the existence of a (single) Mahlo Cardinal, the next cardinal in the hierarchy after the inaccessible cardinal.

2. 0/1-*measures and ultrafilters.* An *ultrafilter* $\mathcal{U}$ on a set X is a set of subsets of X such that

   a. $X \in \mathcal{U}$ and $\emptyset \notin \mathcal{U}$;
   b. if $U \in \mathcal{U}$ and $U \subseteq V$, then $V \in \mathcal{U}$;
   c. if $U \in \mathcal{U}$ and $V \in \mathcal{U}$, then $U \cap V \in \mathcal{U}$;
   d. either $U \in \mathcal{U}$ or $X \setminus U \in \mathcal{U}$,

   for every $U, V \subseteq X$.

   A *trivial ultrafilter* is an ultrafilter such that $\{a\} \in \mathcal{U}$ for some $a \in X$. In such an ultrafilter $U \in \mathcal{U}$ if and only if $a \in U$. The existence of a nontrivial ultrafilter is easily provable (using the Axiom of Choice).

   A nontrivial ultrafilter can be equivalently defined as the set of subsets of measure 1 of a finitely additive 0/1-measure on set X.

   More generally, an ultrafilter is κ*-complete* if every set $\mathcal{A}$ of cardinality less than κ, consisting of pairwise disjoint sets and such that $\bigcup \mathcal{A} = X$ contains an element of $\mathcal{U}$. Nontrivial κ-complete ultrafilters are sets of measure 1 of κ-additive 0/1-measures. Thus a cardinal κ is a measurable cardinal if and only if there exists a nontrivial κ-complete ultrafilter on κ.

   Ultrafilters are important in model theory and set theory. We will encounter them again in the sequel.

3. *The Ramsey Cardinal.*

Definition 5 A cardinal λ is *Ramsey*, if for every coloring of *all* finite subsets of λ, there exists a monochromatic subset X of cardinality λ, which means that for every number k, all k-element subsets of X have the same color (but subsets of different size may have different colors).

Note that ω does not satisfy this property. Ramsey cardinals occur below the first measurable cardinal, but they are much larger than the least weakly compact cardinal.

4. *Large constructive ordinals from large cardinals.* Large cardinals are one of the most abstract objects ever considered. Therefore, it is interesting to see any connections with more concrete concepts. A constructive ordinal has names for all its elements and an algorithm for deciding the order of the elements; hence it is very concrete, though it may be quite complicated. In ordinal analysis of theories one needs very large constructive ordinals and it is a highly nontrivial task to describe them. Using large cardinals this can be done, sometimes, more easily.

 The idea that an ordinal notation can be produced from an uncountable cardinal appeared in a paper by Heinz Bachmann in 1950 [16]. We start with a simple observation that any *countable* set of ordinals, however big the ordinals are, is a well-ordered set whose order type is a *countable ordinal.* The first idea is to take an uncountable ordinal Ω and a few operations. Then we take the closure under these operations, i.e., ordinals generated from Ω using the operations. The set of ordinals that is generated from Ω using the operations is countable, hence the order type of this set is a countable ordinal. If we are lucky, with some additional work we may be able to prove that it is a constructive ordinal.

 To be a little bit more explicit, let us take $\Omega = \omega_1$. As the operation we take plus, times and exponentiation (the two argument function $\xi, \zeta \mapsto \zeta^{\xi}$). The ordinals generated from 0 and 1 are exactly ordinals below ε_0. With Ω included among the initial ordinals, we get more. In particular Ω will be the ε_0th element in the generated ordering. Since the set is closed under exponentiation, its order type must be at least the next ε-number, which is ε_1. This is not a big deal, so we need another idea. The next idea is that when working with countable subsets of ordinals that contain uncountable ordinals, we have a lot of gaps, so we can also generate new ordinals by functions that map large ordinals to small ones. This produces a loop in which ordinals are mapped back and forth, which amplifies the production of new ordinals very much.

 The key operation is called the *collapsing function* and is denoted by ψ. In order to understand what this function does we should interpret Ω and ψ in two ways:

 a. as an ordinal, namely, ω_1, and as a partial function defined on ordinals;
 b. as a mere symbol and as a function that assigns specific countable ordinals to terms obtained from Ω and symbols for the operations and the function ψ.

 The second interpretation will be used to gradually assign names to ordinals that do not have any. The names will be terms in the language $\{0, 1, \Omega, +, \times, \exp, \psi\}$. The first interpretation enable us to define the ordering on these terms. We will first generate ordinals from 0, 1 and Ω using the operations $+, \times, \exp$. When we cannot produce more ordinals in this way, we take the least ordinal that in not in the currently generated set of ordinals X, say γ, and define $\psi(\delta) = \gamma$ for a suitable $\delta \in X$. Then we generate a new set from X and α. We repeat this process transfinitely many times as long as we have "available names", i.e., ordinals for which we can define ψ. The great power of this construction is caused be the fact that as we increase the set X we also get new available names.

Here is the precise definition of ψ. Define sets of ordinals X_α and partial functions ψ_α (approximations to ψ) by transfinite recursion as follows.

a. Let X_0 be the set of ordinals that can be constructed from 0, 1 and Ω using the operations $+, \times, \exp$, and let ψ_0 be undefined everywhere.
b. If $\alpha = \beta + 1$, let γ be the least ordinal not in X_β and let δ be the least ordinal in X_β that is larger than all ordinals for which ψ_β is defined. Then extend ψ_β to ψ_α by defining $\psi_\alpha(\delta) := \gamma$, add γ to X_β and close it off under the operations $+, \times, \exp$ to obtain X_α.
c. If λ is a limit ordinal, then put $X_\lambda = \bigcup_{\alpha<\lambda} X_\alpha$ and $\psi_\lambda = \bigcup_{\alpha<\lambda} \psi_\alpha$.

This process stops when ψ is defined on all ordinals of X_α. The function ψ is the union of all these ψ_α and the Bachmann–Howard ordinal is the least ordinal which is not in any of these sets. Equivalently, the Bachmann–Howard ordinal is the limit of all countable ordinals that appear in this process.

Let us have a closer look at some initial stages of this process. At the beginning Ω plays no role. As we already know, X_0 is the set of all ordinals below ε_0. Hence $\psi(0) = \varepsilon_0$. In general, $\psi(\alpha) = \varepsilon_\alpha$ for all α below the fixed-pint of the function $\xi \to \varepsilon_\xi$, which is denoted by $\phi_2(0)$ using the Veblen function. At this point, Ω kicks in: $\phi_2(0)$ is the lest ordinal not in the currently constructed set X and Ω is the least ordinal in X that is larger than all ordinals for which ψ_β is defined; so we define $\psi(\Omega) := \phi_2(0)$.

Note that for ordinals α, $\phi_2(0) \le \alpha < \Omega$, the function ψ is undefined at this stage, hence it remains undefined throughout the entire process. The reason for having this gap is that we would have to put $\psi(\alpha) = \psi(\Omega)$ for all these ordinals because we want ψ to be order-preserving. This would be useless since we already have a name for this ordinal. (Nevertheless, some authors prefer ψ to be defined on an initial segment of ordinals and do extend the definition in this way.) There are many such gaps because ψ is defined only for countably many values whereas Ω is uncountable.

Another thing to note is that at the stage when we define $\psi(\Omega)$ we have more ordinals between Ω and $\Omega \cdot 2$ than we had originally: we have all ordinals $\Omega + \alpha$ for $\alpha < \phi_2(0)$. Many more will be added before we use $\Omega \cdot 2$. One can compute that we get $\psi(\Omega \cdot 2) = \phi_2(1)$. Further we have $\psi(\Omega^2) = \phi_3(0)$, and more generally, $\psi(\Omega^\alpha) = \phi_{1+\alpha}(0)$ for all $\alpha < \Gamma_0$ (the Feferman-Schütte ordinal), whence we get $\psi(\Omega^\Omega) = \Gamma_0$.

Let us now look at the end of this process. Already at the initial stage we have the ordinals $\Omega, \Omega^\Omega, \Omega^{\Omega^\Omega}, \dots$. Although we will add a lot of new ordinals in the construction, each of them will be bounded by one of these. Since $\Omega = \omega_1$, we have $\omega^\Omega = \Omega$, which means that Ω is an ε-number; in fact $\Omega = \varepsilon_\Omega$. Therefore the limit of $\Omega, \Omega^\Omega, \Omega^{\Omega^\Omega}, \dots$ is $\varepsilon_{\Omega+1}$. If we extended the definition of the collapsing function ψ in a natural way, then the Bachmann–Howard ordinal would be $\psi(\varepsilon_{\Omega+1})$. This term is actually used to denote the Bachmann-Howard ordinal.

The sketch given above is not a proof that the resulting ordinal is constructive. One has to work out representation of these ordinals by terms using the arithmetic operations and ψ. Then one must show that the ordering is computable from the terms representing the ordinals.

Do we need to use an uncountable ordinal for the construction? Theoretically no; it would suffice to take a sufficiently large countable ordinal, but there is no reason why we should do it. To prove that there exists a sufficiently large countable ordinal we would use the fact that the number of countable ordinals is uncountable. So why not to take an uncountable one?

Once we see that one uncountable ordinal is useful for constructing ordinal notations, we may use more uncountable cardinals. It is clear from the example above that doing the same construction with two initial cardinals, say ω_1, ω_2, will produce a larger constructive ordinal. The gain is not very big, so we need to get more cardinals involved. A way to get substantially larger constructive ordinals is to generate larger and larger *cardinals* in the process. Again it may look like a very nonconstructive process. But remember, the only thing that we needed from Ω above, was that it was not accessible by any countable process from below. Eventually Ω was only used as a name. When introducing larger cardinals in the process, the reason is simply to have an object that is far enough from the things generated before. Recall that when constructing Γ_0, the use of the indexing function gave significant strength to the process. The new idea now is that in the construction we can use the indexing function of cardinals, more precisely, the indexing function of a special class of cardinals.

Let us first try something simple. Let us take the construction above and add the cardinal indexing function. Thus we close off under addition, exponentiation, ψ_Ω and the function $\xi \mapsto \omega_\xi$. The ordinal that we get is fairly big, but still not sufficiently big for an ordinal analysis of some theories. This simple construction does not need large cardinals (those whose existence is not provable in Zermelo-Fraenkel Set Theory). To get a really large ordinal we need a new idea. In the construction above, Ω helped us to run the process much longer than we would be able to achieve by only working from below. So let us take a cardinal that is safely above any cardinal that we would generate from below using the operations mentioned, including the cardinal indexing function. This will be our new Ω and then we run the same process (with the cardinal indexing function $\xi \mapsto \omega_\xi$). The ordinals above Ω will keep injecting new *cardinals* below, thus the process will be repeated for a "very long time". How big must this Ω be? It is the same type of question, as we asked before when discussing whether Ω must be uncountable. A cardinal that will easily ensure that we will never reach it in such a process is the first inaccessible cardinal.

Several cardinals larger than the first inaccessible one have been successfully used to produce ordinal notations for large constructive ordinals; in particular the weakly compact cardinal and some larger ones. For more information, see Rathjen [236].

5. *A nonmeasurable set.* I will show how to construct a nonmeasurable subset of a circle. Using a circle we will not lose much generality and the proof will be more transparent. A circle can be mapped on a segment of a line so that rotations correspond to translations. A similar idea will be used in the proof of the Banach-Tarski paradox, which we will discuss shortly.

   Let a measure on a circle C be fixed. The assumptions are that the measure is rotationally invariant, σ-additive and the measure of C is nonzero and finite, say 1.

   Let ρ be the rotation by an angle α such that α is not a rational fraction of π. For every point x on the circle, the points $\rho^i(x)$, $i = \dots, -2, -1, 0, 1, 2, \dots$, obtained by applying ρ i-times forward or backward, are distinct, since π/α is irrational. The set of these points is called the *orbit* of x with respect to the rotation ρ. The orbits form a partition of C, which means that they are disjoint and cover C. Pick one point from every orbit, 'a representative' of this orbit, and let A be the set of these representatives. To this end we must use the Axiom of Choice. The rotations of A by an integer multiple of ρ, namely, the sets $\rho^i(A)$, for $i = \dots, -2, -1, 0, 1, 2, \dots$, form another partition of C. This is a partition into countably many sets, all congruent to A. If A were measurable with measure 0, then, by σ-additivity the measure of C would also be 0. If A had a positive measure, then the measure of C would be infinite. Thus the assumption that A is measurable leads to a contradiction, hence A is not measurable.

   Note that the assumption of σ-additivity is essential. Stefan Banach showed that *finitely additive* translationally invariant measures on 1- and 2-dimensional Euclidean spaces do exist. But this is rather an exception, in three dimensions there is no translationally invariant finitely additive measure (see the Banach-Tarski paradoxical decomposition of a ball, page 218).
6. *How do we compare large cardinals?* Given two definitions of large cardinals, let κ denote the least cardinal of the first kind and λ the first cardinal of the second kind. Suppose the first kind of cardinals should be smaller than the second kind. The first thing that we want to establish is $\kappa \leq \lambda$. In most cases this is done by showing that all cardinals that satisfy the first definition also satisfy the second one. In particular, all large cardinals (so far considered) are inaccessible, thus they are at least as big as the smallest inaccessible cardinal. To separate κ from λ one can use the second incompleteness theorem. If we prove that, assuming the existence of λ in *ZFC*, the *ZFC* set theory with an axiom asserting the existence of κ is consistent, then we get that $\kappa \neq \lambda$ is consistent with *ZFC*, provided that, of course, λ is consistent with *ZFC*. Usually one can show much more—that λ is *much* larger than κ. This done by proving that cardinals of the first type are in some sense abundant below λ. The weakest requirement is that there are λ many cardinals of the first type below λ.
7. *Large cardinals and elementary embeddings.* The strongest known hypotheses are stated using the concept of an elementary embedding which is defined as follows. Let M and N be two structures of the same type, for example, a set with a binary relation. We say that a function j from M to N is an *elementary embedding* if

a. j is one-to-one, and
b. for every formula $\phi(x_1, \ldots, x_n)$ (in the language of the structures) and for every $a_1, \ldots, a_n$ in M, $\phi(a_1, \ldots, a_n)$ holds in M if and only if $\phi(j(a_1), \ldots, j(a_n))$ holds in N.

A special case is when M is a substructure of N and the condition holds for the identity embedding of M into N. Then we say that M is an *elementary substructure* (or *elementary submodel*) of N and N is an *elementary extension* of M.

We will say that an embedding j is *proper* if j is not onto, i.e., there exist elements in the second structure that are not images of elements of the first structure.

One of the strongest large cardinal axioms is the following statement.

I3 *There exist an ordinal* α *and a proper elementary embedding of* $(V_\alpha, \in)$ into $(V_\alpha, \in)$.

Let j be such an embedding. Since j is elementary, it maps ordinals on ordinals and cardinals on cardinals. One can also show that the condition that j is proper implies that there exists a cardinal in V_α that is not in the image of j. The least such cardinal κ is called the *critical cardinal of the embedding*. Such a κ is the large cardinal associated with this axiom. Except for a few large cardinals defined by similar axioms, all others exist below such a κ. This large cardinal does not have a name yet. Perhaps, it is because people are still too afraid that it is inconsistent to assume that it exists. Indeed, small modifications lead to a contradiction. For example, it is known that α cannot be an ordinal of the form $\beta + 2$ (α can only be a limit cardinal or a successor of such a cardinal).

The first cardinal that was characterized by elementary embeddings was the measurable cardinal. The connection between the existence of a measure and an elementary embedding is via an important construction used in model theory called the *ultrapower.* Given a structure M, the ultrapower construction provides a proper elementary extension of M. I defer the description of this construction to page 252. Here we only need to know that it is based on nontrivial ultrafilters. One can apply the ultrapower construction to the class structure $(V, \in)$, where V is the universal class. In general the resulting model does not have nice properties. What we would like to get is a structure in which the membership is the usual $\in$ and which is a model isomorphic to $(M, \in)$ for some class M. The stronger condition in the definition of a measurable cardinal, the σ-additivity, ensures that the ultrapower is isomorphic to such a model. Thus a cardinal is measurable if and only if it is a critical cardinal of an elementary embedding j of $(V, \in)$ into some $(M, \in)$. Various larger cardinals can be defined by postulating that j preserves more of the structure.

8. *The operation of application defined on elementary embeddings.* The operation of application of an elementary embedding j to k, where $j, k : V_\lambda \to V_\lambda$ is defined for λ a limit ordinal. To get the idea, suppose first that $k : V_\alpha \to V_\alpha$ for $\alpha < \lambda$. Then $k \in V_{\alpha+3} \subseteq V_\lambda$, thus k is in the domain of the mapping j. Let us see what is $j(k)$. Since j is elementary, it must preserve the properties of k, in particular, $j(k) : V_{j(\alpha)} \to V_{j(\alpha)}$ is also an elementary embedding. In general,

$k : V_\lambda \to V_\lambda$, k is not in the domain of j and we cannot apply j to k directly, but we can still do it somehow, which is remarkable. To define the application, consider approximations to k, apply j to the approximations and take the union. This operation is denoted by $j[k]$ and it is formally defined by

$$j[k] = \bigcup_{\alpha<\lambda} j(k|_{V_\alpha}).$$

It is easy to check that $j[k]$ is an elementary embedding.

9. *Vopěnka's Principle.* This principle is usually stated in theories with classes:

   For any language L, and any proper class X of structures in L, there are two different structures $M_1, M_2 \in X$ such that there is an elementary embedding of M_1 into M_2.

   Vopěnka's cardinals are those cardinals κ for which $(V_\kappa \cup \mathcal{P}(V_\kappa); \in)$ satisfies Vopěnka's Principle.

10. *The strongest axiom ever proposed.* It may be interesting to see the strongest axiom that has ever been proposed, even without any explanation. By saying '*the strongest*' I mean '*the strongest axiom that is not known to be inconsistent*' because, formally, the strongest axiom is any axiom that implies a contradiction.

    For a set x let $def(x)$ be all subsets of x that are definable in the structure $(x, \in)$ by first order formulas in the language with one binary predicate $\in$ and constants for all elements of x. For a set x, define by transfinite recursion,

$$L_0(x) = x, \qquad L_{\alpha+1}(x) = def\big(L_\alpha(x)\big),$$
$$L_\lambda(x) = \bigcup_{\alpha<\lambda} L_\alpha(x), \quad \text{for } \lambda \text{ a limit ordinal,}$$

    and put

$$L(x) = \bigcup_{\alpha} L_\alpha(x).$$

    The strongest axiom is the following statement proposed by W. Hugh Woodin in 1984:

    I0 *For some ordinal α there exists a proper elementary embedding of $(L(V_{\alpha+1}), \in)$ into $(L(V_{\alpha+1}), \in)$.*

    The critical cardinal of such an embedding is the largest ever considered cardinal (that was not found inconsistent). It is interesting that since 1984 nobody has been able to come up with a larger cardinal.

11. *The measurable cardinal decides a statement related to the Generalized Continuum Hypothesis.* Put $L = L(\emptyset)$. This class was defined by Gödel and he proved that $(L, \in)$ satisfies the Zermelo-Fraenkel axioms and the Continuum Hypothesis (in fact the Generalized Continuum Hypothesis). In this manner he proved that the Continuum Hypothesis is consistent with *ZFC*. Hence, if the Continuum Hypothesis is not true, we must have $V \neq L$. The converse is not true, it is possible that $V \neq L$, but the Continuum Hypothesis still holds true.

Table 3.1 A left self-distributive system with four elements

*	1	2	3	4
1	2	4	2	4
2	3	4	3	4
3	4	4	4	4
4	1	2	3	4

In 1961 D. Scott showed that the existence of a measurable cardinal implies $V \neq L$ [260]. A little later Solovay proved that certain subsets of reals have more of the desirable properties under the assumption of the measurable cardinal than we can prove without it (namely, more sets of reals have the perfect subset property and the Baire property). This was considered to be evidence that large cardinals may decide statements about "small" sets. The current prevailing view is that large cardinal axioms alone cannot decide the Continuum Hypothesis and other sentences that were proved independent by forcing. I will say more about it in Chap. 7.

12. *Laver tables.* Working on elementary embeddings of V_λ to itself, Richard Laver proved a theorem about finite left self-distributive systems [180]. To prove it he used the axiom I3 and up to the present no proof without this assumption is known. Unfortunately, there is no result saying that the theorem needs large cardinals either. If the theorem really needed I3, it would be a remarkable fact, since the combinatorial result is a type of problem on which algebraists work.

Consider Table 3.1. This is a multiplication table of a left self-distributive system with four elements $1, 2, 3, 4$. Look at the first column of the table, which defines multiplication by 1 from the right. Notice that it defines a mapping $x \mapsto x * 1$, which is the cyclic shift $1 \to 2 \to 3 \to 4 \to 1$. One can show that this property of the table and the left self-distributivity uniquely determines all other entries of the table. This holds generally: for every n that is a power of 2, $n = 2^k$, there exists a unique left self-distributive system on the set $\{1, \dots, n\}$ in which $x \mapsto x * 1$ defines the cyclic shift. This fact is provable without any strong axioms.

Another peculiar thing in the table above is that except for the last row, the rows are periodic with a nontrivial period. The periodicity of the rows is the property to which Laver's theorem refers.

Theorem 8 *The periods of the first rows in these $2^k \times 2^k$ tables go to infinity as k goes to infinity.*

Does it seem to you that this theorem should require any strong axioms?

3.4 Controversial Axioms

The Axiom of Choice

In a lot of proofs you can find sentences such as '*Let x be an arbitrary element of the set Y.*' or '*Choose an element x which satisfies property Y.*', etc. The freedom of choice of an element seems to be one of the basic freedoms of mathematicians. But this fact is given by logic, not by set theory. Assuming there exists an element satisfying a property Y, we should be able to speak about such an element, otherwise we could not make any nontrivial logical deductions. The problem which arises in set theory is whether one can continue the selection process for an infinite number of steps. This is outside of the scope of pure logic as it requires the concept of an infinite set. Let us state it precisely in the form of an axiom.

The Axiom of Choice *Let A be a set of disjoint nonempty sets. Then there exists a set C such that for every element B of A, the set C contains exactly one element from B.*

A real world version of it is the following. Suppose you have many boxes each containing at least one thing. The task is to choose *exactly* one thing from every box. Let me stress that there is no hidden restriction; for example, we do not ask to choose a random element, as is the case when doing statistics. So it is a completely trivial task for a human. Now think of a computer. If you have experience with programming, you know that it is not difficult to program a computer to do it, but in most programming languages it is rather annoying that it does require some work. Namely, you must tell the computer which particular elements to choose. You may employ a random number generator, but it would be only a complication, as we do not ask for a random choice.

To make it simpler, let us forget about random number generators. In order to program a computer for such a task, you have to find a rule how to choose one element from a given variety of elements. In computers data are represented by numbers, thus a very simple rule is at hand: take the first number. To break the symmetry, we use the available structure, the linear ordering, on numbers.

Let us return to set theory. The first important observation is that the Axiom of Choice is provable for *finite* sets A. So the validity of the Axiom of Choice in general is again the question of how many properties of finite sets can be transferred to infinite sets.

Examples 1. Let A be a set of disjoint two-element sets of real numbers. Then we can use the same idea as with the natural numbers: we pick from every pair the first element. Of course, we have to show explicitly that the existence of such a set follows from the axioms of set theory, we cannot rely on intuition. Assuming that we have already proved that there is a linear ordering on the real numbers, it is quite easy. In this particular case the elements of the choice set C can be specified explicitly by the formula saying that an element of C is the first element of an element B of A. Formally, we apply the comprehension schema to this formula. Practically, it means that whenever we can define explicitly the set C, there is no problem with the axiom.

2. Next consider a set A which contains disjoint pairs of *subsets* of reals. Why don't we take a linear ordering of subsets of reals again? There must be many such things. Well, there is no *explicit* way of defining ordering on subsets of reals. In

fact, without the Axiom of Choice, it is not possible to prove that there is such an ordering.

All right then, why don't we simply accept the axiom? What is controversial about it? To see why it is controversial, we need some reasons for accepting it and some reasons against. The supporting reasons are consequences that we like, that are in accord with our intuition of what should hold in set theory. The reasons against are consequences that seem counterintuitive. Before presenting such examples, let me mention some general properties. The general negative feature of the Axiom of Choice is its nonconstructive nature. Recall that the set existence axioms of Zermelo-Fraenkel set theory considered so far are special cases of the comprehension schema. This means that we postulate the existence of sets whose elements are determined by some formula. In particular, each of these sets is unique. The Axiom of Choice is different. It postulates the existence of a set which is not uniquely determined, of which we know only some properties. Thus we should expect consequences which are also nonconstructive; proofs of existence of structures without having their explicit definitions. On the other hand, a comforting thing is that the axiom does not introduce large sets, it follows the philosophy of building the universe from the bottom up. Hence we may hope that it will not introduce inconsistency.

The Axiom of Choice was formulated by Zermelo in a paper published in 1904 [317]. He used this axiom to prove that on every set there exists a well-ordering. This fact greatly simplifies the structure of cardinal numbers. In particular every infinite cardinal number κ is $\aleph_\alpha$, for some ordinal α. We have been using this fact throughout this chapter. Without the Axiom of Choice there may exist cardinal numbers which are incomparable. Even more striking is that without the Axiom of Choice the set of all reals can be[13] the union of countably many countable sets.

Having a nice theory of cardinal numbers would not be a sufficient argument for most mathematicians to accept such an axiom, but the Axiom of Choice has many important applications in various fields of mathematics, in particular in algebra, topology and functional analysis. I will only mention three basic results for which this axiom is needed.

1. *Every vector space has a basis.*
2. *Every field has an algebraic closure.*
3. *Every subgroup of a free group is a free group* (*Nielsen-Schreier Theorem*).

Let us consider the first of these three theorems. To construct a basis of a finite-dimensional vector space is easy. We pick vectors one by one so that the next one is independent from the previous ones. When we cannot go on, the construction is finished, since a basis is a maximal set of independent vectors. The same is used for infinite-dimensional vector spaces. The only problem is that again there is no canonical choice of the next vector. Vector spaces have a nice structure, but they are too symmetric. In order to show that this infinite process exists, we have to use the Axiom of Choice.

[13] More precisely, it is consistent with *ZFC* to assume that it is.

The main negative consequence of the Axiom of Choice is the result, which we have already considered, that there is a nonmeasurable set of reals. This is unpleasant, but it is not in clear contradiction with our intuition. It is conceivable that we cannot measure a set which is a sort of a fuzzy collection of random points. But there is a really striking example of pathological subsets of three-dimensional space. It is the famous theorem of Polish mathematicians Stefan Banach (1892–1945) and Alfred Tarski [12]:

Theorem 9 (The paradoxical decomposition of a ball) *Assuming the Axiom of Choice, a ball can be decomposed into ten pieces which can be rearranged into two balls of the same size as the original one.*

This is really difficult to swallow. One thing that may help you is to realize that all the pieces cannot be measurable. Once we have accepted the possibility that a set is not measurable, we can also accept the fact that a measurable set, like a ball, can be decomposed into nonmeasurable ones and then we can assemble it into a set with a different measure. But it is amazing that the new set can be a pair of balls. And this is not the end yet. In fact, one can transform a body of any shape into a body of any other shape and size by using perhaps more pieces, but still only a finite number of them. For example, it is possible to transform a ball into a cube of an arbitrary size in such a way. (What I mean by a body here is a very general set of points where the only restrictions are that it must be of a finite diameter and it must be possible to cut a ball out of it.)

Why isn't it enough for rejecting the Axiom of Choice? As I have mentioned, the concept of an arbitrary set is similar to the concept of an arbitrary function. In fact, it is possible to develop foundations based on the concept of a function instead of a set. (Such a theory was proposed by von Neumann in 1925.) The history of the concept of a function is much longer than that of a set. In the course of making the concept of a function precise there were similar episodes as the one we have just seen. The most remarkable example is the existence of a continuous function which is nowhere differentiable, mentioned in Chap. 1. By the time set theory was discovered, mathematicians had got used to such pathologies in the theory of functions. They accepted as a fact that it is necessary to distinguish between nice smooth functions, which occur in nature, and wild artificial ones, which usually do not have practical applications. Therefore, they accepted the necessity of having to distinguish between nice sets of points in space, which are measurable and have other pleasing properties, and pathological sets, such as those occurring in the Banach-Tarski paradox.

Thus the problem was not to accept the existence of such examples, but the fact that they are caused by a new axiom. We should recall that the pathological functions, such as the one mentioned above, were constructed *without* the Axiom of Choice or any other special axiom. The positive consequences of the Axiom of Choice, in spite of being in accord with intuition, actually supported the suspicion that it is inconsistent. If things work too well in the sense that you are able to prove many theorems you should be careful because it may be caused by a contradiction in your theory. Because once you have a contradiction, all statements follow immediately as theorems. But logical investigations soon showed that the Axiom of Choice is not dangerous in this sense. In 1938 Gödel proved that if the Zermelo-Fraenkel theory without the Axiom of Choice is consistent, then it remains consistent when the Axiom of Choice is included [97]. He proved it by constructing a model of Zermelo-Fraenkel Set Theory in which the Axiom of Choice is true.[14] This was done by selecting only certain sets from the universe of all sets. The selection was based on the philosophy that the set universe grows from below, that is, starting from the empty set gradually extends to larger and larger sets. He called the sets that he chose *constructible*; they were obtained by such a process of adding more sets one by one. The crucial thing was the use of ordinal numbers for enumerating this process. Thinking of ordinals as a time scale, a constructible set is a set which occurs at some moment in the process. Then every constructible set can be naturally associated with an ordinal, namely, the one on which it first appears. Such an indexing by ordinals easily gives the choice sets, since ordinals are well-ordered. For instance, if you have to decide between two sets, take the one which has the smaller index.

The heated discussions about the Axiom of Choice at the beginning of the 20th century are almost forgotten by now. Today the Axiom of Choice is used as a true fact and it is mentioned only if one wants to point to the fact that a proof has a nonconstructive nature.

The Axiom of Determinacy

Philosophically minded people may be attracted by this title, but this axiom has nothing to do with the question of whether or not the world is deterministic. This axiom is connected with mathematical game theory. There is a variety of games, but we will only be concerned with a special type of them. We consider *two player games* with no randomness involved and perfect information. Such games are chess, checkers, *go*, tick-tack-toe (naughts and crosses) and other board games, while card games use, as a rule, the random element of shuffling the cards and the players do not have the same information; this also excludes games in which players toss a dice. The two players are usually called simply *Player 1* and *Player 2*. When the game has a small number of possible positions, we can analyze it completely and find a *winning strategy* for one of the players, provided a tie is not possible and the number of moves is limited. This can be done, for example, for some restricted versions of tick-tack-toe. For most commonly played games, however, the number of possible positions is so enormous that it is physically impossible to construct a computer that would manage to search all positions before the end of our solar system. Still, it is an easy mathematical theorem, attributed to Zermelo, that a winning strategy for one of the players can always be found. When a tie is possible, or when one can

[14]More precisely, he defined an interpretation of the axioms of *ZFC* in the theory without the Axiom of Choice.

repeat the same position infinitely many times, this is not true, but there is at least a strategy to tie.

In particular this is the case for chess. In chess one of the players certainly has a strategy that guarantees at least a tie.[15] However, in spite of the accumulated experience and developed theory, nobody expects that such a strategy will ever be found. It is interesting that it may be not only the complexity of finding the strategy, but just the complexity of the best description that prevents us from using it.

Similarly as for other set-theoretical principles, people wondered whether the theorem that every finite game has a winning strategy for one of the players holds for infinite games as well. A possible way to visualize an infinite game is to think of a game such as *go* played on an infinite board with infinitely many stones. The rules of such a game, however, should not allow removing a stone once it was put on the board. The game ends after playing infinitely many moves. Then we look at the board and decide who wins. Even for the ordinary game of *go* it is sometimes difficult to evaluate the final position, thus it may be really difficult to give a precise rule for the infinite board. But I do not want to define a particular game, I need the general concept of an infinite game. Thus every possible rule defines a game, the only condition being that it classifies all possible final positions into those in which Player 1 wins and those in which Player 2 wins.

Now I can state the axiom.

The Axiom of Determinacy *In every infinite game one of the players has a winning strategy.*

The name 'determinacy' is used because we say that a game is *determined* if one of the players has a winning strategy. Then the axiom can be stated simply as follows: *Every infinite game is determined.* In order to be quite precise, we should add that by 'infinite' we mean 'countably infinite', we do not consider larger infinite games. In order to state the axiom formally, one does not have to talk about a board or stones; moves are simply represented by natural numbers. Note that we do not lose any generality by this representation, since for any game where there is only a finite or a countable number of choices in every move, we can systematically enumerate them. Thus during the game players construct an infinite sequence of numbers. Player 1 plays numbers on odd positions, Player 2 plays numbers on even positions. The game is specified by a set X of sequences of numbers. Player 1 wins the game if the resulting sequence belongs to the set X, otherwise Player 2 wins. When the infinite game is presented in this form, it is clearer that the Axiom of Determinacy has much to do with sets of infinite sequences. As infinite sequences of numbers can be coded by real numbers, the axiom is, in fact, about the sets of the real numbers.

The first reason for studying the axiom was probably sheer curiosity whether the theorem about finite games can be generalized to infinite games. Then it turned

---

[15] More precisely, either one player has a winning strategy, or both players have strategies that guarantee them to win or to tie.

out that whenever a concrete game was defined, a winning strategy was found for one of the players. However, soon it was shown that using the Axiom of Choice one can prove the existence of a non-determined game. In light of the Axiom of Choice being generally accepted this looks like the end of the story. But the axiom has always had its appeal. An especially interesting consequence of the axiom is that every set is measurable and it also implies several other nice properties of the set of reals. If we give up the Axiom of Choice, Zermelo-Fraenkel Set Theory with the Axiom of Determinacy seems to be consistent. The Axiom of Determinacy also implies a weak version of the Axiom of Choice. This is important because we would not like to give up the Axiom of Choice completely. Therefore, it was proposed as an alternative for the Axiom of Choice.

Which of the two axioms should one choose? Looking at the statements of the axioms, the Axiom of Choice seems to be a more fundamental principle, as it is simpler and speaks about sets in general not only about countable sequences, etc. On the other hand we know that real numbers are the most important structure for mathematics. So why should we care so much about the arithmetic of infinite cardinal numbers and why should we accept paradoxical statements such as the Banach-Tarski decomposition of the sphere? Shouldn't we rather care about having nice real numbers? We may wonder what would have happened, if Cantor had not become interested in infinite cardinals, continued his research in analysis and discovered the determinacy principle. Imagine that the Axiom of Determinacy had been introduced first, and before the Axiom of Choice was stated the nice consequences of determinacy, such as measurability of all sets, had been proved. Imagine that then someone would come up with the Axiom of Choice and the paradoxical consequences were proved. Wouldn't the situation now be reversed in the sense that the Axiom of Determinacy would be 'the true axiom', while the Axiom of Choice would be just a bizarre alternative?

Formulas and Games

Let us prove that every finite game with no tie has a winning strategy for one of the players. In fact, we only need the *length* of the game to be finite; there may be an infinite number of possible choices in each move. The proof is by induction on the length of the game. The base case is when there is only one move, which means that only the first player plays and then the game is over. Then either there is a move for him to win, in which case this move is his winning strategy, or there is none, in which case the second player always wins, so she has a winning strategy.[16]

Now suppose that every game with at most n moves has a winning strategy for one of the players. We are to prove that every game with at most $n+1$ moves has this property too. Let a game of length $n+1$ be given. Consider the situation after the

[16] I am using *he* for the first player and *she* for the second.

first move of the first player and imagine we start a game from this situation using the same rules. This game has a length of at most n, so by the induction assumption the theorem holds for it. Consider all possible moves of the first player when he plays the first move. Thus we get various games where either he or the other player has a winning strategy. If there is at least one move for the first player such that he (as the second player now) has a winning strategy in the shorter game, then he has a winning strategy. Namely, we take this particular move and combine it with the winning strategy for the shorter game. Otherwise, whatever the first player plays first, the other has a winning strategy, so the second player has a winning strategy for the longer game.

There is another proof of this theorem, which is based on logic. It is worthwhile going through this proof, as it shows a connection of logic with games. For simplicity, consider a game with four rounds. Namely, in the game the players play in the following order: Player 1, Player 2, Player 1, Player 2. Then the game is over and the winner is determined by the moves they played. The fact that Player 1 has a winning strategy can be stated as follows.

> There is *a move for Player* 1 *such that* for every *move of Player* 2, there is *a move for Player* 1 *such that* for every *move of Player* 2, Player 1 wins.

In order to appreciate the logical structure of this statement, we will rewrite it using logical notation. We will use x_1 and x_2 for the moves of Player 1 and y_1 and y_2 for the moves of Player 2. Let Φ abbreviate the statement that a final position is reached where Player 1 wins. Then the above statement reads:

$$\exists x_1 \forall y_1 \exists x_2 \forall y_2\ \Phi.$$

To prove that either Player 1 or Player 2 has a winning strategy I will now apply the law of excluded middle to this formula. Thus either this formula is true, or its negation. This formula expresses that Player 1 has a winning strategy, so we only need to check that the negation says that Player 2 has a winning strategy. Formally, applying De Morgan's laws to quantifiers, we get the following as the negation:

$$\forall x_1 \exists y_1 \forall x_2 \exists y_2\ \neg\Phi.$$

In words:

> For every *move of Player* 1, there is *a move for Player* 2 *such that* for every *move of Player* 1, there is *a move for Player* 2 *such that* Player 1 looses.

This, clearly, expresses that Player 2 has a winning strategy, and we are done.

Thus we proved the existence of a winning strategy for the special type of games with four rounds. In general the proof is the same, one only needs to use as many quantifiers as there are moves. In this proof I have described winning strategies using alternating quantifiers. Quite often this connection is used in the opposite direction. When we have a formula with many alternating quantifiers, it is difficult to imagine what it means. Then it helps to think about it as expressing the existence of a winning strategy in a game. Sometimes such a game-theoretical interpretation is used even for simple formulas. For instance, when we are trying to prove a formula we identify ourselves with the player who corresponds to existential quantifiers because we should construct the objects that are claimed to exist. The universal quantifiers are identified with an opponent. They provide parameters that we cannot control, so we should be able to handle the problem when the opponent chooses for us the worst case.

The reason why we have a good capacity for games (and why we, in fact, like to play them) is that we are all the time involved in playing games with other people, institutions or with nature. This concerns not only humans, but all intelligent beings. Nature does not require us to understand formulas, but it does require us to be successful players.

Determinacy and Large Cardinals

In spite of its paradoxical consequences the Axiom of Choice is not dangerous. According to the result of Gödel, we cannot introduce inconsistency by adding the Axiom of Choice to the other axioms of Zermelo-Fraenkel Set Theory. This is true not only for Zermelo-Fraenkel Set Theory, but also for various weaker and stronger systems. So this is a clear advantage of the Axiom of Choice: several things become simpler and we do not jeopardize the consistency of the system.

The Axiom of Determinacy is different. The axiom was proposed by Jan Mycielski and Hugo Steinhaus in 1962 [204]. At that time it was a very bold proposal. Being inconsistent with the Axiom of Choice and having strong consequences suggested that it should be inconsistent even if we drop the Axiom of Choice. To explain its role I need to talk about Zermelo-Fraenkel Set Theory without the Axiom of Choice. It would very cumbersome to use such a long name for this theory. So recall that Zermelo-Fraenkel Set Theory (with the Axiom of Choice) is abbreviated by *ZFC*, and *ZF* stands for the theory that we need now, that is, without the Axiom of Choice. For several decades all the attempts to prove its relative consistency with respect to *ZF* failed. It became clear very soon that it was not going to be as simple as with the Axiom of Choice. For example, Solovay proved that the Axiom of Determinacy implies that $\aleph_1$ is a measurable cardinal. In the presence of the Axiom of Choice the existence of a measurable cardinal is a very strong assumption, but how strong this assumption is without the Axiom of Choice is not immediately clear; $\aleph_1$ is the second smallest infinite cardinal, so we are not talking about large cardinals. Nevertheless, this assumption is as strong as the assumption that *ZFC* is consistent with the existence of a measurable cardinal.

In 1984 Woodin proved the relative consistency of *ZF* with the Axiom of Determinacy with respect to *ZFC* with a large cardinal assumption. This means that *ZF* with the Axiom of Determinacy is consistent provided that *ZFC* with the large cardinal assumption is. This is one of the deepest results obtained in set theory. Later, Woodin was not only able to determine which large cardinals suffice to prove it, but he determined exactly the strength of the assumption that *ZF* with the Axiom of Determinacy is consistent. To this end he introduced new large cardinals, called

Woodin cardinals. This result has vindicated the large cardinal theory again, and spurred its further development.

If compared with those we have mentioned, the least Woodin cardinal lies between the least measurable cardinal and the least Vopěnka cardinal, so Woodin cardinals are fairly large. Thus the Axiom of Determinacy is not so safe with respect to a possible contradiction as the Axiom of Choice, but now we know precisely how safe it is. If somebody succeeds in deriving a contradiction from a large cardinal, we will know immediately, if this affects the Axiom of Determinacy. For example, if such an inconsistent cardinal is higher in the hierarchy, we can still use *ZF* with the Axiom of Determinacy.

Solovay's Model

We see that there are two possibilities for set theories each having its own advantages, but they are incompatible. The one that is commonly accepted nowadays is *ZFC* (Zermelo-Fraenkel Set Theory with the Axiom of Choice). It has a nice general set theory, but reals are rather awkward. The other is *ZF* with the Axiom of Determinacy. The theory with the Axiom of Determinacy has other drawbacks, the main problem being that we are not quite sure that it is consistent. To prove that the Axiom of Determinacy is consistent we have to use a very strong assumption. Maybe, after all, we should abandon our hopes for having all subsets of reals be Lebesgue measurable? No, there is another option. What is appealing about the Axiom of Determinacy is not so much the axiom itself, but its consequences such as the measurability of all subsets of reals. If we only want to get those, we do not need Woodin cardinals, it suffices to use a very mild extension of *ZFC*, namely, *ZFC* with an axiom postulating the existence of an inaccessible cardinal. This was shown by Solovay already in 1964;[17] the paper appeared in 1970 [278]. More precisely, he showed:

Theorem 10 *If ZFC with the axiom saying that an inaccessible cardinal exists*[18] *is consistent, then so is the theory with the following axioms*:

1. *all axioms of ZF (Zermelo-Fraenkel Set Theory without the Axiom of Choice)*;
2. *the Axiom of Dependent Choices (a practical version of the Axiom of Choice which is weaker than the unrestricted Axiom of Choice, but which implies the Axiom of Choice for countable sets)*;
3. *all sets of real numbers are Lebesgue measurable*;
4. *every set of reals has the Baire property and the perfect subset property.*

The last two are topological properties. It is not important what they precisely mean; the point is, as with measurability, that sets of reals behave nicely.

[17] See [147], page 132.

[18] S. Shelah proved that even if we only want the Lebesgue measurability, the inaccessible cardinal is needed [266].

This was the first step in the series of results that eventually led to the proof of the consistency of the Axiom of Determinacy with respect to a large-cardinal assumption. In fact, determinacy implies the last two conditions and the countable Axiom of Choice (for subsets of reals). Furthermore, relative to the large-cardinal assumption, the statements above are all consistent with determinacy. So the advantage of this result is that it only needs the existence of an unaccessible cardinal, which is a very small extension of *ZFC*. Theoretically it is possible that *ZFC* is consistent and *ZFC* with an inaccessible cardinal is not, but people will not trust *ZFC*, if it turns out that inaccessible cardinals are not consistent.

Solovay's model might be a nice world to live in if ones concern is analysis, measure theory and alike. But there are other branches of mathematics that do not depend on the real numbers and where many results have been derived using the Axiom of Choice. Therefore, before moving to this 'new nice world' we would have to see what we need to abandon because of loosing the general Axiom of Choice, or rather, what we can substitute for the lost results.

Notes

1. *The Banach-Tarski Paradox.* There are many expositions of this result; the proof below is based on Laczkovich [175].

   To prove the theorem we will first sketch a proof of the following often used result (usually attributed to Cantor and Bernstein, sometimes also to Banach and Schröder). It should be noted that the proof of this theorem is fully constructive, in particular, it does not require the Axiom of Choice.

   Theorem 11 *Given a one-to-one mapping f from a set X into a set Y and a one-to-one mapping g from Y to X, it is possible to construct a bijection between X and Y.*

   Proof To simplify the proof we can assume without loss of generality that Y is a subset of X and g is the identity mapping. To visualize the proof think of X as a disc in the plane and Y as a square contained in it and f simply a contraction of X to a smaller disc Z contained in Y (see Fig. 3.8). Our task is to map the disc X onto the square Y so that no two points go to one point. First we observe that what is between the square Y and the disc Z does not have to move. The points outside Y have to move, so we use f to move them inside. The image of this part, say U is the disc Z with the image of the square of Y cut out. The points in U have to move inside, as we have already mapped something on them. But the situation is exactly the same as at the beginning. So we will continue on and on in this way, see Fig. 3.9. □

   What is actually going on in the proof is that we decompose the disc X into two parts. Then one part is kept, namely, we use the identity, while the other is

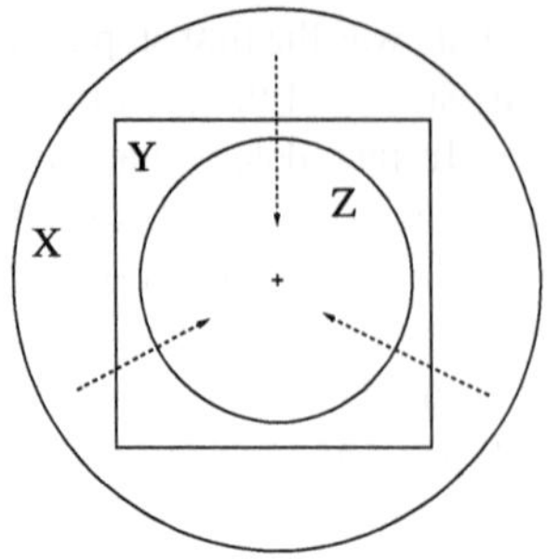

Fig. 3.8 The *square Y* is embedded into the *disc X* by the identity mapping; the *disc X* is embedded into the *square Y* by the mapping shown by the *arrows*; the smaller *disc Z* is the image of the bigger disc

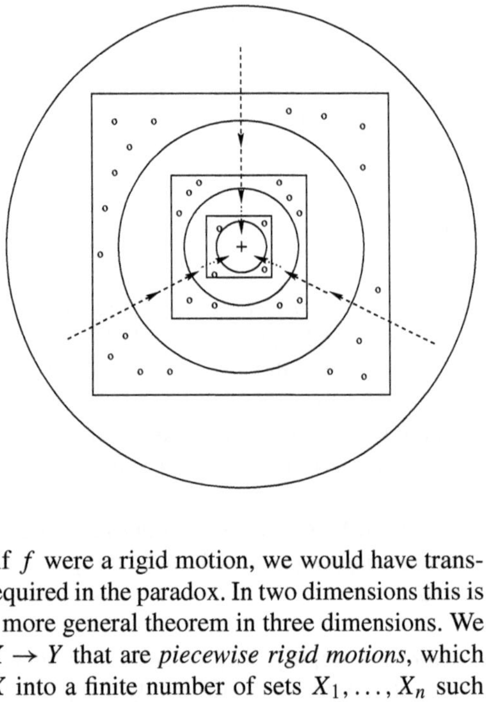

Fig. 3.9 The discs with squares cut off are mapped on the same shapes of the next smaller size (indicated by *arrows*); the squares with discs cut off are mapped on themselves (indicated by *loops*)

moved by the mapping f. Thus if f were a rigid motion, we would have transformed the disc into a square as required in the paradox. In two dimensions this is not possible, but one can prove a more general theorem in three dimensions. We need to consider mappings $f : X \to Y$ that are *piecewise rigid motions*, which means that we can decompose X into a finite number of sets $X_1, \dots, X_n$ such that f is a rigid motion on each set X_i.

Theorem 12 *Let X and Y be subsets of* $\mathbb{R}^3$. *Let* $f : X \to Y$ *and* $g : Y \to X$ *be one-to one mappings which are piecewise rigid motions. Then it is possible to construct a bijection between X and Y which is also a piecewise rigid motion.*

Having this theorem, our task looks much simpler.

Free algebras were mentioned in Chap. 2 (see page 91). Here we need one concrete example of such a structure, the free group F_2 generated by two generators. Let the generators be a and b. The group can be represented by the set of expressions of the form

$$a^{i_1}b^{j_1}a^{i_2}b^{j_2}\dots a^{i_n}b^{j_n},$$

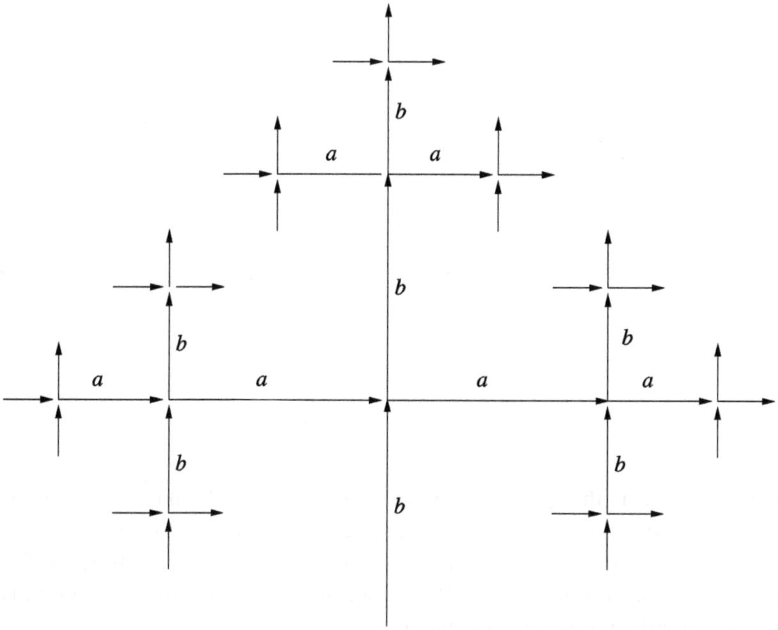

Fig. 3.10 Cayley diagram of the free group F_2. The decreasing size of the *arrows* indicates that it is a process going to infinity, but when embedded on the sphere, the lengths of the arcs will be the same

where n is a natural number and i_k and j_k are integers different from 0, except possibly for i_1 and j_n. If $n = 0$, the expression is empty, so we rather denote it by 1, the unit of the group (see Fig. 3.10). F_2 contains as a subgroup the free group with one generator F_1; for example, the powers of a is such a subgroup. F_1 is isomorphic to the additive group of integers. We used it in the construction of a nonmeasurable set, where we embedded F_1 into the group of rotations of the circle, by assigning the rotation ρ to a. Now we will use an embedding of F_2 into the group of rotation of the sphere S. (We cannot embed it into the group of rotations of a circle, since that group is commutative, while F_2 is not. That's why the paradox cannot be done in two dimensions.)

To embed F_2 we take two orthogonal rotations by the same angle α such that $\cos\alpha$ is a transcendental number; we will use the same letters a and b also for the chosen rotations. It can be proved that this ensures that different elements of F_2 are represented by different rotations. Let G be the group of the rotations that represent F_2. Let x be a point on the sphere S. Again we take the orbit of x as all points to which x can be moved using a rotation from G. A typical orbit of x can be visualized as a drawing of the Cayley graph of F_2 on the sphere. All the arrows have the same length and at every vertex of the graph the arrows labeled a are orthogonal to the arrows labeled b. There are also some exceptional orbits, namely, the orbits of points which are fixpoints of some rotations from G. For every rotation, there are exactly two fixpoints, the points where the axis of the

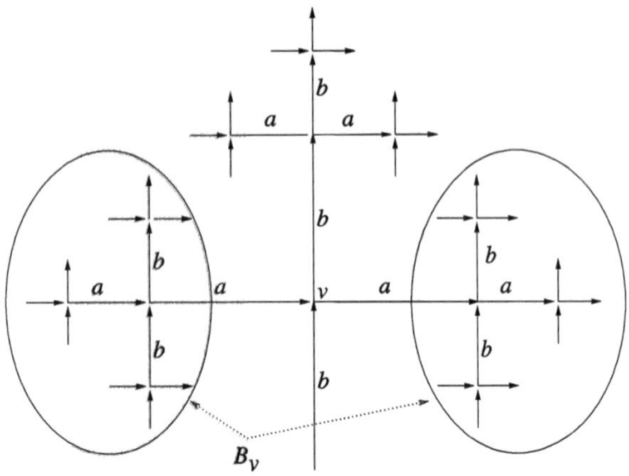

Fig. 3.11 Sets B_v

rotation intersects the sphere. As there are only countably many elements in the free group, there are only countably many such fixpoints, whence there are countably many exceptional orbits and countably many points on them. Let C be the union of the exceptional orbits. The presence of exceptional orbits complicates the proof a little, but it is not essential.

The next step is similar to the construction of a nonmeasurable set on a circle. We choose a representative from every unexceptional orbit. As noted before, this is the step where the Axiom of Choice is needed. You may be a little confused, as the Cayley graph in the picture seems to contain a distinguished vertex, the vertex in the middle, but this is only because of the drawing. In reality all the points on the orbit are equivalent, there is no way to distinguish between them. We know this positively, since the Banach-Tarski paradox cannot be done without the Axiom of Choice.

Consider an unexceptional orbit O and its representative v. Viewed as the Cayley graph of F_2, the orbit consists of four parts connected to v. Let B_v be the union of the two parts connected by arrows labeled a to v (see Fig. 3.11). First observe that by repeated application of rotation b we get infinitely many *disjoint* copies of B_v. Secondly, by shifting it only once by a we cover all the remaining points of O, thus $O \subseteq B_v \cup a(B_v)$. The chief point of the construction is that we can do these operations in parallel in all unexceptional orbits. Namely, let B be the union of all B_v's. Then by rotating by multiples of b *we can embed infinitely many copies of B in the sphere.* On the other hand, the sphere can be covered by two copies of B and the countable set C. As there are only countably many distances between points in C, there exists a rotation which moves $S \setminus C$ so that it covers C. Thus we can conclude, that S can be covered by the union of 4 copies of B. This implies that *we can cut S into 4 pieces, each of which fits into B.*

The two emphasized relations between S and B enable us to construct the decomposition of the ball K surrounded by the sphere S. Let B' be the section of the ball K determined by B with the center of K excluded. Given two copies of

K we can cut them into 8 pieces and two points, such that each piece fits into 8 copies of B'. But we can cut out infinitely many copies of B' from a single ball, so we certainly have 8 copies plus a lot of points left.

If we construct the two new balls from one in this way, there will still be some material left from the original ball. To get a decomposition where nothing is left we finally apply the Cantor-Bernstein construction.

2. *Two Solovay's theorems on measures.* To avoid confusion, I repeat the two results of Solovay mentioned above.

   The first theorem talks about general measures on the set of real numbers, measures that do not have to be translationally invariant and about *ZFC*, Zermelo-Fraenkel Set Theory with the Axiom of Choice.

Theorem 13 *The existence of a measure on the real numbers is consistent with ZFC if and only if the existence of a measurable cardinal is.*

The second theorem talks about a very special measure, the Lebesgue measure and about *ZF*, Zermelo-Fraenkel Set Theory *without* the Axiom of Choice.

Theorem 14 *If the existence of an inaccessible cardinal is consistent with ZFC, then ZF is consistent with the following statements*:

a. *all subsets of the real numbers are Lebesgue measurable,*
b. *the Axiom of Dependent Choices,*
c. *every set of reals has the Baire property and the perfect subset property.*

The Axiom of Dependent Choices says that if R is a binary relation on a nonempty set such that for every x, there exists y such that xRy, then there exists an infinite sequence $x_1Rx_2Rx_3Rx_4\ldots$. This axiom is weaker than the general Axiom of Choice, but it suffices for most proofs in calculus. A set is *perfect* if it is nonempty, closed and has no isolated points. A set has the *perfect subset property* if it is either countable or it contains a perfect subset. Any perfect set is of cardinality $\mathfrak{c}$ (the cardinality of the real numbers). Hence if every set of reals has the perfect subset property, then every set of reals is either small, meaning countable, or large, meaning of cardinality $\mathfrak{c}$. So in this sense the Continuum Hypothesis holds in Solovay's model. The more common way of stating the Continuum Hypothesis is $\mathfrak{c} = \aleph_1$, but this statement is false in the model because the real numbers cannot be well-ordered. (We will not define the Baire property.)

3. *Woodin cardinals.* Woodin cardinals are defined using elementary embeddings of V into transitive class models M. (Models that are proper classes are also called *inner models*.) A Woodin cardinal is a cardinal κ such that for every function $f : \kappa \to \kappa$, there exists a cardinal $\lambda < \kappa$ that satisfies the following two conditions:

   a. for all $\beta < \lambda$, $f(\beta) < \lambda$;
   b. there exist an M and an elementary embedding $j : V \to M$ with a critical point λ such that $V_{j(\lambda)} \subseteq M$.

The least Woodin cardinal is below a superstrong cardinal and above a measurable cardinal.

4. *The consistency of the Axiom of Determinacy.* Soon after people started working on the consistency of determinacy it was proposed to prove that the $(L(\mathbb{R}), \in)$ satisfies the Axiom of Determinacy assuming a large cardinal. It turned out to be the right approach. Indeed, in 1984 Woodin proved that assuming the existence of I0, the Axiom of Determinacy holds in $(L(\mathbb{R}), \in)$ (see [147], page 449). His subsequent work enabled him to reduce the large-cardinal assumption to a substantially weaker one and eventually he determined precisely the strength of the statement that *ZF* with the Axiom of Determinacy is consistent [309].

Theorem 15 *The following theories are equiconsistent* (*meaning that one is consistent if and only if the other is*):

a. *ZF with the Axiom of Determinacy*;
b. *ZFC with the assumption that there exist infinitely many Woodin cardinals.*

To prove that the Axiom of Determinacy holds in $(L(\mathbb{R}), \in)$, one needs a stronger assumption: there exists a measurable cardinal with infinitely many Woodin cardinals below it. If $(L(\mathbb{R}), \in)$ satisfies the Axiom of Determinacy, then we also obtain other nice properties of the real numbers, as mentioned above.

To give an example of an application of the Axiom of Determinacy, we will show that the perfect subset property for the Cantor discontinuum follows from the axiom. The Cantor discontinuum will be represented by the set of all countable strings of zeros and ones $\{0, 1\}^{\omega}$. We will use the modification of the game in which Player 1 plays arbitrary strings of zeros and ones (including the empty string) and Player 2 plays only one bit—zero or one. Let A be an arbitrary subset of $\{0, 1\}^{\omega}$. We will think of A as the winning positions for Player 1, so A defines a game. According to the Axiom of Determinacy we have two possibilities, either Player 1 has a winning strategy, or Player 2 has a winning strategy. We will show that any winning strategy for Player 1 determines a perfect subset of A and any winning strategy for Player 2 determines a countable enumeration of A.

The first statement follows immediately from the definition of a winning strategy. Consider all the infinite strings that can appear when Player 1 uses his strategy. All these strings must be in A, since it is a winning strategy for Player 1. Now look at the structure of these strings. They form an infinite binary tree whose vertices (the places where the strings branch) are places where Player 2 plays. The tree must branch there because Player 2 can choose any of the two values. The fact that this is a perfect subset is just the property that any branch of the tree is in A.

Now suppose that Player 2 has a winning strategy σ. Let $a \in A$ and let us play (simulating Player 1) against the strategy σ as follows. Our first move will be the shortest initial segment of a such that when Player 2 extends it by one bit determined by the strategy σ, the resulting string will still be an initial segment of a. If no such initial string exists, we stop. Otherwise we continue in the same way: we play the shortest string so that the extension determined by σ will still be in a, and repeat it as long as such strings exist. This has to stop after finitely many steps because if we could play in this manner for infinitely many steps, the resulting sequence would be outside of A, since σ is a winning strategy for Player 2, but, at the same time the resulting sequence is $a \in A$. So let us take the finite initial part that we have produced from A in this way. Let $f(a)$ be the set of indices of the bits played by Player 2. It is a finite set of natural numbers, hence if we show that these sets are different for different elements of A, we will have a one-to-one mapping f from A into a countable set. Indeed, let $a, b \in A$ be two different strings. Let n be the first index of a bit on which they differ. Then, clearly, n is exactly in one of the sets $f(a)$ or $f(b)$, thus $f(a)$ is different from $f(b)$. Thus if Player 2 has a winning strategy, A is countable.

3.5 Alternative Set-Theoretical Foundations

We can view the history of Zermelo-Fraenkel Set Theory either as the history of discovering the true principles of set theory, or as the history of one particular axiomatic system that became the standard because of pragmatic reasons. Many alternative systems have been proposed, but it does not make much sense to survey them all without explanation. Therefore, we will rather focus on two and explain them in more detail. They are not real competition to Zermelo-Fraenkel Set Theory, nevertheless, they are quite interesting and everybody seriously interested in the foundations of mathematics should know about them. These systems are based on classical logic. There have been more radical alternatives proposed based on different logics, but I am not going to talk about those.

It is good that mathematicians are united in their opinion of what are the foundations and they accept Zermelo-Fraenkel Set Theory as the foundations. But we have witnessed above that the things are not that bright. Zermelo-Fraenkel Set Theory, as any reasonable axiomatic system, is incomplete and there is no common agreement about how to decide some independent sentences such as the Continuum Hypothesis. Mathematicians left the decision to logicians; logicians usually only give vague arguments that set theory in which the Continuum Hypothesis fails is richer, but no definite answer. The Axiom of Choice is nowadays accepted, but there still remain some doubts about it because there are better alternatives for studying subsets of the real numbers. Only the large cardinal theory, at least so far, does not lead to such dilemmas. Thus our inability to decide the Continuum Hypothesis and the presence of alternatives to the Axiom of Choice should be a warning that we should not be so sure about being right in choosing Zermelo-Fraenkel Set Theory. Studying alternatives may be fruitful, as it was within Zermelo-Fraenkel Set Theory, where the Axiom of Determinacy, an alternative to the Axiom of Choice, opened new areas of descriptive set theory and the large cardinal theory. Moreover, studying alternatives to Zermelo-Fraenkel Set Theory may help us to justify the position of Zermelo-Fraenkel Set Theory, when it turns out that other theories have too many negative properties.

Mathematicians should be informed about alternative approaches to the foundations, but it is naive to think that they will easily switch from Zermelo-Fraenkel Set Theory to something else. It is not their conservatism, or their dislike of logic, there are deeper reason. The most important one is that Zermelo-Fraenkel Set Theory is a very strong theory. In this theory it is possible to model essentially all known situations very efficiently. Formally, it means that we can construct structures with given properties, provided that the requirements are consistent.

Quine's New Foundations

New Foundations is an axiomatic system for set theory conceived by Willard Van Orman Quine (1908–2000) in the 1930s. The name comes from the title of the paper *New Foundations for Mathematical Logic* [231] in which it first appeared. It can be briefly characterized as an attempt to make the Theory of Types attractive to mathematicians while preserving its apparent consistency; unfortunately, it has not quite succeeded in either of its goals. Though more recently the concept of typing has got much attention in programming, for ordinary mathematicians it is rather artificial. Once there is a system, namely Zermelo-Fraenkel Set Theory, which works well without this burden, the systems using types must offer some additional advantages. So the idea is to give up types, but keep the property that seems to ensure the consistency of the theory. According to Russell, paradoxes are caused by 'the vicious circle' and he suggested to solve this problem by introducing the doctrine of types. Quine realized that the problem may only occur when we apply the comprehension axiom, thus only then we have to be careful. So he proposed to get rid of types and instead to restrict the comprehension axioms so that they would only allow the form of formulas that one can use in the Theory of Types.

Formally, the theory can be presented as having the Axiom of Extensionality and a Comprehension Axiom Schema restricted to certain formulas. Roughly speaking, an instance of a comprehension axiom schema is admissible, if we can assign types to all variables occurring in the axiom so that it becomes an instance of such an axiom in the Theory of Types. Practically it amounts to assigning integer indices to the variables of the formula in such a way that whenever two variables occur in the formula in relation $\in$, the index on the right-hand side must be larger by one and if they occur in an equality the indices must be same. Such formulas are called *stratified.*

Examples 1. The formula $x = x$ is stratified because we can assign type 1 to x and then we have the same type on both sides of the inequality. According to the Comprehension Axiom Schema restricted to stratified formulas there exists a set V such that

$$x \in V \equiv x = x,$$

which is the universal set (the set of all sets). Similarly, the stratified formula $x \neq x$ gives us the empty set.

2. The formulas $x \in x$ and $x \notin x$ are not stratified because for any assignment we have the same type on both sides of $\in$. Therefore we cannot reproduce Russell's paradox.

It should be noted that non-stratified formulas are not disallowed in general. One can use any formulas in proofs. Furthermore, some sets which have only non-stratified definitions are proved to exist by indirect arguments.

Example The formula $x \in y \vee x = z$ is stratified (assign 1 to x and z, and 2 to y). Hence for every y and z, there exists the set $y \cup \{z\}$. This does not exclude $y = z$, so we have $y \cup \{y\}$ for every y, although $x \in y \vee x = y$ is not stratified.

The main objection to this system is that the restriction of the comprehension schema is not given by a general set theoretical principle (such as 'small sets are consistent', used in Zermelo-Fraenkel Set Theory), but a syntactical condition. So to say, you cannot teach set theory based on this system without talking about logic. Recall that the universe of the Theory of Types and the universe of Cantorian set theories such as Zermelo-Fraenkel Set Theory are not very different. The universe of Zermelo-Fraenkel Set Theory is also divided into levels (namely V_α); the difference is only that the hierarchy is transfinite and cumulative (level V_α contains V_β for all $\beta < \alpha$). So typing is not unnatural even from the point of view of Cantorian set theories. Therefore, the condition used to restrict the comprehension schema is also not as unnatural as it may seem at first glance. The main argument for New Foundations is the fact, proved by the Swiss mathematician Ernst Specker (1920–2011), that New Foundations are equivalent to *the Simple Type Theory with an additional axiom saying that all types are isomorphic* [283]. A model of this theory is a structure that we can easily imagine, but, unfortunately, are not able to construct.

If we disregard the objection about the syntactical nature of the restriction on the Axiom of Comprehension, the theory looks quite appealing. It has some means that we are lacking in Zermelo-Fraenkel Set Theory. We have the universal set, we can define the number n to be the set of all sets of cardinality n, we can prove the existence of the set of natural numbers without augmenting the theory with the Axiom of Infinity, the Power-Set Axiom is an instance of the schema and so on, while in Zermelo-Fraenkel Set Theory we had to add several other axioms on top of the restricted comprehension schema. Furthermore, New Foundations is finitely axiomatizable. J.B. Rosser has shown that mathematics can be founded on this system very well [250].

Clearly, there must be aspects of the theory in which it deviates from Cantorian set theories. One such an anomaly is the fact that, in general, the natural bijection that to every element x of a set A assigns the set $\{x\}$ (the set containing x as the unique element) cannot be represented by a set, unless New Foundations is inconsistent. This is to be expected because such a set would allow us to circumvent the restriction to stratified formulas. One can show that among sets that are in some sense small such anomalies do not arise. One can also add axioms that guarantee that some concrete sets are small in this sense. However, this cannot be used as an argument for accepting New Foundations; on the contrary, it shows that it is better to stick to *ZFC* which uses the paradigm of small sets explicitly.

The most intriguing problem is the consistency of New Foundations. The Theory of Types is quite weak and its consistency is readily provable in *ZFC*. New Foundations looks like an extension of the Simple Type Theory, so one would expect that it is not much stronger. But so far nobody has succeeded in proving its consistency in *ZFC* or its extension. The canonical way of proving the consistency of a theory in *ZFC* is to construct a model of the theory, a structure that satisfies the axioms of the theory. Because of the rather strange (or just unusual) way in which sets behave in New Foundations, to construct such models directly seems hard. Instead one can use Specker's reduction to the construction of a model of New Foundations to a model of the Simple Type Theory with an isomorphism between the types. Having this reduction, the task of proving the consistency of New Foundations does not look so hopeless; nevertheless, the problem has remained open since 1937!

The fact that we are not able to prove the consistency of New Foundations relative to *ZFC* and its extensions is rather alarming. Quine was hoping that the theory would be safe against antinomies because it uses the Theory of Types as a paradigm, but suddenly we are confronted with a theory that we cannot show to be consistent using means that suffice for all other theories. Furthermore, the advantages that it offers are not so great. While, for example, the Axiom of Determinacy has dramatic consequences for subsets of reals, no such results are known about New Foundations. Moreover, almost nobody uses the theory, so the practical experience is close to zero. Put otherwise, in New Foundations sets behave quite differently than we are used to, hence there may exist simple proofs that we are overlooking, and there may even exist such a simple proof of contradiction. T. Jech [137] suggested that in such a case it may be better to let a computer look for inconsistency and he actually performed such experiments. That he did not find a contradiction does not mean much, as the current theorem provers are still very inefficient compared to a human. The possibility of New Foundations being inconsistent is still very real.

New Foundations is often considered to be only a nice curiosity in logic with no practical relevance to mathematics. If a contradiction is eventually found, it will probably be forgotten completely. If a model is found in *ZFC*, then it may encourage more people to study it, and some may even use it for doing mathematics, but in that case it would only strengthen the general belief that *ZFC* is omnipotent. From the point of view of foundational studies, the most interesting case would be if New Foundations were consistent and there were no proof of it in *ZFC* or its natural extensions. This is as it appears now, but, unfortunately, if this is really the case, we will never be sure that it is so. According to Gödel's second incompleteness theorem we can prove the consistency of New Foundations only relative to another theory, but if we cannot prove the consistency in *ZFC*, then what else can we do?

Robinson's Nonstandard Analysis and Vopěnka's Alternative Set Theory

'New Foundations' and 'Alternative Set Theory' may sound similar, like catchy names for new products, but the philosophy behind them is completely different. Vopěnka claims that, in fact, he did not invent a completely new theory, but simply reconstructed mathematicians' past views on the foundations of mathematics, mainly the views of Leibniz, who introduced differential calculus using infinitely small quantities. Infinitely small quantities helped the development of calculus and mathematicians have been using them in an intuitive way all the time. Though stimulating, the concepts of infinitely small and infinitely large led often to nonsensical results. When the formal foundations of calculus were being developed, no consistent formalization of infinitely small and infinitely large was available, therefore these concepts were completely eliminated. Instead an approach based on the concept of a limit was accepted. The usual definition of the limit starts with '*for every* $\varepsilon > 0$ *there exists* $\delta > 0\ldots$', thus this method is also called the ε–δ formalization. I prefer to use the latter term, since the limit itself can be defined by infinitely small quantities.

As an example, consider the definition of the limit of a function $f(x)$ at x_0. Roughly speaking, y_0 is the limit of $f(x)$ at x_0, if $f(x)$ tends to y_0 when x tends to x_0. Here are two possible definitions:

1. Using Infinitely Small Quantities *y_0 is the limit of $f(x)$ at x_0, if $f(x)$ is infinitely close to y_0 whenever x is infinitely close to x_0.*

2. ε–δ Definition *y_0 is the limit of $f(x)$ at x_0, if for every $\varepsilon > 0$, there exists $\delta > 0$ such that whenever the distance of x from x_0 is less than δ then the distance of $f(x)$ from y_0 is less than ε.*

It is interesting to note that the first definition talks about infinity directly, while in the second one it is only implicit. It is like using actual infinity in the first definition and potential infinity in the second. Although actual infinity is generally accepted, infinitely small quantities are not.

The first definition is clearer and it is the one that mathematicians originally used, but the problem is how to interpret the term 'infinitely close'. Contemporary mathematicians are so much used to the second approach that they would argue that the second one is clearer and that, in fact, if you try to make the first one precise, you end up with the second one.

For some period of time, the general opinion was that the only rigorous definition was the second one. When logic, in particular its branch model theory, developed into a modern mathematical area it turned out that it was possible to speak about infinitely small and infinitely large quantities in a way that does not lead to a contradiction. As envisioned by Leibniz, it is possible to extend the structure of real numbers so that it contains numbers that we can interpret as infinitely small and infinitely large numbers.

The basic idea is not difficult. First, realize that infinitely small quantities are reciprocals of infinitely large quantities and vice versa, thus only one of the two needs to be explained. We will use infinitely large quantities because then it suffices to speak about natural numbers. A positive real number is infinitely small, if it is less than $\frac{1}{n}$ for an infinitely large natural number n, and a real number is infinitely large, if it is larger than an infinitely large natural number.

To get infinitely large numbers we will consider a *nonstandard model* M of natural numbers. Recall that in such a model we have elements that interpret the standard natural numbers $0, 1, 2, \dots$, and furthermore we have others, which necessarily must be larger than the standard ones (see page 87). In spite of being different from the standard model, M can be constructed so that exactly the same sentences are true in M as in the standard model. The apparent paradox that the standard model and M, which is different from the standard model, satisfy the same axioms is explained by the limited possibility of the language that we consider. The properties that distinguish the two models are not expressible in the language. In particular, no formula can define the standard numbers in M. We say that *inside* of M this set is not definable. However, when studying the model, we do not have to restrict ourselves to a particular language. In particular we can speak about arbitrary subsets, including those that are not definable inside of M. So to say, looking at it from *outside* we see more. The standard numbers in M are also called *finite numbers;* the others are called *nonstandard numbers*, or *infinite numbers*, and will be used as the *infinitely large* numbers in definitions such as the one above.

Thus in a nonstandard model we can distinguish finite and infinite integers. This enables us to use the terms infinitely small and infinitely large in the sense that these concepts were used in the past. In particular, in the first definition we only need to define what it means to be infinitely close which is now clear: it means that the difference must be less than $\frac{1}{n}$ for some *infinite* number n. Recall that we cannot make the distinction between finite and infinite numbers using the original language of the model M; we say that these concepts are *external*, as opposed to *internal* concepts that are those that can be defined in the language of M. In order to talk about external concepts, we must use a richer language. This presents no problem, it suffices to add a predicate symbol whose interpretation is the set of finite numbers. Let us call this predicate *FN*. We are, of course, prepared for some complications, as nothing is for free. The main complication is that the concepts defined using *FN* behave differently than those defined using the original language. To see that just realize that the principle of mathematical induction fails for *FN* (it contains 0, it is closed under the successor function, but it does not contain all numbers). Therefore, one has to be careful and distinguish when only inner language is used and when an external concept is involved. In the second case we are not allowed to apply the principle of induction. I should stress that this restriction does not affect the internal concepts at all; we keep the same true statements as they are in the standard natural numbers. We have learned that in order to avoid some paradoxes, it is necessary to distinguish between the object language and the language that we use to describe the object. Thus the necessity to use a hierarchy of two languages here should not seem so alien to us.

So far we have only talked about natural numbers, which may have given the impression that we have to restrict ourselves to a particular language that uses only arithmetical operations. But this approach is very general. The model M can be a model of set theory, the real numbers, or anything you like. In such a case we do not have a canonical model, thus we use the distinction 'standard' and 'nonstandard' according to what the natural numbers in M look like. In order to be able to introduce the nonstandard concepts such as infinitely small numbers, we need a model M in which natural numbers form a nonstandard model of arithmetic. The construction of such a model does not present any problems.

Several logicians (including C.S. Pierce and A.A. Fraenkel) considered the problem of designing a structure that would represent infinitely small quantities. The problem was solved in 1960 by Abraham Robinson (1918–1974). He coined his method *nonstandard analysis*. The name is derived from the application of nonstandard models to "standard" problems, mostly in analysis. The most amazing thing is that it really works perfectly. One can prove, quite easily, that definition 1. is equivalent to definition 2. The same is true for all the main concepts concerning functional analysis, metric spaces and some parts of the measure theory. Robinson showed in his book devoted to this subject [247] that one can introduce the concepts of analysis and prove basic theorems using exclusively infinitely small numbers. The most impressive was his solution of a problem of P.R. Halmos and K.T. Smith, an open problem about infinite-dimensional Hilbert spaces. Robinson solved this problem using his nonstandard analysis. The basic idea is that the problem has a positive solution in finite-dimensional spaces. In a nonstandard model '*finite dimension*' is interpreted as '*dimension n, for some natural number n*', but then n can be also infinitely large. Hence the theorem is also true for spaces with dimensions finite in the model but in reality infinite. Let me stress that this is only a rough idea and considerable ingenuity was needed to obtain the result. (An exposition of the proof is in Notes.) Nonstandard analysis does not give an automatic process of reducing theorems on infinitely large entities to theorems about finite entities, but it gives us a different angle from which we can view problems and thus we can get more inspiration.

Robinson successfully applied his method in a number of other fields of mathematics ranging from number theory to mathematical economics, in which he also solved an open problem. He found many followers and a number of striking applications were found later. A notable example of these is J. Hirschfeld's nonstandard proof of Hilbert's fifth problem [131].

Vopěnka's work [301] can be viewed as a part of nonstandard analysis, but there is an essential difference between his approach and Robinson's. While Robinson thought of nonstandard analysis as a tool for proving theorems in the classical theory, Vopěnka proposed to use nonstandard analysis as the foundations of mathematics. Instead of studying nonstandard models in Zermelo-Fraenkel theory, he suggested stating the principles of nonstandard analysis as axioms and working in such an axiomatic system. Once we start in this way we are not dependent on Zermelo-Fraenkel Set Theory and we have the freedom to rethink all axioms of set theory. The aim is to develop a theory that would be close to the intuition we use in calculus and enable us to reason with infinitely small numbers. Since, according to Vopěnka, the intuition of contemporary mathematicians is greatly distorted by being educated exclusively in the ε–δ formalism, he studied the ideas of mathematicians of the past and his theory is inspired by the views that were common before calculus was formalized. His explanation of the name of the theory is as follows. When mathematics progressed to the point that it was necessary to set precise foundations for mathematical analysis, there were two main possibilities: one was to use the ε–δ formalization, the other was to develop a consistent theory of infinitesimals. The first one was chosen and the second one was completely abandoned. In Alternative Set Theory we are exploring the second *alternative*.

Here are the main principles of Alternative Set Theory.

The Axiom of Finiteness *Every set is finite.*

It should be noted that the word '*finite*' is used in Alternative Set Theory with a different meaning, and hence also the axiom is stated in a different way. This needs an explanation, but let us first look at Vopěnka's justification of this axiom.

In the real world we observe only finite entities, infinite sets are only abstractions. We use infinite sets in mathematics because without infinity we would loose most parts of it. In Alternative Set Theory we do not loose those parts because infinity is present there, but it is introduced in a different way, using nonstandard concepts. This is the crucial discovery, so let me spell it out:

> It is possible to introduce infinity in a totally different way than it is done in the classical Cantorian set theory.

According to Vopěnka, if mathematicians knew this possibility at the time when concepts in calculus were formalized, then perhaps Cantor's theory of infinite cardinals would not have been introduced, as large infinite cardinalities are artificial and they do not occur in the physical world.

The Axiom of Finiteness alone, of course, does not suffice to develop a theory of finite sets; we need some construction axioms. One possibility is to accept all axioms of Zermelo-Fraenkel Set Theory, except for the Axiom of Infinity. This is not in contradiction with what I said about being independent of Zermelo-Fraenkel Set Theory. As far as finite set theory is concerned, there is good agreement on what axioms should be in a basic theory of finite sets and Zermelo-Fraenkel Set Theory less the Axiom of Infinity is just one possible set of axioms that defines the natural theory that we call Finite Set Theory (see page 116). Finite Set Theory can be axiomatized in a simpler way, which is what Vopěnka did. In particular, instead of the Replacement Schema, he took the Axiom Schema of Induction (in a form adapted for finite sets, see Notes).

Alternative Set Theory contains also classes. As usual, it is more convenient to consider sets as being a special type of classes. Hence, all objects of the theory are classes and a class is a set, if it is an element of a class, otherwise it is called a *proper class*. In contrast to other theories a proper class does not have to be large. Vopěnka defined that a class X is a *semiset*, if it is a subset of a set. A class is a *proper semiset*, if it is a semiset, but it is not a set. (The concept of a semiset goes back to his earlier research, where he studied extensions of Zermelo-Fraenkel Set Theory allowing semisets. The purpose of that was to have a theory in which one could prove independence results without referring to models of Zermelo-Fraenkel Set Theory.) The existence of a proper semiset is another basic axiom.

The Axiom of a Proper Semiset *There exists a proper semiset.*

There are sets that do not contain a proper subsemiset. A trivial example is the empty set whose only subclass is the empty set itself. More generally, any concrete finite set does not contain a proper semiset. This suggests that sets that do not contain a proper semiset are small and only large sets can contain a proper semiset.

To distinguish small and large sets we will change our terminology. Since all sets are finite, the adjective 'finite' is superfluous in connection with sets. Therefore, from now on we will only use it for small sets. Thus we call sets that do not contain a proper subsemiset *finite* and those that do contain a proper semiset *infinite*. After this change, one must also state the Axiom of Finiteness in a different way.

The new terminology is in accord with the usage of these terms in nonstandard analysis.[19] The class of finite numbers is denoted by *FN*. This is a proper semiset, in particular it does not contain all numbers. One can easily show that a set is finite if and only if its cardinality is a finite number. Having numbers divided into finite and infinite ones, one can develop analysis in a similar way to Robinson's.

Vopěnka took pains to explain all axioms. The Axiom of a Proper Semiset needs special care, as this is the one which distinguishes the theory from all others. When justifying the Axiom of Finiteness, we claimed that actually observable sets are finite. Now we are introducing some sort of infinity, so we must demonstrate that it also occurs in the real world. An interpretation of proper semisets is given by some vague, or not precisely specified concepts. Such concepts are excluded from classical mathematics, but they are frequently used in ordinary speech. For example, a pile of sand, forest, etc. How many grains of sand constitute a pile? How many trees do you need to call it a forest? They concern finite, but big numbers. Others refer to discrete, but gradual changes. For instance, from which generation on should we call our ancestors *homo sapiens*? Such examples are almost ubiquitous. Thus sometimes Alternative Set Theory is promoted as a theory that is able to handle vagueness. I think this is not correct. We can use these examples as explanations, but the theory cannot be applied to practical situations where we need to handle vagueness. As we have observed, every concrete number that we write down is finite from the point of view of Alternative Set Theory and we can write down an upper bound on the number of grains of sand on the whole Earth (as already Archimedes did), thus there is no proper semiset of grains of sand on the Earth. In fact, we can never observe a proper semiset in the sense of the theory. So the relation to vague concepts cannot be taken literally.

---

[19]Note, however, that nonstandard analysis uses models, whereas here we are talking about a theory.

Another argument that Vopěnka gave to justify the Axiom of a Proper Semiset is that proper semisets can be used to explain some paradoxes, namely, Berry's paradox. For this paradox, the explanation is: there is no first number that cannot be defined by an English sentence with at most one hundred letters because *the class of numbers that can be defined in such a way is a proper semiset* and the induction principle fails for semisets. Unfortunately, there is the same problem. The set in which this semiset should be included, is a concrete set, its size is 27^{100}. Such sets do not contain proper subsemisets. So either we separate the language of the theory from the language that we use to argue about it and the paradox disappears in the same way, as in other theories, or we get a contradiction anyway.

The Prolongation Axiom *Every function defined on FN can be extended to a function which is a set.*

This is a rather delicate statement. Think of a function defined on the finite numbers as a procedure that is doing something useful for us. Then the axiom says, roughly speaking, that we can run this procedure successfully not only on finite numbers, but, in fact, on some initial segment of numbers $[0, n]$ where n is an infinite number. In spite of looking very technical, this axiom has a very natural explanation. Look at the milestones along a road going to the horizon, or at ties on a railway. You expect that the same pattern continues, at least some distance, beyond the horizon. What we see are the pieces enumerated by the semiset *FN*, the finite numbers. *FN* is a proper semiset, while roads and trucks are real objects. As real objects are sets, they cannot be restricted to *FN*, therefore roads and tracks must continue beyond our horizon, at least a little bit. Again, we cannot take this explanation literally. It is more persuasive, if instead of watching a landscape we talk about measurements performed in physics. When we investigate the micro-structure of space and matter, or aim our telescopes into the most distant parts of the universe, we are always limited by the instruments available. The horizon of our observations is not fixed, but due to the progress in science and technology it encompasses more and more. This brings new discoveries, new phenomena, but in most cases we assume that a small increase of our observational horizon will not require changing our basic physical theories.[20] The same can be said about some parts of mathematics. Consider number theory and the particular branch studying natural numbers. Most of the research concerns general laws, but recently a field called experimental number theory became popular because computers enable us to do experiments that were impossible before.[21] The part of the set of natural numbers that humankind, or any intelligent beings, will ever explore is small. Nevertheless, we firmly believe, as in other sciences, that what we have learned on a finite part of $\mathbb{N}$ extends beyond

[20] An important exception is high energy physics. Physicist hope to discover new particles by using more powerful accelerators.

[21] However, experimental computations with natural numbers are perhaps as old as mathematics itself.

these limits; in particular, the basic arithmetical laws should hold true for arbitrarily large numbers.

An important consequence of this axiom is that there are at least two infinite cardinalities: one is the cardinality of FN, another one is the cardinality of N. They correspond to $\aleph_0$, the countable cardinality, and $2^{\aleph_0}$, the cardinality of the continuum. Note that the existence of two infinite cardinalities is proved in an essentially different way than in Cantorian set theory.

The complete list of axioms is in Notes, page 250.

Most mathematicians are sceptical about the idea that an extensive use of nonstandard analysis may bring a revolution in some fields of mathematics. They agree that reformulating a problem using different concepts often helps, but when you are confronted with a really difficult problem, you need much more than that. What seems more promising is proving theorems that do not have a standard counterpart, but which have a lot of consequences in the standard world of mathematics.

Notes

1. *Categorical foundations.* As the name suggests, this is an approach where one uses the theory of categories instead of set theory. The key concept there is a *topos*. The idea of using topoi as a foundation of mathematics is due to F.W. Lawvere [181]. Topoi are a class of categories that are very universal in the sense that a lot of concepts can be formalized in them. In particular one can interpret Zermelo Set Theory in a topos satisfying some additional axioms (a *well-pointed topos*). In this setting it is also possible to reproduce the proofs of the independence of the Axiom of Choice and the Continuum Hypothesis [189].

   In principle, one can interpret the whole Zermelo-Fraenkel Set Theory and its extensions in topoi satisfying more axioms. However, it would be more interesting to find *natural* categorical axioms that transcend axioms proposed in set theory.
2. *The axioms of New Foundations.* The language has $\in$ as the only nonlogical symbol. The axioms are:

   The Axiom of Extensionality
   $$x = y \equiv \forall z(z \in x \equiv z \in y).$$

   Stratified Comprehension Axiom Schema
   $$\exists x \forall y\big(y \in x \equiv \varphi(y)\big),$$
   where φ is stratified and x does not occur in it.

   A formula is stratified, if it is possible to assign integer indices to its variables (all occurrences of a variable get the same index) so that for every subformula of the form $x = y$, the variables x and y have the same index and for every subformula of the form $x \in y$, the index of y is equal to the index of x plus 1.

3. *The relation of New Foundations to the Simple Type Theory.* Consider a model of the Simple Type Theory $(U_0, U_1, U_2, \ldots, \in)$ together with a function f which is a bijection between U_i and U_{i+1} and satisfies $x \in y \equiv f(x) \in f(y)$. In other words, f determines an isomorphism between every pair of structures $(U_i, U_{i+1}, \in)$ and $(U_{i+1}, U_{i+2}, \in)$ (hence iterations of f give an isomorphism between every $(U_i, U_{i+1}, \in)$ and $(U_j, U_{j+1}, \in)$). Such a model exists if and only if there exists a model of New Foundations.

   To prove the equivalence, first let M be a model of New Foundations. Let U_i be copies of M and f the natural bijection between the consecutive sets. To be more explicit, take $U_i = M \times \{i\}$ and $f((a, i)) = (a, i + 1)$. Then define the membership relation by setting $(a, i) \in (b, i+1)$ in the new model, if $a \in b$ holds true in M. The resulting structure is a model of the Simple Type Theory because every formula corresponds to a stratified formula in M, hence the comprehension axioms follow from the comprehension axioms for stratified formulas in M.

   Conversely, given a model of the Simple Type Theory $(U_0, U_1, U_2, \ldots, \in)$ with such an f, one defines a model of New Foundations by taking U_0 as the universe and setting $x \in y$ in this model, if $x \in f(y)$. Then stratified formulas in the new model are translated to well formed formulas of the model of the Simple Type Theory.

   To see how it works consider the following theorem of New Foundations: $\exists x(x \in x)$. It holds because the set of all sets contains itself. This is translated as $\exists x(x \in f(x))$ in the Simple Type Theory with an isomorphism. There it is true because the universal set of type i is mapped by f on the universal set of type $i + 1$ and the latter contains the former.

   Specker showed (using model theory) that this model-theoretical relation can be replaced by a syntactical relation. Namely, New Foundations are consistent if and only if the Simple Type Theory is consistent when we add a rule saying that all types are equivalent. More precisely the rule means that for any sentence φ, we add $\varphi \equiv \varphi^+$ as an axiom, where φ^+ is φ with all type indices increased by one.
4. *Jensen's model with urelements.* Of the variants of New Foundations that were proved to be consistent the most interesting is the one considered by R. Jensen [139]. It is also a nice example of an application of Ramsey's theorem.

   Suppose we want to construct a model of the Simple Type Theory with the additional Specker's axioms using the cumulative hierarchy $\{V_\alpha\}_{\alpha \in ON}$. It can easily be seen that we cannot take $V_\alpha, V_{\alpha+1}, V_{\alpha+2}, \ldots,$ as the ordinals $\alpha, \alpha + 1, \alpha + 2, \ldots$ are definable in the corresponding levels and they are distinguishable. However, it is possible to take $V_{\alpha_0}, V_{\alpha_1}, V_{\alpha_2}, \ldots,$ where $\alpha_0, \alpha_1, \alpha_2, \ldots$ are not consecutive ordinals. There is only one problem: there are sets in $V_{\alpha_{n+1}}$ that are not subsets of V_{α_n}, as they may also contain other elements which appear on the intermediate levels. If we follow the construction of a model of New Foundations of the previous paragraph, the only thing that breaks down is the extensionality axiom. But it does not break in too bad a way. The elements that appeared on the intermediate levels enter into the model as *urelements*. Recall that these are elements that do not contain other elements, in other words, they are not sets, they can only be elements of sets. This, of course, violates the extensionality, since the extensionality would make them all equal to the empty set. For other elements of the model, namely those which contain at least one element, the Axiom of Extensionality remains valid. Having urelements is not so unnatural. Things in the real world are urelements—we do not think of them as sets of atoms (if we did, then atoms would be urelements, etc.). Furthermore, urelements naturally occur, for instance, in the Theory of Types in which the bottom type consists of urelements.

Constructing the model in the form $V_{\alpha_0}, V_{\alpha_1}, V_{\alpha_2}, \dots$ would need large cardinals, but this is overkill; to prove the consistency, one needs relatively little. Let us take $V_\omega, V_{\omega+1}, V_{\omega+2}, \dots$ and let ϕ be a sentence of the Simple Type Theory which mentions types up to k. Now color every k-tuple $(n_1, \dots, n_k)$, $n_1 < \dots < n_k$ red, if ϕ holds in $V_{\omega+n_1}, \dots, V_{\omega+n_k}$, otherwise color the k-tuple blue. By Ramsey's theorem we can find an increasing sequence $i_1, i_1, i_2, \dots$ such that either the formula holds for all choices of k-tuples from this sequence, or it does not hold for any. Thus we have achieved for ϕ what we need for all sentences. By repeated applications of Ramsey's theorem we get it for any finite set of sentences. Once we have it for arbitrary finite sets, we have the consistency of all sentences by compactness.

Jensen's New Foundations with urelements seems to be close to the original system, but there are facts that indicate otherwise. In the modified system it is consistent to assume the existence of a well-ordering of V, which implies the Axiom of Choice, while in the original theory the Axiom of Choice has been disproved. Jensen's construction has the property that many sentences valid in *ZFC* are also true in this model; the Axiom of Choice is only one of them. If we want to construct a model of New Foundations along these lines in *ZFC*, we have to start with something that already captures the differences between *ZFC* and New Foundations. Hence a substantially new idea is needed.

The relation of the type-free λ-calculus to its typed version is similar to the relation of New Foundations to the Simple Type Theory. So the ideas used to construct models of the λ-calculus, perhaps, may be used to construct a model of New Foundations. Attempts of this kind have already been made.

5. *Hao Wang's system Σ.* One of the many systems based on the Theory of Types that have been proposed is the system Σ, due to the Chinese American logician Hao Wang (1921–1995) [302] (see also [305], Chaps. XXIII and XXIV). The details are not important, what I want to mention is one particular idea, which possibly can be used not only in connection with the Theory of Types. When proposing a system for the foundations of mathematics, our goal is to present it as a *formal system* so that the correctness of syntax (formulas, axioms, proofs, etc.) can be efficiently tested. The general consensus is that the right framework for formal systems is finite combinatorics, or if we want to be more specific, the combinatorics of finite sequences of symbols. This is also apparent from the rather obsolete term *symbolic logic*, for which we nowadays use the term *mathematical logic*. But one can question this restriction: why should elementary combinatorics of finite sequences be more basic than other mathematical concepts?

Consider, say, ordinals. The finite ones are natural numbers and they are always accepted. Now take the ordinals $\omega, \omega+1, \omega+2, \ldots$, the infinite ordinals less than $\omega+\omega$. They are considered bad, as they are infinite objects; already ω has infinitely many ordinals before it. Nevertheless, as a structure, where you do not analyze the substance of the elements, $\omega+\omega$ is as transparent as ω. Take a symbol representing ω, then all $\omega+n$ are only finite sequences. Surely, you cannot do the same for all countable ordinals, but whenever you have a system of notation for some segment of ordinals, you can think of those ordinals as finite objects. The best example is the segment of ordinals below ε_0. These ordinals can be represented in the Cantor normal form (see page 193) which is essentially a finite tree.

The cumulative hierarchy of sets in Zermelo-Fraenkel Set Theory suggests considering theories of types with transfinite types indexed by ordinals. If one has a system of notation for a segment of ordinals and restricts the types to this segment, the resulting system is perfectly formal. The advantage of using transfinite types is that we obtain stronger systems: the longer segment of ordinals we take, the stronger system we get. But how strong a system do we need and how big a segment should we take? For practical purposes, doing ordinary mathematics, we do not need much: Russell's Theory of Types, having only finite types, amply suffices. The appeal of this approach is not in that it enables us to create a sufficiently strong system, but in the possibility of avoiding the incompleteness phenomenon. Unlike in other systems, here, it seems, there exists a clear way of extending the system indefinitely by taking larger and larger constructive ordinals. Wang's proposal was to use all constructive ordinals as types. As a result the entire system is not constructive, but each part determined by the types up to a constructive ordinal is a constructive system. Thus one should rather think of it as an evolving system, in which we take larger and larger constructive ordinals, but never work with the entire system.

How strong is this system? Wang claimed that the consistency of Σ_α, the system restricted to types below the ordinal α, is provable in $\Sigma_{\alpha+2}$ and that the system avoids the basic problems of formal systems such as the possibility to be easily extended "*to a new system which at the same time contains new theorems and yet can be proved to be no less reliable than the original*", [305], pages 564–565.

All this looks like a wonderful thing, almost like achieving the ultimate goal of foundations, but optimism is not warranted. By using constructive ordinals we are only sweeping the problems under the rug. Taking larger constructive ordinals is not an automatic process. It requires a high degree of ingenuity to find a system of notations for higher constructive ordinals. Once we have it we are not finished, we have to *prove* that it determines a well-ordering. Where should we prove it? Moreover, there is an even more fundamental problem: what is a natural ordinal notation? We cannot use an arbitrary one because this would lead to trivialities. We will learn more about it in Chap. 7.

6. *Basic concepts of nonstandard analysis.* Let M be a mathematical structure. M is associated with a logical language L that contains the necessary means for

expressing statements true in M. In particular, if M is a structure of order n then so is the language. But we do not have to use higher order languages, since we can assume that M is a model of a theory that contains the concept of a set, hence sets are the usual first order elements of the structure.

Definition 6 An *enlargement* (also called *nonstandard extension*) of M is a structure $^*M$ and a mapping $x \mapsto {}^*x$ which satisfy the two conditions below.

1. $x \mapsto {}^*x$ is an elementary embedding[22] of M into $^*M$, in particular, the same sentences of L are true in $^*M$ as in M.
2. Let R_i, $i \in \omega$ be sets or relations of the same type definable in $^*M$ (however, we do not require that the *sequence* $\{R_i\}_{i\in\omega}$ is definable in $^*M$). Suppose that $\bigcap_{i=0}^{n} R_i \neq \emptyset$, for every $n \in \omega$. Then $\bigcap_{i\in\omega} R_i \neq \emptyset$.

Various modifications of condition 2. are used in the literature; here we are using the strongest form. In model theory this condition is called $\aleph_1$*-saturation;* in Alternative Set Theory the Prolongation Axiom is used instead of it.

For a set or a relation (of arbitrary type) R definable in M, we will denote by $^*R$ the relation defined in $^*M$ by the same formula that defines R in M. Thus the mapping $x \mapsto {}^*x$ is naturally extended to all entities of M.

When having an embedding it is often a good idea to identify elements with their images, but here it is better to do so only for some elements. In particular, I will assume $n = {}^*n$ for all standard integers, but not for real numbers. The general rule is that we put $x = {}^*x$ for elements that have finite structure and otherwise not. The explication of having finite structure is that the element is represented by a hereditarily finite set.[23]

In nonstandard analysis we use the pair $(M, {}^*M)$ to prove theorems about structure M. To this end we do not have to restrict ourselves to language L, on the contrary, in order to be able to distinguish the substructure M in $^*M$, we have to use more. We use L only when we want to deduce that a particular sentence holds in $^*M$. The concepts definable in L are called *internal*, all others are called *external.* According to condition 1., $^*M$ satisfies exactly those sentences of L that are true in M. We call elements of M *standard elements*, and elements in $^*M \setminus M$ *nonstandard elements.* As regards natural numbers, we also use terms *finite numbers* and *infinite numbers.*

For some elements one can define *the standard part.* In particular this can be done for finite real numbers. A real number in $^*M$ is *finite*, if its absolute value is less than a finite natural number. Let us use decimal representations of real numbers. Thus a real number $r \geq 0$ of $^*M$ is a sequence $\{r_n\}_{n\in{}^*\mathbb{N}}$, where r_n are numbers between 0 and 9 with one exception which is the decimal point. Then r is finite, if the point occurs on a finite position. For such an r, the finite part is simply the standard number represented by $\{r_n\}_{n\in\mathbb{N}}$.

[22] See page 212.

[23] See page 175.

Let me express it ‘pictorially’. Let

3.1415926... ...222090721... ...17012003...

be a finite nonstandard number. Then its standard part is

3.1415926...[24]

An infinitely small real number is a number whose absolute value is less than $1/n$ for every finite positive integer. Clearly, a real number is infinitely small, if and only if it is finite and its standard part is 0. I will say that r *is infinitely close to* s, and write $r \approx s$, if $r - s$ is infinitely small.

Let us consider some examples. For a standard function f and a standard real number a, we want to define the derivative of f at a. We define $f'(a)$ to be the standard part of

$$\frac{{}^*f({}^*a + \eta) - {}^*f({}^*a)}{\eta}$$

for $\eta > 0$ infinitely small, *provided that this number does not depend on the particular choice of* η. Thus the expression above defines a number which is infinitely close to ${}^*(f(a)')$. Let us prove the formula for differentiation of the product of two functions.

$$\begin{aligned}
&{}^*\big((f(a)g(a))'\big)\\
&\approx \frac{{}^*f({}^*a+\eta){}^*g({}^*a+\eta) - {}^*f({}^*a){}^*g({}^*a)}{\eta}\\
&= \frac{{}^*f({}^*a+\eta){}^*g({}^*a+\eta) - {}^*f({}^*a){}^*g({}^*a+\eta) + {}^*f({}^*a){}^*g({}^*a+\eta) - {}^*f({}^*a){}^*g({}^*a)}{\eta}\\
&\approx {}^*\big(f(a)'\big){}^*g\big({}^*a+\eta\big) + {}^*f\big({}^*a\big){}^*\big(g(a)'\big)\\
&\approx {}^*\big(f(a)'g(a) + f(a)g(a)'\big).
\end{aligned}$$

In the last step I am using $g(a + \eta) \approx g(a)$ which is the continuity of g at a (a consequence of the existence of the derivative at a). To get the derivatives we take the standard parts. Since the standard parts of infinitely close numbers are the same, the formula is proved.

Next consider infinite sums. We want to define the sum of a series of standard real numbers $A = \{a_n\}_{n=0}^{\infty}$. We define $\sum_{n=0}^{\infty} a_n$ to be the standard part of

$$\sum_{n=0}^{\kappa} {}^*a_n$$

for κ an infinitely large integer, *provided that this number does not depend on the particular choice of* κ. Thus the expression above defines a number which is infinitely close to ${}^*(\sum_{n=0}^{\infty} a_n)$.

---

[24] It is not clear that the two numbers can be ${}^*\pi$ and π; to this end one would have to prove that the strings 222090721 and 17012003 occur infinitely often in the decimal expansion of π.

The classical paradox of summation is based on the following series:

$$1-1+1-1+1-1+1-1+1-1\ldots$$

By distributing parentheses in two ways we get two contradictory conclusions,

$$(1-1)+(1-1)+(1-1)+(1-1)+(1-1)\ldots=0,$$
$$1+(-1+1)+(-1+1)+(-1+1)+(-1+1)\ldots=1.$$

Using our definition we also get these two values, depending on the parity of κ, but because of the proviso, it means that the sum does not exist.

Notice that in the example above we spoke about the parity of an infinite number. Apparently this is the thing that confused mathematicians who had considered the problem of using infinitesimals prior to Robinson. They knew the basic property of infinite cardinals, which is $x+1=x$, so how can one speak about the parity of x? In an enlargement the cardinality of sets $\{0,1,2,\ldots,\kappa-1\}$ may be the same for all infinite κ, but this cardinality and the position of κ in the structure are different things. The cardinality of $\{0,1,2,\ldots,\kappa-1\}$ is not definable in the enlargement $^*M$, it is an external concept, whereas parity is definable because it is an internal concept. We used a very simple property of integers (parity) for showing that the series above does not converge, in general, one may need much more complex properties.

A simple model in which one can talk about infinitesimals was known already in the 19th century. An ordered field is called *archimedian* if for every number x, there exists a natural number n such that $|x|\leq\bar{n}$, where $\bar{n}$ denotes the sum of n ones of the field. The first nonarchimedian field was constructed using formal power series. In a nonarchimedian field we have numbers that are finite and those that are infinite according to the aforementioned definition, thus one can develop some rudiments of nonstandard analysis. However, such a field does not have to contain an enlargement of the natural numbers, thus we cannot use such fields for infinite sums and similar more complex concepts. The crucial insight of Robinson was that infinite entities should have the same properties as finite ones. But to achieve this, one has to distinguish the languages that one uses to speak about them because infinite objects are different from finite ones.

7. *A theorem that was first proved by means of Nonstandard Analysis.*

Theorem 16 *Let T be a bounded linear operator on an infinite-dimensional Hilbert space H such that $p(T)$ is compact, for some a nonzero polynomial p. Then T has a nontrivial closed invariant subspace.*

Recall that a linear operator is *compact*, if it maps the unit ball onto a set whose closure is compact. A subspace L is *invariant* for T, if T maps L into itself, i.e., $T(L)\subseteq L$. A subspace is called *nontrivial*, or *proper*, if it is different from $\{0\}$ and from the entire space H.

This theorem was proved for $p(z)=z^2$ by Robinson which solved a well-known open problem. The proof was generalized to arbitrary nonzero polynomials by A.R. Bernstein and then published in a joint paper [15]. I will try to give a brief overview of the proof below.

First we recall a classical result of Schur.

Theorem 17 *Every complex square matrix A can be expressed in the form*

$$A = UTU^{-1},$$

where U is unitary and T is upper triangular.

Notice that this entails that in a finite-dimensional space *any* linear operator has a lot of nontrivial invariant subspaces. Indeed, the application of the unitary matrices means only an orthonormal change of the basis, thus we can assume M already is upper triangular. For such a matrix, the space of vectors $(c_1, c_2, \ldots, c_i, 0, 0, \ldots, 0)$ is an invariant subspace for every fixed i.

It is clear that we cannot use Schur's theorem to directly obtain the theorem about infinite-dimensional spaces because it does not mention compactness or polynomials. However, we can get something else that seems to be close to the solution. To this end let us make some preliminary considerations. Take σ a nonzero element of H and consider the set $\{\sigma, T\sigma, T^2\sigma, \ldots\}$. If the set is not linearly independent, then we get a nonzero finite invariant subspace. If it does not generate H (in the sense of Hilbert spaces, which means by infinite sums), then its closure is a proper invariant subspace. Thus we can assume in the rest of the proof that neither of the two holds, so the set is independent and generates H. Apply the Gram-Schmidt orthonormalization to this set and denote by $Z = \{\eta_1, \eta_2, \ldots\}$ the resulting orthonormal basis (in the sense of Hilbert spaces). For every n, the sets $\{\sigma, T\sigma, \ldots, T^{n-1}\sigma\}$ and $\{\eta_1, \eta_2, \ldots, \eta_n\}$ span the same subspace, whence $T\sigma, T^2\sigma, \ldots, T^n\sigma \in \{\eta_1, \eta_2, \ldots, \eta_{n+1}\}$. This means that if we take the infinite matrix $A = (a_{ij})_{i,j=1}^{\infty}$ that defines T in the basis Z, then A is *almost upper triangular*, meaning that $a_{ij} = 0$ for $i > j + 1$. Now I can state the key lemma.

Lemma 2 *There exists κ such that $a_{\kappa+1,\kappa}$ is infinitely small.*

The point is that if instead $a_{\kappa+1,\kappa}$ were zero, we would be done. The invariant subspace would be again the space of all vectors $(c_1, c_2, \ldots, c_\kappa, 0, 0, \ldots, 0)$. So we are closer again, but more work is needed, as one cannot eliminate $a_{\kappa+1,\kappa}$ completely (otherwise we would get a finite invariant subspace, which does not exist in general).

To state this lemma precisely and to prove it, we have to talk about enlargements. I will assume that an enlargement $^*H$ of H is given and that H is a subset of $^*H$. The latter assumption is perhaps not quite natural, but simplifies notation considerably. Also we have enlargements $^*\mathbb{N}$ of the natural numbers $\mathbb{N}$ and $^*\mathbb{C}$ of complex numbers $\mathbb{C}$. Elements of $^*H$ are represented by infinite sequences $\{c_n\}_{n\in{}^*\mathbb{N}}$ of complex numbers $c_n \in {}^*\mathbb{C}$.

Lemma 3 *Let $(b_{ij})_{i,j\in{}^*\mathbb{N}}$ be the matrix of a compact linear operator S. Then for every fixed i, b_{ij} is infinitesimal for every infinite j.*

If D is a compact subset of H, then for every $\varepsilon > 0$, there exists a finite subset $K_\varepsilon \subseteq D$ such that:

every element of D is in distance less than ε *from* K_ε.

The last statement is stated in the internal language, thus the same must hold in the enlargement. Since K_ε is finite we get:

every element of $^*D$ *is in distance less than* ε *from* K_ε.

Hence every element of $^*D$ is arbitrarily close to an element of H, in other words, every element of $^*D$ is infinitely close to an element of H. Now let $^*D$ be the closure of the set to which the unit ball is mapped by S. By definition $^*D$ is compact. Take a basis element η_j. This element is in the ball, so it must be mapped to an element that is infinitely close to an element of H. But all coordinates with infinite indices of an element of H are infinitely small. Thus also $S\eta_j$ has all coordinates with infinite indices infinitely small, which proves Lemma 3.

Now we can finish the proof of Lemma 2 very quickly. I will only do it for $p(T) = T^2$ leaving the easy generalization to arbitrary nonzero polynomials to the reader. Let (b_{ij}) be the matrix of T^2. Clearly, $b_{i+2,i} = a_{i+1,i}a_{i+2,i+1}$ because A is almost upper triangular. If i is infinite, then by Lemma 3, $b_{i+2,i}$ is infinitely small, whence either $a_{i+1,i}$, or $a_{i+2,i+1}$ is infinitely small.

The plan of the rest of the proof is as follows. Fix a number κ such that $a_{\kappa,\kappa+1}$ is infinitely small. Let E_κ be the subspace of elements whose coordinates with indices larger than κ are zero. Then T "almost leaves E_κ invariant"; more precisely, if $\xi \in E_\kappa$ is an element with finite norm, then $T\xi$ has $\kappa + 1$st coordinate infinitely small and all coordinates with larger indices are zero. We take an operator T' that leaves E_κ genuinely invariant and that is "infinitely close" to T. Referring to Schur's Theorem, we take invariant subspaces of T'

$$E_0 \subseteq E_1 \subseteq E_2 \subseteq \cdots \subseteq E_\kappa$$

such that $\dim(E_n) = n$, for $n = 0, 1, \ldots, \kappa$. We define $^\circ E_n$ to be the subspace of H of elements that are infinitely close to $E_n \cap H$. Then we show that every $^\circ E_n$ is invariant under T and at least one of them is nontrivial.

Here are more details. Let P denote the projection operator on E_κ. We define $T' = PTP$. Let $E_0, E_1, E_2, \ldots, E_\kappa$ be as above, and let P_n be the projection operators on E_n, for $n = 0, 1, \ldots, \kappa$. The following facts can be proved easily:

1. $^\circ E_n$, as defined above, are invariant under T;
2. $^\circ E_0 = \{0\}$ and $^\circ E_\kappa = H$;
3. $\dim(^\circ E_n/^\circ E_{n-1}) \leq 1$, for $n = 1, 2, \ldots, \kappa$;
4. if $\xi \in E_\kappa$ has a finite norm, then $p(T)\xi$ is infinitely close to $p(T')\xi$.

Here it may seem that the first three facts should be enough, but be careful. The mapping $E_n \mapsto {}^\circ E_n$ is not definable internally, as we speak about infinitely small elements in the definition! Therefore, we cannot guarantee that there exists the smallest n such that $^\circ E_n = H$.

Thus we need a different argument. Fix an element $\xi \in H$ of a finite norm, say $\|\xi\| = 1$, and consider the norms

$$r_n = \| p(T')\xi - p(T')P_n\xi \|,$$

for $n = 0, 1, \ldots, \kappa$. Then $r_0 = \|p(T')\xi\| > 0$ because if $p(T')\xi = 0$ then, by 4., $p(T)\xi$ would be infinitely small, hence also equal to zero because T and ξ are standard. But we assume that $\xi, T\xi, T^2\xi, \ldots$ are linearly independent, so it is not possible. Since ξ is standard, $\|\xi - P\xi\|$ is infinitely small (all coordinates of ξ with infinite indices are infinitely small). Hence

$$r_\kappa \leq \| p(T')\| \, \|\xi - P_\kappa\xi\| = \| p(T')\| \, \|\xi - P\xi\|,$$

is also infinitely small (T is standard and $\|P\| \leq 1$, so $\|p(T')\|$ is finite). Thus we can take the least n such that $r_n < r_0/2$. Now such an n exists because the numbers r_i were defined in the internal language of the expansion, i.e., the definition does not mention any infinitely small or large quantities.

We will show that $°E_n \neq \{0\}$ and $°E_{n-1} \neq H$. This immediately implies that either $°E_{n-1}$ or $°E_n$ is a proper subspace of H. If it were not, then the only remaining possibility would be that $°E_{n-1} = \{0\}$ and $°E_n = H$. But that is excluded by 3.

So suppose that $°E_n = \{0\}$. Let $\zeta = p(T')P_n\xi$. As $p(T')$ leaves E_n invariant, $\zeta \in E_n$. Furthermore, $\zeta \approx p(T)P_n\xi$ by 4. Since $p(T)$ is compact, there is a standard element $°\zeta \approx p(T)P_n\xi \approx \zeta$. According to $°E_n = \{0\}$ we have $°\zeta = 0$, thus ζ is infinitely small. But

$$r_n \geq \| p(T')\xi \| - \| p(T')P_n\xi \| = r_0 - \|\zeta\|,$$

which is in contradiction with $r_n < r_0/2$.

Now suppose that $°E_{n-1} = H$. Then $\|\xi - P_{n-1}\xi\|$ would be infinitely small. But $r_{n-1} \geq r_0/2$ is not infinitely small and $r_{n-1} \leq \|p(T')\| \, \|\xi - P_{n-1}\xi\|$. So this is also not possible. This finishes the proof.

Hilbert's fifth problem was solved by A. Gleason and by D. Montgomery and L. Zippin in 1952. The answer was given by proving the following theorem.

Theorem 18 *Every locally Euclidean group is a Lie group.*

In 1990 J. Hirschfeld published a proof of this theorem based on nonstandard analysis [131].

8. *The axioms of Alternative Set Theory.* The theory has one binary relation $\in$ denoting the membership. The objects of the theory are called *classes*. Those classes that are elements of other classes are called *sets*. As usual, in order to simplify notation, we distinguish sets from classes by using upper case letters for classes and lower case letters for sets. Note that Alternative Set Theory was proposed as an open system to which more axioms can be added. Thus the list of axioms below consists only of basic axioms and stronger systems can be found in the literature.

The Axiom of Extensionality *(as usual)*

The Axiom of the Empty Set $\exists x \forall Y (Y \notin x)$; *this set is denoted by* $\emptyset$.

The Axiom of the Set-Successors $\forall x, y \exists z \forall u (u \in z \equiv (u \in x \vee u = y))$; *this set is denoted by* $x \cup \{y\}$.

The Axiom Schema of Induction *For every formula* $\phi(x)$ *in which only sets are quantified,*

$$\big(\phi(\emptyset) \wedge \forall x, y\big(\phi(x) \to \phi\big(x \cup \{y\}\big)\big)\big) \to \forall x \phi(x).$$

Note that the Comprehension Schema for sets and the Axiom of Foundations are derivable from this schema.

The Axiom Schema of Comprehension for Classes *For every formula* $\phi(x)$ *which does not contain the variable* Y,

$$\exists Y \forall x \big(x \in Y \equiv \phi(x)\big).$$

A *semiset* is a class that is a subclass of a set. A *proper semiset* is a semiset that is not a set.

The Axiom of the Existence of a Proper Semiset *There exists a proper semiset.*

A class is *finite*, if it does not contain a proper semiset. *FN* is the class of all finite numbers. One can prove that *FN* is a semiset, hence there exist infinite numbers. A set is finite if and only if it is equinumerous to a finite number.

The Prolongation Axiom *Every function defined on FN can be extended to a function which is a set.*

The Axiom of Cardinalities *Every class is either finite, or equinumerous to FN, or equinumerous to the universal class.*

Vopěnka does not consider this axiom to be essential and is willing to replace it by a an axiom postulating a larger number of infinite cardinalities if this turns out to be useful.

Collections of classes can be coded by binary relations: a relation R codes the classes $\{x; R(x, y)\}$, for $y \in V$. Such a coding is said to be *extensional*, if no two rows are equal.

The Axiom Schema of Extensional Coding *Every codable collection is extensionally codable, i.e., for every relation R, there exists a relation S that codes the same classes and is extensional.*

This axiom is only given as an example of a variety of axioms on classes by which the theory can be expanded.

9. *Models of nonstandard analysis and Alternative Set Theory.* Enlargements and models of Alternative Set Theory are best constructed using ultrapowers. I have briefly mentioned this construction in connection with measurable cardinals (see

page 213). Let me now describe the construction in more detail. Let M be an arbitrary model. Let $\mathcal{U}$ be a nontrivial ultrafilter on ω. Let M^ω be the set of all functions $f : \omega \to M$. For two functions, we define an equivalence relation by

$$f \sim_{\mathcal{U}} g \equiv \{n;\ f(n) = g(n)\} \in \mathcal{U}.$$

The universe of $^*M$ is $M^\omega / \sim_{\mathcal{U}}$, the equivalence classes of $\sim_{\mathcal{U}}$. Given a function f, we will denote by $[f]$ its equivalence class. For a k-ary relation R of the structure M, define $^*R$ by

$$^*R([f_1], \dots, [f_k]) \equiv \{n;\ R(f_1(n), \dots, f_k(n))\} \in \mathcal{U}.$$

To define operations in the ultrapower we treat them as relations with one more argument.

The most important information about the ultrapower is given by *Łos's Theorem* which states that the above formula is true for *all definable* relations in M. The proof is a simple argument using induction on the complexity of the formula defining R. The embedding $x \mapsto {}^*x$ is given by $^*a = [c_a]$, where c_a is a function constantly equal to a. The fact that it is an elementary embedding is an immediate consequence of Łos's Theorem.

The proof of the condition 2. of the definition of an enlargement (the $\aleph_1$-saturation) is also easy. Let $^*S_i$, $i \in \omega$ be sets definable in $^*M$. So S_i, $i \in \omega$ are the corresponding sets defined by the same formulas in M. Suppose that $\bigcap_{i=0}^n {}^*S_i$ is nonempty for every $n \in \omega$. Since the mapping $x \mapsto {}^*x$ is an elementary embedding, this implies that also $\bigcap_{i=0}^n S_i$ is nonempty for every $n \in \omega$. It is because each condition $\bigcap_{i=0}^n S_i \neq \emptyset$ speaks about only a finite number of definable sets. For the sake of simplifying the proof, let us assume that $^*S_i$, and hence also S_i, $i \in \omega$ are sets. For every n, pick an element $a_n \in \bigcap_{i=0}^n S_i$ and let $f : \omega \to M$ be defined by $f(n) = a_n$. We will show that $[f] \in \bigcap_{i \in \omega} {}^*S_i$. Again it follows immediately from Łos's Theorem. Let n be given. Then $[f] \in {}^*S_n$ if and only if $\{m;\ f(m) \in S_n\}$ is in $\mathcal{U}$. But $\{m;\ f(m) \in S_n\} \supseteq \{m;\ m \geq n\}$, so indeed it is in $\mathcal{U}$.

A model of Alternative Set Theory is constructed by taking an ultrapower of the structure $(V_\omega, \in)$, where V_ω denotes the set of all hereditarily finite sets. Sets in the model are elements of $^*V_\omega$, classes are subsets of $^*V_\omega$, but we have to identify (or simply remove) those that are extensional with sets. The semiset FN is $\mathbb{N}$. The Prolongation Axiom is proved in the same way as condition 2. of Nonstandard Analysis. (In fact, because $FN = \mathbb{N}$ and we take all subsets as classes, the Prolongation Axiom is equivalent to $\aleph_1$-saturation.) The only axiom that we do not get automatically is the Axiom of Cardinalities. For this axiom to hold in the model described, we need the Continuum Hypothesis. But this is also not a problem. There are models of Zermelo-Fraenkel Set Theory in which the Continuum Hypothesis holds true, thus we can do the ultrapower construction inside of such a model and we get all axioms of Alternative Set Theory.

Main Points of the Chapter

- The two major axiomatic systems for set theory are Bertrand Russell's Theory of Types and Zermelo-Fraenkel Set Theory. Today the latter one is accepted by most mathematicians as a theory describing the true universe of sets.
- Though based on different ideas, the Theory of Types and Zermelo Set Theory, a basic part of Zermelo-Fraenkel Set Theory, are essentially the same. Zermelo-Fraenkel set theory is conceptually simpler, it is stronger, and it can be naturally extended to much stronger systems.
- Infinite sets are defined as sets satisfying a certain property that is not satisfied by finite sets.
- In Zermelo-Fraenkel Set Theory there are infinitely many different infinite cardinalities and there is no largest one.
- The cardinality of the real numbers is bigger than the cardinality of natural numbers, but we do not know, and, maybe, never will, whether there are other infinite cardinalities in between.
- Large cardinal axioms are new axioms that serve to make Zermelo-Fraenkel Set Theory stronger. Using these axioms we can decide some important statements that are undecidable in pure Zermelo-Fraenkel Set Theory.
- The problem with such axioms is that the stronger they are the greater the danger is that they are inconsistent. There is no way to secure their consistency; we can only rely on experience that a lot of research has been done and no contradiction has been found.
- The Axiom of Choice has some consequences which look paradoxical, but it does not introduce an inconsistency into Zermelo-Fraenkel Set Theory.
- It is possible to replace the Axiom of Choice by axioms that make sets of real numbers look better. In particular, it is consistent to assume that all subsets of the real numbers are measurable, provided that we abandon the unrestricted Axiom of Choice and use only its weaker version.
- The particular set of axioms that we are using may be the result of a historical accident. It is conceivable that if the Axiom of Determinacy had been discovered before the Axiom of Choice, it may have become the preferred one.
- Various other axiom systems for set theory have been proposed, but only the Zermelo-Fraenkel system has been accepted by working mathematicians. A likely reason is that it is the strongest available theory.
- New Foundations is an interesting system because we still do not know whether it is consistent. More precisely, we do not have proof of contradiction in the system, nor are we able to prove its consistency using Zermelo-Fraenkel Set Theory (even with the help of large cardinal axioms).
- Nonstandard analysis offers a different way of developing the foundations of calculus and in some cases has helped solve open problems.

Chapter 4
Proofs of Impossibility

"He is right," Watson nodded. "The essential thing about mathematics is that it gives aesthetic pleasure without coming through the senses."

Rudy Rucker, *A New Golden Age*

To prove that something is impossible is usually much harder than the opposite task. To find the proverbial needle in a haystack is hard, but it is much harder to prove that it is not there. The point is that you may find the needle by chance and then you are sure that it is there, but to be completely sure that it is not there, you have to systematically search every piece of the haystack. In court it is the prosecution who must prove allegations, because it is much more difficult for the (presumably innocent) defendant to disprove them. The same can be observed in mathematics. More often it is harder to prove a negative statement, a theorem saying that something does not exist, than a positive one, which shows the existence of a particular mathematical structure. Not only the proofs of non-existence are as a rule more difficult, they also require greater ingenuity on the part of authors. In such proofs one must use some general properties of all possible entities; in other words, it is necessary to develop a theory.

There are many such negative results in mathematics—results saying that an equation does not have a solution, a structure with certain properties does not exist, a proof of a given sentence does not exist etc. However deep and interesting these results may be, I am going to consider only a special kind of negative results in mathematics, those that have something to do with foundations. I call these results *proofs of impossibility,* but the name does not quite convey the intended meaning. A little better, but rather long description would be '*negative results somewhat related to the foundations of mathematics*'. These mathematical results often had been very difficult problems, open for a long time. As there were no mathematical tools available for solving them, they attracted the attention not only of professional mathematicians, but also amateurs. (In some cases, such as the trisection of the angle and the incompleteness theorem, amateurs have not given up, in spite of the existing impossibility proofs, and they keep annoying mathematics departments with their meaningless writings.) Some of these problems go back to the ancient Greeks, some appeared in 20th century mathematics when studying foundations. There are

P. Pudlák, *Logical Foundations of Mathematics and Computational Complexity*, Springer Monographs in Mathematics, DOI 10.1007/978-3-319-00119-7_4,

also problems in contemporary mathematics that ask for proofs of impossibility. (As they are still open, we are not sure that the answer is negative, thus it may turn out after all that they do not fall into this category.) These problems come from the new branch of mathematics and theoretical computer science called *complexity theory*. I will argue in the chapters to follow that these problems may be relevant to foundations. Facing these new problems now, we have no idea what to use to solve them, exactly as our ancestors who wanted to solve the old ones. It may be therefore useful to go back to the old impossibility proofs and look at how new ideas emerged and how they led to the solutions.

4.1 Impossibility Proofs in Geometry and Algebra

I will start with problems that are very old and nowadays may seem completely unimportant. Who might ever care about using a ruler and a compass to draw geometrical figures, when a computer can do it in a split second? Who could be so naive as to try to compute solutions of algebraic equations using cube, fourth and higher roots, which you can hardly compute without a computer or an advanced calculator anyway? The problems are completely unimportant from the point of view of practical computations, no doubt, and that was the case already before they were solved. But they were a big challenge—do we have enough means to answer such questions? Once the means had been developed, they turned out to be very useful for a lot of other problems in mathematics. It would certainly be very interesting to trace the consequences through the history of mathematics up to some contemporary applications, but that is a topic for a different book. This book is about the foundations of mathematics, so I want only to offer you a glimpse of how one can prove such impossibility results.

Irrational Numbers

Proofs of impossibility are almost as old as mathematics. Mathematicians knew that some problems have solutions and others don't. This is not a special feature of mathematics, everyday tasks may be realizable, or may be not. But mathematicians have had the advantage of being able, at least in some cases, *to prove* that something is impossible. The first historically recorded proof of impossibility can be stated as the insolubility of a certain equation. This may seem an unimportant event, but it had a big impact on the understanding of the concept of a number. It was the proof of the existence of an irrational number. Recall that *rational numbers* are those that can be expressed as fractions n/m with n, m integers, $m \neq 0$. Numbers that are not rational are called *irrational*. Rational numbers are very concrete things, as concrete as integers. We need only two integers to represent a rational number. If we want to represent natural numbers by some real things, we may identify them with symbols representing them. It does not matter whether we represent five by 5, or by 101, or by |||||; what is important is that we can do it, in principle, for every natural number. The same is true for rational numbers; we can use pairs of symbols that represent natural numbers (and a symbol for minus). Unfortunately, we cannot interpret all numbers in this way as there are irrational numbers. This was discovered already by the ancient Greeks. Nowadays we feel quite comfortable with irrational numbers, but it wasn't always so. According to one legend, Pythagoreans could not bear this disharmony in the world of numbers and drowned the fellow who found it. One reason why this result was discomforting for ancient Greeks was that they could not avoid such numbers in geometry. We get $\sqrt{2}$ as the length of the diagonal in the unit square (by the Pythagorean Theorem).

The proof that $\sqrt{2}$ is irrational is very easy. As is typical for negative results, the proof is by way of contradiction. Suppose $\sqrt{2} = n/m$ for some integers n, $m > 0$. Squaring the equation we get $2 = n^2/m^2$ and then

$$2m^2 = n^2.$$

We have reduced the problem to the insolubility of the above equation in the domain of integers. Now we need some basic number theory, namely, the result that every positive integer has a unique factorization into the product of powers of prime numbers. In fact we need only to know that every positive integer can be uniquely written as $2^k p$ with p odd. So let $n = 2^k p$ and $m = 2^l q$, with p, q odd. Substituting these into the equation above we get $2(2^l q)^2 = (2^k p)$, which is

$$2^{2l+1} q^2 = 2^{2k} p^2.$$

Now we have on the left hand side a number that is the product of an *odd* power of 2 and an odd number, while on the right hand side we have a number that is the product of an *even* power of two and an odd number. By the uniqueness of such representations, these two number cannot be equal. This contradiction shows that the equation above does not have integer solution and thus $\sqrt{2}$ is irrational.

But this is not the end of the story. $\sqrt{2}$ is a solution of an equation with rational (in fact integer) coefficients, the equation $x^2 = 2$. There are numbers that do not have even this property, they are not solutions of any equations with rational coefficients. Such numbers are called *transcendental*. One of these numbers was well known in the old ages, the π, but it took more than two millennia before a proof of its transcendence was found. The numbers that are solutions of equations with rational coefficients are called *algebraic*; thus $\sqrt{2}$ is algebraic, whereas π is not, it is transcendental.

The proof of the existence of transcendental numbers was not a surprise. Mathematicians had suspected that π and e are transcendental, but it was difficult to prove it. Even today we are not able to prove that, for example, π^e is transcendental (or disprove it).[1] But let us look at what consequences it has for the concept of a number. If we need to denote the positive solution of an equation $x^n = b$, for b positive rational and $n \geq 2$ natural, we use the symbol $\sqrt[n]{b}$. But for transcendental

[1]Note, however, that e^π is known to be transcendental.

numbers we need new symbols. We may introduce symbols for the transcendental numbers that we frequently use, and we may try to introduce some system of notation for some classes of transcendental numbers, but it is impossible to name every number. This follows from Cantor's theorem that the set of real numbers is not countable and from the fact that any system of notation can give names only to countably many elements. A consequence for foundations is that in order to define real numbers, we need the concept of an infinite set or an infinite sequence. We write $\pi = 3.1415926\ldots$, but this is not a definition; the dots give us no clue how to continue. We may identify real numbers with *infinite decimal expansions*, but the infinite expansion are not things that we could write down.

Cantor's theorem can be used to prove that there exist transcendental numbers. By Cantor's theorem, there are uncountably many real numbers, thus to prove the existence of a transcendental number it suffices to show that there are only countably many algebraic numbers. We can enumerate algebraic numbers by enumerating equations and their roots. In this way we surely overcount, as an algebraic number is a root of many equations, but this does not matter as long as we show a countable upper bound. Every equation with rational coefficients is determined by its coefficients, hence it is given by a finite number of rational numbers, hence finite number of integers. The number of finite sets of integers is countable. Thus there are only countably many equations with rational coefficients. An equation of degree d has at most d solutions, according to a well-known result in algebra. Thus we have countably many equations, each having finitely many solutions. Therefore, there are only countably many algebraic numbers.

The three classes of numbers give us a scale by which we can classify real numbers according to their complexity:

1. rational numbers—simple;
2. algebraic numbers that are not rational—moderately difficult;
3. transcendental numbers—difficult.

One would expect that the simplest task should be to prove that the two extremes are distinct, a *difficult* thing is not *simple*, and that it should be harder to show that *simple* is not *moderately difficult*, or *moderately difficult* is not *difficult*. Thus we would expect that the first result was a proof that a transcendental number was not rational. But this is not the case. Looking at the dates[2] when particular results were proved we see clearly that it isn't so:

1. the proof of the irrationality of $\sqrt{2}$—approximately 430 BCE (probably by the Pythagorean school);
2. the proof of the irrationality of π—1761 (by J.H. Lambert);
3. the first example of a transcendental number with a proof—1844 (by J. Liouville);
4. the proof that e is transcendental—1873 (by C. Hermite);
5. the proof that π is transcendental—1882 (by F. von Lindemann).

[2]See [150, 287].

The explanation is that a number such as $\sqrt{2}$ has a very simple definition, so we can work with it more easily, while merely to define a number such as π we have to use infinitesimal calculus. In the next chapter we will observe the same phenomenon in complexity theory, but there we have only reached a very early stage: for very simple functions we can prove that they are somewhat complex.

Cantor's result that implies the existence of transcendental numbers dates 1874, but it is not the most difficult one, on the contrary, as we have seen in the previous chapter, it is quite simple. The reason why it was discovered so late is not the complexity of the proof, but the lack of precise concepts of sets and infinity in the pre-Cantorian mathematics. Another reason why this proof had not been discovered earlier is its non-constructive character. Cantor's proof shows that almost every number is transcendental, but it does not provide a single concrete example of a transcendental number. We will also watch a parallel of Cantor's non-constructive proof with non-constructive proofs in complexity theory in the next chapter.

Constructions with a Ruler and a Compass

The ancient Greeks achieved great ingenuity in solving geometrical problems using a ruler and a compass. The part of geometry in which constructions of geometrical figures were studied used to be an important field of applied science. It wasn't so long ago that engineers still used these instruments. A ruler and a compass were not important for the foundations of geometry. Nobody was so naive as to define the circle as the form obtained by using a compass. But these two instruments were important for developing the concept of an *algorithm*. To define a concept of algorithm, as the method to construct something from given data, you need to say what are the elementary operations that are allowed. Performing an algorithm means doing such elementary operations one after the other. As the ancient geometers were so successful in using rulers and compasses, it was only natural to suggest that the basic elementary operations were those that can be done using a ruler and a compass and that using these operations one could construct everything.

Though everybody has learned constructions with a ruler and a compass in the school, it is worthwhile to write down explicitly what are the rules of this game.[3]

- *We are given some points, circles and straight lines in the plane and we are to construct certain other points, lines and circles. Thus input and output data are a finite number of points, circles and lines.*
- *We can use two operations*:
  1. *given two points we can draw the line through the points*;
  2. *given two points we can draw the circle with the center at the first point and passing through the second point*;

[3] Another reason for stating the rules explicitly is that it has been shown that using these instruments in a special way one *can* solve the trisection of the angle.

3. *it is tacitly assumed that whenever circles and lines intersect, we get also the intersection points as the result of the operations above.*

In practice one is also allowed to choose a "random" point on a line on a circle or outside of them. This operation can be avoided, though it may be a little cumbersome. It is important for us not to use this option, as we want to know what can be constructed using the above operations. A random choice may result in an accidental solution, however unlikely it may be.

It turned out that there are problems that the ancient Greeks could not solve using these constructions. Three of them became especially famous:

1. *duplication of the cube*, (*given a line segment x construct a line segment y such that the cube with a side of length y has double volume as the cube with side x*);
2. *quadrature of the circle*, (*given a circle construct a square with the same area*);
3. *trisection of the angle*, (*divide a given angle into thirds*).

It has been proved that these problems are not solvable using a ruler and a compass in the way defined above. The *quadrature of the circle* is even used as a synonym for an unsolvable problem.

To show that these problems are unsolvable using a ruler and a compass, the first step is to transform them into algebraic problems. Using elementary analytic geometry one gets that 1. asks for constructing a line segment of length $\sqrt[3]{2}$ given a unit line segment and 2. asks for constructing a line segment of length $1/\sqrt{\pi}$ given a unit line segment. It is not difficult to show that given $1/\sqrt{\pi}$, one can construct π and vice versa, so the problem is equivalent to constructing π. The last one transforms, using a little trigonometry, to constructing a solution of the equation

$$4x^3 + 3x - a = 0$$

given a unit line segment and a line segment of length a.

The next question is to find out what lengths can be constructed from the unit line segment. Again, only elementary considerations are needed to show that the lengths that can be constructed from a unit line segment are exactly those that can be expressed using the basic arithmetical operations and the square root operation. For example,

$$\frac{5 + 4\sqrt{8 - \sqrt{7 + \sqrt{2}}}}{9 - \sqrt{23}}$$

is a number that can be obtained as the length of a line segment constructed from a unit line segment. The form of these numbers is given by the equations of the geometrical figures that one uses. Straight lines are given by linear equations and circles are given by the familiar quadratic equations. The coordinates of the intersections are obtained by solving systems of such equations. For solving linear equations, we only need the arithmetical operations, for solving quadratic equations we also need square roots. The distance of two points can be computed from their coordinates using the Pythagorean theorem, which, again, only needs a square root.

Let us focus on the first problem, which is the simplest one. What remains to do now is to prove that $\sqrt[3]{2}$ cannot be expressed only using square roots. This is, of course, the hardest part. The key tool is *number fields*. Several number fields are familiar to us: rational numbers, real numbers and complex numbers. In general a number field is a substructure of complex numbers that contains 0 and 1, in which addition, subtraction and multiplication is defined for all elements and in which it is possible to divide by a nonzero element. Shortly stated, it is a substructure that shares the basic properties with the structure of rational numbers. Then the idea is to assign a number field to a particular type of numbers and study the properties of these fields. Using these properties one shows that the numbers that are expressed in a specified way cannot be expressed in another specified way. In particular the number $\sqrt[3]{2}$ that is expressed using the cube root, cannot be expressed using square roots.

The introduction of number fields for solving such problems is an important twist. The original problem is very much combinatorial. We have expressions of certain form, namely expressions with square roots. We need to show that a cube root cannot be expressed in this way. But instead of analyzing these expressions we introduce new structures. These structures are more complex objects; according to logical classification, they are *higher order* structures, but in some sense, they are much better. Furthermore, they provide us with information that is not apparent in the original objects. This is very typical for contemporary mathematics. So as to introduce order into the problem that we are studying, we embed the studied objects into larger structures. Then we see that the apparently accidental relations that we observed initially are, in fact, a special case of a general law.

Logic provides a general explanation why applying higher order structures is often successful, though particular cases may have its own special features that need different explanations. The explanation is that using higher order structures also means using larger sets. We know from set theory that the larger cardinalities we introduce the stronger theory we get.

Example Let us look at the idea of the proof that $\sqrt[3]{2}$ cannot be expressed using a single square root, say $\sqrt{2}$. It is ridiculous to use number fields for such a purpose, the statement has an easy direct proof, but it will give us an idea of what happens in more difficult cases. In our little example one can easily show that

1. the field generated by $\sqrt{2}$ consists of all numbers of the form
$$a+b\sqrt{2},$$
a, b rational numbers; moreover each element of the field has a unique representation of this form;
2. the field generated by $\sqrt[3]{2}$ consists of all numbers of the form
$$a+b\sqrt[3]{2}+c(\sqrt[3]{2})^2,$$
a, b, c rational numbers; moreover each element of the field has a unique representation of this form.

Now the argument continues as follows. Suppose $\sqrt[3]{2}$ is expressible using $\sqrt{2}$. This means that $\sqrt[3]{2}$ is in the first field. Then the whole second field must be contained in the first field, since every element of the second field can be obtained from $\sqrt[3]{2}$. So it remains to show that the second field cannot be contained in the first one. The reason for that is size. The first field is parameterized by two parameters, the second by three. Therefore, the second field is larger, hence cannot be a part of the first one.

You see that the crucial property, the number of parameters, the *degree,* is a very natural invariant of the fields, but it only appeared when the new structure was associated with the particular number.

Solutions of Equations by Radicals

We have seen that solving geometrical problems using a compass and a ruler is equivalent to finding expressions with square roots for certain numbers. Now we extend our tools to all roots: cube roots, fourth roots, etc. These roots are also called *radicals*, thus we speak about solving problems by radicals. We know that there are transcendental numbers and such numbers cannot be expressed using radicals. Hence the only numbers that we can hope to be expressible by radicals are solutions of algebraic equations. For quadratic equations, there is a well known formula. For cubic equations the formula is more complicated. If we consider the following special form of the cubic equation

$$x^3 + px + q + 0,$$

to which any general one can be easily transformed, then one of the solutions is given by

$$x = \sqrt[3]{-\frac{q}{2} + \sqrt[2]{\left(\frac{q}{2}\right)^2 + \left(\frac{p}{3}\right)^3}} + \sqrt[3]{-\frac{q}{2} - \sqrt[2]{\left(\frac{q}{2}\right)^2 + \left(\frac{p}{3}\right)^3}}. \tag{4.1}$$

For quartic (degree 4) equations, the formula is quite involved, but there is one. For quintic (degree 5) equations

$$x^5 + ax^4 + bx^3 + cx^2 + dx + f = 0, \tag{4.2}$$

one would expect some horrible formula, but, surprisingly, there is no formula at all; such equations cannot be solved by radicals. More precisely this means: there is no general formula that expresses a root of quintic equation (4.2) in terms of a, b, c, d, f. In fact, an even stronger theorem is true: there are explicit quintic equations with integer coefficients none of which solutions can be expressed using rational numbers and roots of arbitrary degree. For example, the following equation, which has three real roots, cannot be solved by radicals:

$$x^5 - 4x + 2 = 0. \tag{4.3}$$

The spirit of this problem is the same as the one concerning constructions with a ruler and a compass. The essential difference, however, is that the solution of this problem led to a new deep theory. This new theory is based on the concept of a group, something that was not used before. The most important aspect of it is that the theory needs to study arbitrary finite groups. Thus it works not with a few given structures, as was the case before, but with a whole class of structures. This has influenced the development of mathematics towards the structural approach that we use nowadays. Contemporary mathematics studies not only numbers, geometrical figures and functions, but also various structures and classes of structures. Instead of numbers, number fields become more important, instead of function, function spaces dominate and instead of symmetries, mathematicians study groups.

This topic is connected with three important names, Paolo Ruffini (1765–1822), Niels Henrik Abel (1802–1829) and Évariste Galois (1811–1832). These three mathematicians worked on the solvability of algebraic equations by radicals, in particular on the problem of the solvability of quintic and higher degree equations. Ruffini got very close to solving this problem. His proof that quintic equations are not solvable by radicals contained a gap that he might have corrected, had anybody cared to point it out. Independently, a correct proof was given by Abel.[4] Galois developed a theory that now bears his name and that plays a key role in algebra. He also introduced the concept of a group into this field.

Interestingly, these mathematicians had a hard time persuading others about the importance of their work. Ruffini sent a number of papers to Lagrange, Galois sent his papers to Cauchy and Fourier, but in both cases either the papers were lost, or no response was given, or the work was proclaimed incomprehensible. It took some time before the mathematical community recognized that the theory developed for solving this problem surpasses the importance of the problem itself and that there was a new type of mathematics emerging from it—mathematics that was not in the main stream of infinitesimal calculus, but that was as interesting and as important as calculus. One should never underestimate results only because they do not belong to the main-stream research.

Galois Theory

I will sketch some ideas that are used in the proof that quintic and higher degree equations are not solvable by radicals. These results are part of what is called now *Galois theory*, an important research field in algebra.

We will need the concept of *field extensions*. We have used it already implicitly when considering the constructions with a ruler and a compass. In general, it is a pair of fields K and L such that K is a *subfield* of field L; we also say that L is an *extension* of K, or simply write L/K. We can view this pair as a composed structure.

[4]Note that Gauss claimed that he also had a proof.

The key idea is that one can get a lot of important information only knowing the symmetries of this structure.

The most familiar extension is $\mathbb{C}/\mathbb{R}$, the extension of real numbers $\mathbb{R}$ to complex numbers $\mathbb{C}$. Every complex number can be uniquely expressed as

$$a + ib$$

with a, b reals. If you know how to compute with reals, then all you need to compute with complex numbers is the identity $\mathrm{i}^2 = -1$. For example, the inverse of $1 + \mathrm{i}$ is computed as follows:

$$\frac{1}{1+\mathrm{i}} = \frac{1}{1+\mathrm{i}} \cdot \frac{1-\mathrm{i}}{1-\mathrm{i}} = \frac{1-\mathrm{i}}{1-\mathrm{i}^2} = \frac{1-\mathrm{i}}{1+1} = \frac{1}{2} + \frac{\mathrm{i}}{2}.$$

We also have the identity $(-\mathrm{i})^2 = -1$, which means that $-\mathrm{i}$ satisfies the same identity as i. Hence, if we have a computation with complex numbers (expressed using reals in the form above), then we can replace systematically all occurrences of i by $-\mathrm{i}$ (in particular, $-\mathrm{i}$ will be replaced by $-(-\mathrm{i})$ which is i) and we get a valid computation again. Thus there is no way to distinguish i from $-\mathrm{i}$; it is only a matter of notation that we denote one by i and the other by $-\mathrm{i}$.

The extension $\mathbb{C}/\mathbb{R}$ does not have other symmetries, except for the trivial one that does not do anything. So we have two symmetries, the trivial one, which we will denote by I and the one that transposes $a + ib$ and $a - ib$, which we will denote by σ. Saying 'symmetry' is not quite precise; the technical term is *automorphism*. An automorphism is a permutation of the set $\mathbb{C}$ that preserve the arithmetical operations. Both I and σ are trivial on $\mathbb{R}$, as they do not move any element of $\mathbb{R}$. Permutations can be composed, which means that given two permutations we apply the first one and then the second one. In the case of I, σ, I is the identity and $\sigma\sigma = I$, so we get a very simple group with the following multiplication table:

	I	σ
I	I	σ
σ	σ	I

But what does this have to do with equations? The above extension is connected with the equation

$$x^2 + 1 = 0.$$

Namely, the two critical elements i and $-\mathrm{i}$ are the two roots of this equation. We think of extending $\mathbb{R}$ to $\mathbb{C}$ as a process of adding the roots of this equation to $\mathbb{R}$. Since the extension is determined by the roots of the equation, also the automorphisms are determined completely by their actions on the roots of the equation. Hence we may view the group as the group of permutations of the roots.

Thus we have learned that a group is associated to each field extension. This group is called *the Galois group* of the extension. Let us look at another example, the equation

$$x^2 - 2 = 0. \tag{4.4}$$

Now we will concentrate on extensions of the field of rational numbers $\mathbb{Q}$, which is what we need for the problem of solubility of equations by radicals. We extend $\mathbb{Q}$ to the field that contains the roots of the equation (4.4). We will denote this field by $\mathbb{Q}(\sqrt{2})$. This is an extension that we have considered on page 261. Now it seems that the two roots can be distinguished, as one of them, $\sqrt{2}$, is positive and the other, $-\sqrt{2}$, is negative. But in Galois theory we only take the arithmetical structure of the fields, but we do not include the ordering of reals. The two roots are indistinguishable with respect to the arithmetical structure. Again, all the information that we have about them is that their squares equal 2. Thus we can swap the two roots as we did in the previous extension. The two extension we have considered, though different, look very similar. Indeed, their Galois groups are isomorphic.

If we extend a field by adding one root of a quadratic equation, we get automatically the second root. This is not the case in general. It is possible to add some roots without adding others. Those extensions that are obtained by adding all roots of an equation play a special role in the theory and they have a special name: *Galois extensions*.[5] A nice property of Galois extensions is that their Galois groups say a lot about them. The main theorem of Galois theory states, among other things, that there is a unique correspondence (called *Galois correspondence*) between subgroups of the Galois group and intermediate fields. In particular, knowing only the Galois group one can determine whether or not the equation is soluble by radicals.

So far everything has been 'Galois' and nothing 'Abel'. But Abel will also get some credit. The commutative law does not hold in all groups; those that satisfy it form an important class of groups. Such groups are called either *commutative,* or, more often, *abelian*. Naturally, when the Galois group of an extension is abelian, we call such an extension abelian. Let $p(x) = 0$ be an equation with coefficients in a field K and let L be the Galois extension obtained by adding all roots of $p(x) = 0$. If L/K is an abelian extension, then the roots can be expressed as radicals of elements of K.

But being abelian does not characterize the solubility by radicals; there are equations that give non-abelian extensions, but still can be solved by radicals. Consider the following situation. Let $p(x) = 0$ and $q(x) = 0$ be two polynomial equations with rational coefficients; let K and L be their Galois extensions. Suppose that L is an extension of K, L/K is abelian and $K/\mathbb{Q}$ is also abelian. Yet it may happen that $L/\mathbb{Q}$ is not abelian. On the other hand, the roots of $p(x) = 0$ are radicals of rational numbers, hence all numbers in K are radicals of rational numbers. The roots of $q(x) = 0$ are radicals of the numbers contained in K, hence also radicals of rational numbers.

Fortunately this is the only complication and we can state the characterization.

Theorem 19 *Let $p(x) = 0$ be a polynomial equation with rational coefficients and let L be the extension of $\mathbb{Q}$ by adding the roots of $p(x) = 0$. Then the equation is solvable by radicals if and only if there exists a chain of abelian Galois extensions*

[5]This is not a precise definition, but it suffices for this rough description of the theory.

starting at $\mathbb{Q}$ *and ending at* L *(i.e.,* $\mathbb{Q} \subseteq K_1 \subseteq K_2 \subseteq \ldots \subseteq K_n \subseteq L$, *with* $K_1/\mathbb{Q}$ *abelian Galois,* K_1/K_2 *abelian Galois, etc.).*

The main theorem of the Galois theory, the Galois correspondence, enables us to give similar characterization in terms of subgroups of the Galois group. Finding the Galois group may be a difficult task, but once we have it, is not difficult to check this property. The Galois groups of equations are finite, as they can be represented by groups of permutations of the roots of the equation. Thus, in principle, we can search them completely and determine all their properties.

Of course, one needs much more theory to explain the group theoretical condition of solubility and why the abelian extensions are soluble and so on. I will present the theory in a bit greater detail in Notes. To conclude this part I will only explain why the degree five is crucial. Given a degree d equation, the group can be represented as a group of permutations of d elements. The largest such group is the group of all permutations of the d elements; it has $d!$ elements. If d is at most four, the groups are quite small and therefore all have the property that ensures solubility. The first permutation group that does not have the property occurs on five elements. One such group is the group of all permutations of the five elements. Then one can also show that there are equations with such a Galois group. One such equation is (4.3) on page 262.

Notes

1. *A transcendental number.* Liuville's proof is based on the theory of approximations of real numbers. Consider the following two numbers in decimal representation:

$$\xi = .001000000100000000000000000001\ldots = \sum_{n=1}^{\infty} 10^{-3^n},$$

$$1/99 = .0101010101\ldots.$$

In a certain sense ξ is easier to approximate than $1/99$. Namely, .001 is a 3 digit approximation of ξ with a precision of 8 digits, .001000001 is a 9 digit approximation of ξ with a precision of 27 digits, etc., while $1/99$ can be approximated with n digits only to a precision of $n+1$ digits. In general, rational numbers that are not decimal fractions are difficult to approximate by decimal fractions. Liouville proved a general theorem about approximation of algebraic numbers. The special case of the theorem for decimal representations says that an irrational number that is a root of a rational equation of degree d cannot be approximated by n digits to a precision bigger than $(d+\varepsilon)n$ for infinitely many n's for any $\varepsilon > 0$. A corollary of this theorem is that the following number is transcendental

$$\sum_{n=1}^{\infty} 10^{-n!}.$$

Note that the best theorem about approximations, Roth's theorem, eliminates the dependence of the approximation on the degree of the polynomial (one consequence is that the number denoted by ξ above is also transcendental). But to prove the transcendence of e and π a completely different approach is needed.

2. *The number fields for the ruler and compass constructions.* To make the argument that $\sqrt[3]{2}$ cannot be expressed using $\sqrt{2}$ precise, we define the degree of the filed over the field of rationals $\mathbb{Q}$. If K is a filed that extends rationals, we can think of K as a vector space over $\mathbb{Q}$. So elements of K are thought of as vectors and elements of $\mathbb{Q}$ are scalars, i.e., *only* elements of $\mathbb{Q}$ are scalars. Then the *degree* of K over $\mathbb{Q}$ is simply the dimension of this vector space. In the same way we can define the degree of L over K for an arbitrary field K and its extension L.

   The example above that shows that $\sqrt[3]{2}$ cannot be expressed using $\sqrt{2}$ is simple, but not very persuasive because of two reasons. First, the degree of the extension is the same as the degree of the roots in the two fields considered. Secondly, by taking more square roots we easily get a field of degree larger than 2, thus we get no contradiction. Before going to theory, let us consider a slightly bigger example. Suppose we have two square roots, say $\sqrt{2}$ and $\sqrt{3}$. Then the associated number field is neither of degree 2, as you may try to deduce from having only square roots, nor of degree 3, as one may guess because of having one more term, but 4. Namely, every element in the field is uniquely expressible in the form $a+b\sqrt{2}+c\sqrt{3}+d\sqrt{2}\cdot\sqrt{3}$, with a, b, c, d rationals. Furthermore, we will see shortly that every subfield of such a field has degree 1, 2 or 4, i.e., only the divisors of 4. Hence the field generated by $\sqrt[3]{2}$ is not a subfield of this field. The reason is that 3 does not divide 4.

   Let me now mention a few basic results in field theory from which the insolubility of the duplication of the cube follows. Above we only considered number fields which are subfields of the field of complex numbers. The elementary results we need are true for general fields. We will denote by $[L:K]$ the degree of an extension L of a field K. The degree may be an infinite cardinal, but here we will need only finite extensions.

   The first result that we need is the following. Let L be an extension of K and let M be an extension of L then

$$[M:K]=[M:L][L:K]. \tag{4.5}$$

   A corollary of the formula is that $[L:K]$ *divides* $[M:K]$. The proof of this formula, as well as of the second result, is easy.

   The second result is about extensions obtained by adding a root of a polynomial. Let L be an extension of K again. Let $p(x)$ be a polynomial with coefficients in K and suppose that p has a root α in L, but it does not have one in K. Then there exists a minimal extension of K contained in L that contains α. It is denoted by $K(\alpha)$. The result that we need is:

Proposition 5 *If p is irreducible over K (cannot be factored into a product of two non-constant polynomials) and the degree of p is d, then*

$$\big[K(\alpha):K\big]=d.$$

In the case of the constructions with a ruler and a compass, we start with the field of rational numbers $\mathbb{Q}$ and we are only adding roots of quadratic polynomials. Consequently, we get, by the two formulas, that degrees of such extensions over $\mathbb{Q}$ are always powers of 2. On the other hand, the degree of $\mathbb{Q}(\sqrt[3]{2})$ over $\mathbb{Q}$ is 3. Since 3 does not divide any power of 2, $\sqrt[3]{2}$ is not contained in any extension obtained by adding square roots. Hence the cube cannot be duplicated.

(To apply the results we need also to show that the polynomial $x^3 - 2$ is irreducible over $\mathbb{Q}$. This is also easy. Suppose it is not irreducible. Then one of the factors must be linear. This means that one of the roots of the polynomial is in $\mathbb{Q}$. But this can be disproved in the same manner as one shows that $\sqrt{2}$ is not in $\mathbb{Q}$.)

To show the insolubility of the trisection of the angle one needs only to find a rational number a such that the polynomial $4x^3 + 3x - a$ is irreducible over $\mathbb{Q}$, which is not a difficult task. Thus in principle it is the same argument as for the duplication of the cube. To show the insolubility of the quadrature of the circle, one can use another basic fact that says that $[K(\alpha) : K]$ is infinite if α is not a root of any (nonzero) polynomial with coefficients in K. Then one can use the nontrivial result that π is not algebraic. Hence $[\mathbb{Q}(\pi) : \mathbb{Q}]$ is infinite, while adding finitely many quadratic roots (in fact, adding any roots) we only get finite extensions.

3. *Some Galois theory.* Even though this is just a brief sketch of the theory needed for the insolubility by radicals, a fair amount of concepts have to be defined. To simplify it, we will only consider fields that contain $\mathbb{Q}$ and every extension will be of finite degree. I will not give proofs, but where a proof is simple, I will give at least a hint.

Given a field extension L/K, we will denote by $G(L/K)$ the Galois group of the extension. It is the group of automorphisms of L that do not move any element of K. An extension L/K is a *Galois extension* if every polynomial $p(x)$ that has coefficients in K is irreducible in K and has a root in L has all its roots in L. We say that $p(x)$ *splits completely* because then it can be factored into linear terms of the form $(x - \alpha)$, α a root of $p(x)$. The extension of K generated by adding the roots of $p(x)$ is called *the splitting field* of $p(x)$.

We will need the following easy fact:

Lemma 4 *If M/K is a Galois extension and L is an intermediate field, then M/L is also a Galois extension.*

The hint is that the irreducible polynomial over L of an element $a \in M \setminus L$ divides its irreducible polynomial over K.

A subgroup N of a group G is called *normal* if for every $x \in G$ and $y \in N$, we have $xyx^{-1} \in N$. If N is a normal subgroup of G, then we can form a group G/N, the factor group. There is a group homomorphism from G onto G/N whose kernel is N. Also any kernel of a group homomorphism is a normal group.

Let M/K be a Galois extension. The *Galois correspondence* is the following relation between the subgroups of $G(M/K)$ and the fields intermediate between K and M. Given a field L between K and M, we assign to it $G(M/L)$; given a subgroup H of $G(M/K)$, we assign to it $F(H)$, *the fixed field of* H, that is the intermediate field of all elements of M that are not moved by any of the automorphisms from H. In this situation a number of useful conditions are satisfied. Here are some of them:

1. *the degree of* M/K *equals the order of* $G(M/K)$;
2. *the mappings* $L \mapsto G(M/L)$ *and* $H \mapsto F(H)$ *are inverse each to the other, thus* $F(G(M/L)) = L$ *and* $G(M/F(H)) = H$;
3. *the mappings reverse the inclusion relation: if* $L_1 \subseteq L_2$, *then* $G(L_1/K) \supseteq G(L_2/K)$; *if* $H_1 \subseteq H_2$, *then* $F(H_1) \supseteq F(H_2)$;
4. L/K *is a Galois extension if and only if* $G(M/L)$ *is a normal subgroup of* $G(M/K)$;
5. *if* $G(M/L)$ *is normal, then* $G(L/K) = G(M/K)/G(M/L)$.

M/K is called a *radical* extension if it is obtained by successively adding nth roots (for various n's). More precisely, it means that there exists a chain of extensions

$$K = L_0 \subseteq L_1 \subseteq \cdots \subseteq L_k = M$$

such that each L_{i+1} is the splitting field of some polynomial of the form $x^n - a$ for $a \in L_i$. It seems that we are asking for more than needed, as one may prefer to take only a, which is a positive real and add only the positive real root of $x^n - a$. However, it not difficult to see that solubility in this more general sense is equivalent to the solubility in the restricted sense.

Example Consider the equation $x^3 + px + q = 0$ with $p, q \in \mathbb{Q}$. Suppose it does not have rational roots and it has only one real root. To get a radical extension that contains roots of this polynomial, we first take the extension of $\mathbb{Q}$ to the splitting field L of the polynomial

$$x^2 - \left(\frac{q}{2}\right)^2 - \left(\frac{p}{3}\right)^3.$$

Let us denote, as usual, by $\sqrt{(\frac{q}{2})^2 + (\frac{p}{3})^3}$ the positive real root of this polynomial. (The assumptions ensure that we take the square root of a positive number.) Thus

$$L = \mathbb{Q}\left(\sqrt{\left(\frac{q}{2}\right)^2 + \left(\frac{p}{3}\right)^3}\right).$$

Then we extend L to the splitting field M of the polynomial

$$x^3 + \frac{q}{2} - \sqrt{\left(\frac{q}{2}\right)^2 + \left(\frac{p}{3}\right)^3}.$$

Thus we get the first term of the expression for a root of $x^3 + px + q = 0$

$$\sqrt[3]{-\frac{q}{2} + \sqrt{\left(\frac{q}{2}\right)^2 + \left(\frac{p}{3}\right)^3}}.$$

But the second term of (4.1), page 262, is also in M, hence M is a radical extension that contains the roots of $x^3+px+q=0$. (The reason why we get the second term automatically is that the degree of the extension $M/\mathbb{Q}$ is already 6 which is the degree of the splitting field of x^3+px+q.)

Theorem 19 (page 265) gives a characterization of solubility by radicals in terms of field extensions. (It was only stated for extensions of $\mathbb{Q}$, but it holds in general.) The corresponding group-theoretical concept is the *soluble group*. A group G is soluble if there exists a chain of subgroups starting with the trivial group end ending with G, say $\{1\}=G_0, G_1, G_2, \ldots, G_{n-1}, G_n=G$, such that G_i is a normal subgroup of G_{i+1} and G_{i+1}/G_i is an abelian group. The equivalence follows from the Galois correspondence. Consider a Galois extension M/K and let

$$K=L_0\subseteq L_1\subseteq L_2\subseteq L_3\subseteq\cdots\subseteq L_k=M,$$
$$G(M/K)=G_0\supseteq G_1\supseteq G_2\supseteq G_3\supseteq\cdots\supseteq G_k=\{1\}$$

be chains of intermediate fields and the corresponding groups. According to property 4. of the Galois correspondence, L_1/L_0 is a Galois extension if and only if G_1 is normal in G_0. Since M/L_1 is also a Galois extension (by Lemma 4), we can continue and consider the shorter chains starting with L_1 and G_1 respectively. Then we get that L_2/L_1 is a Galois extension if and only if G_2 is normal in G_1 and so on. If the consecutive extensions are Galois, we get from the Galois correspondence that L_{i+1}/L_i is an abelian extension if and only if the group G_{i+1}/G_i is abelian. Thus the two conditions are equivalent. Let me state it explicitly.

Theorem 20 *An extension is radical if and only if its Galois group is soluble.*

I will only explain why a radical extension has a soluble Galois group, the other direction is not needed for the insolubility. First we need to show that by adding nth roots we get soluble extensions. It suffices to prove it for n prime, as a general root can be obtained by successively applying prime roots.

Lemma 5 *Let K be a field, $a\in K$ and p a prime number. Then the Galois group of the splitting field of x^p-a over K is soluble.*

To prove it, we construct the splitting field of x^p-a in two stages. First we construct L, the splitting field of x^p-1 over K, and then M the splitting field of x^p-a over L. Thus we have a chain of extensions $K\subseteq L\subseteq M$. It will be clear in a moment that M is also the splitting field of x^p-a over K. Then we will prove that both L/K and M/L are abelian which implies that $G(M/K)$ is soluble. Note that the Galois group $G(M/K)$ is not abelian, except for $p=2$, but it has an abelian normal subgroup $G(M/L)$ such that $G(M/K)/G(M/L)$ is abelian.

Viewed geometrically, the roots of x^p-1 are the vertices of a regular p-gon in the plain of complex numbers with vertices on the unit circle and one vertex being 1. Namely, the roots are $1, \xi, \xi^2, \ldots, \xi^{p-1}$, $\xi = \cos(2\pi/p) + \mathrm{i}\sin(2\pi/p)$. The automorphisms of L over K are determined by their actions on the roots, but, in fact, we only need to know the image of the generator ξ. If $\xi \mapsto \xi^n$, then the automorphism maps every root y to y^n. Since $(y^n)^m = (y^m)^n$, automorphisms of L/K commute, thus $G(L/K)$ is abelian.

To prove that $G(M/L)$ is abelian is also easy. The roots of $x^p - a$ are

$$\sqrt[p]{a},\ \xi\sqrt[p]{a},\ \xi^2\sqrt[p]{a},\ \ldots,\ \xi^{p-1}\sqrt[p]{a}.$$

Since $\xi, \xi^2, \ldots, \xi^{p-1}$ are in L, an automorphism of M over L is determined by its action on $\sqrt[p]{a}$. If it maps $\sqrt[p]{a}$ to $\xi^n\sqrt[p]{a}$, then it maps every $\xi^i\sqrt[p]{a}$ to $\xi^i\xi^n\sqrt[p]{a}$. Hence the action of an automorphism on the roots is simply multiplication by some ξ^n. As multiplication of numbers is commutative, the Galois group $G(M/L)$ is commutative.

Now the part of Theorem 20 that we are interested in follows easily. If we have a chain of extension where each consecutive extension is the splitting field of some polynomial $x^p - a$, we can refine it to get a chain of abelian Galois extensions (it will be twice as long).

4. *The insolubility of the quintic equation.* We need a quintic polynomial with integer coefficients whose Galois group (more precisely, the Galois group of its splitting field) is not soluble. Such a polynomial is the polynomial $x^5 - 4x + 2$ from page 262, but in fact any polynomial with the following two properties will do:

   1. it has three simple real roots (and two complex roots that are not real) and
   2. it is irreducible over $\mathbb{Q}$.

   To see 1, simply look at the graph of the function $y = x^5 - 4x + 2$. This is not a proof, but using elementary calculus it is easy to verify that the graph crosses the x-axis exactly three times. Proving that a polynomial with integer coefficients is irreducible over $\mathbb{Q}$ is harder, but it is facilitated by the *Gauss lemma.* This simple result says that if such a polynomial is not irreducible over $\mathbb{Q}$, then it can be factored into a product of polynomials with integer coefficients. In general, the factorization problem for rational polynomials is algorithmically solvable. To show that our polynomial is irreducible, it suffices to consider the divisibility by powers of 2 of coefficients in an alleged factorization of the polynomial. This is similar to showing that $\sqrt{2}$ is irrational.

   Now we need an insoluble group. In Chap. 1 we have defined simple groups as those that have only trivial homomorphic images (a one-element group or an isomorphic group). An equivalent condition is that the group does not have a proper normal subgroup. Thus we get examples of insoluble groups by taking non-abelian simple groups. Such a group that is a permutation group on a five element set, called *alternating group* and denoted by A_5.[6] The elements of this

[6]Finite groups are often presented as groups of symmetries of some structures; when introducing groups I mentioned the group of symmetries of the regular triangle. A_5 can be presented as the group of symmetries of the regular icosahedron; therefore it is also called *the icosahedral group.*

group are all *even permutations*, permutations that have an even number of inversions; equivalently, the permutations that have an even number of even cycles. To show that it is not abelian just take two suitable even permutations. To prove that A_5 is simple requires considering several cases, but we do not need to use essentially any theory.

Using elementary group theory one shows that any group that contains an insoluble group is also insoluble. In particular S_5, the group of *all* permutations of five elements is insoluble. Now we will show that the splitting field of any degree 5 polynomial satisfying 1 and 2 above has S_5 as the Galois group. To prove it we will show that the group contains a 2-cycle (a transposition) and a 5-cycle. Then we will refer to the easy and well-known fact that a 2-cycle and an n-cycle generate whole S_n (the group of all permutations on an n-element set).

To show the 2-cycle in the Galois group of the splitting field of our polynomial is very easy: the complex conjugation swaps the two complex roots while keeping the real ones fixed. To show the 5-cycle, take any root α of the equation and consider the extension $\mathbb{Q}(\alpha)$ obtained by adding only this root. The degree of this extension is 5 by Proposition 2 above. Unfortunately, it is not a Galois extension, so we cannot argue that a 5-cycle is in its Galois group. We have to use a less direct argument. Since $\mathbb{Q}(\alpha)$ is an intermediate field in the extension that we consider, we get, by the multiplication formula for degrees (see (4.5) on page 267), that the degree of the splitting field is divisible by 5. Hence also the size of the Galois group is divisible by 5. According to a basic result in group theory (attributed to Cauchy), if the size of a group is divisible by a prime p, then it has an element g of the order p, which means that $g^p = 1$ and $g^i \neq 1$ for $1 < i < p$. Thus the splitting field contains an element of order 5. The only type of a permutation on five elements that has such an order is a 5-cycle. Thus we are done.

This shows that a *concrete* quintic equations with integer coefficients is not solvable by radicals, which implies that there is *no general formula* for quintic equations such as the formulas for equations of degrees less than five. The latter fact is weaker and is also a little easier to prove because to this end, it suffices to find an insoluble quintic equation with coefficients in some field, not necessarily the field of rational numbers.

4.2 The Incompleteness Theorems

The incompleteness theorems are the most important results in the foundations of mathematics. These theorems are well known, but still many mathematicians do not fully realize their consequences. It is, perhaps, because the theorems do not have much to do with the everyday work of a mathematician. But for those who study the foundations of mathematics, those who wonder what mathematical reality is and what mathematical truth is, this is the most important information. The incompleteness phenomenon ruins the hope that we could build a formal system for the whole of mathematics that would be complete and for which we could prove the consistency. Instead, we are facing a much more complex situation. In spite of the many decades that have passed since this breakthrough result appeared, we have not been able to find satisfactory answers to all the problems raised by the incompleteness theorems.

In this section we will be mainly concerned with the proofs of these theorems. They are a kind of results that can be easily explained on an informal level. But one has to be very careful when using imprecise descriptions of mathematical results and such descriptions must not be confused with real proofs. An informal argument used instead of proof may easily lead to false statements, especially in this case. Such wrong deductions, unfortunately, have been used to derive "consequences" of Gödel's theorems that have little to do with reality. (I will mention some in Chap. 7.)

Let us recall the theorems.

Gödel's Incompleteness Theorems

1. *Any consistent formal theory T able to formalize a certain part of arithmetic is incomplete. More precisely, there is an arithmetical sentence ϕ such that neither ϕ nor it negation $\neg\phi$ is provable in T.*
2. *If T is a consistent formal theory able to formalize certain part of arithmetic, then T does not prove its own consistency.*

Self-reference

One of the key ingredients of Gödel's proof is what is often called *self-reference*. This is a situation in which some subject expresses facts about itself. More generally, we may include also situations in which a program processes its own code and similar ones. Why self-reference is important is clear—the proof is based on the liar paradox and for that we need a kind of circular statement. Let us look at the following statement.

This sentence has 24 letters.

The sentence refers to itself by the word '*this*'. In natural languages we do not have problems stating such sentences, but in formal systems the language is restricted and it is not clear if one can make such statements. What precisely does '*this*' mean in the sentence above? We know that it refers to the sentence itself, but we deduce it from the context (there is nothing else to which the word might point). Let us take another sentence.

The following sentence has 24 letters "It's a beautiful day today.".

(The sentence is false, but this is of no concern now.) Here we point to a sentence by words '*the following*'. This way of referring to a sentence is much less ambiguous. It is customary to refer to a sentence by putting it between quotation marks, so that the sentence is determined uniquely. Furthermore, the sentence to which we refer is given more explicitly. In the first case, we had to finish reading and then we had to go back to read the sentence that is referred to. This is not the usual way one reads the text. Returning back to the beginning of the sentence is not the main problem. What matters more is that the second time we read it we do not read the sentence in order to get a message, but we take it as plain text, as a sequence of symbols. While in the first case we may doubt whether we should allow such constructions in formal systems because it leads to paradoxical sentences such as:

This sentence is false.

the second kind of referring to sentences seems quite innocent. Often the first thing that you learn when studying a programming language is how to make a computer to print a sentence on the screen, for example:

```
printf("Hello world!")
```

This is another example of a direct reference; we are not greeting the computer, we are asking the computer to greet. In fact, even if we wanted we could not forbid such a direct reference in programming languages. A sentence is merely a sequence of letters; we cannot prohibit talking about sequences of concrete symbols. Input data for a computer are such sequences, numbers represented in decimal or binary forms are such sequences, etc.

So let us assume that direct reference is possible in a formal language that we are using in logic. This can be easily checked for concrete versions of logical calculi, but we do not have to do it, as we know that a system without this property would be very limited. Contrary to what the above examples may suggest *direct reference* alone enables us to write down *self-referential* sentences. Of course, we need a little trick. To discover this trick let us try to do it directly. Our first attempt would look like this:

The following sentence has 32 letters "The following sentence has 32 letters.".

This is true, but the sentence does not refer to itself. It refers to a sentence that is only a part of it. Doing it in such a straightforward manner we will never succeed, for the whole sentence must always be as long as the sentence that we refer to. The sentence above is twice as long as the sentence that it refers to. But there is something even more conspicuous in that sentence: it consists of two copies of the same sentence, some quotation marks and periods. Perhaps, we can make use of this property? Yes we can! Look at this:

The following sentence written once with no modification and then once again between the quotation marks has 100 letters "The following sentence written once with no modification and then once again between the quotation marks has 100 letters"

This is a sentence that speaks about itself. (To make it simpler I left out the periods, thus it is not quite grammatically correct, but this is only a minor point.) It is clear that this trick works in general and we can say whatever we want about the sentence. It is worthwhile to repeat it with the liar paradox sentence.

The following sentence written once with no modification and then once again between the quotation marks is false "The following sentence written once with no modification and then once again between the quotation marks is false"

Self-reference is a very interesting phenomenon, but what we need for the proof of Gödel's Theorem is merely a simple technical trick based on saying that the sentence should be written twice. So do not confuse the incompleteness, which is the impossibility of proving all true sentences, with the impossibility of referring to itself. The latter is false—sentences *can* refer to themselves! So can also axiomatic theories express things about themselves (but for this we even do not need any special tricks). The reason why people often confuse these two things is the following. In order to resolve the liar paradox, one has two possibilities: either to say that it is caused by self-reference, or by the impossibility of defining truth for every sentence. We have seen that it is impossible to forbid self-reference without severely restricting the language. Hence the problem is in the definition of truth. For some people, this seems less acceptable, therefore they mistakenly think that self-reference is the problem.

To conclude this section, here is a simple exercise in programming in which you can check the possibility of self-reference in a computer language.

Exercise Write a program (in your favorite language) that prints its own code. (The program should run on any machine, so you cannot cheat by telling the program the place where the code is stored. Check the Internet for the shortest known such program!)

Arithmetization of Syntax

When trying to learn Gödel's theorems you may have come across such a rather strange title. Maybe, that was the reason why you gave up reading on. In fact, there is nothing deep behind it, and my aim in this section is to show you that this is not such an essential part of the argument as often believed to be.

Let me start with *formalization of syntax*. By this we mean giving a precise mathematical definition of the concept of a proof and proving the basic properties of this concept. From Chap. 2 we know that proofs can be rigorously defined. Furthermore, we know that essentially all mathematics can be formalized in Zermelo-Fraenkel Set Theory. Thus we have a formal definition of axiomatic systems, and we have one axiomatic system in which we can formalize axiomatic systems. In particular in this system we are able to reason about the system itself. This suffices for the proof of the First Incompleteness Theorem. In particular, it enables us to prove the First Incompleteness Theorem for Zermelo-Fraenkel Set Theory. But notice that we have only used that Zermelo-Fraenkel Set Theory is strong enough to prove basic properties about the syntax of logic. Hence, if we take a stronger theory, we will be able to prove its incompleteness again. On the other hand, if we take a weaker theory, then it is trivially incomplete, since it does not prove all the axioms of Zermelo-Fraenkel Set Theory.

Thus as long as we are only interested in proving the incompleteness of set theories compatible with Zermelo-Fraenkel Set Theory, we can simply refer to the fact that logic can be formalized. When we want to prove the second incompleteness theorem, we have to be more careful. In order to prove that the consistency of a theory T is not provable in T, we have to formalize syntax in T. It does not suffice to do it in an extension of T. So the question of what is the minimal strength of axioms which suffice for formalization is relevant. One can show that we do not need such a strong theory as Zermelo-Fraenkel Set Theory; it suffices to use Finite Set Theory (which is Zermelo-Fraenkel Set Theory less the Axiom of Infinity). The reason is that syntactical entities are finite, thus we only need finite sets. Hence the Second Incompleteness Theorem can be proved for all extensions of Finite Set Theory.

Syntax can be formalized not only in set theories, but also in arithmetical theories. When we formalize the concept of a proof in arithmetic, we call it *arithmetization of syntax.* We do not get such a formalization automatically as in set theory. In arithmetic we do not have the concept of a sequence of symbols, hence we have to simulate it by a number-theoretical concept.

Gödel first proved the First Incompleteness Theorem for the Theory of Types, the system introduced by Russell and Whitehead in *Principia Mathematica.* In this theory a logical calculus can be represented directly, which means that the definitions of formulas and proofs can be literary used—in the same way as in Zermelo-Fraenkel Set Theory. Gödel used such a natural formalization in his proof—numbers to represent symbols, sequences of numbers to represent formulas and sequences of sequences of numbers to represent proofs. (The reason for choosing the Theory of Types instead of Zermelo-Fraenkel Set Theory was simply the fact that the latter was not as well established as it is now.) So Gödel did not use arithmetization of syntax in his first proof and the resulting independent sentence was a sentence of set theory. After presenting his proof in Königsberg in 1930, von Neumann asked him whether he could construct an independent sentence that would be purely arithmetical. It was a challenge, since such sentences would concern number theory, the heart of mathematics, rather than axiomatic set theory, an invention of logicians. Gödel succeeded in finding such sentences shortly after that. To this end, he had to represent all syntactical objects, not only symbols, by numbers. In other words, he had to arithmetize syntax.

In retrospect, this does not seem like a very difficult task. We are now used to the fact that all finite structures can be encoded by strings of bits and strings of bits can readily be interpreted as digits of numbers. But that was more than a decade before the first digital computers appeared.

Let me briefly sketch the main ideas of the arithmetization of syntax. The key observation is that syntactical objects can be viewed as texts, which are sequences of symbols. We start with identifying the finitely many symbols of logic with some numbers. Say, if we have 26 symbols, we can identify them with numbers $1, 2, \ldots, 26$. Then we need to code sequences of symbols, which means to code sequences of numbers. The most popular way is to use exponents in prime factorizations of numbers. In this encoding the number 1500 represents the sequence $(2, 1, 3)$, since $1500 = 2^2 \cdot 3^1 \cdot 5^3$. Since Peano Arithmetic does not have a symbol for the operation of exponentiation, we have to show that exponentiation can be defined. This is a little tricky and Gödel invented an ingenious way of defining sequences without exponentiation (see page 293). Anyway, it is only a technical detail, since if we include exponentiation as a basic operation, we can avoid this complication.

Formalization does not mean only to assign certain sets or numbers to syntactical entities. We need also to prove that the representations behave as they should. Furthermore, we need to formalize syntactical concepts by formulas of low quantifier complexity. This is the reason the incompleteness theorems require that the set of axioms of the theory be decidable.

To summarize the preceding paragraphs, we need formalization of syntax for the incompleteness theorems, but it does not have to be arithmetization. If we want to prove the theorems for set theories, we can use standard formalizations, if we want to prove them for theories of arithmetic, we have to use formalization based on numbers, called arithmetization, and if we wanted to prove them for some other theory, we would need a formalization based on the concepts of that particular theory.

This brings us to the natural question: which are the theories to which incompleteness applies? Recall that there are two kinds of theories. The first kind are theories that define a class of structures and the aim is to define a large variety of structures of a certain type. For example, we define groups by a set of equations, but we do not want to decide properties, such as the commutativity of the group operation. We want to study commutative groups as well as non-commutative ones. Thus such a system of axioms is designed to be incomplete.

The incompleteness of the type that appears in Gödel's Theorem is different. It applies to theories that attempt to describe a single structure, such as the universe of sets, the algebraic structure of the natural numbers or real numbers. There we would like to get a complete theory, but sometimes it is impossible. It is this second kind of theories for which we use Gödel's theorems, but not all such theories are incomplete. There are important mathematical structures that can be completely axiomatized.[7]

It is difficult to state exactly for which theories T one can use Gödel's argument, since the proof can be generalized in various ways. Essentially, we need three conditions.

1. *The theory T is consistent.*

In the proof below we will use a stronger condition:

[7] An example of an important structure with a decidable set of axioms is $(\mathbb{R}; +, \cdot, \leq)$, the structure of real numbers with operations $+$ and $\cdot$ and the relation $\leq$, see the Theory of real closed fields on page 91.

1′. *The theory T is sound.*

This means that all sentences provable in T are true. Condition 1′ clearly implies Condition 1 (indeed, if T is inconsistent, then it proves $0 = 1$, which is a false arithmetical sentence). The reason for using this condition is twofold. First, the proof is simpler and, second, I can explain the original proof of Gödel, which will also enable me to explain the proof of the Second Incompleteness Theorem. Since only sentences of certain logical complexity play a role in the proof, one can weaken the condition and only require that provable sentences of that complexity are true. This is what Gödel actually did by using the ω-consistency. (Defining this concept here may obscure the essence of the proof, so I prefer to do it in Notes.)

2. *The set of axioms of the theory T is decidable.*

I have already explained this condition in Chap. 2, so let me only briefly remind you its meaning: there exists an algorithm for deciding whether a sentence is an axiom or not. In this book I am only considering such theories and using the term '*a formal theory*' to stress this fact when needed. Other authors use the term '*a formal system*' meaning essentially the same thing.

3. *T is compatible with a theory in which it is possible to formalize syntax.*

This means that it is possible to extend T to a consistent theory in which it is possible to formalize syntax. This suffices for the First Incompleteness Theorem, but for the Second Theorem we need more:

3′. *It is possible to formalize syntax in T.*

Conditions 3 and 3′ are stated informally, since I have not explicated what a formalization of syntax means. This can be made precise by saying that there is a formalization of the provability predicate that satisfies certain *derivability conditions* (see page 297). Condition 3 is not quite convenient for applications; it is much better to give some minimal requirements about what should be provable in theory T. One can show, in particular, that it suffices that T proves some very basic properties of arithmetical operations, or in the case of set theories, some very simple theorems about sets. (For more, detail see Notes.)

The Proof of the First Incompleteness Theorem

Now that we have all the ingredients for the proofs of the incompleteness theorems, we can start with the proof of the First Incompleteness Theorem. As explained above, in order to make the proof simpler, we will assume that T is a *sound theory.*

We need to construct a sentence that is not provable in T. It will be a self-referential sentence, which will be called γ_T. Informally, the sentence expresses the following:

This sentence does not have a proof in T.

Stated more precisely, γ_T is a sentence such that the following is provable in T:

γ_T is true if and only if γ_T does not have a proof in T.

This is the sentence that Gödel used in his proof. In the preceding pages I have explained why it is possible to write down such a sentence in a formal logical language.

First I will prove that γ_T is true. This is proved by arguing by contradiction as follows. Suppose that γ_T is provable in T. Then γ_T is false, since γ_T asserts that γ_T is not provable. On the other hand T can only prove true sentences, hence assuming that γ_T is provable, γ_T must be true. So, assuming that T proves γ_T, we have proved that γ_T is false and true at the same time, which is a contradiction. Thus γ_T is not provable.

Once we know that γ_T is not provable, we are done because we have:

1. γ_T is unprovable, and
2. γ_T is true because it expresses the statement above.

This finishes the proof of the First Incompleteness Theorem. In order to fully understand what is going on in the proof it may be helpful to consider all of the possibilities concerning the provability and truth of the Gödel sentence. These four cases are listed in the table below.

	γ_T is true	$\neg\gamma_T$ is true
γ_T is provable in T	**impossible:** γ_T asserts that γ_T is unprovable in T	**impossible:** only true sentences are provable in T
γ_T is unprovable in T	the only **possible** case	**impossible:** $\neg\gamma_T$ asserts that γ_T is provable in T

The two possibilities on the diagonal (left upper rectangle—right lower rectangle) are impossible because of the self-referential nature of the sentence. The possibility in the right upper rectangle is excluded due to the soundness of the theory T.

The proof resembles very much the liar paradox, more precisely, the paradoxical sentence '*This sentence is not true.*' The difference is in using the predicate '*provable in*' T instead of '*true*'. The concept of a proof in a formal theory is a precise mathematical concept. Moreover, it is a combinatorial concept, a concept that does not require infinite sets. On the other hand, the concept of truth is not a syntactical concept and thus cannot be defined in combinatorial terms. We will see in the sequel that actually this paradoxical sentence proves limitation on how truth can be defined.

The Proof of the Second Incompleteness Theorem

The Second Incompleteness Theorem asserts that, assuming that T is a sufficiently strong theory, T cannot prove the sentence expressing that T is consistent. Hence, if

T is sound, then the Second Incompleteness Theorem extends the first one by giving an explicit statement that is unprovable. Furthermore, the unprovable sentence is a very important one, since it concerns the problem of consistency. Note that the proof of the first incompleteness theorem does give quite a concrete independent sentence, but it is a rather strange sentence.

Let us denote by Con_T the statement that T is consistent. We want to prove that T does not prove this sentence. The idea of the proof is to show that Con_T implies the Gödel sentence γ_T. Then Con_T cannot be provable in T because otherwise also γ_T would be provable and we know that the latter is not the case. To prove this implication, we do not need any technicalities. This is done simply by analyzing the proof of the First Incompleteness Theorem. However, instead of the First Incompleteness Theorem, we will use the following lemma.

Lemma 6 *If T is a consistent formal theory able to formalize certain part of arithmetic then T does not prove γ_T.*

The difference is that we do not assume the soundness of T and that we do not claim that γ_T is independent. When showing the unprovability of γ_T we used soundness only to exclude the case that γ_T is provable, but not true. So we need to show that it suffices to assume the consistency of T to exclude this case.[8]

By way of contradiction, suppose that γ_T is provable in T. We want to prove that T is inconsistent. Let d be a proof of γ_T in T. Depending on how we formalize syntax, d is a concrete finite set, number, or a finite string. Checking that d is a proof can be done by a simple procedure. Assuming T is a sufficiently strong theory, we can formalize the steps of the procedure applied to d, and thus we show that T can verify that d is a proof of γ_T. Whence it is provable in T that

γ_T is provable in T.

But this statement is equivalent to $\neg\gamma_T$ in T. Hence $\neg\gamma_T$ is provable in T. Putting it together with our assumption that T proves γ_T, we get that T is inconsistent, which finishes the proof of the lemma.

Now we can finish the proof of the Second Incompleteness Theorem. Recall that γ_T says that "*γ_T is not provable in T*". Hence the statement of Lemma 6 can be simplified to

If T is consistent, then γ_T.

Thus the sentence Con_T implies the sentence γ_T. This is true in the "real world", but this does not suffice, as we need to prove this implication in T. However, look at the argument above—it is completely elementary. Hence any theory that is able to formalize basic things about finite sets can also prove this implication. To complete the proof of the Second Incompleteness Theorem, it remains to combine the two facts that we have obtained:

---

[8]We used the soundness in the proof above because the argument is simpler when this assumption is present.

1. T proves $Con_T \to \gamma_T$ (which is the formalization of Lemma 6), and
2. T does not prove γ_T (which is exactly Lemma 6).

These two sentences clearly imply that T does not prove Con_T.

Although we do not need it for the proof, it is worthwhile noting that γ_T is *equivalent* to Con_T. This means that each formula implies the other and this is provable in T. We have already shown that Con_T implies γ_T. The converse implication is even simpler. The sentence γ_T asserts that something (namely γ_T) is not provable in T. However, if some sentence is not provable, the theory must be consistent, for in an inconsistent theory everything is provable. Hence γ_T entails Con_T.

In general, the Second Incompleteness theorem does not imply that theories are incomplete. The proofs above show that γ_T and Con_T are not provable in T, provided that T is consistent, but we cannot conclude that T is incomplete, for it can happen that $\neg\gamma_T$ and $\neg Con_T$ are provable. Indeed, take a theory S to which Gödel's Theorems apply and extend it by the axiom $\neg Con_S$. This theory is consistent because its consistency is equivalent to the statement that S does not prove Con_S (the negation of the added axiom), which is the content of the Second Incompleteness Theorem. Let T be the theory S extended with $\neg Con_S$. Since the inconsistency of S entails the inconsistency of T, T proves its inconsistency $\neg Con_T$. Now, Con_T is equivalent to γ_T, so $\neg\gamma_T$ is also provable in T.

Thus Gödel's Theorems do not exclude the possibility that one could find an extension, say of Peano Arithmetic, which would be a consistent complete theory (for every sentence ϕ, either ϕ or $\neg\phi$ would be provable in the theory). The theorems only imply that such an extension cannot be a theory that only proves true sentences. Nevertheless, one can construct just a slightly more sophisticated self-referential sentence ρ_T such that neither ρ_T nor $\neg\rho_T$ are provable in any sufficiently strong consistent T (see *Rosser's sentence*, page 291).

The Undefinability of Truth

Gödel showed the impossibility to realize Hilbert's program by his second incompleteness theorem. Interestingly enough, he discovered it while working in the direction that Hilbert proposed. Gödel wanted to show that assuming that a first order theory of arithmetic, is consistent, then also a second order theory of arithmetic is consistent. The way Gödel arrived at the Incompleteness Theorem is briefly described in Wang's article [306]. Unfortunately, Wang was very terse and we can only guess the details. So, let us try to reconstruct what could have been Gödel's train of thought.

Suppose we want to reduce the proof of consistency of Second Order Arithmetic (see page 295 for the list of axioms) to Peano Arithmetic. Second Order Arithmetic differs from Peano Arithmetic in that it has two sorts of objects: numbers and sets of numbers. The most important axioms are instances of the Comprehension Axiom Schema that says that every formula with one free variable defines a set. The natural idea that Gödel tried to use was to represent sets by formulas. In this way one can automatically satisfy the Comprehension Axiom Schema—given a formula $\phi(x)$, the set that it defines will be represented by $\phi(x)$. Furthermore, a formula can be represented by a number if we use a suitable coding. There is, however, one essential problem: in order to define the membership relation '*n is in a set represented by* $\phi(x)$' we need to be able to define the relation '*n satisfies* $\phi(x)$'. This is possible for a single formula, or for some restricted classes of formulas, but, as we know now, not in general.

It is likely that Gödel realized at this point that if a definition of satisfiability were possible, one would be able to reproduce the liar paradox and thus prove the inconsistency of the theory. Hence, if the theory is consistent, there is no definition of satisfiability.

Now, suppose that Second Order Arithmetic is sound and complete. Then we can replace satisfiability by provability because a sentence is true if and only if it is provable. Provability, unlike satisfiability, has low logical complexity and can be formalized in Peano Arithmetic. However, we know that this cannot work, since satisfiability is not definable. So the assumption that Second Order Arithmetic is sound and complete must fail. Hence if it is sound, it must be incomplete.

We are talking about Second Order Arithmetic, but by now it is already clear that the argument is quite general; it holds for any T containing Peano Arithmetic. Thus we have obtained the Incompleteness Theorem.

The next question is: can we give an example of a sentence that is true, but unprovable? If we look at the proof of incompleteness above, we notice that it uses the sentence '*I am false*' from the Liar Paradox, but where '*false*' is replaces by '*unprovable*', which is exactly Gödel's sentence γ_T. As we have seen, it is possible to eliminate the concepts of satisfiability and truth from the proof completely.

In Gödel's paper the fact about the undefinability of truth does not appear, but is clear that he was aware of this fact. The theorem on the undefinability of truth was published by Tarski in 1936 [290].

Before stating the theorem, let us recall some basic facts about first-order structures. A structure, which is called a *model* in logic, consists of a universe X, some relations $R_1, \dots, R_k$ and some functions $f_1, \dots, f_l$ defined on X. Let us denote this structure by M. The relations and the functions determine the *language of the structure,* which will be denoted by L. In this language we have a relation symbol for every relation and a function symbol for every function. Formulas are constructed from these symbols, connectives, quantifiers and brackets. It is possible to define, for every formula ϕ with n free variables and every string of n elements $a_1, \dots, a_n$ of M, if the formula is satisfied by these elements. In particular, if ϕ is a sentence (no free variables), then we say that ϕ is *true* in M if it is satisfied.

The question is whether or not the property 'ϕ *is true in* M' (or more generally, the relation 'ϕ *is satisfied by elements* $a_1, \dots, a_n$') is definable in the language L. However, this question is meaningful only if we have a natural representation of formulas of L by elements of M. Hence the class of structures for which we can state and prove the theorem corresponds to the class of theories to which the First Incompleteness Theorem is applicable. I will state the theorem using Peano Arithmetic, but, in fact, for both the First Incompleteness Theorem and the theorem below, a much weaker fragment of arithmetic, Robinson Arithmetic (see page 116), suffices.

The Gödel-Tarski Theorem *Let M be a structure in which a model of Peano Arithmetic can be defined. Let L be the language of M. Then there is no formula $Tr(x)$ in L such that for every sentence ϕ in L,*

$$\phi \equiv Tr(\ulcorner \phi \urcorner)$$

holds true in $\mathcal{M}$.

The symbol $\ulcorner \phi \urcorner$ denotes the element representing ϕ in M.

The proof is a straightforward application of self-reference. Suppose one can define truth in M by some formula $Tr(x)$. Construct a formula ϕ that says '*I am not true*'. Then we get a contradiction as in the liar paradox. If we do it formally, the property of ϕ is that it satisfies

$$\phi \equiv \neg Tr(\ulcorner \phi \urcorner),$$

which is in contradiction with the formula in the theorem.

The proof of the undefinability of truth also explains in which context the definition is possible and where it is not. The proof requires self-reference, in order to formalize the liar paradox. If we talk about a structure from outside, there is no self-reference. See also Fig. 4.1 for a schematic explanation of the levels of discourse.

Rather than having deep consequences, this theorem provides explanations. Firstly, it explains semantic paradoxes such as the liar paradox, (explanation being that truth is not definable) and Berry's paradox (explanation being that 'definable' is not definable). Secondly, it helps us understand why a theory cannot prove its own consistency. The most common way of proving the consistency of a theory is to provide a model of it. To use the model to prove that the theory is consistent we need a definition of truth for the model. In set theory, if the model is a set, there is no problem. We can also define class models, but for such models truth is not definable because "classes are too big". In particular, the universe of all sets can be viewed as a class model. If we could define truth for it, we could use this class model to prove the consistency of the theory in itself. So the explanation that the theorem provides is roughly the following. The universe in which we argue is too big to be described by itself, hence there is no way to prove that it is consistent.

This may suggest that the theorem about the undefinability of truth says that, after all, a certain type of self-reference is impossible, namely, in a language L we cannot define the truth about L. It is disputable whether this should be called self-reference, but words are not important. The difference between the self-reference used in the proofs of Gödel's theorems and the undefinability of truth is that the former concerns syntax whereas the latter concerns semantics. In a given language we may be able to define its syntax, but not its semantics.

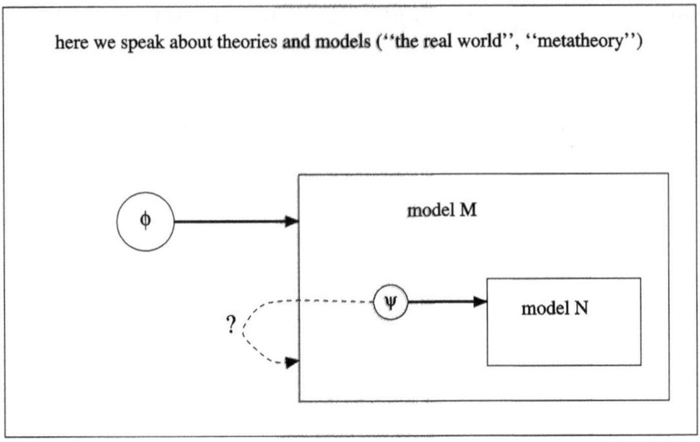

Fig. 4.1 A picture showing the hierarchy of systems in which we speak about concepts of logic. The largest box is the highest level. This is the level on which we communicate; here we use a natural language, such as English, and we assume the most common mathematical principles, such as mathematical induction (however, it is possible to use a formal system, for example, a programming language, also on this level). At this level we define when a sentence ϕ is true in a model M. If M has sufficiently rich structure, we can define logical concepts inside of M; in particular, we can define when a sentence ψ is true in a model N which is defined in M. But in M we do not have a general definition of the truth of sentences in M itself

Interlude—Free Will and the Hierarchy of Observers

In order to explain the necessity of distinguishing the two levels of discourse, we will make a short digression to an apparently unrelated topic. A classical problem in philosophy is whether free will is compatible with the idea that everything is determined by the laws of physics. The problem is complicated by the fact that according to the laws of quantum mechanics randomness is an inherent factor in the basic laws. So let us consider a simplified problem, a thought experiment, whether or not free will is compatible with a world based on completely deterministic laws. '*Deterministic*' means that the future state of the universe is uniquely determined by the present state (or the state at any given moment). The well-known *Laplace daemon*, an entity that knows the positions of all atoms, can determine all events in the future. Such a deterministic world seems clearly in contradiction with free will: a person does not have the possibility to choose his or her action, as all actions are completely determined by past events. Nevertheless, many people believe that this is only an apparent contradiction. The German physicist Max Planck gave an explanation of this paradox based on the role of observers [218]. According to Planck, free will is a subjective feeling caused by the fact that we are not able to predict our own decisions. We can, of course, observe ourselves, we can realize what we are thinking about, recall what we remember, but in observing our mind we are limited; we cannot observe completely the state of our mind, thus we miss many details that may be important for the decision that we are making. Therefore, we can only partly predict our future actions. On the other hand, such limitations do not apply to an external observer. An external observer may have, at least theoretically, the complete information about the state of our mind, and thus the observer may be able to predict our next action.

Let us consider an even simpler situation. Let us imagine a computer trying to predict its own actions in the future. (Incidentally, the problem of free will is also important in artificial intelligence.) Suppose it does it in the middle of a difficult computation for which it needs essentially all its power. Clearly, to predict the outcome of the computation before it is actually performed, it would have to find a more efficient way to solve the given task, but, maybe, there isn't any. Then the only way to find the truth is to go on and compute. On the other hand, the computer can be built so that we can stop it at any moment and copy the state of all logical gates and the content of the memory. As the computer is stopped, we have plenty of time to compute what it will do next when we let it go on. Thus we can know what it will do *before* it will actually do it. Whether you are willing to accept that computers may eventually have free will, or not, you have to admit that this is quite a general argument and it applies to any information processing devices, including human brains.

We are studying mathematical structures and logical theories and they certainly do not have free will. Yet the role of observers is similar. The difference is in that we do not talk about predicting the future, but about the possibility that a structure or a logical theory contains information about sentences that are true in it.

Some Peculiarities of Incompleteness

People are often baffled by the fact that for an arithmetical formula $\varphi(x)$ it can happen that the sentence

$$\textit{for every number } x, \varphi(x)$$

is unprovable, but all sentences

$$\varphi(0),\ \varphi(1),\ \varphi(2),\ \varphi(3)\ \dots$$

are provable. We are talking, of course, about provability in a fixed theory T. When writing $\varphi(0), \varphi(1), \varphi(2), \varphi(3)\dots$, we assume that either T contains explicitly all numerals $0, 1, 2, \dots$, or we have some representation of them by suitable terms in T. If, for instance, T contains constants for 0 and 1 and the operation of addition, then we can use $0, 1, 1+1, 1+1+1, \dots$; thus we get

$$\varphi(0),\ \varphi(1),\ \varphi(1+1),\ \varphi(1+1+1)\ \dots$$

The first thing one should realize is that the first expression is a single sentence Φ, whereas the second one is an infinite set of sentences. Clearly, using basic logical rules we can derive every sentence of the set from Φ, but there is no rule in logic that allows us to do the opposite. Since all proofs in logic are finite, any proof can use only a finite number of premises. Hence, if we had a proof of the sentence Φ from the infinite set $\{\varphi(0), \varphi(1), \varphi(2), \ldots\}$, it would use only a finite number of sentences of the form $\varphi(n)$, for n a numeral. Clearly, a statement about all numbers does not logically follow from a finite number of special cases.

Suppose Φ expresses the fact that an algebraic equation does not have a solution in the domain of natural numbers. We know that it may be very difficult to prove such a sentence. For example, to prove Fermat's Last Theorem for the exponent 3,

$$(x+1)^3 + (y+1)^3 \neq (z+1)^3,$$

is a nontrivial task. However, to prove that a given concrete number is not a solution is trivial and everybody can do it. To this end you simply need to evaluate the terms of the equation and compare them. Such a computation constitutes a proof, in fact, it is the most basic kind of proof.

To be quite precise, mathematicians accept computations as proofs, or parts of proofs, but computations are not explicitly included in the logical calculus. In order to be able to use computations as formal proofs, we have to assume as axioms some basic properties of the operations. In the case of arithmetic computations with integers we need a few simple axioms that are present in any reasonable axiomatization of the arithmetic of natural numbers. In particular the axioms of Peano Arithmetic, even without the axioms of induction, suffice for this purpose.

It is not difficult to find sentences $\varphi(x)$ with the property stated at the beginning of this subsection. As a matter of fact, every unprovable true sentence which is a universal finite formula provides us with such an example. Let us consider Gödel's formula Con_T which expresses that a theory T is consistent. It is a statement of the form

Every string of symbols is not a proof of contradiction from the axioms of T.

Suppose T is consistent. Then by Gödel's Theorem this sentence is unprovable in T, but for every given string of symbols, we can verify that it is not a proof. This verification can again be written as a formal proof in T (provided that T is sufficiently strong to prove basic properties of strings of symbols).

Flexible Formulas and Unpredictable Programs

In the literature there are a variety of results inspired by Gödel's incompleteness theorem. One of these which strikes me most is the *flexible formula* of the Polish logician Andrzei Mostowski (1913–1975), [201]. Let T be a sufficiently strong theory that contains arithmetic. Then it is possible to construct a formula $\mu_T(x)$ with one free variable x with the following property. Not only is $\mu_T(n)$ independent from T for every numeral n, but also these sentences are *mutually independent*. This means that the axioms of T together with sentences of the form $\mu_T(m)$ or $\neg\mu_T(m)$ for $m \neq n$ are still unable to decide if $\mu_T(n)$ holds. Formally, it is stated by saying that if we choose arbitrarily $\mu_T(n)$ or $\neg\mu_T(n)$, but not both, for every n, then this set of sentences together with T is consistent for every such choice. For example,

$$T \cup \{\mu_T(0), \mu_T(1), \neg\mu_T(2), \neg\mu_T(3), \mu_T(4), \neg\mu_T(5), \mu_T(6), \neg\mu_T(7), \mu_T(8), \dots\}$$

is a consistent theory, whatever the continuation of the sequence of formulas is.

The unpredictable program is an example of the same flavor. We already know that it is algorithmically undecidable whether a given program stops. When we are not able to decide if the program stops, can we at least prove something about what the program will do? For instance, can we at least prove that the program is not a virus that will damage our computer? The answer is again no, but again, as in all independence results, the unpredictability depends on the theory T in which we argue about the program.

Here is a program which, from the point of view of T, can print any string. More precisely, we cannot prove in T that the program will not print y for any given string y.

Program $Unpr_T$.

1. `systematically search all proofs of` T;
2. `if you find a` T`-proof of the sentence` "*$Unpr_T$ does not print y*", `for some string of symbols` y, `then print` y.

This is not an explicit definition of a program because it refers to itself, but we already know how to replace self-reference by direct reference. So the above program can be written explicitly in every reasonable programming language.

Let us show that assuming that T proves only true sentences, it is consistent with T that $Unpr_T$ prints an arbitrary string y. Arguing by contradiction, let y be an arbitrary string and suppose that T does prove that $Unpr_T$ never prints y. Let d be this proof. It may happen that there are actually several proofs of this statement, moreover there may exist such proofs for other strings too. Thus we consider the order in which $Unpr_T$ searches proofs and take the first such proof d_0. Let y_0 be the string to which this proof refers to. Then, by definition, $Unpr_T$ must print y_0. But this is impossible because T proves only true sentences. We got a contradiction, hence the assumption that T proves that $Unpr_T$ never prints the string y is false.

Having the unpredictable program, it is easy to construct the flexible formula. Define

$\mu_T(n)$ is true if and only if $Unpr_T$ prints a string w of 0s and 1s of length at least $n+1$ such that the $n+1$-st element of w is 1.

If we are given an infinite sequence $(a_0, a_1, a_2, \dots)$ of 0s and 1s, then for every finite initial segment $(a_0, a_1, a_2, \dots, a_n)$ of it, it is consistent that $Unpr_T$ prints $(a_0, a_1, a_2, \dots, a_n)$. Thus

$$T \cup \{\mu_T(a_0), \mu_T(a_1), \mu_T(a_2), \dots, \mu_T(a_n)\}$$

is consistent for every n, whence T with the infinite sequence of sentences,

$$T \cup \{\mu_T(a_0), \mu_T(a_1), \mu_T(a_2), \dots\},$$

is also consistent.

Looking at $Unpr_T$ from a higher perspective we can show that it will never stop, so it will not print any string. But notice that this requires the assumption that T is consistent. Indeed, if T is inconsistent, then T will stop and print some string because if T is inconsistent, there are proofs of all sentences in T and it is only the matter of which string y will occur in these proofs first. You may counter by saying that we can prove in T that $Unpr_T$ is quite innocent—it will either run for ever, or print a string and stop. So look at the following modification.

Program $Unpr_T^+$.

1. `run` $Unpr_T$;
2. `if` $Unpr_T$ `prints a string` y `which is a code of a program, then run` y.

Now you cannot prove using only T that this program will not destroy your computer.

Notes

1. *Doing things formally...* may obscure the main ideas, but once the essence of the proof is clear, seeing more precise arguments will certainly be useful. The only technical part in the proofs of incompleteness theorems is constructing self-referential sentences, so we will concentrate on it.

   As I said, writing correct formulas which express facts about formulas requires representation of formulas by terms. Suppose we have a theory T that contains arithmetic of natural numbers and we want to speak about a concrete number, such as 3 or 2001. It is possible that T contains decimal numerals as constants and there is no problem, but it does not have to be the case. In such a case we have to describe the number somehow. When the theory contains constants and function symbols, we may use terms. For example, using the constant 1 and the binary function symbol $+$ we write 3 as $1+1+1$. Some theories do not have constants or function symbols. Then we can still describe a concrete number (for example, 0 is the least number, 1 is the next after the least, etc.). Also in programming languages some data can be expressed directly (certainly, numbers) and some have to be described. When talking about syntactical objects we have to be more careful. We cannot assume that each syntactical object can be used also as the name of itself. This is not possible even in natural written language. Consider the sentences:

   I am talking about you.

   and

   I am talking about "you".

We have to use quotation marks to distinguish *you* as a person from *you* as a word. To make this distinction in formal languages people use various notations. Here we will use $\ulcorner \ldots \urcorner$ as such quotation marks. Thus $\ulcorner \phi \urcorner$ is a term that represents the formula. In the case of a theory that only speaks about numbers, we first represent formulas as numbers, and then $\ulcorner \phi \urcorner$ will be a term representing the number assigned to the formula ϕ. (Should it sound confusing, think of $\ulcorner \ldots \urcorner$ simply as quotation marks.)

The next thing is to represent functions. Suppose, for instance, that we want to talk about squaring numbers. If we have $\times$ in T, then we can use the term $x \times x$. Having only a small number of function symbols and constants we cannot hope to represent all functions. There will be a lot of functions that we can describe by formulas, but not by terms. For presenting the proof, it is, however, very convenient to assume that we can represent one particular function by a term. Namely, we need the *doubling* function

$$\sigma(x) \mapsto \sigma(\ulcorner \sigma \urcorner). \tag{4.6}$$

This is the function that given a formula $\sigma(x)$ with a free variable, substitutes the term representing $\sigma(x)$ for the variable x. We will denote this function by d and assume that it can be defined by a term in T (and use the same symbol for the term). Function d is the function '*write the following sentence twice...*' from the verbal exposition of self-reference in the previous section. Formally, we need the following to be provable in T for every formula $\sigma(x)$,

$$d(\ulcorner \sigma(x) \urcorner) = \ulcorner \sigma(\ulcorner \sigma(x) \urcorner) \urcorner. \tag{4.7}$$

This is a formalization of the fact that d represents the function defined by (4.6). Now let $\phi(x)$ be an arbitrary formula. We will prove an important lemma which is a formal statement of self-reference. The lemma is called the *Diagonal Lemma* or the *Fixed Point Theorem*. The idea is due to Gödel, but in an explicit form, it first appeared in Carnap [38].

Lemma 7 *Given a formula $\phi(x)$, it is possible to construct a formula ψ such that the following equivalence is provable in T*

$$\psi \equiv \phi(\ulcorner \psi \urcorner).$$

The meaning of ψ is the self-referential statement

This formula satisfies condition ϕ.

We already know how to do it:

The following formula written twice satisfies condition ϕ: "The following formula written twice satisfies condition ϕ".

(To be precise, one has to add that the second occurrence is after a colon and in the scope of quotation marks.)

To do it formally, we use the doubling function d. Thus ψ is

$$\phi\big(d\big(\ulcorner\phi(d(x))\urcorner\big)\big). \tag{4.8}$$

It is worth checking that the formula expresses the same as the sentence above: $\phi(\ldots)$ stands for *satisfies condition* ϕ, $d(\ldots)$ for *written twice* and $\ulcorner\phi(d(x))\urcorner$ is the direct reference "*The following formula written twice... satisfies condition* ϕ".

Now we will show that the sentence expresses what it should. Notice that ψ has the form $\phi(t)$, where t is the term $d(\ulcorner\phi(d(x))\urcorner)$. Thus we need to prove in T that $t = \ulcorner\psi\urcorner$. Indeed, by (4.7) and (4.8),

$$d\big(\ulcorner\phi(d(x))\urcorner\big) = \ulcorner\phi\big(d\big(\ulcorner\phi(d(x))\urcorner\big)\big)\urcorner = \ulcorner\psi\urcorner.$$

Hence ψ is equivalent to $\phi(\ulcorner\psi\urcorner)$.

Note that this proof is almost identical with the proof of the Fixed Point Theorem of λ-calculus (see page 149). This is not surprising, especially for those who solved the problem on page 275. That exercise asks for a fixed point of the function computed by the compiler for the programming language—giving the compiler the program it simply prints the program.

For the proof of the incompleteness theorem we take a special formula for $\phi(x)$, the formula saying that x has no proof in T. Let us denote this formula $\neg\pi_T(x)$. Then there exists a formula γ_T such that in T

$$\gamma_T \equiv \neg\pi_T\big(\ulcorner\gamma_T\urcorner\big).$$

To finish the proof we need one more fact: *if a sentence ϕ is provable in T, then $\pi_T(\ulcorner\phi\urcorner)$ is also provable in T*. This is a special case of the following general principle, called *Σ-completeness*:[9]

If ψ is a true Σ_1 sentence, then ψ is provable in T.

So if a true sentence is sufficiently simple, then it is provable. Here is the idea of why it is true. Consider the following very simple true sentence.

$$(1+1)\cdot(1+1) = 1+\big(1+(1+1)\big).$$

This equality can be easily derived using the distributive and associative laws. We have such laws in any reasonable arithmetical theory. Now consider the following Σ_1 sentence (a sentence asserting that 4 is even).

$$\exists x\big(x\cdot(1+1) = 1+\big(1+(1+1)\big)\big).$$

An important observation is that to derive this sentence from the one above, we need only logic. So the same theory that proves the first sentence, proves the second one. In Σ_1 sentences we may also use bounded universal quantifiers, so we may have a sentence of the form

$$(\forall x \leq 1+1)\alpha(x).$$

[9]It would be more appropriate to call it Σ_1 completeness, but I will stick to the traditional notation.

Proving such a sentence reduces, using only some basic axioms of arithmetic, to proving three sentences

$$\alpha(0),\ \alpha(1),\ \alpha(1+1).$$

In such a way we can gradually reduce proving a Σ_1 sentence to proving simple equations and inequalities of the type above with no free variables. The proofs of such equations and inequalities are essentially numerical evaluations of terms.

Now we need to check that the concept of provability in T is Σ_1. This is the part of the proof that uses the fact that the set of axioms of T is decidable. This condition on T is, in fact, stronger than what is needed for the proof; it suffices only to assume that the axioms of T form a recursively enumerable set (a Σ_1 set). However, axiom systems with a recursively enumerable set of axioms never occur in practice, and, furthermore, any such set of axioms can easily be replaced by recursive set of equivalent axioms. So let us assume that the set of axioms is decidable. The statement that a sentence ϕ is provable has the form:

> *there exists a sequence of formulas such that every formula in the sequence is either an axiom of logic, or an axiom of the theory T or it follows from some previous formulas by a logical rule, and the last formula in the sequence is* ϕ.

Clearly, the only quantifier is '*there exists a sequence*'; the remaining part is a statement that can be decided by an algorithm. This is exactly the form of Σ_1 sentences.

Now we can finish the proof quickly. Suppose that γ_T is provable in T. Then, by Σ-completeness, also $\pi_T(\ulcorner\gamma_T\urcorner)$ is provable in T. But γ_T is equivalent to $\neg\pi_T(\ulcorner\gamma_T\urcorner)$. Hence we have both $\pi_T(\ulcorner\gamma_T\urcorner)$ and $\neg\pi_T(\ulcorner\gamma_T\urcorner)$, which means that T is inconsistent. Thus, assuming that T is consistent, γ_T is not provable in T. Furthermore, γ_T is true because γ_T is equivalent to $\neg\pi_T(\ulcorner\gamma_T\urcorner)$ which expresses the true fact that γ_T is not provable in T.

2. *The General Diagonal Lemma.* Often a more general version of the Diagonal Lemma is needed in which a parameter is allowed.

Lemma 8 *Given a formula $\phi(x)$, it is possible to construct a formula $\psi(y)$ such that the following sentence is provable in T*

$$\forall y(\psi(y) \equiv \phi(\ulcorner\psi(\bar{y})\urcorner)).$$

The expression $\ulcorner\psi(\bar{y})\urcorner$ is the Gödel number of the formula $\psi(\bar{y})$, where $\bar{y}$ is the *numeral* y, which is the term $SS\ldots S(0)$ with y occurrences of S.

3. *Rosser's sentence.* In the exposition above, we used the assumption that the theory T was sound. Gödel stated his theorems in greater generality by using the weaker assumption of *ω-consistency.* This property is the restriction of the

assumption that T is sound (proves only true sentences) to the class of Σ_1 sentences. Thus ω-consistency means that if a Σ_1 sentence ψ is provable in T, then ψ is true. Equivalently we can say that if $\exists x\phi(x)$ is provable in T for a Σ_0 formula $\phi(x)$, then ϕ holds for a concrete number. (One can also view this property as the converse of Σ-completeness.) The reason for introducing this concept was not only to get a more general theorem, but also because the concept of a sound theory is rather vague. By restricting this property to Σ_1 sentences about natural numbers, he was able to avoid such ambiguities.

Thus Gödel's results left open the possibility that some ω-inconsistent theories were complete. However, it is not difficult to rule out also this strange possibility. This was done by J.B. Rosser [249] shortly after Gödel. Rosser defined the following self-referential sentence ρ_T:

For every proof of ρ_T, there exists a shorter proof of $\neg\rho_T$.

The idea behind this construction is to obtain a sentence that is symmetric in the sense that the sentence and its negation behave in the same way. The sentence ρ_T is, clearly, not quite symmetric. It says that "*if there exists* a proof of ρ_T, then...", whereas its negation says "*there exists* a proof of $\neg\rho_T$...". However we get symmetry if we assume that *either there exists a proof of ρ_T or there exists a proof of $\neg\rho_T$*. Under this assumption, ρ_T is equivalent to:

the shortest among the proofs of ρ_T and $\neg\rho_T$ is a proof of $\neg\rho_T$,

and $\neg\rho_T$ is equivalent to:

the shortest among the proofs of ρ_T and $\neg\rho_T$ is a proof of ρ_T.

After these preliminary considerations, the proof is easy. Assume that ρ_T is not independent. Well, this is exactly the assumption that guarantees the symmetry. Furthermore, according to the Σ-completeness this statement is also provable in T.

So it suffices to consider one of the two (non-exclusive) possibilities, say that $\neg\rho_T$ is provable in T. Let d be such a proof of $\neg\rho_T$. Since T is consistent, there is no proof of ρ_T in T, in particular, there is no such proof below d. The sentence "*d is a proof of $\neg\rho_T$ and there exists no proof of ρ_T below d*" is a Σ_1 sentence (in fact even Σ_0), thus, by Σ-completeness, it is provable in T. But this sentence implies ρ_T, hence T is not consistent. Thus we got a contradiction, which shows that $\neg\rho_T$ is not provable in T.

4. *Some attempts to get round the second incompleteness theorem.* Gödel's Theorem excludes the possibility for a theory to prove its consistency. To check that you understand this theorem and the concepts involved, it is instructive to look at some attempts to extend a given theory T so that the extension proves its consistency. In the sequel we will write $Con(T)$ instead of Con_T in order to avoid indices with indices, and use the expression $T+\phi$ to denote the theory obtained from a theory T by adding a sentence ϕ as an additional axiom.

   (1) Define $T_0 = T$, $T_1 = T_0 + Con(T_0)$, $T_2 = T_1 + Con(T_1)$ and so on. Let S be the union of theories $T_0, T_1, T_2, \ldots$. Assume that T is a sound theory. Then,

not only is T consistent, but also T_1 because $Con(T)$ is sound. Moreover, T_1 is also sound. In the same manner we can prove by induction that all T_n are sound.

Here is a *wrong argument* that S proves its consistency.

> *Suppose that a contradiction can be derived from S. Then it must be derived from some finite part, say T_n. Now, $Con(T_n)$ is provable in S because it is already provable in T_{n+1}.*

What is wrong in this argument? We want to prove $Con(S)$ in S. Let us see what is the sentence $Con(S)$ that we want to prove. It says that S is consistent, hence it must imply that every part of S is also consistent. Thus in S the following implication is provable $Con(S) \to \forall x\ Con(T_x)$. One can prove in S also the converse implication. In fact, if we formalize the above wrong argument we get a correct proof of $\forall x\ Con(T_x) \to Con(S)$ in S. So the problem with the above argument is that we are trying to use concrete instances $Con(T_0), Con(T_1), Con(T_2), \ldots$ instead of the sentence $\forall x\ Con(T_x)$.

(2) Since the above construction did not produce what we wanted, one may try to construct a kind of limit of the process of adding consistencies. The Fixed Point Theorem mentioned above seems the right tool for this purpose, since usually a limit is a fixed point. So what we want is a sentence δ such that

$$\delta \equiv T \text{ together with } \delta \text{ is consistent,}$$

or, using the *Con* notation

$$\delta \equiv Con(T + \delta).$$

Our hope is that $T + \delta$ is consistent and proves its own consistency. The second condition follows immediately from the construction of δ. Hence, by Gödel's Theorem, the other condition must fail, so $T + \delta$ is not consistent. As the proof of the Second Incompleteness Theorem is not difficult, we can find where the contradiction appears by following its proof for this particular theory line by line, but an intuitive explanation is more useful. If we add consistency statements one by one as in (1), we get stronger and stronger theories. In $T + \delta$ we add a sort of loop that generates automatically all such extensions. Because of that the strength of the theory goes beyond any limit and thus becomes too strong—inconsistent.

(3) Another process that has been studied is to extend the process of adding consistencies transfinitely. I will have to say more about it in Chap. 7.

5. *Arithmetization of syntax in Peano Arithmetic.* Peano Arithmetic has only 0, S, $+$ and $\times$ as primitives. Therefore, arithmetization of syntax in Peano Arithmetic is not quite easy. The problem is that all natural ways of coding sequences of numbers by numbers are based on exponentiation which is not among primitives. Exponentiation can be defined, but again it is hard to do it without having a way of coding sequences.

Gödel devised a tricky way of coding sequences that avoids exponentiation, his *β-function:*

$$\beta(u,d,i) = \text{Rem}(u, id + 1),$$

where $\text{Rem}(x, y)$ denotes the remainder of x after division by y. Using this function one can code arbitrary finite sequences of natural numbers. Given a sequence $a_1, \dots, a_n$, the code is a triple (u, d, n) such that for all $i = 1, \dots, n$, $a_i = \beta(u, d, i)$. We can find such a code as follows. Let $d = (\max_i a_i n)!$. Then the numbers $d + 1, 2d + 1, \dots, nd + 1$ are pairwise relatively prime, thus, by the Chinese Remainder Theorem, there exists some u such that

$$u \equiv a_i \mod (id + 1), \quad \text{for } i = 1, \dots, n.$$

Since $a_i < id + 1$, we have $a_i = \text{Rem}(u, id + 1) = \beta(u, d, i)$. When proving properties of the β function, we cannot assume that we have the factorial function, but we can easily prove by induction that for every x, there exists a number y divisible by all the numbers $z \leq y$, which suffices for our purpose.

Furthermore, β-function has a simple arithmetical definition:

$$\beta(u,d,i) = a \equiv \exists z \exists w \big(u = (id + 1)z + a \wedge a + w = id\big).$$

Several other methods of coding finite sequences have been found. For example, it is possible to first define exponentiation and then use definitions of sequences based on it.

Peano Arithmetic is certainly not the weakest theory for which one can prove the incompleteness theorems. One can show that the first incompleteness theorem holds true for any consistent theory that contains *Robinson Arithmetic*, which is the theory axiomatized by the basic seven axioms of Peano Arithmetic, without the schema of induction (see page 116).

For such a weak theory, it is not clear whether it makes sense at all to ask whether the second incompleteness theorem holds in it. Proving the second incompleteness theorem means proving that it is consistent to assume in T that there exists a proof of contradiction in T. However, if we cannot prove the basic properties of the formalized syntax, the proof of contradiction would only be some number satisfying some strange formula.

Thus it is a bit surprising that one can prove a meaningful version of the second incompleteness theorem even for such weak theories. The point is that one can define an interpretation of a much stronger theory in Robinson Arithmetic, a theory that is not as strong as PA, but strong enough to properly formalize syntax. The interpretation is defined by a suitable formula $\nu(x)$. The numbers that satisfy ν form an initial segment of the whole universe of numbers of Robinson Arithmetic, a segment closed under addition and multiplication. The interpretation is the restriction to ν. Then one can prove that it is consistent to assume that there exists a proof of contradiction already in the segment defined by ν. Since we have all the basic properties on the segment, such a contradiction is represented by a number that indeed encodes a proof of contradiction.

6. *Second-Order Arithmetic.* The theories I am going to describe are axiomatized by sentences that are true in the structure

$$\big(\mathbb{N}, \mathcal{P}(\mathbb{N}); 0, S, +, \times, \leq, \in\big).$$

This structure has two sorts of objects: numbers and sets of numbers. Correspondingly, the language of these theories L_2 has the constant 0, the arithmetical operations (defined only on numbers), the relation $\leq$ (defined only on numbers) and the membership relation $x \in Y$ that expresses that a number x is in a set Y. I will use lower case letters for numbers and upper case letters for sets of number.

The role that Peano Arithmetic plays for the structure of natural numbers $(\mathbb{N}; 0, S, +, \times, \leq)$ is now played by *Second Order Arithmetic,* which is usually denoted by Z_2. The axioms of Z_2 are the axioms of Robinson Arithmetic, the Schema of Comprehension and the Axiom of Induction.

The Schema of Comprehension *For every formula $\phi(x)$ of L_2 (possibly with other free variables treated as parameters)*,

$$\exists Y \forall x\ \big(x \in Y \equiv \phi(x)\big),$$

where $\phi(x)$ may contain parameters. This schema enables us to define a set of numbers by any formula in the language of Z_2.

The Axiom of Induction

$$\forall Y\big(0 \in Y \wedge \forall x\big(x \in Y \to S(x) \in Y\big) \to \forall x\ x \in Y\big).$$

The schema of induction for all formulas $\phi(x)$ of L_2,

$$\phi(0) \wedge \forall x\big(\phi(x) \to \phi\big(S(x)\big)\big) \to \forall x \phi(x),$$

is a consequence of the Schema of Comprehension and the Axiom of Induction.

7. *Proofs of the impossibility of impossibility.* Sometimes people ask if it possible to prove some kind of higher level incompleteness in the sense that the statement of unprovability itself is unprovable. A trivial example of this kind is the question of whether a contradiction, say $\phi \wedge \neg\phi$ is provable in a theory T. By Gödel, T is unable to decide if it is consistent, so it is unable to decide if $\phi \wedge \neg\phi$ is provable in T or not. To get something a little bit more interesting, suppose we want to have theories S and T and a sentence ϕ such that

   1. it is provable in S that T is consistent;
   2. it is consistent with S to assume that ϕ is provable in T;
   3. it is consistent with S to assume that ϕ is *not* provable in T.

   Again, the existence of such theories and such a sentence is a straightforward consequence of the second incompleteness theorem. Take T a sound theory that is strong enough for Gödel's Theorem to be applicable. Let S be $T + Con(T)$, thus 1. is satisfied. Let ϕ be $\neg Con(T)$. By Gödel's Theorem, we have both S is consistent with $\neg Con(S)$ and S is consistent with $Con(S)$. Note that $\neg Con(S)$ is $\neg Con(T + Con(T))$, which means that T proves $\neg Con(T)$ which is ϕ. Thus we have 2. Furthermore, $Con(S)$ is $Con(T + Con(T))$, which means that $\neg Con(T)$ is not provable in T, thus we get also 3.

A more interesting question is the following. Is it possible that we prove that ϕ is undecidable in some theory without knowing whether ϕ is true or false? The Continuum Hypothesis is probably an example. However, the question whether it is true, or false is rather delicate and it is even not clear that one can answer this question. So we would prefer an arithmetical sentence. The least arithmetical complexity for which such independence can occur is Π_2 and Σ_2 (see Proposition 13). In a short unpublished note Petr Hájek described a contrived example of such an independence result. Here is a brief sketch of a version of his result. Let $P_{PA}(x)$, respectively $P_{ZFC}(x)$, denote formulas expressing the provability predicates of Peano Arithmetic, respectively Zermelo-Fraenkel set theory. Applying the Diagonal Lemma construct a sentence δ such that *PA* proves

$$\delta \equiv P_{ZFC}(\ulcorner\delta\urcorner) \to P_{PA}(\ulcorner\neg\delta\urcorner). \tag{4.9}$$

Since *ZFC* proves all theorems of PA, the equivalence above is also provable in *ZFC*. We will show that

a. it is provable in *ZFC* that δ is independent from *PA* and
b. δ is independent from *ZFC*.

(For the second part of the statement we need to assume that *ZFC* is consistent and does not prove that it is inconsistent; needless to say, these assumptions are accepted as true by almost all mathematicians.)

First we will argue in *ZFC* and prove the independence from *PA*. We will use the fact that for every arithmetical formula ψ, we can prove in *ZFC*

$$P_{PA}(\ulcorner\psi\urcorner) \to \psi.$$

This is called the *reflection principle for PA* and it is an assumption stronger than the statement that *PA* is consistent.

Still arguing in *ZFC*, suppose $PA \vdash \delta$, then we get $PA \vdash P_{PA}(\ulcorner\delta\urcorner)$, by Σ-completeness, and also $PA \vdash P_{ZFC}(\ulcorner\delta\urcorner)$, as *PA* can see that *ZFC* is at least as strong as *PA*. From $PA \vdash \delta$ and $PA \vdash P_{ZFC}(\ulcorner\delta\urcorner)$ we get $PA \vdash P_{PA}(\ulcorner\neg\delta\urcorner)$ using the definition of δ and modus ponens. Thus *PA* proves that it is inconsistent. By the reflection principle, we also have that *PA* is inconsistent, but in *ZFC* we know that *PA* is consistent. Thus *PA* does not prove δ.

Now suppose $PA \vdash \neg\delta$. Then $PA \vdash P_{PA}(\ulcorner\neg\delta\urcorner)$. Note that $\neg\delta$ is $P_{ZFC}(\ulcorner\delta\urcorner) \wedge \neg P_{PA}(\ulcorner\neg\delta\urcorner)$, so we also have $PA \vdash \neg P_{PA}(\ulcorner\delta\urcorner)$. Thus *PA* is inconsistent, which is not true. Hence we have shown in *ZFC* that δ is independent from *PA*.

Now we prove that δ is independent from *ZFC*.

Suppose $ZFC \vdash \delta$. Then $ZFC \vdash P_{ZFC}(\ulcorner\delta\urcorner)$, hence $ZFC \vdash P_{PA}(\ulcorner\neg\delta\urcorner)$. By reflection, $ZFC \vdash \neg\delta$, thus *ZFC* would be inconsistent. So $ZFC \nvdash \delta$.

Suppose $ZFC \vdash \neg\delta$. Then $ZFC \vdash P_{ZFC}(\ulcorner\neg\delta\urcorner)$ and $ZFC \vdash P_{ZFC}(\ulcorner\delta\urcorner) \wedge \neg P_{PA}(\ulcorner\delta\urcorner)$, by the definition (4.9) of δ. Whence $ZFC \vdash P_{ZFC}(\ulcorner\neg\delta\urcorner) \wedge P_{ZFC}(\ulcorner\delta\urcorner)$. Thus *ZFC* proves that it is not consistent, which is not true. Hence δ is independent from *ZFC*.

8. *Modal logic of provability.* Let $P_{PA}(x)$ denote, as above, that the formula with the Gödel number x is provable in *PA*. In particular it defines an operator that from a given formula ϕ produces another formula $P_{PA}(\ulcorner\phi\urcorner)$. It is natural to think of this operator as a modality, as the meaning of $P_{PA}(\ulcorner\phi\urcorner)$ is that ϕ holds in some stronger sense, namely, we have a proof of it. So let us use $\Box$ for it and let us see what kind of modal logic it gives. Again, we will only be interested in the propositional fragment of such a logic. We will only consider the provability in *PA*, but the results are valid for a large class of theories. Nevertheless, you should keep in mind that the properties of this modality may depend on the theory whose provability we consider.

   It has been shown that this gives rise to a modal logic which can be axiomatized by a few simple axiom schemata and rules. This logic is called *Provability Logic*. It is not very difficult to prove that the provability predicate for *PA* satisfies all axioms and rules of the Provability Logic. The difficult part of the characterization is to show that all true statements about provability in *PA* expressible in modal logic can be proved in Provability Logic. This was proved by Solovay in 1976 [280].

   Provability Logic, PRL, is axiomatized by the standard axioms and rules of the classical propositional calculus and the following modal ones: the generalization rule

   (1) *from* ϕ, *derive* $\Box\phi$,

   and three axiom schemas

   (2) $\Box(\phi \to \psi) \to (\Box\phi \to \Box\psi)$,
   (3) $\Box\phi \to \Box\Box\phi$,
   (4) $\Box(\Box\phi \to \phi) \to \Box\phi$.

   The rule (1) and the axioms (2) and (3) are satisfied by the usual provability predicate of Peano Arithmetic. They are also called *the derivability conditions*, as they axiomatize which properties the provability predicate must satisfy so that we can prove Gödel's Theorem. In this book we have always assumed that the provability predicate is formalized in a natural way. This means that we express the syntactical concepts needed in the formalization in the most direct way. However, it is natural to ask for which other formalizations Gödel's Theorem still holds true. One possible answer to this question is that these are all predicates that satisfy conditions (1), (2) and (3) above.

   The axiom (4) is the opposite of the axiom $\Box\phi \to \phi$ of S_4, which is not valid in *PRL*. It says, in terms of provability in *PA*, that the only way one can prove $P_{PA}(\ulcorner\phi\urcorner) \to \phi$ is to prove ϕ. This statement is known as *Löb's Theorem*, thus the axiom is also called *Löb's axiom*.

   G. Sambin and D. de Jongh independently proved that in *PRL* one can define explicitly fixed points. This is the following schema that corresponds to the Diagonal Lemma.

   (5) *Given a modal formula* $\phi(x)$ *in which the propositional variable* x *occurs only in the scope of the modality* $\Box$, *one can find a modal formula* δ *such that* $\delta \equiv \phi(\delta)$ *is provable in PRL.*

We will prove Gödel's theorems in this formalism. By (5) we get a formula δ such that

$$\delta \equiv \neg\Box\delta. \tag{4.10}$$

Then, in particular, we have

$$\delta \rightarrow \neg\Box\delta.$$

Applying generalization (1), axiom (2) and modus ponens we get

$$\Box\delta \rightarrow \Box\neg\Box\delta.$$

By (3) we also have

$$\Box\delta \rightarrow \Box\Box\delta.$$

The last two give together

$$\Box\delta \rightarrow \Box\Box\delta \wedge \Box\neg\Box\delta.$$

Using classical rules and (2) one can easily show $\Box\Box\delta \wedge \Box\neg\Box\delta \rightarrow \Box(\Box\delta \wedge \neg\Box\delta)$ which gives

$$\Box\delta \rightarrow \Box(\Box\delta \wedge \neg\Box\delta).$$

Let us abbreviate $\Box\delta$ by ϕ. Thus the last formula is

$$\Box\delta \rightarrow \Box(\phi \wedge \neg\phi).$$

Now the meaning of the consequent is clear: it says that a contradiction is provable. (The contradiction is represented by $\phi \wedge \neg\phi$, but we know that all contradictions are equivalent.) Let us take the contrapositive implication

$$\neg\Box(\phi \wedge \neg\phi) \rightarrow \neg\Box\delta. \tag{4.11}$$

The meaning is that if a contradiction is not provable then δ is not provable. But if δ is not provable then, by (4.10), δ is true. Thus we have obtained the First Incompleteness Theorem. The Second Incompleteness Theorem follows immediately by observing that the meaning of $\neg\Box(\phi \wedge \neg\phi)$ is the consistency of *PA* and that it implies δ.

9. *A definition of a provability predicate for which Gödel's Theorem fails.* Let T be an arbitrary (recursively axiomatized) theory. Let $P_T(x)$ be the natural definition of the provability predicate for T. Define

$$P_T^*(x) \equiv P_T(x) \wedge Con(T).$$

Then we can prove $\neg P_T^*(\ulcorner 0 = 1 \urcorner)$, the consistency expressed using P_T^*, without knowing almost anything about T. The argument is: either $Con(T)$ is true and then we cannot derive a contradiction from the axioms of T, or it is false, in which case the provability predicate $P_T^*(x)$ defines an empty set.

If T is consistent, then $P_T(x)$ and $P_T^*(x)$ define the same set of sentences provable in T. Thus, for example, $P_{PA}^*(x)$ defines (in $\mathbb{N}$) the sentences provable in Peano Arithmetic, and Peano Arithmetic proves $\neg P_{PA}^*(\ulcorner 0 = 1 \urcorner)$.

This definition is, clearly, not natural. In particular it has higher quantifier complexity than is needed in the proof of Gödel theorems and thus it does not satisfy the provability condition (3).

10. *A decreasing sequence of theories.* S. Feferman and H. Friedman [275] proved that it is possible to construct a sequence of theories with decreasing consistency strength—a sequence of recursively axiomatized theories $S_0, S_1, S_2, \dots$ such that S_n proves the consistency of S_{n+1} and every S_n is consistent. Equivalently, the strict ordering of theories, defined by $T < S$ if S proves the consistency of T, is not well-founded.

    The basic idea can be explained as follows. Take an *increasing* sequence of consistent recursively axiomatized theories $T_0, T_1, T_2, \dots$, where T_{n+1} proves the consistency of T_n. We know how to construct such a sequence: given a theory T, define $T_0 = T$ and, for $n > 0$, $T_{n+1} := T_n + Con(T_n)$. Now consider a nonstandard model of arithmetic M. In this model we have theories T_ν also for infinitely large numbers ν. Thus in M we can take an infinite number ν and define $S_n = T_{\nu-n}$ for *all finite* n. What remains to do is to simulate this construction without referring to a nonstandard model. The trick is that, due to the incompleteness theorem, for every recursively axiomatized extension of Peano Arithmetic T, we can define a number which either does not exist, or is infinitely large, and it is consistent with T that the number exists. Namely, take ν to be the least number that encodes a proof of contradiction in T and for $n \geq 0$, put $S_n = T_{\nu-n}$ if ν exists, otherwise $S_n = T$.

    To prove that the consistency of S_{n+1} is provable in S_n, we argue in S_n as follows. Either T is consistent and $S_{n+1} = T$, so S_{n+1} is consistent. Or there exists ν and then $S_n = S_{n+1} + Con(S_{n+1})$, hence we have $Con(S_{n+1})$ as an axiom. Let us observe that the theories S_n are not only consistent, but also sound provided that we start with a sound T.
11. *The undefinability of truth and complexity hierarchies.* The undefinability of truth in a language L can also be explained using the concept of complexity: the definition of truth for L has higher complexity than any relation that can be defined in L.

    This is best seen on the example of the natural numbers. Consider the structure with the usual arithmetical operations and the relation of inequality $(\mathbb{N}; 0, 1, +, \times, \leq)$. The sets definable in this structure are classified into the arithmetical hierarchy consisting of classes Σ_i and Π_i (see page 141). We know that the hierarchy of classes Σ_i is strictly increasing and every definable set and relation is in Σ_i for some i. Using a definition of truth, one can define every definable set without using additional quantifiers. If the definition of truth were in some Σ_i, it would not be able to define sets in Σ_{i+1}. So the complexity of the truth predicate in $(\mathbb{N}; 0, 1, +, \times, \leq)$ is higher than all Σ_i and Π_i.
12. *A curiosity.* I am indebted to A. Visser and A.C. Franco for the following observation.

Theorem 21 *For every sufficiently strong finitely axiomatized arithmetical theory T, one can construct a sentence α_T such that*

a. *Peano Arithmetic proves α_T,*
b. *T proves $\alpha_T \equiv Con_T$.*

'Sufficiently strong' means that T is sequential (see page 574) and proves the cut-elimination theorem; 'T is arithmetical' means that it is formalized in the standard language of arithmetic; Con_T is the standard formalization of the consistency of T.

We will use the following result from [222].

Lemma 9 *For every finitely axiomatized sequential theory T, there exists a formula $\nu(x)$ such that T proves*

a. $\nu(0) \wedge \forall x\ (\nu(x) \rightarrow \nu(x+1))$,
b. *there is no cut-free proof of contradiction in T whose* Gödel *number satisfies ν.*

If T is arithmetical, then $\nu(x)$ is also an arithmetical formula. Let

$$\beta := \big(\nu(0) \wedge \forall x\ \big(\nu(x) \rightarrow \nu(x+1)\big)\big) \rightarrow \forall x\ \nu(x).$$

This is just an instance of the induction schema, hence it is an axiom of *PA*. By the lemma, it is provable in T that β implies that there is no cut-free proof of contradiction in T. Since T proves the cut-elimination theorem, this implies Con_T. So $\beta \rightarrow Con_T$ in T. To make it equivalent, put $\alpha_T := \beta \vee Con_T$.

To see that this theorem has nothing to do with Hilbert's Program, just notice that there is no assumption about the consistency of T in the theorem. So even if T is inconsistent, *PA* proves α_T.

4.3 Algorithmically Unsolvable Problems

We will now consider problems that have an infinite number of instances. The instances are determined by parameters. We would like to know whether a uniform way of solving all instances of a given problem exists. More precisely, we ask whether there is an *algorithm* that, for every parameter, finds a solution of the problem. When there is no such algorithm, we say that the problem is algorithmically unsolvable. Note that this does not exclude the possibility to solve the problem for some instances. We will focus on *decision problems,* the problems for which the answer is *yes* or *no.*

This question is, certainly, relevant for practice. When the problem is algorithmically unsolvable, we cannot program it. However, the question is also very important for foundations. We will see connections between algorithmic unsolvability and unprovability.

The Halting Problem

The most basic algorithmically unsolvable problem is the *halting problem.* The problem and its algorithmic unsolvability is formally stated in the following theorem due to Turing [293].

Theorem 22 *The following problem is algorithmically undecidable: Given a code of a program P and input data D, to decide whether or not P will halt after a finite number of steps when run on data D.*

We are talking about programs, but one may replace programs by Turing machines (as was in Turing's paper), or any other concept that formalizes computations. In the sequel we will simply say that "*a program P halts on data D*" assuming implicitly that the program will compute for a finite number of steps and then it will halt.

Here is a sketch a proof of this theorem. In fact, we will prove a stronger theorem: *it is undecidable if a program P will halt when run on its own code.* The proof is easy and it goes by contradiction. Suppose the problem is decidable. Let Q be a program that decides it. Thus Q does the following:

1. Q always outputs 0 or 1;
2. if $Q(P) = 0$, then P does not halt on its own code;
3. if $Q(P) = 1$, then P halts on its own code.

Now we modify Q so that instead of printing 1, it will run for ever. To obtain the modified program Q' we only need to replace the command `'print 1'` by a few lines that will define a loop from which there is no escape; for example

`while` $n > 0$ `do` $n := n + 1$

Let us see now what happens if we run Q' on its own code. To this end we first run Q on the code of Q' and when $Q(Q') = 1$ we apply the modification. Thus there are two possibilities.

1. If $Q(Q') = 0$, then also $Q'(Q') = 0$, in particular, Q' halts. This is a contradiction because, if Q does what we assume, $Q(Q') = 0$ implies that Q' should not halt on its own code.
2. Now assume $Q(Q') = 1$. then Q' will use the modification and run into an infinite loop. This is again a contradiction because according to $Q(Q') = 1$, Q' should halt on its own code.

   As there are no other possibilities, we have shown that the assumption that there exists a decision algorithm for the halting problem is false. Hence the theorem is proved.

Clearly, the proof uses the same type of self-reference as Russell's paradox and Gödel's Theorem. While the two concern rather abstract areas—set theory and proof theory—here we are dealing with a very concrete problem. In spite of that the proof is weird: who would ever want to run a program on its own code? Thus one may propose to forbid running programs on their codes and hope the undecidability will be eliminated. This should be taken seriously because somebody may conclude that only computers may do stupid things such as running a program on its own code and, therefore, computers are inferior to humans in their computing abilities (if we disregard their speed). This conclusion is totally wrong. To explain why, let us assume that we have two programming languages L_1 and L_2. Let us now change the problem a little and ask if a program P written in L_1 halts on the code of the same algorithm written in L_2. The proof above remains correct with this modification. Hence it does not matter how we encode the program before we give it as an input, as long as the encoding is computable. Now suppose that we get a program P and *arbitrary* data D. It is always possible that D encodes P in some, perhaps strange, way. So the restriction to programs running on their own codes is only a means to prove the undecidability, but the result is really about the general halting problem.

Furthermore, an easy consequence of the result is that we can find *one* program for which it is undecidable if it halts on given data. Such a program is any universal program (see page 130). A universal program U simulates what an arbitrary given program does with given input data. In particular, if we apply U to an input that encodes a program P and data D, then the computation of U on this input will halt if and only if the computation of P will halt on D. Hence, if we could solve the halting problem for U, we could also solve the halting problem for P.

The undecidability of the halting problem is a basic result that is used to prove a number of other undecidability results. In fact all undecidable problems of a certain form (those that define recursively enumerable sets, see page 140) are derived from it. This is done by encoding computations by the mathematical structures appearing in those problems. I will describe several such concrete undecidability results in the following sections. One of these is the problem about tiling the plane with a given set of tiles. In this particular case the reduction is very complicated, but it is easy to imagine what is going on. We want to represent the computation process of a Turing machine. What the machine does in one computation step is not so far from putting a tile next to others. If we allow sufficiently complex shapes or color patterns of tiles, we can encode by them the rules that the machine uses to move its head and to write symbols on the tape. For other problems, the reduction may be even more difficult; in particular, the undecidability of Diophantine equations is one of the major results obtained in the 20th century, but again, the whole point is how to encode computations.

Tilings of the Plane

Of all the concrete computationally unsolvable problems the problem of tiling the plane has always attracted the most attention. The statement of the problem is: given a finite set of shapes of tiles, decide whether the entire plane can be tiled only using tiles of these shapes. A more general problem is with tiles having colored patterns and with the additional requirement that the patterns must match.

If the problem were to tile a given finite region of the plane, the problem could be solved by exhaustive search, but we are asked to tile the whole infinite plane. In some cases we may be sure that it is possible; in particular this is the case when the set of tiles can be used to tile the plane in a periodic way (see Fig. 4.2). When trying to construct such a tiling, as soon as we see that the pattern is repeating, we conclude that we can continue in the same way *ad infinitum*. But, assuming that a given set of tiles can be used to tile the whole plane, is it always possible to find a periodic tiling with the same tiles?

Fig. 4.2 A periodic tiling of the plane by three Wang tiles

At first glance, one necessarily gets the impression that it is a typical example of a completely useless problem that mathematicians study only for their pleasure. However, the contrary is true. The problem originated in connection with a question about decidability of a certain part of first order logic. Thus the tiling problem is related to one of the fundamental problems of mathematics. Subsequent research led to further interesting results, even with applications in physics.

The problem was first posed by Hao Wang in 1961 [304]. He considered square tiles with colored edges. In an admissible tiling the colors of neighboring tiles must match. Such tilings are also called *domino tilings*, since the rule which tiles match is essentially the same as in dominos; however, Wang tiles are not allowed to be rotated or flipped, they can only be translated. Hao Wang became interested in the problem because he was able to reduce this problem to a problem in logic. He proved that the existence of such domino tilings of the plane can be expressed by sentences of a certain type. Hence, if the former problem was undecidable, then so was the problem of deciding if a sentence of this type is true. He also noticed that if every domino tiling of the plane were periodic, then the problem would be decidable.

The domino tiling problem was solved by R. Berger in 1966 [21]. Berger proved that there is no algorithm for deciding if a given finite set of tiles can tile the plane. This implies that there must exist finite sets of tiles which admit only non-periodic tilings. Berger's proof gives a concrete set of tiles by which the plane can only be tiled in aperiodic ways, however, this set is very large (20,426 tiles). Subsequently R.M. Robinson reduced the number to only six tiles and eventually R. Penrose to two. Penrose found two such pairs, neither of which consists of Wang tiles. The Penrose tiles are not squares; they can be represented by simple shapes with more complex color patterns, or colorless tiles with more complicated shapes. Penrose obtained a US patent for his tiles in 1979. It is an open problem whether or not there exists a single "aperiodic" tile.

By definition, aperiodic tilings do not possess symmetries which are (nontrivial) translations, but they may be symmetric with respect to rotations. Penrose tilings are especially interesting. They may have rotational symmetry, but those that do not have such a symmetry possess at least *quasisymmetries* which means, roughly speaking, that when they are suitably translated and rotated by an angle $2\pi/5$ they match very well, although not completely. A matter formed from atoms or molecules

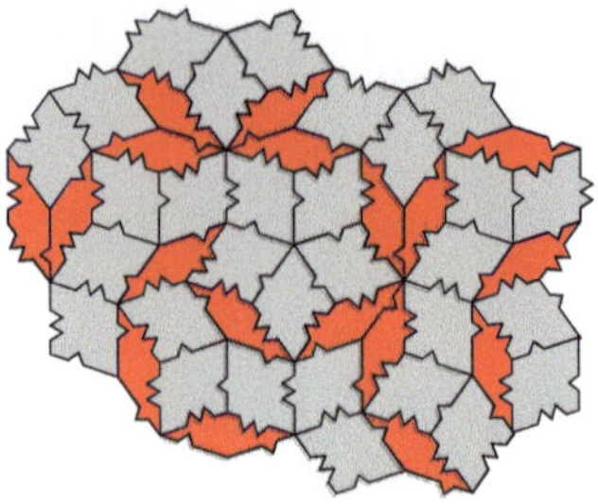

Fig. 4.3 A section of an aperiodic tiling by the Penrose rhombi. There are other tilings of the plane that use the same two tiles, but all are aperiodic. Every angle in the two quadrangles is a multiple of $\pi/5$. Consequently, any two edges in a tiling of the plane determine an angle that is a multiple of $\pi/5$. The projections and indents ensure that the rules about which edges of the tiles and how they are matched are satisfied

that would behave like these tiles would exhibit a *fivefold* symmetry on the macroscopic level. Such symmetries are impossible in true crystals, therefore such materials were coined *quasicrystals*. First quasicrystals were synthesized from aluminium-manganese alloy in 1984. In 2011 the Nobel Prize in chemistry was awarded to Dan Shechtman for the discovery of quasicrystals. Thus a problem in mathematical logic is connected with a Nobel Prize in chemistry!

The growth of quasicrystals does not have a simple explanation such as for ordinary crystals. It might be a phenomenon that cannot be efficiently simulated by classical computers and thus might require quantum computers. (Quantum computations will be treated in the next chapter.)

Algorithmically Unsolvable Problems in Number Theory

Undecidable problems can be found in almost every field of mathematics. Number theory is particularly interesting because it is one of the oldest fields. The most important problem among the unsolvable problems in number theory is the problem of deciding whether or not a given Diophantine equation has a solution. Recall that a Diophantine equation is an equation with integral coefficients and we are only looking for integral solutions (see page 56). This problem was among the problems that David Hilbert selected for his famous list in 1900; it was problem number 10:

> "Given a Diophantine equation with any number of unknown quantities and with rational integral numerical coefficients: *To devise a process according to which it can be determined by a finite number of operations whether the equation is solvable in rational integers*."[10]

[10]See [125], page 458. '*Rational integers*' are the usual whole numbers. Since the concept of integers can also be defined in number fields which are different from the field of rational numbers, he used the specification '*rational*'.

Hilbert did not speak about algorithms as the concept had not been precisely defined and the wording also suggests that he thought that an algorithm could be found. At that time it was hard to imagine that such a concrete finite problem could not be solved by an algorithm. The progress in number theory, to which Hilbert contributed a lot, was evidence that the solution should be positive. For more and more classes of Diophantine equations, theories were developed and this gave decision procedures for many kinds of equations. Yet the general problem is unsolvable. Let me stress again that this fact is more important for theory than for practice. The Diophantine equations that we encounter in practical problems are usually solvable; the problem is only that the algorithms are not efficient enough for being used in practice. The consequence for number theory is more relevant: in contrast with the special classes of equations that we now understand well, it is not possible to design a theory of general Diophantine equations that would give us a decision procedure.

The algorithmic unsolvability of the Diophantine problem was proved by Yuri Matiyasevich in 1970 [193]. As substantial work on the problem had been done before, part of the credit should also be given to Martin Davis, Julia Robinson and Hillary Putnam whose fundamental results were used in Matiyasevich's solution of the tenth Hilbert problem.

It is always good to see a concrete example. Below I present a compact set of 18 Diophantine equations found by J.P. Jones [143].

$$
\begin{aligned}
q &= b^{5^{60}}\\
l &= \mathbf{A} + t\theta\\
e &= \mathbf{C} + m\theta\\
n &= q^{16}\\
r &= \left[g + eq^3 + lq^5 + \left(2(e - \mathbf{D}\lambda)\left(1 + \mathbf{B}b^5 + g\right)^4 + \lambda b^5 + \lambda b^5 q^4\right)q^4\right]\left(n^2\right.\\
&\qquad \left.- n\right) + \left(q^3 - bl + l + \theta\lambda q^3 + \left(b^5 - 2\right)q^5\right)\left(n^2 - 1\right)\\
p &= 2ws^2r^2n^2\\
k &= r + 1 + hp - h\\
a &= \left(wn^2 + 1\right)rsn^2\\
c &= 2r + 1 + \phi\\
d &= bw + ca - 2c + 4a\gamma - 5\gamma\\
\tau^2 &= p^2k^2 - k^2 + 1\\
k^2 &= 4\left(c - ksn^2\right)^2 + \eta\\
d^2 &= \left(a^2 - 1\right)c^2 + 1\\
f^2 &= \left(a^2 - 1\right)i^2c^4 + 1\\
b^5 &= \theta + 2\mathbf{D}\\
\lambda + q^4 &= 1 + \lambda b^5\\
elg^2 + \alpha &= (b - \mathbf{BC})q^2\\
(d + of)^2 &= \left(\left(a + f^2\left(d^2 - a\right)\right)^2 - 1\right)(2r + 1 + jc)^2 + 1.
\end{aligned}
$$

In these equations **A**, **B**, **C** and **D** are parameters and all remaining letters are unknowns. Jones proved that the problem to determine, for given natural numbers **A**, **B**, **C**, **D**, whether the system has a solution in the domain of natural numbers is undecidable. Notice that one can eliminate the first 10 equations by substituting the right hand side expressions for $q, l, e, \dots, d$. The reason why Jones did not do it was that he aimed at the most compact presentation—the substitutions would increase the size of equations substantially. This set of equations is not exactly what Hilbert had in mind; his original problem was about single equations. But there is a simple way to transform a finite set of Diophantine equations into one that has exactly the same set of solutions. For every equation from the set, we subtract the left hand side from the right hand side and take the square. If some numbers satisfy the equation then we get zero, otherwise we get a positive number. Hence if we sum all these squares we get zero if and only if all equations are satisfied. This transformation produces the following equation from the 18 equations above:

$$\begin{aligned}&\left[q - b^{5^{60}}\right]^2 + [l - \mathbf{A} - t\theta]^2 + [e - \mathbf{C} - m\theta]^2 + \cdots \\ &\quad + \left[(d + of)^2 - \left((a + f^2(d^2 - a))^2 - 1\right)(2r + 1 + jc)^2 - 1\right]^2 = 0. \quad (4.12)\end{aligned}$$

Thus we have a single equation such that it is undecidable for which parameters **A**, **B**, **C**, **D** it has a solution. This is similar to the halting problem. We have observed that the halting problem is undecidable for a universal program (or Turing machine). The equation above is also in a sense universal: taking suitable parameters we can encode every Diophantine equation.

Unprovability from Undecidability

The most important algorithmically unsolvable problem from the point of view of the foundations is the problem to decide whether or not a given sentence is provable in first order logic. By the Completeness Theorem, this is the same as asking to decide the logical validity of the sentence. The history of this problem, *Entscheidungsproblem,* was mentioned in Chap. 2.

Having algorithmically undecidable problems, it is a simple matter to show that the provability in first order logic is also undecidable. The essence is that in first order logic we are able to speak about programs, Turing machines, tiles, and, of course, Diophantine equations. So we are able to express statements such as '*program P halts at data D*', '*a set S of tiles can be used to tile the whole plane*', etc. If provability in logic were decidable, then we could also decide these statements, which we know is not possible.

Here is a description of this argument in more detail. The undecidable problem that best suits this purpose is the undecidability of the Diophantine problem. (Note, however, that one does not need such a strong result for this proof.) The undecidability of the Diophantine problem means that it is not possible to find an algorithm that would decide for a given polynomial equation

$$p(x_1, \dots, x_n) = q(x_1, \dots, x_n),$$

where p and q polynomials with integral coefficients if there exists an integral solution. This is the same as asking whether or not a sentence of the form

$$\exists x_1 \dots \exists x_n \; p(x_1, \dots, x_n) = q(x_1, \dots, x_n)$$

is true. Thus the *truth* of such sentences is undecidable. However, we need to show that the *provability* of some sentences is undecidable. The provability in first order logic and the truth of an arithmetical sentence are different things, so we have to use provability in some theory. We need a finite set of axioms, which is the same as asking for one sentence α (the conjunction of the axioms), such that in the theory axiomatized by α every sentence of the form above is provable if and only if it is true.

By the incompleteness theorem, we already know that there are no theories that would satisfy this property for *every arithmetical* sentence, but the sentences that we consider here are very special. Suppose such a sentence is true; how should we prove it? If the sentence is true, the polynomial equation has a solution. Now, given a solution to a polynomial equation it is very easy to verify it—simply evaluate the polynomials (see the example on page 290). Computations are not formal proofs, but it is clear that there must be a simple theory that formalizes such computations. Such a theory is the theory of rings (see page 20). Hence if α is the conjunction of the axioms of rings,

$$\alpha \rightarrow \exists x_1 \dots \exists x_n \; p(x_1, \dots, x_n) = q(x_1, \dots, x_n)$$

is true if and only if it is provable in first order logic. This is exactly what we need for reducing the undecidability of the Diophantine problem to the undecidability of first order logic: since there is no algorithm for deciding if a Diophantine equation has a solution, there is also no algorithm for deciding the provability of such sentences in first order logic.

In a similar vein we can, furthermore, derive Gödel's incompleteness theorem. Arguing by contradiction, suppose that we have a complete axiomatization of arithmetic by a decidable set of axioms. Recall that this means that the set of axioms is decidable. Then we claim that there is also an algorithm to decide if a sentence is provable in this system. The point is that we can systematically generate all proofs in T. Hence if we want to decide a particular sentence ϕ, we generate proofs until we find a proof of ϕ or a proof of $\neg\phi$. The completeness of T, the fact that always either ϕ is provable, or $\neg\phi$ is provable, guarantees that this procedure always halts. We do not know how many proofs we need to generate, but we will always find a proof of ϕ or a proof of $\neg\phi$.

Example Let ϕ be a sentence that expresses that a program P halts on data D. If we had a complete axiomatization of arithmetic, then we would always be able to find a proof either of ϕ or of $\neg\phi$, and thus decide the halting problem. Thus there is no complete axiom system for arithmetic (whose axioms could be systematically generated).

This proof does not point to an explicit sentence not provable in T, but more detailed analysis of this argument does give a procedure for constructing such an independent sentence. It can also be shown that we do not get anything new—the sentence that we obtain is equivalent to the Gödel sentence for T. So, clearly, the two proofs, one showing the incompleteness, the other showing the undecidability, share a common principle, a principle appearing in various disguises such as the liar paradox, Cantor's diagonal method, and self-reference.

Using this connection we can produce a lot of "concrete" independence results. Here are three such results stated in the form of a theorem.

Theorem 23 *Let T be a consistent theory*[11] *in which it is possible to express facts about finite structures. Then one can find*

1. *a program P and data D such that P does not halt on D,*
2. *a finite set of tiles S that can tile the whole plane,*
3. *a Diophantine equation $p = q$ that does not have a solution,*

such that none of these three sentences is provable in T.

These sentences are very explicit, yet they have an unpleasant property, which is, paradoxically, the universal applicability of this theorem. This is best seen in the following version of sentence 3:

3′. *one can find four natural numbers such that the Diophantine equation* (4.12) *with these numbers substituted for parameters* **A, B, C, D** *is unsolvable, and this fact is not provable in T.*

Hence the unprovable sentences obtained in this way are all the same except for the four numerical parameters. We would like to see a relation between the theory T and the unprovable sentence, but all this is hidden in the four numbers. If number theorists would like to understand such equations, they would have to analyze the structure of the numerical parameters, since the overall structure of the sentence does not provide relevant information. The same concerns the logicians who would like to show the unprovability of a well-known problem in Diophantine theory by reducing it to such a sentence. Therefore, such sentences do not count as truly concrete. In Sect. 4.4 we will see independence results that do not suffer this drawback.

We can also use the connection between unprovability and undecidability to explain why there are algorithmically undecidable problems. Many problems have the following form: given a computable relation $R(x, y)$, we are to decide, for every x, whether there exists y satisfying the reaction. Suppose a is an element for which there is no such y. This is expressed by the formula

$$\neg\exists y\ R(a, y). \tag{4.13}$$

The question is: how difficult is it to prove this sentence? One can show that if the problem is algorithmically unsolvable, then there is no bound on how hard is to prove such sentences—whatever theory T one takes, there are always instances that

[11] Recall that in this book a theory is always axiomatized by a decidable set of axioms.

are unprovable in T. Moreover, this characterizes precisely when such a problem is algorithmically unsolvable.

This is summarized in the following theorem (for the proof, see Notes).

Theorem 24 *Let $R(x, y)$ be a computable relation. Let P be the problem: for a given x, decide whether there exists a y such that $R(x, y)$. Then the following are equivalent.*

1. *The problem is algorithmically undecidable.*
2. *There is no theory T such that for every a, the sentence* (4.13) *is true if and only if it is provable in T.*

The Complexity of Ramsey's Theorem

Instead of continuing the list of concrete undecidable problems I will turn now to one of the recurring motifs of the book, which is Ramsey's Theorem. The result that I am going to present here is also a good example of a different type of result connected with algorithmic undecidability, a result whose nature is *impossibility* rather than *undecidability*. We already know that the Ramsey property that is proved to hold for every subset of the natural numbers is so complex that if applied to higher cardinalities it defines extremely large cardinals. Thus we will not be surprised that there is something computationally hard in the theorem.

Let us start with a very simple example. Suppose we are given a computable function f on the natural numbers which has two values 0 and 1. The function is given as a program to compute f. Suppose our task is to determine a value a such that $f(n) = a$ for infinitely many natural numbers. Is this a decidable problem?

The answer is no, and the proof is easy. It suffices to reduce the halting problem to it. Let a program P and data D be given. Let f be the function defined by $f(n) = 0$ if P does not halt during the first n steps of the computation, otherwise $f(n) = 1$. Then $f(n)$ is always 0 or it is 0 only for a finite number of natural numbers n according to whether or not P halts on D. As we cannot decide the latter, we also cannot decide the former.

In Ramsey's theorem we consider a function f that gives one of the two values (usually we call it *colors*) to each k-element subset of integers. The claim is that there exists an infinite subset such that all its k-element subsets get the same value. We call such sets *homogeneous.* Now we can ask again: given a program for f, can we compute a value a such that there exists an infinite homogeneous set of this color? Since the above toy problem is a special version with $k = 1$, the answer is no; in fact, for every k, it is undecidable.

As the undecidability arises already at the trivial level $k = 1$, this is not interesting. What is an interesting problem is this: *how difficult is it to describe an infinite homogeneous set?* There always exist uncountably many homogeneous sets, so there are certainly very difficult ones, but what the problem is about is to find the simplest ones.

In particular we can ask the following.

Assume that f is a computable function, does there exist an infinite homogeneous set X such that X is decidable?

Now if $k = 1$, then trivially such an X exists—take the color a that occurs infinitely many times and take all numbers such that $f(n) = a$. The algorithm for computing f can also be used to decide if a number is in this set. However, for $k \geq 2$ it is different. The proof of Ramsey's Theorem does not produce a decidable set because when constructing a homogeneous set we have to decide an infinite number of questions of the form: is a particular set finite or infinite. This suggests (but it does not prove!) that for some computable f there exists no infinite decidable homogeneous set. This intuition was eventually confirmed by a theorem proved by C.G. Jockusch [141].

Theorem 25 *There exists a computable partition of pairs of natural numbers such that there exists no computable infinite homogeneous set.*

This theorem does not have direct consequences of the form of unprovable true sentences, but the techniques used in the proof are closely related to those that are used in the unprovable version of Ramsey's Theorem, which we will consider shortly.

Notes

1. *Proof of Theorem* 24. Suppose that the problem P is decidable. Then the set of true sentences of the form $\neg\exists y R(a, y)$ is decidable. So we can simply take this set of sentences as T.

   Now suppose that T proves $\neg\exists y R(a, y)$ if and only if this sentence is true. Then, for a given a, we can decide the truth of $\exists y R(a, y)$ as follows. We systematically check $R(a, y)$ for all elements y and in parallel we also check all proofs, looking for a proof of $\neg\exists y R(a, y)$. One of these must always succeed.

   Note that Theorem 23 is a corollary of Theorem 24.
2. *Computability theory*, also called *recursion theory*, is an important field in mathematical logic. Some basic concepts of this theory were introduced in Chap. 2. Here we will continue with a few more facts. We will start with repeating the definitions of the most important concepts.

   The most important concept is the concept of an *algorithmically decidable set,* also called simply *decidable set*, or *recursive set*. These are sets for which there exists an algorithm for deciding if a given element is in the set. By the elements we mean numbers, or strings of symbols in a finite alphabet, or another natural class of finite structures. The second most important concept is a *computable function*, also called a *recursive function*. This is a function f for which there exists an algorithm to compute $f(x)$ from a given element x. Thus we can identify recursive sets with recursive functions that have two values 0 and 1.

The next most important concept is a *recursively enumerable set*. A set X is recursively enumerable if there exists an algorithm that halts exactly on the elements of X. A related concept is a *partial recursive function*, which is a function computable by an algorithm, but it does not have to be defined for all inputs.

The following are basic facts concerning these concepts:

- *A set X is recursive if and only if both X and the complement of X are recursively enumerable.*

The proof is trivial: given an x run the two algorithms, one for X and the other for the complement. One of them must stop, and then we know the answer. Notice, however, that this principle has non-constructive nature: we have no a priori bound on the time we need to run the two algorithms. We have used this fact to show that a complete theory is decidable.

- *A set is recursively enumerable if and only if it is the domain of a partial recursive function.*
- *A set is recursively enumerable if and only if it is the range of a partial recursive function.*
- *A set is recursively enumerable if and only if it is the range of a recursive function.*

The last fact justifies the name *recursively enumerable*.

The new concept that we will consider here is the reducibility of one set to another one. By this we want to express the intuitive notion that sometimes we can reduce one decision problem to another one. There are several nonequivalent ways one can define it. The following is the most useful one.

A set X is reducible to a set Y if there exists a computable function f such that

$$x \in X \quad \text{if and only if} \quad f(x) \in X.$$

Let us denote by H the *halting set*, which is the set of programs that will halt when run on their own code. Clearly H is recursively enumerable.

Theorem 26 *Every recursively enumerable set X is reducible to H.*

Thus H is the most complex set among recursively enumerable sets. We say that H is *complete* in the class of recursively enumerable sets. On the other hand, all natural undecidable problems that are recursively enumerable are proved to be undecidable by reducing H to them, hence all such sets are complete. It is natural then to ask if there are other recursively enumerable undecidable sets. That it is so, is a theorem of R.M. Friedberg [80] and A.A. Mučnik [202]. They also proved that there are pairs of recursively enumerable sets such that they are incomparable with respect to the relation of reducibility. Their proof furthermore gives that for every two recursively enumerable sets X and Y such that X is reducible to Y, there exists a recursively enumerable set Z which is strictly in between. To state these results, it is better first to factorize the set of all recursively enumerable sets by the relation of mutual reducibility, and talk about the resulting ordering. Then the class of complete sets becomes the largest element of the ordering and the two properties can be expressed by saying that this ordering is not linear and it is dense.

Friedberg and Mučnik introduced a proof technique called the *priority method*, which soon became the main tool in recursion theory. It can be viewed as an essential extension of the diagonal method. When using the diagonal method to show that a set X is not in a class C, we run an infinite process which at stage n ensures that X is different from the nth set of C. Many problems in recursion theory, including the problem solved by Friedberg and Mučnik, can be presented in a similar way: we need to construct a set that satisfies an infinite list of requirements. However, the requirements may be conflicting, thus by satisfying a requirement n we may injure an already satisfied requirement m. Therefore, it is much harder to define a process that would accomplish it. Essentially, it amounts to carefully define which requirements have higher and which lower priorities to be satisfied. The concrete applications of this method cannot be presented without introducing more concepts from recursion theory and proving several auxiliary results, which is outside of the scope of this book.

3. *Tilings and formulas.* Once we know that first order logic is undecidable, it is natural to ask what is the smallest complexity of formulas for which there is no decision procedure. The answer, of course, depends on the parameters by which we measure the complexity of formulas. The basic classification is by the number and the type of quantifiers. We consider sentences in the prenex form, i.e., with all quantifiers in front of the formula and classify them by the type of prefix. From this point of view the simplest undecidable class consists of sentences of the form

$$\forall x \exists y \forall z\ \phi(x, y, z).$$

Recall that a sentence is a formula that does not have any free variables. Hence formulas of this type have only the three explicitly mentioned variables. Furthermore, we assume that the sentences contain binary relations, but no constants or function symbols. (The reason for disallowing constants and function symbols is that they could simulate quantifiers.) The undecidability of this class of sentences was proved by A.S. Kahr, E.F. Moore and H. Wang [146]. Below we will sketch main ideas of a reduction of the domino tiling problem to the problem of logical validity of sentences of this class.

A set of Wang tiles can be defined as follows. S will be the set of tiles, H will be a binary relation that specifies which tiles can be placed next to each other horizontally, and V is the relation that specifies which tiles can be placed next to each other vertically. A tiling of the plane T is an assignment which for every pair of integers, determines the tile with these coordinates and which respects the relations H and V.

Having such a formal description it is routine to write down a formula that expresses the existence of a tiling. However, we need a very special formula:

it must not use any function symbols or constants and it may use only three quantifiers (of a particular type). To express the existence of a tiling by such a simple formula is quite tricky. First we need a different formalization. We will imagine a tiling to be a structure whose universe are integers and the presence of a tile t with coordinates (i, j) will be expressed by a binary relation $R_t(i, j)$. Here is the formula:

$$\forall x \exists y \forall z \Big(\bigwedge_{t \neq t'} (\neg R_t(x, z) \vee \neg R_{t'}(x, z)) \wedge \bigvee_{(t,t') \in H} (R_t(x, z) \wedge R_{t'}(y, z)) \wedge \bigvee_{(t,t') \in V} (R_t(z, x) \wedge R_{t'}(z, y)) \Big).$$

Let us denote this formula $\forall x \exists y \forall z\, \tau(x, y, z)$. Our goal is to prove that $\forall x \exists y \forall z\ \tau(x, y, z)$ has a model if and only if there exists a tiling of the plane by tiles defined by (S, H, V). Since the tiling problem is undecidable, we will obtain that it is undecidable if formulas of the form $\forall x \exists y \forall z\ \phi(x, y, z)$ have a model, which is equivalent to the undecidability of the problem if formulas of the form $\exists x \forall y \exists z\ \psi(x, y, z)$ are provable (ϕ and ψ are quantifier-free).

Given a tiling by (S, H, V), it is very easy to check that the tiling is a model of $\forall x \exists y \forall z\ \tau(x, y, z)$. Interpret y as $x + 1$, then the first part of the formula says that for every pair of coordinates (x, z), there is at most one tile with these coordinates; the second part says that next to the right from (x, z) there is a tile that satisfies the relation H; the third part says that next upwards from (z, x) there is a tile that satisfies V.

The opposite direction is more difficult. Given a model M of $\forall x \exists y \forall z\ \tau(x, y, z)$, we should prove that there exists a tiling. We cannot assume that M itself is a tiling—if the formula has a model then it has infinitely many different models. Thus we will only show that M has a *submodel* which is a tiling, albeit a tiling only of one quadrant. This suffices for proving that there is a tiling of the entire plane. To construct the submodel we recall an idea of Skolem (see page 89). We enlarge M to a model M' by adding a unary function f so that M' satisfies

$$\forall x \forall z\ \tau\big(x, f(x), z\big). \tag{4.14}$$

For every x, the value $f(x)$ is simply one of the elements y such that $\forall z\ \tau(x, y, z)$. Pick an arbitrary element a_0 of M' and let $a_1 = f(a_0)$, $a_2 = f(a_1), \ldots$. Take the submodel M'' of M' induced by the universe $\{a_0, a_1, a_2, \ldots\}$. We claim that if we interpret elements a_n as coordinates n, then M'' is a tiling of the quadrant of the plain whose points have nonnegative coordinates. To see that we observe that formula (4.14) is a universally quantified conjunction of three formulas, hence it is equivalent to the conjunction of the following three formulas:

$$\forall x \forall z \Big(\bigwedge_{t \neq t'} (\neg R_t(x, z) \vee \neg R_{t'}(x, z)) \Big),$$

$$\forall x \forall z \Big(\bigvee_{(t,t') \in H} \big(R_t(x,z) \wedge R_{t'}(f(x),z)\big)\Big),$$

$$\forall x \forall z \Big(\bigvee_{(t,t') \in V} \big(R_t(z,x) \wedge R_{t'}(z,f(x))\big)\Big).$$

These three sentences express precisely what we need for M'' to be a tiling of the nonnegative quadrant. (Note that if the set $\{a_0, a_1, a_2, \dots\}$ is finite, we get a periodic tiling.)

Concerning the proof that the tiling problem is undecidable, let me only mention the main difficulty that one has to overcome. The basic idea is to encode Turing machine computations by tilings, so that we can reduce the halting problem to the nonexistence of a tiling. In Chap. 2 we introduced the matrix model of computation (see page 137), which is almost a reduction to tilings. In the matrix model the dependence of the entries in the matrix is vertical and diagonal. But this is only a minor difference; it is a simple task to replace diagonal relations by vertical and horizontal relations. What is a problem is that in the matrix model (as in any other straightforward simulation of Turing machine computations) we have a special part, the first row, that corresponds to the initial configuration of the simulated machine. We would like to reserve a special configuration of tiles for the initial configuration, but the following fact is an obstacle.

Proposition 6 *If a finite configuration of tiles occurs in a tiling of the plane only finitely many times, then there exists another tiling of the plane in which it does not occur at all.*

This can easily be shown using König's Lemma.

Consequently, the initial configuration cannot be encoded on one particular place of the tiling, *it must be encoded on infinitely many places in the tiling.*

4. *Aperiodic sets of tiles.* It is always more difficult to work with objects that are irregular than regular ones. Thus the task of proving that a particular set of tiles can produce only aperiodic tilings seems quite intriguing. But in reality it is not very difficult. The point is that in spite of the fact that aperiodic tilings (such as in Fig. 4.3) look very chaotic, some sort of regularity is always present. The most straightforward way of proving aperiodicity is to prove that there is a unique way of tiling the plane and to describe explicitly this tiling, or at least some key properties of it. However, this is not always possible; in particular the Penrose tiles (both types shown above) produce an infinite number of different tilings of the plane. The basic idea of proofs of the aperiodicity for such sets of tiles is as follows.

   Let S be a set of tiles. For the sake of simplicity, suppose that the tiles do not have any pattern or colors, so only the shape matters. First we need to show that there is at least one tiling of the plane. Suppose it is possible to reduce the size of the tiles uniformly so that the set of reduced tiles S' can be used to tile each tile of S. Then, of course, we can reduce S' to even smaller tiles S'' so that each tile of S' can be tiled by tiles from S'' and so on. Now pick an arbitrary tile t from S and apply this process of decomposition into smaller tiles for several steps. Thus we obtain a tiling of t by small tiles that have the same shape as the original ones, except that they are reduced. Blow up this tiling so that the small tiles become tiles of the original set S. Then we obtain a tiling of a large region by tiles S. Hence, we can tile arbitrarily large regions of the plane. This implies that the whole plane can be tiled.

The idea of proving that only aperiodic tilings are possible is based on a reverse process. Let a tiling T of the plane by a set of tiles S be given. Suppose that it is possible to connect groups of tiles in such a way that we obtain a tiling T' by a set S' of larger tiles. Now we have to suppose moreover that this can be done in a *unique way*. Then we can conclude that every symmetry of T must also be a symmetry of T' because otherwise it would be possible to enlarge T in at least two possible ways to a tiling by tiles S'. If such a symmetry is a nontrivial translation, then it must be a translation by at least the distance equal to the inner diameter of the tiles of S' (the maximal diameter of a circle that can fit in every tile of S'). Hence if we can enlarge T (say, by repeating the same process) to a tiling by arbitrarily large tiles (more precisely, by tiles with arbitrarily large diameter) in a *unique way*, then no translation preserves T, which means that T is aperiodic.

To apply these ideas to Penrose tilings some modifications are needed. The main trick is to split each of the two rhombi into two triangles. Thus, for example, the two rhombi are replaced by four triangles—two pairs of congruent triangles with different color patterns. Whereas it is not possible to tile a Penrose rhombus by smaller Penrose rhombi because one has to preserve the color patterns on the edges, there is a way to combine pairs of triangles into larger triangles in a unique way. It is necessary to check a number of details, but no higher mathematics is involved in this proof.

The mathematics of aperiodic tilings is very interesting. For example, it is well-known that any configuration that occurs in a tiling by Penrose rhombi must occur infinitely many times in every tiling by Penrose rhombi; furthermore, in different tilings these configurations may occur with different frequencies. Alain Connes proposed to study Penrose tilings using his new theory *Noncommutative Geometry*.

5. *Matiyasevich's theorem.* Discussing details of the proof of Matiyasevich's theorem would take us too much into number theory, therefore, I will mention only a few ideas, mainly those that concern logic. Several different proofs were discovered later; here we will follow, more or less, the original approach.

We will use polynomials with integer coefficients, positive and negative. However, we will consider only *nonnegative* solutions to polynomial equations. We will call a relation $R(x_1, \dots, x_k)$ on the natural numbers *Diophantine* if for some polynomial equation the relation can be represented as the set of parameters for which the equation has solutions. Formally, it means that $R(x_1, \dots, x_k)$ has a representation of the form

$$\exists y_1 \dots \exists y_n \big(p(x_1, \dots, x_k, y_1, \dots, y_n) = q(x_1, \dots, x_k, y_1, \dots, y_n)\big),$$

for some polynomials p and q with integral coefficients and the quantifiers range over natural numbers. We will call $x_1, \ldots, x_k$ *parameters,* and $y_1, \ldots, y_n$ *unknowns.* Hilbert posed the question about solutions in the ring of integers, but one can easily prove that the algorithmic solvabilities of these two versions are equivalent. Here we only need to prove that the unsolvability of the Diophantine problem in the domain of the natural numbers implies the unsolvability of this problem in the domain of the integers. This follows from Lagrange's Theorem: *Every natural number is the sum of four squares.* Hence, if we substitute for every unknown y_i the sum $y_{i,1}^2 + y_{i,2}^2 + y_{i,3}^2 + y_{i,4}^2$, then the equation has a solution in the natural numbers if and only if it has a solution in the integers.

The first step of the proof is to show that every recursively enumerable relation can be defined by an arithmetical formula in a prenex form where all universal quantifiers are bounded, a Σ_1 formula. Once we know that there exists a schema of coding finite sequences by numbers which can be described by an arithmetical formula (for example, Gödel's β-function, see page 294), this is a routine though rather tedious task. A computation can be represented by an $n \times m$ matrix and the matrix can be coded as a string of length nm. We know that everything is finite, except that we do not know how long the computation is. That is why we need at least one unbounded existential quantifier.

The next step is to show that we can use arithmetical formulas of a very special form. The result is due to Davis.

Lemma 10 *Every recursively enumerable relation $R(x_1, \ldots, x_k)$ on natural numbers can be defined by a formula of the form*

$$\exists y \forall z \leq y \exists y_1 \leq y \ldots \exists y_n \leq y \; p(x_1, \ldots, x_k, y, z, y_1, \ldots, y_n) = 0,$$

where p is a polynomial with integral coefficients.

This is very close to Matiyasevich's theorem, the only problem is the universal quantifier. If it were not there, we would only need to get rid of the bounds in $\exists y_i \leq y$, which is very easy. But this apparently small kink is an essential obstacle that requires ingenious tricks to be overcome.

To prove the lemma we need to do two things with the Σ_1 formula: (1) replace the general quantifier-free arithmetical formula by an equation, (2) put the quantifier prefix into the special form.

We can assume that the quantifier-free formula is in the CNF form, which means that it is a conjunctions of disjunctions of equalities and nonequalities. A nonequality $p \neq q$ can be replaced by

$$\exists y \ldots (p + y + 1 = q \vee p = q + y + 1).$$

The dots indicate that we put the quantifier in front of the whole formula. To eliminate disjunctions and conjunctions we use the following transformations:

$$p_1 = q_1 \vee p_2 = q_2 \mapsto (p_1 - q_1)(p_2 - q_2) = 0,$$
$$p_1 = q_1 \wedge p_2 = q_2 \mapsto (p_1 - q_1)^2 + (p_2 - q_2)^2 = 0.$$

The solution of (2) is based on two transformations:

(a) switching bounded universal quantifies with an existential quantifier;
(b) contracting two quantifiers of the same type into one.

I will explain (a) in an example. Suppose we have a formula

$$\forall u \leq v \exists w\, \alpha(v, u, w).$$

As was observed by Skolem, such a formula is equivalent to

$$\exists f \forall u \leq v\, \alpha\big(v, u, f(u)\big),$$

where f is a variable for a function. Since $u \leq v$, f needs to be defined only on a finite domain, in other words, f is a finite sequence. Thus we can apply the β-function and get

$$\exists p, q \forall u \leq v\, \alpha\big(v, u, \beta(p, q, u)\big),$$

where now p and q are numbers. We still have to describe the β-function by an arithmetical formula. This will result in new existential quantifiers after $\forall u \leq v$, so apparently we have not gained anything, but we have. The point is that we only need to collect all universal quantifiers at one place of the prefix and we do not have to put the existential quantifiers from the definition of β-function immediately after $\forall u \leq v$. Suppose $\alpha(v, u, w)$ is of the form $\mathbf{Q}\ \gamma(v, u, w, \bar{z})$, where $\mathbf{Q}$ denotes a quantifier prefix. Then we can put the existential quantifiers of the definition of the β-function between $\mathbf{Q}$ and $\gamma(v, u, w, \bar{z})$. Thus we can move all bounded universal quantifiers to one location in the quantifier prefix.

To contract quantifiers one uses a pairing function defined by a polynomial, for example, Cantor's pairing function

$$(x, y) \mapsto \big((x + y)^2 + 3x + y\big)/2.$$

Since we rather need the two inverse functions, this also introduces new existential quantifiers, but again, we can put them at the end of the prefix.

It still remains to show that we can bound all quantifiers uniformly by y. We leave out this part and pass immediately to the next step of the proof Matiyasevich's theorem. This is a result of Davis, Putnam and Robinson [57].

Lemma 11 *Every recursively enumerable relation $R(x_1, \ldots, x_k)$ on natural numbers can be defined by a formula of the form*

$$\exists y_1 \ldots \exists y_n \exists z_1 \ldots \exists z_m\ p\big(x_1, \ldots, x_k, y_1, \ldots, y_n, 2^{z_1}, \ldots, 2^{z_n}\big) = 0,$$

where p is a polynomial with integral coefficients.

Due to the previous lemma we only need to get rid of one bounded universal quantifier and we can use exponential terms. Let us consider a slightly simplified situation. Let a polynomial of two variables $q(x, y)$ be given and suppose we want to remove the universal quantifier from the formula

$$\forall y \leq x\ q(x, y) = 0. \tag{4.15}$$

We will use the Chinese Remainder Theorem and modular arithmetic. Let x be fixed and let $m_0, m_1, \ldots, m_x$ be pairwise relatively prime numbers, all larger than $\max_{y \le x} q(x, y)$. Then, trivially, for $y \le x$, $q(x, y) = 0$ if and only if m_y divides $q(x, y)$. Take u such that

$$u \equiv y \mod m_y, \qquad \text{for all } y \le x.$$

Then the equation (4.15) is equivalent to

$$q(x, u) \equiv 0 \mod (m_0 m_1 \dots m_x),$$

which is the same as saying that $m_0 m_1 \dots m_x$ divides $q(x, u)$. The relation "a divides b" has a simple Diophantine representation $\exists c(b = ac)$. Hence we only need to define the product of a sequence of sufficiently large pairwise relatively prime numbers and a u satisfying the relation above. To define the product of a general sequence is a hard problem, but we can take a very simple sequence, essentially the same as for the β-function:

$$m_y = (y + 1)d + 1,$$

where $d = k!$ for a sufficiently large k. (A definition of the factorial function is explained below.) The product $m_0 m_1 \dots m_x = \prod_{y \le x}((y + 1)d + 1)$ is congruent to 1 modulo d, hence

$$\prod_{y \le x} \big((y + 1)d + 1\big) = (u + 1)d + 1,$$

for some u. Coincidentally, for every $z \le x$,

$$u \equiv y \mod \big((z + 1)d + 1\big).$$

Indeed,

$$(u + 1)d + 1 = \prod_{y \le x} \big((y + 1)d + 1\big) \equiv 0 \equiv (z + 1)d + 1 \mod \big((z + 1)d + 1\big).$$

We subtract 1 from both sides, then divide by d (which is relatively prime with the modulus) and subtract 1 again, to get the congruence above. Hence (4.15) is equivalent to

$$\prod_{y \le x} \big((y + 1)d + 1\big) = (u + 1)d + 1 \wedge q(x, u) \equiv 0 \mod \left(\prod_{y \le x} \big((y + 1)d + 1\big)\right),$$

plus a condition that d is sufficiently large, which we omit for the sake of brevity.

The definition of the product would require two more pages, so let me only give a hint. The binomial coefficient is defined as the fraction of two products

$$\binom{n}{m} = \frac{n(n-1)\dots(n-m+1)}{m(m-1)\dots 1} = \frac{n(n-1)\dots(n-m+1)}{m!}.$$

Thus if we could define $\binom{n}{m}$ and $m!$, we would also get the product $n(n - 1)\dots(n - m + 1)$, which is the product of an arithmetic progression with difference 1. The product of a general arithmetic progression, which is what we need, can also be obtained from a binomial coefficient, a factorial and an exponential term, but it is more complicated. The binomial coefficient $\binom{n}{m}$ can be defined as follows. Take $A > 2^n$, and consider

$$(A+1)^n = \sum_{i=0}^{n} \binom{n}{i} A^i = \sum_{i=0}^{m-1} \binom{n}{i} A^i + \binom{n}{m} A^m + \sum_{i=m+1}^{n} \binom{n}{i} A^i.$$

The three terms in this decomposition are determined by these conditions: the first is less than A^m, the second is xA^m for some $x < A$, the third is divisible by A^{m+1}. So the binomial coefficient is uniquely determined as the x.

The definition of the factorial $m!$ is based on the formula

$$m! = \lim_{n\to\infty} n^m \binom{n}{m}^{-1}.$$

Let us skip to the last part of the proof, the part that is due to Matiyasevich. What remains is to find a Diophantine representation of the binary relation $y = 2^x$. It had been known before the problem was solved that it sufficed to find a Diophantine representation of the graph of a function that is *approximately* exponential. Originally Matiyasevich used the Fibonacci numbers defined by

$$F(0) = 0, \qquad F(1) = 1, \qquad F(n+2) = F(n) + F(n+1).$$

Later proofs used solutions of the Pell equations of the special form

$$x^2 - (a^2 - 1)y^2 = 1.$$

The solutions are some pairs (x_1, y_1), (x_2, y_2), ... If we enumerate the sequences x_n and y_n in increasing order, then they grow exponentially ($(2a-1)^n/2a < x_n \leq (2a)^{n-1}$, $(2a-1)^n \leq y_n \leq (2a)^n$). These numbers are solutions of a Diophantine equation, but this not very important. The really difficult thing is to show that the relation '*x is the nth solution*' is Diophantine. It would take us too far afield to discuss this part of the proof.

6. *More on the complexity of Ramsey's Theorem.* Jockusch proved a more general theorem [141].

Theorem 27 *Let $k \geq 2$. Then*

a. *for every computable partition of k-element subsets of $\mathbb{N}$, there exists an infinite homogeneous set in the class Π_k;*
b. *there exists a computable partition of k-element subsets of $\mathbb{N}$, such that there is no infinite homogeneous set in the class Σ_k.*

4.4 Concrete Independence

By Gödel's Theorem, we know that for every sound theory, there are true statements that are not provable in it. We also know that the axioms that mathematicians use can be collected into one theory because we have set theoretical foundations of all present mathematics. Hence there are sentences that cannot be proved using currently accepted postulates. Contrary to this, the experience of working mathematicians is that every problem, however difficult it is, is eventually solved, and the difficulty lies not in discovering new axioms, but it is only the complexity of a proof that makes the problem hard. The only exception is set theory in which we have many mathematically interesting independence results. It is not quite clear whether it is because set theory is inherently incomplete, or because we possess a powerful technique of proving independence. However, all these independent sentences concern properties of infinite sets. When we focus our attention on finite problems, problems in finite combinatorics and number theory, problems about finite algebras, etc., the situation is dramatically different. Until the late 1970s all independent sentences of this type were only of the type discussed in the previous section. Our ultimate goal is to be able to prove that some well-known mathematical problems are not solvable in particular theories. Sentences that look the same for every theory can hardly be used for this purpose.

In the mid-1970s, J.B. Paris came up with a very nice idea: to use concepts from the theory of large cardinals to study models of Peano Arithmetic. Peano Arithmetic is essentially the same theory as Finite Set Theory, thus there are no infinite cardinals, let alone large cardinals, definable in it. However, a nonstandard model of Peano Arithmetic (a model that is not isomorphic to the natural numbers) is an ordered set with a very rich structure. Though a nonstandard model of Peano Arithmetic is never well-ordered, still it resembles infinite ordinal numbers. The complexity of these models suggested looking for properties similar to the properties of large cardinals. Paris managed to define initial segments of nonstandard models that satisfy properties analogous to those used to define large cardinals [213]. Then he realized that since the models are essentially the models of Finite Set Theory, there must be some *finite* concepts associated with these initial segments. This eventually led to the construction of very explicit sentences that are true but not provable in Peano Arithmetic (and Finite Set Theory). His method was later used to find a number of other concrete independence results.

Fast Growing Functions

Let us consider the game of playing the largest number again, but now let us be fair and disallow infinite numbers. Is this a trivial matter once we can use only natural numbers? No, in fact, playing such a game we face similar problems as when we played it with infinite numbers. It is because we do not restrict the way a number is defined. There may be a simple clever definition of a very large number, but the problem may arise whether or not the definition is correct, that is, whether such a number actually exists, similarly as it was with large infinite cardinal numbers. Let us try to define some big numbers. Already arithmetic operations enable us to define huge numbers. In the times of Archimedes mathematicians did not use exponentiation as we do now, so Archimedes could not simply write $10^{8\cdot 10^{16}}$. We get large numbers using exponentiation once, or twice, but a much more efficient way is to use it iteratively. Thus

$$2^{2^{2^{2^{2}}}}$$

is much larger than 2^{2222}, though it uses the same number of the same characters. Iterated exponentiation of 2 is, sometimes, called *superexponentiation* and denoted *supexp*. The value of *supexp* on 1000 (the stack of 1000 twos) is an unimaginably large number, but we can go on and iterate superexponentiation to get another function and so on.

Iterating functions in the manner described above we get larger and larger numbers, but there is no problem with their existence. However large the numbers are, we are sure that they exist, although we cannot evaluate them in decimal representation nor can we run the computer for so many steps. It is a different thing, however, if we define a number implicitly, say, as the solution of an equation, or as the number obtained by some algorithm. Then the existence of the numbers is not obvious at all.

A very nice example was given by R.L. Goodstein in 1944 [105]. He defined, for every number n, a sequence, which we will call *the Goodstein sequence of* n. First we consider a simplified version of such sequences ("Goodstein sequences for beginners"). The first term of the sequence is n. Then we extend the sequence by applying an operation to the last term. To apply the operation on the m-th term of the sequence we first represent the number in base $m+1$, then we change the base to $m+2$ while keeping the representation. Finally we subtract one. The sequence stops when it reaches 0 (but we do not know yet whether it ever reaches 0).

Example Start with 5, which is in binary 101_2. The next term is obtained by taking 101_3, that is, interpreting the number in ternary and subtracting 1, which gives $100_3 = 9$. The next number in the sequence is $100_4 - 1 = 33_4 = 15$.

Though this was only a simplified version, it may also be an interesting concept. The original Goodstein sequence is more complicated. Note that the binary representation of 5 can be written using arithmetic operations as $2^2 + 1$. Now 2 occurs also in the exponent, thus it should also be changed to 3. The general rule is to write also the exponents as the sum of powers of the basis considered at a particular step, then the exponents of the exponents, etc. We add, though it may seem ridiculous at this point, that the sequence stops when it reaches 0. Here is an example of the beginning of a genuine Goodstein sequence starting at 21:

$$2^{2^2} + 2^2 + 1,$$
$$3^{3^3} + 3^3,$$
$$4^{4^4} + 4^4 - 1 = 4^{4^4} + 3 \cdot 4^3 + 3 \cdot 4^2 + 3 \cdot 4 + 3.$$

Another example is in the middle column of Table 4.1; it is the Goodstein sequence starting at 4. The numbers increase very rapidly. If we look only on their numerical values, then they show no tendency to eventually stop increasing and decrease. The remarkable fact is, however, that no matter which number we start with, the

Table 4.1 The Goodstein sequence starting at 4. The number n of an element of the Goodstein sequence is in the first column; the value $G(n)$ is in the second column and it is represented in base $n+1$; the corresponding ordinal numbers are in the third column. The maximal value of $G(n)$ is attained for n such that the corresponding ordinal is 2ω, then it stagnates until one step after ω, from which point on it decreases

n	$G(n)$	*Ordinal*
1	2^2	ω^ω
2	$2 \cdot 3^2 + 2 \cdot 3 + 2$	$2 \cdot \omega^2 + 2 \cdot \omega + 2$
3	$2 \cdot 4^2 + 2 \cdot 4 + 1$	$2 \cdot \omega^2 + 2 \cdot \omega + 1$
4	$2 \cdot 5^2 + 2 \cdot 5$	$2 \cdot \omega^2 + 2 \cdot \omega$
5	$2 \cdot 6^2 + 6 + 5$	$2 \cdot \omega^2 + \omega + 5$
$\vdots$	$\vdots$	$\vdots$
10	$2 \cdot 11^2 + 11$	$2 \cdot \omega^2 + \omega$
11	$2 \cdot 12^2 + 11$	$2 \cdot \omega^2 + 11$
$\vdots$	$\vdots$	$\vdots$
22	$2 \cdot 23^2$	$2 \cdot \omega^2$
23	$24^2 + 23 \cdot 24 + 23$	$\omega^2 + 23 \cdot \omega + 23$
24	$25^2 + 23 \cdot 25 + 22$	$\omega^2 + 23 \cdot \omega + 22$
$\vdots$	$\vdots$	$\vdots$
$\vdots$	$\vdots$	$\vdots$
$3 \cdot 2^{402653209} - 3$	$2 \cdot (3 \cdot 2^{402653209} - 2) + 1$	$2 \cdot \omega + 1$
$3 \cdot 2^{402653209} - 2$	$2 \cdot (3 \cdot 2^{402653209} - 1)$	$2 \cdot \omega$
$3 \cdot 2^{402653209} - 1$	$(3 \cdot 2^{402653209}) + 3 \cdot 2^{402653209} - 1$	$\omega + 3 \cdot 2^{402653209} - 1$
$3 \cdot 2^{402653209}$	$(3 \cdot 2^{402653209} + 1) + 3 \cdot 2^{402653209} - 2$	$\omega + 3 \cdot 2^{402653209} - 2$
$\vdots$	$\vdots$	$\vdots$
$\vdots$	$\vdots$	$\vdots$
$3 \cdot 2^{402653210} - 3$	$(3 \cdot 2^{402653210} - 2) + 1$	$\omega + 1$
$3 \cdot 2^{402653210} - 2$	$(3 \cdot 2^{402653210} - 1)$	ω
$3 \cdot 2^{402653210} - 1$	$3 \cdot 2^{402653210} - 1$	$3 \cdot 2^{402653210} - 1$
$3 \cdot 2^{402653210}$	$3 \cdot 2^{402653210} - 2$	$3 \cdot 2^{402653210} - 2$
$\vdots$	$\vdots$	$\vdots$
$\vdots$	$\vdots$	$\vdots$
$3 \cdot 2^{402653211} - 4$	2	2
$3 \cdot 2^{402653211} - 3$	1	1
$3 \cdot 2^{402653211} - 2$	0	0

Goodstein sequence eventually does reach 0. This seems very unlikely, since the operation of changing the base of the representation of the number increases the number so much that subtracting 1 afterwards apparently cannot help, but it is true.

To see why a Goodstein sequence has to terminate look at the example in Table 4.1. In the last column there is a sequence that follows the pattern of the numbers, but the base is replaced by the symbol ω. Think of ω as an infinite number. Doing the proof formally we would take the first infinite ordinal number, but we only need a few basic properties, so one can follow the proof without being familiar with infinite ordinals. If you do not like infinite numbers, think of ω simply as a huge natural number, bigger than every number that we may ever need during the computation. Interpreting ω in this way the sequence in the third column *decreases from the very beginning*! If ω is a large number, then surely ω^ω is much larger than $2 \cdot \omega^2 + 2 \cdot \omega + 2$.

Let us look more closely at what is going on in the third column. When going from 2 to 3 or 3 to 4, it is simple, we decrease the finite term by one. It more interesting what happens when going from step 1 to 2, or 4 to 5 or 10 to 11. Let us consider the last one. The smallest term in 10 is ω; in 11 we replace it by 11 (we first replace it by 12 and then subtract 1). Thus the infinite number is replaced by a finite number, hence the number decreases. If we interpret ω as a large natural number, then this large number is replaced by a small number.

The argument with taking ω a very large finite number looks quite persuasive, but there is a loophole in it. For that argument, we must take ω larger than any other number that may occur in the computation. If the computation were finite, then we could certainly find such a number, but the finiteness of the computation is exactly what we want to prove. Thus we only know *ex post* that taking a finite number works as well. Therefore, we rather interpret ω as an infinite number and then there is no problem with showing that the infinite numbers in the third column decrease. Still, the fact that they decrease does not automatically guarantee that they eventually reach 0. To this end we have to apply a property of infinite ordinal numbers, the property of being *well-ordered,* which is exactly what we need: every decreasing sequence must get to 0 after finitely many steps. Notice that the numbers in the third column copy the process in the second column. In particular, when we get a finite number in the third column it is the same as in the second column. Thus the two sequences reach 0 after the same number of steps. We can conclude that every Goodstein sequence reaches 0 eventually.

This argument uses a principle (that a particular set of ordinals is well-ordered) which is not difficult but it is intrinsically based on infinite sets and thus is unprovable in Finite Set Theory. Hence the termination of Goodstein sequences cannot be proved in Finite Set Theory using this argument. That a particular proof cannot be formalized in some theory does not imply that the theorem is not provable in this theory. However, for this particular theorem, L. Kirby and J.B. Paris proved that there is no proof of it in Finite Set Theory [152]. Let us denote by $G(n)$ the number of elements of the Goodstein sequence starting at n. The reason why we cannot prove that every Goodstein sequence terminates is the very rapid growth of the function $G(n)$. (It is difficult to watch this growth by computing the values for some small numbers, since the numbers are extremely large already starting with $G(4)$.) Intuitively, the sequence grows so rapidly that Finite Set Theory cannot prove that such large numbers exist. (More precisely, it can prove for every given number n that $G(n)$ exists, but it cannot prove the general statement that $G(x)$ exists *for every* x.)

One can show that every sound theory is able to manage only some limited growth. If a function grows faster than this bound, the theory is not able to verify that the function is properly defined, which means that it is consistent with the theory that for some argument n the value of the function is undefined. $G(n)$ is such a function for Finite Set Theory. Constructions of such functions for Finite Set Theory based on other combinatorial principles and for other theories have been given; we will consider some of these results below.

Unprovability of the existence of rapidly growing functions is an important aspect of the incompleteness, but there are also unprovable true sentences that have nothing to do with any rapid growth. In particular, Gödel's sentences are such.

Interlude—Hercules and Hydra

Here is an even more entertaining problem. In the old legend Hydra had the miraculous ability to grow two heads for each head chopped off. Hercules had to use fire to defeat it. We will see that in fact he did not need to use fire; mere chopping off heads would suffice—at least if the rules were set suitably. However, this would take him a tremendously long time to win.

The idea of presenting an unprovable sentence as a game with Hercules and Hydra is due to L. Kirby and J.B. Paris [152]. They defined Hydras to be rooted finite trees. A tree is simply an acyclic connected graph; '*rooted*' means that one node is distinguished (and called the root). A *head* is an end node, a *leaf,* with the arc connecting it to the rest of the tree. At each stage Hercules chops off one head. If the head is attached to the root, nothing happens. If the head is attached to a node N which is not a root, then the node N splits into several ones and the subtrees above it are replicated. The number of nodes into which it splits depends on the stage: in the first stage it doubles, in the second it triples and so on. An example is shown in Fig. 4.4.

So the position of Hercules seems to be even worse than in the legend. However, notice the minor modification that the new parts of Hydra do not grow from the scar, but from the next place below, if there is any. This seemingly innocent change gives Hercules an advantage. A big advantage—Hercules wins whatever he does! Any strategy he uses to chop off heads succeeds; everything will be removed, only the root will remain.

The counterintuitive character of this theorem suggests that it should be difficult to prove it, or at least that it should require some trick. The trick is the same as used for Goodstein sequences. Assign ordinals to Hydras and show that the ordinals always decrease (while the size of the tree may temporarily increase). An intuitive explanation is that the number of heads at first increases, but always a complex head is replaced by (several) simpler ones. This proof is fairly simple, but it requires

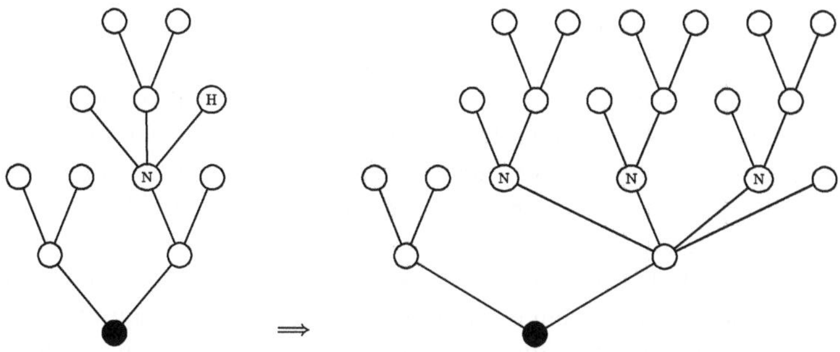

Fig. 4.4 After Hercules chops off the head H at stage 2, the remaining part above N will triple. Note that each of the three subtrees is smaller than the original subtree above the node N because it does not contain the head H. This is the reason why Hercules always wins

means not available in Finite Set Theory. The statement that Hercules always wins is not provable in this theory. Furthermore, as in the case of Goodstein sequences, one can show that the reason for unprovability is, again, the big growth rate of the number of steps needed to reduce the tree to a single node.

The Collatz Problem

Goodstein's Theorem had been proved long before it was shown that it is unprovable in Finite Set Theory. So it is not an ad hoc manufactured sentence. Later we will consider the Continuum Hypothesis, a famous open problem in set theory that, instead of being solved, was shown independent of Zermelo-Fraenkel Set Theory. In this section we are interested in combinatorial or number theoretical independent sentences, sentences that do not speak about infinite sets. There is nothing like the Continuum Hypothesis if we restrict our attention to such statements: there is no open problem in combinatorics or number theory that we can show to be unprovable from some reasonable set of axioms.

Here is an *open problem* that is somewhat reminiscent of Goodstein's theorem. The problem was posed by L. Collatz in 1937 and is known under a half a dozen names; most frequently it is called *the Collatz Problem* and $3x + 1$ *conjecture*. As we did above, we start with a number and define a sequence by applying a certain operation to get the next term of the sequence. The operation is now defined by the following simple rule:

1. *if the number is even, then divide it by two*;
2. *otherwise multiply it by three and add one.*

Example Starting with 9 we get

$$9, 28, 14, 7, 22, 11, 34, 17, 52, 26, 13, 40, 20, 10, 5, 16, 8, 4, 2, 1.$$

After reaching 1 we always stop because then the sequence would repeat 4, 2, 1 for ever.

The open problem is whether or not the same happens for every number, namely, if for every positive number, the sequence starting with this number will eventually reach 1. With the advent of computers this problem attracted also a lot of computer scientists because it is very suitable for experimental testing. So far experiments confirm the conjecture that every Collatz sequence reaches 1. At the time of writing these lines, the conjecture was confirmed for numbers up to 20×2^{58}.

A weaker conjecture is that there are no finite cycles defined by this rule, except for the trivial one $1, 4, 2, 1$. We only know that any such cycle would be at least 272,500,658 long. This problem is equivalent to whether or not starting with some number the sequence goes to infinity. Indeed, if the sequence does not go to infinity, then, necessarily, it must hit the same number twice. Once it hits the same number twice, it has to repeat. Thus we can rephrase the weaker problem as follows. Is it possible to give a bound on the numbers occurring in the sequence in terms of the initial number?

We know that intuition may be misleading, but the process of discovering a proof always starts with some vague ideas, which sometimes lead to a solution, sometimes not. Thus let us try some *heuristic* arguments, arguments based on not completely justified assumptions. The first reaction to the problem is that the process is not balanced: we multiply by 3 but divide only by 2, hence we are more increasing than decreasing the numbers. Moreover we are adding 1 after multiplying by 3. Hence the sequence should typically go to infinity. This argument is definitely wrong, as it ignores an important fact: after multiplying by 3 (and adding 1) we always divide by 2, while after dividing by 2 we may sometimes divide by 2 again. Sometimes there may be many divisions by two before we hit an odd number. It turns out that if we could assume that we get a *completely random* even number, then division by two would, in fact, prevail. Let us do a simple calculation. Think of the number as written in binary, then the number of divisions by 2 is the number of zeros at the end of its binary representation. A random even number has at the end

1 zero with probability $1/2$,
2 zeros with probability $1/4$,
3 zeros with probability $1/8$,
...

Hence, the expected number of zeros at the end of a random even number written in binary is

$$1 \cdot \frac{1}{2} + 2 \cdot \frac{1}{4} + 3 \cdot \frac{1}{8} + \cdots = 2.$$

Thus on the average we divide by 4 before we apply the $3x + 1$ rule again. Hence, if we could assume that the even numbers occurring in the process are random, we could view it as alternations of steps in which we multiply by 3 and add 1 and steps in which we divide by four. Each pair of steps would decrease the number approximately by a factor $3/4$. Consequently, such a sequence would converge very rapidly to 1.

Summing up, if we could somehow justify the assumption that the even numbers that occur in the sequences are generated randomly, we would have a proof of the conjecture. However, that is impossible: the sequences are given by rules that determine the numbers uniquely, hence there is no randomness in it. Yet it is possible to compare what we get by computing with what we should get if the sequences were random. The Collatz problem is an ideal object for experimental mathematics, since the computation of the sequence is very simple and only rarely it is necessary to do many iterations. In particular we can compute the average number of iterations needed to obtain 1 for very large sets of natural numbers. The results of these computations are in good agreement with what the heuristic argument above predicts. For some other quantities, the experiments were even more successful.

Though the analogy with random numbers does not give a proof of the conjecture it can be used to prove some partial results. A more sophisticated argument was used to prove that the conjecture is true for almost all numbers. Thus if you pick randomly a natural number n, then with high probability the sequence started at n will reach 1. In other words, either the conjecture is true, or the exceptions are very rare.

The similarity with Goodstein sequences is rather superficial and it does not seem likely that one can prove the independence of Collatz's problem from the axioms of Finite Set Theory using methods similar to those used for Goodstein sequences. The main problem is that there does not seem to be any fast growing function involved. If Collatz's sequences behave like random ones, then each sequence reaches a cycle quite soon. So the problem is rather to prove that they do behave like random sequences. What is possible, however, is to prove an independence result for a problem similar to the $3x + 1$ conjecture.

Let us first state the $3x + 1$ conjecture more formally. We will call the function C defined by

$$C(x) = \begin{cases} x/2 & \text{if } x \text{ is even,} \\ 3x + 1 & \text{if } x \text{ is odd,} \end{cases}$$

the *Collatz function*. The problem is then what happens if we iterate this function starting with a natural number n: does it eventually reach 1? We can ask the same question about the convergence of iterations for other functions. A natural generalization, however, must preserve the spirit of the problem. J. Conway came up with a class of functions that generalize the Collatz function. A function in this class is defined by considering the remainders of x when divided by a number k, and defining the value of the function to be a linear function in each case. One of such functions, different from the Collatz function, which has been studied is:

$$U(x) = \begin{cases} 7x + 3 & \text{if } x \equiv 0 (\text{mod } 3), \\ (7x + 2)/3 & \text{if } x \equiv 1 (\text{mod } 3), \\ (x - 2)/3 & \text{if } x \equiv 2 (\text{mod } 3). \end{cases}$$

Conway proved that it is an undecidable problem to determine for a given function from the class if the analogue of the $3x + 1$ conjecture holds [47]. So this is another number-theoretical undecidable problem. Since the undecidability is again proved by a reduction of the halting problem, it is a universal problem in the sense explained in the previous section. Consequently, also here we can derive independence results:

Theorem 28 *For every theory T, it is possible to find a function f in the class defined by Conway such that it is undecidable in T whether for every n, the sequence $n, f(n), f(f(n)), \dots$ contains* 1.

This gives us evidence that *problems of this type* are difficult, but it does not say anything about *the* $3x + 1$ *conjecture itself.*

An Unprovable Version of the Finite Ramsey Theorem

The most interesting unprovable combinatorial sentence was found by J.B. Paris and L. Harrington in 1977 [214]. This is a sentence that is true, but unprovable in Finite Set Theory and it is a formally minor modification of the well-known and important Ramsey theorem. Recall that the Finite Ramsey Theorem says that, for every r, if we color k element subsets of a sufficiently big set by two colors in an arbitrary way, then one can find a subset of size r on which all the k-element subsets have the same color. (For the sake of simplicity, we consider only two colors.) The precise statement is as follows.

The Finite Ramsey Theorem *For every k and r positive natural numbers, there exists a natural number n such that for every coloring of k-element subsets of the set $\{1, 2, \dots, n\}$ by two colors, there exists a subset $X \subseteq \{1, 2, \dots, n\}$, X having size r such that all k-element subsets of X have the same color.*

The set X is called *monochromatic*.

In this theorem it is irrelevant what the underlying set is; the only thing that matters is its size n. So we take $\{1, 2, \dots, n\}$ as a suitable representative of an n-element set. However, it will be important to talk about subsets of natural numbers when we consider the unprovable modification. To this end we also need to define a special property of finite subsets of natural numbers. We say that a finite set X of natural numbers is *relatively large* if its size is greater or equal to the least element of X. Here is the modification invented by Paris and Harrington.

The Paris-Harrington Theorem *For every k and m positive natural numbers, there exists a natural number n such that for every coloring of k-element subsets of the set $\{m, m + 1, \dots, n\}$ by two colors, there exists a* relatively large *subset $X \subseteq \{m, m + 1, \dots, n\}$ such that all k-element subsets of X have the same color.*

Let us look at what the modification does. Since we now consider subsets X of $\{m, m + 1, \dots, n\}$, the least element of X is at least m. Hence, since X is relatively large, it has a size at least m. This only serves us to take off from the ground; the seemingly innocent requirement of being relatively large has a much more dramatic effect. Here is why. We can easily deduce from the Finite Ramsey Theorem that there exists an m-element monochromatic subset of $\{m, m+1, \ldots, n\}$ if n is sufficiently large, but it can happen that none of these sets contains m. Hence the minimal elements of these sets are at least $m+1$. Thus we have to look for monochromatic sets of a size of at least $m+1$. Again, taking n is sufficiently large, we can ensure that there are monochromatic subsets of size $m+1$, but we have to take n larger than in the previous case. Now, again, the Finite Ramsey Theorem only says that there are such monochromatic sets, but does not guarantee that one of them contains $m+1$. So we may be forced to look for even larger monochromatic sets. Thus it seems that it may happen that we never ensure the existence of a relatively large monochromatic subset, however big n we take, because the bigger monochromatic sets we take the bigger their least elements will be. This is not true, but the positive feedback has the same effect as in the combinatorial sentences that I described before. If we define a function $PH(k, m) = n$ as the least n that satisfies the condition of the Paris-Harrington theorem, the function has again a tremendous rate of growth. As it is not quite clear what is the growth of a function of two variables, we should rather say that the "diagonal" function $PH(x, x)$ grows very rapidly, so rapidly that one cannot prove in Finite Set Theory that it is defined for all arguments.

First we have to prove, say in Zermelo-Fraenkel Set Theory, that the function is properly defined, which means simply proving the Paris-Harrington theorem. If infinite sets are allowed in the proofs, the proof is quite easy. In fact, essentially the same proof that shows that the Infinite Ramsey Theorem implies its finite version (see page 24) gives the Paris-Harrington theorem. I can explain it without going into the details of that proof. Fix k, the size of colored subsets. We have argued by contradiction and assumed that the Finite Ramsey Theorem fails for some r. Namely, however large n we take, there is some coloring of k element subsets of $\{1, 2, \ldots, n\}$ without a monochromatic set of size r. Then we have combined these counterexamples into a single counterexample to the Infinite Ramsey Theorem. Assuming the infinite Ramsey Theorem we get an *infinite* homogeneous subset X. Taking r elements from it, we showed that one of the counterexamples could not have been a counterexample for the Finite Ramsey Theorem. So the Infinite Ramsey Theorem gave us more: instead of just an r element set, an infinite set. Now the point is that finding a relatively large subset of an infinite set is as easy as finding a finite set of a given size. Namely, let s be the least element of X, then taking s along with $s-1$ other elements of X we get a relatively large set. To recap it in a single statement: Paris-Harrington theorem holds true because every infinite set is relatively large.

How does one prove the unprovability of such sentences in Finite Set Theory? One possible approach would be to study the growth of the functions that can be defined in Finite Set Theory and show that a function such as $PH(x, x)$ grows faster. This is possible, but it is rather tedious work [151]. Instead, Paris and Harrington used a model theoretical argument, which is best explained by using models of Peano Arithmetic. Take a nonstandard countable model M of Peano Arithmetic. If it does not satisfy the Paris-Harrington sentence, then we are done because this implies the sentence is not provable in Peano Arithmetic. Otherwise, construct a new model M' from M such that M' is still a model of Peano Arithmetic, but it no longer satisfies the Paris-Harrington sentence. Thus one gets again that the sentence is not provable in Peano Arithmetic.

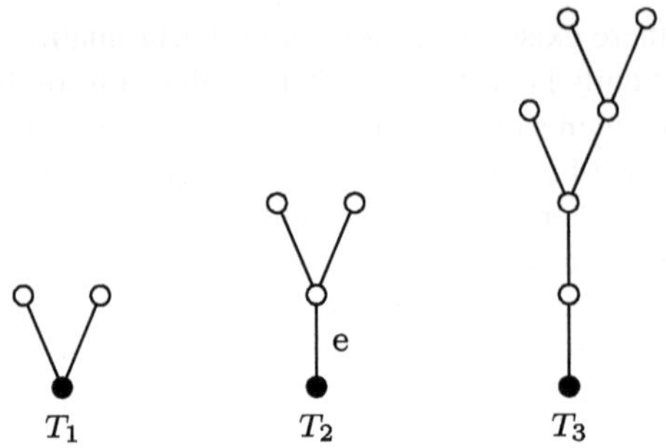

Fig. 4.5 There is a homomorphic embedding of T_2 into T_3 because we can cut off the two top vertices from T_3 and split the edge e of T_2 into two edges in order to get isomorphic trees. There is no homomorphic embedding of T_1 into T_2 or T_3

The key part of the proof is, of course, the construction of M'. This model is not constructed from scratch; it is a submodel of M, in fact, an initial segment of M. As said before, initial segments of a nonstandard model of arithmetic resemble infinite ordinals, though they are in many respects completely different. An initial segment closed under the successor is called *cut.* To construct a submodel we, of course, need that cuts be closed also under addition and multiplication.

Paris and his student Kirby defined several types of cuts. The properties by which they were defined were originally motivated by large cardinals, but eventually they used properties which are satisfied also by $\aleph_0$. The most important type of cuts are *strong cuts* which are defined using a condition related to Ramsey's Theorem. Every strong cut is a model of Peano Arithmetic and this was used to prove independence results (see Notes).

Independence Beyond Finite Set Theory

Explicit independent combinatorial sentences have been found for stronger theories. The most interesting one is a modification of a well-known theorem of J. Kruskal. This is also a statement about finite rooted trees, but it is even simpler than the Hercules and Hydra sentence. We only need to define homomorphic embeddings. Rooted trees come with a natural ordering. A *homomorphic embedding* of a tree T_1 into a tree T_2 is a one-to-one order preserving mapping that maps the root of T_1 onto the root of T_2 and preserves infima. A homomorphic embedding of T_1 into T_2 exists, roughly speaking, if we can prune T_2 and stretch T_1 so that we get identical trees (see an example in Fig. 4.5). The following is a weaker form of Kruskal's theorem.

Theorem 29 (Kruskal [171]) *For every infinite sequence $\{T_n\}_{n=1}^{\infty}$ of finite trees, there are two numbers $i < j$ such that there exists a homomorphic embedding of T_i into T_j.*

This is not a statement that speaks only about finite sets, as it contains the concept of an infinite sequence. Thus it does not make sense to talk about provability in

Finite Set Theory. Harvey Friedman found an ingenious way to transform it into a finite statement (a Π_2 sentence). To explain Friedman's finite version, consider the following simple modification of Kruskal's Theorem.

Theorem 30 *For every function f, there exists a natural number n, such that for every sequence $\{T_i\}_{i=1}^n$ of finite trees such that the size of each T_i is at most $f(i)$, there are two numbers $i < j \leq n$ such that there exists a homomorphic embedding of T_i into T_j.*

Clearly, the original theorem follows from this version (indeed, given a sequence $\{T_n\}_{n=1}^\infty$, take $f(n)$ to be the size of T_n). The advantage of the new form is that if we fix a particular function f, then the statement speaks only about finite sequences. Surprisingly, the statement is strong already for very simple functions. To obtain an unprovable sentence, it suffices to take the simple linear functions $x + c$ for c a constant. Thus one gets the following sentence.

Theorem 31 (Friedman's miniaturization [269, 274]) *For every c, there exists an n such that for every sequence of trees $\{T_i\}_{i=1}^n$, where the size of T_i is at most $i + c$ for $i = 1, \ldots, n$, there are two numbers $i < j \leq n$ such that there exists a homomorphic embedding of T_i into T_j.*

This sentence is unprovable in ATR_0, a theory essentially stronger than Finite Set Theory and Peano Arithmetic.[12] A version of this theorem with labeled trees is unprovable in $\Pi_1^1 - CA_0$, an even stronger theory. Though stronger than Finite Set Theory, these theories are still very weak compared to Zermelo-Fraenkel Set Theory. Friedman has found other combinatorial sentences which are independent of Zermelo-Fraenkel set theory and some extensions of this theory by large cardinals. These sentences are fairly easy to state, but it is almost impossible to grasp their mathematical meaning.

Unprovable Sentences that Are not Associated with Fast Growing Functions

An extreme rate of growth is only one of the possible reasons for unprovability. There are sentences that are unprovable in a theory T and which are not connected with any fast growing function. Such a sentence is the Gödel sentence expressing the consistency of the theory. The sentence says that every sequence of symbols has some property; the property is that it is not a proof of contradiction from the axioms of T. We can distinguish such sentences from those that are connected with fast growing functions by their logical complexity. In Chap. 2 (page 139) we have introduced quantifier complexity of sentences. According to this classification *universal*

[12] See page 643 for the definition of ATR_0.

finite sentences, sentences that have one universal quantifier are Π_1. An unprovable (in a theory T) sentence of this form is the Gödel sentence (for T). Sentences that have one universal quantifier followed by one existential quantifier are classified as Π_2. Note that what matters are only unbounded quantifiers; if the range of a quantifier is restricted to a finite domain, we do not count it. Furthermore, if we have several same quantifiers, they can be combined into one. Thus a more precise definition of Π_2 is: a block of universal quantifiers, followed by a block of existential quantifiers, followed by a formula with only bounded quantifiers. The Finite Ramsey Theorem and the Paris-Harrington Theorem are Π_2.

Every sentence of the form '*for every* x, *there exists* y *such that* $\phi(x, y)$' is associated with a function because y depends on x. We can express the same by introducing a function f that describes this dependence explicitly. So the sentence reduces '*for every* x, $\phi(x, f(x))$'. It may happen that a sentence that is syntactically Π_2 is equivalent to a simpler sentence which is Π_1. The equivalence of sentences is with respect to a theory T that we investigate. So there are Π_2 sentences that can be reduced to Π_1 and those that cannot, which are intrinsically Π_2. The latter ones are exactly those that are associated with fast growing functions. The Finite Ramsey Theorem is provable in Finite Set Theory, so it is equivalent to a tautology such as $0=0$; Paris-Harrington Theorem is intrinsically Π_2.

Our interest in unprovable Π_1 sentences has deeper reasons. Eventually we would like to prove independence of statements that are open problems. The independent combinatorial sentences that we have so far were never open problems; their truth was always almost obvious and they are Π_2. The most interesting open problems are typically Π_1. Sometimes it is not trivial to prove that a problem is Π_1 (for example, for the Riemann Hypothesis), sometimes we do not know if a problem can be represented in this way, but often the most interesting special instances of the problem are Π_1 (for example, problems in the computational complexity theory). I surmise that Collatz's problem is also Π_1. Lacking techniques to prove independence of any combinatorial Π_1 sentence (assuming we exclude Gödel sentences because they talk about logic), we can hardly hope to prove the independence of such open problems.

Notes

1. *Hercules, Hydra and ordinals.* The assignment of ordinals to the trees representing Hydras is defined inductively on the depth of the trees. We assign the ordinal 0 to a tree with a single vertex. Suppose we have already assigned ordinals to $\alpha_1, \alpha_2, \ldots, \alpha_n$ the maximal proper subtrees of a tree T, then we order the ordinals so that $\alpha_{i_1} \geq \alpha_{i_2} \geq \cdots \geq \alpha_{i_n}$ and assign
$$\omega^{\alpha_{i_1}} + \omega^{\alpha_{i_2}} + \cdots + \omega^{\alpha_{i_n}}$$
to T. Thus the ordinal assigned to the first tree in Fig. 4.4 is:
$$\omega^{\omega^{\omega^2+2}+1} + \omega^2.$$

The ordinal assigned to the second tree (the tree of Hydra after cutting a head) is:

$$\omega^{(\omega^{\omega^2+1})\cdot 3+1} + \omega^2,$$

which is clearly smaller.

2. *Cuts in models of Peano Arithmetic.* Let M be a countable nonstandard model of *PA*. A subset I of M is called a *cut* if it is closed downwards (formally, $x \in I$ and $y < x$ implies $y \in I$) and it does not have the largest element ($x \in I$ implies $x + 1 \in I$). The model M itself is a cut; the set of standard numbers N of M is a cut; given a nonstandard element a, the set $\{x \in M\ ;\ x \leq a+n,\ \text{for some standard } n\}$ is a cut. The last cut is not a substructure of M because it is not closed under addition and multiplication. Let

$$I = \left\{x \in M;\ x \leq a^n,\ \text{for some standard } n\right\}.$$

Then I is a substructure, but it is not a model of PA because a^a is not defined in it. Cuts that are also substructures will be our main object of study.

Let I be a cut and $X \subseteq I$. We will say that X is $*$*-definable* in I if there exists a set Y definable in M such that $X = Y \cap I$. The concepts of $*$-definable relations and functions are defined in a similar fashion. Using $*$-definability one can define interesting properties of cuts.

A cut I is *regular* if no $*$-definable sequence $a_0, a_1, \ldots$ indexed by numbers $i \leq b$, for some $b \in I$, is unbounded in I. This definition was, clearly, inspired by regular cardinals.

A cut I is *strong* if it is regular and for every $*$-definable coloring of triples (u, v, w), $u, v, w \in I$, by two colors, there exists a monochromatic $*$-definable set $X \subseteq I$, unbounded in I. In plain words, in a strong cut the Ramsey Theorem holds for $*$-definable colorings of triples with $*$-definable monochromatic sets. One can show that in a strong cut the Ramsey Theorem holds true also for $*$-definable colorings of k-element sets by a colors for $k, a \in I$.[13]

3. *The unprovability of the Paris-Harrington sentence in Peano Arithmetic.* Our goal is to construct a cut I which is a model of PA, but it is not closed under the fast growing function $PH(x, x)$ associated with the Paris-Harrington theorem. This will immediately imply the independence result we are after. The following lemma formalizes our goal more precisely.

Lemma 12 *Let a be a nonstandard element of M and $b = PH(a, a)$. Then there exists a cut M' which is a model of PA and $a \in M'$, but $b \notin M'$.*

To derive the independence, consider two cases. First, if Paris-Harrington Theorem does not hold in M, we are done. Otherwise, take an arbitrary nonstandard element a. Since Paris-Harrington Theorem holds in M, there exists

[13]In set theory there is a concept of *strong cardinals,* but this has nothing in common with strong cuts. Strong cuts are rather related to weakly compact cardinals.

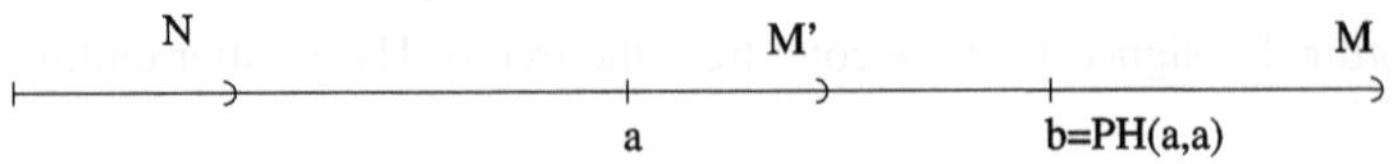

Fig. 4.6 A diagram of the model M. Elements are drawn as |, cuts as)

a $b = PH(a, a)$ in M. Then we can construct a model of *PA* in which Paris-Harrington Theorem fails by the lemma.

It remains to prove the lemma. This is based on the following two facts.

a. There exists a strong cut between any a and b such that $b = PH(a, a)$.
b. Every strong cut is a model of Peano Arithmetic.

The situation is schematically shown in Fig. 4.6.

The construction of a strong cut between a and b is rather tedious, but the basic idea can be demonstrated on the following simple lemma.

Lemma 13 *For every $k \le a$ and every definable coloring of k-element subsets of $[0, b]$ by two colors, there exists a monochromatic definable subset $X \subseteq [0, b]$ and a cut I between a and b such that $X \cap I$ is unbounded in I. In particular, X is $*$-definable in I.*

The assumption $b = PH(a, a)$ and $k \le a$ implies $b \ge PH(k, a)$. Hence there exists a relatively large monochromatic subset X. In particular the size of X is at least a, which is a nonstandard number. Let $X = \{x_1, x_2, \ldots\}$ in increasing order. Take $I = \{y \in M; y \le x_n$, for some standard number $n\}$. This proves the lemma.

We now consider the second step of the proof which shows that a strong cut is a model of PA.

First we need to show that the cut is closed under addition and multiplication. This can be ensured by taking a special coloring and using the fact that the monochromatic set is relatively large in an essential way (not just to get a sufficiently big set) to prove that the sequence $x_1, x_2, \ldots$ increases at least as rapidly as $a, a^2, a^3, \ldots$.

One can easily check that the basic axioms are satisfied by every cut closed under addition and multiplication. So it remains to prove the induction axioms in I. This follows from the next lemma.

Lemma 14 *Every definable set in (the structure induced by) I is $*$-definable.*

To see that this lemma suffices, recall that induction is equivalent to the least number principle. Let $\phi(x)$ be a formula that defines a nonempty subset X of I. We should show that X has the minimal element. According to the lemma, there exists a Y definable in M such that $X = Y \cap I$. Since M satisfies the least number principle, Y has the least number, say y. Clearly, y is also the least number of X (because I is an initial segment).

The lemma is proved by induction on the quantifier complexity of formulas. The base case concerns bounded formulas. These formulas are absolute, so the definability in I, $*$-definability, and definability in M are all the same.

To prove the induction step, suppose $\phi(x)$ is a formula of the form $\exists y\, \psi(x, y)$, where ψ is a $*$-definable relation. Define a function f on the interval $[a, b]$ by

$$f(x) = \begin{cases} \min\{y;\ y \leq b \text{ and } \psi(x, y)\}, & \text{if there exists any such } y, \\ 0 & \text{otherwise.} \end{cases}$$

Define a coloring of triples of elements (u, v, w), $a \leq u < v < w \leq b$, by the following rule.

- If for some $x < u - 3$, $v < f(x) \leq w$, then color (u, v, w) blue;
- if for all $x < u - 3$, $f(x) \leq v$ or $f(x) > w$, then color (u, v, w) red.

An interesting property of this coloring is that every unbounded monochromatic $*$-definable set is red, none is blue. Suppose $U = \{u_1, u_2 \ldots\}$ is a blue unbounded monochromatic $*$-definable set. Consider the following triples (u_1, u_2, u_3), (u_1, u_3, u_4), $(u_1, u_4, u_5), \ldots$. Since all these are blue, we have $x_1, x_2, x_3, \ldots < u_1 - 3$ such that $f(x_1) \in (u_2, u_3]$, $f(x_2) \in (u_3, u_4]$, $f(x_3) \in (u_4, u_5], \ldots$. This is impossible because $x_1, x_2, x_3, \ldots$ is unbounded in I and I is regular. So, indeed, there are only red monochromatic relatively large sets.

To prove that $X = \{x \in I;\ \exists y \in I\ \psi(x, y)\}$ is $*$-definable, let $c \in X$. Since U is unbounded in I, there is a $u_t \in U \cap I$ such that $c < u_t - 3$. Since for some $d \in I$, we have $\psi(c, d)$, we also have $\psi(c, f(c))$ and $f(c) \in I$. Let r be such that $f(c) \leq u_r$ and $r > t + 1$. Now consider the triple (u_t, u_{t+1}, u_r). Since it is colored red, we have either $f(c) \leq u_{t+1}$ or $f(c) > u_r$, but only the first possibility can be true. So $f(c) \leq u_{t+1}$. This shows that for every $c \leq u_t - 3$,

$$\exists y \in I\ \psi(c, y) \quad \text{if and only if} \quad \exists y \leq u_{t+1} \psi(c, y).$$

Since for every c, the bound u_{t+1} is definable in M, we have shown that X is $*$-definable.

4. *Indicators of models of Peano Arithmetic.* Define $PH(m, n)$ to be the largest k such that for every coloring of k-element subsets of the interval $[m, n]$ by two colors, there exists a relatively large monochromatic subset of $[m, n]$ (it is a sort of inverse function to $PH(k, m)$). Paris called such a function an *indicator* of models of *PA* because of the following theorem that he proved [212].

Theorem 32 *There exists a cut between a and b which is a model of PA if and only if* $PH(a, b)$ *is a nonstandard number.*

5. *Unprovable combinatorial statements equivalent to a reflection principle.* In their seminal paper Paris and Harrington observed that their sentence, though expressed as a combinatorial statement related to the Ramsey Theorem, is equivalent to a strengthening of the Gödel sentence, a restricted form of the reflection principle (called Σ_1*-Reflection Principle*) for Peano Arithmetic. It is the following sentence:

For all Σ_1 *sentences* ψ*, if PA proves* ψ*, then* ψ *is true.*

Essentially, all known independent combinatorial sentences are equivalent to the reflection principle for the corresponding theory. Should we be disappointed because the allegedly combinatorial sentences are, in fact, sentences about logic? Not at all, one can show that many unrelated theorems are equivalent in weak theories. (In strong theories they are provable, hence trivially equivalent.)

We will discuss reflection principles in Chap. 7.

6. *Measuring the growth rate by ordinals.* To measure the growth rate of functions defined on natural numbers one defines a sequence of functions that serve as a scale. The functions are indexed by constructive ordinals. The history of this research goes back to the work of Kleene [155].

There are several definitions of such scales, called *hierarchies of functions.* Here is one definition of such a hierarchy, the *fast growing hierarchy*. We start with a simple function, which is usually the successor function. So we put

$$f_0(n) = n+1.$$

Then we progress by transfinite recursion as follows. If β is not a limit ordinal, thus $\beta = \alpha + 1$, for some ordinal α, then define

$$f_\beta(n) = f_{\alpha+1}(n) = \underbrace{f_\alpha(\dots(f_\alpha(f_\alpha(n)))\dots)}_{n\text{-times}}.$$

If β is a limit ordinal, then we need a sequence of smaller ordinals that converges to β. These sequences are called *fundamental sequences*. Let $\beta_0, \beta_1, \beta_2, \dots$ be a fundamental sequence for β. Then we define

$$f_\beta(n) = f_{\beta_n}(n).$$

So we take a sort of diagonal of the smaller functions. Clearly, the definition depends very much on the choice of fundamental sequences. Usually there is one natural choice for the fundamental sequence. For example, for ω we take, of course, the sequence $0, 1, 2, \dots$, for ω^2, we take $1, \omega, \omega^2, \omega^3, \dots$, for ε_0, we take $1, \omega, \omega^\omega, \omega^{\omega^\omega}, \dots$.

Let us compute some functions in the hierarchy. Clearly we have $f_1(n) = 2n$, $f_2(n) = 2^n n$,

$$f_3(n) \geq supexp(n) = \underbrace{2^{2^{\cdot^{\cdot^{2^n}}}}}_{n \text{ twos}}.$$

f_4 is larger than iterated superexponentiation and so on. f_ω is the diagonal of these and it is a function that grows extremely fast. A version of this function (in the sense of having approximately the same growth rate) is the *Ackermann function.* Yet this is only the beginning of the hierarchy. It is clear that we cannot observe the growth of such functions empirically.

Using this hierarchy one can determine the possible growth of definable functions in some theories. Namely, if one can show in Finite Set Theory that a function f is defined for all natural numbers, then for some $\alpha < \varepsilon_0$ and some n_0, $f(n) < f_\alpha(n)$ for all $n > n_0$. This result is tight, since all f_α, for $\alpha < \varepsilon_0$

are definable in Finite Set Theory. For every recursively axiomatized theory, there is a constructive ordinal associated with it in this way. These ordinals have been determined for moderately strong theories. For strong theories, such as the Zermelo-Fraenkel set theory, it is beyond current means to describe these ordinals.

7. *Measuring the "size" of finite sets by infinite ordinals.* An interesting spin-off of this theory is a way of measuring finite subsets of natural numbers by countable ordinals. For finite ordinals, let *n-large mean* having at least n elements. Then we define ω-large to mean relatively large. We continue by defining X is $\omega+1$-large if X is still ω-large after deleting the least element, X is $\omega+2$-large if X is still $\omega+1$-large after deleting the least element, etc. In general, define

$$X \text{ is } \alpha+1\text{-large} \quad \text{if } X\setminus\{\min X\} \text{ is } \alpha\text{-large},$$

and, for a limit ordinal λ

$$X \text{ is } \lambda\text{-large} \quad \text{if } X \text{ is } \lambda_{\min X}\text{-large},$$

where $\lambda_0, \lambda_1, \ldots$ is the chosen fundamental sequence for λ.

Since the arithmetic of ordinals does not have as good properties as the arithmetic of natural numbers, we cannot expect that this concept will have as good properties as the cardinality of finite sets. But at least in some cases some properties are preserved. For example, if $\beta \leq \alpha$ then a set X is $\omega^\alpha + \omega^\beta$-large if and only if it can be divided into two sets X_1 and X_2 with all elements of X_1 less than all elements of X_2 such that X_1 is β-large and X_2 is α-large. Using a little bit more sophisticated definition of α-largeness, we can have a similar relation also for products.

J. Ketonen and R. Solovay introduced this concept in order to analyze the Paris-Harrington Theorem and to give an alternative proof of its independence from Peano Arithmetic [151]. This approach can also be used to determine the proof-theoretical ordinals of theories (the least constructive ordinals for which theories prove transfinite induction) using model-theoretical means, see [10].

8. *Why is Kruskal's theorem stronger?* Hydras correspond to ordinals less then ε_0, and in Kruskal's theorem we also have rooted trees, so why is the latter stronger? The point is that the orderings of rooted trees are different in the two problems. To see that one gets more using the quasiordering by homomorphic embeddings, we will show an order preserving mapping onto all ordinals below Γ_0 (see page 194 for the definition of this ordinal). Given a tree, the associated ordinal will be denoted by $\alpha(T)$. This mapping is defined by induction on the height of the tree. If the height is 0, which means that the tree only consists of the root, $\alpha(T)=0$. Otherwise we distinguish four cases according to the number of subtrees attached to the root.

   a. If there is only one such subtree T_1, then put $\alpha(T)=\alpha(T_1)$.
   b. If there are two, T_1 and T_2, then, assuming without loss of generality $\alpha(T_1)\geq\alpha(T_2)$, define $\alpha(T)=\alpha(T_1)+\alpha(T_2)$.
   c. If there are three, T_1, T_2, T_3, then, assuming without loss of generality $\alpha(T_1)\geq\alpha(T_2)\geq\alpha(T_3)$, define $\alpha(T)=\phi(\alpha(T_2),\alpha(T_1))$ (where ϕ is the Veblen function, see page 195).

d. If there are more than three, $T_1, T_2, \ldots$, then, assuming without loss of generality $\alpha(T_1) \geq \alpha(T_2) \geq \cdots$, define $\alpha(T) = \phi(\alpha(T_1), \alpha(T_2))$.

One can check that this mapping preserves ordering, which means that if there exists a homomorphic embedding of T_1 into T_2, then $\alpha(T_1) \leq \alpha(T_2)$. The reason why we get all ordinals below Γ_0 is the following normal form. Every ordinal $\alpha < \Gamma_0$ can be presented as

$$\alpha = \phi(\alpha_1, \beta_1) + \cdots + \phi(\alpha_n, \beta_n),$$

with $\alpha_1, \ldots, \alpha_n,\ \beta_1, \ldots, \beta_n < \alpha,\ n \in \omega$. Applying this fact recursively to $\alpha_1, \ldots, \alpha_n,\ \beta_1, \ldots, \beta_n$ and so on, we obtain a term based only on symbols ϕ, $+$ and 0.

Thus one can reduce the principle of the transfinite induction up to Γ_0 (more precisely, the statement that the ordinal notation based on the Veblen function defines a well-ordering) to Kruskal's theorem. As this is stronger than the transfinite induction to ε_0, also Kruskal's theorem is stronger. Γ_0 is not the largest such ordinal. Every well-quasiordering is associated with an ordinal. The ordinal associated with trees and the quasiordering by homomorphic embeddings has been determined and is larger than Γ_0.

The fact that Kruskal's theorem is so strong, however, does not imply that Friedman's "miniaturization" of the theorem is also a strong statement. Kruskal's theorem has higher logical complexity, in particular, it is an inherently infinite statement, while Friedman's one is finite, a Π_2 sentence; hence Friedman's miniaturization of Kruskal's theorem *cannot* be equivalent to the original theorem. So we have to scrutinize it more closely.

The above interpretation of ordinals by trees shows that Kruskal's theorem implies the transfinite induction over Γ_0. The principle of transfinite induction can be stated as the statement that there is no infinite decreasing sequence of ordinals. Equivalently, in every infinite sequence of ordinals there is a pair of consecutive elements α_n, α_{n+1} such that $\alpha_n \leq \alpha_{n+1}$. The relation of being less than or equal to is interpreted as an embedding of the corresponding trees. In Friedman's miniaturization the modification is that we allow only special sequences, those in which the trees increase slowly. In terms of ordinals this means that we only say that there are no infinite decreasing sequences of ordinals in which the size of the expressions for the ordinals grows slowly. We know that this weakened version of the principle of the transfinite induction cannot capture the general principle, but one can show that it is enough to prove the consistency of ATR_0.

Here is the basic idea. To prove the consistency of ATR_0 we need to assume the condition about infinite sequences only for sequences where the growth of the representations of ordinals α_n is bounded by a primitive recursive function, which in terms of the hierarchy f_α means bounded by some $f_k(x)+l$ for $k, l \in \omega$. This ensures that if we replace $\alpha_0, \alpha_1, \alpha_2, \ldots$ by the sequence

$$\omega^\omega \cdot \alpha_0,\ \omega^\omega \cdot \alpha_1,\ \omega^\omega \cdot \alpha_2,\ \ldots,$$

the gaps between the consecutive ordinals of the new sequence will be so large that we will be able to insert long decreasing sequences between each pair and make the growth of the sizes of the ordinals linear. This translates to a linear growth of the corresponding trees.

9. *Combinatorial sentences provable only using large cardinals.* While proof-theoretical ordinals are known only for fairly weak fragments of set theory, combinatorial sentences that require very strong axioms about sets have been found. The model-theoretical method of Paris and Harrington, which was sketched above, can be extended to set theories. This line of research has been pursued mainly by H. Friedman. For many years, he has been improving his results by showing independence from stronger large-cardinal axioms and by making the sentences more natural.

   One of his sentences is related to the Finite Ramsey Theorem. To see the connection, we will follow Friedman's exposition in first presenting some well-known versions of the Infinite Ramsey Theorem. Recall that in the Infinite Ramsey Theorem (see page 25) it is important that we color k-element subsets of natural numbers by a *finite* number of colors. However there are versions of the theorem where the number of colors can be infinite. Here is a typical result of this type. Call a function F from k-element subsets of $\mathbb{N}$ to $\mathbb{N}$ *regressive,* if for every k-element set of numbers x, $F(x) = 0$ if $\min(x) = 0$ and $F(x) < \min(x)$ otherwise.

Theorem 33 *For every regressive function F from k-element subsets of $\mathbb{N}$ to $\mathbb{N}$, there exists an infinite subset of natural numbers X, such that for k-element subsets x of X, the value $F(x)$ depends only on the least element of x.*

The straightforward finite version of this theorem has the same status as the Paris-Harrington Theorem: it is true but unprovable in Finite Set Theory. It is interesting that in the finite version we do not have to add anything like 'relatively large' [149].

In the theorem above, it can happen that the set X is homogeneous for some F, but in general it is not true. A counterexample is F defined by $F(x) = 0$ if $\min(x) = 0$ and $F(x) = \min(x) - 1$ otherwise. Nonetheless, this is possible if we replace $\mathbb{N}$ by a suitable large cardinal. Recall that cardinals are represented by some ordinals, thus they are naturally equipped with a well-ordering. Hence the concept of a regressive function makes sense also for them. In fact, such a property defines particular large cardinals.

Definition 7 For $k > 0$, an infinite cardinal κ is called $(k-1)$*-almost ineffable* if for every regressive function F from k-element subsets of κ to κ, there exists $X \subseteq \kappa$ that is homogeneous for F and has cardinality κ.

It is a miniaturization of this definition which Friedman showed to be provable only using large cardinals. Since $\mathbb{N}$ does not have the property of the almost ineffable cardinals, it is clear that the miniaturization has to be nontrivial. Thus instead of one coloring function, it is necessary to take a family of functions satisfying certain properties. We will consider mappings that to a given set $A \subseteq$

$\{0, 1, \ldots, n\}^k$, assign a function $F_A : A \to A$. Following Friedman, we define that such a family of functions is *#-decreasing,* if for every $A \subseteq \{0, 1, \ldots, n\}^k$ and $b \in \{0, 1, \ldots, n\}^k$,

a. either $F_{A\cup\{b\}}$ extends F_A, which means $F_A(a) = F_{A\cup\{b\}}(a)$, for every $a \in A$,
b. or for some $a \in A$, $\max(b) < \min(a)$ and $F_A(a) > F_{A\cup\{b\}}(a)$.

Theorem 34 (Friedman [82]) *For every k and r, there exists n such that for every family $\{F_A\}_{A\subseteq\{0,1,\ldots,n\}^k}$ of #-decreasing functions, there exists an $A \subseteq \{0, 1, \ldots, n\}^k$ and an $E \subseteq \{0, 1, \ldots, n\}$ such that E has r elements, $E^k \subseteq A$ and there are at most k^k elements $b \in A$ such that for some $a \in E^k$, $\max(b) < \min(a)$ and $F_A(a) = b$.*

This needs some explanation. First, instead of restricting the class of functions F_A to regressive functions, we take all, but then we count the number of the values on arguments a on which they behave as regressive functions. Secondly, we are now considering functions defined on k-element sequences of numbers, instead of k-element sets. Such sequences can be easily distinguished by their order type. For example (a_1, a_2), a sequence of length two, has one of the three order types: $a_1 < a_2$, $a_1 > a_2$ and $a_1 = a_2$. Thus in general it is not possible to reduce the number of colors below the number of order types. The number k^k is a (generous) upper bound on the number of order types.

Friedman proved:

a. Theorem 34 is provable in *ZFC* augmented with the axiom saying that for every $k > 0$, there exists a k-subtle cardinal;
b. for every natural number $k > 0$, Finite Set Theory augmented with Theorem 34 as an additional axiom proves the consistency of *ZFC* plus the axiom that there exists a k-subtle cardinal.

These results are stated for a more familiar kind of cardinals, but they hold true also for k-almost ineffable cardinals because the hierarchies of these two types of cardinals are mutually interleaved.

4.5 The Independent Sentences of Set Theory

Reading your proof had a similarly pleasant effect on me as seeing a really good play.

Kurt Gödel on Cohen's proof

Finally, I will turn to one of the most exciting topics in the foundations of mathematics—independence results in set theory. The reason why these results are so valuable is that they not only give very concrete independent sentences, they are also solutions of problems that had been open for decades and that were very important for the development of set theory. When speaking about independence in set theory, I mean Zermelo-Fraenkel Set Theory, the most important axiomatic system for set theory. Some independence results had been obtained for weaker systems before the problem of the Continuum Hypothesis was solved, but the golden era of set theory started only after Paul Cohen introduced the method of *forcing*.

Let us recall some basic facts. The Continuum Hypothesis is the conjecture that the cardinality of the set of all subsets of natural numbers is the first uncountable cardinality; in mathematical symbols $2^{\aleph_0} = \aleph_1$. It can also be phrased as the statement that there is no intermediate cardinality between the cardinality of the set of natural numbers and the cardinality of the set of real numbers. This statement is independent of Zermelo-Fraenkel Set Theory, which means that it is neither refutable in this theory nor provable. The first part, which can also be expressed as the consistency of the Continuum Hypothesis with the axioms of Zermelo-Fraenkel Set Theory, was proved by Gödel in 1938 [97]. The second part, the consistency of the negation of the Continuum Hypothesis, was proved by Cohen in 1963 [46]. For this result he was awarded the Fields Medal of the International Mathematical Union in 1966. (Until now, he has been the only mathematician who received this prize for a result in logic.) Immediately after that, other researchers started generalizing and improving his method and applying it to other problems. Several longstanding open problems in infinite combinatorics and set-theoretical topology were solved soon after. Forcing has become a standard tool in modern set theory.

The amount of independence results proved in set theory and the flexibility of the forcing method is quite astounding. It shows that Zermelo-Fraenkel Set Theory, in spite of being so strong and apparently containing all natural axioms that we can think of, is not able to decide problems of a certain type. Unfortunately, there does not seem to exist a natural way how to resolve these independences by adding more axioms. New axioms studied in contemporary set theory are almost invariably large cardinal axioms. Axioms of that type have the advantage of not being controversial because when we decide whether to add larger cardinals or to prohibit them, we prefer to add cardinals, since it makes the theory stronger than adding the opposite axiom. However, large cardinal axioms do not help us to decide problems such as the Continuum Hypothesis. Certainly, it is possible to add other axioms ad hoc (for instance, we can simply postulate the Continuum Hypothesis), but set-theorists are not able to agree on such axioms. In fact, most of them prefer to leave such statements undecided, as it enables them to study a variety of different models of set theory. It is like in other fields of mathematics where mathematicians prefer to have the freedom of using different structures and they generalize concepts rather than restrict them to a single one. For example, we do not want to postulate that real numbers is the only *field* which should be studied; on the contrary, we want to have the possibility to study other structures that satisfy the axioms of a field, the complex numbers, quaternions, number fields, finite fields, etc.

It should be noted that all the independence results in set theory only concern sentences about *infinite* sets. Forcing and other methods used in set theory fail to prove the independence of any sentence about numbers or other finite structures. In the previous section I described a method that works in the context of finite problems, but it is only applicable to problems associated with very fast growing functions. Otherwise we only have Gödel's Theorem, which is universally applicable, but from which we are not able to extract mathematically interesting sentences.

Maybe the difficulty is in that natural numbers are very much determined by basic principles such as induction. Maybe there are deeper reasons why we lack such methods. But maybe we just have to work harder to develop theory and to find a method that would be as powerful as forcing and will work also for finite problems. Then, hopefully, we will be able to show that some persistent open problems in number theory, or in complexity theory are independent (say, from the axioms of Finite Set Theory). To develop such methods is one of the principal goals of the research in proof complexity.

The Consistency of the Continuum Hypothesis

Rather than surveying independence results in set theory I am going to focus on the methods that are used to prove the independence result. I think this is more interesting and reveals more about the essence of the incompleteness phenomenon. We will start with the method by which Gödel proved the consistency of the Continuum Hypothesis.

As I am going to describe fairly difficult results, it is good to start slowly. Therefore, let us first briefly recall what one has to do prove the independence of a sentence ϕ from a theory T. We need do two things: to prove that ϕ is not provable in T and to prove that the negation of ϕ is not provable in T. These two conditions are equivalent to $\neg\phi$ being consistent with T and, respectively, ϕ being consistent with T. By the completeness theorem for first order logic, $\neg\phi$ is consistent with T if and only if there exists a model of T in which ϕ is false. Likewise, ϕ is consistent with T if and only if there exists a model of T in which ϕ is true. Thus the task of proving the independence reduces to constructing two models. There are methods for proving independence not based on models, but in the case of Zermelo-Fraenkel Set Theory it the preferred choice.

It is difficult to construct a model of a strong theory from scratch. But notice that if we are to prove that T is consistent with some sentence ϕ, then in particular T must be consistent. If we do not assume that T is consistent, we cannot do anything. When T is consistent, then by the Completeness Theorem, we also know that there exists a model of T. Then we can start with such a model and modify it to obtain a model of T in which ϕ is true.

Thus our goal now is to construct a model of Zermelo-Fraenkel Set Theory in which the Continuum Hypothesis is true from a general model M of Zermelo-Fraenkel Set Theory. Gödel's idea was to define a submodel of M which is in some sense minimal. His hope was that if he picked as few elements of M as possible, then M would be very "thin", in particular, there would exist few subsets of the set of natural numbers. With a bit of luck, there would be only $\aleph_1$ such subsets, hence the consistency of the Continuum Hypothesis would be established. His intuition was correct: this approach really does what he intended to achieve.

The main problem that he had to solve was how to pick elements so that the resulting structure would satisfy the axioms of Zermelo-Fraenkel Set Theory and, at the same time, would have few elements. Gödel did it by choosing the constructible sets. I mentioned briefly this concept in Chap. 3. Let me now say just a little bit more about it and give some details of this construction Notes.

Recall that in the Zermelo-Fraenkel universe every set is, in a way, constructed from the empty set. We start with the empty set and then using transfinite recursion over all ordinal numbers we add power sets. These sets are denoted by

$$V_0 \subseteq V_1 \subseteq V_2 \subseteq \cdots \subseteq V_\alpha \subseteq \cdots,$$

and called the cumulative hierarchy (see page 168). For example, at stage ω (the first infinite ordinal) we obtain all hereditarily finite sets (the sets whose elements are finite and elements of elements are finite, etc.).

Gödel's approach can be explained as taking this principle seriously and really *constructing* all sets from below. In the cumulative hierarchy we simply take all subsets of the set from the previous stage. What Gödel did instead was to take only definable subsets. Furthermore, the definability is meant in a restricted sense. In this way, Gödel defined the *constructible hierarchy*

$$L_0 \subseteq L_1 \subseteq L_2 \subseteq \cdots \subseteq L_\alpha \subseteq \cdots.$$

In contrast to the cumulative hierarchy, these sets may not cover all sets. The union of these sets is a proper class denoted by L and its elements are called *constructible sets*.

Having a set x, we put into our model all sets that are definable in the structure with the universe x and the membership relation $\in$. Hence, when defining new sets at this stage, we can only refer to sets contained in x and we are not allowed to mention sets that have not been constructed yet. This is called a predicative definition. Because of this constructive nature of the definition, Gödel used the term 'constructible'. However, one should not think of constructible sets as something very explicit. Since we want L to be a model of Zermelo-Fraenkel Set Theory, the process has to be done by transfinite recursion over *all ordinals*. Thus we assume that we can take ordinals for granted, or in other words, that every ordinal is constructive. One should not confuse this kind of constructibility with the concept based on algorithms and Turing machines.

Given a model M of Zermelo-Fraenkel Set Theory, we will denote the submodel formed by constructible sets by L^M. It is the minimal model among submodels of M that contain all ordinals of M. Intuitively, L^M is a model of Zermelo-Fraenkel Set Theory because essentially all axioms are instances of the Comprehension Schema and we are adding all definable sets. The Replacement Axiom Schema also should not be a problem because we take all cardinals from M. To prove it formally is a nontrivial task and it is not possible to discuss it without going into rather technical details. (Some more details are in Notes.)

Gödel used L^M also to prove the consistency of the Axiom of Choice. In order to define constructible sets, the Axiom of Choice is not needed. Thus one can start with a model M which does not satisfy this axiom. Then taking L^M in this model we obtain a model in which the Axiom of Choice is true. This proves that if Zermelo-Fraenkel Set Theory is consistent without the Axiom of Choice, then it is also consistent with it. This is a very important result because it dismisses doubts about the consistency of the Axioms of Choice, the doubts caused by its paradoxical consequences.

To prove that L^M satisfies the Axiom of Choice, even if M does not, is not difficult. The process of defining constructible sets can be used to define an enumeration of all constructible sets by ordinals. Thus every constructible set is associated with an ordinal. Now suppose that we should pick one element from a nonempty set x. Then there is a canonical way how to do it: pick the element associated with the smallest ordinal. Since we have a systematic way of choosing elements from nonempty sets, the Axiom of Choice is true.

How to Enlarge a Model of Set Theory

Our goal now is to construct a model in which the Continuum Hypothesis is false. We will again start with a model M of Zermelo-Fraenkel Set Theory, but we will moreover assume that M is countable and the natural numbers in M are the actual natural numbers. (We know that if a theory is consistent, then it has a countable model; for the second assumption later see Notes.) If the Continuum Hypothesis fails in M we are done. So let the Continuum Hypothesis be true in M. This gives us some information about M, but in order to be able to manipulate with M, we would like to know more. By the result above, we can take the submodel consisting of constructible sets of M. So we can assume without loss of generality that already in M all sets are constructible. This, of course, goes in the wrong direction, as we want to have many subsets of natural numbers. We will not use this information in an essential way, but it should help us to create a mental picture of what is going on. Notice that so far we cannot exclude that in all models there are only constructible sets.

We want to enlarge M so that it contains lots of subsets of natural numbers. Since M is countable, also $\mathcal{P}(\omega)^M$, the set that represents the set of all subsets of integers in M, is countable. On the other hand, since in fact the number of subsets of natural numbers is uncountable, we know positively that there are some subsets of integers which are not in $\mathcal{P}(\omega)^M$. Thus it seems that there are sets that we can use, but we have no guarantee that they can be added in a consistent way. Obviously, we cannot only add one new subset of integers. We need to obtain a model of Zermelo-Fraenkel's axioms, hence once we add a new set X, we must also add a lot of other sets, in particular, all those that are definable from X. Yet it may happen that whatever we add to X, we will not be able to round it off to a model of Zermelo-Fraenkel Set Theory.

Thus we will start with the simplest problem: to prove that we can add at least one non-constructible subset. Knowing the concept of constructible sets, it is natural to try the following. Let us construct sets as Gödel did, except that at some stage we inject a subset of integers that is not in M. Let X be a such a subset of integers. We may simply start with X and then follow the constructive process without any further changes hoping that the modification of the construction is so small that we will still be able to verify Zermelo-Fraenkel's axioms as we did without X. This answers the question what should we do when adding a set X, but it does not solve the problem which sets X can be used.

In order to motivate the next step, we will make a short digression to algebraic structures. Given an algebraic structure A, it is very easy to enlarge it by adding one or more elements. This is because there is a very special way of doing it. We can add elements in the *free* way. This means that we will not assume any special properties, except those implied by the axioms of the particular kind of algebras. As an example consider the field of real numbers $\mathbb{R}$. To add an element x in the free way means to enlarge $\mathbb{R}$ into the field of rational functions with indeterminate x. (Rational functions are fractions $p(x)/q(x)$, where $p(x)$ and $q(x)$ are polynomials and $q(x)$ is not the zero polynomial.) This field is denoted by $\mathbb{R}(x)$. It is the minimal enlargement that contains x and in which x does not satisfy any nontrivial equation.

What we attempted to do with models, also resulted in a uniquely determined, and in a sense minimal structure. Let $M[X]$ denote this minimal extension of M obtained by adding a set X. Why is it sometimes not a model of Zermelo-Fraenkel's axioms, whereas $\mathbb{R}(x)$ is always a field? The reason why algebras behave better than general models is that they are defined by equations and equations are all that we are interested in. Equations are positive statements and combining positive statements cannot lead to a contradiction because the negation sign does not occur in them. Furthermore, if we only have equations, we can easily compare the sets of formulas that an element satisfies; the more equations we impose on an element the more special the element is. Thus the most general situation, the situation in which the element is *free*, corresponds to the empty set of equations.[14] In the case of general structures, there is no preferred set of formulas and we have to treat positive and negative statements in the same way, hence a contradiction can easily occur. For example, should $M[X]$ be larger than M, X must not be empty or the whole set of natural numbers, as these sets are already present in M. Thus X must satisfy some positive statements of the form $n \in X$ and some negative statements $n \notin X$ as well. If we add a subset X of natural numbers, we must also add its complement. There is a kind of symmetry between positive and negative statements and we have to break it.

At this point one may fear that the analogy between algebras and models is misleading and it is pointless to try this approach, but that is a wrong conclusion. Let me draw another parallel. In geometry we often talk about *general positions*. If, for example, we have a finite set of points, then a line in a general position does not hit any of the points, it is not parallel with any of the lines determined by pairs of points, etc. In this example the idea is to avoid all positive relations, but there is an essential difference. Whereas the free extensions by one element are all isomorphic, in geometry there are always infinitely many general positions; in fact, most positions are general.

---

[14]When we talk about fields, we prefer to say '*transcendental*', instead of '*free*'.

Cohen's surprising discovery is that it is possible to define a concept in set theory that is similar to freeness in algebra and general position in geometry. He coined this notion *a generic set*. As the name suggests, a generic set G should look like a typical set X such that $M[X]$ is a model of Zermelo-Fraenkel Set Theory. We cannot use this property as the definition because so far we are no able to prove that there exists any X that we can add consistently to M. Thus one has first to give a rather technical definition of being generic and then show that it has this property. I will give a precise definition in Notes; here I will only try to convey some intuition about this concept.

It will often be convenient to think about a generic set G as determined by an infinite sequence $g = g(0), g(1), g(2), \ldots$ of zeros and ones: $g(n) = 0$ says that $n \notin G$ and $g(n) = 1$ says that $n \in G$. Thus we will also speak about generic sequences.

Obviously, whether or not G is generic does not depend on whether or not a concrete number, such as 5, is in G. Clearly, it should also not depend on the relation of G with any fixed finite set of numbers. In terms of zero-one sequences, it means that any finite string of zeros and ones can be an initial segment of a generic sequence. Thus a generic sequence may start in an arbitrary way and we can determine if it is generic only after we know all the infinitely many values. It may then seem rather paradoxical that all properties of the extension generated by a generic sequence are determined by its *finite* segments. But the point is that all generic extensions will be "essentially the same", so they will only differ in "inessential details" such as whether or not G contains 5. Hence it is reasonable to assume that every "inessential detail" should be determined by a few values of the generic sequence.

To get intuition it helps to think about generic sequences as constructed in infinitely many steps using the following rule of thumb.

The Generic Set Rule *If something can happen, then it will happen.*

This informal rule enables us to derive various facts about G. We will start with proving that both G and its complement are infinite. Indeed, suppose that we have already constructed $g(0), g(1), g(2), \ldots, g(n)$ and we wonder if there will be some $m > n$ such that $g(m) = 1$. We know that the property of being generic does not depend on initial segments of g, hence there is nothing that prevents us from defining $g(m) = 1$ for $m = n + 1$ or any larger number. Thus *it can happen that* $g(m) = 1$ *for some* $m > n$, hence by the rule $g(m)$ *must be equal to 1 for some* $m > n$. Now we can repeat this argument again and again, whence we conclude that there must be infinitely many numbers m such that $g(m) = 1$. By the same token, there must be infinitely many numbers m such that $g(m) = 0$. Hence both G and its complement are infinite.

What about prime numbers, does G contain infinitely many primes? Exactly the same argument shows that G does contain infinitely many primes. Furthermore, the number of primes that G does not contain is also infinite. In particular, G is never equal to the set of prime numbers P. Notice that the only property of P that was used in this argument was that P was an infinite subset of natural numbers in the model M. Hence we can apply the same argument to any infinite subset of the natural numbers in M. Since G is infinite, this proves that G is not in the model M. Thus a generic set for a model M is never present in the model.

Let us try some other properties. Does a generic G contain two consecutive numbers? It surely does. Given an initial segment $g(0), g(1), g(2), \dots, g(n)$ we may define $g(m) = 1$ and $g(m+1) = 1$ for any $m > n$ and still this sequence can be completed to a generic one. Hence there must be an infinite number of such pairs. By the same token, there infinitely many segments in G which are equal to the binary file of this book. This may suggest that we should imagine a generic sequence s as a random sequence. It is true that g shares some properties with random infinite sequences of zeros and ones (for example, there are also infinitely many copies of this book in every random sequence), but g is not random. Consider the frequency of ones in $r(0), r(1), r(2), \dots, r(n)$ for a random sequence r. As n goes to infinity this frequency very quickly approaches $1/2$. In contrast to that the frequency of ones in $g(0), g(1), g(2), \dots, g(n)$ unpredictably oscillates. To see that consider the following property of m:

$$g(m) = 0 \quad \text{and} \quad g(m+1) = 0 \quad \text{and} \quad g(m+2) = 0 \quad \text{and} \quad \dots \quad \text{and} \quad g(10m) = 0.$$

This property implies that the frequency of ones on the segment $[0, 10m]$ is less than $1/10$. Using again the same argument, one can prove that the number of such numbers m is infinite. Thus the frequency will drop below $1/10$ infinitely often and, by symmetry, also jump over $9/10$ infinitely often.

Assuming that we can prove the existence of a generic set G, we are almost done with the problem of extending the model M to a model that contains a non-constructible set. We know that G is not in M, but this does not automatically guarantee that G is not constructible in the extension $M[G]$. We have to prove that the constructible sets of $M[G]$ are the same as the constructible sets of M. Intuitively this is clear because the ordinals are the same in both models and when defining new constructible sets we refer only to those that we already have constructed. Hence the set that is outside this process keeps staying outside.

I have said very little about how to do these things rigorously. I am going to give more details below; for now, let me only explain one thing that may puzzle you after this informal exposition: *How to interpret the Generic Set Rule in a consistent way?*

Let us look at a couple of possible conflicts in applying the rule. The simplest conflict arises when we take the statement $n \in G$ for a fixed number n. Since both $n \in G$ and $n \notin G$ are possible, the rule dictates that both must be true. This is resolved by disallowing applying the rule to such statements. The intuition is that such single statements have nothing to do with being generic and we can decide them in an arbitrary way. Hence we do not need to use any rule for them.

The second example concerns the proof that a generic set G is infinite. Apparently the same argument shows that G is finite. If we have already constructed $g(0), g(1), g(2), \dots, g(n)$ then nobody tells us that we must not continue with zeros to infinity. Since $g(m)$ *can be* equal to 0 for all $m > n$, then, according to the rule, it *must be*. Thus we conclude that G must be finite.

The usual method to resolve conflicts of rules is to introduce a hierarchy. Then if two rules are in conflict, the rule that is higher in the hierarchy gets the preference.

We can apply this method also here. In our case the hierarchy will be given by the complexity of the formulas—the simpler a formula is, the higher in the hierarchy it is. Furthermore, the existential quantifiers have the priority over the universal quantifiers. In the previous example we had a formula saying that there *exists* an $m > n$ which is in G and the formula saying that *for all* $m > n$, m is not in G; so the first formula gets the preference. Hence the proof that G is infinite is correct, whereas the proof of the converse is wrong.

A Model in Which the Continuum Hypothesis Fails

Now we want to construct a model in which $2^{\aleph_0} = \aleph_2$. To this end we need to enlarge M so that we get $\aleph_2^M$ subsets of natural numbers. Here $\aleph_2^M$ denotes $\aleph_2$ in the sense of the model M. The basic idea is to add a sequence of $\aleph_2^M$ generic sets at once. The sequence of $\aleph_2^M$ generic sets will be denoted by $\{G_\alpha\}_{\alpha<\aleph_2^M}$.

In fact, we do not only want every single G_α to be generic, but, instead, the whole sequence as a single object should be generic. This can be stated more precisely by saying that we need a generic function $g(x, y)$ whose first argument is a natural number, the second argument is an ordinal in M below $\aleph_2^M$ and whose values are 0 and 1. The value $g(n, \alpha)$ tells us whether or not $n \in G_\alpha$. The reason for having the whole sequence generic is twofold. Firstly, we have to ensure that $M[\{G_\alpha\}_{\alpha<\aleph_2^M}]$ is a model of Zermelo-Fraenkel's axioms. Secondly, if the sequence is generic, we can prove that the sets in the sequence are pairwise distinct, which is the heart of the matter. It is very easy to prove the latter fact using the Generic Set Rule. Let $\alpha < \beta < \aleph_2^M$. Suppose that a finite number of values of g is fixed. Then some $g(n, \alpha)$ and $g(n, \beta)$ is not defined yet, and we can put one of them equal to 0 and the other equal to 1. Since this can happen, it must happen for some n. Hence $G_\alpha \neq G_\beta$.

This may give the impression that proving independence results in set theory is an easy job. Indeed, once the main theorems about the method of forcing are proved, a lot of independence results can be shown without much effort. The method is very powerful, but we always must be careful and prove things properly. For example, the above argument is not complete. We have not proved that after enlarging M the cardinal $\aleph_2^M$ is the second uncountable cardinal also in the new model. It could happen that one of the sets that we added along with generic sets G_α was a one-to-one mapping from $\aleph_2^M$ to $\aleph_1^M$. (Recall that M is a countable model, so the actual cardinalities of $\aleph_2^M$ and $\aleph_1^M$ are the same.) This is not the case in this particular construction, so $M[\{G_\alpha\}_{\alpha<\aleph_2^M}]$ is indeed a model in which the Continuum Hypothesis fails.

An equivalent way of constructing this model is to simply construct a generic subset of $\aleph_2^M$. Such a subset automatically encodes $\aleph_2^M$ distinct subsets of natural numbers. This is more elegant, but less intuitive.

Note that exactly the same argument works for many other uncountable cardinal numbers, for example, all $\aleph_n$ for n a natural number.

Forcing

The modern approach to Cohen's forcing method is to first define the concept of a generic set and then define forcing using this concept. The advantage of this approach is that one gets the definition of both concepts very easily, but it is somewhat deceiving, as eventually one has to use the clauses that define forcing syntactically anyway. Another reason for not taking this route is that it is more interesting to get an idea how Cohen conceived these formidable results than learning technicalities. Therefore, I will try to follow his path to the concept of forcing.

I will explain forcing on the example with which we started. Recall that the goal was to add one new subset of natural numbers to a model M. The idea behind forcing is to mimic the proof of the completeness theorem for first order logic. In that proof the aim was to construct a model of a theory T. The main idea was to go over all sentences and for each sentence to decide to add the sentence or its negation to the theory. (It was also necessary to add constants that would represent objects claimed to exist, but we can ignore this aspect for the moment.) Now instead of the theory we have a model M and we want to add a set X of natural numbers to it. We will think of X as a formal constant and try to form a list of sentences that will be all true sentences about it in some extension $M[G]$, where G is an interpretation of the constant X.

The problem is that we have only a rough idea what G should be, so we are not able to decide, for a general sentence ϕ, whether or not we should add it on our list. The only thing that we know for sure is that we may add sentences of the form $n \in X$ and $n \notin X$, for n a natural number, at our will. The problem would be solved if already these simple sentences decided all the rest. To be more precise, we would like, for every sentence ϕ, to be decided by a *finite number* of such sentences. Now it is clear that the crucial thing is what '*decided*' means. If we interpreted it simply as logical consequences, then it would not work. A finite set of such sentences implies very little; for example, it never implies that X is infinite, or that it is finite. So we need to use the fact that we are talking about a generic set. The special way in which the finite sets of basic sentences decide other sentences is the forcing.

To define forcing we first have to define the *forcing conditions*. In our special case, these are the finite sets of basic statements mentioned in the preceding paragraph. The nature of the problem that we are concerned with here, allows us to take sets of a special form, namely, sets that decide whether or not $n \in X$ for all $n = 0, 1, 2, \ldots, m$, for some m. We represent these sets by strings $(g(0), g(1), g(2), \ldots, g(m))$, where each $g(n)$ is either 0 or 1. It is convenient to consider also the empty sequence $()$. Thus forcing conditions are such strings. We say that a forcing condition p extends a forcing condition q, and write $p \supseteq q$ if p is an extension of q. Furthermore, we will need constants for every set in the extension of the model M by a set G. In forcing it is customary to call these constants *names* (the reason for not using the standard term 'constant' is, perhaps, that one set will typically have many names). The names can be defined *a priori,* before an extension is constructed. There is a canonical way to generate the sets of an extension $M[G]$

from the set G and sets of M, thus we can assign a name to every stage of this process. We will use X as the name for the subset of natural numbers that we want to add to the model. So X is a name for a generic set G.

The following is the obvious beginning of the definition of forcing.

1. *a forcing condition* $(g(0), g(1), g(2), \ldots, g(m))$ *forces* $n \in X$ *if* $n \leq m$ *and* $g(n) = 1$;
2. *a forcing condition* $(g(0), g(1), g(2), \ldots, g(m))$ *forces* $n \notin X$ *if* $n \leq m$ *and* $g(n) = 0$.

Then definition continues in the way you expect:

3. *a forcing condition* p *forces* $\phi \vee \psi$ *if* p *forces* ϕ *or* p *forces* ψ;
4. *a forcing condition* p *forces* $\exists x\, \phi(x)$ *if, for some* c, p *forces* $\phi(c)$.

But the definition of forcing a negated sentence is probably different from what you would guess:

5. *a forcing condition* p *forces* $\neg\phi$ *if no forcing condition* q, $q \supseteq p$, *forces* ϕ.

This, at first glance strange clause, is the gist of the definition of forcing.

Example Take the forcing condition $p = (101)$. This condition forces $0 \in X$, $\neg(1 \in X)$, $2 \in X$. Let us check that it also forces $\neg\neg(0 \in X)$. Indeed, every forcing condition q that forces $\neg(0 \in X)$ starts with 0, but such strings are not extensions of (1). Since no extension of p forces $\neg(0 \in X)$, according to 5., p forces $\neg\neg(0 \in X)$.

The definition of forcing the negation of a sentence ensures the following two key properties. The first one is the consistency.

Proposition 7 *For every forcing conditions* p *and every sentences* ϕ, *it is not true that* p *forces both* ϕ *and* $\neg\phi$.

This is an immediate consequence of the definition. For if p forces ϕ, then an extension of p, namely p itself forces p, hence by definition, p does not force $\neg\phi$.

The second property ensures that we can extend any condition so that any sentence is decided.

Proposition 8 *For every forcing conditions* p *and every sentences* ϕ, *there exists an extension* $q \supseteq p$ *such that either* q *forces* ϕ, *or* q *forces* $\neg\phi$.

This is also an immediate consequence of 5., since according to the definition, either there exists an extension $q \supseteq p$ such that q forces ϕ, or already p forces $\neg\phi$.

The last proposition enables us to accomplish our partial goal of mimicking the proof of the completeness theorem. Let $\phi_1, \phi_2, \ldots$ be an enumeration of all sentences that speak about the extension of M. (Since M is countable, the number of names that we need is countable, hence also the number of all such sentences is countable.) We will construct an infinite sequence of zeros and ones

$(g(0), g(1), g(2), \ldots)$ as follows. First we pick $(g(0), g(1), g(2), \ldots, g(n_1))$ so that this forcing condition either forces ϕ_1 or forces $\neg\phi_1$, then we extend it to $(g(0), g(1), g(2), \ldots, g(n_1), g(n_1+1), g(n_1+2), \ldots, g(n_2))$ so that this forcing condition either forces ϕ_2 or $\neg\phi_2$, and so on. The resulting sequence has the remarkable property that *every sentence is decided by an initial segment of* g. This property is taken as the definition of being generic. Let us state it explicitly using the set G instead of the sequence g.

Definition 8 A set G is generic if for every sentence ϕ, there exists a forcing condition determined by a finite part of X which either forces ϕ or it forces $\neg\phi$.

Example Let us check that in our example the condition that X is infinite is forced. Let k be an arbitrary natural number. Consider the sentences

$$\exists n\ (n \geq k \wedge n \in X) \quad \text{and} \quad \forall n\ (n \geq k \rightarrow n \notin X).$$

For every forcing condition p, we can extend p to q, $q \supseteq p$, so that q forces the first sentence. So the first sentence can be forced. The second sentence has to be first expressed using the existential quantifier and negation: $\neg\exists n\ (n \geq k \rightarrow n \notin X)$. Hence in order to force this sentence, we need a condition p such that for all $q \supseteq p$ and all $n \geq k$, q forces $n \notin X$, which is impossible. Since every sentence is decided by a generic set, it is the first sentence that is always forced.

This demonstrates how the special condition 5. breaks the symmetry between these two sentences.

Sets satisfying this definition seem to be interesting objects of study, but it is not clear that they will help us to achieve our main goal, which is to obtain an extension $M[G]$ which is a model of Zermelo-Fraenkel's axioms. Here is a theorem that fully justifies this definition.

Theorem 35 *If G is generic, then for every sentence ϕ, the sentence ϕ holds true in $M[G]$ if and only if there exists a forcing condition determined by a finite part of G which forces ϕ.*

This theorem reduces the task of proving independence results to forcing. To prove that the extension satisfies an axiom, we now only need to prove that this axiom is forced by some forcing condition. In the previous sections I sketched how one can prove such results using the informal Generic Set Rule. The formal proofs based on forcing follow essentially the same lines. Another consequence of this theorem is that the sentences that are true in all generic extensions are precisely those that are forced by the empty forcing condition.

To prove that all Zermelo-Fraenkel's axioms are forced is a rather tedious task. It requires precisely defining the set of names that represent elements of the extension and then writing down a rather technical definition of forcing the atomic sentences $c_1 \in c_2$ and $c_1 = c_2$. (I only defined forcing of $c_1 \in c_2$ in the special case of c_1 representing a natural number and c_2 representing G.) Fortunately, Cohen proved a general theorem that Zermelo-Fraenkel's axioms are always forced, whatever type of forcing conditions one chooses. Once we know this, we can fully concentrate on the sentences that we want to prove to be consistent with Zermelo-Fraenkel's axioms.

For the sake of simplicity, I have considered a concrete example of forcing conditions. Using these conditions one can construct a model which contains a non-constructible subset of natural numbers. If we need to prove other independence results we have to use other forcing conditions.

Example If we want to show the unprovability of the Continuum Hypothesis by proving that it is consistent to assume that $2^{\aleph_0} = \aleph_2$, we can take forcing conditions of the form

$$\big((\alpha_1, n_1, a_1), (\alpha_2, n_2, a_2), \ldots, (\alpha_k, n_k, a_k)\big),$$

where $k, n_1, n_2, \ldots, n_k$ are natural numbers, $\alpha_1, \alpha_2, \ldots, \alpha_k < \aleph_2^M$, $a_1, a_2, \ldots, a_k \in \{0, 1\}$ and $(\alpha_i, n_i) \neq (\alpha_j, n_j)$ for $i \neq j$. A condition (α_i, n_i, a_i) determines whether or not n_i is in the generic set G_{α_i}.

One can prove various independence results by choosing suitable sets of forcing conditions. Thus the forcing method is the art of designing forcing conditions.

The Independence of the Axiom of Choice

After Gödel had proved that the Axiom of Choice is consistent with the rest of Zermelo-Fraenkel's axioms (provided that Zermelo-Fraenkel's axioms without the Axiom of Choice are consistent), it remained to show that the same is true about the negation of the Axiom of Choice. The method of forcing enabled Cohen to prove also this. He showed how, from a given model, to construct a larger model in which the Axiom of Choice fails. I will sketch the basic idea of this construction.

Let M be a countable model of Zermelo-Fraenkel's axioms. Take forcing conditions that enable us to obtain an infinite set of generic sets. This is done in the same way as in the construction that was used to show that the negation of the Continuum Hypothesis is consistent, but we only need a countable set of generic sets. Let these generic sets be denoted by $G_0, G_1, G_2, \ldots$. Now we have to do something different because the previous construction produced a model in which the Axiom of Choice was true. The idea is to add the sets $G_0, G_1, G_2, \ldots$ to M in such a way that they are in a certain sense indistinguishable. In particular, it will not be possible to determine the index n from the set G_n. It seems plausible that such a construction is possible because our intuition about generic sets is that they do not have any special properties, hence it should be hard to tell them apart.

More precisely, the goal is to enlarge M by $G_0, G_1, G_2, \ldots$ so that M contains the set $\Gamma = \{G_0, G_1, G_2, \ldots\}$, and such that for every A, a subset of Γ, either A is finite, or it contains all elements of Γ except possibly for a finite number of them.

Once we have such a model, we will be done because this will be in contradiction with the Axiom of Choice, as one can easily derive from the Axiom of Choice that every infinite set has a subset which is infinite and whose complement is also infinite.

Let us see what the extension $M[\Gamma]$ must contain and what it must not. First, it has to contain all finite subsets of Γ and all subsets of Γ whose complement is finite. So far it is simple, but if we go higher in the hierarchy of sets, it is not so clear. For example, $M[\Gamma]$ has to contain the set of all finite subsets of Γ, but what about the set $\{\emptyset, \{G_0\}, \{G_0, G_1\}, \{G_0, G_1, G_2\}, \dots\}$? We must not add this set because from it we would be able to determine the indices of the generic sets and having the indices we would be able to take the set of sets G_n with n even. That would spoil the construction.

As a matter of fact, the condition that determines which sets are allowed and which are not is quite simple. The condition is:

A set A can be added if and only if there is some number n such that all sets G_m with $m > n$ are indistinguishable from the point of view of A.

The indistinguishability means that for every $m_1, m_2 > n$, we can switch G_{m_1} with G_{m_2} without changing A.

Obviously, one has to check a lot of details to prove that it works, but the main idea is quite simple. Every forcing condition is finite, thus it mentions only a finite number of the names representing the generic sets $G_0, G_1, G_2, \dots$. If, for some n, a forcing condition p does not contain names for G_m with $m > n$, then those sets are indistinguishable for p. Hence no forcing condition can force the existence of a set that does not satisfy the condition above.

Notes

1. *Models of ZFC*. The best way to explain forcing is to use countable transitive models of *ZFC*. A *transitive set* is a set x such that for all y and z, if $y \in x$ and $z \in y$ then $z \in x$. In words, with every element y that x contains, x contains also all elements of y. Notice that this implies that it also contains all elements of z, etc. A *transitive model* of *ZFC* is a model M whose universe is a transitive set and in which the interpretation of the membership relation is the usual membership relation restricted to the universe (formally, $M \models y \in x$ if and only if $y \in x$).

   The advantage of transitive models is that many concepts are *absolute* in them. We say that a relation defined by a formula $\phi(x_1, \dots, x_k)$ is absolute in M if for every $a_1, \dots, a_k \in M$, $M \models \phi[a_1, \dots, a_k]$ if and only if $\phi(a_1, \dots, a_k)$ is true. The most important relation that is absolute (by the definition of transitive models) is the membership relation $\in$. Furthermore, the properties of being a natural number and being an ordinal are also absolute. (For this reason such models are sometimes called *standard*, but this is rather misleading because if there exists a transitive model, then there are many different transitive models.) What is not absolute is the concept of being a cardinal number. For example, if M is countable, then the only absolute cardinal numbers are the finite ones and $\aleph_0$.

According to the completeness theorem the existence of a model of *ZFC* is equivalent to the consistency of *ZFC*. The statement that there exists a transitive model is stronger and it cannot be derived from mere consistency of *ZFC*. But it is not a strong axiom; in fact, it is weaker than the axioms postulating the existence of a strongly inaccessible cardinal. If κ is a strongly inaccessible cardinal, then V_κ is a transitive model. However, if there exists a transitive model, then there exists also a countable transitive model. This is proved by applying the Löwenheim-Skolem Theorem first to get a countable submodel and then recursively deleting all subsets that spoil transitivity staring from those of the lowest rank and going upwards. (The latter procedure is called the *Mostowski collapse.*)

2. *Interpretations and inner models.* When we speak about consistency, we always mean relative consistency with respect to some assumption. The minimal assumption that one can use when studying *ZFC* is the consistency of *ZFC* because adding more axioms can only make things worse. In fact this is also all that is needed in this independence result. Thus the precise statement of Cohen's independence result is that if *ZFC* is consistent, then the Continuum Hypothesis is independent. This is in contrast with results such as the consistency of the Axiom of Determinacy with ZF (*ZFC* without the Axiom of Choice) which require large cardinals.

   The approach based on countable transitive models helps readers to get an intuition, but it requires an assumption stronger than the consistency of *ZFC*. If we want only to use the consistency of *ZFC*, it is best to use *interpretations* of theories (in the sense defined on page 117). Interpretations enable us to prove independence results using only syntactical concepts, which are finite structures. Thus implications such as

$$Con(ZFC) \to Con\big(ZFC + 2^{\aleph_0} = \aleph_1\big) \quad \text{and}$$
$$Con(ZFC) \to Con\big(ZFC + 2^{\aleph_0} > \aleph_1\big)$$

   are provable already in Finite Set Theory. In set theory interpretations of *ZFC* are called *inner models* of *ZFC*.

   Gödel's proof of the consistency of the Continuum Hypothesis is a construction of an inner model in which the relation of membership is translated identically and the domain of the interpretation consists of constructible sets. Constructible sets are defined by a formula Λ in the language of set theory. Thus this interpretation is the syntactical object $(\in, \Lambda)$. To prove that it is an interpretation, one has to show that all axioms of *ZFC* are provable if we restrict the range of quantification to Λ. As we want to get the consistency of the Continuum Hypothesis, thus we have to prove also the Continuum Hypothesis relativized to Λ. To show that forcing constructions can be presented as inner models, I have to change the definition of forcing. What I sketched on the preceding pages is the original concept of forcing which has the strange feature that it does not preserve logical validity. For example, $\phi \vee \neg\phi$ is a tautology for every sentence ϕ, but there are sentences for which such a sentence is not forced. This can easily be rectified by redefining that a forcing condition p forces a sentence ψ in the new sense if p forces $\neg\neg\psi$ in the old sense. Then we can define an interpretation by taking the names of potential sets in the extension as the domain, and for two names a and b, translating the membership relation $a \in b$ by the formula '*the empty condition forces* $a \in b$'.

We need also to define a translation of the equality relation, which I omit.

3. *More about constructible sets.* Let us first recall the definition of L (on page 214 we defined a more general concept $L(x)$). Sets L_α are defined by transfinite recursion for all ordinals as follows. $L_0 = \emptyset$, $L_{\alpha+1} = def(L_\alpha)$, and $L_\lambda = \bigcup_{\alpha<\lambda} L_\alpha$, λ a limit ordinal. Then we put $L = \bigcup_\alpha L_\alpha$. We denote by $def(x)$ the set of sets definable in the structure $(x, \in)$ by first order formulas with parameters from x. The sequence $\{L_\alpha\}$ indexed by all ordinals is called the constructive hierarchy.

To prove that L satisfies the axioms of *ZFC* is not a trivial task. Superficially it may seem that the Axiom Schema of Comprehension is easy because it says that a collection of elements of a set defined by a formula forms a set. But the definability in the schema is the definability in the whole universe of sets, which is in our case L, so it is not restricted to the elements of a given set. Thus one has to prove that it is possible to replace L by a part of it which is a set. More precisely, if x is a subset of y defined in L by formula $\phi(z)$ and $y \in L_\alpha$, then it is possible to find β (which may be larger than α) such that $x \in def(L_\beta)$. To this end, we need to replace the *class* model $(L, x, \in)$ by a *set* model $(L_\beta, x, \in)$. We cannot work with class models as with set models because it is not possible to define the truth for all formulas in a class model. But since we only need formulas of limited complexity, namely, up to the complexity of ϕ, it suffices to apply the idea of the Löwenheim-Skolem Theorem and obtain a set from a class.

The proof of the Power-Set Axiom is also not trivial. The problem is that, for a given set x, its subsets can be defined on arbitrary large stages L_α. However, from some ordinal α on, all these definitions define sets that have appeared before. To see that, for every subset $y \subseteq x$, assign to y the least ordinal α such that $y \in L_\alpha$. By the Replacement Axiom Schema the range of this assignment is a set. Then there exists an upper bound β on these ordinals because a set of ordinals has always an upper bound (their union is such an ordinal).

To prove that in L the Continuum Hypothesis is true is the most difficult and also the most interesting part of Gödel's proof. One can easily show that the set of natural numbers ω occurs at a stage indexed with a countable ordinal. We want to prove that every subset x of ω which appears in some L_α, appears already in some L_β for some *countable* β. Let us first show that this entails the Continuum Hypothesis. Since there are $\aleph_1$ countable ordinals and for every countable ordinal, we define only a countable number of subsets of ω, there are at most $\aleph_1$ such sets. But there is a snag: $\aleph_1^L$, the first uncountable ordinal from the point of view of L, can be smaller than the genuine $\aleph_1$. Fortunately, this problem can be avoided by doing the proof not in the whole universe, but only in L. This is not a real complication because we assume that we have already proved the axioms of *ZFC* in L, thus L is as good as any other universe of sets.

Thus the problem is reduced to showing that if all sets are constructible then the Continuum Hypothesis is true. This assumption is called the Axiom of Constructibility and is abbreviated by $V = L$. Though we use classes to state this axiom concisely, it is an axiom that can be stated in *ZFC*.

To prove that every subset of ω occurs in some L_β with $\beta < \aleph_1^L$ one uses essentially the same idea as to prove that there exists a countable transitive model of *ZFC* if there exists an inaccessible cardinal. Instead of transitive models we will use transitive sets that are models of a weaker theory. We need a finite subtheory of *ZFC*, say axiomatized by an axiom Θ, with the following property. For every transitive set z, if the structure $(z; \in)$ is a model of Θ, α is an ordinal in z and $x \in z$, then $x \in L_\alpha$ if and only if this sentence is true in $(z; \in)$. In other words, in transitive sets satisfying Θ the concept of a constructible set is absolute. The next step is to show that if $x \in L_\alpha$, then there exists a transitive set which contains α, and x, and satisfies Θ. One can take L_γ for a suitable $\gamma \geq \alpha$. The last step is to take a countable submodel of (L_γ, α, x) and collapse it to a countable transitive set. This collapse moves α to a countable ordinal β and it preserves the property that x is constructed at that stage. Since the relation of being constructed at stage β is absolute, we obtain $x \in L_\beta$.

I only used it for the Continuum Hypothesis, but the argument is quite general and one can prove the Generalized Continuum Hypothesis, which is $2^{\aleph_\alpha} = \aleph_{\alpha+1}$ for every α.

4. *More about generic sets and forcing.* A generic extension of a transitive countable model M is determined by a set of forcing conditions P. A set of *forcing conditions* P is a partially ordered set with the largest element. In the examples above the largest element was the empty string and the ordering was given by the reverse inclusion. We will denote the ordering on P by $\leq$ and the top element by 1. We say that a subset D of P is *dense* if for every $p \in P$, there exists a $q \in D$ such that $q \leq p$. A subset $G \subseteq P$ is called *generic* if the following two conditions are satisfied:

   a. G is a filter, which means that for every $p, q \in G$, there exists $r \in G$ such that $r \leq p$ and $r \leq q$, and for every $s \in G$ and $s \leq t$, we have $t \in G$;
   b. for every set D which is dense and which is in M, $G \cap D \neq \emptyset$.

   It is crucial for the forcing construction that the partially ordered set is in M and we consider only dense sets in M, whereas G does not have to be in M. In fact, in all nontrivial cases there is no generic set in M (which is what we need in order to enlarge M). The fact that there exists a generic set (outside of M) is easy. Since M is countable the number of dense sets in M is also countable. Hence we can enumerate them $D_1, D_2, \ldots$, and we can inductively choose a sequence

   $$p_1 \in D_1, \quad p_2 \leq p_1, \quad p_2 \in D_2, \quad p_3 \leq p_2, \quad p_3 \in D_3, \quad \ldots$$

   Then we take $G = \{p \in P;\ \exists n p_n \leq p\}$, the filter generated by the sequence.

   In order to define forcing, we need constants representing every possible set in the extension that we are going to construct. These constants are called *names*.

This is an important part of the definition which I neglected in the exposition in the informal description. Indeed, without knowing the structure of names it seems to be a miracle that forcing conditions can force all axioms of *ZFC*. The names, in fact, almost completely describe the extension of M by a generic set G. They form sort of semi-finished model of *ZFC*, where one has only to decide about details (such as which particular n is in G, in the case of the first example). This is quite apparent in the Boolean valued models, which I am going to describe shortly. An important thing that we must keep in mind is that forcing, hence also names, should be definable in M.

Cohen's original construction was based on an extension of the concept of constructible sets. Soon it turned out that was unnecessarily complicated. I will motivate the contemporary treatment of forcing names by an example. Let C be a set of forcing names in M and suppose we have a mapping $f : P \to C$ in M which assigns a name from C to every forcing condition $p \in P$. Since in an extension $M[G]$ we have the mapping f and the set G, we also have $f(G)$, the f image of G. Thus given f and a set of names C, we need a name for $f(G)$. More generally, if we have a relation $R \subseteq P \times C$, we need a name for $R(G) = \{c \in G;\ \exists p\ (p, c) \in R\}$. This is clearly a necessary condition which the set of names must satisfy, but in fact it is also a sufficient condition.

A natural way to construct such a class of forcing names is to define by transfinite recursion:

a. C_0 is an arbitrary class containing a name for every set in M,
b. $C_{\alpha+1}$ contains names for every relation $R \subseteq P \times C_\alpha$, and
c. $C_\lambda = \bigcup_{\alpha<\lambda} C_\alpha$ for limit ordinals λ.

A closer look at the construction reveals that the first step is actually not needed, namely, we can put $C_0 = \emptyset$ (which has the advantage that C_α are sets). Let $C = \bigcup_\alpha C_\alpha$, where the union is over all ordinal numbers α.

Now we need to define forcing of atomic statements. Suppose c_1 is a name assigned to a relation $R \subseteq P \times C$, let (p, c_2) be in the relation R. Then we, certainly, want to define forcing so that all q, $q \leq p$ force the sentence $c_2 \in c_1$ because if p is in G, then $c_2 \in c_1$ is true in $M[G]$. Furthermore, we have to define forcing of atomic sentences of the form $c_1 = c_2$ because different names may denote the same set in the extension. Thus we define

6. *p forces $c_2 \in c_1$ if c_1 is a name associated with a relation R and $(p, c_2) \in R$;*
7. *p forces $c_1 = c_2$ if it forces the sentence $\forall x(x \in c_1 \equiv x \in c_2)$.*

This will ensure the Axiom of Extensionality.

The way we defined forcing is close to the original definition of Cohen. We have already noticed that this definition is not closed under logical consequences and that this complication can be avoided by applying the above definition to sentences to which we add double negations. However, we also have to take a suitable modification of the definition of forcing of atomic sentences. Thus in 7. we require p to force $\neg\neg\forall x(x \in c_1 \equiv x \in c_2)$ and instead of 6. we take the following clause

6'. *p forces* $c_2 \in c_1$ *if* c_1 *is a name associated with a relation* R, *and for some* c_3 *such that* $(p, c_3) \in R$, *p forces the sentence* $\neg\neg(c_3 = c_2)$.

This looks like a circular definition, as we define forcing of atomic sentences with the membership relation using forcing of atomic sentences with the equality, and the other way round. But notice that we always refer to names from some previous stage, hence formally, the definition is by transfinite recursion.

5. *More about the independence of the Axiom of Choice.* The axioms of *ZFC*, which include the Axiom of Choice, hold true in every generic extension $M[G]$ of a countable transitive model M. Thus to construct a model in which the Axiom of Choice fails, it is necessary to change the construction. The change is in the names used in the construction; the concept of forcing is the same.

   Recall that we want to add infinitely many generic sets $G_1, G_2, \ldots$, but we do not want to add them as a sequence. Thus our aim is to extend M only by sets that can distinguish only a finite number of the sets $G_1, G_2, \ldots$ (to distinguish in the sense of the indistinguishability condition on page 353). Therefore, we will use only names that satisfy this condition, hoping that it will guarantee that in the extension all names will be represented by sets satisfying the condition. That is true, but it must be proved.

   Let us demonstrate it on a name c of the following form:

$$\{(p_1, X_1), (p_2, X_2), \ldots\},$$

   where $p_1, p_2, \ldots$ are forcing conditions and $X_1, X_2, \ldots$ are names for $G_1, G_2, \ldots$. We have to prove that the set corresponding to this name, which we will denote by A, is either finite, or it contains all sets $G_1, G_2, \ldots$ except for finitely many. The assumption is that for some n, the name c does not distinguish names X_i with $i \geq n$. Suppose that A is infinite. Then for some forcing condition p which determines the generic extension, (p, X_i) must be in c for some $i \geq n$. According to the condition, (p, X_j) is in c for all $j \geq n$. Hence p forces $X_j \in c$ for all $j \geq n$, which proves that A contains all G_j except for at most $n - 1$ of them.

   Once we know that all sets in the generic extension satisfy the condition, we are done.

6. *Collapsing cardinals.* The concept of a cardinal number is not absolute in transitive models and in fact for each uncountable cardinal κ in M, it is possible that κ is not a cardinal in a generic extension $M[G]$. This can happen because in $M[G]$ a one-to-one mapping from a smaller cardinal λ onto κ may be added. We say that κ collapses to λ.

   To construct an extension in which κ collapses to ω, define forcing conditions as follows: p is a forcing condition if for some finite subset A of ω, p is a one-to-one mapping from A to λ. Then it is easy to check that every generic set G is a mapping from ω onto κ.

   I still owe you the proof that in Cohen's construction of a model of *ZFC* with the negation of the Continuum Hypothesis cardinals do not collapse, so let us do it now. The proof is based on a simple criterion for extensions without collapsing cardinals. We say that two forcing conditions p and q are *incompatible* if there is no forcing condition r such that $r \leq p$ and $r \leq q$.

Proposition 9 *If in M there exists no uncountable set of mutually incompatible forcing conditions, then no cardinal collapses in $M[G]$.*

The assumption of the theorem is known as the *countable chain condition*.

The proof of this proposition is simple and instructive, hence it is worthwhile doing it in detail. Arguing by contradiction, suppose the proposition is false. By what we know about forcing, it means that there exists a forcing condition p and a forcing name c such that p forces '*c is a mapping from λ onto κ*' where κ and λ are infinite cardinals and λ is smaller than κ. Take an element $\alpha \in \lambda$ and consider all β such that some q, $q \leq p$, forces $c(\alpha) = \beta$. For every such β, choose one such q and denote it by q_β.

We claim that for every pair $\beta \neq \beta'$, the forcing conditions q_β and $q_{\beta'}$ are incompatible. Indeed, if $r \leq q_\beta, q_{\beta'}$, then r forces $c(\alpha) = \beta$ and $c(\alpha) = \beta'$ which is inconsistent with the fact that c is a mapping.

Since there are at most a countable number of mutually incompatible conditions, there are at most a countable number of such elements β. Remember that forcing is definable in M, hence the set of these β is a set in M. Thus we have proved that for every α, there is a countable set of B_α such that in every generic extension $c(\alpha) \in B_\alpha$. The assignment $\alpha \mapsto B_\alpha$ is definable, hence also $\bigcup_{\alpha \in \lambda} B_\alpha$ is a set in M. It follows that in $M[G]$, the range of the mapping which is the interpretation of the name c is contained in $\bigcup_{\alpha \in \lambda} B_\alpha$. This set has cardinality at most $\aleph_0 \cdot \lambda = \lambda < \kappa$, thus the range of the mapping cannot be the whole κ. This contradiction finishes the proof of the proposition.

It remains to show that the particular set of forcing conditions used by Cohen has the property stated in the proposition, which I leave to the reader with the following hint: consider the smallest natural number k such that there exists an uncountable set of mutually incompatible forcing conditions of length k.

The formidable universality of the forcing method can be illustrated by the fact that one can use it also to prove the consistency of the Continuum Hypothesis. Let M be a countable transitive model in which the Continuum Hypothesis fails. Extend M by a generic set G which is a one-to-one mapping from ω_1^M onto $\mathcal{P}^M(\omega)$, the set of all subsets of ω in the sense of the model M. (This is the same as adding such a mapping from ω_1^M onto $(2^{\omega_0})^M$.) This does not automatically ensure the equality $\aleph_1 = 2^{\aleph_0}$, since these cardinals may be different in $M[G]$. To keep $\aleph_1$ intact, one uses a generalization of the above proposition. To ensure that $2^{\aleph_0}$ remains the same, it suffices not to add any new subset of ω. Forcing conditions that produce such an extension are all one-to-one mappings from countable subsets of ω_1 into 2^{ω_0}.

7. *Boolean valued models*. This is an alternative approach to constructions of generic extensions introduced by Scott and Solovay [262], and by Vopěnka [300]. It has the appeal of algebraic constructions and for a model theorist is more natural than Cohen's construction. However, these two constructions are completely equivalent, thus one cannot obtain more results by using Boolean valued models. Furthermore, when we need to prove nontrivial properties of a generic extension, we have to resort to forcing anyway.

Boolean algebras were defined on page 21. A Boolean algebra is *complete* if every set S of elements has the least upper bound, which entails that every set has the largest lower bound. These two elements are denoted by $\bigvee S$ and $\bigwedge S$ respectively. An *atom* is an element different from 0 below which there is only 0.

Example 1 Let X be an arbitrary set. Let the elements of the Boolean algebra B_1 be all subsets of X, with operations $\cap$, $\cup$, the complement in X and the constants $\emptyset$ and X. This is an important example of a complete Boolean algebra, but for Boolean valued models it is not good because it has atoms (the one-element sets).

Example 2 This example requires some basic knowledge of topology. We start with the Cantor discontinuum C. This topological space can be represented by the set of all countably infinite strings of zeros and ones. The basis of open sets consists of sets which contain all infinite extensions of a finite string. Elements of B_2 are all subsets of points $S \subseteq C$ that satisfy the following equality

$$S = Int(Cl(S)), \tag{4.16}$$

where *Int* denotes the interior and *Cl* denotes the closure. The operations are defined by

$$S \wedge T = S \cap T, \quad S \vee T = Int\big(Cl(S \cup T)\big), \quad S' = Int(C \setminus S).$$

It is not difficult to prove that B_2 does not contain any atoms.

Remarks 1. The strange formula (4.16) has a simple explanation. Suppose we use all open sets instead of set satisfying (4.16). Then such an algebra satisfies weaker axioms, namely those that come from intuitionistic logic. When we need to embed classical logic into intuitionistic logic, we use double negation and this is exactly what happens here—we pick the elements that satisfy $S = (S')'$.

2. In this way we can define a Boolean algebra for every topological space. Such Boolean algebras should not be confused with the Stone representation theorem. That theorem says that every Boolean algebra is isomorphic to an algebra of sets whose operations are the set operations of intersection, union and complement.

The idea behind Boolean valued models is to replace the two truth values *true* and *false* by values in an infinite complete Boolean algebra B. The constants 1 and 0 of B represent *true* and *false*, the other elements represent "intermediate truth values". One can study Boolean valued models for every first order theory. For simplicity, suppose that the only nonlogical symbol is R, a symbol for a binary relation. Then a structure with Boolean values in B consists of a universe A and a mapping $v : A \times A \to B$ which assigns a value in B to every pair of elements of A. For $a, b \in A$, $v(a, b)$ is the truth value of the sentence $R(a, b)$. Then we define the truth values of all first order sentences a natural way. For example, we define the value of $\exists x\ \phi(x)$ by $v(\exists x\ \phi(x)) = \bigvee_{a \in A} v(\phi(a))$ (assuming that we have already defined the values of $\phi(a)$). This valuation preserves logical validity in the following sense:

a. if ϕ is logically true, then in every Boolean valued model $v(\phi) = 1$;
b. if ϕ is a logical consequence of a set of sentences Ψ, then $\bigwedge_{\psi \in \Psi} v(\psi) \leq v(\phi)$.

Hence Boolean valued models behave very much like ordinary ones.

If we focus on models of *ZFC*, then we can use some specific properties. Suppose we have a model M and a complete Boolean algebra B in M, which is not complete in the metatheory. This can happen because in M there are only some subsets of B and we require that meets and joins exist only for sets in M. Then we can still construct Boolean valued models with values in B whose valuations preserve logical validity. Namely, if we define a B-valued model fully in M, then the only infinite meets and joins that we will need are those that are present in M.

The second specific property of *ZFC* is that given a model M of *ZFC* and a complete Boolean algebra B in M, there is an almost canonical way to define a Boolean valued model that extends M. I will explain it in a simple motivating example. Let x be a set and $f : x \to B$ a mapping from x to B. Then we can treat f as a "Boolean valued subset of x". For $y \in x$, the function gives the truth value $f(y)$ to the statement that a y is in this subset. This is very much like a *fuzzy subset* of x, which is a mapping from x to the unit interval $[0, 1]$. Now having Boolean valued sets, we can define Boolean valued sets of Boolean valued sets and so on. Formally, we define by transfinite recursion $M_0^B = \emptyset$, $M_{\alpha+1}^B$ is the set of all functions with domain in M_α^B and range in B, and $M_\lambda^B = \bigcup_{\alpha<\lambda} M_\alpha^B$, for λ a limit ordinal. Let $M^B = \bigcup_\alpha M_\alpha^B$. You may have noticed the similarity of this construction and the construction of forcing names. This is not accidental: the elements of M^B indeed correspond to forcing names.

We need to equip M^B with a valuation v in B. Every element of M^B is a function f from the set $dom(f)$, the domain of f, into the Boolean algebra B. Contrary to what one may expect the value of the sentence $x \in f$ is, in general, not $f(x)$. The reason is that M^B with such a valuation would not satisfy the Axiom of Extensionality. In order to remedy it, we have to treat the equality relation also as a Boolean valued relation. But once we change equality, we have to change also the membership relation because the axioms of equality for the membership relation would not hold true. The following is the right way to define a Boolean valuation v on M^B (using recursion on the rank of elements):

$$v(x \in f) = \bigvee_{y \in dom(f)} f(y) \wedge v(y = x);$$
$$v(f = g) = \left(\bigwedge_{x \in dom(f)} f(x)' \vee v(x \in g)\right) \wedge \left(\bigwedge_{x \in dom(g)} g(x)' \vee v(x \in f)\right).$$

Recall that $a' \vee b$ represents the implication $a \Rightarrow b$.

Now we need to identify the element of M^B that plays the role of the generic set. It is the identity function on B, which we will denote ι_B. We can easily compute what the model M^B thinks about ι_B because $v(x \in \iota_B)$ is 0 if $x \notin B$ and it is x if $x \in B$. (By saying 'M^B *thinks* ϕ' we mean that the Boolean value of ϕ is 1.) Thus M^B thinks that ι_B is a subset of B. Furthermore, it thinks that

(1) $0 \notin \iota_B$,

because $v(0 \in \iota_B) = 0$. Let x and y be in B. Then we have

$$v\big((x \in \iota_B \wedge y \in \iota_B) \Rightarrow (x \wedge y \in \iota_B)\big) = \big(\iota_B(x) \wedge \iota_B(y)\big)' \vee \iota_B(x \wedge y)$$
$$= (x \wedge y)' \vee (x \wedge y) = 1.$$

In words, M^B thinks that

(2) *if x and y are in ι_B, then also $x \wedge y$ is in ι_B.*

You can furthermore check that M^B thinks that

(3) *if x is in ι_B and $x \leq y$, then y is in ι_B*, and
(4) *if S is set in M and $\bigvee_{x \in S} = 1$, then for some $x \in S$, x is in ι_B.*

The conditions (1)–(4) define *complete ultrafilters* except that (4) is stated only for sets in M. Thus from the point of view of M^B, if the element ι_B were in M, then it would be a complete ultrafilter on B. However, a complete ultrafilter is always trivial—it is the set of all elements above an atom of the Boolean algebra. Hence if B does not contain any atoms, then M^B thinks that ι_B is a set *outside* of M.

If M is countable, then there exists a subset G of B (a genuine set, not a Boolean valued one) that satisfies conditions (1)–(4). Recall that for an ultrafilter, it suffices to have condition (4) only for finite sets S. Thus G satisfies a stronger condition, a condition which depends on M, therefore it is called an M *generic ultrafilter*. G determines a mapping, in fact a homomorphism of Boolean algebras, from B to the two-element Boolean algebra: map an element $x \in B$ to 1 if $x \in G$, and to 0 if $x \notin G$. Using this homomorphism we can transform a B-valued model into a two-valued model. A two-valued model is almost an ordinary model, we only have to identify elements a and b whenever the value of $a = b$ is 1. Thus we obtain the generic extension $M[G]$.

It only remains to find out what is the relation of forcing conditions to complete Boolean algebras. Call an ordering $(P; \leq)$ *separative* if it satisfies the following condition: if $p \nleq q$ then there exists an r such that $r \leq p$ and q and r are incompatible. We say that a subset P of a Boolean algebra B is dense if it does not contain 0 and for every nonzero element $x \in B$, there exists $p \in P$ such that $p \leq x$. One can prove that for every separative ordering $(P; \leq)$ there exists a complete Boolean algebra such that P is a dense subset of B, and the ordering of elements of P in B is the same as in $(P; \leq)$. This Boolean algebra is unique up to the isomorphism. On the other hand, we can find a dense separative set in every Boolean algebra; it is simply the set of all nonzero elements.

Thus forcing conditions can be thought of as a concise way of representing a complete Boolean algebra.

Example 2, continued The Boolean algebra B_2 is determined by the dense set of all finite strings of zeros and ones. A finite string s corresponds to the open set of all infinite continuations of s. Thus B_2 is the Boolean algebra associated with the forcing conditions of the very first example of forcing (page 349).

8. *Random generic sets.* When explaining generic sets, I stressed the fact that they do not have properties typical for random sets. This is not quite precise and it concerns only the original constructions of Cohen. There are generic extensions by sets that look random. To this end we do not have to develop a new kind of forcing; we only have to take a suitable set of forcing conditions, or, which is equivalent, to take a suitable complete Boolean algebra.

   Let us demonstrate it by the problem of adding a non-constructible subset of natural numbers. To this end Cohen used forcing conditions that lead to the Boolean algebra B_2 defined above. In order to obtain a randomly looking generic subset of natural numbers, we take a different Boolean algebra, which we will denote by B_3. To define B_3 we start with the same set as we did with B_2, the set of countably infinite sequences of zeros and ones $\{0, 1\}^{\omega}$, but instead of using topology, we will use measure. The measure that we need is the natural measure such that the whole space has measure 1, the set of sequences starting with 0 (respectively 1) has measure $1/2$, etc., (that is, the measure of the set of infinite sequences that extend a fixed sequence of length n has measure $1/2^n$). The details of how this is precisely defined are not important, but let me stress the fact that measure is defined only for some subsets of $\{0, 1\}^{\omega}$, which are called *measurable sets*. To define the elements of B_3 we identify measurable sets whose difference is a set of measure 0. For example, all countable sets have measure 0 (but not only those), hence, in particular, two sets that differ in a countable number of points will be identified. Formally, the elements of B_3 are classes of measurable sets that differ by sets of measure 0. The operations are defined by taking representatives from the classes, applying the corresponding Boolean operation, and taking the class containing the result.

   Thus Boolean algebra B_3 produces a generic extension $M[F]$ in which F is not constructible. The disadvantage of this construction is that whereas B_2 has a succinct description by a countable set of forcing conditions (finite strings of zeros and ones), B_3 does not have such a representation: B_3 is not generated by a countable set of forcing conditions.

   Let us now compare $M[F]$ with a generic extension $M[G]$ produced by Cohen's forcing conditions. We know that F is different from G (for example, the average number of zeros in initial segments of F converges to $1/2$), but this is not enough to prove that the two models are different. We know that such extensions contain a lot of other subsets of natural numbers that are not present in M. Thus G could be among the sets generated from F. To prove that it is not so, we have to find some property that distinguishes these two generic extensions. An interesting property that does it is the following. In $M[F]$ every function on natural numbers is bounded from above by a function from M, while in $M[G]$ there are functions that grow faster than any function in M. Though models constructible by such '*random forcing*' are different from those produced by '*Cohen forcing*', many results, including the unprovability of the Continuum Hypothesis, can be reproved using random forcing.
9. *Martin's Axiom.* Researchers in set theory prefer to assume the negation of the Continuum Hypothesis, since the universe of sets satisfying this axiom is richer.

If one needs to use the Continuum Hypothesis to prove a theorem, one can say that this theorem is true for constructible sets. However, there are many consequences of the Continuum Hypothesis which are consistent with the negation of the Continuum Hypothesis. For such theorems, there is a better alternative. In 1970 Martin and Solovay proposed an axiom which is consistent with the negation of the Continuum Hypothesis and which proves a number of consequences of the (non-negated) Continuum Hypothesis [192].

This axiom is known as Martin's Axiom. Unfortunately, it is a rather technical statement that does not look like a natural universal principle. It has several equivalent versions; I will state the most elementary one which is based on concepts from forcing. This is not a coincidence, Martin's Axiom was discovered when studying independence proofs. We already have definitions of all concepts needed to state it.

Martin's Axiom *Let* $(P, \leq)$ *be a partially ordered set with the countable chain condition. Let* $\mathcal{D}$ *be a set of dense subsets of* $(P, \leq)$ *and suppose that the cardinality of* $\mathcal{D}$ *is strictly less than* 2^{ω}*. Then there exists a filter* F *that meets all elements of* $\mathcal{D}$*.*

Main Points of the Chapter

- Many important results in mathematics have the form of a proof of impossibility.
- Very often proofs of impossibility are difficult and require the use of abstract concepts, even though the problems themselves may have elementary formulations.
- The proofs of Gödel's incompleteness theorems are based on a formula resembling the liar paradox. It uses the ability of the language of logic to express self-referential sentences. The unprovable sentence is equivalent to the consistency of the theory.
- A similar argument is used to prove that some problems are algorithmically unsolvable. One can also deduce unprovability of some sentences from proofs of undecidability of related problems.
- Algorithmically unsolvable problems can be found in various branches of mathematics including number theory and geometry.
- One can show unprovability of some concrete mathematical problems in Finite Set Theory. The proof method, however, requires that the problems encode very fast growing functions.
- There is a very efficient method, the method of forcing invented by Paul Cohen, using which one can prove many independence results in set theory, including the independence of the Continuum Hypothesis. The method is applicable only to sentences about infinite sets.

Chapter 5
The Complexity of Computations

Hiding in the alternating patterns of digits, deep inside the transcendental number, was a perfect circle, its form traced out by unities in a field of naughts.

Carl Sagan, *Contact*

COMPLEXITY is a notion about which we do not learn in schools, but which is very familiar to us. Our generation has witnessed a tremendous increase of complexity in various parts of our life. It is not only the complexity of industrial products that we use. The world economy is a much more complex system now than it used to be; the same is true about transportation, laws and so on. Computers help us to cope with it, but they also enhance the process of making our lives more complex. The progress in science reveals more and more about the complexity of nature. This concerns not only biology and physics, but also mathematics. In spite of the great role that it plays in our lives, complexity has become an object of mathematical research only recently. More precisely, the word complexity had not been used until about the 1960s, but many parameters introduced long before can be thought of as some sort of complexity measures. Already the words used for these parameters suggest that they are used to classify concepts according to their complexity: *degree, rank, dimension*, etc. The most important instantiation of the notion of complexity is in computability theory, which is the subject of this chapter.

Originally the motivation for studying computational complexity was to understand which algorithms can be used in practice. It had been known that some problems, although algorithmically solvable, require so large a number of steps that they never can be used. It was, therefore, necessary to develop a theory for classifying problems according to their feasibility. When theoretical studies began, it turned out that there are fundamental problems concerning computational complexity. Moreover, some of these problems appeared to be very difficult. We now appreciate their difficulty because only a few of them have been solved after many years.

These problems concern the relationship of the basic resources used by algorithms: time, space, nondeterminism and randomness. Our inability to make any substantial progress in solving them suggests that there may be fundamental obstacles that prevent us from solving them. It is conceivable that these problems not only need new methods, but may need new axioms. This seems to be a rather bold

P. Pudlák, *Logical Foundations of Mathematics and Computational Complexity*, Springer Monographs in Mathematics, DOI 10.1007/978-3-319-00119-7_5,

conjecture, but recall the history of Diophantine equations. The problem appeared to be just a difficult number-theoretical problem and Hilbert even assumed that it was algorithmically solvable. Now we know that this is not the case: there is no theory that would suffice to prove the unsolvability of every unsolvable equation. History may repeat itself in computational complexity and we may need mathematical logic to solve the fundamental problems of computational complexity theory.

In the next chapter, we will see slightly more explicit connections of computational complexity with logic and the foundations of mathematics, mediated by proof complexity.

5.1 What Is Complexity?

From our daily experience we know that there are easy tasks and there are difficult ones. Everybody knows that it is more difficult to multiply two numbers than to add them. Those who use computers more extensively also know that they are able to solve certain problems fast, while some other problems require a long time. But we also know that some people are faster than others, that we can solve a task more easily if we know more about it and that some programs are slow for a given problem, but sometimes a sophisticated program can solve the same problem very efficiently. Thus it is not clear whether there is a particular property of problems that prevents us (and computers) from solving some problems quickly, or if it is just the question of knowing how to solve a particular problem fast.

Therefore the first thing to learn is that, indeed, there is a quantity associated with every problem, which we call the complexity of the problem, that determines how efficiently the problem can be solved. This quantity is represented by a *natural number*. When studying computational complexity, we always consider only algorithmically solvable problems, problems solvable using a finite amount of computational resources. Since algorithms make discrete steps, also the resources can be measured in discrete units. The amount of computational resources needed to solve a particular instance of a problem is this number. In fact there is not only one, but several such quantities corresponding to the type of resources that we study. Furthermore, each one depends on the particular model of computation that we use.

Let us start with the most important type of complexity, which is called *time*. If we use the classical model of computation, Turing machines, then the time complexity of a problem is the minimal number of steps that a Turing machine needs to solve the problem. However, the time complexity of computations cannot be defined for a single input. Recall that when we considered the concept of decidability, it was important to have an infinite *set of instances of a given problem*. Typically, we asked if a property of natural numbers was decidable. For a finite set, there always exists an algorithm—a look up table. So the same is true about complexity; it only makes sense, if we have an infinite, or at least very large set of inputs.

Suppose, for example, that the problem is to decide if a given number N has an even number of prime divisors. The problem is, clearly, decidable: we can enumerate all primes less than N and try to divide N by each of them. This is certainly not the best way to solve this problem, but it seems that the problem is difficult if we do not know anything special about N. But suppose that we somehow determine that the number of prime divisors of N is, say, even. Then we cannot say anymore that, for this particular N, it is difficult to decide this property, because we know the answer. Therefore talking about the complexity of such a problem only makes sense if we consider a large set of numbers. To be more specific, consider a Turing machine M that correctly decides the above property for every number. Since M has to work with arbitrarily large numbers, the number of steps that M needs to answer will vary with the input; in general, it will increase. If M is a formalization of an efficient algorithm, then it will not increase very fast. If, however, the problem is hard, it will increase fast for every Turing machine.

The idea of defining the time complexity of a problem itself, not just with respect to a particular machine, is to take the least number of steps that a machine needs for solving the problem. Ideally, we would like to prove that there exists a machine which is the fastest one and define its time complexity to be the time complexity of the problem. This is surely not possible, since we know, for example, that for every fixed number there exists a procedure that solves the problem almost immediately. (If we use Turing machines it will take some nontrivial time for the machine to compare the input number written in binary with the one it has in its lookup table, but it will be fairly short.) Therefore we content ourselves with *asymptotic estimates* on the possible time complexities of Turing machines solving the problem.

For every nontrivial problem, we naturally expect that the time will increase with the size of the input data. The time that an algorithm needs varies not only with the size, but also if the input size is fixed, it may need a different number of steps for different data. In order to be able to decide if we can use an algorithm for data of a given length, we need an upper bound on the time needed for all such data. Thus we define the *time complexity* of an algorithm (or of a Turing machine, etc.) to be the function $t(n)$ such that $t(n)$ is the maximal time the algorithm needs on inputs of size n. The size of an input is usually the length of a string which encodes the data using a finite alphabet; we call it the *input length*. This approach is called *worst case complexity*, since we classify algorithms by how they behave in the worst case on data of a given input size. (In practice one may prefer to use *average case complexity*, but I will not deal with this concept here.)

Hence the computational complexity of a problem is not measured by a simple object such as a number, but rather by a function that depends on the input length. In most cases it seems very difficult even to estimate these functions; in fact, the task of determining the complexity of particular problems is so difficult that often we are happy to get *any* estimates of the time bounds. Therefore we usually content ourselves with asymptotic bounds.

The Three Types of Numbers

Rather than talking about asymptotic bounds, it seems better to start with a more pedestrian point of view. The three types of numbers mentioned in the title are *small, medium* and *large* natural numbers. Such a classification makes sense only if we specify what we want to do with numbers. Let us assume that we want to do elementary computations, more specifically, we want to use the basic arithmetical operations: addition, subtraction, multiplication and division. Then we can ask, how large numbers can be added, multiplied etc. This, of course, depends on whether we want to do the computation with a pencil and paper, or with a computer (and what kind of computer), and how much time we are willing to spend. If needed, we can do such computations without a computer with numbers that have dozens of digits. If we use a computer, we can easily handle numbers that have millions of digits, and with some effort we can, perhaps, go as far as 10^{20} digits. So these should be the medium size numbers.

On the other hand, we estimate that 10^{200} digits cannot be physically represented in the visible part of the universe. Hence numbers with so many digits are large. When a number is large, it does not mean that we cannot represent it at all and that we cannot compute with it. For example, $10^{10^{100}}$ is such a number and we see immediately that it is divisible by 5, and with a little bit of math we can show that it is not divisible by 7. But notice that we need to apply mathematics to prove these assertions; we cannot simply do the divisions and see what the remainder is.

Small numbers are specified by means of algorithms based on the *brute-force search.* These are algorithms that search for the solution in a very simplistic way: they just check all possible values that can be a solution of the problem.[1] The oldest problem to which people have applied such algorithms is the *integer factoring* problem. This is the problem, for a given natural number N, to find a proper divisor of N, called a *factor*. A factor is a number which divides N, and it is not a trivial divisor, which means, it is different from 1 and N. The simplest factoring algorithm is to take numbers from 2 to $N-1$ one by one and try to divide n by them. We can save a lot of work if we realize that we only need to test numbers less than or equal to $\sqrt{N}$. This is because if N has a nontrivial divisor, then it can be factored as ab with a, b different from 1 and N and then either a is less than or equal to $\sqrt{N}$ or b is less than or equal to $\sqrt{N}$. Though it is an improvement, such an algorithm is still not applicable to typical medium size numbers.

Suppose, for instance, that we want to factor a medium size composite number[2]

$$114381625757888867669235779976146612010218296721242362562561842935706935245733897830597123563958705058989075147599290026879543541$$

with 129 decimal digits. If we systematically tried all numbers starting with 2, then we would use about 10^{64} divisions until we found the first factor that has 64 digits:

$$3490529510847650949147849619903898133417764638493387843990820577$$

Each division can be computed fairly easily; with enough patience, we can even do it using only paper and pencil. But the number of divisions that we have to do is so huge that we cannot do them all even with a powerful computer.

[1]In Russian a single word, *perebor*, which means *picking over*, is used for this type of algorithms.

[2]This number is called RSA-129.

(The second factor has 65 digits:

32769132993266709549961988190834461413177642967992942539798288533

These factors are primes, hence they are the only proper divisors.)

Small, medium size and large are not mathematical concepts. They are rather vague concepts concerning our present or future ability to perform certain type of algorithms. Nevertheless, there are "mathematical" relations between them. Observe that if n is a small number, then a number with n digits is medium size. In other words, if n is small, then 10^n is medium size. Also, if N is medium size, but not small, then 10^N is large. Notice that these relations are based on the exponential function, which plays an important role in complexity.

Often we also compute with data that are not numbers. In such a case we encode them by strings in a finite alphabet, usually strings of zeros and ones. This is not much different from numbers, as we can always imagine numbers as written in binary representation. To get a more general description of the three sizes, we can speak about elementary operations with strings.

A Field Full of Open Problems

In spite of the huge amount of results produced in mathematics, what we know seems to be still just a small fraction of what we would like to. That there are more unsolved problems than results, can be said about every field of science. Typically, when a big problem is solved, it raises more new questions than it gives answers to. However, there are differences; in some fields we have a lot of fundamental results and we just need to get deeper knowledge, in others the fundamental questions are still open. Complexity theory is an example of the latter kind. It seems that what we can prove now are only basic facts, while the truly interesting facts are still out of our reach. We can make conjectures about the fundamental relations, but we do not have means to prove them.

The problem that is the simplest to explain is:

Is multiplication more difficult than addition?

Everybody "knows" that it takes more time to multiply two large natural numbers than to add them. Therefore children start with addition and learn multiplication later. Circuit designers also "know" that circuits for multiplication are more complex than circuits for addition. But then why are we not able to prove it? The point is that most people only know the usual school algorithm for multiplication in which we have to, among other things, multiply every digit of the first number with every digit of the second one. This, of course, takes more time than only adding the digits on the same places. Thus this particular algorithm for multiplication needs more time than the usual algorithm for addition. However, the question is whether *every* algorithm for multiplication needs more time than the usual algorithm for addition. The most naive approach to this problem would be to show that the school algorithm is the fastest possible. However, that is not true: we do have algorithms that run much faster than the school algorithm when the numbers are large.

If the answer to this problem is as we expect (that the multiplication of integers is a more complex operation than addition), then it is a typical impossibility problem. It would mean that *it is impossible* to find an algorithm for multiplication that is as fast as the algorithm for addition. This is essentially the form of all big problems in complexity theory.

What is not quite clear is why we are not able to solve these problems. As we already know, impossibility problems are usually hard, which is one explanation. Another reason may be that since the field is young, the theory and the proof methods are not sufficiently developed. But there may be more fundamental reasons. We will see in Chap. 6 that the problems in complexity theory are connected with problems in foundations of mathematics. Thus it is conceivable that we may need new mathematical axioms to solve them.

Let us look at the most important of these problems.

The **P** *Versus* **NP** *Problem*

The **P** versus **NP** problem is the main open problem in complexity theory. It is not by accident that it was the first of the deep problems asked in this field; it is because it is a really fundamental question, and at the same time it is of practical interest. In plain words it can be very roughly stated as the question:

> *Can we always replace the brute-force search by an essentially more efficient algorithm?*

Suppose we are looking for a solution of a problem P and we have an efficient way how to determine what is a solution and what isn't. Then we can find a solution, if there is any, by searching the entire space of possible solutions. The question is then whether we are able to find a better algorithm which does not need so much time.

It is important to realize that here the brute-force algorithms are used only to define a certain class of problems. It does not mean that we are interested in such algorithms; on the contrary, we would like to avoid them. There are other ways to define the same class. One of them is based on the concept of guessing. If, for example we are searching a factor of a number, we may simply try to guess it. If the number is of medium size, the probability that we succeed is usually extremely small and we cannot use it in practice, but in theoretical research we can use this property as the definition of a certain type of problems. Thus we can restate the **P** vs. **NP** problem, again very roughly, as follows.

> *Suppose we know that we can find a solution by guessing. Can we then find a solution by a fast algorithm?*

It should be stressed that the two descriptions above are only attempts to describe the problem succinctly and in plain words. I have also assumed that most readers have some experience with programming, that's why I presented it in the form of replacing one type of algorithm by another, but people with different backgrounds may prefer different descriptions. A logician, for instance, would see the essence of the problem rather in the classical question of replacing existence by construction in the context of efficient computations. I will shortly give a more precise definition, which will eliminate possible ambiguities in the interpretation.

Intuitively the answer to the problem seems to be clear: there is no reason why such an algorithm should always be possible. This intuition is based on our everyday experience: if I absent-mindedly put some paper in a random place, then next time I need the paper I have to search all drawers to find it because the information of where it is lost. However, this argument is wrong. In the **P** vs. **NP** problem the crucial point is that we ask about *mathematical* properties. Hence the information about the solution is present; there is no uncertainty about where the solution is.

Consider the problem of factoring integers, which is one of the situations to which **P** vs. **NP** problem refers. For given numbers N and M, we can easily determine if M is a factor of N, but we do not know how to find M without trying a lot of numbers. If I forget where I put the paper, then from my point of view it can be anywhere and I cannot use reasoning to determine its place. On the other hand, factors of a given number N are uniquely determined by N and I can use mathematics to find one. There are several ingenious algorithms for integer factoring based on non-trivial mathematical results which perform much better than the trivial brute force search (but they are still not fast enough).

From the point of view of foundations, the most important search problem is the *proof search*. It is the problem to find a proof for a given formula assuming that we know that there is a medium size proof.

Here is an example of a problem coming from practice. Suppose an agent needs to visit certain towns. There are airline connections between some pairs of towns but not between every pair. Can he travel so that he lands in every town exactly once and return to the town where they started? Assuming that every flight costs the same this would optimize the cost of his task.[3] Formally, it is a problem about graphs. It is the question if a given graph has a *Hamiltonian Cycle*, where a Hamiltonian cycle is a cycle that goes through every vertex of the graph and passes every vertex exactly once.

Example The graph of the cube is Hamiltonian, as apparent in the second drawing in Fig. 1.1 on page 6.

This problem is simply called *Hamiltonian Cycle*. Again the trivial algorithm for this problem is to try all possible ways to go along the edges of the graph and return to the same vertex. The number of these attempts to find a Hamiltonian cycle can be extremely large even for fairly small graphs. The question is: can we do it essentially better?

---

[3]This is a simplified version of the *Traveling Salesperson Problem* in which we may have different costs associated with different connections.

A lot of problems of this type can be presented as a problem of finding a solution of one or a system of equations in some limited range. Integer factoring is the problem of finding a solution to the following simple equation

$$x \cdot y = N$$

for a given natural number N with the constraint $x, y > 1$. Solving polynomial equations in finite fields is another important class of such problems. Yet another very important problem is called *Integer Linear Programming*. It is the problem of solving systems of linear inequalities in the domain of integers. One can find such problems in every branch of mathematics, as soon as one starts looking for algorithms.

Polynomial Time and Nondeterministic Polynomial Time

I will now state the **P** versus **NP** problem more precisely in order to show that it is a concrete mathematical problem.

First we have to replace the vague concepts 'small' and 'large' by a precise one. To this end we have to return to considering all infinitely many inputs and the dependence of the time on the length of inputs. Then we will specify a class of problems according to the asymptotic behavior of the functions that bound the time and say that these problems are easy. This will be the class **P**, *Polynomial Time*.

Let us recall the concept of a *decision problem*. We usually think of a decision problem as a condition that specifies certain numbers, strings, or other finite structures. But since in mathematics we use set theoretical approach that identifies the set with the problem, a decision problem is simply a set of some finite structures. If we assume the standard computational model, the Turing machine, the inputs will be finite strings of symbols from a finite alphabet. So a decision problem is formally a set of strings and when we talk about sets of numbers, or a finite structure of some type, we are assuming that they are encoded by strings.

The class of functions that we use to define **P** are polynomials. Thus **P** is the class of decision problems that can be computed using at most polynomially many steps. Here is a formal definition.

Definition 9 **P** is the class of sets of strings such that A is in **P** if and only if there exists a Turing machine M and a polynomial p such that

1. M stops on every input, and for inputs of length n, it uses at most $p(n)$ steps before it stops;
2. M accepts the set A (which means that it prints 0) if and only if the input belongs to A.

Computations that use only a polynomial number of steps play a central role in complexity theory. So we will use this concept also in the context of functions. We say that a function f is *computable in polynomial time* if there exists a Turing machine that computes the function f within such a polynomial bound.

In general we can consider all polynomial functions (for example, $3x^2 - 2x + 9$) but what really matters is their asymptotic growth. The asymptotic growth of a polynomial is fully determined by the leading term (which is $3x^2$ in the example). Hence we could use a simpler class of function instead of polynomials, say, the class of functions of the form ax^b, where a and b are positive constants.

We call sets such as **P** *complexity classes* because they define sets of decision problems of certain complexity. **P** is a mathematical approximation of decision problems that can be practically solved, the "easy" problems. In computability theory we say that a problem is decidable, or recursive, if there is a Turing machine that decides the problem in a *finite number* of steps. In complexity theory we have a better approximation: a problem is in **P** if it is decidable in a *polynomial number* of steps.

How good is this approximation? One can immediately give examples showing that it has little to do with practical solvability. The first example is when the constant at the leading term is large. Say, we have the polynomial function $10^{1000}x$. An algorithm requiring so much time is practically useless. If the exponent is large, say we have the polynomial function x^{1000}, it does not work either. We can solve some small instances with such an algorithm, but the large growth of this function prevents us from using it for just a little larger data.

Yet, in general, **P** works pretty well. For one thing, if we show that a decision problem is in **P** by finding a polynomial algorithm, then the constants in the polynomials are, as a rule, very small. For another, this definition has a very desirable property: if we combine several polynomial time algorithms, we obtain again a polynomial algorithm. Naturally, when combining a few simple things, we expect the result to be simple too. In fact, the class of functions bounded by polynomials is the smallest natural class that ensures this property.

Next we need to define the complexity class **NP**. In plain words, a set A is in **NP** if the membership in A can be characterized as follows. An element x is in A if there exists a witness of bounded size y that testifies that x is in A. To make this more precise, we require that

1. the size of y is *at most polynomially larger* than x, and
2. given an x and y, one can decide *in polynomial time* whether or not y is a witness.

Here is a formal definition.

Definition 10 **NP** is the class of sets A that can be defined as follows. For some binary relation $R \in \mathbf{P}$ and a polynomial p,

$$x \in A \quad \text{if and only if} \quad \text{there exists a } y \text{ such that } |y| \leq p(|x|) \text{ and } (x, y) \in R. \tag{5.1}$$

Here $|x|$, $|y|$ denote the lengths of the strings x and y. A binary relation R on strings is in **P** if its natural encoding R' by a set of strings is in **P**. One natural encoding is obtained by taking an extra separating symbol # and defining $R' = \{x\#y;\ (x, y) \in R\}$.

Let A be in **NP** and let p be a polynomial that limits the size of witnesses. Given an input string x of length n, we can decide whether x is in A by trying all y such that $|y| \leq p(|x|)$. We can test each y in polynomial time—this is the condition that the relation R is in **P**. However, the number of potential witnesses that we need to test is huge: $2^{p(n)}$. In concrete examples, for every input string x, there is a natural set of potential witnesses B_x. Although the lengths of the strings in a typical B_x are slightly smaller than the length of x, the size of B_x is still too large and we cannot apply brute force search. If one wants to prove that a problem in **NP** is actually in **P**, then trying to reduce the search space is not a good strategy.

Let us consider some examples.

Examples 1. The first example is the problem to determine if a natural number is prime. This problem is mathematically represented by the set *Primes*, the set of all prime numbers. More precisely, *Primes* is the set of 0–1 strings that represent prime numbers in binary notation. It is easy to show that the complement of this set, the set *Composites*, is in **NP**. Given a number N, a witness of N being composite is a number M such that $1 < M < N$ and M divides N. Since the division can be computed in polynomial time, the latter condition defines a binary relation in **P**. The bounding polynomial $p(x)$ is simply x.

Note that the input length is $n \approx \log_2 N$. Hence the number of potential witnesses, the numbers between 1 and N, is exponentially large. Nevertheless, there is an ingenious algorithm that decides primality and runs in polynomial time. This is a result of M. Agrawal, N. Kayal, N. Saksena [2]. So this is an example of a problem that is in **NP** by definition, but in fact it is even in **P**.

2. The problem *Hamiltonian Cycle* is represented by the set of all graphs that have Hamiltonian cycles, the *Hamiltonian graphs*. In this example, for a graph G, a natural set of potential witnesses is the set of all cycles C on the set of vertices of G. The relation R is: C is a Hamiltonian cycle in G. For given G and C, it is very easy to check whether C is a Hamiltonian cycle in G; in particular, one can do it in polynomial time.

The **P** vs. **NP** is the problem whether or not these two classes are the same, which can be written as the simple equation:

$$\mathbf{P} = \mathbf{NP}?$$

P is clearly a subclass of **NP** (use a dummy witness to prove it). Thus the open problem is whether **NP** is larger than **P**.

As stated, the **P** vs. **NP** problem is to determine, for every set in **NP**, whether it is in **P**. But this pair of complexity classes has a remarkable property that reduces this question to a single **NP** set. There is a set A in **NP** such that $\mathbf{P} = \mathbf{NP}$ if and only if A is in **P**. In fact, hundreds of such sets have been found; one of them is *Hamiltonian Cycle*. Such results are proved using *polynomial reductions* between sets. A *polynomial reduction* of a set X to a set Y is a polynomial time algorithm that reduces the decision problem for X to the decision problem for Y. A set X is called **NP**-*complete*, if it is in **NP** and every set in **NP** can be reduced to it. **NP**-complete sets have the property mentioned above.

Proving **NP**-completeness is also important from a practical point of view. If we know that a set A is **NP**-complete, then it still may have a polynomial time algorithm, but since we know that the problem **P** vs. **NP** is open, we know that nobody knows a polynomial algorithm for the set A. This has become a standard way of showing that a problem is hard, though formally it is hard only if $\mathbf{P} \neq \mathbf{NP}$. Thus in practice we use $\mathbf{P} \neq \mathbf{NP}$ as an axiom.

When talking about algorithms all the time, one may get the impression that such things are not relevant for classical parts of mathematics, which is not true. Imagine, for example, that you are a mathematician working in finite combinatorics and your favorite topic is Hamiltonian graphs. Then most likely the theorem that you would like to prove is a characterization of the graphs that are Hamiltonian. The theorem should say that a graph is Hamiltonian if and only if some condition (simpler than the one by which they are defined) is satisfied by the graph. To give an example of such a condition, consider *Euler graphs*. These are like Hamiltonian graphs, but instead of having a cycle that goes exactly once through every vertex, they contain a cycle that goes exactly once through every *edge* (and it may visit vertices repeatedly). You surely recall puzzles in which you should draw a given picture without lifting your pen and ending on the same point on which you started. The well-known characterization of these graphs is that in such a graph every vertex is incident with an even number of edges. (This condition is, clearly, necessary; the nontrivial part, though not very difficult, is to prove the converse.) There is no such theorem for Hamiltonian graphs. If **P** were not equal to **NP**, we would have an explanation for the absence of such a theorem. If, moreover, we specified precisely what type of characterization we wanted, we might be able to prove that there is no such characterization. Our experience confirms that there is a relation between computational complexity and characterizations: **NP**-complete problems typically do not have nice characterizations.

The problem **P** vs. **NP** was first explicitly stated by Stephen A. Cook in a paper published in 1971 [48]. The title of the paper is *The Complexity of Theorem-proving Procedures*, which hints that the paper concerns logic. The main result of the paper was the theorem that the problem of satisfiability of propositional formulas is an **NP**-complete set. Independently and approximately at the same time the same result was proved by Leonid Levin [182]. Though nobody stated the problem explicitly earlier, some researchers considered related questions before Cook and Levin. Often quoted is Gödel's letter to von Neumann sent in 1956 in which he asked how difficult it is to decide, for a given first order formula ϕ and a number n, whether there exists a proof of ϕ of length at most n [100]. Specifically, he asked if the time complexity of this problem can be bounded by the function cx or cx^2 for a constant c. This is the proof-search problem mentioned above and we know that this is an **NP**-complete set. Another problem that he mentioned in his letter was the complexity of *Primes*.

The reason for Cook's choosing **P** is obvious: it stands for *polynomial*. The **NP** comes from *nondeterministic polynomial*. A *nondeterministic Turing machine* is a modification of Turing machines in which the machine in some states can do one of several actions. Hence, for a given input string, the way how the machine computes is not uniquely determined, which means that several computations are possible. We say that a nondeterministic Turing machine accepts an input string if at least one of the possible computations ends in the accepting state. Using nondeterministic Turing machines we define **NP** as the class of sets that are accepted by nondeterministic

Turing machines in polynomial time. Notice that the **P** vs. **NP** problem is a question about the possibility of eliminating an existential quantifier whose range is bounded. The concept of nondeterministic machines is just another way of expressing such an existential quantification.

The **P** vs. **NP** problem immediately drew attention of researchers in computer science. Many attempted to solve it, but soon it became clear that there were no mathematical means that one could use. A number of results related to this problem have been proved, but the general feeling is that we have not progressed very much during those more than 40 years. Gradually the **P** vs. **NP** problem became well-known in the whole mathematical community. Bets about when it will be solved and prizes for the solution have been proposed. In 2000 the scientific board of the Clay Mathematics Institute included this problem on the list of seven most important problems in mathematics. The institute offers the prize of one million US dollars for the solution of each of these problems. (In Chap. 1 I mentioned another problem that is on the list, the Riemann Hypothesis.)

Complements of NP *Sets*

The existential quantifier plays a crucial role in the definition of the class **NP**. What happens if we replace it by the universal quantifier? We know that negation inverts the quantifiers; specifically, if we negate an existentially quantified formula, it becomes equivalent to the universally quantified negated formula. In symbols it is:

$$\neg \exists x\ \phi \equiv \forall x\ \neg \phi.$$

If we translate this relation into the set-theoretical language, we will see that replacing the existential quantifier by the universal quantifier in the definition of **NP** results in replacing the sets in the class by their complements. So we define **coNP**, *co-Nondeterministic Polynomial Time*, as the class of complements of sets $X \in$ **NP**.

This duality can be extended to other concepts. For example, we can define **coNP**-complete sets. The basic **NP**-complete set is *SAT*, the set of satisfiable boolean formulas. By the duality, we obtain that *TAUT*, the set of propositional tautologies, is **coNP**-complete.

Since **P** is closed under complements, **P** = **coNP** if and only if **P** = **NP**. The question whether **NP** = **coNP** is different. One can easily see that **P** = **NP** implies **P** = **coNP**, but we do not know if the opposite implication holds true. So we cannot exclude that **NP** = **coNP** but **P** $\neq$ **NP**. The problem **NP** versus **coNP** plays the central role in proof complexity.

Time Versus Space

When we define Polynomial Time, the complexity class **P**, it does not matter which computational model we use. Time measured by all reasonable models differs at most by a polynomial. When we want to have more precise estimates on time we must specify the model. In complexity theory we use Turing machines, which may seem not quite adequate. Indeed, the original Turing machines which have a single tape are not very fast. For instance, if such a machine has to determine if two strings on the tape are the same, it has to go back and forth between the two strings many times. Therefore we rather use *multitape Turing machines*, Turing machines with several tapes. In such machines every tape is equipped with one read-write head. The task in the above example is easy for a two tape machine: the machines first copies the first string from the input tape to the second tape by moving its heads simultaneously; then it puts the input head on the first letter of the second string and the head on the second tape on first letter of the string on the second tape. Then it can compare the two strings by moving both heads simultaneously.

Computers do not work like a multitape machine. A theoretical model that describes computers better is the *random access machine*. By the random access we mean the possibility to access any place in the memory directly. This means that when a machine reads (or writes to) a memory unit i and next it needs to read (or write to) the memory unit j, it does not have to pass through the intermediate positions between i and j and can jump directly to j. It is also much easier to program a random access machine than a multitape Turing machine, but surprisingly, they are not essentially faster. Thus in theoretical research we prefer the simpler concept, the multitape Turing machine.

Once we have specified the computation model we can measure the time complexity very precisely. One of the early results in complexity theory (from 1960s) was the *Time Hierarchy Theorem* proved by J. Hartmanis and R.E. Stearns [114]. It states that for every reasonable function bounding the time of Turing machines, there exists a set whose complexity is very close to this function. Thus we have a whole range of possible time complexities.

In order to state this result more precisely, one should describe the set of functions that we can use as time bounds in this theorem, the *time constructible functions*. Since this is a rather technical point I will postpone it to the Notes. Let me only mention that the class includes all polynomials and other functions defined by the commonly used functions, such as the exponential function.

Theorem 36 (Time Hierarchy) *Let $f(x)$ and $g(x)$ be time constructible functions. Assume that $f(x)$ grows faster than $g(x)$ in the following sense:*

$$\lim_{x\to\infty} \frac{g(x)\log g(x)}{f(x)} = 0.$$

Then there exists a set that can be computed in time $f(x)$, but cannot be computed in time $g(x)$.

The proof of this theorem is, essentially, an adaptation of the undecidability of the halting problem (see page 301). As it is an important application of the diagonalization method, it is worthwhile to describe the proof in more detail. Let Time(g) denote the class of sets that are computable in time $g(x)$. Our goal is to construct a set A such that

1. A is outside Time(g), and
2. A is computable in slightly larger time (the better upper bound we get on the time complexity of A, the finer hierarchy we obtain).

To construct a set A that is not in Time(g) by diagonalization is easy. We pick an input string w_B for every set B in Time(g). We put w_B in A if and only if w_B is not in B. This guarantees that A is not in Time(g). Since we need A to be computable in limited time, we have to be more careful; we cannot simply enumerate all sets in Time(g). We also cannot enumerate all Turing machines that run in time $g(x)$, since this property is (for most g) undecidable. So what we do is to enumerate *all* Turing machines. For every M, we pick an input string w_M, say the code of M. Then we decide whether or not to put w_M as follows. We simulate the computation of M on w_M for $g(n)$ steps, where n is the length of w_M. If the machine does not stop within this time bound, we do not care, since it does not define a set in Time(g). If it stops, we put w_M to A if and only if M does not accept this string.

Thus A is not in Time(g) and it remains to determine the time complexity of the set A. It turns out that we need just a little more than $g(n)\log g(n)$ to simulate $g(n)$ steps of a given machine. The reason is that we have to do it on a machine with k tapes, where k is a fixed constant, whereas machines that we need to simulate have arbitrarily large (finite) number of tapes. Machines with more tapes are faster.

Let us consider a couple of applications of the Time Hierarchy Theorem. According to this theorem there are sets computable in quadratic time (with time bound cx^2, c constant), but not in linear time (with time bound cx). Let **EXPTIME** be the class of sets computable in time $\exp(x^c)$ for c constant. Clearly, $\mathbf{P} \subseteq \mathbf{EXP}$. It is a simple corollary of the theorem that **EXP** contains more sets than **P**; thus $\mathbf{P} \neq \mathbf{EXP}$.

The space complexity is informally defined as the amount of memory that is needed for computation. Again, we measure it as a function of the length of the input data. Formally, we say that a Turing machine uses space s on a given input, if its heads visit s squares on the tapes during the computation. In analogy with time one defines *Polynomial Space*, denoted by **PSPACE**, as the class of sets that can be computed with space bounded by a polynomial.

For space complexity, we have a similar hierarchy theorem. The Space Hierarchy Theorem can be proved in a form which is stronger than we have for time. It suffices to assume $\lim_{x\to\infty} g(x)/f(x) = 0$ in order to prove that there exists a set computable in space $f(x)$ and not in space $g(x)$.

The hierarchy theorems give us almost optimal information about time classes and space classes when time and space is considered separately. Problems start when we want to compare time classes with space classes. All we can do is to prove a couple of simple relations. The first one is that if a set is computable in time $f(x)$, then it is also computable in space $cf(x)+c$, for some constant c. This is easy: if a Turing machine with c tapes makes only $f(n)$ moves, then it can visit at most $cf(n)+c$ squares. Thus, in particular, we have $\mathbf{P} \subseteq \mathbf{PSPACE}$. It is also not difficult to show that $\mathbf{NP} \subseteq \mathbf{PSPACE}$. Indeed, brute-force search needs only polynomial space. Thus $\mathbf{P} \neq \mathbf{PSPACE}$ would follow from $\mathbf{P} \neq \mathbf{NP}$, but we also conjecture that $\mathbf{NP} \neq \mathbf{PSPACE}$. The class **PSPACE** also possesses complete problems; typical

PSPACE-complete problems are to decide who has a winning strategy in a combinatorial two-player game of bounded length. This is an explanation of why we do not have winning strategies for games such as chess and go. However, it is only a supporting evidence about the hardness of these concrete problems; we are not able to prove any nontrivial lower bounds on the complexity of these games.

Examples 1. We may never find out whether or not white has a winning strategy in chess. Because of the 50-move rule, one could, in principle, analyze the complete tree of plays using 50 chessboards and determine who has a winning strategy or whether there is a forced draw, but the number of plays is enormous. So the problem is not *space*, but *time*.

2. The game of Hex is simpler and thus more amenable to be analyzed. A simple argument shows that the first player has a winning strategy, but winning strategies are known only for boards smaller than the standard one. The general problem to determine, for a position on an $n \times n$ board, who has a winning strategy is **PSPACE**-complete, which means that the existence of an efficient algorithm for this problem is very unlikely [244].

To bound space by time we need to know how many steps a machine can make if it uses space $f(x)$. One can easily estimate this number by $\exp(c \cdot f(x))$, for some constant c. This is the number of all possible configurations of the machine and the tapes that can appear if space is restricted to $f(x)$. This is also an upper bound on how many steps the machine can make, since if the machine ran longer, it would go into an infinite loop, but we assume that it always stops and produces the correct answer. This, in particular, yields **PSPACE** $\subseteq$ **EXP**.

There is a small improvement of the simulation of time by space, but except for that nothing else is known.[4] It is possible that these simulations are the best possible, but we are not able to prove it. The best way to state these open problems is in terms of complexity classes. We have

$$\mathbf{P} \subseteq \mathbf{PSPACE} \subseteq \mathbf{EXP}.$$

Thus **PSPACE** is somewhere between **P** and **EXP**, and *this is essentially all we know about it!* We believe that both inclusions are strict, but we are neither able to prove that $\mathbf{P} \neq \mathbf{PSPACE}$, nor to prove that $\mathbf{PSPACE} \neq \mathbf{EXP}$. It is interesting, however, that we know that at least one of them is strict. This follows from what we have observed above that $\mathbf{P} \neq \mathbf{EXP}$.

For researchers working in computational complexity theory, it is important to know that at least one of the above inclusions is not equality. People from outside of this field sometimes say: "*What if somebody finds a polynomial algorithm for an* **NP**-*complete problem and we thus get* $\mathbf{P} = \mathbf{NP}$? *Then all the talk about deep problems will turn out to be only humbug!*" It is true that finding such an algorithm would probably be the simplest way to solve the **P** vs. **NP** problem and it may not require developing a deep theory. This can happen essentially with any of the open

[4] Assuming f is time constructible, sets computable in time $f(n)$ are computable in space $f(n)/\log f(n)$, see [132].

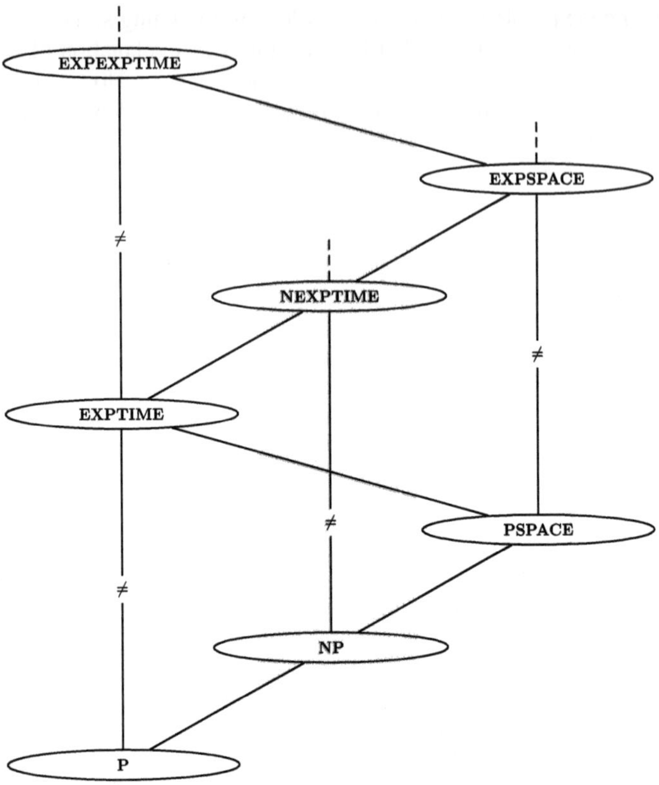

Fig. 5.1 A diagram of some basic complexity classes. The *upward lines* show strict inclusions. Since each of these columns corresponds to one type of resource, inequalities in the columns can be proved by diagonalization. The *right-upward lines* show inclusions that correspond to stronger resources; all these inclusions may be equalities. Equalities propagate upwards; for example, if **P** = **NP**, then **EXP** = **NEXP**. The *left-upward lines* are exponential simulations; for them, it is also open whether they are strict inclusions or equalities. **EXPEXPTIME** is doubly exponential time, **EXPSPACE** is exponential space, and **NEXP** is nondeterministic exponential time

problems about complexity classes, but we do have reasons to believe that it cannot happen for all. In the pair of problems **P** = **PSPACE**? and **PSPACE** = **EXP**? at least one has the negative solution. We may be able to prove, for example, **P** ≠ **PSPACE** by proving **PSPACE** = **EXP**, but there are more complexity classes between **P** and **EXP** and it is not likely that each collapses to **P** or to **EXP**. So, very likely, some inequalities between complexity classes cannot be proved by proving equalities.

Some basic complexity classes and relations between them are shown in Fig. 5.1.

Circuits

In computational complexity we use another important model of computation—circuits. In circuits functions are decomposed into a combination of elementary

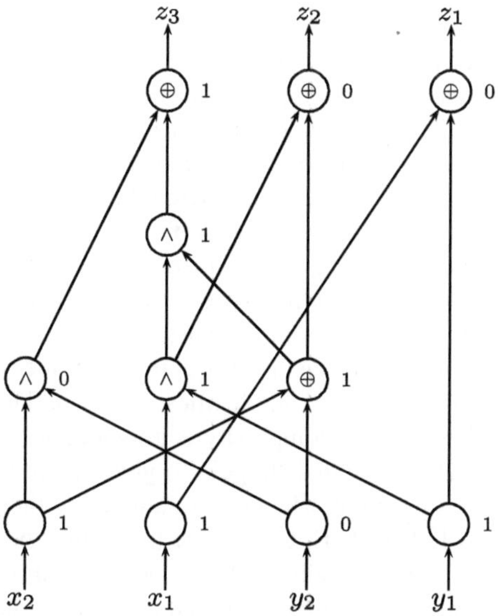

Fig. 5.2 A Boolean circuit for computing the sum of two two-bit numbers x_2x_1 and y_2y_1 yielding a three-bit number $z_3z_2z_1$. The symbols $\wedge$ and $\oplus$ denote the operations AND and XOR (multiplication and addition modulo 2). The computation proceeds from the *bottom* to the *top*. The 0's and 1's at the gates show the computation of $11 + 1 = 100$

functions. Circuits are mainly used to compute Boolean functions $f : \{0, 1\}^n \to \{0, 1\}^m$. This means that the task is for an input string of zeros and ones of length n to compute the string of length m assigned to it by the function. In *Boolean circuits* the elementary functions are some simple Boolean functions, usually unary and binary Boolean functions. Such elementary Boolean functions are also called logical connectives; therefore some authors use the term *logical circuits*. We are now interested in complexity, so we do not care about the interpretation of the Boolean functions in logic; the main thing is that they define simple operations. Also the fact that we use two values is not important, it is just because this is the smallest value that we can use. In electronic circuits two values is also the standard, but the reason for that is the suitability for production, reliability etc. In theory we also use circuits computing with various other sets of values, in particular, algebraic circuits which compute with numbers.

Formally, a Boolean circuit is an acyclic graph in which some nodes are labeled by the input variables and the other nodes are labeled by elementary Boolean functions. Furthermore, some nodes are also labeled by the output variables. This is very much like in real circuits, thus we also often call the edges of the graphs *wires* and the nodes *logic gates*. Given an input string the computation proceeds as follows. First we assign the values to the input nodes and then we gradually compute the values on the nodes labeled by elementary Boolean functions. Eventually we read the output bits on the nodes labeled by the output variables. An example of a Boolean circuit is in Fig. 5.2.

The complexity of the circuit is the number of nodes. The circuit complexity of a Boolean function is the minimal complexity of a circuit that computes the Boolean function. Having a single number as the complexity of a Boolean function corresponds better to our intuition about what complexity is. However, soon I will talk about infinite sequences of Boolean functions and the asymptotic growth of the complexity, which is not much different from the time and space complexities defined by means of Turing machines.

The set of elementary functions is assumed to be finite and complete in the sense that it must be possible to express any Boolean function using the elementary ones. We call such a set a complete set of connectives, or simply basis. Circuit complexity depends on bases, but for different bases the value is the same up to a multiplicative factor. Hence, if we are interested in the asymptotic growth, the basis does not matter. The same is true if we replace the two values by some finite set of values; again the difference is only a constant factor.

In order to justify this complexity measure, we must show that there are functions of various complexities, small and large. Again, it is natural to call a function whose circuit complexity is polynomial 'small' and those whose circuit complexity is exponential 'large'. To give examples of functions with small complexity is not a problem. The circuit complexity theory was founded by Claude Shannon in the 1940s [264]. One of the first things he observed was that there exist functions of complexity almost 2^n, in fact, the majority of all functions of n variables are so complex. The proof is based on a simple idea. Count the number of Boolean function of n variables (this is 2^{2^n}) and count the number of circuits of size at most s. If the number of circuits is smaller, then there must be a function which cannot be computed by such circuits. One can compute that the number of circuits smaller than $2^n/cn$, for some constant c, is less than 2^{2^n}, hence there are functions whose complexity is larger than $2^n/cn$.

Note that this is a nonconstructive proof, a proof which does not provide any explicit example of such a function. Seventy years later we are still unable to give a constructive proof. In fact, we are unable to prove a *nonlinear lower bound* for an explicitly defined function!

Such a statement may give the impression that nobody has worked on circuit complexity since Shannon. This is not true; there are numerous results in circuit complexity. Very efficient circuits have been found for many functions, lower bounds have been proved for some restricted classes of circuits and circuits play an important role in other parts of complexity theory. Lower bounds for general circuits is the only part where no progress has been achieved.

One may suggest that Turing machine complexity is related to the complexity of software and circuit complexity is related to the complexity of computer hardware. The truth is rather that these two concepts are just two facets of the same thing. Let me first give a brief intuitive explanation, based on real computers, why Turing machine and circuit complexities are closely related. Suppose we want to have a device for efficiently computing a function F. We can either construct a processor for F, or program a computer to compute F. Given an electronic circuit, we can program a computer to simulate the circuit; thus the algorithmic complexity is not larger than the hardware complexity. *Vice versa*, given a program for computing F, we can design a circuit by assigning a gate to each elementary operation needed to execute the program; thus we obtain a circuit computing the function F whose size is bounded by the time complexity of the program. So these two complexities are the same up to some factor.

I will now explain it in more detail using the matrix model of computation introduced in Chap. 2, page 137. On page 144 I described a transformation of a computation of a Turing machine T into a matrix M for inputs of a given length n. Recall that the rows of the matrix M correspond to the steps of the computation of the machine T and the columns of M correspond to the squares on the tape of T. (For the sake of simplicity, I will assume that T has only one tape.) The entry in the matrix should furthermore encode a bit of information about whether the head of T is present on this square or not, and about the state of the machine. Thus the entries in the matrix M are from some finite alphabet A which is big enough to encode a symbol on the tape, a bit marking the position of the head and a state of T. The symbols in the first row are determined by the input string. For $i > 1$, the symbol in the row i and column j is uniquely determined by the symbols on positions $(i-1, j-1), (i-1, j), (i-1, j+1)$, the three adjacent symbols in the row above. More precisely, if j is the first or the last column, then we consider only $(i-1, j), (i-1, j+1)$, or $(i-1, j-1), (i-1, j)$, the two adjacent symbols in the row above. In other words, the entry on (i, j) is a *function* of the previous two or three entries.

In this way the matrix can be viewed as a circuit C. The circuit computes with values in the set A and uses binary and ternary functions defined on A as the elementary functions. It has a rectangular form, in which one dimension corresponds to the time of the machine T and the other to space. We only take as many rows as is the maximal number of steps that T makes on inputs of length n and the number of columns is the maximal number of squares that T visits in some computation on inputs of length n.

Now suppose that T operates only on bits. Then for the fixed input size n, T computes a Boolean function. So, instead of the circuit C operating with symbols in the finite alphabet A, we may want to have a Boolean circuit which uses only binary Boolean function as the basis. To obtain such a circuit we encode the elements of A by binary strings. Thus the binary and ternary functions defined on A become some Boolean functions. For each such function g, we choose a Boolean circuit and replace every occurrence of the function g in C by this Boolean circuit. By this transformation, we increase the size of the circuit only by a constant factor. This shows that if the time complexity of T is $t(x)$ and the space complexity is $s(x)$, we get a circuit of size at most $ct(n)s(n)$, for some constant c. If we are only interested in time, we can upper-bound the size of the circuit by $ct(n)^2$, since we know that always $s(x) \leq t(x)$. In particular, if the time of T is bounded by a polynomial, then the size of the Boolean circuit is also bounded by a polynomial.

It is clear that the above transformation produces only circuits of a special form. In order to understand the relation of Turing machines to circuits, we should examine what is the difference between a general circuit and a circuit obtained from a Turing machine. To simplify this problem, let us only consider the circuits that were obtained in the first part of the transformation. In the notation used, it is the circuit C which uses values from the set A. Let us simplify it further and compare C only with circuits that have the same rectangular form with the same connections between the

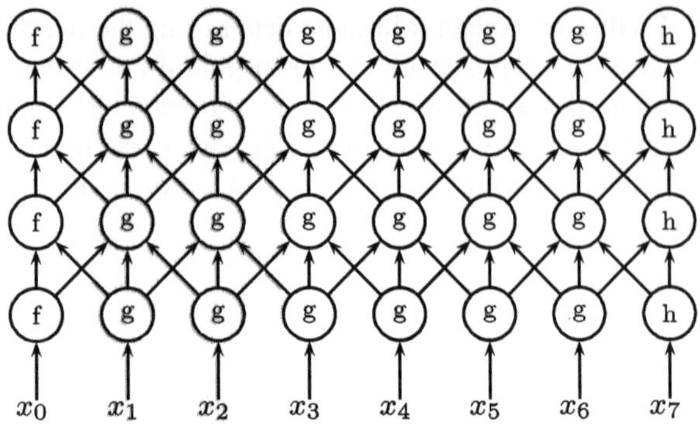

Fig. 5.3 A uniform circuit

gates[5] and which use the same set of values A. Let D be such a general rectangular circuit. As the two circuits have the same underlying graph, the difference between them is only in what functions at which nodes of the graph are used. One can show that in C there are only three different functions: one binary function assigned to all nodes that correspond to the first row, one ternary function assigned to all nodes that are neither in the first column nor in the last one, and one binary function assigned to all nodes corresponding to the last column. In D, however, various functions can be assigned to the nodes in an arbitrary manner.

Furthermore, for the Turing machine T, the three functions are not only assigned to a given input size n, but the same functions are also used in the circuits for all input sizes. Hence the circuits obtained from Turing machines have a very regular form; we say that they are *uniform*, in contrast with general circuits that are *nonuniform* (see Fig. 5.3). This naturally leads to the concept of *nonuniform complexity classes* and to the question what properties do they share with their uniform counterparts. I will define only the **nonuniformP**, the nonuniform version of **P**.[6] This is the class of all sets of 0–1 strings with the following property. A set A is in **nonuniform-P** if and only if for some polynomial $p(x)$, for every input length n, there exists a circuit C_n of size at most $p(n)$ which accepts exactly those strings of length n that are in A. In other words, we have a sequence of circuits of polynomial size that compute the sections of the set of A.

The transformation of a Turing machine computation into a circuit shows that **P** is contained in **nonuniform-P**. To determine how much **nonuniform-P** is different from **P** is another fundamental problem of complexity theory. We know for sure that the two classes are different. This follows from the fact that **nonuniform-P** contains sets that are not computable. The reason is that since we are free to choose a circuit for each n in an arbitrary way, we can take some trivial circuits that accept all inputs of length n for some numbers n, and take some circuits that reject all inputs of length n for the other numbers. In this manner we can encode an arbitrary subset of natural numbers into a set in **nonuniform-P**.

---

[5] One can prove that this is an inessential restriction.

[6] I deviate from the standard notation which is **P/poly**.

We seemingly deviate from our original goal to restrict the class of computable sets to a subclass which is closer to what is practically computable, but our intuition tells us that it is all right. In practice we consider only some small range of the input sizes, the 'medium size' inputs, and we do not care what happens for inputs whose size is 'large'. Thus we may expend a lot of effort to design a special purpose circuit for a given task and then use it for efficient computation. From the theoretical point of view, there is no distinction between circuits and Turing machines if we have a fixed input size. We can equivalently say that we may need a lot of effort to produce a fairly complex program, but once we have it, it runs fast. For instance, such a program may use precomputed tables of numbers, tables that need a lot of time to be produced. To express it shortly, **P** is a class of problems efficiently computable by *small programs*, whereas **nonuniform-P** is a class of problems efficiently computable using *programs that may be very hard to produce*.

The main problem concerning nonuniform complexity classes is how big uniform classes are contained in **nonuniform-P**. We believe that none of the reasonable uniform extensions of **P** is contained in it. In particular we conjecture that **NP** is not a subset of **nonuniform-P**. That said, we are even unable to prove that **EXP** is not contained in **nonuniform-P**. We will see later that this is an important problem.

Another important application, actually the first one, of the transformation of Turing machine computations to circuits is in the **NP**-completeness theory. I will say more about it in Notes.

How to Prove that P $\neq$ NP?

The first idea that comes to mind is to use the same method as is used in proving that some problems are algorithmically unsolvable. This is the method that we call *diagonalization*, whose origin goes back to Cantor's proof that the set of real numbers (or equivalently, the set of all subsets of natural numbers) has larger cardinality than the set of natural numbers. This method works well, as we have seen, when one needs to separate two classes defined by restricting the same computational resource. Thus one can separate pairs of time complexity classes and pairs of space complexity classes. But when we need to separate two classes of different a type, it does not work. We even have an argument that shows that diagonalization cannot solve problems such as **P** vs. **NP**, which I am going to sketch now. (But one can never exclude that some very unusual application of the method will work anyway.)

If you look at a typical proof based on diagonalization, such as the proof of the Time Hierarchy Theorem, you will notice that it uses very little information about how Turing machines compute. In particular the proof does not use the fact that the steps of the computation are very simple operations (reading the input symbol, rewriting it and moving the head to an adjacent square). If instead of these simple operations, more precisely, if, on top of these, we also allow some complex ones, the proof goes through without a problem. So let us consider an alternative world in which Turing machines can use a certain complex operation. You can imagine it as if some company would succeed in producing a special purpose processor that computed some hard function. Then on computers equipped with this processor one could compute many things that otherwise would be too hard. Everything would be the same, except that when programming one would be allowed to use this special function like the usual functions present in all programming languages.

Let us now consider the classes **P** and **NP** in such a world. We call classes modified in this way *relativized*. We are not able to solve the **P** vs. **NP** for the original, unrelativized classes, but it is possible to decide the corresponding question for some relativized ones. T. Baker, J. Gill and R. Solovay found (already in 1975) relativizations for which $\mathbf{P} = \mathbf{NP}$ and others for which $\mathbf{P} \neq \mathbf{NP}$, [17]. This proves that:

It is not possible to solve the **P** *vs.* **NP** *problem without using the fact that the basic operations are the simple ones.*

In particular it demonstrates that:

No direct application of diagonalization can prove $\mathbf{P} \neq \mathbf{NP}$.

So what else can we use? When looking for a method that essentially uses the fact that a single step of computation may only use an elementary function, we are naturally led to circuits. We know that in order to prove $\mathbf{P} \neq \mathbf{NP}$, it suffices to find a set in **NP** whose sections cannot be computed by polynomial size circuits. Thus the computation-theoretical problem is reduced to a combinatorial problem. Given an explicit Boolean function, such as the function that computes whether the input string of length n is an encoding of a Hamiltonian graph, we need to prove that every circuit that computes the function must have a size bigger than $f(n)$ for some function f which grows faster than any polynomial. Hamiltonian graphs are typical combinatorial concepts and the nature of circuits is also combinatorial. So the first reaction is that all this is just finite combinators, maybe a bit more difficult, but nothing more. Yet many problems, including very difficult ones, can be stated as problems in finite combinatorics; one should not judge the problem only by its appearance.

Our negative experience suggests that proving superpolynomial lower bounds on explicitly presented Boolean functions is such a difficult problem. We also have some mathematical results that support this presumption. The form of these results is that a certain method of proving such lower bounds on the circuit size, a method that we consider to be a natural way of solving such a task, in fact cannot work. These results are mainly due to A.A. Razborov. In the 1980s Razborov proved several important lower bounds on the size of circuits of a special type [238, 239]. This created a great excitement; we hoped that the new methods that he introduced could be used to prove $\mathbf{P} \neq \mathbf{NP}$. However, nobody was able to apply these methods to general circuits. Later Razborov proved that his methods, unfortunately, cannot be applied to general circuits [240]. (Again it does not exclude the possibility that some particular generalizations of the method can still work.)

Let us fix some terminology, before explaining the first type of negative result. In a circuit an elementary Boolean function, a gate, is assigned to every node of the underlying graph. We will assume that these functions are arbitrary binary Boolean functions (put otherwise, the basis is fixed to be the set of all binary Boolean functions). Furthermore, we associate every node of the circuit with a Boolean function computed at this node. This is the function that expresses the dependence of the output bit of the node on the input variables of the circuit. If the node is the output node, then the associated function is the function that the circuit computes. In the sequel we will assume that we want to prove a lower bound on the size of circuits computing some Boolean function f.

The first result concerns lower-bound methods based on the idea of *a progress in computing* f. This is a very natural idea: if f has large complexity and every step of computation makes only small progress (which means that it can increase the complexity of the computed functions only a little bit), then there must be many steps in every computation of f. The computations that we have in mind are circuits. As each node of a circuit computes some function, this schema makes sense. Indeed, at each step the circuit complexity of the computed functions can increase only very little.

In this form the method is just a reformulation of the task of proving a lower bound; we have to replace the circuit complexity by something more specific. The most natural specification of this method, which is also what Razborov used in his lower bounds on monotone circuits, is to consider the *distance* from f as the measure of progress. Whereas the circuit complexity of a function is a concept that we are not able to handle, the distance is a very simple property. It is just the number of inputs on which a given function differs from f.

However natural this approach seems, the method with this specification does not work (in the case of general circuits). The essence of the argument showing that it does not work is in the following equation.

$$f = g \oplus (f \oplus g).$$

Here $\oplus$ denotes the Boolean function *exclusive or*, also called *XOR*, which can be interpreted as the addition modulo two. To evaluate such an expression it is better to use the latter interpretation. As we count modulo two, we have $g \oplus g = 2g = 0$, which proves the equation. (The order of summands and parentheses are irrelevant for the computation, but they are important for the argument that I am going to describe.)

Suppose we have a circuit C that has two parts, one computing some function g and the other computing some function $f \oplus g$, and the output node computes the exclusive or of these two circuits. Hence C computes f according to the above equality. Thus if our lower bound method worked, one of the functions g or $f \oplus g$ must be close to f. But g can be an arbitrary function, hence should the method work for f, we must have, for every g: either g is close to f, or $f \oplus g$ is close to f. Since the pairs $\{g, f \oplus g\}$ are disjoint for distinct functions g, it follows that half of all Boolean functions must be close to f, which is impossible.

I have sketched only one basic idea of Razborov's results. It shows that it is difficult to base a lower-bound method on studying what functions are computed locally at single nodes. One can use the progress in computing a function f for lower bounds, but it is necessary to treat it as a global property. Every single node, except for the output node, can compute a function that is completely unrelated to f.

The next result is, in a sense, more general, but it is based on an unproved assumption from complexity theory. As it may seem rather strange to use such an assumption to prove unprovability of another assumption, I will first explain why it is reasonable to use such an assumption. Let us assume the hypothetical situation that we can prove that a certain method $\mathcal{M}$ cannot be applied to prove $\mathbf{P} \neq \mathbf{NP}$. Furthermore assume that for this very proof, we need to use the assumption $\mathbf{P} \neq \mathbf{NP}$. In other words, we can prove the following statement:

If $\mathbf{P} \neq \mathbf{NP}$, *then it is not possible to prove* $\mathbf{P} \neq \mathbf{NP}$ *using method* $\mathcal{M}$.

But if $\mathbf{P} = \mathbf{NP}$, then no method can give a proof of the opposite. Thus in such a case we can eliminate the assumption $\mathbf{P} \neq \mathbf{NP}$ and we can conclude:

It is not possible to prove $\mathbf{P} \neq \mathbf{NP}$ *using method* $\mathcal{M}$.

Now suppose that instead of $\mathbf{P} \neq \mathbf{NP}$, we use a stronger assumption $\mathcal{A}$. Then we cannot remove it, but such a result is still interesting; for example, we can conclude that $\mathcal{A}$ itself is not provable using method $\mathcal{M}$, without using any unproved statement.

We will now consider a method of proving lower bounds based on the following idea. First characterize the Boolean functions that have large circuit complexity using a property P and then prove that a particular Boolean function, if possible a function from **NP**, satisfies P. This kind of proof is often used in mathematics; many important theorems are either explicitly stated as a characterization of some concept, or can be interpreted in such a way. What we consider to be a nice characterization depends on the particular field of mathematics, and on personal taste; there is no precise definition of it. Having an efficient algorithm to decide property P is not exactly what we mean by a nice characterization, but it is often a consequence of it, and conversely, if there is no such algorithm, it strongly suggests that a nice characterization is impossible.

Thus for this method, the crucial question is: *can we decide in polynomial time if a given Boolean function f has complexity larger than S?* When one speaks about computing properties of Boolean functions, one should make clear, how the functions are given. Here we assume that a Boolean function is given by its truth table, represented by a string of zeros and ones of length 2^n, where n is the number of variables of the function. The above question is more precisely:

Problem 2 Is it possible to decide in polynomial time which strings of length 2^n encode Boolean functions of complexity larger than S?[7]

Now I can state a simplified version of a result concerning the above method. It uses an unproved conjecture about the existence of pseudorandom generators, which I will abbreviate as the *PRG-Conjecture*. I will discuss pseudorandom generators in the next section, for now it is not very important what exactly the conjecture says (for the exact statement see page 435). It suffices to say that although it is essentially stronger than $\mathbf{P} \neq \mathbf{NP}$, it is considered very plausible.

[7] Note that the input size is 2^n, hence polynomial time means 2^{cn} for some constant c.

Theorem 37 *The PRG-Conjecture implies that the answer to the above question is no, i.e., it cannot be decided in polynomial time whether a string encodes a Boolean function of complexity larger than S.*

Thus assuming the PRG-Conjecture, it is unlikely that a suitable characterization of hard Boolean functions can be found. Therefore it is unlikely that we could first characterize hard Boolean function by a polynomial time property and then prove that a given function has the property. Such a method of proving that a Boolean function has large complexity is unlikely to work.

This is a weaker form of a result of A. Razborov and S. Rudich [243]. I will now describe their result in more detail. Observe that for proving a lower bound we do not need a precise characterization of functions whose complexity is larger than a given bound S. What we only need is to characterize a subset of hard Boolean functions and to show that our function belongs to this subset. The essence of the result can be briefly stated as follows.

> *Assuming the PRG-Conjecture, only small subsets of binary strings encoding hard Boolean functions can be computed by polynomial time algorithms.*

Hence we cannot prove circuit lower bounds by first characterizing a *large* subset of hard functions and then showing that our function belongs to this set. This still leaves open the possibility of characterizing a small subset of hard functions. Why cannot we first characterize a *small* subset of hard functions and then prove that our function is in this set? Theoretically this is, of course, possible; theoretically our function may be the unique function with such a property, but then we can hardly speak about a *method*. The idea here is that a method must be based on a *general* combinatorial property, general meaning that *many* functions have it.

Razborov and Rudich proved more than just the above result. They analyzed all important lower bounds in circuit complexity (there is a number of such results for restricted classes of circuits) and they showed that all these proofs are based on characterizing a subset of functions that are hard for the class of circuits involved and the subset is both computable in polynomial time and large. (For restricted classes of circuits this does not contradict the PRG-Conjecture.) Thus the assumptions about the form of lower-bound proofs, including the largeness, are natural; we do not have other proofs. Therefore Razborov and Rudich proposed to call lower-bound proofs of this form *natural proofs*.

There is another argument that justifies the largeness condition. As we saw before, if we try to use the idea of the progress in computing a function f, then half of the functions are "close" to f. Hence this type of method also leads to a characterization of large sets of hard functions.

To recapitulate it, Razborov and Rudich defined a precise notion of a natural proof for a given class of circuits, and they proved:

1. *Essentially all known lower bound proofs are natural.*
2. *Assuming a plausible conjecture, there is no natural proof of an exponential lower bound for general Boolean circuits.*

We therefore have to look for a proof that is essentially different from the proofs that have been found so far.

The last remark about natural proofs concerns the possibility of removing the unproved assumption in the way I mentioned before. At least in one case this is indeed possible. A. Wigderson showed that one can prove *without any unproven assumption* that there is no natural proof of the hardness of the *discrete logarithm.* This function is conjectured to be hard; in particular, it should not have polynomial size Boolean circuits.

The Problem of Proving Lower Bounds

To show that $\mathbf{P} \neq \mathbf{NP}$ we need to prove that for some set A in **NP**, the question "*is a given x in A?*" cannot be decided using a polynomial time algorithm. In other words, we need to prove that any algorithm for this problem must run in time $t(n)$ that cannot be bounded by a polynomial in n. Thus we need to prove a *lower bound* on $t(n)$. Essentially all problems about separating complexity can be stated as problems of proving certain lower bounds.

Proving lower bounds on the complexity of computations is also important if one needs to precisely determine the complexity of particular problems. When we have an algorithm A for some problem P, it is usually not very difficult to estimate its running time $t(n)$ (or the space $s(n)$) it requires. Thus an algorithm gives us an *upper bound* on the complexity of P. Then we wonder whether the problem can be solved by a better algorithm. To show that there is no better algorithm requires proving a lower bound. The ideal situation would be when, for every problem used in practice, complexity theory were able to exactly determine the computational complexity of the problem. Presently this is only wish as one can prove precise bounds only in a few very special cases.

As usual, when a problem cannot be solved at once, people try to prove at least partial results. Let us focus on the problem of proving lower bounds on the size of Boolean circuits. There are two ways how one can make this problem simpler. First, one can try to prove just some nontrivial lower bounds. Second, one can restrict the class of circuits.

The first approach is the most frustrating area in computational complexity. One can say that almost nothing has been achieved here. At the time of writing these lines, the largest lower bound on circuits with n variables computing an explicitly defined Boolean function is only $3n - o(n)$,[8] see [27]. Furthermore, these proofs use only elementary arguments, so they do not provide any insight into what the essence of computational complexity is.

In contrast to this, the area of lower bounds on circuits from restricted classes is one of the most interesting parts of theoretical computer science. I will describe

---

[8]This is for circuits in the basis of all binary connectives; $o(n)$ denotes a term of a lower order than n.

one basic result of this kind. Consider the Boolean function $PARITY(x_1, x_2, \ldots, x_n)$ which computes the parity of the number of ones in the input. So it outputs 0 if the number of ones among $x_1, \ldots, x_n$ is even, otherwise it will output 1. Now suppose we want to compute it as a disjunction of conjunctions of variables and negated variables. Formally, we want to express it as

$$PARITY(x_1, x_2, \ldots, x_n) = \bigvee_i \bigwedge_j z_{ij}$$

where each z_{ij} is a variable or negated variable (x_k or $\neg x_k$ for some $k = 1, \ldots, n$). If some conjunction $\bigwedge_j z_{ij}$ contained less than n variables, then the right hand side would accept inputs of both parities, so this is not possible. If a conjunction contains all variables, then it accepts exactly one input (assuming every variable occurs in it only once). Hence the number of conjunctions must be 2^{n-1}, the number of odd inputs. This is a very easy lower bound. Now consider circuits of depth 3, which means expressing *PARITY* as

$$PARITY(x_1, x_2, \ldots, x_n) = \bigvee_i \bigwedge_j \bigvee_k z_{ijk}$$

where each z_{ijk} is a variable or negated variable. Then one can still prove an exponential lower bound $2^{\sqrt{n}}$, but the proof is not trivial. More generally, one can prove exponential lower bounds for any fixed depth. The proof is based on the random restriction method, which is briefly described in Notes.

It would take us to far afield to discuss the importance of this result and its relation to others. Let me only draw your attention to the striking simplicity of the function used in this lower bound. Naively, one would expect that it should be easier to prove such lower bounds for functions that are hard, rather than such a simple one. The point is that the lower bound proof uses the simple property that *PARITY* changes its value if we flip any of the input bits. This function and its negation are the unique functions that have this property.

This is one of the first nontrivial lower bounds. In more recent results a bit more complex properties are used, but they are still fairly simple. Our inability to use complex properties of hard functions shows that the field is still not mature enough. The state of affairs is comparable with number theory at the time when only the irrationality of $\sqrt{2}$ was known.

Existence and Construction

When we prove that there exists a mathematical entity (a number, a mathematical structure, etc.) with some property, we usually give an explicit description of it. Sometimes, however, one can only prove that it exists without being able to exhibit a specific example of such a thing. The first type of proof is called *constructive*, the second type *nonconstructive* or *purely existential*. Purely existential proofs appeared in mathematics not so long ago. At first they were looked upon with great suspicion and some mathematicians even rejected them. Some mathematicians thought that they were the source of paradoxes and therefore in intuitionistic mathematics such proofs are not allowed. In 1889, Hilbert proved a fundamental result that the number of invariants of an n-form is finite [123], which generalized a former proof of Paul Gordan for binary forms. Hilbert used an indirect argument, and since it was one of the first purely existential proofs, Gordan's reaction was: "*this is not mathematics, it is theology*". Another well known nonconstructive result is Roth's Theorem about approximations of irrational numbers by integers. It states that for a certain type of precision, there are only finitely many rational numbers that approximate a given rational number with that precision. (Weaker versions of this theorem were proved by Liouville, Thue and Siegel.)

Gradually mathematicians got used to it and in some fields there are many purely existential proofs. The reason is not that we like them, on the contrary, we always prefer a constructive proof, but sometimes this is the only way to prove a theorem. Quite often, an existential proof is easy and a constructive proof of the theorem is difficult and therefore found much later. But even nowadays we usually do not consider a purely existential proof to be a complete solution of a problem. Typically, when looking for a solution of equations, *proving the existence* of a solution and *solving* the equations are always treated as two separate things; solving means to describe a solution explicitly.

In the chapter about set theory we noticed that the Axiom of Choice has a very nonconstructive nature. Among its consequences, there are results stating the existence of very counterintuitive objects, such as nonmeasurable sets of reals, the paradoxical transformation of a ball into two balls, etc. But existential proofs are also used in finite mathematics where we do not need the Axiom of Choice. They were introduced to finite graph theory by Erdős in the 1940s. Probably the first paper in which his *probabilistic method* was used concerned estimates of the Ramsey numbers [67].

Let us briefly recall the Finite Ramsey theorem and the Ramsey numbers (defined on page 16, see also page 328). The theorem states that for every natural number n, there exists a number R such that, for every graph on R vertices, there exists a set of n vertices such that either every pair of them is connected by an edge, or no pair is. We call such sets of vertices monochromatic. The minimal R such that this holds is the Ramsey number $R(n)$. We do not have a precise formula for the numerical function $R(n)$, but already back then Erdős established that it grows exponentially.

To prove such a result one has to show an exponential upper bound and an exponential lower bound. The upper bound $R(n) \leq U$ means that every graph with U vertices contains a monochromatic set of size n. It was proved constructively (for $U = \binom{2n-2}{n-1}$), in the sense that, for every graph of size U, a monochromatic subset of size n was constructed by an efficient algorithm. To prove the lower bound $L < R(n)$ one has to prove that *there exists* a graph on L vertices which does not contain a monochromatic set of size n. This is where Erdős used a purely existential proof (for $L = 2^{(n-1)/2}n/e$).

His argument can be explained in two equivalent ways. (1) He counted the number of all graphs on L vertices that contain a monochromatic subset of size n and showed that it is smaller than the number of all graphs on L vertices. (2) He counted the probability that a random graph on L vertices contains a monochromatic subset of size L and showed that the probability is less then 1. Since the probability that a random graph has a property P is the number of graphs having property P divided by the number of all graphs, (1) and (2) are the same, except that they use different language. In both cases we immediately conclude that there exists a graph on L vertices that does not contain a monochromatic subset of size n, without being able to exhibit a single example of such a graph. More than sixty years after Erdős asked this question, we are still unable to construct explicitly a *Ramsey graph*, a graph that would show an exponential lower bound on $R(n)$.[9]

Problems of this type are clearly related to the **P** vs. **NP** problem. Since the number of graphs on L vertices is finite and the condition that we are interested in is algorithmically decidable, we can search all such graphs and find one. The problem is that for medium size L the number of such graphs is too large, hence this algorithm is too slow. If **P** were equal to **NP** we could construct such a graph in time bounded by a polynomial in L. What is more important is that the brute-force algorithm above does not yield any additional knowledge about such graphs. Why is an explicit construction better? Consider the problem of finding more precise estimates on $R(n)$. Erdős's lower bound is asymptotically $2^{n/2}$ and his upper bound is 2^{2n}. These bounds have been improved only in lower order terms, so the situation now is essentially the same as sixty years ago. A likely explanation of our failure to close the gap is the following: the actual value of $R(n)$ is close to the upper bound 2^{2n}, whereas the value for random graphs is close to $2^{n/2}$ and the probabilistic method is only able to determine the value for random graphs. Apparently we need to find explicit constructions in order to be able to determine the Ramsey numbers. This is not a sheer speculation; in Notes I will mention two combinatorial problems that were solved by means of explicit constructions.

Probabilistic proofs show the existence of graphs with typical values of parameters which is bad for problems such as the problem of computing the Ramsey numbers. On the other hand, this approach opened a completely new area of research, the study of properties of "random graphs", or more generally any "random structures". Using the word '*random*' is only a façon de parler about properties that are shared by most structures of a given type. Saying that '*a random graph has property P*' is just a convenient abbreviation of the longer '*if we randomly choose a graph with N vertices, then the probability that it has property P goes to 1 as N goes to infinity*'. That said, I should stress that a random graph is not abstract nonsense that we can never observe in reality. On the contrary, we can very easily produce such a graph: for each pair of vertices decide whether to draw or not to draw an edge between them by tossing a coin. Such a graph will be for all practical purposes random. At the end of this chapter I will mention a possibility how to define formally a random graph as a specific graph.

[9]To give an explicit construction of a Ramsey graph is one of the many famous Erdős problems for which he offered money prizes.

Shannon's proof that there exist Boolean functions whose circuit complexity is exponential appeared only two years after Erdős's paper. I do not know if Shannon was aware of Erdős's result, but his proof is based on the same idea—comparing the number of all n variable Boolean functions with the number of circuits of some size S, which bounds the number of Boolean function of circuit complexity S. Thus we know that "random Boolean functions" have complexity of the order $2^n/n$, but we are even not able to construct functions that have nonlinear complexity. The words '*construct*' and '*explicitly define*' are not precisely defined. In complexity theory we, of course, would interpret them as giving an algorithm of certain low complexity. If we succeed in constructing a function with, say, nonlinear circuit complexity, then depending on how strong constructibility condition we ensure we get a correspondingly strong separation result. One of the best would be to find a sequence of functions that compute sections of an **NP** problem A and which have superpolynomial circuit complexity. Superpolynomial circuit complexity means that A is not in **nonuniform-P**. So this would give us **NP** $\not\subseteq$ **nonuniform-P**, from which **P** $\neq$ **NP** follows. Another natural interpretation of being explicit is to require that explicit Boolean functions should be constructed in polynomial time in 2^n (which is the size of the truth tables). This is much weaker but it would still give the interesting consequences, for example, the separation **EXP** $\not\subseteq$ **nonuniform-P**.

Another example of a structure whose existence can be proved easily, but it is nontrivial to construct it explicitly are expander graphs (see page 431). These are graphs that, roughly speaking, have few edges, but where every set of vertices has many neighbors (unless the set is too large). Explicitly defined expander graphs are very useful components of many constructions and derandomization techniques.

Let us pause and compare the problem of constructive proofs in finite combinatorics and computational complexity with similar problems in other fields of mathematics. There are numerous examples of problems for which we have only existential proofs. Typically one has a criterion for solvability of equations of some type, but the result says nothing about how to find an explicit solution.

I have already mentioned that integer factoring is a problem of this type. A particular instance is the famous problem to determine which Mersenne numbers are primes. A Mersenne number is a number of the form $2^p - 1$, where p is a prime. The well-known Lucas-Lehmer test can be used to prove primality and non-primality of very large Mersenne numbers. If such a number is not a prime, it means that $xy = 2^p - 1$ has a solution with $x, y > 1$, but finding the factors is much harder.

Likewise, theorems about the existence of solutions of differential equations can be applied to many types of equations, but only in some special cases it is possible to describe a solution explicitly. In such problems a solution is an infinite object, thus we do not have a precise definition of being explicit. If the solution is unique, we can often compute it with arbitrary precision, but this is not enough.

It is also interesting to compare the methods used to prove such results. Cantor proved that the cardinality of the set of all real numbers is bigger than the cardinality of the set of all algebraic numbers. As we have observed in Chap. 4, a simple corollary of this result is that there exist non-algebraic, i.e., transcendental numbers. It had been proved before that particular numbers are transcendental, but Cantor's proof is much simpler. In circuit complexity we use finite cardinalities to prove the existence of Boolean functions with large complexity, but unfortunately, we still do not have methods to give specific examples of such functions. The method used in these two cases is the same, both proofs are fairly easy and both do not provide specific examples.[10]

The lack of progress in solving fundamental problems of the complexity theory suggests that we must first learn how to use deeper mathematical methods. Recent development shows that explicit constructions of finite structures with certain useful properties are often the key to making progress where we had been stuck for a long time. Finding explicit constructions where there are only existential proofs is a very interesting topic. However, we should not expect that every existential proof can be replaced by explicit constructions. For example, it is not excluded that there are no polynomial time constructions of Ramsey graphs.

The Complexity of Algebraic Computations

Computing with digital devices requires encoding numbers by strings of symbols, say zeros and ones. What if we ignore this problem and assume arithmetic operations to be elementary? Given an algebraic function, which is a function defined by a polynomial, it is natural to ask how many basic algebraic operations we need to compute the function. This is *algebraic complexity*.

It should be stressed that if we determine the algebraic complexity of an algebraic function, we still do not know how difficult it is to compute it using a general computational model such as the Turing machine, or Boolean circuits. In the algebraic setting the cost of computing the product of two numbers is one because it requires only one step, but when the numbers are encoded in binary, it is a nontrivial task (and we do not know the precise cost). What is more important is the opposite: for some algebraic functions an algebraic program may require much more time than a Turing machine. The point is that a Turing machine can compute basic arithmetic operations fairly fast, but on top of that it can also compute a lot of operations that are not algebraic. For example, it can flip the binary string that represents the number. However, algebraic programs have the big advantage of being more versatile. We can often use the same algorithm for various fields, sometimes even for rings.

[10] In fact one can extract a specific transcendental number from this proof. We can compute solutions of algebraic equations with rational coefficients to arbitrary precision and we can enumerate them (possibly with repetitions). Thus applying the diagonal trick we can define digits of a transcendental number by an algorithm. Similarly, we can enumerate all Boolean functions of n variables and taking the first one whose complexity exceeds a given bound. We do not consider such definitions explicit, as the defined entity is chosen by a process that has little to do with the property that we need to ensure.

One of the central problems in algebraic complexity is the complexity of matrix multiplication. Let $A = (a_{ij})$ and $B = (b_{ij})$ be two $n \times n$ matrices. Their product is defined by

$$AB = \Big(\sum_k a_{ik} b_{kj}\Big).$$

If we use this formula to compute the product we need n^3 multiplications and $n^2(n-1)$ additions. It is tempting to conjecture that this is the best possible, but it isn't.

The fact that it suffices to use essentially fewer operations was discovered by Volker Strassen. Sometimes around 1969 he decided to determine the complexity of matrix multiplication in the simplest possible case. It was the case of two by two matrices in the two element field. The two by two is the smallest nontrivial dimension of matrices and the two element field $\mathbb{F}_2$ is the simplest field since it consists only of 0 and 1. Thus in this field the number of possible terms is very limited. Strassen found a very remarkable way of computing such a product [286].

The product of two such matrices is

$$\begin{pmatrix} a_{11} & a_{12} \\ a_{21} & a_{22} \end{pmatrix} \begin{pmatrix} b_{11} & b_{12} \\ b_{21} & b_{22} \end{pmatrix} = \begin{pmatrix} c_{11} & c_{12} \\ c_{21} & c_{22} \end{pmatrix}$$

where the numbers c_{ij} are, according to the definition,

$$\begin{aligned} c_{11} &= a_{11}b_{11} + a_{12}b_{21} \\ c_{12} &= a_{11}b_{12} + a_{12}b_{22} \\ c_{21} &= a_{21}b_{11} + a_{22}b_{21} \\ c_{22} &= a_{21}b_{12} + a_{22}b_{22}. \end{aligned}$$

Thus the task is to compute these four bilinear forms in $a_{ij}, b_{k,l}$. Using this definition we can compute them with 8 multiplications and 4 additions. Strassen found the following algorithm.

$$\begin{aligned} d_1 &= (a_{11} + a_{22})(b_{11} + b_{22}) \\ d_2 &= (a_{21} + a_{22})b_{11} \\ d_3 &= a_{11}(b_{12} - b_{22}) \\ d_4 &= a_{22}(-b_{11} + b_{21}) \\ d_5 &= (a_{11} + a_{12})b_{22} \\ d_6 &= (-a_{11} + a_{21})(b_{11} + b_{12}) \\ d_7 &= (a_{12} - a_{22})(b_{21} + b_{22}) \\ c_{11} &= d_1 + d_4 - d_5 + d_7 \\ c_{21} &= d_2 + d_4 \\ c_{12} &= d_3 + d_5 \\ c_{22} &= d_1 + d_3 - d_2 + d_6. \end{aligned} \tag{5.2}$$

This enables one to compute the product with 7 multiplications and 18 additions. As it is often the case, it turned out that what he learned in the special case could be widely generalized. Namely:

1. Although Strassen's algorithm uses 25 operations whereas the definition gives only 12, the important thing is that the number of multiplications has been reduced. This is crucial in the recursive application of this formula, which enabled him to show a better *asymptotic* upper bound on the number of operations when the dimension of matrices goes to infinity.
2. The algorithm works not only in the two element field; it works generally in every field (in fact, it also works in rings, which is needed for recursive applications to higher-dimensional matrices).

It is not difficult to explain the generalization to matrices of any dimension. An important ring is the ring of $n \times n$ matrices. One can apply the above formula to this ring as follows. Take two $2n \times 2n$ matrices A and B and divide them into $n \times n$ blocks A_{ij}, B_{ij}, $i, j = 1, 2$. It is not difficult to see that the product AB can be computed using the decomposition into blocks, as if we had two 2×2 matrices with entries A_{ij}, B_{ij}, $i, j = 1, 2$. In this computation the operations of additions and multiplication are the addition and multiplication in the ring of $n \times n$ matrices. Hence using formulas (5.2) we can reduce the computation of the product AB to 7 products and 18 additions of matrices of twice smaller dimensions. Thus if we can multiply $n \times n$ matrices using $\mu(n)$ multiplications and $\alpha(n)$ additions, we can multiply $2n \times 2n$ matrices using $7\mu(n)$ multiplications and $7\alpha(n) + 18(2n)^2$ additions.

If n is a power of 2, $n = 2^k$, then the recursive application of this reduction yields an algorithm with at most $c \cdot 7^k$ multiplications and additions, for some constant c. This is asymptotically $n^{\log_2 7} = n^{2.8073\dots}$, which is better than the asymptotic complexity of the defining formula, which is n^3. The current best algorithm gives $n^{2.3727}$.

The lesson from these results is that our initial intuition about complexity may be very poor. The defining formulas of the matrix multiplication are esthetically much more pleasing than algorithms such as Strassen's, so we tend to conjecture that they should be the optimal ones. But the "natural" way of computing a certain problem may be far from the optimal one.

The experts on matrix multiplication conjecture that it should be possible to reduce the exponent in the upper bound arbitrarily close to 2. Still it seems unlikely that the number of operations needed for matrix multiplication is linear in n^2, more likely it is $cn^2 \log n$ or higher, but we are not able to prove it (notice that here n^2 is the input size). As in Boolean circuit complexity, also in algebraic circuit complexity we are not able to prove nonlinear lower bounds on *any* explicitly defined algebraic function.

Notes

1. *Time constructible functions.* Time constructible functions are a special concept that is needed for the Time Hierarchy Theorem. A function f defined on natural

numbers is *time constructible* if there exists a Turing machine M which for every n, stops after exactly $f(n)$ steps on every input word of length n.

Here is why we need such functions. When we diagonalize the sets computable in time $f(x)$, we have to consider all Turing machines, since it is undecidable if a machine stops within such a limit. Therefore we have to truncate the simulated computation after $f(n)$ steps. If $f(x)$ were not time constructible, we could not do it.

All the functions that naturally appear as time bounds are time constructible. However, for some artificially defined functions the Time Hierarchy Theorem fails.

Everything can be very easily adapted to space, so I will not discuss space constructible functions.

2. *The complexity of multiplication.* In order to add two n-bit numbers, we need, for every position i, to add the ith bit and a carry, which requires a constant number of bit operations. Hence the asymptotic time complexity of addition is linear: cn for some constant c.

   The school algorithm for multiplication of two n-bit numbers is based on computing the table of the products of the ith bit of the first number and jth bit of the second one, for all $i, j = 1, \dots, n$, then one adds diagonals of the table. Thus we get a quadratic bound, dn^2 for some constant d. In 1971 A. Schönhage and V. Strassen found an algorithm that needs time only $d'n \log n \log\log n$, for some constant d', [257]. This algorithm is not used in computer hardware, as d' is fairly large and $d'n \log n \log\log n$ beats dn^2 only for fairly large numbers, but it is useful in experimental mathematics and cryptography.
3. *Factoring numbers and testing primality.* Deciding if a given number is prime and finding its nontrivial factors are closely related problems, but it seems that they have different complexities. We know that it is possible to decide in polynomial time if a number is prime, which is formally expressed by $\mathit{Primes} \in \mathbf{P}$, but most researchers believe that finding factors is much harder.

   For testing primality, several probabilistic polynomial time algorithms were found in the 1970s. More recently, in 2001, M. Agrawal, N. Kayal and N. Saxena found a deterministic polynomial time algorithm [2].

   The fastest factoring algorithms are probabilistic. The *number field sieve algorithm* seems to run in time bounded by

$$e^{cn^{1/3}(\log n)^{2/3}},$$

   where n is the number of digits of the number to be factored and c is a constant. Thus it runs approximately in time exponential in the third root of the length of the number, which is much more than polynomial. This bound has not been proved formally; it is only based on a heuristic argument. The best bound proved formally is only exponential in the second root

$$e^{(1+o(1))(n \log n)^{1/2}},$$

   where $o(1)$ denotes a function such that $o(1) \to 0$ as $n \to \infty$. See [156] for a presentation of these algorithms.

4. *Search problems.* We started the informal description of the **P** vs. **NP** problem with search problems, whereas the definition speaks only about decision problems. It is not difficult to prove that if $\mathbf{P} = \mathbf{NP}$, then these problems also have polynomial time algorithms. I will show it with the example of the integer factoring problem. Consider the following decision problem derived from integer factoring: for a given number N and a string of zeros and ones w, decide if there exists a proper divisor M of N whose binary representation starts with the string w. This is clearly a decision problem and it is in **NP**, since we can guess such a divisor and easily verify the correctness. If it were $\mathbf{P} = \mathbf{NP}$ then we would have a polynomial time algorithm A for this problem and we could apply it to factor a composite number N as follows. First use A to determine the first bit of a proper divisor of N. When we already know that there exists a proper divisor that starts with w, we can determine if there exists a proper divisor starting with $0w$ or $1w$ by a single application of A. Hence we can find a divisor by repeating this k times, where k denotes the number of binary digits of N.

   Say that a binary relation $B(x, y)$ defines an **NP** search problem, if $B \in \mathbf{NP}$ and for some b, for every x and y, if $B(x, y)$, then $|y| \leq |x|^b$. It is easy to generalize the example above and prove that $\mathbf{P} = \mathbf{NP}$ if and only if **NP** search problems possess polynomial time algorithms.
5. *Sets, languages and Boolean functions.* We often speak about the complexity of sets, but to be quite precise we should consider only sets of strings in a finite alphabet. Recall that in computer science we call such sets *languages*. So for example, when I spoke about the complexity of the set of all Hamiltonian graphs, I implicitly assumed that we take a natural encoding of all graphs by strings in a finite alphabet. It is difficult to define what are natural encodings. Intuitively an encoding should be compact, i.e., should not use excessively more bits than needed and should not contain information about the encoded entity that is not easily computable. For example, an encoding of graphs in which the first bits determines whether the graph is Hamiltonian is not natural. Apparently the only way to do it formally is to define one particular encoding and say that a natural encoding is an encoding that is equivalent to a fixed one in the sense that one can be computed from the other in polynomial time. In particular, we encode numbers by their binary representation and graphs by their incidence matrices.

   If we want to study the relation between languages and Boolean function, it is best to focus on languages in the two element alphabet $\{0, 1\}$. For a language L, we consider its sections $L \cap \{0, 1\}^n$ for $n = 0, 1, 2, \ldots$. Each such section is associated with the Boolean function $f_n : \{0, 1\}^n \to \{0, 1\}$ defined by

$$f_n(x) = 1 \quad \text{if and only if} \quad x \in L,$$

   for $x \in \{0, 1\}^n$. Thus L defines uniquely the sequence of Boolean functions $\{f_n\}$, and *vice versa* any sequence of Boolean functions $\{f_n\}$ such that $f_n : \{0, 1\}^n \to \{0, 1\}$ defines uniquely a language L in the alphabet $\{0, 1\}$.

   In the subsection about circuits I mentioned that Turing machine computations can be simulated by circuits whose size is not much larger than the time complexity of the machine. A consequence of that can be stated formally as follows.

Theorem 38 [48, 182] *If $L \in \mathbf{P}$, then for the corresponding sequence of Boolean functions $\{f_n\}$ there exists a sequence of circuits $\{C_n\}$ such that C_n computes f_n and the sizes of C_n are bounded by a polynomial in n.*

One can prove a quantitatively more precise theorem saying that if the time complexity of L is bounded by a function $t(x)$, then the size of the circuits can be bounded by $c \cdot t(n) \log t(n)$ for some constant c depending on L.

6. **NP**-*completeness.* A set A is *polynomially reducible* to set B if there exists a function f computable in polynomial time such that for every x,

$$x \in A \quad \text{if and only if} \quad f(x) \in B.$$

A set B is **NP**-*hard* (intuitively, at least as hard as every set in **NP**) if every set $A \in \mathbf{NP}$ is polynomially reducible to B.

A set B is **NP**-*complete*, if it is in **NP** and it is **NP**-hard.

Call a circuit C satisfiable, if there exists an assignment to the input variables a such that $C(a) = 1$.

Theorem 39 *The set of all satisfiable circuits is an* **NP**-*complete set.*

Proof-sketch The fact that the set is in **NP** is easy: we can guess the satisfying assignment and check it in polynomial time by simulating the computation of C.

Now let A be an **NP** set. Suppose A is defined by the formula (5.1) on page 373. Consider an inputs x of length n. The witnesses are bounded by $p(n)$ for some polynomial p. Without loss of generality, we can assume that all have the same length $m \leq P(n)$. Let M be a Turing machine computing the relation R in polynomial time. By transforming M into a sequence of circuits we get a circuit $C(x, y)$ with x representing n input bits and y representing y input bits, with the following property. If u has length n, then $u \in A$ if and only if there exists w such that $C(u, w) = 1$. Given u, let $C_u(y)$ be the circuit C with the inputs x fixed to u. Then we have $u \in A$ if and only if there exists w such that $C_u(w) = 1$, which is by definition if and only if C_u is satisfiable. Thus the mapping $u \mapsto C_u$ reduces the question if $u \in A$ to the question if C_u is satisfiable. One can check that construction of C_u can be done in polynomial time, hence the function is a polynomial reduction. This finishes the proof. □

This is the basic **NP**-complete set from which the **NP**-completeness of all other known sets has been derived. When we have one **NP**-complete set, we can prove **NP** hardness by defining polynomial time reductions from this set. This is conceptually much simpler, since we only need to find an algorithm. Nevertheless, many polynomial reductions are quite difficult to define and to prove their correctness.

In the same way one can define completeness for other classes. We know, for example, that there are **PSPACE**-complete problems and **EXP**-complete problems.

7. *The Polynomial Hierarchy.* We have already met the Arithmetical Hierarchy (see page 141). Let us recall that it is a hierarchy of arithmetically defined subsets of natural numbers defined as follows. The level Σ_n are the sets definable by formulas with n alternations between existential and universal quantifiers, where the formula starts with an existential quantifier. The level Π_n is defined in the same way, except that we require the defining formulas to start with a universal quantifier.

The Polynomial Hierarchy is a feasible version of the arithmetical hierarchy where we only allow quantifications limited to finite domains. The finite domains consist of numbers, or of strings of polynomial length. The levels of the Polynomial Hierarchy are denoted by Σ_n^p and Π_n^p (the superscript p standing for 'polynomial').

Let us consider a couple of examples.

1. The class Σ_1^p is the familiar class **NP**. We know that a set $A \in \mathbf{NP}$ can be defined by a formula of the following form.

$$x \in A \textit{ if and only if } \exists y(|y| \leq p(|x|) \wedge B(x, y)).$$

In this formula, $|\cdots|$ denotes the length and p denotes some polynomial, so the first part says that the length of y is polynomially bounded by the length of x. The binary relation $B(x, y)$ is assumed to be computable in polynomial time.

2. The class Π_1^b is defined by formulas of the form

$$x \in A \textit{ if and only if } \forall y(|y| \leq p(|x|) \rightarrow B(x, y)).$$

This class is dual to the class **NP**, in the sense that it is the class of complements of sets in **NP**. Therefore, it is also denoted by **coNP**.

3. The class Σ_2^p is the class of all sets A that can be defined as follows.

$$x \in A \textit{ if and only if } (\exists y_1, |y_1| \leq p_1(|x|))(\forall y_2, |y_2| \leq |p_2(x)|)C(x, y_1, y_2),$$

where p_1 and p_2 are polynomials, and C is a ternary relation computable in polynomial time. Here I am using a shorter notation for bounded quantifiers.

An example of a class in Σ_2^p is the set of all pairs (G, k) such that the clique number of G *equals to* k. To define such pairs we must say that *there exists* a clique of size k and *for every* subset X of vertices of size $k + 1$, X is not a clique in G. (Since we can switch the quantifiers in this example, the set is also in Π_2^p.)

If we view **NP** as an extension of **P** by adding one existential quantifier, the Polynomial Hierarchy is the hierarchy obtained by further extensions of these classes by adding more quantifiers. It has been conjectured that all the classes **P**, Σ_n^p and Π_n^p, for $n = 1, 2, \ldots,$ are distinct, that is, the Polynomial Hierarchy is a proper hierarchy. One can show that if $\Sigma_n^p = \Pi_n^p$, then all the classes with higher indices collapse to Σ_n^p.

One reason for believing that the Polynomial Hierarchy does not collapse is that the analogous relations do in fact hold for the class of recursive sets **Rec** and the levels Σ_n and Π_n of the Arithmetical Hierarchy. This is a rather weak argument, because some analogies apparently fail. While $\mathbf{Rec} = \Sigma_1 \cap \Pi_1$, we conjecture that **P** is a proper subclass of **NP** $\cap$ **coNP**. If **P** $=$ **NP** $\cap$ **coNP** were true, one would be able to invert any polynomial time computable length preserving bijection.

Several conjectures can be reduced to the conjecture that the Polynomial Hierarchy does not collapse. In particular, if **NP** $\subseteq$ **nonuniform-P**, then $\Sigma_2^p = \Pi_2^p$. Hence, if Polynomial Hierarchy does not collapse, then **NP** $\nsubseteq$ **nonuniform-P**.

8. *Nondeterministic space.* Using nondeterministic Turing machines we can define nondeterministic space classes. The relations between deterministic and nondeterministic space classes are different from the corresponding relations between time classes. One can prove that if a set is computable by a nondeterministic Turing machine in space $s(x)$, then it is computable by a deterministic Turing machine in space $s(x)^2$. Thus if we denote the class of sets computable in nondeterministic polynomial space by **NPSPACE**, then we have

$$\mathbf{PSPACE} = \mathbf{NPSPACE}.$$

But this does not mean that we know all about the relation of nondeterministic and deterministic space classes. In fact the general feeling is that the above equation is not the one that corresponds to **P** vs. **NP**. The essential question, which is open like all essential questions in complexity theory, is whether more than linear increase of space is needed to eliminate nondeterminism.

9. *Proving disjunctions of conjectures.* It is an interesting phenomenon that in complexity theory we are able to prove several disjunctions of statements that we conjecture to be true. I have already shown examples based on sequences of inclusion where we know that the extreme terms are distinct. But that is not the only way to prove such disjunctions. Here is one based on a different argument.

$$\mathbf{P} \neq \mathbf{NP} \quad \text{or} \quad \mathbf{EXP} \nsubseteq \mathbf{nonuniform\text{-}P}.$$

We believe that both are true, but we are not able to prove either of the two (though the second one seems much easier than the first one).

Here is the idea of the proof. We know that there are Boolean functions whose circuit complexity is exponential. Suppose we could, for every n, construct a truth table of such a function f_n of n variables in time 2^{cn} for some constant c. (As $2^{cn} = (2^n)^c$, it is in polynomial time in the size of the truth table.) Then we could define a language in **EXP** which is not in **nonuniform-P** as follows. For an input word w of length n, first compute f_n and then accept w if and only if $f_n(w) = 1$.

So we only need to show that one can get such functions if **P** $=$ **NP**. Let $S(n) = 2^{n/2}$. This function grows faster than all polynomials and one can show that there are Boolean functions whose circuit complexity is larger. Given a truth table of a function f, to decide if f has circuit complexity less than $S(n)$ is a problem in **NP**. Indeed, the input length is 2^n, a circuit of size $\leq S(n)$ can be encoded by a string of length $\leq 2^n$ and to check that such a circuit computes f we only need to evaluate it on 2^n inputs and compare it with the truth table. Now consider the problem of finding such a function. This is an **NP** search problem.

Thus if $\mathbf{P} = \mathbf{NP}$ we can find f polynomial time, which is 2^{cn}. This proves the disjunction. (In fact, the proof shows more than promised: either $\mathbf{P} = \mathbf{NP}$ or there is a language computable in time 2^{cn} whose circuit complexity is as large as $2^{n/2}$.)

10. *Some lower bounds methods for restricted classes of circuits.* In the quest for proving $\mathbf{P} \neq \mathbf{NP}$ and showing other separations of complexity classes many interesting circuit lower bounds have been proved. But all the methods introduced to date have only a restricted range of applications. They are incapable of proving nonlinear lower bounds on general Boolean circuits. So is it worth taking time to survey them? In set theory all the major problems had been widely open until the forcing method was discovered. Then everything dramatically changed and since then proving independence became a matter of routine. This may well happen in complexity theory too and then the current weak results will be forgotten as they were in set theory. Yet there are some basic ideas that may play an important role in future lower bound techniques and, after all, the main purpose of this paragraph is to show you what a lower bound proof can look like.

(i) The *method of random restrictions* was introduced independently by M. Ajtai [1] and by M. Furst, J. Saxe and M. Sipser [83] in the early 1980s. It is based on the following idea. Suppose a Boolean function f is computed by a small circuit C. Pick randomly a small subset of the input variables and assign randomly zeros and ones to the rest. What we want to achieve is that after this substitution we can reduce C to a substantially simpler circuit C'. At the same time we want the restricted function f' to be still hard. If C' is very simple, we can see that it cannot compute f', thus we get a contradiction with our assumption that f can be computed by a small circuit. One possible realization of this idea is to show that C' computes a constant function (0 or 1) whereas f' is not constant.

The reason why some circuits tend to shrink when we apply random restriction is that if we have a conjunction, then it suffices to have one of the inputs to be fixed to 0 and it becomes 0. The same is true about disjunction and the value 1. The hope is that fixing some inputs creates a chain reaction resulting in trivializing a lot of gates in the circuit.

This indeed works very well if the circuit uses conjunctions, disjunctions and negations and the number of alternations between different operations is small. But if the basis contains also the parity, this method completely fails. It is because fixing one input in $x \oplus y$, say x, does not fix the gate to a constant; the bit y remains intact or is flipped. This method also does not work if the depth of the circuits is not bounded.

(ii) The most useful is the approximation method introduced by Razborov in the mid-1980s. Let $\mathcal{F}_n$ denote the set of all n variable Boolean functions. Let $\mathcal{S}$ be a proper subset of $\mathcal{F}_n$; I will call $\mathcal{S}$ *simple functions*. The idea of the approximation method is to choose $\mathcal{S}$ so that

a. it contains the initial Boolean functions;
b. the Boolean operations used by circuits can be "well approximated" by some operations on the set $\mathcal{S}$;

c. the function f for which we are proving a lower bound has only "poor approximations" in $\mathcal{S}$.

Then one can prove a lower bound as follows. Let C be a circuit for f. We can use the same circuit to compute in the domain $\mathcal{S}$ using the approximate operations. The initial functions are the same and at each step of computation the error of the approximation changes very little. Then there must be many steps in the computation because the output functions of the computation in $\mathcal{S}$ approximates f poorly.

To describe the method in more detail, we need some more notation. The *initial functions* are the functions computed at initial nodes of circuits. If an initial node is labeled by a variable x_i, it is the function that outputs the ith bit of the input string. We can use other simple functions as the initial ones; one often uses $\neg x_i$. For every operation o from the basis of operations K that the circuit uses, we have an operation $\bar{o}$ defined on $\mathcal{S}$. For the sake of simplicity we will assume that K contains only binary operations. A subtle point is how to measure how good an approximation is. For $f \in \mathcal{F}_n$ and $g \in \mathcal{S}$, we define the error

$$\delta(f, g) = \{x;\ f(x) \neq g(x)\}.$$

So the error is the *set* of input assignments for which the two functions disagree. For every pair of functions $g_1, g_2 \in \mathcal{S}$ and every operation $o \in K$, we define

$$\delta_o(g_1, g_2) = \delta(g_1 o g_2, g_1 \bar{o} g_2),$$

the error produced when we use $\bar{o}$ instead of the operation o. Let

$$\Delta = \{\delta_o(g_1, g_2);\ o \in K, g_1, g_2 \in \mathcal{S}\},$$

the set of all errors that can occur on the operations $o \in K$ and elements of $\mathcal{S}$. Finally define

$$\rho(f) = \min\left\{t;\ \delta(f, g) \subseteq \bigcup_{i=1}^{t} \delta_i \text{ for some } \delta_1, \dots, \delta_t \in \Delta\right\},$$

the *distance of f from $\mathcal{S}$*. The following simple proposition is a formalization of the argument sketched above:

Proposition 10 *The circuit complexity of f is at least $\rho(f)$.*

The general framework is simple, but applications of this method require ingenious choices of the components. Also I did not present it in the most general form. Some applications require to generalize it further, but it is not difficult to imagine such extensions.

The method was first applied to circuits with the basis consisting of $\wedge, \vee$. Such circuits compute only some Boolean functions, namely, they compute all *monotone* Boolean functions. A function is monotone if $x_1 \leq y_1, \dots, x_n \leq y_n$ implies $f(x_1, \dots, x_n) \leq f(y_1, \dots, y_n)$. Razborov first proved a superpolynomial lower bound on an **NP**-complete function. It was very exciting, but soon he published another paper in which he proved such a bound also for a function in **P**. This proved that monotone circuits (circuits with only $\wedge, \vee$) are weaker than general circuits. So it is necessary to use a different proof.

The status of this method is that although theoretically it is a universal method by which it is possible to prove exponential lower bounds on the size of general circuits, such proofs must be rather unnatural. The natural way to apply the method is the following. Let e be the minimum of the cardinalities of $\delta(f, g)$ over all $g \in \mathcal{S}$, let d be the maximum over all cardinalities of δ over all $\delta \in \Delta$. Then e/d is clearly a lower bound on $\rho(f)$, hence also a lower bound on the circuit complexity. But Razborov showed that, unlike in the monotone case, d is always large if we have a complete basis. Thus applying the method in this way we even cannot get a nonlinear lower bound.

(iii) If we only want to prove more than linear lower bounds, the variety of methods is even larger. One of the methods that was used to prove nonlinear lower bounds for a class of circuits uses the following idea. Since we know that there are hard functions, we can find one such function with a small number of variables by brute force. Then we will define our hard function by taking several independent copies of the small hard function. In typical implementation of this idea the first $n/2$ input bits are interpreted as a truth table of a Boolean function of $\log(n/2)$ variables. The remaining $n/2$ bits are split into blocks of size $\log(n/2)$ and we compute the function defined by the first $n/2$ on every block. Among the assignments to the first $n/2$ there are surely those that are the truth tables of the hardest $\log(n/2)$ variable functions.

The reason why this does not work for general circuits is rather counterintuitive (I mentioned this "paradox" on page 42). For this method to work, we need to show that when computing independent copies of the same function the complexity adds up, but one can show that this is not always true. There are Boolean functions $f(x)$ such that computing simultaneously $f(x)$ and $f(y)$ for two independent inputs x and y needs only a little bit more than computing $f(x)$ for a single input (see [295]).

(iv) The *communication complexity* method uses the following idea. Given a circuit C for a Boolean function f, cut the circuit into two parts each containing one half of the input bits and see how much information must be exchanged between these two parts. If we can prove that a lot of communication must be done in order to compute the function, we get a lower bound on the number of edges of the circuit that we cut into halves.

Formally the communication complexity of a function for a given division of the input bits is defined by means of a two player game. In this game the two players cooperate in order to compute the function. Each player knows only his half of the input bits. The communication complexity is the number of bits that they exchange in the worst case when they use the best possible strategy. Notice that we completely ignore the complexity of computing the exchanged bits; we are only interested in the amount of information exchanged.

The advantage of communication complexity is that it eliminates circuits and reduces lower bounds to a combinatorial property. Unfortunately the combinatorial problems that one needs to solve in order to prove interesting separations of complexity classes are still too difficult. Before we solve the problem of explicit construction of Ramsey graphs we have little chance to solve these problems, which have the same nature, but are more difficult.

Furthermore, also in communication complexity counterintuitive things happen, as exemplified on page 42.

11. *Probabilistic proofs in finite combinatorics and explicit constructions.* We will start with Erdős's lower bound on the Ramsey numbers [67], which is a prototype of all probabilistic proofs in finite combinatorics.

Consider graphs on R vertices. The number of all graphs is $2^{\binom{R}{2}}$ because there are $\binom{R}{2}$ pairs of vertices and for every pair, we have two possibilities. Let X be a subset of vertices of size n. Then the number of graphs in which X is monochromatic is $2^{\binom{R}{2}-\binom{n}{2}+1}$ because on X we have only two possibilities, either to take all edges, or none. Since the number of subsets of size n is $\binom{R}{n}$, we can estimate from above the number of graphs that contain a monochromatic subset of size n by

$$\binom{R}{n}2^{\binom{R}{2}-\binom{n}{2}+1}.$$

Hence if

$$2^{\binom{R}{2}} > \binom{R}{n}2^{\binom{R}{2}-\binom{n}{2}+1},$$

then there exists a graph on R vertices with no monochromatic subset of size n. Using the Stirling formula, one can show that the last inequality is implied by

$$R > 2^{(n-1)/2}n/e.$$

Hence if this inequality is satisfied, such a graph exists.

I will now describe two results in which explicit constructions improved the previous bounds obtained by nonconstructive means. The first one is closely related to the Ramsey number problem. The problem is: given numbers n, s and t, how many edges can a graph on n vertices have without containing two disjoint sets S and T of sizes s and t with all the edges between S and T. This is a typical problem from the *extremal graph theory*. In this branch of graph theory we often ask how many vertices or edges can a graph have without containing certain prohibited configurations. In the Ramsey number problem the prohibited configurations are complete and empty graphs of a given size.

It was proved that for constants $s \le t$ such graphs cannot have more than $cn^{2-1/s}$ edges, for some constant c depending on s and t. Using a probabilistic argument, it was also shown that there exist such graphs with $c'n^{2-2/s}$ edges, for some constant $c' > 0$. This is the best one can get by the probabilistic method, since for random graphs the true value is around $n^{2-2/s}$. Much later Kollár, Rónyai and Szabó found an explicit construction for $t = s! + 1$ which have asymptotically $n^{2-1/s}$ vertices [159]. So they were not only able to give an explicit construction for these values of s and t, but their construction also matches the upper bound up to a multiplicative constant. Their proof is highly nontrivial and uses results from algebraic geometry. (The extremal problem for $t = s$ is still open.)

The second example is from a different field, the theory of error correcting codes. A *code* C is, by definition, a subset of strings of length n with elements in a finite set A; n is the *length of the code* and A is the *alphabet*. The elements of C are called *codewords*. If M is the number of codewords of C, then we can use C to code at most M different messages. The purpose of encoding is to introduce some redundancy which can be used to recover corrupted messages. It would take us too far afield to explain how this works, so I will introduce only the concepts necessary to describe the mathematical result.

The main two parameters of a code are the *rate* and the *minimal distance*. The rate is $\rho(C) = \log_q M$, where q denotes the size of the alphabet. The distance of two codewords is the number of positions in the strings in which they differ. The relative minimal distance $\delta(C)$ is the minimal distance d of pairs of distinct codewords divided by n. The basic observation is that if an arbitrary word w has distance less than $n\delta(C)/2$ from some codeword u, then u is the unique codeword with this property. This enables us to uniquely decode messages in which less than $n\delta(C)/2$ letters are wrong.

It is desirable to have both the rate and the relative minimal distance as large as possible because a large rate enables one to send more messages and a large relative minimal distance is good for correcting large errors. But there is a tradeoff between the two parameters: if one is large the other cannot be. A central problems of this field is to determine this dependence precisely. More specifically, we are interested in the maximal rate that can be achieved for large code lengths as the function of the relative minimal distance.

To prove a lower bound on this function means to prove the existence of codes with such parameters. A classical result, the *Gilbert-Varshamov lower bound*, is proved nonconstructively and, again, it is a fairly easy counting argument. We can interpret it also as giving an estimate on *random codes*. Several upper bounds have been proved, but they do not match the lower bounds in any part of the range of parameters (except for the extreme points 0 and 1) for any alphabet size. In 1981 V.D. Goppa introduced *algebraic geometry codes* and a little later M.A. Tsfasman, S.G. Vlăduţ and T. Zink constructed certain algebraic geometric codes that beat the Gilbert-Varshamov bound for certain alphabet sizes [292]. They used an advanced part of algebraic geometry, the theory of modular curves. Different constructions were found later, but they also use a fair amount of theory.

So this is another example of proving the existence of structures with parameters better than the parameters of random ones. In both cases this was achieved by explicit constructions. In both case, however the problems are not completely solved. The most tantalizing question about codes is if it is possible to break the Gilbert-Varshamov bound for the alphabet size two (the smallest size for which this is known is 49).

12. *Algebraic complexity classes.* Algorithms such as Strassen's matrix multiplication can be presented as *algebraic circuits.* They are a natural model of nonuniform algebraic complexity. It is interesting that there is also a natural model of uniform algebraic complexity. This approach was pioneered by L. Blum, M. Shub and S. Smale [25]. In fact, they define a computation model not only for algebraic structures (such as fields and rings), but for any (first order) structure. Let $\mathbf{A} = (A; F_1, \dots, F_n, R_1, \dots, R_n)$ be a structure, where A is the universe, $F_1, \dots, F_n$ are some functions and $R_1, \dots, R_m$ are some relations defined on the set A. We can associate a class of machines with $\mathbf{A}$ as follows. The machines are like Turing machines that work with elements of A instead of a finite set of symbols. A machine has a finite number of registers that can store elements of A and a tape for writing and reading elements of A. The program that controls the machine makes decisions using the relations of $\mathbf{A}$ and computes new elements using the functions of $\mathbf{A}$. When the universe of the structure A is infinite, some nontrivial algorithms can be performed even without the tape.

For example, the machines that one obtains from the structure $(\mathbb{R}; +, \cdot, <)$ can be used to formalize several basic algorithms working with real numbers. The difference between this model and Turing machines is that a Turing machine uses only a finite number of symbols, hence we can only approximate real numbers. In the Blum-Shub-Smale model, one can add, multiply and compare real numbers with infinite precision. The input and output data for such a machine are finite lists of real numbers. Thus the machine computes a function from a finite Cartesian product of real numbers into another finite Cartesian product of real numbers (provided that the machine always halts).

The Blum-Shub-Smale model is also a proper generalization of the original Turing machine. We get the original Turing machine if we take a sufficiently complex finite structure, e.g., the two-element field.

Furthermore, for every structure $\mathbf{A}$, one can define the complexity classes $\mathbf{P}_\mathbf{A}$ and $\mathbf{NP}_\mathbf{A}$ that correspond to the classical $\mathbf{P}$ and $\mathbf{NP}$. The question whether $\mathbf{P}_\mathbf{A} = \mathbf{NP}_\mathbf{A}$ has been resolved only for some simple structures. For the most interesting structures, $\mathbb{R}$ and $\mathbb{C}$, it is still open. It is interesting that these versions of the $\mathbf{P}$ vs. $\mathbf{NP}$ problem seem to be very much related to some classical problems in number theory. Blum, Cucker, Shub and Smale stated the following conjecture [26].

Conjecture 1 *Let $f(x)$ be a nonzero polynomial in one variable x with integral coefficients and suppose that it can be computed by an algebraic circuit of size t. Then the number of integral zeros of $f(x)$ is at most $(t+1)^c$, where c is a universal constant.*

In other words, the conjecture says that the number of integral zeros is polynomially bounded by the circuit complexity of the polynomial function. They proved that the conjecture implies $\mathbf{P}_\mathbb{C} \neq \mathbf{NP}_\mathbb{C}$.

Mentioning circuits in the conjecture gives the impression that it is more related to computational complexity than classical problems in number theory.

But there are indications that the relation to number theory is much tighter than it may seem at first glance. One of such results is due to Qi Cheng [40]. Cheng considered a related conjecture and proved that it implies a classical deep result, the *Strong Torsion Theorem* for elliptic curves qi-cheng. Presently we only have circumstantial evidence that such versions of the **P** vs. **NP** problem must be difficult. We have implications, but we do not have a proof of an equivalence with a difficult number-theoretical problem. Thus it may still happen that the problem will turn out to be easy. Most researchers believe, however, that both the original **P** vs. **NP** problem and the algebraic versions are difficult.

13. *Geometric complexity theory.* In the late 1970s, L. Valiant proposed another algebraic version of the **P** vs. **NP** problem. He showed that his version of the problem can be reduced to a conjecture that can be stated in purely algebraic terms [297]. If the conjecture is true, then Valiant's versions of **P** and **NP** are different.

    Let Det_n and Per_n denote the determinant, respectively, the permanent, of $n \times n$ matrices, where we view Det_n and Per_n as homogeneous polynomials in n^2 variables.

Conjecture 2 (Valiant) *There exists no polynomial p such that for all m,*

$$Per_n(y_{ij}) = Det_{p(m)}\big(\sigma(x_{ij})\big),$$

where σ is a projection.

A projection is a substitution of variables and constants. This problem is still open; the best lower bound on $p(m)$ is only quadratic.

More recently, K. Mulmuley proposed an approach to Valiant's conjecture which in principle could also lead to the proof of $\mathbf{P} \neq \mathbf{NP}$ in the original formulation [203]. He calls his approach *geometric complexity theory*. The basic idea is to replace projections by invertible linear mappings. These mappings form an important group, the *general linear group* GL_n. Thus substitutions are replaced by actions of the elements of this group on the polynomial Det_n and one can use group representation theory.

In order to transform the problem in this way, one has to make two modifications. First we need polynomials of the same dimension. To this end, it is sufficient to replace Per_m by $z^{n-m}Per_m$, where $n = p(m)$. The second obstacle is that, although projections are linear, they are not invertible in general. This is solved, roughly speaking, by computing the permanent only approximately. In terms of representation theory it means that we ask whether $z^{n-m}Per_m$ is in the closure of the orbit (with respect to GL_n) of Det_n. Thus the following conjecture implies Valiant's conjecture.

Conjecture 3 (Mulmuley) *There exists no polynomial p such that for all m and $n = p(m)$, $z^{n-m}Per_m$ is in the closure of the GL_n orbit of Det_n.*

This conjecture has attracted a number of mathematicians working in group representation theory and related fields. Mulmuley and other mathematicians working on this project have proved theorems that apparently go in the right direction. Unfortunately, none of the results proved so far can be interpreted as a theorem about algebraic complexity classes.

14. *The "simplest" problem in algebraic complexity theory.* One can easily prove the following simple fact.

Proposition 11 *For every $n \geq 1$, there exists a real $n \times n$ matrix A such that for every factorization $A = XY$ into a product of two $n \times n$ real matrices X and Y, the total number of nonzero elements in X and Y is at least n^2.*

The argument used in the proof (based on elementary facts from algebraic geometry) is that one cannot parameterize the variety of $n \times n$ matrices by less than n^2 numbers.

The open problem is to *define such matrices explicitly.* The problem is simple to state, but it seems very difficult to solve. To see how little we are able to prove, let me mention that the best lower bound that one can prove for explicitly defined matrices is only of the form $c \cdot n(\log n/\log\log n)^2$, for some constant $c > 0$, [87].

15. *A message encoded in π.* What if a picture of a circle is indeed encoded by the digits of π (as in Carl Sagan's novel *Contact*)? What if some other message (of a super-civilization that can alter laws of mathematics) is encoded there? Can we find it?

The answer to this hypothetical question depends on where the message is supposed to be. If it is right at the beginning, encoded by segment of 'small' digits, we will certainly notice it. We can also compute all digits of any short segment in the medium range, since the digits of π can be computed very efficiently, but we must know where the segment with the message is; we cannot search the whole medium range, or we must be extremely lucky. Thus such an experimental approach will almost certainly fail. Since the digits of π are not random numbers, it is not excluded that we can determine such a segment indirectly, using mathematical reasoning. This may be very hard, especially if we did not know the content of the message. If the message is encoded by digits that are on large number positions, we can only do it using theory; no experimental search can help.

5.2 Randomness, Interaction and Cryptography

When writing about time and space of computations I presented it as *limitations:* the computation is constrained by the fact that it has to finish in certain time and must use only given space. But one can present it also positively as using *computational resources*. When talking about randomness, it is quite natural to think about it as a resource. A randomized computation uses random numbers, or random bits, which we can visualize as coin-flips. While in the case of space, we can reuse the same memory locations several times because we can erase the information that we do not need anymore, random bits cannot be used repeatedly. Once a random bit is used, it is no more random because the data that we process may *depend* on this bit.

Randomness is one computational resource that I am going to deal with in this section; another is interaction. It is much more difficult to classify and quantify interaction, as it has many forms. Interaction is present whenever a person performs computations on a powerful computer. The person gives the data to the computer and, after the computer processes it, receives the output. This is a completely trivial example, nevertheless, already in this simple case we can ask: How does the person know that they received the right answer? Can he verify the answer, and if so, why did he have to use the computer to get the answer? etc.

This example also shows the typical assumption that one of the interacting entities has small computational power and the other has a big one. We think of these two entities as players who cooperatively perform computation in order to determine whether a given input belongs to a given set, or to determine the value of a function on a given input. The one with little computation power is usually called *Verifier;* the one with strong computation power, sometimes even unlimited, is called *Prover*. The computation is done according to some rules called the *protocol*. The best way to understand it is to imagine it as a game in which the goal of Prover is to persuade Verifier about a correct answer. We assume that Prover may cheat and the protocol must be designed so that Verifier can detect the lie. (Therefore Prover must cooperate.)

The complexity class **NP** can be presented as an extension of **P** by a basic form of interaction. The idea is to interpret what I called 'guessing' by an action of Prover. Recall that a set A is in **NP**, if there is an associated binary relation R such that x is in A if and only if there is a y such that $R(x, y)$. Thus we can compute A using the following simple protocol:

1. Verifier and Prover get the input string x;
2. Prover produces a string y;
3. Verifier checks if $R(x, y)$;
4. if so Verifier declares x to belong to A.

To get **NP**, one has also to specify that the string y can be only polynomially longer than x and that Verifier is, in fact, a polynomial time algorithm. On the other hand, we do not limit Prover in any way.

This is, of course, a trivial reformulation of the definition of **NP**, but viewing it from this perspective, generalizations of this concept come immediately to mind. For instance, what happens if we allow more rounds of interaction? Well, if we just allow more rounds in the protocol above, nothing happens; we get **NP** again. The reason is simple: since Verifier is a deterministic algorithm, Prover can compute all communication before it starts. Thus instead of communicating in rounds it can send all its answers at once. If, however, Verifier can toss coins, Prover cannot predict the communication anymore and then things become much more interesting.

This is just one case that shows that randomness is important in interactive computations. Another area of communication protocols where randomness is essential is cryptography—secure transmission of data. One of the basic tasks in cryptography is to choose keys that cannot be predicted. Without randomness it is impossible.

Cryptography is the field of applied science that is most tightly connected with complexity theory. Users of an ordinary program are happy with the information that the program is based on the fastest known algorithm for the given task. They do not care whether or not it has been *proved* that no better algorithm exists. In contrast to this, the information that so far nobody has broken the protocol is not satisfactory for users of a cryptographic protocol. They would very much like to have a *mathematical proof* that the protocol is secure. This amounts to proving that the task of breaking the protocol is computationally infeasible, which means that it requires so much time that nobody can solve it. Unfortunately, this is what theory is still not able to provide them with. Still, theory is very useful for cryptography: we cannot prove security, but we can determine reasonable assumptions which imply it. The most common of these assumptions is the conjecture that integer factoring is a computationally hard problem.

Theoretical cryptography is an extremely interesting field, but this is not the main reason for including it in this book. It turns out that the concepts and results of cryptography are also very important for complexity theory itself. In particular they are relevant for the question whether randomness in computations is helpful.

How Can Randomness Be Helpful?

We view computers as instruments that are able to achieve formidable results due to their high speed and extreme precision. Computers are indeed very precise; if programmed correctly, they almost never err. Randomness seems to be in conflict with precision. We associate randomness with disorganized behavior which is apparently not good for solving hard problems. So, how can one make any good use of it?

One explanation is that randomness enables us to easily construct complex structures. Consider for example a finite random graph. It is the result of a random process, where for each pair of vertices, we decide randomly (we flip a coin) whether or not they are connected by an edge. The resulting graph has no regular structure; it is complex. Such a structure may have properties that we are not able to ensure using a deterministic algorithm. The structure may be a solution to our problem, or it may be a tool that we can use to solve the problem.

If we want to obtain a structure with a particular property by a random process, such structures must be abundant, otherwise the probability of finding one would be small and we can find it only by many trials. Fortunately, quite often the structure that we need to find occurs with high probability.

An example of this are quadratic nonresidues. Let $p > 2$ be a prime. An integer r, $0 < r < p$, is called a *quadratic residue modulo p* if it is the remainder of a square (a number of the form x^2) divided by p. Formally, it means that the equation

$$x^2 \equiv r \mod p$$

has a solution. An integer n, $0 < n < p$, that does not satisfy this condition is called a *quadratic nonresidue modulo* p. It is an easy exercise to prove that half of the numbers between 0 and p are quadratic residues and half are quadratic nonresidues. It is very easy to find residues: clearly, $1^2 \equiv 1 \mod p$ is a residue, and we can generate others by taking an arbitrary number x and computing the remainder of x^2 divided by p. To find a quadratic nonresidue is more difficult. There is an algorithm that given a number y, $0 < y < p$ decides in polynomial time whether or not y is a quadratic residue. The algorithm is based on computing the *Legendre-Jacobi symbol*

$$\left(\frac{y}{p}\right),$$

which, for p prime, is equal to 1 if y is a quadratic residue and -1 if it is a quadratic nonresidue. Thus we have a simple probabilistic algorithm for finding a quadratic nonresidue.

1. `choose randomly` y, $1 < y < p$;
2. `compute` $(\frac{y}{p})$;
3. `if` $(\frac{y}{p}) = -1$ `output` y, `otherwise go to 1.`

This algorithm finds a quadratic nonresidue in the first round with probability $1/2$; in general, the probability that it will need more than m rounds is $1/2^m$. So the algorithm finds a quadratic nonresidue after a few rounds with very high probability.

What about finding a quadratic nonresidue by a deterministic algorithm? (I am now using the word 'deterministic' to stress that the algorithm does not use randomness.) A straightforward adaptation of the above algorithm is to replace the random choice of y by the systematic search $y = 2, 3, 4, \ldots$. In this way we certainly find a quadratic nonresidue; the question is only when. If p is of medium size, we cannot search all numbers between 1 and p; there are too many of them. If we use the concept of polynomial time, instead of small, medium and large numbers, then an algorithm that searches all numbers up to p is not a polynomial time algorithm. As the length of the input is $\log p$, a polynomial time algorithm must run in time polynomial in $\log p$. So for this algorithm to be polynomial time, we would need to prove that there is always a quadratic nonresidue below $f(\log p)$, where $f(x)$ is some polynomial. It is quite likely that this is true, but there is no proof of it. We only know that a certain conjecture, the *Extended Riemann Hypothesis*, implies the stronger statement that there exists a constant c such that for every prime p, there exists a quadratic nonresidue below $c \log p$. Thus if the Extended Riemann Hypothesis is true, we can find a quadratic nonresidue in polynomial time, but the hypothesis may be false. (The Extended Riemann Hypothesis is, as the name suggests, a certain generalization of the Riemann Hypothesis.)

So in spite of the fact that half of the numbers are quadratic nonresidues, to find them by a deterministic polynomial time algorithm is an open problem.

In the next example of a problem solvable by a probabilistic algorithm in polynomial time we even do not have a conjecture how to solve it deterministically. Let $s(x_1, \ldots, x_n)$ and $t(x_1, \ldots, x_n)$ be two algebraic formulas and suppose that we want to decide whether

$$s(x_1, \ldots, x_n) = t(x_1, \ldots, x_n) \tag{5.3}$$

is a true identity in integers. The standard algorithm that mathematicians use is to rewrite the expressions into a sum of monomials. Then it is an identity if and only if we get the same expressions on both sides. However, this procedure may result in an exponential blow up, so it is not a polynomial time algorithm. We can also test the identity by choosing some numbers $a_1, \dots, a_n$, evaluating both sides and checking if they give the same value. If we do not get the same value, the identity is refuted, but to prove that it is an identity we would need to use, in general, exponentially many samples. If, however, we are satisfied with showing that it is an identity with high probability, we can use an efficient randomized algorithm.

The idea of the randomized algorithm for the identity testing is based on the observation that if the equation (5.3) is not an identity, then it has a bounded number of solutions [259, 320]. The bound is exponential, but it is good enough for our purpose. It is possible to take M that is not much larger than the size of the formulas such that if we choose $a_1, \dots, a_n$ randomly from the interval $[0, M]$, the probability that the two formulas evaluate to the same number will be less than $1/2$. If we repeat the test several times and it will always pass, we will know with high probability that it is an identity.

Holographic Proofs

Holographic proofs are a nice example of how randomness can help in other situations, not just for computing. It concerns the problem of verifying the correctness of documents. Signing a contract without reading it through is very dangerous—a single sentence that you miss may have disastrous consequences for you. The same concerns mathematical proofs. A referee of a mathematical paper is supposed to check the proofs. If only one step in the proof is wrong the proof is invalid. Verifying documents and proofs is time consuming and often boring, so we wish there were a way to do it more efficiently without compromising too much precision.

There exists, indeed, a method that at least in theory can achieve this goal. The essence is to use encoding of the documents in a certain way which spreads any possible error in the document over a large part of it. Then it is possible to test the correctness by looking at a constant number of randomly chosen bits of the document. The correctness is verified, of course, with some error, but the error decreases exponentially with the number of bits tested; hence one can get very high certainty with a relatively small number of bits.

Obviously, more details are needed to explain this vague idea. First, it should be stressed that the task is not only to check that the format of the document is correct, but we also need to check that it is a correct document for a given purpose. It is best to think of it as a proof P of a theorem ϕ. Then the theorem ϕ is given and we need to check that P is a proof *of* ϕ (by reading a constant number of bits of P), not only that P is *some* proof. Another important thing one should realize is that there may be many proofs of ϕ and the constant number of bits that we get does not tell us which of them we have checked. Thus it is more correct to say that we are testing the *existence* of a proof of ϕ.

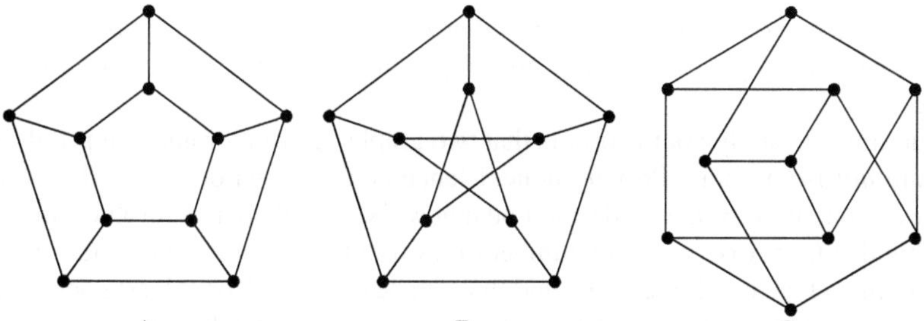

Fig. 5.4 Which graphs are isomorphic? (See page 436)

Holography is a method of making images of 3-dimensional objects. It uses usual light-sensitive films, but it needs coherent monochromatic light, the best source of which are lasers. The image is encoded in the two dimensions available on the film and needs a similar arrangement of coherent light in order to create a 3-dimensional image of the object visible to the human eye. The relation with proofs has nothing to do with dimensions. In calling the proofs holographic people refer to the fact that in a holographic photograph the details of the objects are not encoded in localized regions. So one can take a part of the photo and it will still produce the same 3-dimensional image of the whole object, only the quality will decrease. It should be pointed out that the concept of a holographic proof is more complex; it is not just a robust way of encoding. If we only needed to be able to reconstruct the information from each sufficiently large part, we could use a much simpler means, the error correcting codes.

The existence of holographic proofs is one of the most remarkable results in complexity theory. It has important applications in proving that some **NP**-complete problems not only cannot be solved in polynomial time, but it is impossible even to solve them approximately (provided that $\mathbf{P} \neq \mathbf{NP}$). The precise statement of the result is rather technical, so I defer it to the section Notes.

Interactive Proofs

Interactive protocols are used to define new complexity classes and give an alternative definition of some standard classes. In the two examples below we will see how one can compute more by combining polynomial time computations with interaction than one can do without it.

The first example concerns the *Graph Isomorphism Problem*. It is the problem to decide whether two given graphs are isomorphic. As isomorphic graphs must have the same number of vertices, we can assume that the two given graphs have the same set of vertices. The problem then is whether there is a permutation of vertices that transforms one graph onto the other (see Fig. 5.4). This is, clearly, a problem in **NP** for we can guess the permutation and check if it works. Note that we can ask the same problem about isomorphism for other kinds of structures. We take graphs as the canonical example because they are simple and other structures can be coded by them.

Our aim is to study not the proofs that two graphs are isomorphic, but proofs that they are *not* isomorphic. Proving nonexistence is usually harder, as we know, and here we want, moreover, to find a method that works for all nonisomorphic pairs. It is possible that there exists a small number of invariants that completely determine the isomorphism type of a finite graph and that can be computed, or at least guessed and checked in polynomial time. Then such sets of invariants can be considered proofs of non-isomorphism. Such sets, however are known only for restricted classes of graphs, so we have to use something else, which is interaction.

The two players that appear in the interactive protocol introduced by L. Babai [11] are called Arthur and Merlin (the king and his wizard). Arthur is a man without any supernatural abilities, thus he represents an entity that can only compute in polynomial time, but he does have an advantage: he can toss coins. Merlin, in contrast, is omnipotent, which means that his computational power is unlimited. In particular, he can easily recognize whether or not two graphs are isomorphic. Unfortunately, he cannot always be trusted (after all, he is a son of a devil) and that is the problem. Thus Arthur can ask Merlin any question, but he can learn something only if he can somehow verify Merlin's answer.

Once Merlin comes to the king with two graphs G and H and wants to persuade him that they are not isomorphic. As Arthur does not trust him, they need a schema how to do it. Here is how it works. After receiving the two graphs, Arthur sends Merlin away. He tosses a coin to choose one of them and then tosses the coin again several times to randomly permute the vertices of the chosen graph. Let the resulting graph be called F. Then Arthur calls Merlin back, presents F, and asks Merlin to tell which of the two graphs, G or H, he used to construct F. His reasoning is: if G is not isomorphic to H and as Merlin can recognize isomorphic graphs, he can easily answer this question correctly; on the other hand, if Merlin is trying to cheat him and the graphs are isomorphic, then F is a permutation of both graphs G and H, hence the probability that he answers correctly will be $1/2$. The last probability can be reduced to $(1/2)^k$ if Arthur presents k such graphs. Thus Arthur can easily and with very high probability verify that the graphs are not isomorphic.

The Graph Isomorphism Problem is a rare example of a set in **NP** which is neither known to be in **P** nor to be **NP**-complete. Although a general theorem says that if $\mathbf{P} \neq \mathbf{NP}$, then there are many such sets [176], it seems hard to find such a set that would be defined in a natural way. The protocol for non-isomorphism presented above is evidence that the Graph Isomorphism Problem is not **NP**-complete. Now we only need evidence that it is not in **P**.

Proofs that Convey no Information

A proof conveys the fact that the theorem is true. Can this be the only knowledge that it reveals? Consider a very familiar situation: you went to a meeting to discuss a problem, you talked with people, you spent an hour or more, but after the meeting you had the feeling it was completely useless—you did not learn anything that you hadn't known before. From your point of view it was "zero-knowledge" communication. But was it really useless? Maybe you do not realize it, but your opinions about the subjects discussed at the meeting have changed and maybe this was exactly the purpose of the meeting.

What we are actually interested in is a more concrete situation. Suppose you succeed in finding a solution to an important practical problem and you want to sell your solution. To this end you must persuade a potential buyer that you really have such a solution. But if you show the solution to somebody, you risk that they steal the idea without paying anything. Therefore, you need to be able to present some evidence that you have a solution, but the evidence should not reveal any details about the solution. Are such *zero-knowledge proofs* possible?

Let us start with a more basic question: what does 'zero knowledge' exactly mean? Surprisingly, this can be defined quite easily. Suppose A and B communicate by sending some strings of bits. When they finish, we say that it was zero-knowledge communication for A if all the strings that A obtained A could have computed without the help of B. This is fine, but why should A bother at all to talk with B in such a case? The point is that in spite of being able to compute all that A gets from B, A does learn something: *A learns that B is able to compute these strings.* So it is better to interpret such communication as *A testing B*. Think of B as a student taking an oral exam with professor A. Then A, of course, knows the answers to his questions; what A does not know is if B knows the answers.

The protocol for proving non-isomorphism of graphs described above is an example of a zero-knowledge protocol. In that protocol Arthur is only testing Merlin whether he indeed knows that the graphs are not isomorphic. This gives such a protocol for the *complement* of one set in **NP**. There is a general theorem saying that the complement of every set in **NP** has such a protocol. This theorem is proved by presenting a zero-knowledge protocol for the **NP**-complete problem of coloring graphs by 3 colors and then referring to the fact that other problems in **NP** are polynomially reducible to it. I will sketch the protocol.

Recall that a graph G is 3-colorable if one can assign one of three colors to every vertex so that no two vertices connected by an edge have the same color. The idea of the zero-knowledge protocol is that Merlin encodes the colors of a 3-coloring so that they can be determined only using secret keys; every vertex has a different key. After receiving this code, Arthur selects randomly two vertices that are connected by an edge and asks Merlin for the keys. Thus Arthur can verify that the two vertices have different colors. They repeat this protocol many times, but each time Merlin permutes randomly the colors and uses new keys. Thus the colors that Arthur learns in different rounds have nothing in common. Hence the only information he obtains is what he expects: that Merlin knows a proper 3-coloring.

In order to prove this theorem, one needs to have the means to produce secure encoding. The existence of such encodings has not been proved formally, but we believe that they do exist. It is the basic assumption used in cryptography.

Cryptography

Cryptography is the science of secure information transmission. It studies means to efficiently transmit information to authorized subjects while preventing an unauthorized subject from learning even small parts of the transmitted messages. This field of science studies many practical problems as well as theoretical ones. The theory of cryptography is closely related with computational complexity and can be considered part of it.

Preventing an unauthorized person from reading a message can be done by arranging physically that the message can only reach the authorized people. The essence of cryptography is, however, to arrange it so that even if an unauthorized person gets hold of the message, they should not be able to get the *information* that the message contains. There are basically two ways how to do it. One is to prevent an eavesdropper from decoding the message by hiding the information needed to decode it. This means that without some specific knowledge, a *key*, it is impossible to decode the message, or it is possible to do it only with extremely small probability. Such protocols are studied using *information theory.* The second way is to make it impossible to decode the message using the available computers and limited time. Thus the barrier is the *computational complexity* of the task that an eavesdropper has to solve. In practice one has to combine both approaches.

Thus cryptography bears a special relation to computational complexity. In all other applications, large computational complexity of a problem is viewed negatively, whereas cryptography without computationally hard problems is impossible. Furthermore, for cryptography it is very desirable to actually *prove* the hardness of particular functions. Ideally, a cryptographic protocol should always be accompanied with a proof of security, a proof that the task of breaking the code is computationally infeasible. We know that proving computationally hardness of specific Boolean functions is still beyond the reach of our mathematical means; so we are very far from this ideal. The special nature of functions needed for secure encoding makes the task even harder. We are not able to reduce these problems to the standard conjectures of computational complexity such as $\mathbf{P} \neq \mathbf{NP}$. The security of all known protocols is based on (apparently) much stronger assumptions.

As it happens so often in science, the concepts introduced in cryptography for practical reasons turned out also to be important for theory. They play an important role in the study of fundamental problems in complexity theory, for example, in the study of randomized computations. These concepts turned out to be very useful also in the more remote area of proof complexity, which is the subject of the next chapter. Let us have a look at the most important ones.

One-Way Functions

We know very well that there are things that are very easy to do but very difficult, or impossible to undo. In physics there is a law that explains this phenomenon. It is the *Second Law of Thermodynamics* that asserts that *the entropy in an isolated system can only increase with time*. In plain words, if we need to restore order in some system, we must pay for it by taking away order from another system.

In mathematics a one-way function is, roughly speaking, a function f which can be efficiently computed, but its inverse cannot. The connection with thermodynamics is not superficial. Should the computed function be one-way, we must erase information during the computation of the function and that is only possible by dissipating heat. If we erase some information, then it looks very likely that it should be difficult to reverse the computation, but in fact it may be very difficult to prove it formally. Let us try a bit more precise definition using polynomial time computability.

A possible definition is that f is one-way-function if $x \mapsto f(x)$ is computable in polynomial time, but there is no polynomial time algorithm which from a given y computes some x such that $f(x) = y$, provided such an x exists. In order to avoid trivialities, we will assume that the lengths of x and y are polynomially related—one can bound the length of x (respectively of y) by a polynomial depending on the length of y (respectively of x).

The existence of such functions follows easily from the (unproved) assumption $\mathbf{P} \neq \mathbf{NP}$. Indeed, if $\mathbf{P} \neq \mathbf{NP}$, then there is a search problem $R(x, y)$ such that finding y for a given x is not possible in polynomial time. (I am tacitly assuming the same conditions that were used to define the class **NP**: R is computable in polynomial time and the length of y is bounded by a polynomial in the length of x). Define $f(x, y) = x$ if $R(x, y)$ and $f(x, y) = 0$ otherwise. Clearly, if we were able to recover x, y from the value of the function $f(x, y)$ in polynomial time, then we also would be able to solve the search problem in polynomial time.

The definition above is not very useful. Therefore the term '*one-way function*' has been reserved to a more useful, but also more difficult concept. To understand the motivation behind this concept, we have to look at the applications of one-way functions in cryptography. In the basic model one has an encoding function F which, given a *encoding key* e and a message x called a *plaintext*, produces the encoded message $y = F(e, x)$ called the *ciphertext*. There is also a decoding function G which from the ciphertext y and the decoding key d associated to e produces the original plaintext $x = G(d, y)$. The *secret-key* systems are systems where both keys have to be secret. In such systems often the decoding and the encoding keys are the same. In *public-key* systems, the encoding key is public because it is assumed that the decoding key cannot be computed from it.

If one could use keys of the same length as the messages and each time use a new pair of keys, everything would be simple. In such a case there are simple coding and decoding functions that are provably secure. Their security is based only on information-theoretical arguments, which is good because we do not have to use any unproven assumptions, but such systems are highly impractical. Therefore the main problem in cryptography is to design encoding and decoding functions with short keys that can be reused many times.

When researching the security of a code, one has to take into account various possible scenarios. For us, it suffices to consider the most basic type of attack. This is an attempt to break the code using pairs of several plaintexts and the corresponding ciphertexts, say, $(x_1, y_1), (x_2, y_2), \ldots, (x_n, y_n)$. One can show that if these pairs are randomly chosen, then already for a small n both keys of the system are uniquely determined.[11] In practice no messages are random; they are messages in some natural language about things currently happening. Nevertheless, what is not random for us may still look random from the point of view of the coding system. In any case, we cannot exclude that the keys are determined from a small sequence of samples. Now, since the information about the keys is present in the sequence $(x_1, y_1), (x_2, y_2), \ldots, (x_n, y_n)$, the only way that the system can be resistant against this attack is if it is *computationally infeasible* to extract this information.

The task of computing the keys from samples of plaintext and ciphertext is a typical search problem. Thus one may naively propose to justify the security of a system by the commonly accepted conjecture $\mathbf{P} \neq \mathbf{NP}$. Why is it naive? Firstly, we cannot use any asymptotic statements, such as $\mathbf{P} \neq \mathbf{NP}$, to make claims about algorithms that only work with inputs up to some concrete finite size. Secondly, and this is the gist of this subsection, $\mathbf{P} \neq \mathbf{NP}$ is a conjecture about the *worst-case* complexity, which does not suffice for cryptographic security. $\mathbf{P} \neq \mathbf{NP}$ means that some X in **NP** is not in **P**. But if $X \notin \mathbf{P}$, we only know that for every polynomial algorithm A, A is wrong on at least *one* input. Indeed, for many **NP**-complete sets, which are the hardest sets in **NP**, the decision problem is simple for *random* inputs. Thus such sets are easy *on average*, while they still may be hard *in the worst case*. What is needed for good cryptographic functions is that they are hard *almost always*; this means that for every algorithm A running in short time, the probability that it solves the search problem for a random input is negligible.

Another important thing to realize is that enemies may try to use arbitrary means to break the code. Here we are only interested in the computational aspects, so the question is what kind of computations they can use. Presently we only know about one additional means that may possibly extend the class of computations in limited time—randomness. It is possible that some problems may be solvable using randomized computations, whereas every deterministic algorithm for them would run for too long. Therefore we have to assume that the attacks on the code will be done using randomized algorithms.

After this digression to cryptography the definition of one-way functions makes much more sense. It is a theoretical concept, so one uses asymptotic bounds: polynomial upper bounds on the time of the machines and bounds on the probability of the form one over a polynomial. It is worth stating the definition quite formally, since the concept is rather subtle.

Let $p_1, p_2, \ldots$ be an infinite sequence of positive reals. We will say that they are *negligible*, if there exists a function $\gamma(n)$ that grows faster than any polynomial such that $0 \leq p_n \leq 1/\gamma(n)$ for every n.

Definition 11 Let f be a function mapping binary strings into binary strings such that the lengths of x and $f(x)$ are polynomially related (as explained above). Then we say that f is a *one-way function*, if

---

[11] In general, they are only determined up to the functional equivalence; I will ignore this subtlety.

1. f is computable in polynomial time (without using randomness);
2. for every randomized Turing machine M running in polynomial time, the probability, for a random x of length n, of the following event is negligible: *given* $y = f(x)$ *as the input, M finds a pre-image of y (an* x' *such that* $f(x') = y$*).*

The complicated condition 2. formally expresses that any polynomial time randomized algorithm inverts function f with negligible probability.

It should be noted that although the motivation presented above was from cryptography, the concept is important also in other parts of complexity theory.

A Complexity Class Defined Using Randomness

When studying the relation of randomized computations to non-randomized ones (deterministic computations) it is good to have a benchmark to which one could refer. As usual in complexity theory, it should be a complexity class. Probabilistic complexity classes are almost exclusively defined using the worst-case complexity, like the non-probabilistic ones, which means that we require that a probabilistic Turing machine must use limited time or space on *each* input. Further, for *each* input we specify with which probability it should be accepted.

A probabilistic Turing machine is simply a nondeterministic Turing machine (as defined on page 375); what is different is only the way we interpret it. Whereas when defining that a nondeterministic Turing machine accepts an input we only ask if there is an accepting computation, in the case of probabilistic machines we count the probability that they accept a given input.

The best is to give an example. It is the most important probabilistic class, the *Bounded error Probabilistic Polynomial time*, **BPP**.

Definition 12 A set X is in **BPP** if there exists a probabilistic Turing machine M that stops on every input after polynomially many steps and such that for every input a,

1. if $a \in X$, then M accepts a with probability at least $2/3$;
2. if $a \notin X$ then M accepts a with probability at most $1/3$.

If we have a set X and a machine M satisfying the definition, then M can make errors on both kinds of inputs, those that are in X and those that are not in X. The machine accepts elements of X with higher probability than non-elements. It is important that there is a gap between the two probabilities. This gap enables us to show that **BPP** is very close to **P**.

The argument is very simple. Given an input a, run M on a several times, say m-times. Then accept a using "the majority vote", meaning that we accept a if at least in $m/2$ cases M accepted a. Then the claim is that we will accept every $a \in X$ with probability at least $1 - \varepsilon$ and reject every $a \notin X$ with probability at least $1 - \varepsilon$ where $\varepsilon > 0$ is exponentially small (with the exponential function only depending on m). Indeed, suppose first $a \in X$. Let p_a be the probability that M accepts a. By definition, $p_a \geq 2/3$. The expected number of times that M accepts a is $p_a m$. By the law of large numbers, the distribution of this random variable is sharply concentrated around the expected value (most likely it will be $p_a m \pm \sqrt{m}$ and the probability drops exponentially when we go outside this region). Thus, in particular, the probability that M will accept less than $m/2$-times is exponentially small. Whence the majority vote will be correct with probability exponentially close to 1. The proof for $a \notin X$ is completely symmetric.

The bottom line is that for all practical purposes **BPP** is as good as **P** because by using only a few repetitions we can make ε so small that an error will never occur in practice. Hence *the class of feasibly computable sets is not* **P**, *but* **BPP**! That said, most likely the following is true:

Conjecture **BPP** = **P**.

However, if the conjecture is true, it does not mean that randomness is completely useless. The conjecture only says that we can eliminate randomness by at most polynomially increasing the time. It is likely that some problems need, say, quadratic time if computed deterministically, but only linear time when computed probabilistically. In large scale computations done in practice this can make a big difference.

To understand the reasons why most researchers believe in this conjecture, we must learn about the ways how to eliminate randomness from computations.

Pseudorandomness—Imitation of Randomness

Consider a problem for which we only have a probabilistic polynomial time algorithm. For such a problem, there is a good chance that one can find a deterministic polynomial time algorithm because it has happened so in many cases. But the deterministic algorithms are as a rule based on different ideas than original probabilistic algorithms. For example, the recently discovered polynomial time algorithm for testing primality uses an idea different from all previously found probabilistic algorithms. So our experience does not suggest any general method of eliminating randomness. There is, however, a simple idea that potentially may work in general. It is based on the assumption that it is possible to replace random bits by bits generated deterministically in a suitable way, by *pseudorandom* bits.[12]

In fact, if you program your computer to run a probabilistic algorithm, it will do almost exactly that: it will use the current time as a random seed, but then it will compute the required random numbers deterministically. For most applications, it

[12] A related concept had been studied in the research area *algorithmic randomness* before the mathematical definition of pseudorandom generators was introduced. In that area the aim was to define and study countably infinite sequences that share properties with typical random sequences.

works well, but it cannot be used if security is at stake. The random number generators used in most implementations are very simple; they pass simple statistical tests, but a bit more sophisticated test can easily discover that the sequence of numbers is not random.

The basic question is whether pseudorandomness is a mathematical concept, i.e., whether there is a precise mathematical definition of this concept, or it is only a vague intuitive concept. If we consider sequences of n zeros and ones, then probability theory does not distinguish between them—the sequence of n zeros is as random as any other. When we talk about *one* random sequence in probability theory, we in fact refer to all sequences. Thus the first important thing to learn is that *using complexity theory one can define pseudorandomness.*

Before stating the definition, I will present some arguments that more or less uniquely lead to the definition. Let us focus on sequences of n bits. First we observe that we need at least a little bit of true randomness. If we produced a sequence of bits without any random seed, it would always be the same sequence, thus one would easily recognize that it is not a random sequence. Therefore we do not talk about pseudorandom sequences of bits, but rather pseudorandom *generators*. A pseudorandom generator produces a sequence of n bits from a sequence of m bits for some $n > m$. We say that a generator stretches m random bits to n pseudorandom bits. I will explain later how much a generator should stretch a given sequence of random bits.

The main idea of the definition of pseudorandom generators is to replace simple statistical tests by all tests that can be computed in polynomial time. Thus a test is any polynomial time algorithm that outputs zero or one. When a test is simple, we know what we should obtain when testing a pseudorandom generator. For example, if we just test how often a given subsequence of k bits occurs in the sequence produced by the generator, we know that the frequency should be $1/2^k$. But what should we do with a test that is based on an algorithm that we do not know? The solution is simple: we do not need to be able to compute the probability that the test accepts a random sequence of length n, we just test pseudorandom sequences *and random sequences*, and then we compare the frequencies that we have obtained. If we have a good pseudorandom generator, the frequencies should be very close. It is quite natural to use polynomial time algorithms as tests, but for certain technical reasons it is better to define pseudorandom generators using polynomial size Boolean circuits.

Having defined tests we can now define that P is a pseudorandom generator, if it is computable in polynomial time and it passes all tests. To pass a test defined by a circuit C means that there is a negligible difference between the probabilities with which C accepts the outputs of P and with which it accepts random sequences. Here is a slightly more formal definition.

Definition 13 A *pseudorandom generator* is a function P computable by a deterministic polynomial time Turing machine such that for every m,

1. P maps the set of zero-one strings of length m into the set of zero-one strings of length n, for some $m < n$;

2. for every polynomial size circuit C, the difference between the probability that $C(P(x)) = 1$ for a random string x of length m and the probability that $C(y) = 1$ for random string y of length n is negligible.

(The meaning of 'negligible' is the same as in Definition 11.)

Let us play with "our new toy". Let $X \in \mathbf{BPP}$ and let M be a probabilistic Turing machine that accepts X as required by the definition of **BPP**. In general, M gets random bits ('tosses a coin') during the computations, but it is not difficult to see that equivalently, M can get all random bits at the beginning, store them on the tape and then use them, one by one, as needed during the computation. Thus M is in fact a Turing machine that gets two inputs, a and the random bits. Suppose $a \in X$ and M needs n random bits for the computation on a. By definition, the probability that M accepts a is $p \geq 2/3$.

Now suppose that we have a pseudorandom generator P that stretches m bits to n bits. Then we can try to run M using the pseudorandom bits produced by P. If we think of a being fixed, then we can think of M as a test, a test that checks pseudorandomness of the bits. If p' is the probability that M accepts a with pseudorandom bits from P, then p' must be very close to p, when n goes to infinity. Thus $p' \geq 2/3 - \varepsilon$, where $\varepsilon \to 0$ when $n \to \infty$. The argument is completely symmetric for an input $b \notin X$. So, if we denote by q' the probability that M accepts b using pseudorandom bits, then $q' \leq 1/3 + \varepsilon$, where $\varepsilon \to 0$ when $n \to \infty$. For n sufficiently large, we get $q' \leq 1/3 + \varepsilon < 2/3 - \varepsilon \leq p'$; so we have a constant size gap between q' and p'. *Hence pseudorandom bits are as good as random ones!*

But is it of any practical interest? According to the definition, we only know $m < n$, so it can be just $m = n + 1$. In such a case we have saved only *one* random bit—not a big deal! If we want to get rid of random bits, the crucial question is how much can pseudorandom generators stretch the input bits. Fortunately, there is a very simple way to get larger stretching: to compose the generators. If we compose k pseudorandom generators, each stretching only by one bit, we get a generator that stretches the input by k bits. This works as long as k is bounded by a polynomial in m. In this way we can save a substantial part of random bits.

If we want to show that randomness can be eliminated, we have to get rid of all random bits. This means that eventually we have to get rid also of the random seed of the generators. Here the idea is also simple: run the algorithm for all random seeds. To see how it works, consider again the same situation as above, where $a \in X$ and we used a pseudorandom generator P stretching m bits to the number of random bits that the machine M needed. We concluded that M accepts a with probability $p' \geq 2/3 - \varepsilon$. This means that of all the 2^m strings produced by P, at least $(2/3 - \varepsilon)2^m$ will let M accept. Similarly, if $b \notin X$, the number of strings that will let M accept b will be at most $(1/3 + \varepsilon)2^m$. Hence the simple majority vote rule will decide whether we should accept the input.

In order to obtain a polynomial time algorithm, we would need to have 2^m to be bounded by a polynomial in the input length. Let n denote the input length, then it means that m should be bounded by $c \log n$, for a constant c. A probabilistic polynomial time algorithm on such inputs may need polynomially many random bits. Thus we need a generator that stretches the input exponentially, from $c \log n$ to $p(n)$, for some polynomial p. Such generators do not satisfy our definition of pseudorandom generators. The problem is that we measure the time as the function of the input length. Then no function that stretches the length exponentially can be polynomial time computable. Nevertheless, this approach still makes sense; only the definition is too restrictive. For the purpose of derandomization, it suffices to bound the time by a polynomial in the output length. After we change the definition in this way, we have to make another change—we must bound the size of circuits used as a test by a specific polynomial.

The technical details of the more general definition of pseudorandomness are not important. It suffices that we know that it can be done and that such generators can derandomize **BPP**. Thus, in particular, the problem of proving $\mathbf{P} = \mathbf{BPP}$ can be reduced to a construction of pseudorandom generators of a certain type.

Derandomization and Proving Lower Bounds

We do not know if pseudorandom generators exist. A necessary condition for their existence is that $\mathbf{P} \neq \mathbf{NP}$, but we also do not know if the existence of pseudorandom generators can be proved from this conjecture. It seems that the conjecture that they exist is stronger than the conjecture that $\mathbf{P} \neq \mathbf{NP}$. But we do know some interesting facts about them.

The first remarkable fact is that the existence of pseudorandom generators is equivalent to the existence of one-way functions (in the sense of Definitions 11 and 13). One of the implications is very simple: every pseudorandom generator that stretches m bits to $2m$ bits *is* a one-way function. Here is a sketch of a proof. Suppose the contrary, that a pseudorandom generator P is not a one-way function. It means that we can invert it with non-negligible probability. Stated formally, one can compute in polynomial time a function g such that

$$P\big(g\big(P(x)\big)\big) = P(x)$$

is true with non-negligible probability. But if we take y a random string of $2m$ bits, then the probability that it is in the range of P is at most 2^{-m}, whence the probability that

$$P\big(g(y)\big) = y \tag{5.4}$$

is also at most 2^{-m}. Thus (5.4) defines a test that can distinguish between random strings y and $P(x)$; so P is not a pseudorandom generator.

To construct a pseudorandom generator P from a one-way function f is much more difficult, so I will consider only a special case, where the one-way function satisfies a couple of additional conditions. The first condition is that f is a one-to-one mapping from the set of n bit strings into itself. In other words, f is a permutation of the set $\{0, 1\}^n$. The second condition is that for $x \in \{0, 1\}^n$, the first bit of x, denoted by x_1, is not predictable from $f(x)$. (This can be defined precisely, but I will omit this definition in the informal presentation here.) If these two conditions are satisfied, we can define a pseudorandom generator by

$$x_1 x_2 \ldots x_n \mapsto x_1 y_1 y_2 \ldots y_n,$$

where $y_1 y_2 \ldots y_n = f(x_1 x_2 \ldots x_n)$. So it is a pseudorandom generator that stretches n bits to $n+1$ bits. The assumption about unpredictability of the first bit is quite natural. A cryptographically good one-way function should hide as much as possible about the input, in particular, all input bits should be unpredictable from the output.

The fact that the existence of pseudorandom generators is equivalent to the existence of one-way functions shows that these concepts are good and that it is natural to conjecture that they exist. Yet, we would prefer to find a relation to a conjecture about basic complexity classes. A lot of progress in this direction has been achieved in the last two decades. I will describe what I think is the most interesting one of these results. This result gives a statement that seems very plausible and which implies that **BPP** = **P**. Pseudorandom generators are not explicitly mentioned, but they play the crucial role in the proof.

Recall that we view circuit complexity as the nonuniform version of Turing machine time complexity. We know that sets in **P** can be computed by polynomial size circuits. We have also observed that in general, if A can be computed in time $t(n)$, then it can be computed using circuits of size $c \cdot t(n)^2$, where c is a constant. It seems that this relation cannot be substantially improved. The result says that if this is indeed the case, then **BPP** = **P**. The precise statement is in the following theorem of R. Impagliazzo and A. Wigderson [136].

Theorem 40 *If there exists a set A of 0–1 strings such that*

1. *A is computable in time 2^{cn}, where c is a constant, and,*
2. *for every n, the circuit complexity of the set $A \cap \{0,1\}^n$ is at least $2^{\delta n}$, where $\delta > 0$ is a constant,*

then **BPP** = **P**.

We can view this theorem also from the perspective of proving lower bounds. Most research on lower bounds focuses on circuit complexity of Boolean functions. Since we know that there are Boolean functions that have exponential circuit complexity, the problem is to prove it for *explicitly defined* Boolean functions. Explicitly defined is an intuitive concept that we can interpret in various ways, but it always means some restriction on the complexity. If, for instance, we interpret it as **NP**, then proving superpolynomial lower bounds would imply **P** ≠ **NP**. If we interpret 'explicitly defined' as 'computable in time 2^{cn}', then proving exponential lower bounds on an explicitly defined function would give **BPP** = **P**.

Thus this result shows what most researchers agree on: proving lower bounds on the circuit complexity of Boolean functions is the central problem in complexity theory. Some recent results confirm it even more. Although the converse to the implication in the theorem above is not known, there are some partial results that show that **BPP** = **P** implies certain lower bounds on circuit complexity [145]. Thus it is impossible to derandomize **BPP** without proving some nontrivial lower bounds.

Two Important Functions

The methods developed for derandomization enable us to construct one-way functions and pseudorandom generators from any sufficiently hard functions. Although we cannot prove their hardness formally, to find suitable candidates is not a problem. For example, a natural candidate for a set A that satisfies the conditions of Theorem 40 is the well-known **NP**-complete set *SAT*, the set of satisfiable Boolean formulas. That said, this is completely useless for practical applications, especially for cryptography. When a construction is based on a series of complicated reductions, its desirable properties will manifest only for very large input lengths, which occur rarely in practice. What is needed are functions that exhibit these properties for inputs of fairly small lengths. There are two functions that are prominent among all proposed candidates of one-way functions. They are *multiplication of integers* and *exponentiation modulo a prime*.

The inverse problem to multiplication is *factoring*, which I have already mentioned several times. For multiplication to be a one-way function, we need not only to know that the factoring problem is hard, but also that it is hard for almost all inputs. This is clearly not true: every other number is even, hence the product of two random numbers of a given length n is even with probability $3/4$; to find a factor of an even number is trivial. Therefore, it has been proposed to restrict the domain to pairs of prime numbers. Let an input length n be given, then one considers the function:

`input:` p, q `primes of length` n
`output:` pq

In all known algorithms the most difficult numbers to factor are those that are products of a small number of large primes whose differences are also large. In particular, the most difficult seems to be to factor the product of two large primes p and q, with $|p-q|$ large. It is possible that multiplication restricted to pairs of primes is indeed a one-way function.

The second function also comes from number theory. If p be a prime number, then there exists a number g such that every number $1 \leq y \leq p-1$ can be uniquely expressed as the remainder of g^x when divided by p for some $0 \leq x \leq p-2$; we write $y = g^x \mod p$. Such a g is called a *generator of the multiplicative group modulo* p. Since this is a one-to-one mapping from $\{0, 1, \ldots, p-2\}$ onto $\{1, 2, \ldots, p-1\}$, the inverse function is defined, and since it is the inverse to exponentiation, it is called the *discrete logarithm;* we write $x = \log_g y$.

It is not quite obvious that exponentiation modulo a prime is computable in polynomial time. One cannot do it by first computing g^x and then computing the remainder. It is necessary to use modular arithmetic, namely, to take the remainder after each step of computation because otherwise one would have to use exponentially long numbers. Also we cannot perform x multiplications; one has to use the trick of 'repeated squaring' (see page 431). Finding a generator g is another interesting problem that one has to solve in practical applications; unfortunately, it would take us too far afield to discuss it.

What is more important is the complexity of the inverse problem, the discrete logarithm. Again, some algorithms are known, but all need exponential time. Since in number theory it is a well-known problem, studied for a long time, and we still do not have efficient algorithms, it is reasonable to conjecture that there is no polynomial time algorithm.

The most interesting property of the discrete logarithm function is that when it is hard in the worst case, then it is hard also on average. This is an extremely desirable property and in all other proposed candidates it is difficult to find a good justification for the hardness on average. In the case of the discrete logarithm *we can actually prove it*, provided we have hardness in the worst case, moreover the proof is very easy. Here is a sketch of the proof.

Suppose M is an algorithm that solves the discrete logarithm for random inputs. Thus for a *randomly* chosen y, $1 \leq y \leq p-1$, it computes $\log_g y$ with probability at least n^{-c}, where n is the input length (the number of bits of p) and c is a constant. We define M' which solves with the same probability the problem for *every* given y', $1 \leq y \leq p-1$. M' first takes a random r, $0 \leq r \leq p-1$ and computes $y = y'g^r$. Then applies M to y; let x the number computed by M. Then M' outputs $x - r$.

To show that M' computes $\log_g y'$ with probability at least n^{-c}, suppose that in the subroutine M succeeds to find $x = \log_g y$. Then $x = \log_g y'g^r = \log_g y' + \log_g g^r = \log_g y' + r$. Thus, indeed, M' finds $\log_g y'$. To get a probability arbitrarily close to 1, we only need to repeat the subroutine M polynomially many times.

Our only reason to believe that factoring and the discrete logarithm are hard functions is the lack of efficient algorithms in spite of the deep insight of number theorists into these problems. Unlike in the case of **NP**-complete problems, where we know that they are the most difficult ones in the class **NP**, we do not have any result of this kind for these two functions. Some classes, in particular complexity classes of randomized computations, do not seem to have complete problems; the same is probably true for one-way functions and pseudorandom generators.

Notes

1. *Probabilistic primality tests.* As the polynomial bound in the Agrawal-Kayal-Saxena test is still quite large, in practice it is better to use faster probabilistic tests discovered before. One of these tests is due to Solovay and Strassen [282]. Given an N which is to be tested, we randomly choose an a, $1 \leq a < N$ and test

   a. $(a, N) = 1$ (are a and N coprime)?
   b. $a^{(N-1)/2} \equiv (\frac{a}{N}) \bmod N$?

   If N passes both tests then it is a prime with probability at least $1/2$. The precise meaning of this statement is: if N is *not* a prime, then at most $1/2$ of all numbers $1 \leq a < N$ satisfy both equations, and if N is a prime, then all a satisfy the two conditions. In this test the only random part is the choice of a, the expressions in the equations can be computed in polynomial time. (For computing the Legendre-Jacobi symbol it suffices to recursively use the well-known relations, one of which is the Quadratic Reciprocity Law.)

If we repeat the test twice we get:

a. if N is a prime then it is accepted with probability 1;
b. if N is not a prime, then it is accepted with probability at most $1/4$.

This formally proves that the set of primes is in the class **BPP** (we know that it is, in fact, in **P** which is contained in **BPP**. Notice that the first condition is much stronger than it is required in the definition of **BPP** because in the first case there is zero error. Many concrete problems that are in **BPP** satisfy this stronger condition.

Similarly, as in the case of finding quadratic nonresidua, one can turn some probabilistic primality tests into deterministic polynomial time algorithms if certain number-theoretical conjectures are true. Such a test based on the Extended Riemann's Hypothesis was proposed by G.L. Miller [198]. The proof of the Extended Riemann's Hypothesis, if ever found, will surely be much more difficult than the proof of the correctness of the Agrawal-Kayal-Saxena algorithm.

2. *Probabilistic Turing machines and hardness on average.* An alternative way of defining probabilistic Turing machines is to equip deterministic machines with an additional tape for random bits. Then we count the probability of acceptance of a given input when the tape contains random bits.

   Sometimes people confuse this model with a different problem. Suppose a set A and a deterministic machine M is given. Determine, for a *random input*, the probability that the machine correctly decides if the input is in A. The same problem is more often studied for Boolean circuits and Boolean functions. We say that a function $f : \{0, 1\}^n \to \{0, 1\}$ is *hard on average*, if for every circuit C of subexponential complexity, the probability that C accepts or rejects a random input correctly is less than $1/2 + \varepsilon$, where $\varepsilon \to 0$ as $n \to \infty$.
3. *Randomized circuits.* It is also possible to study randomized circuits. Such circuit C has input variables of two types. One serves as the ordinary input variables $x_1, x_2, \ldots, x_n$, the others are for random bits $r_1, r_2, \ldots, r_m$. For a given assignment to values to $x_1, x_2, \ldots, x_n$, we count what is the probability that C accepts (= computes 1) for a random string of $r_1, r_2, \ldots, r_m$. If C computes a Boolean function f with bounded error (as in the definition of **BPP**: if $f(x) = 1$ then $C(x, r) = 1$ with probability $\geq 2/3$ and if $f(x) = 0$ then $C(x, r) = 1$ with probability $\leq 1/3$), then we can amplify the probability exponentially by taking several circuits and computing the majority vote. In this way one can construct a circuit whose probability of error is less than 2^{-n}. When the error is so small, there exists one string $r_1, r_2, \ldots, r_m$ with which C computes correctly for all 2^n inputs. Thus we can eliminate randomness by only polynomially increasing the size of the circuit.

   Therefore, randomness is not interesting in the context of non-uniform complexity.
4. *More on holographic proofs.* Here is the definition of a *holographic* or *polynomially checkable* proofs. I continue with the example of proofs and formulas. A system of holographic proofs consists of a suitable definition of proofs and an algorithm A. The algorithm is probabilistic and has two parameters: a constant

$0 < \alpha < 1$ and an integer $c \geq 1$. Given a formula ϕ and a text P, A computes c bit positions in P and reads these bits, say $b_1, b_2, \ldots, b_c$. Using only these bits and formula ϕ, it either accepts or rejects. Furthermore,

(a) if P is a correct proof, then A always accepts P;
(b) if there exist no proof of ϕ, then A rejects P with probability at least α.

Such proofs can be constructed not only for the particular case of formulas and their proofs, but for any problem in **NP**. Recall that a set A is in **NP** if it is defined using a search problem, namely, $x \in A$ if the associated search problem has a solution. Testing the existence of a certain object is the heart of the matter. So one first proves that an **NP**-complete problem has polynomially checkable proofs and then uses reductions to prove that all **NP** problems have such proofs. The fact that every set in **NP** has polynomially checkable proofs is the famous *PCP Theorem* [7, 9].

5. *Public key cryptography.* The most interesting concept in cryptography is a public key system. These systems are not only very interesting from the point of view of theory, but they are also the most useful applications of cryptography.

   Let us recall the standard setting. We have an encoding function $F(e, x)$ and a decoding function $G(d, y)$, where x is the plaintext, y is the ciphertext, e an encoding key and d is the decoding key corresponding to e. It is assumed that the functions are publicly known. In the case of a public key system, Alice knows both e and d and publicly announces the encoding key e. Then everybody can send messages to Alice, but only she can decode them. Thus for a fixed e, the function $x \mapsto F(e, x)$ is not one-way because it has an easy inverse $G(d, y)$, but it is hard to find the inverse, unless one knows d. Such functions are called *trapdoor functions.*

   The most common public key system is the RSA, invented by R.I. Rivest, A. Shamir and L. Adleman [246]. In RSA Alice needs a sufficiently large composite number N and its factorization. The best seems to take two large primes p and q whose difference is also large and let $N = pq$. The number N is public, but p and q must be kept secret. An encoding key is a randomly chosen e, $1 < e < \phi(N)$, where $\phi(N) = (p-1)(q-1)$ (the Euler function of N). Knowing $\phi(N)$, Alice can compute the decoding key $1 < d < \phi(N)$ such that $ed \equiv 1 \mod \phi(N)$, (this is done using the well-known Euclid algorithm).[13] The encoding function is

$$F(e, x) = x^e \mod N;$$

   the decoding function is, in fact, the same

$$G(d, y) = y^d \mod N.$$

   In this way we can encrypt any number from 1 to $N-1$ except for p and q. If N is sufficiently large and messages are more or less random, the probability that a message will be p or q is negligible. Using the well-known Euler's Theorem

[13] Strictly speaking, the encoding and decoding keys are not only the numbers e and d, but the pairs (e, N) and (d, N).

which says that $z^{\phi(N)} \equiv 1 \mod n$, we see immediately that d is the decoding key for e:

$$\left(x^e\right)^d = x^{ed} = x^{1+c\phi(N)} = x \cdot \left(x^{\phi(N)}\right)^c \equiv x \cdot 1^c = x \mod N.$$

If one can factor N, then one would be able to compute the decoding key and thus break the system. It seems that this is the only way how one can successfully attack it, but there is no proof of it. There is, however, a public key system, for which one can prove that it is secure if and only if it is hard to factor a randomly chosen product of two primes. Such a system was designed by M.O. Rabin [234]. The encoding function in his system is simply $x \mapsto x^2 \mod N$, where N is as in RSA.

However, RSA has some additional good properties which make it more attractive than other systems. In particular, the keys commute: if we need to apply two keys then it does not matter in which order we apply them. This has a number of useful applications.

6. *Modular exponentiation* is needed in many cryptographic protocols. Suppose we want to compute A^B modulo N. The input size is the sum of the lengths of the numbers A, B, N. Since B can be exponential in the input size, we cannot simply multiply AB-times. Therefore, we use the trick of repeated squaring. It is best seen in an example. Let $B = 1011001$ in binary. First express B as follows

$$1011001 = 1 + 2(0 + 2(0 + 2(1 + 2(1 + 2(0 + 2 \cdot 1))))).$$

(Notice that 0's and 1's occur in the reverse order on the right-hand side.) Of course, we can delete all 0's and the last 1,

$$1011001 = 1 + 2 \cdot 2 \cdot 2(1 + 2(1 + 2 \cdot 2)).$$

Now we can compute A^B efficiently:

$$A^{1011001} = (((((A^2)^2 \cdot A)^2 \cdot A)^2)^2)^2 \cdot A.$$

Furthermore, we must not do the operations in $\mathbb{Z}$ because the sizes of numbers would be exponentially large. Instead, we compute in $\mathbb{Z}_N$; technically, it means that after each operation we replace the result by its remainder modulo N.

7. *Expander graphs.* Explicitly constructed expander graphs are probably the most useful structures in complexity theory. Expander graphs are graphs satisfying a certain property which is typical for random graphs. Since we have an explicit construction, which means that we can construct them efficiently without using randomness, we can use them to simulate randomness in some specific situations.

The graphs that are of interest for us are graphs with bounded degree. This means that the degree of every vertex (the number of edges incident with the vertex) is bounded by a constant d. Often we can assume the stronger condition that the degree of every vertex is equal to the constant d. Such graphs are called *d-regular*.

The property by which expander graphs are defined is, roughly speaking, that every subset of vertices X is connected with many vertices outside of X. Obviously, if X is the set of all vertices or it contains almost all vertices, then there are not many vertices left. Therefore one has to state the condition so that this case is avoided. Here is the definition.

Definition 14 A finite d-regular graph G on a set of vertices V is called an ε-expander, for $\varepsilon > 0$, if for every subset of vertices X that contains at most one half of all vertices, the number of vertices outside of X connected with vertices in X is at least ε times the size of X.

The name expander is used here because the set of all vertices that are in X or connected to a vertex in X has size at least $(1+\varepsilon)$ times the size of X. Thus every X which is not too large "expands" at least by factor $1+\varepsilon$. In applications we need infinite families of graphs that are d-regular ε-expanders for some constants d and constant $\varepsilon > 0$. To prove that such graphs exist by non-constructive means is easy—a random d-regular graph is an ε-expander for some $\varepsilon > 0$ with high probability. To construct a family of explicit expanders is, however, a non-trivial task. The graphs may be quite simple to describe, but to prove that they are expanders is difficult. Although the condition by which expanders are defined is purely combinatorial, the best way to prove nontrivial bounds on the expansion rate is to apply algebra (to show a gap between the largest and the second largest absolute values of eigenvalues of the adjacency matrix of the graph).

Example The first explicit construction of a family of expanders was given by Margulis [191]. For every $n > 0$, he defined a graph on the set of vertices $\mathbb{Z}_n \times Z_n$ in which every vertex (x, y) is connected by an edge with $(x+y, y)$, $(x-y, y)$, $(x, x+y)$, $(x, x-y)$, $(x+y+1, y)$, $(x-y+1, y)$, $(x, x+y+1)$, $(x, x-y+1)$ where addition is modulo 2. Some edges my actually be loops. This is in order to formally satisfy the condition that the graph is 8-regular.

Let us now consider a simple application of expanders. Let G be a d-regular expander on a set of vertices V of size N. A *random walk* on G means that starting in a random vertex v_0, we randomly choose v_1 to be one of the d neighbors of v_0 and go to v_1; then we choose v_2 to be a random neighbor of v_1 and so on. One can show that the sequence of vertices chosen in this way behaves like a genuinely random sequence of vertices. Namely, suppose that we need to find a vertex in some subset U of vertices whose size is $1/2$ of the size of V. For example, let V be numbers $1, 2, \ldots, p-1$ for some prime p, and our task is to find a quadratic nonresidue. Then the probability that we hit a nonresidue will increase exponentially with the length of the walk.

What is the advantage of using a random walk instead of random vertices? If we pick a random vertex, we have N possibilities, so we need $\log_2 N$ random bits. In a random walk on a d-regular graph we need only $\log_2 d$, i.e., a constant number of random bits, to get the next vertex. Thus random walks on expanders enable us to save many random bits when we need to amplify the probability in problems such as finding a quadratic nonresiduum.

There are numerous other applications of expander graphs. To mention at least one, expanders can be used to construct error correcting codes with very good parameters.

8. *Derandomization by means of hard functions.* I will sketch two main ideas of the proof of Theorem 40. The first one is the *Nisan-Wigderson* generator of N. Nisan and A. Wigderson [209]. This is a construction that is used in many proofs in this field of research.

Let $0 < \alpha, \beta, \gamma, \delta < 1$ be such that $\beta + \gamma < \delta < \alpha$. Let n (an input size) be given. In order not to overload notation with additional symbols I will assume that αn and γn are integers. To construct a Nisan-Wigderson generator we need furthermore a suitable set system $\mathcal{S}$ and a hard function f.

The set system $\mathcal{S}$ contains $2^{\gamma n}$ subsets of the set of variables $\{x_1, x_2, \dots, x_n\}$; each subset is of size αn and the intersection of every pair of different elements of $\mathcal{S}$ has size at most βn.

The Boolean function f maps $\{0, 1\}^{\alpha n} \to \{0, 1\}$ and we will assume that the circuit complexity of f is at least $2^{\delta n}$.

The set system $\mathcal{S}$ and the function f determine a Nisan-Wigderson generator—the function $G(x_1, x_2, \dots, x_n)$ defined by

$$G(x_1, x_2, \dots, x_n) =_{def} \{f(Y)\}_{Y \in \mathcal{S}}.$$

On the right hand side, $f(Y)$ denotes f computed on a subset of variables $Y \subseteq \{x_1, x_2, \dots, x_n\}$. Thus G maps bit strings of length n on bit strings of length $m = 2^{\gamma n}$, which gives exponential stretching, as needed for derandomization.

In order to get intuition about how it works, assume that the bound βn on the sizes of the intersections of subsets is much smaller than their size αn. Then we can say that they are "almost disjoint". If the sets in the set system $\mathcal{S}$ were indeed disjoint, the output bits would be independent and we would get truly random bits, but no stretching. We hope that if they are "almost disjoint", then the output bits will be "almost independent". Furthermore, there are families of almost disjoint sets that have exponentially many members.

Recall that we need to show that the generator produces bits that look random to *circuits of small size*. From the *statistical* point of view the output bits are very dependent, so any formal proof has to focus on the complexity of computing some properties of the output bits. One can show that everything boils down to proving that any of the output bits cannot be computed from the remaining ones by a small circuit. More precisely, we need to show that the bit cannot be predicted by a small circuit with non-negligible probability.

Let Y_0 be an arbitrary element of the set system $\mathcal{S}$. Let us simplify the task and only prove that $f(Y_0)$ cannot be computed by a small circuit from the values $f(Y)$, where $Y \in \mathcal{S}$ and $Y \neq Y_0$. So we will assume that C is a circuit that computes $f(Y_0)$ from these values and prove a lower bound on its size. We can express our assumption by the following equality:

$$f(Y_0) = C(\{f(Y)\}_{Y \in \mathcal{S}, Y \neq Y_0}).$$

Let a be an arbitrary assignment to the variables outside of Y_0. For $Y \in \mathcal{S}$, we will denote by $Y|_a$ the string in which the variables of $Y \setminus Y_0$ are set according to a, and the variables in $Y \cap Y_0$ remain unchanged. Since the equation above holds generally, it must also hold true after this substitution. We thus get

$$f(Y_0) = C\big(\{f(Y|_a)\}_{Y \in \mathcal{S}, Y \neq Y_0}\big).$$

Since $Y \cap Y_0$ has size at most βn, the restricted function $f(Y|_a)$ depends on at most βn variables. Therefore it can be computed by a circuit of size at most $2^{\beta n}$. Combining all these circuits with C we get a circuit of size at most

$$2^{\beta n} 2^{\gamma n} + |C|,$$

(where $|C|$ denotes the size of C) and C computes $f(Y_0)$. Since the circuit complexity of $f(Y_0)$ is $\geq 2^{\delta n}$ and the term $2^{\beta n} 2^{\gamma n} = 2^{(\beta+\gamma)n}$ is asymptotically smaller, the size of C satisfies the inequality

$$|C| \geq (1-\varepsilon) 2^{\delta n} = (1-\varepsilon) m^{\delta/\gamma},$$

where $\varepsilon \to 0$ as $n \to \infty$.

We do not get a superpolynomial lower bound on the size of C, but one can show that it is possible to choose parameters and construct the set system so that the exponent δ/γ is an arbitrary large constant. In other words, we can beat any polynomial bound.

On the other hand, we can also compute the Nisan-Wigderson generator G in time that is polynomial in m. To this end we need the assumption that f is computable in exponential time; more precisely, f is a restriction to $\{0,1\}^{\alpha n}$ of a function computable in time $2^{c\alpha n}$. In terms of m the bound is $m^{c\alpha/\gamma}$, i.e., polynomial in m.

The above analysis demonstrates the basic idea of Nisan-Wigderson generators, but the property that we have shown does not suffice for derandomization of **BPP**. To get the stronger property of the generator (that the bits cannot be predicted with non-negligible probability from the remaining ones) we also need a stronger property of the function f used there. Instead of only assuming the *worst-case* complexity of f to be large, we need the *average-case* complexity to be large. We need that for some $\varepsilon > 0$, every circuit of size at most $2^{\delta n}$ disagrees with f on the fraction ε of all inputs. Since there is a way to construct a function hard in the average from a function that is only hard in the worst case, Theorem 40 only assumes the existence of functions that are hard in the worst case. So the second idea, which I am going to explain now, is how to get average-case complexity from worst-case complexity.

I described error correcting codes a few pages back (page 407). Let Γ be a binary code with the minimal distance d. Recall that if u is a codeword and we change less than $d/2$ bits then we still can uniquely decode the resulting word. This suggests the following idea.

Think of a Boolean function f of n variables as a binary string w of length $N = 2^n$ where the bits of the string are the values of f for all possible inputs. Using a suitable error-correcting code Γ, encode w by a longer string $u \in \Gamma$ of length $M = 2^m$. Now we can interpret u as a Boolean function g of m variables. Let εM be the minimal distance of Γ, for some $\varepsilon > 0$. Let D be a Boolean circuit with m variables which approximately computes g. If D outputs a wrong value on less than $\frac{\varepsilon}{2}M$ inputs, then it defines a string that is in distance less than $\frac{\varepsilon}{2}M$ from u. Thus in such a case we can completely recover g and f, in spite of the many possible errors that D makes. Put differently, given such a circuit D we can compute $f(x)$ for *every input* x, in particular also for the "hard inputs".

It is natural to expect g to be hard on average, but in general this is not the case. It can happen that whereas f is hard, g can be very simple. To make the idea work, we must use coding that preserves complexity. More specifically, we need a code that has the following properties:

a. It is possible to decode *locally* and *efficiently*. This means that to compute a value of $f(x)$, we only need a few values $g(y_1), g(y_2), \dots, g(y_k)$ and we need a small circuit for computing $f(x)$ from these values; moreover this circuit should give the correct value of $f(x)$ also if part of the values are wrong.
b. It should be possible to choose the inputs $y_1, y_2, \dots, y_k$ *randomly* so that if we use $D(y_1), D(y_2), \dots, D(y_k)$ instead of $g(y_1), g(y_2), \dots, g(y_k)$, most of the values are correct.

Having such a circuit C for decoding Γ, we can combine it with circuit D and we get a circuit computing f precisely. If C were small, then the resulting circuit would also be small. But we assume that such a circuit does not exist. Hence it is not possible to compute g by a small circuit even on average.

Constructions of locally decodable codes are very interesting, but it would take us too far afield. Let me only say that they are based on interpolation of low degree polynomials.

9. *Pseudorandom generators and natural proofs.* The PRG-Conjecture mentioned on page 388 says that there exist pseudorandom generators that satisfy a stronger requirement than the one stated in Definition 13.

The PRG-Conjecture *There exist an $\varepsilon > 0$ and a polynomial time computable function P such that for every n,*

a. *P maps the set of bit strings of length n into the set of bit strings of length $2n$;*
b. *for every circuit C of size at most 2^{n^ε},*
$$\left|Prob\big(C(P(x)) = 1\big) - Prob\big(C(y) = 1\big)\right| < 2^{-n^\varepsilon},$$
where $Prob$ is probability with respect to the uniform distribution on bit strings of length n.

This means that P is a very strong pseudorandom generator—it cannot be distinguished from a truly random source by exponentially large circuits even with exponentially small probability.

Such a pseudorandom generator P is then used to construct a pseudorandom generator Q that stretches n bits to 2^m bits, where $m = n^\delta$ for some positive constant δ. This is similar to what one needs for derandomization of **BPP**, but now we need a stronger condition on the computability of Q, namely, we need that for a given M, $1 \leq M \leq 2^m$, it is possible to compute the Mth bit of the output of $Q(x_1, x_2, \dots, x_n)$ in time polynomial in n. The generator Q is constructed by suitably composing P with itself.

Let us denote the function that computes the Mth bit of the generator Q by $F(x_1, x_2, \dots, x_n, y_1, y_2, \dots, y_m)$, where $x_1, x_2, \dots, x_n$ are the seed of the generator Q and $y_1, y_2, \dots, y_m$ are the bits encoding the number M. So this function is computable in polynomial time. Such a function is called a *pseudo-random function generator* F because it can be used to imitate random functions. If we randomly choose an assignment of zeros and ones $a_1, a_2, \dots, a_n$ to the variables $x_1, x_2, \dots, x_n$, then $F(a_1, a_2, \dots, a_n, y_1, y_2, \dots, y_m)$ is computationally indistinguishable from a truly random function of m variables.

Recall that the nonexistence of natural proofs is essentially the statement that one cannot define a large set of hard Boolean functions by a polynomial time computable condition. (For explaining the argument it is not necessary to specify what *hard* and *large* mean; you can think of it as *superpolynomial circuit complexity* and *a positive fraction of all functions*, respectively.) Assuming we have a pseudorandom generator described above, it is easy to prove it. Let A be a polynomial time algorithm that defines a large subset of hard Boolean functions of m variables. Think of this algorithm as a test of the pseudorandom generator Q. As Q is pseudorandom, A should also accept a large subset of all possible outputs of Q (what we actually need is that it accepts at least one). Since $F(a_1, a_2, \dots, a_n, y_1, y_2, \dots, y_m)$ is computable in polynomial time, every string produced by Q codes a Boolean function that has a polynomial size circuit. This proves that A also accepts a string computable by a polynomial size circuit, which is a contradiction. In this way the Razborov-Rudich result is proved.

10. *The game morra.* If you think that generating random numbers is easy, try morra. Morra is an ancient game that is still played in Italy and Spain. Two players use their voices and each one hand. They simultaneously show numbers, 1–5, on their hands and shout a number that is a guess of the sum the shown numbers. If one player guesses correctly the sum shown on the hands, he scores a point; if both guess wrong numbers or both guess correctly, none gets a point.

    The game seems trivial from the point of view of classical game theory—the best strategy is to show a random number n uniformly distributed between 1 and 5 and shout a random number m uniformly distributed between $n+1$ and $n+5$. But when the game is actually played, due to high rhythm, players have very little time to think up a number that the opponent cannot predict. It requires skill and practice to be good at morra. One has to be able to produce unpredictable numbers and recognize regularities in the opponents play. Furthermore one has to do all that in a short interval between the rounds, including the addition of his number with the predicted number of the opponent.

11. *Which graphs are isomorphic in Fig. 5.4? B* is isomorphic to C; they are drawings of the *Petersen graph.*

5.3 Parallel Computations

When computing difficult problems we can often speed up the computation by using several processors which compute certain parts of the computational task simultaneously. We call it *parallel computations* in contrast to *sequential computations* in which we perform only one action at a time. There are problems whose computation can be sped up by using parallelism and there are others for which parallelism does not help. This phenomenon is well known in manufacturing. If a product consists of many components, one can produce the components independently of each other and then quickly assemble them to the final product. But if we are to build, say, a house, we must first build the foundations, then the first floor, and so on, the roof coming last.

A prime example of a mathematical problem that is amenable to parallelization is the problem of factoring a natural number N. The most trivial algorithm, based on trying to divide N by all numbers up to $\sqrt{N}$, can easily be split into independent tasks. For example, we can divide the interval $[1, \sqrt{N}]$ into as many segments as we have processors, and let each processor try the numbers in its segment. Other, more sophisticated and faster algorithms have this property too. One of the first spectacular applications of the Internet for solving a large scale problem was the project of factoring the RSA-129 number. This 129 decimal digit composite number was proposed in 1977 by Rivest, Shamir and Adleman, the authors of the most popular cryptographic protocol. They challenged everyone to factor this number. Their aim was to test the security of their system for keys of this length. In summer 1988 Arjen Lenstra launched a project to factor this number using computers connected to the Internet. The following April they announced the solution. In course of computation more than 600 people took part in solving some of the tasks to which the problem was split on their computers. (The RSA-129 number and its factors are on page 368.)

Since then not only have computers became much faster, but also new algorithms have been discovered and the old ones have been improved. In 2005 a 200 decimal digit number RSA 200 was factorized using many computers working in parallel. The computation ran for one and a half years. The authors estimated that it would take about 55 years if it were done on a single machine. More recently, in 2009, a number with 232 decimal digits was factored. (Check the Internet for the present integer factorization record.)

The Ideal Parallel Computer

In many situations already reducing the time by a factor of two may be important. For example, in weather forecasting the computations are so time consuming that sometimes they would be finished only after the time for which we needed the prediction. In theoretical investigation we rather study the distinction between polynomial and exponential. To this end we need a mathematical model of an idealized parallel computer. Such a computer looks formally very much like an ordinary personal computer running modern operating systems such as Linux. It can run several processes in parallel and every process can start another new process. In a personal computer, however, all this is done by a single processor which alternates between all the processes.[14] Thus it only seems that the processes run in parallel, but in fact at each moment the computation progresses only in one of the started process. Truly parallel machines have been built, but the number of processors in them is always limited. In the ideal parallel machine we imagine having an unlimited number of independent processors that can work simultaneously on the processes assigned to them.

Having an infinite number of processors does not give us infinite computational power. Recall that a Turing machine also has an infinite tape, but still it can only compute some functions (the computable functions) because at each step of computation it only uses a finite part of the tape. Similarly, a parallel machine can only use a finite number of processes at each step. The advantage of the parallel machine is that the number of processors can increase exponentially. For instance, if at each step every process starts another new process and no processes terminate, we will have 2^t running processes at time t.

With exponentially many processors at our disposal, we should be able to compute more in polynomial time than we can on sequential machines. Let us see what we can compute on a parallel machine in polynomial time. One can easily check that we can compute all **NP** problems: every **NP** problem is associated with a search problem and to solve the search problem on a parallel machine in polynomial time, we can activate an exponential number of processors, each processor working on a single item of the search space. But it is possible to do more than **NP**. One can prove that the sets accepted by parallel machines in polynomial time are exactly the sets computable by Turing machines in polynomial space. We can express it as an equality between complexity classes by

$$\mathbf{parallelP} = \mathbf{PSPACE}.$$

But, alas, whether or not $\mathbf{P} = \mathbf{PSPACE}$ is one of the big open problems in complexity theory! Hence it is also an open problem whether parallel machines are essentially more powerful than sequential machines. Since $\mathbf{P} \neq \mathbf{NP}$ implies that $\mathbf{P} \neq \mathbf{PSPACE}$, it seems very likely that they really are, but we are not able to prove it.

It should be stressed that parallel machines could compute more problems in polynomial time than sequential machines only because we allow them to use more than a polynomial number of processors. A parallel machine with a polynomial number of processors can be simulated sequentially with only a polynomial increase in time. From the practical point of view it is not realistic to assume more than a polynomial number of processors. Even if we could build a parallel machine with infinitely many processors, we would still be able to use only polynomially many of them in polynomial time because the universe has three dimensions. The number of processors that such a machine could use in time n would be of the order of n^3 and thus we could simulate such a machine in polynomial time. Thus **parallelP** is not a realistic model of what is efficiently computable.

[14] Recently chips that integrate several cores have been introduced. Computers equipped with such processors can run some processes in parallel.

This is the theoretical point of view; in practice parallel machines are useful because even a relatively small speed-up may play a significant role.

Interlude 1—Parallel Computations in the Brain

I have made a lot of digressions from the main topic, so there is no need for giving an apology for the next one. Indeed, the main reason for including the next two subsections is to show some applications of the concept of parallel computing in life sciences. On the other hand, when talking about the foundations of mathematics we have to take into account human abilities. I strongly believe that mathematics describes some fundamental principles of nature that are independent of human beings. But as we are limited beings, only some of these principles are within our reach and we have to represent them in a way comprehensible to us. The limitations are of various natures; for instance, we have a built in framework for using languages which may limit the way we use logical deduction. The most important restriction is, however, the limited computational power and limited memory. Complexity theory is the theory that enables us to quantify and compare this type of limitations.

In this subsection I want to make two claims. The first is:

In order to perform complex tasks, our brains have to work like parallel machines.[15]

This is a generally accepted fact, nevertheless, it is worthwhile to consider arguments that support this claim.

It seems to us that at each instant we are thinking only about a single idea, but in fact a lot things are going on at the same time in our brains. When we are speaking, we are pronouncing a sentence, but we are already thinking about the next one. Not only are we thinking about the next one, we are also contemplating how to go on afterwards, we are looking for suitable arguments to make our point and so on. At the same time we watch our audience to see their reactions, or drive a car; some people are even able to talk about one thing and type an unrelated message on the computer.

That people can do several things at the same time is a well-known fact. What is a more intricate question is whether our brain uses parallelism when solving problems. There are several observations that suggest that the brain is working sequentially when solving problems. One of these is the fact that we present the solution as a sequence of words, either spoken or written. In particular mathematical proofs and computer programs are given as text. But often one can present solutions of problems in the form of diagrams more efficiently, so this is not a very strong argument. Another argument is based on watching eye movements. When we let people solve a problem that involves a picture or a geometric configuration, such as a chess

[15] I will only be concerned with the parts of the brain that are responsible for cognitive processes or motor actions.

problem, we can get some information about the way they solve the problem by observing their eyes. The rapid movement of eyes from one place on the picture to another suggests that there is a long sequence of deductions made one after another. If the problem is simple enough, one can explain the problem solving process by a series of steps. But if it is difficult, it seems that there is not enough time to do it in this way. For example, in a typical situation chess players have to consider such a large number of possible moves that it is not possible for them to do it in the short time that is available by contemplating one alternative after another.

With open eyes our brain processes a huge amount of information at every moment. So let us close our eyes and observe what is going on in our mind. What we would like to find out is what happens in a single moment, namely, whether we are thinking about one idea, or there are more. I cannot speak for others, but it is natural to expect that what I observe is the same as others do. When I try to recall my state of mind, I find out that there is one leading idea, but there are also several ideas around that are less stressed and less clear. I feel that those ideas around are somehow connected with more ideas, but I cannot tell if the latter are really present. Thus I surmise that our consciousness has a hierarchical structure: there is one leading idea and a hierarchy of ideas which we realize less and less. Probably, we do not realize at all the ideas in the lowest part of the hierarchy, or at least we are not able to report about them later. Of course, more elaborate experiments are needed to find out what is really going on. Also there is an obvious problem involved that the brain puts only certain things to memory; thus we are never sure that we report everything that we experienced. But at least one thing we can confirm: when thinking about a certain subject, we are able to recall related things faster, than if we were asked to recall them without preparation. Our brain, sort of, has always the related data at hand.

When comparing brains with computers the most striking fact is how slow the components of brains are compared to the electronic components of computers. A neuron can fire at most about one thousand times per second, which is 1 kHz in terms of frequency. Present computers run on frequencies of several GHz, thus they are more than several million times faster. Yet people, animals and birds can outperform computers in many tasks. Imagine the computations done by the brains of a table tennis player or a swallow catching insects in flight. In order for the computations to be done on time, the information has to go through a very small number of layers of neurons. The study of how visual information is processed shows that some abstract concepts already appear after a few layers of neurons. The slowness of neurons implies that when working on tasks that require complex and fast reactions the brain must use massive parallelism; most likely it uses a lot of parallelism all the time.

My second claim is that parallel processing is also sufficient for an explanation of the astounding abilities of human brains. More specifically:

> *It is consistent with our present knowledge that the information processing in human brains can be approximated by the parallel computational model of threshold circuits.*

A *threshold circuit* is a network of elements that compute *threshold functions*. Threshold functions are Boolean functions of several input variables and one output variable. A threshold function t of n variables is determined by n real numbers $a_1, a_2, \ldots, a_n$, called the *weights*, and one real number b, called the *threshold.* The function is defined as follows. Given input bits $x_1, x_2, \ldots, x_n$,

$$t(x_1, x_2, \ldots, x_n) = 1 \quad \text{if and only if} \quad \sum_{i=1}^{n} a_i x_i \geq b.$$

In the simplest case all the weights are the same and equal to 1 and the threshold is a natural number b. Then t outputs 1 if and only if at least b input bits are ones. In particular the AND_n, the *and-function* of n variables, is the threshold function with unit weights and the threshold n, and similarly, the OR_n, the *or-function*, is the threshold function with unit weights and the threshold 1. Another important threshold function is MAJ_n, the *majority function*, defined by unit weights and $b = n/2$. In general threshold functions use arbitrary real weights and the weights can also be negative.

A threshold function is a (very simplified) model of the electric activity of a neuron. We distinguish two states of a neuron: the inactive state, represented by 0, and firing of the neuron, represented by 1. A neuron receives signals from other neurons (via dendrites) and according to the weighted sum of these signals it decides to fire or not to fire. When it fires, this signal is sent to other neurons (via the *axon*). Negative weights are possible because some connections between neurons (*synapses*) use inhibitory neurotransmitters.

Formally, a threshold circuit is a Boolean circuit in which each gate computes some threshold function. Such circuits are rather different from those that we considered before. The circuits that I dealt with before used only a few very simple gates. Now I am allowing more complicated gates and they may have a large number of inputs. In threshold circuits we usually do not bound the number of inputs of a gate (a real neuron may be connected to several tens of thousands other neurons by synaptic connections). Instead we usually impose another restriction: we bound the depth of the circuits. The depth is the length of the longest path from inputs to the outputs of the circuit. We speak about *bounded depth* circuits when the depth is bounded by a constant. This, of course, makes sense only when considering infinite families of circuits. When we have a single circuit, such as the brain, we should think of the depth bound to be a very small number. Imagine a circuit consisting of a few layers of threshold gates between the inputs and outputs.

The bounded depth restriction corresponds to what we know about the anatomy of the brain. We also have an indirect evidence that the signal processing takes place only on a small number of layers of neurons: it is the fact that people are able to react very fast in spite of the relative slowness of neurons. It is also interesting to compare the volume occupied by *somas*, the central parts of neurons on the one side and by axons, the threads that transmit electric signals, on the other side. Somas are in the thin layer of the *gray matter* on the surface of the brain, whereas most volume is occupied by the inner part, the *white matter*, through which neurons are connected by axons. This shows that the large number of connections between neurons must play a very important role. These connections make it possible to spread information to a large number of neurons in a small number of steps.

Bounded depth threshold circuits have been studied extensively, so we have some idea how strong they are, which may help us to assess whether or not this model of the brain is oversimplified. Bounded depth and polynomial size threshold circuits have been constructed for several nontrivial functions. We have such circuits in particular for arithmetic operations and various analytical functions. There are some functions computable in polynomial time for which we do not have such circuits, but in general bounded depth polynomial time threshold circuits seem fairly strong. One indirect evidence of the strength of bounded depth circuits is that we are not able to prove for any explicitly defined function that it cannot be computed by such circuits. In other words, our lower bound methods fail for such circuits even though they seem much more restricted than general Boolean circuits. We cannot exclude that threshold circuits of depth 3 and polynomial size can compute the same class as general polynomial size Boolean circuits (the class **nonuniform-P**).

What consequences for the brain can we draw from what we know about threshold circuits? Let us look at how much faster threshold circuits can be than the ordinary circuits that use gates with only two inputs. One can show that a single threshold function can be computed by a polynomial size[16] logarithmic depth Boolean circuit. It is also obvious that for a nontrivial threshold function the depth of such a circuit must be at least $\log_2 n$, where n is the number of input variables. Hence, if a threshold gate needed a unit time for its computation, then we could compute some functions in constant time on threshold circuits, while Boolean circuits would require at least logarithmic time. It is, of course, unrealistic to assume that a device with a large number of inputs would work as fast as a simple device with only two inputs. However, for adding many electric potentials we do not need more time than for adding two. Thus the delays would be caused rather by the larger distances from which we need to get the electric charges. It seems likely that a living cell cannot produce signals with very high frequency and there must be some intrinsic limitation. When it was not possible to produce components of higher speed, nature decided to use components that have more inputs. Having a component with more inputs certainly helps to compute some functions faster, but the speed-up factor is not very big; in the case of the brain it could be in the order of tens, which is far from several millions, the factor by which current transistors are faster than neurons. Thus the likely reason why our brains still surpass computers in some tasks is rather in the ability of the brain to get large numbers of neurons involved into computations, in other words, to use massive parallelism.

Another indication that this kind of circuit is quite powerful is that it is as powerful as its probabilistic version. From the previous section we know that randomness often helps to get faster and simpler algorithms. Given a randomized bounded depth threshold circuit, one can construct a deterministic circuit that computes the same function with no error, has size only polynomially larger and depth only larger by one.

Apart from being a fairly strong computational device, bounded depth threshold circuits have an important practical advantage: it is possible to design fairly reliable

[16]In fact, $cn \log n$ size, for c a constant.

circuits from somewhat unreliable elements. This may be another reason why the brain uses elements similar to threshold functions.

Some researchers suggested that the great power of the human brain cannot be explained by such simple models as threshold circuits. They say that the extraordinary abilities of the brain can only be explained if we assume that single neurons are able to perform complex operations. I agree that a cell is a very complex system and more research should be done concerning the complexity of the tasks that single neurons can do. But I think that so far we do not have a reason to doubt that threshold circuits of the size and the speed of a brain can actually compute what the brain does.

Finally, one should not forget that the brain evolved during hundreds of millions of years. Thus the complexity of the brain is not determined only by the components and their number, but also by its design, "the computer architecture". When neuroscience progresses to the stage that we will be able to decode some algorithms used by the brain, we may be quite surprised by their intricacy. From the point of view of complexity theory, we should view the brain as a nonuniform device with the extra power that non-uniformity gives.

So let us now look at the ways how evolution could have produced such complex structures.

Interlude 2—Computation and Life

It is not a coincidence that computers and brains use electric charges to encode bits of information. If we need components that are fast and simple to produce, it is probably the best solution to use components that process and send electric charges. But we should not conclude that computations are possible only in systems composed of small devices sending electric impulses. When high speed is not needed, it is possible to use other means. A human body uses a lot of chemical signals to control organs. It is much slower than using nerves, but in many cases it suffices and it is an efficient way to send the signal to all parts of the body.[17] Biochemical processes in a living organism are so complex and interwoven that we can also consider them as kind of computation.

This kind of complexity is present already on the level of a single cell. A cell controls its chemical processes by producing proteins that are either directly involved in the processes or function as enzymes. Proteins are produced by transforming the information about a protein encoded on DNA into an actual protein. The pieces of the DNA string that code a protein are called *genes*. Usually, only some genes are active, those that the cell actually needs. The control of a gene may be simple—a lack

[17]Also signals between neurons are transmitted chemically. There is a tiny gap between a synapse and another neuron, the synaptic cleft. An electric signal arriving to a synapse causes release of neurotransmitters into the gap, which in turn triggers, or inhibits, an electric signal in the adjacent neuron. As the gaps are very small, this does not cause much time delay.

of a certain molecule triggers production of an enzyme that enables the production of this molecule. In some cases, however, the mechanism is very complex. In particular, when a cell of a developing multicellular organism differentiates, some genes must be switched off and some must be turned on, and this has to be done without much external influence. Such a complex control is achieved by means of *regulatory genes*. The role of a regulatory gene is not to produce an enzyme that will be used directly, but to enhance or inhibit the activity of another gene. One gene can be controlled by several other genes. In such a case the activation of the gene is a Boolean function of the activation of the controlling genes.

Several different Boolean functions were found in such regulatory systems. This includes negation (one gene suppresses the activation of another), conjunction (a gene is activated if two other genes are active) and disjunction. Conjunction and negation (or disjunction and negation) is a basis of all Boolean functions. Thus we have components to build, in principle, an arbitrary Boolean circuit. A number of such genetic regulatory circuits have been described (e.g., the circuit of the *bacteriophage lambda*). These circuits are not exactly Boolean circuits as I have defined them. The difference is that regulatory circuit may have a large number of feedbacks. This makes the description of the computation more complicated and they may be unstable and may oscillate. However, from the point of view of complexity of computations, they have the same computational power as circuits without feedback.

Computations can also occur on a much higher level. Consider the process of adaptation of organisms of some species to a changing environment. We can view such a group of organisms as a system that receives information from the environment in which the organisms live and reacts to it by adapting the organisms to the particular state of the environment. Such a system, viewed as a computational device, works as follows. It stores information in the strings of DNA which are present in each individual organisms. The information from the environment enters the system by means of removing certain strings from the population (expressed more poetically, Nature tells the system which genomes she likes). The system processes information by editing the strings of DNA.

Let us recall that there are two basic ways of editing genetic information.

1. *Mutations* produce some small random changes of the strings of DNA. They occur with low frequency.
2. In *crossing over*, or *recombination*, pairs of DNA strings exchange some segments.[18] This is the essence of sexual reproduction.

A possible explanation of the role of mutations is that they are useful for producing new genes which are not present in the DNA strings and which are more beneficial for the survival than those that they replace. The role of crossing over is assumed to be in spreading the genes in the population. When crossing-over is

[18]I prefer to use 'crossing-over' since it is less ambiguous than 'recombination'. Recombination often refers to all possible editing operations that ever occur. For example, sometimes a segment of the string is inverted, but this is rather an error, like a mutation, and as such it may be useful, but most often it is detrimental.

present, a beneficial gene spreads quickly and before long almost every organism has this gene. If several new beneficial genes appear in the population, an organism that has them all will appear fairly soon. If there were no crossing-over, it would take a much longer time to select organism with many good genes.

The adaptation is viewed as the process that reduces the number of occurrences of genes that are bad and increases the number of beneficial ones. But now suppose that the expression of a gene is controlled by a complex circuit of control genes. Then the selection mechanism acts on all genes from the circuit. The resulting process is thus more complicated. It is quite possible that this gives much more computational power to the system, but I am not aware of any research done in this direction. In any case this gives at least one advantage: the possibility of switching off the expression of a gene in the population without removing the gene from the DNA strings. This is important because crossing over does not produce new genes and it would take a long time to re-create the gene by mutations once it is lost.

What has been researched is the power of the editing operation of crossing over [226]. Assuming crossing over is not a completely random process, such a system can have very strong computational power. This can be shown in the following computing system motivated by genetics.

The system consists of many strings of the same length, where a particular type of strings may occur several times. It operates in discrete steps; in every step we randomly form pairs of strings and perform the crossing-over operation described below. I will assume that the number of (copies of) strings is even, so that it is possible to match all strings.

In every crossing over the two strings involved are cut in only one spot, the same in both strings. Then two corresponding pieces are switched and reconnected. (Thus the first segment of the first string is connected with the second segment of the second string and the first segment of the second string is connected with the second segment of the first string.) The key assumption is that the spot where the strings are cut is determined by a short context around it. This means that we have a set of rules that determine where the crossing occurs. Such a rule can be, for example:

If the first string contains a segment AAAGGG and the second contains TTTCCC, then cut the first string between AAA and GGG and the second between TTT and CCC. (Then connect AAA to CCC, and TTT to GGG.)

Thus the system is completely determined by the length of strings, the number of strings and a set of crossing-over rules. In order to perform a computation, one sets all strings to a particular form (or forms) which encode the input data. Then we let the system evolve and observe what happens. If the rules are properly chosen and sufficiently many steps are done, the majority of all strings may encode the output value of the function that we wanted to compute.

One can show that in this way the system can simulate Turing machines. But not only that, one can also show that if the number of strings is exponential, then the system can simulate a parallel machine. It is, of course, not realistic to assume populations of exponential size, so the relevant conclusion is rather that the system has the *potential* to work as a parallel machine. This means that if the population is large then it may solve some problems faster than a sequential machine can.

Apparently not much is known about which are the spots where crossing over happens and how random this process is. It has been observed however that crossing over is not completely random. Some short sequences of nucleic bases have been determined that make the possibility of crossing over more likely than crossing over elsewhere. What is much more understood are the transcription processes from DNA to messenger RNA and from messenger RNA to proteins. Since these are complex processes controlled by what is written on the strings of DNA and RNA, it is conceivable that the crossing over process could be a highly complex mechanism as well.

In conclusion I should mention, at least briefly, attempts to use biological elements such as DNA molecules, for computing. Though interesting, I do not find this subject relevant enough to be discussed here at length. The reason why this approach may help compute faster is only because the components used in the devices are smaller than in the current electronic computers. Thus it is only a different way of miniaturization, *not a conceptually new type of computer.* It is likely that further miniaturization of electronic components will produce components comparable in size to such macromolecules.

Notes

1. *Parallel machines.* There is a host of models of parallel computers in the literature. Most of them are (most likely) weaker than what I considered on previous pages, but some are even stronger. The one that I briefly described is essentially the machine proposed by W.J. Savitch and M.J. Stimson [255]. There is agreement among researchers in parallel computations that the true parallelism should enable machines to compute in polynomial *time* what is computed in polynomial *space* by sequential machines (Turing machines). This is called the *Parallel Computation Thesis*.

   Such strong models of parallel computers are, however, interesting only from the theoretical point of view. For designing efficient algorithms, we rather need machines that can solve nontrivial problems in *sublinear time*. For Turing machines, sublinear time is uninteresting—the machine cannot even read the whole input. Thus we need different models. The focus is not on enlarging the class of efficiently computable sets and function, but in finding faster algorithms for what we already know to be computable in polynomial time.

   The standard model used for this purpose is PRAM, the *Parallel Random Access Machine*. I will not introduce this concept, since one can estimate the power of PRAMs by Boolean circuits. If we ignore uniformity (PRAM is a uniform model of computation, circuits are nonuniform), then the time of PRAMs roughly corresponds to the depth of circuits, and the number of processors to the size of circuits (more precisely, size divided by depth). The depth of a circuit is the length of the longest path from an input node to an output node. Thus the problems for which one can gain a substantial reduction of time by using parallel computers are those which can be computed by polynomial size circuits with sublinear depth. The smallest nontrivial depth restriction is logarithmic ($c \cdot \log n$, for c a constant, n the length of the input). We conjecture that there are functions computable by polynomial size circuits, but not computable by polynomial size circuits of *logarithmic depth*. Again, this a widely open problem.

2. *Threshold circuits*. For a lot of functions it has been shown that they can be computed by threshold circuits of constant depth and polynomial size. These functions include the common numerical functions, such as addition, multiplication, division, sine, exponentiation, square roots and others. When the output should be a real number, it is computed with exponential precision. Furthermore, such circuits also compute some functions that are conjectured to be pseudorandom. Thus an exponential lower bound on the size of threshold circuits computing an explicitly defined function seems a very difficult problem, as it would either refute some conjecture about pseudorandomness, or it would be a non-natural proof in the sense of Razborov-Rudich. There are also some fairly simple functions that we do not know how to compute by bounded depth and polynomial size threshold circuits; in particular, it is the connectivity of graphs.

   If we want to explain the function of the brain using threshold circuits we have to take into account the problem of precision. We cannot expect any actual device to be able to compute with precise real numbers. Let us see what precision is needed in threshold circuits.

   The first observation is that given a definition of a threshold function, we can replace real weights and a real threshold by rational weights and a rational threshold. To see it, imagine the 2^n input strings of the threshold function as the vertices of the n-dimensional cube in $\mathbb{R}^n$. If the function is defined by

$$\sum_{i=1}^{n} a_i x_i \geq b, \tag{5.5}$$

   then think of the function as separating the accepted inputs from the rejected inputs by the hyperplane $\sum_{i=1}^{n} a_i x_i = b$. Thus we only need to move the hyperplane slightly so that it still separates the same vertices and has rational coefficients. Once we have rational coefficients, we can multiply the inequality (5.5) by a suitable natural number and we get all coefficients integral. Having integral coefficients, we can measure the precision required by the gate by the size of the coefficients (their absolute values).

   In general the coefficients have to be exponentially large. It is unrealistic to assume that neurons can compute with such a precision. However, one can show that such high precision threshold gates can be avoided. One can show that a given circuit with general threshold gates can be transformed into a circuit with threshold gates that use only coefficients of polynomial size and this transformation increases the size of the circuit only polynomially and the depth only by one. If we allow a slightly larger increase of depth, by a constant factor, then it suffices to use very special gates. These gates are the majority functions of possibly negated inputs. In terms of coefficients it means that we take $a_i = \pm 1$, for $i = 1, \ldots, n$, and $b = n/2$.

Let me now address the question of reliability, which is another thing that could negatively influence the power of actual threshold circuits. I will show that with threshold circuits we can also cope with this problem very well.

It impossible to construct circuits that could tolerate arbitrary errors; one has to make some assumptions about the nature of errors. One reasonable assumption is that a threshold gate makes an error with probability at most $1/4$ and if the inputs are such that they sum to value far from the threshold, it makes an error with very small probability. This means that if gate g is defined by the inequality (5.5), then it operates reliably on inputs $(x_1, \ldots, x_n)$ for which either $\sum_i a_i x_i \ll b$ or $\sum_i a_i x_i \gg b$.

In such a case we can replace every threshold gate by a small circuit that operates always with high reliability. The circuit is simply

$$MAJ_m(g_1, \ldots, g_m),$$

where MAJ_m is the majority gate and $g_1, \ldots, g_m$ are copies of the simulated gate g. Suppose g makes an error with probability ε, $0 < \varepsilon < 1/2$. Then if we have an input $(x_1, \ldots, x_n)$ for which g should be 0, then in the average εm of the gates $g_1, \ldots, g_m$ will output 0 and the rest will output 1. If we have an input $(x_1, \ldots, x_n)$ for which g should be 1, then it will be the other way around. More importantly, the law of large numbers ensures that in the first case the probability that $m/2$ or more gates $g_1, \ldots, g_m$ will output 1 will be exponentially small, and symmetrically in the second case the probability that less than $m/2$ gates will output 1 will be exponentially small. (Exponentially means exponentially in m.) Thus if m is sufficiently large, we can replace gate g by this circuit, and it will make errors only with slightly larger probability than the gate MAJ_m itself. The error will add up with more gates to simulate, but starting with a very small probability of error we will still get a reasonably reliable circuit.

3. *Neural networks.* Threshold circuits can be viewed as a very special kind of neural networks. In neural networks the gates are, as a rule, smooth approximations of threshold functions. The research into neural networks focuses on learning algorithms, which are algorithms that adapt the weights of the gates using examples of input and output values of an unknown function f in order to train the network to compute f. From the point of view of complexity theory, they are not more powerful than threshold circuits, unless one uses very special functions as the gates.

5.4 Quantum Computations

Let us recall that the Church-Turing thesis is the conjecture that every computable function is computable on a Turing machine (or an equivalent device). The concept of computability used in the thesis is not a precise mathematical notion; it refers to the intuitive concept of computability, which we usually associate with a process that can be physically realized at least in principle. So let us now consider the thesis from the point of view of physics. If we imagine that computations are done by mechanical devices, the thesis seems to be very likely true. But physics has much more to offer than mechanics. The traditional concept of an algorithm is based on the idea that an algorithm performs a finite number of elementary operations. This is a very 'mechanistic' approach. Let us abandon such presumptions and simply ask which functions can arise in physical processes. Then a computation can be viewed as a physical experiment in which we set the initial conditions according to the input data and get the output as the result of the experiment. At the early stages of computer science such devices were proposed. They were called *analog computers* in contrast to *digital computers*. For example, we can compute addition of two real numbers by combining the corresponding voltages. These devices suffered from very poor precision and therefore they were soon abandoned (a mechanical device of such a kind that survived the longest was the slide rule). As modern computers developed, more computations were done also in physics. Then the problem reappeared again, but from the opposite side. The question was whether there are physical processes that we cannot simulate using computers. In 1982 the American physicist Richard Feynman brought up this problem in connection with quantum physics. There are several reasons why quantum physics should be studied from this point of view. First, in quantum physics one can perform experiments with extremely high precision. The second general reason is that the quantum world is so much different from our experience. The main reason, however, why quantum computing is so popular nowadays is that it seems that it can really help us to compute things that we are not able to compute with classical means.

In spite of sometimes looking very bizarre, quantum physics does not contain phenomena that are not computable by ordinary Turing machines. Thus from the point of view of pure computability quantum physics does not give us new concepts. However, if we take into account the complexity of computations it seems that we can really gain something. In his seminal paper Feynman gave arguments why computer simulations of quantum phenomena may need exponential time and thus may be infeasible [73]. He also suggested that we may be able to construct quantum computers which would be able to perform such simulations, hence they would be more powerful than classical computers. (The word 'classical' is used to distinguish concepts not based on quantum physics from those that use quantum physics.)

The basic principles needed for quantum computations can be explained without much mathematics. In quantum physics it is possible to form combinations of states. Formally, if $S_1, \dots, S_k$ are possible states of a given system $\mathcal{S}$ and $a_1, \dots, a_k$ are nonzero complex numbers satisfying a certain restriction, the system can also be in the compound state $a_1|S_1\rangle + \dots + a_k|S_k\rangle$. This is called a *linear superposition* of the states $S_1, \dots, S_k$ and the numbers $a_1, \dots, a_k$ are called *amplitudes*. The meaning of the linear superposition is, roughly speaking, that the system $\mathcal{S}$ is, in a way, in all the states $S_1, \dots, S_k$ at the same time. To distinguish the states S_i from nontrivial superpositions, I will call them *basis states* (it should be noted that this is not an absolute concept, it depends on the way we describe the system). What makes the simulation of quantum phenomena difficult is that k, the number of possible states, can be very large. For example, suppose the system consists of n memory registers, each registers holding one bit. Then the number of all possible states is 2^n. Thus to simulate a quantum computation with n quantum bits we would need to keep track of 2^n complex amplitudes.

The possibility of having a linear superposition of an exponential number of states suggests that quantum computers could compute like truly parallel machines. Instead of having exponentially many processors, one for every process, we could use one processor to work on exponentially many tasks at the same time. But it is not so simple. Suppose that we want to solve a search problem, for instance, to find a factor of a composite number N. A naive approach would be to put the memory of a computer into the superposition that includes all numbers M, $1 < M < N$ and let the computer try to divide N by M. Thus we obtain a superposition of the states of the computer such that at least one of the states contains the solution. This can be done, at least theoretically, but the problem is how to get the solution from the superposition? As I will explain below, the amplitudes determine the probability that we can get a particular state from the superposition. In such a superposition they are exponentially small. In fact, in this way we would not get more than if we just picked M at random and tested if it divides N.

In order to obtain more than one can get by using only randomized algorithms, one has to use the fact that the amplitudes do not have to be positive real numbers. The crucial property is that the amplitudes can sometimes cancel out, which results in increasing the others. A quantum algorithm that is able to achieve more than a probabilistic one, has to be tricky: it has to reduce the amplitudes of unwanted results while increasing the wanted ones. This has been accomplished only in a few cases so far. Thus quantum computations are able to use parallelism to some extent, but it seems that only in some limited way. In particular, we are not able to show that all **NP** problems are solvable in polynomial time on quantum computers, and we rather conjecture that this is not the case.

The two most important problems that are solvable on quantum computers in polynomial time but no polynomial time classical algorithm is known for them are factoring of composite numbers and computing discrete logarithms. Both algorithms were found by Peter Shor in 1996 [268]. These problems play a key role in cryptography, where one uses their apparent hardness. Hence constructing a quantum computer would have rather destructive consequences—we would have to abandon the RSA protocol and others. (But keep in mind that we are not able to prove the hardness of these functions anyway; thus it is not excluded that we can break these protocols even using classical computers.)

Another quantum algorithm, found by L.K. Grover in 1998, is not polynomial time but is very general [109]. This is an algorithm for solving search problems. It has the remarkable property that to find a solution in a search space of size N, the algorithm needs only $c\sqrt{N}$ steps, for some constant c. For problems that we are able to solve only by the brute force search, this is a significant speed up. Since the search space is typically exponential in the size of inputs, this means that a quantum computer of the same speed as a classical computer would be able to handle almost twice as large input data.

In the following decade a number of generalizations of these algorithms and some new ones have been found. I will not discuss them here because they are rather tech-

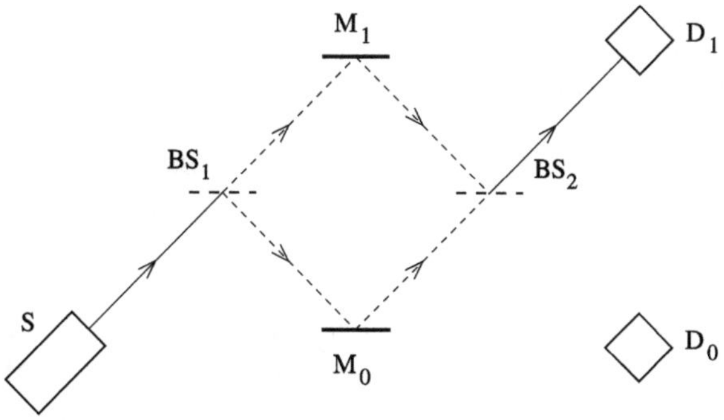

Fig. 5.5 The Mach-Zehnder interferometer

nical results and none of them has presented a breakthrough comparable to Shor's algorithm. A lot of work has been done in related areas of quantum information and quantum cryptography, which also will not be treated in this book (except for a brief remark in Notes).

A Brief Visit in the Quantum World

Several important features of quantum physics can be explained on a simple experimental set-up called the *Mach-Zehnder interferometer*. It consists of a source of coherent light, such as a laser, which sends a beam to a beam splitter, say a half-silvered mirror. After the beam splits into two, the two beams are reflected to the same point at another beam splitter. There are two detectors behind the second beam splitter placed in the direction of the two beams, see Fig. 5.5. Each beam should split into two on the second beam splitter, thus one may expect that both detectors should detect light. However, if all the components are accurately placed, only detector D_1 records light. If the splitters and mirrors did not absorb any light, then the intensity of the beam reaching detector D_1 would be the same as at the source.

This can easily be explained using the fact that light is a special type of electromagnetic wave. Each of the beams $M_0 \to BS_2$ and $M_1 \to BS_2$ splits on the beam splitter BS_2 into a pair of beams $BS_2 \to D_0$ and $BS_2 \to D_1$. The resulting two beams $BS_2 \to D_1$ have the same phase, so they add up, whereas the two beams $BS_2 \to D_0$ have opposite phases, so they cancel out. This is called respectively *positive interference* and *negative interference*.

Now assume that we gradually decrease the intensity of the light emitted from the source, say, using filters that absorb light. If light was just waves, we should detect smaller and smaller intensity at D_1. What happens in reality is, however, different. If the detector is sensitive enough, at some point it will not register decreasing intensity of light, but instead it will record pulses whose frequency will decrease with the decreasing intensity of the emitted light. This is not caused by the interferometer; it would be the same if we aimed the source directly to the detector. The explanation is that light consists of *quanta*, which we call *photons*. For each wavelength, this quantum is uniquely determined, in other words, the energy of the photon is deter-

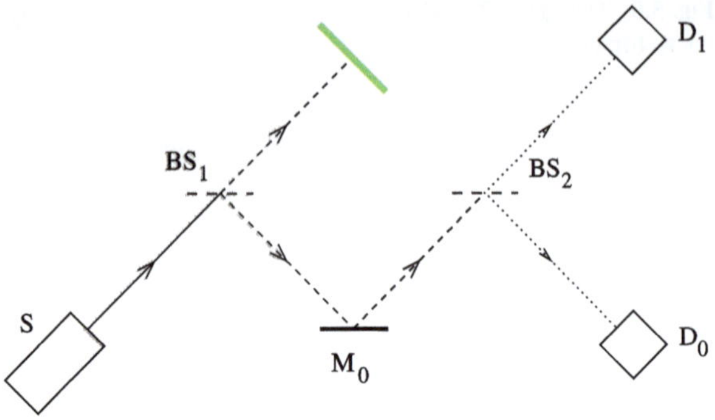

Fig. 5.6 The Mach-Zehnder interferometer with one mirror removed

mined by the wavelength (it is inversely proportional to it). For a given frequency, it is not possible to send a smaller amount of energy than the energy of the photon of that wavelength.

This is the wave-particle duality. We can explain light by waves, as well as the kinematics of particles, but if a phenomenon has features of both, we need a new theory. Indeed, if the source sends a single particle to the interferometer, then, according to classical notions, the particle has to choose which way it passes the interferometer. We can confirm it by temporarily putting two detectors in place of the two mirrors (or at any place on the two paths). Then we register a photon always at only one detector, which suggests that the particle did not split into two. But how can we have interference on the second beam splitter if only one particle arrives there?

Suppose we block one of the paths between the two beam splitters, say, we remove mirror M_1. Then, in terms of waves, there is no interference on the second beam splitter, hence both detectors detect light, each detecting one quarter of the original intensity (see Fig. 5.6). But now suppose that the source emits a single photon. With probability $1/2$, it will choose the path that is not blocked and after reaching the second beam splitter it will go to one of the detectors with equal probability. Compare this with the original situation in which the photon always goes to detector D_1. In particular also the photons that go via the lower path. Thus if we think of a photon as a particle that always has a definite position, we get a contradiction. It looks like the photon senses if mirror M_1 is present in spite of going a different route; if the mirror is present it always goes to D_1, otherwise it sometimes goes to D_0.[19]

It has been proposed that in this way we are able to send a signal somewhere without sending any physical object there. In the setting of the Mach-Zehnder interferometer we can send such a "signal" from the position of mirror M_1 to detector D_0. To send a signal we simply remove the mirror. Then we argue as follows. If the mirror is present, no signal reaches D_0, so there is no communication present.

[19]One may suggest that a photon is not just a point, but rather a wave packet, and a part of this packet may touch mirror M_1 while the center of the packet will be at M_0. But this explanation also does not work, since there is no restriction on the distances in the interferometer. Even if the distances were cosmic, it would behave exactly the same way.

If the mirror is removed, only photons that use the lower path reach D_0 and these are photons that did not reach the position of mirror M_1, so we can say that we have nothing to do with them. If, moreover, detector D_0 controls some action that we are not supposed to do, we can claim that we have not caused it.

This is just playing with the interpretation of what is going on in the Mach-Zehnder interferometer. The next example, the *Elitzur-Vaidman bomb test*, however shows something that we can actually do using quantum physics, but not without it [66]. For this example, we need something that is sensitive to light and is destroyed by it. In the original setting it is a bomb connected with a light detector so that the bomb explodes whenever light is detected. If you want to do it experimentally I suggest using a piece of film instead; it will be safe and will serve the purpose equally well.

In the test we assume that at the place of mirror M_1 there is either the original mirror, or the thing that is sensitive to light, say, the film. Our goal is to determine the presence of the film without destroying it. We will do it using detector D_0. We already know what will happen:

1. if the film is not present, we have the standard setting of the Mach-Zehnder interferometer, so D_0 never records a photon;
2. if the film is present, then after the beam splitter BS_1,
   a. with probability $1/2$, the photon will go up, the film will be destroyed, and no detector detects the photon;
   b. with probability $1/2$, it will go down avoiding the film, and after reflecting from M_0 and reaching BS_2 it will,
      i. with probability $1/4$, go up and reach detector D_1, and
      ii. with probability $1/4$, go down and reach detector D_0.

The point is that we only detect the photon at D_0 when the film *is* present, and when this happens the film is *not* destroyed (case 2.(b) ii). Thus, with probability $1/4$, we can determine the presence of the film using light and without destroying it. Using more complicated settings one can detect the film without destroying it with probability arbitrarily close to 1.

Good scientists should not be satisfied with data they cannot fully explain; they should be curious what is really going on. So, is there a way to watch what the photon actually does? Similar experiments can be done with particles that have nonzero mass and travel at speeds lower than the speed of light, thus it is possible to detect the presence of a particle without deviating it too much off its course. What happens then is that whenever we set up the experiment so that we know which way the particle goes, the interference disappears. It is like observing a magician: if we do not know the trick, it is magic, but as soon as we learn the trick, the magic disappears.

The Quantum Bit

In order to understand quantum computations, we do not need any physics. It suffices to learn a few basic rules of the game called quantum computing and then we can play. Moreover, once we understand quantum computing, we can model interesting quantum phenomena on quantum circuits. Such models are so far restricted to thought experiments, since we do not have a physical realization of quantum computers yet.

The best way to view quantum computations is as a generalization of the matrix model introduced in Chap. 1 (see page 137). In that model a column is interpreted as a memory location and rows correspond to discrete time moments. Thus a row holds information about the current content of the memory. A particular matrix model corresponds to an algorithm or a circuit. It is determined by rules that posit how to rewrite a row to the next one. The crucial condition is that the rules must be simple; therefore we allow only rules that change a small number of the entries.

The quantum version of the matrix model is based on replacing the usual bits by quantum bits.[20] What is a quantum bit? A quantum bit (more fashionably called a *qubit*) is simply a linear superposition of the two classical bits, 0 and 1. Formally, it is the expression

$$a|0\rangle + b|1\rangle,$$

where a and b are complex numbers such that

$$|a|^2 + |b|^2 = 1. \tag{5.6}$$

I put 0 and 1 in these strange brackets as it is customary in quantum physics to denote the states in this way. It is a useful notation introduced by Paul Dirac; an expression $|x\rangle$ is called a *ket vector*, where '*ket*' is the second half of the word '*bracket*'.[21] To give some meaning to this expression it is good to state the first rule.

Rule 1 *If we observe a superposition of states* $a_1|S_1\rangle + \cdots + a_k|S_k\rangle$, *then we see state* S_i *with probability* $|a_i|^2$.

This rule implicitly says that we cannot see superpositions; we can only see basis states.[22] The process of looking at a quantum system is usually referred to as *measurement* because in general the result can be a real number. Here we will consider measurements that can give one of a finite set of values. The standard interpretation of what is going on in measurements is that the system suddenly collapses from a superposition to a basis state.

Thus, in particular, if we observe the quantum bit $a|0\rangle + b|1\rangle$, we get 0 with probability $|a|^2$ and 1 with probability $|b|^2$ (where $|z|$ denotes the *absolute value*, the *modulus* of a complex number z). This explains the condition $|a|^2 + |b|^2 = 1$; the total probability must be 1. So far it is not clear why we need complex numbers.

---

[20] We could use alphabets larger than two, but it would be just an unnecessary complication.

[21] We will not need *bra vectors*.

[22] We cannot see the superposition because the measuring apparatus can produce only one of k possible values. But we can set the apparatus differently and then what was previously a superposition may become a basis state. In particular, we can detect in the new setting what was originally a superposition.

This will become clearer after I state the second rule, but before doing so let us talk more about the quantum bit. Whereas the classical bit is determined by two discrete values, the quantum bit is determined by two complex numbers a, b. Thus we interpret it as a vector in the two dimensional complex linear space $\mathbb{C}^2$. The condition on the sum of the squares of the absolute values is simply the condition that the length of this vector is equal to 1.

The next thing that we need to know is how a system changes in time. We envisage that quantum computers will work in discrete steps, like the classical ones, thus we will focus on discrete time. Let us see how a single quantum bit can change. Formally, we ask what transformations of $\mathbb{C}^2$ are those that correspond to physical processes. The answer is *linear transformations*; linearity is the crucial property of quantum physics. There is only one additional restriction: the transformations must preserve the length of vectors. (In fact, we need it only for vectors of length 1, but linearity automatically implies that the lengths of all vectors are preserved.) Linear transformations satisfying this condition are called *unitary*.

Rule 2 *A transition from one superposition to another is a unitary transformation.*

Let us consider a couple of examples. Consider the following unitary transformation H:

$$|0\rangle \mapsto \frac{1}{\sqrt{2}}|0\rangle + \frac{1}{\sqrt{2}}|1\rangle$$

$$|1\rangle \mapsto \frac{1}{\sqrt{2}}|0\rangle - \frac{1}{\sqrt{2}}|1\rangle.$$

Suppose that the initial state is $|0\rangle$. After applying H we obtain

$$\frac{1}{\sqrt{2}}|0\rangle + \frac{1}{\sqrt{2}}|1\rangle.$$

If we observe (perform a measurement on) this state, we obtain 0 or 1 with equal probability. If instead of observing, we apply H again, we obtain the following state

$$\frac{1}{\sqrt{2}}\left(\frac{1}{\sqrt{2}}|0\rangle + \frac{1}{\sqrt{2}}|1\rangle\right) + \frac{1}{\sqrt{2}}\left(\frac{1}{\sqrt{2}}|0\rangle - \frac{1}{\sqrt{2}}|1\rangle\right) = |0\rangle.$$

This is essentially what the Mach-Zehnder interferometer does. The first application of H is the first beam splitter. After a photon passes trough it, it is in the superposition of two states, one taking the upper route the other taking the lower one. After the second beam splitter we get the original basis state. If we observed the state after the first application of H, it would collapse to $|0\rangle$ or to $|1\rangle$, and then the second application of H would put it into a superposition. (A more precise description of the events in Mach-Zehnder interferometer is given in Notes.)

A more concise representation of unitary transformations is by matrices. The above transformation H is represented by the matrix

$$\begin{pmatrix} \frac{1}{\sqrt{2}} & \frac{1}{\sqrt{2}} \\ \frac{1}{\sqrt{2}} & -\frac{1}{\sqrt{2}} \end{pmatrix}. \tag{5.7}$$

An application of the transformation to a vector is simply matrix-vector multiplication. In our example two applications of H give the original state, which can be expressed using matrix multiplication as $HH = I$, with I denoting the unit matrix.

An important property of unitary transformations is that they are invertible. Hence every quantum process can be done in a reverse order. This seemingly makes it impossible to do classical computations on quantum computers, as classical computers use irreversible gates, but in fact it is not a problem; there is an efficient way to transform classical computations into reversible ones. I will get to this later.

As the next example, consider the negation as the only nontrivial reversible operation on one bit. It is defined by the matrix

$$\begin{pmatrix} 0 & 1 \\ 1 & 0 \end{pmatrix}.$$

What is more interesting is that in quantum world we also have a square root of the negation:

$$\begin{pmatrix} \frac{1}{\sqrt{2}} & \frac{1}{\sqrt{2}} \\ -\frac{1}{\sqrt{2}} & \frac{1}{\sqrt{2}} \end{pmatrix}.$$

You can check that this matrix multiplied with itself gives the matrix of the negation.

The next is an example of a unitary transformation that preserves the bits.

$$\begin{pmatrix} 1 & 0 \\ 0 & \mathrm{i} \end{pmatrix}.$$

If we apply it to a quantum bit and observe, we do not see any difference; the probabilities of observing zero and one remain the same. Yet such transformations are important in combinations with others.

A quantum bit is not a mere idealization. There are many experiments that give precisely two possible results. In every such case we can interpret the measured physical quantity as a quantum bit. One that is easiest to visualize is the spin of the electron. Spin in quantum physics is a property of particles which is the quantum version of classical rotation. Let us fix an axis, that is, an oriented line through the electron. Then it can spin around it in two possible ways, the speed being always the same. By convention we say that in one case the spin is *up*, in the other it is *down*. For the given axis, the electron can be in a basis state up, or down, or in any superposition of these two. In the latter case measurements will sometimes give the value of spin up, sometimes down. What is interesting is that given such a superposition we can find an axis for which this is a basis state. Hence we can imagine the spin as an arrow attached to the electron. If we measure in the direction of the arrow we get always value up, if we measure in the opposite direction we get always value down, for other directions, we get values up and down with probabilities depending on the angle between them and the arrow.

A direction in three-dimensional space is determined by two real numbers. This does not seem to fit well with a quantum bit which is determined by two complex numbers. Two complex numbers are given by four real numbers and we have one equation (5.6) that they should satisfy, which gives three degrees of freedom. The explanation is that for a quantum bit $a|0\rangle + b|1\rangle$ only the ratio $a : b$ has an interpretation in the real world. If we have a description of a quantum system, we can multiply everything by a complex unit (a complex number c whose absolute value is 1) and everything will work the same. Thus the mathematical description has an extra parameter that has no physical interpretation.

Quantum Circuits

We need to study a system consisting of more than one quantum bit in order to understand quantum circuits. The generalization from one quantum bit is not difficult, but some caution is necessary. For example, one may be tempted to say that n quantum bits is an element of the direct product of n copies of $\mathbb{C}^2$, thus it is an element of $\mathbb{C}^{2n}$. This is not true, we need a space of much higher dimension.

To get the right concept of n quantum bits, we must realize what the basis states are. If we observe n quantum bits we should see n classical bits. Thus a basis state is a sequence of n classical bits, and a general state is a linear superposition of them. Formally, n quantum bits are the following expression

$$\sum_{e_1,\dots,e_n \in \{0,1\}} a_{e_1,\dots,e_n} |e_1, \dots, e_n\rangle,$$

where $a_{e_1,\dots,e_n}$ are complex numbers satisfying

$$\sum_{e_1,\dots,e_n \in \{0,1\}} |a_{e_1,\dots,e_n}|^2 = 1,$$

and $|e_1, \dots, e_n\rangle$ denotes the basis state where the bits are $e_1, \dots, e_n$. Notice that such a state is determined by 2^n complex numbers, hence we are in the 2^n-dimensional complex linear space $\mathbb{C}^{2^n}$. This is what makes the simulation of quantum circuits by classical computers difficult; if the dimension was only $2n$ it would be easy.

Incidentally, a sequence of n quantum bits is an element of a certain kind of product of n copies of $\mathbb{C}^2$, but it is not the usual direct product, it is the *tensor product* (see Notes).

The transition from one state of n quantum bits to another is again by unitary transformations. Thus every computation step is such a transformation and the function that the quantum circuit computes is the composition of these unitary transformations. If we want to write down such a unitary transformation explicitly, we need a huge matrix, a matrix of dimensions 2^n by 2^n. As in classical computations, we want to decompose a given unitary transformation into a product of some simple elementary ones. The minimal number of elementary unitary transformations needed to express a given unitary transformation U as a product is the complexity of U.

Thus what remains is to say what the elementary unitary operations are. As in classical circuits so also in quantum circuits the elementary operations are defined to be those that operate on a small number of quantum bits. In other words, the operations that change only a constant number of bits. This is a reasonable proposal and it is generally accepted as the right one. However, it is harder to justify it than in classical computations. In quantum systems *very distant* particles can be entangled (the *Einstein-Podolski-Rosen pairs*) and there are systems that are more localized, but consist of *large numbers* of entangles particles (the *Bose-Einstein condensates*); both phenomena have been demonstrated experimentally. Nevertheless, it is reasonable to assume that the elementary operations act only on a small number of bits. How a unitary transformation is applied only to some bits is best seen in examples.

Suppose that the initial state is, say,

$$|00000\rangle .$$

Apply the unitary transformation defined by the matrix H, (see (5.7) above), to the first quantum bit. Then we obtain

$$\frac{1}{\sqrt{2}}|00000\rangle + \frac{1}{\sqrt{2}}|10000\rangle . \tag{5.8}$$

Let us now apply the same transformation to the second quantum bit. We get

$$\tfrac{1}{2}|00000\rangle + \tfrac{1}{2}|01000\rangle + \tfrac{1}{2}|10000\rangle + \tfrac{1}{2}|11000\rangle .$$

Clearly, after applying this to all bits we obtain a superposition of all possible basis states with the same amplitudes.

As an example of a unitary transformation on two bits consider the quantum version of the classical Boolean functions that maps a pair of bits (x, y) to the pair $(x, x \oplus y)$. This gate is called *controlled not* since the second bit is negated if and only if the first bit is 1. I will denote it by *CNOT*. The matrix of this unitary transformation is

$$\begin{pmatrix} 1 & 0 & 0 & 0 \\ 0 & 1 & 0 & 0 \\ 0 & 0 & 0 & 1 \\ 0 & 0 & 1 & 0 \end{pmatrix} . \tag{5.9}$$

It is a permutation matrix, as are all matrices of classical reversible transformations. Suppose we apply it to the first two bits of the state (5.8). We thus get

$$\frac{1}{\sqrt{2}}|00000\rangle + \frac{1}{\sqrt{2}}|11000\rangle .$$

We can draw schemas of quantum circuits in very much the same way as ordinary circuits. The special property of quantum circuits is that the number of wires going into a gate always equals the number of wires going out because the number of bits that it processes has to be preserved throughout the entire computation. But whereas electronic circuits are very much like these diagrams, the physical realizations of quantum circuits are quite different. We cannot send quantum bits by wires, which is only a minor problem compared to others hurdles that researchers in this field have to cope with.

Let us consider a very simple quantum circuit in Fig. 5.7. This circuit works with two quantum bits. We can think of the quantum bits as stored in two memory

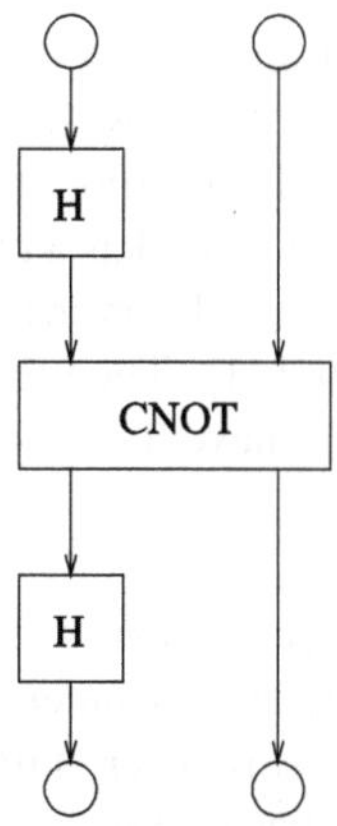

Fig. 5.7 A quantum circuit

registers and certain operations are applied to them. In this circuit we first apply a unary quantum operation to the first register, then a binary quantum operation to the first and the second registers, and finally we apply again a unary quantum operation to the first register. The unary gates are both H, defined by the matrix (5.7), and the binary gate is the *CNOT*, defined by the matrix (5.9). Let the input to the circuit be 00. Then the computation proceeds as follows:

$$|00\rangle$$
$$\downarrow$$
$$\frac{1}{\sqrt{2}}|00\rangle + \frac{1}{\sqrt{2}}|10\rangle$$
$$\downarrow$$
$$\frac{1}{\sqrt{2}}|00\rangle + \frac{1}{\sqrt{2}}|11\rangle$$
$$\downarrow$$
$$\tfrac{1}{2}|00\rangle + \tfrac{1}{2}|10\rangle + \tfrac{1}{2}|01\rangle - \tfrac{1}{2}|11\rangle.$$

After the computation is done, we look at the output bits. According to the Rule 1, we observe one of the four possibilities 00, 10, 01, 11, each with probability $1/4$.

This circuit demonstrates an important phenomenon: the role of entanglement of quantum bits. Notice that the first output bit is 0 or 1, each with probability $1/2$. We have already observed that applying H twice gives the identity mapping. Further, *CNOT* does not change the first bit. Thus we could expect that starting with 0 in the first register, we always get 0 in the first register. The reason why we do not, is that between the two applications H, the first bit gets entangled with the second bit (in the specific situation considered here with the second bit originally being 0, the first bit is simply copied to the second register). You can observe it also on the computation. The terms that would cancel each other if there weren't *CNOT* interposed do not cancel out here because they are distinguished by the second quantum bit.

This example also bears on the fundamental question of what happens when a measurement is done. There are a variety of different explanations; the most commonly accepted one is the following. We assume that quantum laws are universally applicable, hence they apply also to measurements. A measurement is then a process in which the observed entities get entangled with the measurement devices.

If an observation is done by a human, then eventually the state of their brain gets entangled with the observed entities.

In our tiny example we can view the first register as an experimental object and the second as a measuring device. (We can also assume that it is not just a device but a live observer.) First we transform $|0\rangle$ in the first register into a superposition of $|0\rangle$ and $|1\rangle$. Then the application of *CNOT* is a measurement. After the measurement we have the following superposition.

$$\frac{1}{\sqrt{2}}|00\rangle + \frac{1}{\sqrt{2}}|11\rangle.$$

Let us look at it from the point of view of the measuring device. The device can be in two states, but in each of the states of the device, the value of the bit in the first register is unique. So from its point of view the superposition has collapsed to a basis state.

This brings us back to the Mach-Zehnder interferometer. The interposed gate *CNOT* corresponds to a detector that determines which way the photon went. If it is present, we do not obtain interference.

Some people are sceptical about the possibility of constructing quantum computers that would work with a sufficiently large number of quantum bits so that they would be able to solve problems that classical computers cannot. However, no persuasive general argument against quantum computers has been presented so far. One of the objections that has been raised concerns the small amplitudes with which quantum computers will have to work. Indeed, suppose that the computer used a superposition of all 2^{1000} combinations of 1000 bits. If all the amplitudes had the same absolute value, then this absolute value would be 2^{-500}. Then the argument says that no physical instrument can have such a precision. What is wrong with this argument? Firstly, we can easily design experiments in which we have such superpositions—just send a packet of 1000 photons through the Mach-Zehnder interferometer. Secondly, the argument completely misses the point. If we had such a precise instrument we would not need quantum circuit and intricate quantum algorithms. The trick of quantum computing is that *it is possible to obtain something measurable*, that is, a state with a relatively large amplitude, *by combining many states with extremely small amplitudes.* Adding amplitudes is similar to adding probabilities. In efficient probabilistic algorithms exponentially small probabilities add up to large probabilities, such as $1/2$.

The Quantum Computing Thesis

Is there any quantum analog of the Church-Turing Thesis? One can re-state the Church-Turing Thesis using quantum Turing machines, but as I said (already in Chap. 2), as far as computability is concerned we do not get more than we do with classical Turing machines. This is because the advantage of quantum Turing machines is only in their ability to perform some parallel computations. If time and space of computations are not limited, we can simulate parallel computations by sequential ones. So the advantage is apparent only if we consider the complexity of computations. But then we can ask a version of the thesis with limited computational resources: *Are quantum circuits the best possible computational device?* The precise meaning of this question is:

> *Can every physical instrument be simulated by at most a polynomially larger quantum circuit?*

Since we are not aware of any phenomena that would require more complex computations, we conjecture that the answer is positive. This is the *Quantum Computing Thesis*.

It is unlikely that one can answer such a question before we have a unified theory of all physical phenomena. But maybe one can at least show that we cannot get more computational power from quantum theory than we have in quantum circuits. If this is true, then quantum circuits (if they are ever constructed) will not only be a useful computational device, but they will also be useful experimental devices for testing quantum theory.

However, even before quantum circuits with a sufficiently large number quantum bits are constructed, we can at least use the concept of a quantum circuit for thought experiments, as quantum circuits can easily be used to model complex situations. When thinking about them we can focus on information theoretical aspects and we are not distracted by physical phenomena that are specific to physical experiments. Above I have shown a small example of circuits for the problem of measurement, but one can study much larger circuits.

Reversible Computations

Quantum circuits are not generalizations of classical circuits. Recall that quantum circuits are reversible, which property is, in general, not satisfied by classical circuits. Although one can efficiently simulate classical circuits by quantum circuits, it is not a trivial task. Fortunately, reversibility is the only property that we need—a reversible classical circuit can be directly interpreted as a quantum circuit.

What does it precisely mean for a computation to be reversible? I will talk about computations in general, as the essence is the same for circuits and Turing machines. A computation is a sequence of states. A transitions from a given state to the next one is done by an elementary operation. In particular, the next state is uniquely determined by the previous one. But in general, *the previous state* does not have to be determined uniquely from the current state; some information may be lost. (A side effect of the loss of information is the heat that processors have to dissipate.) In a *reversible computation* the previous state must be uniquely determined by the current state. But this is not enough; we want to be able to actually reverse the entire computation—to start with the output and compute backwards to the input. Therefore, we require that, given a state, *the previous state should be computable by an elementary operation.*

It is not difficult to show that every computation can be turned into a reversible one without substantially increasing the running time, or the circuit size. The basic trick is very simple: record all the history of the computation. It works like the editor I am now using. If I erase something by mistake, it is not a disaster; I can simply invoke the command `undo` because what I erased on the screen remains stored somewhere in the memory of the computer.

If we wanted only to show that quantum computers can simulate classical computers, this would suffice. However, what we also need is to know whether we can use well-known classical algorithms as subroutines in quantum algorithms. A reversible algorithm can always be performed by a quantum circuit, but if used inside of a quantum algorithm the additional bits that the algorithm produces may cause problems. In particular, the simulation based on recording the full history has a serious problem: it creates a lot of garbage. In classical computers we can erase all the data that we do not need for the rest of the computation, or we can simply ignore it. In quantum computations we neither can erase, nor ignore it. The garbage data contains information that may result in entanglements at places where we do not want it. Recall that truly quantum phenomena can occur only when some information is not present. (For example, the Mach-Zehnder interferometer does not work if the information about the path of the particle is recorded.) Thus, if possible, the redundant bits should be eliminated from reversible computations.

Therefore, we need a better simulation. Again, another simple trick suffices to show that we can erase all the history except for the input data. More formally, if we have an algorithm for computing

$$x \mapsto f(x),$$

then there is a reversible algorithm for computing

$$x \mapsto \big(x, f(x)\big)$$

whose running time is not essentially longer. The whole point is to erase the history of the computation gradually and in the reverse order. The crucial observation is that if we erase the last item of the history, we can restore it very easily because it is determined by the previous item. Hence the action of erasing the last item is reversible! Thus we can continue until only the input data x remains on the history record.

What about the input bits x, can we get rid of them too? A necessary condition is, clearly, that f is one-to-one, otherwise we would loose information. The answer to this question is, again, not difficult. I will state it in terms of polynomial time algorithms. This theorem is due to C.H. Bennett [19].

Theorem 41 *Given a one-to-one function f, one can compute $x \mapsto f(x)$ reversibly in polynomial time, if and only if one can compute both f and its inverse function f^{-1} in polynomial time.*

The forward direction is trivial. The proof of the opposite direction is based on the fact mentioned before the theorem and a simple idea. For the sake of symmetry, let us denote $f(x)$ by y and express our assumption as follows: we can compute in polynomial time $x \to y$ and $y \to x$. This implies that we can also compute *reversibly* in polynomial time $x \to (x, y)$ and $y \to (y, x)$. Clearly, we can switch x and y in (x, y). Since the computation $y \to (x, y)$ is reversible, we can compute $(x, y) \to y$ reversibly in polynomial time. Combining the computations $x \to (x, y)$ and $(x, y) \to y$, we get a reversible polynomial time computation of $x \to y$.

The theorem, in particular, implies that in order for f to be computable by a polynomial size quantum circuit (without having to store anything but the output value $f(x)$), it suffices that f and its inverse function f^{-1} are computable in polynomial time (on a classical Turing machine). On the other hand, any one-way function is an example of a function for which this is not possible.

Quantum Algorithms

If a quantum computer is ever built, it will most likely be a physical realization of the quantum circuits, as described in previous sections. Hence the best way to define an algorithm is by means of quantum circuits. But an algorithm is not a circuit; it is an idea how to compute a function. Thus when presenting a quantum algorithm, one should first explain the essence by a less formal description and only then give the circuits for the procedures used in the algorithm.

We do not know any *exact* quantum algorithms that are faster than classical. The exactness means that when computing a Boolean function f for a given input x, we obtain the output $f(x)$ with amplitude 1, which means that the measurement of the output gives always the string of bits $f(x)$. Thus it is possible that quantum circuits that compute Boolean functions exactly are not more powerful than reversible classical circuits. The quantum algorithms that apparently are superior to any classical algorithms are not exact; they produce the required output only with a sufficiently large probability. So they behave like probabilistic algorithms, but their essence is different. The output is a linear superposition of strings of bits, one of which is the output value $f(x)$. The amplitude of the string that we need should have a large amplitude, so that we get it with large probability. As in probabilistic algorithms, 'large' means at least $1/2$.

Quantum algorithms are usually presented as a combination of classical and quantum computation. Typically, there is a preprocessing classical phase, then a quantum computation, and finally, a classical postprocessing phase. The main reason for presenting quantum algorithms in this way is that it is easier to prove for a classical algorithm that it is correct (in most cases only well-known algorithms are used in the classical parts of the quantum algorithms, thus one does not have to prove their correctness at all). Also it is good to know which particular part of the algorithm uses quantum effects in an essential way. Another reason is that researchers would like to demonstrate that nontrivial quantum computations are possible. However, the number of quantum bits that the current experimental methods are able to realize is very small. Therefore, it is an advantage to strip a quantum algorithm of everything that does not have to be quantum. That said, it is always possible to perform the entire algorithm on a quantum circuit.

However, not everything that looks classical can be taken out of quantum computations. The typical structure of the quantum phase of quantum algorithms is that we first apply quantum operation to form a linear superposition of many basis states, then we apply a classical reversible algorithm to each of these states, and then we apply again a quantum operation. One may get the impression that the middle part should be easier to realize, since it is classical. But one should always bear in mind that the algorithm is working with a linear superposition of states also in this part of the computation and we need to preserve the quantum nature of the superposition. Thus one cannot use classical electronic components; we need to compute with quantum bits also there.

To get a feel for quantum algorithms, I will briefly sketch the most interesting and the most important quantum algorithm, which is Shor's algorithm for factoring natural numbers. Shor's algorithm allows us to factor numbers in time that is polynomial in the length of the number. Recall that we do not know any polynomial time probabilistic algorithm for factoring and the commonly accepted conjecture is that there exists no such an algorithm. Furthermore, if successfully implemented, it would break many of the currently used cryptographic protocols.

Suppose we want to factor a composite number N. Think of N as a medium size number, which means that number of digits is small enough so that we can compute with it, but N itself is so large that we cannot examine any substantial part of the set of numbers less than N. Several algorithms, running in exponential time, are based on the following simple observation attributed to Fermat. If we find two numbers a and b such that

$$a^2 \equiv b^2 \bmod N \quad \text{and} \quad a \not\equiv \pm b \bmod N,$$

then

$$(a+b)(a-b) \equiv 0 \bmod N \quad \text{and} \quad a+b, a-b \not\equiv 0 \bmod N.$$

Hence both $a+b$ and $a-b$ contain nontrivial factors of N. Thus having such a pair a, b we can find a factor of N by computing the greatest common divisor of $a+b$ and N (or of $a-b$ and N). The greatest common divisor can be computed in polynomial time using the ancient Euclid algorithm. The problem is, however, how to find such a pair a, b efficiently.

One possibility is to find, for some c, its multiplicative period r in the ring of integers modulo N. The period r is the smallest positive integer such that $c^r \equiv 1 \bmod N$. If r is even and if $c^{r/2} \not\equiv -1 \bmod N$, then we can take $a = c^{r/2}$ and $b = 1$. Since r is minimal such that $c^r \equiv 1 \bmod N$, the condition $c^{r/2} \not\equiv 1 \bmod N$, is also guaranteed and we get a factor of N using the above argument. One can show, using elementary number theory, that if we choose $1 < c < N$ at random, then with probability at least $1/2$ either the two conditions above are satisfied, or already c has a common factor with N. Hence, the only essential problem is to find the period.

We do not know how to compute multiplicative periods in polynomial time on classical machines, but it is possible to do it on quantum machines—which is the key component of Shor's algorithm.

Let us forget about the factoring problem for a moment and focus on the problem of computing the period r of a function f defined on integers. The idea of the quantum algorithm for this problem comes from the branch of mathematics called *harmonic analysis*. It is well known that a tone can be decomposed into pure tones, tones whose form is sinusoid. Mathematically this is a decomposition of a periodic function $f(x)$, whose period divides 2π into an infinite sum of the functions $1, \sin x, \cos x, \sin 2x, \cos 2x, \ldots$ multiplied by suitable constants. The constant at a particular function of this set depends on to what degree the function agrees or disagrees with f. In the physical world such a good agreement can be observed as various forms of *resonance* and *interference*. If we want to determine the pitch of a tone, we can play it near a string instrument and watch which string resonates. In particular, if

$$f(x) = a_1 \sin x + b_1 \cos x + a_2 \sin 2x + b_2 \cos 2x \ldots$$

and the only nonzero coefficients are at $\sin \ell x$ and $\cos \ell x$ for which p divides ℓ, then we know that the period of the function is $2\pi/p$.

To explain the main idea of the algorithm, we will consider the following simplified problem. Suppose we have to determine an unknown natural number $p > 2$ which is represented by a regular p-gon with the center at the point $(0, 0)$ and vertices at the unit circle; otherwise its position is completely random. The information that we can get is a randomly chosen vertex of the polygon. Using classical means, we cannot infer anything about p because we do not know how the polygon is rotated, hence what we get is a random point on the unit circle. Now suppose we can get more samples of the vertices, but each time we ask for another vertex, the polygon is rotated by a random angle. Again, the information that we get are simply random points on the unit circle.

However, the quantum version of this problem is solvable. In the quantum setting the vertices of the polygon are not given to us randomly; instead we get a quantum superposition of all vertices of the p-gon, each vertex with the same amplitude $1/\sqrt{p}$. If we do a measurement on this quantum state, we get a random vertex. But we can first transform it to another quantum state and measure the new state. (Instead of applying the transformation we can also view it as doing a *different measurement* on the same state.) Then we can learn relevant information about p.

Mathematically, this means that if the lines through the vertices form angles $k\frac{2\pi}{p} + h$ with the x-axis, for $k = 0, 1, \ldots, p - 1$ and some real number h, then we get the state

$$\sum_{k=0}^{p-1} \frac{1}{\sqrt{p}} |k\tfrac{2\pi}{p} + h\rangle, \tag{5.10}$$

which represents the linear superposition of the vertices of the p-gon. Next we apply a suitable unitary transformation to obtain a state of the form

$$\sum_{\ell \text{ divisible by } p} \alpha_\ell |\ell\rangle, \tag{5.11}$$

where we assume some upper bound M on ℓ in order to get a finite sum (and the sum of the squares of the absolute values of α_ℓ is 1). If all these nonzero amplitudes α_ℓ have the same absolute value, then, by measuring the state, we get a random multiple of p that is less than M. If we get enough samples, we obtain multiples of p from which we will be able to determine p with high probability. The key technical problem is how to compute (5.11) from (5.10). I will come back to it shortly.

The general problem of finding a period of a periodic function f can be reduced to the above special case as follows. Suppose f is defined on integers and has period r. We take a number M that is sufficiently larger than r and consider the function f on the interval $[0, 1, \ldots, M-1]$. Furthermore, I will assume that f takes on r distinct values and M is divisible by r. The latter assumption is not justified; I will explain later how to eliminate it. Now rather than being defined on integers, we can think of f as being defined on M points of the unit circle which form the regular M-gon with one point on the axis x. Let w be one of the r values of f. The points on which the value of f is w form a regular p-gon with $p = M/r$. Thus getting a random sample of an argument and a value of f, $(a, f(a))$, is the same as getting a random vertex from one of these p-gons: the argument a tells us the point on the unit circle and the value $f(a)$ tells us the 'name' of the polygon. Since M is exponentially large, the probability that we obtain vertices from the same polygon is negligible. (This explains why in the simplified problem above we rotate the polygon by a random angle each time we should get a new sample.) Thus using the solution for the problem about polygons, we get p, and then we compute the period of the function $r = M/p$.

This was a high level description of the quantum algorithm for finding periods with a lot of details omitted. Some of these are less important, some are essential. One of the inessential technicalities concerns the divisibility of M by r. Finding a random multiple of r is equivalent to finding r itself, thus we cannot assume that r divides M. But taking M not divisible by r causes only minor problems; we just do not get precise numbers and have to do some rounding to get integers.[23]

What is more important is how the main transformations can be realized by quantum circuits. The computation of the period of the function f starts with computing the linear superposition of all natural numbers a less than M. Since we have the freedom to take any M in a large range (we need $M > 2r^2$ and M must have length bounded by a polynomial in the input N), we take M to be a power of 2. Then such a superposition is the superposition of all binary strings of the length $\log_2 M$. This is fairly easy, as I showed on page 458.

The next step is to compute the linear superposition of all pairs $(a, f(a))$. To this end we only need to compute reversibly $(a, f(a))$ from a. As shown in the previous section, this can be done by a polynomial size reversible circuit if f is

[23]Here is an intuitive explanation. If r does not divide M, the p-gons are not quite regular—one edge is shorter, thus we only know that $\frac{M}{p} < r < \frac{M}{p-1}$. But if $M > 2r^2$, this interval is shorter than 1, hence r is determined uniquely.

computable in polynomial time. The particular function that we use in factoring is $f(x) = c^x \mod N$, which is known to be computable in polynomial time.

The last transformation is called the *discrete quantum Fourier transform*. It is a transformation based on the complex function $\mathrm{e}^{\mathrm{i}x}$ defined, for all real numbers x, by

$$\mathrm{e}^{\mathrm{i}x} = \cos x + \mathrm{i} \sin x.$$

Viewed geometrically in the plane of complex numbers, as x goes from 0 to 2π, the values of $\mathrm{e}^{\mathrm{i}x}$ follow the unit circle (in the counter-clockwise direction). Thus one can easily see that if we sum the values $f(x)$ for $x = k\frac{2\pi}{p} + h$ and $k = 0, 1, \dots, p-1$, we get 0, the center of gravity of the p-gon.

Now consider the function $\mathrm{e}^{\mathrm{i}\ell x}$ for some natural number $\ell > 1$. If ℓ is not divisible by p, and we sum the values of this function for $x = k\frac{2\pi}{p} + h$ and $k = 0, 1, \dots, M-1$, where M is a common multiple of p and ℓ, then, again, one can easily show by computation that the sum is 0. But if ℓ is divisible by p, then we sum the terms of the form

$$\mathrm{e}^{\mathrm{i}(\ell k \frac{2\pi}{p} + h)} = \mathrm{e}^{\mathrm{i}\ell k \frac{2\pi}{p} + \mathrm{i}h} = \mathrm{e}^{\mathrm{i}\ell k \frac{2\pi}{p}} \mathrm{e}^{\mathrm{i}h} = \left(\mathrm{e}^{\mathrm{i}2\pi \frac{\ell}{p}}\right)^k \mathrm{e}^{\mathrm{i}h} = \mathrm{e}^{\mathrm{i}h}$$

because $\frac{\ell}{p}$ is an integer and $\mathrm{e}^{\mathrm{i}x} = 1$ if x is a multiple of 2π. Hence the terms add up to the nonzero value $M\mathrm{e}^{\mathrm{i}h}$. Thus we get a nonzero value if and only if ℓ is divisible by p, which is the resonance principle mentioned above.

Let us now consider Fourier transform applied to a p-gon on the unit circle. The classical Fourier transform consists of the sums of the above form for ℓ in a certain range. Shor's quantum Fourier transform is a unitary transformation that replaces the vertices of the p-gon by the numbers ℓ and their amplitudes are the sums considered above, suitably normalized. Thus we get the quantum state (5.11) which enables us to determine p. Let us just recall that in the factoring algorithm, instead of zeros and non-zeros, we only have small amplitudes and large amplitudes respectively because we cannot assume the divisibility of M by p.

Having a nice formula for a unitary transformation does not guarantee that it can be computed by a polynomial size circuit. Indeed, designing a quantum circuit for discrete quantum Fourier transform is a nontrivial task. I will describe it in Notes.

This amazing algorithm and other applications of the quantum Fourier transform are, unfortunately, good only for very special problems. Ideally, we would like to show that all problems form a complexity class, such as **NP**, can be computed by quantum machines in polynomial time. According to our experience that we have so far, that is very unlikely. It seems that the problems that are solvable by quantum machines in polynomial time must have a very special nature, or they are already efficiently solvable by classical machines (see *The hidden subgroup problem* on page 475).

If we do not insist on having a polynomial time algorithm and are satisfied with any improvement in the running time, then it is different. Grover's algorithm helps us to solve search problems for which we do not have any nontrivial algorithm, only the thorough search of all instances. Such is, for example, the problem of finding a satisfying assignment for a Boolean formula. For this problem, all known algorithms run in time 2^n, (unless the formula is of a special form). Grover showed that using quantum machines we can speed up the search quadratically. Thus in the case of Boolean formulas with n variables, instead of searching all 2^n assignments, we can find a satisfying assignment in time $c\sqrt{2^n}$, for some constant c, which is $c2^{n/2}$. If quantum computers are built and had the same speed as classical ones, this would enable us to solve twice larger instances of this problem.

Another example of a potential application of Grover's algorithm is searching a secret key. Quadratic reduction of the time for this problem, could be a substantial help, but there is a simple remedy: take twice longer keys.

This algorithm is also an example of how we can prove in a particular setting that quantum computations help. The setting is that we use a black box model in which the box acts as an "oracle" that tells us whether the given string of bits is right or not. If we use classical computations, we can only check systematically, or randomly, all inputs, since we do not know how the box computes and we are not allowed to open it. If we search systematically, we need to search all strings in the worst case; if we do it probabilistically, we will check half of them on average. But if we do in the quantum way, the number of steps needed is approximately only the square root of the size of the search space.

Notes

1. *A more precise explanation of the Mach-Zehnder interferometer.* I will denote by $|\nearrow\rangle$ and $|\searrow\rangle$ a photon flying in the corresponding directions. When the photon is reflected from the beam splitter (half-silvered mirror) or from a mirror, then not only $|\nearrow\rangle$ changes to $|\searrow\rangle$ and *vice versa*, but also the amplitude rotates by i. Thus the sequence of states is as follows:

$$|\nearrow\rangle \quad \mapsto \quad \begin{matrix} \frac{1}{\sqrt{2}}|\nearrow\rangle \\ + \\ \frac{1}{\sqrt{2}}\mathrm{i}\,|\searrow\rangle \end{matrix} \quad \mapsto \quad \begin{matrix} \frac{1}{\sqrt{2}}\mathrm{i}\,|\searrow\rangle \\ + \\ -\frac{1}{\sqrt{2}}|\nearrow\rangle \end{matrix} \quad \mapsto \quad -|\nearrow\rangle$$

The three unitary transformations are defined by the following matrices:

$$\begin{pmatrix} \frac{1}{\sqrt{2}} & \frac{1}{\sqrt{2}}\mathrm{i} \\ \frac{1}{\sqrt{2}}\mathrm{i} & \frac{1}{\sqrt{2}} \end{pmatrix} \quad \begin{pmatrix} 0 & \mathrm{i} \\ \mathrm{i} & 0 \end{pmatrix} \quad \begin{pmatrix} \frac{1}{\sqrt{2}} & \frac{1}{\sqrt{2}}\mathrm{i} \\ \frac{1}{\sqrt{2}}\mathrm{i} & \frac{1}{\sqrt{2}} \end{pmatrix}.$$

The state of the photon is not identical after passing through the interferometer; its amplitude is rotated by -1, but the detector cannot distinguish such states.

2. *The mathematics of quantum circuits.* The tensor product is a natural concept defined in every category of structures. Here I will use an explicit definition for vector spaces, which should be more comprehensible for those who have never heard about it. For two complex vector spaces V and W of dimensions c and d, the tensor product is a space Z of dimension cd together with a certain bilinear mapping from the set-theoretical product $V \times W$ into Z. Z is denoted by $V \otimes W$; the mapping is also denoted by $\otimes$. If we choose bases $v_1, \ldots, v_c$ of

V and $w_1, \ldots, w_d$, then the vectors $v_i \otimes w_j$ for $i = 1, \ldots, c$, and $j = 1, \ldots, d$ form a basis of $V \otimes W$. Furthermore, we will assume that the mapping $\otimes$ is chosen so that it preserves unit lengths, that is, the tensor product of two vectors of length 1 has length 1. The tensor product of more than two vector spaces is defined in the same way.

Let us apply this concept to quantum bits. We define a string of n quantum bits to be a vector of length 1 of the tensor product of n copies of the two-dimensional complex space $\mathbb{C}^2$, which is denoted by $(\mathbb{C}^2)^{\otimes n}$. In $\mathbb{C}^2$ we take one orthonormal basis and denote its elements by $|0\rangle, |1\rangle$. This determines the following basis of the tensor product of n such spaces: the set of the vectors $|i_1\rangle \otimes \cdots \otimes |i_n\rangle$, where $i_1, \ldots, i_n \in \{0, 1\}$. In order to simplify notation, we abbreviate $|i_1\rangle \otimes \cdots \otimes |i_n\rangle$ by $|i_1, \ldots, i_n\rangle$.

Given two linear mappings $L : V \to V'$ and $K : W \to W'$, their tensor product $L \otimes K : V \otimes W \to V' \otimes W'$ is defined in a natural way: we define $L \otimes K(u \otimes v) = L(u) \otimes K(v)$ and extend it to the whole space $V \otimes W$ by linearity.

Let $U : \mathbb{C}^2 \otimes \mathbb{C}^2 \to \mathbb{C}^2 \otimes \mathbb{C}^2$ be a unitary transformation; we think of U as representing a binary quantum gate (a quantum gate that works with two quantum bits). Let $n > 2$ and suppose we apply U to the first two quantum bits of the string of n quantum bits. Then, as a unitary transformation of the whole space $(\mathbb{C}^2)^{\otimes n}$, it corresponds to $U \otimes I_{n-2}$, where I_{n-2} denotes the identity mapping on $(\mathbb{C}^2)^{\otimes n-2}$ (which is the tensor product of $n - 2$ copies of the identity mapping on $\mathbb{C}^2$). If this unitary transformation is applied to two non-consecutive quantum bits, then we would have to extend our notation to express it in such a way, but it is clear that such a mapping is the same up to a permutation of the terms of the tensor product $(\mathbb{C}^2)^{\otimes n}$. Similarly, if we have a unitary transformation $U' : C^2 \to C^2$, i.e., a transformation of one quantum bit, the corresponding transformation of the whole space is $U' \otimes I_{n-1}$ (up to permuting the terms in the tensor product).

We will restrict ourselves to the circuits that use at most binary gates; these transformations will be our elementary operation. So given an arbitrary unitary transformation $T : (\mathbb{C}^2)^{\otimes n} \to (\mathbb{C}^2)^{\otimes n}$, we want to know the minimal number k of elementary transformations into which it can be decomposed. This k is the *quantum circuit complexity* of T.

This is a very clean and natural mathematical concept, but what we actually need is a little different. There are two modifications that we have to add.

1. As in classical computations, we should expect that in order to compute efficiently, we will often need more memory bits than just those that store the input bits. Thus when computing some $T : (\mathbb{C}^2)^{\otimes n} \to (\mathbb{C}^2)^{\otimes n}$ we should be allowed to use more than n quantum bits. As usual we will assume that the registers that do not contain input bits are initially set to 0.

2. We should keep in mind that the unitary transformation computed by a quantum circuit only serves to compute a *Boolean function*. The input data that we want to use are basis states, actual strings of zeros and ones, and what we can read from the output is again one of the basis states. It would be natural to

require that the unitary transformation T computed by a quantum circuit must map a basis state to a basis state (thus T would be the quantum extension of a permutation of the set $\{0, 1\}^n$). Unfortunately, this seems also to be a too sever restriction. Presently, we do not have any example of a quantum circuit of this type that would be significantly smaller than known classical circuits computing the same function. Thus we allow the circuit to output a superposition from which we get the required output value with a reasonable probability. Like in the case of probabilistic circuits, it suffices if the probability is at least $1/n^c$, where n is the input size and c is a constant because it enables us to boost the probability to become close to 1 by repeating the computation a polynomial number of times.

3. *Quantum complexity classes.* It seems natural to define *Quantum Polynomial Time*, **QP** as the class of sets computable by quantum Turing machines in polynomial time. (I have not defined quantum Turing machines, but it is not difficult to imagine what they should be.) In the definition of **QP** we require that the input-output behavior of a quantum Turing machine be deterministic. Formally, it means that the machine computes a function $F : \Sigma^* \to \{0, 1\}$ in such a way that for a given input $x \in \Sigma^*$, it outputs $F(x)$ with an amplitude whose absolute value is 1.

   This class is not as natural as one would expect. The problem is that when we require quantum Turing machines to compute precisely, we do not get universal machines. Consider just a unary quantum gate. There are infinitely many such gates and it is impossible to simulate them using a finite number of finitely dimensional unitary transformations with absolute precision. The main reason, however, why researchers do not like this class is that it does not seem to help us to compute faster than using classical deterministic Turing machines. Whether or not **QP** = **P** is an open problem, but it is even hard to conjecture what is true.

   The most important class is the *Bounded Error Quantum Polynomial Time*, **BQP**. This is an analog of the probabilistic class **BPP**, where instead of probabilistic Turing machines we use quantum Turing machines. The condition for accepting an input is the same: if we observe the output of the machine, we see an accepting state with probability at least $2/3$ if the input is in the computed set, and with probability at most $1/3$ if it is not. This can easily be restated in terms of the amplitudes of the accepting and rejecting states.

   BQP contains **BPP** because instead of using r random bits we can take the quantum superposition of all strings of zeros and ones of length r with equal amplitudes. This can be done using r unary gates H (see page 455) and the rest of the computation is done by a reversible circuit. The commonly accepted conjecture is that **BQP** contains more sets than **BPP**. This is supported by the fact that we have polynomial time bounded error quantum algorithms for factoring and the discrete logarithm, whereas no such algorithms are known if we only use probabilistic algorithms. The two problems concern computations of functions, but they can easily be reduced to sets. For example, the problem to determine, for given numbers N and i, if the ith bit of the smallest nontrivial factor of N is 0 is equivalent to factoring integers. This problem is in **NP** as we can guess the complete factorization of N. However, we do not believe that the whole **NP** is contained in **BQP** and we rather conjecture that no **NP**-complete problem is contained in **BQP**.

Concerning the upper bounds, the best upper bound on **BQP** that one can state using the complexity classes defined in this book is **BQP** $\subseteq$ **PSPACE**. Thus we cannot exclude that **BQP** contains sets that are outside **NP**; for all we know, it might even contain **PSPACE**-complete problems.

In quantum computing it is often more convenient to work with circuits rather than Turing machines. Both classes **QP** and **BQP** can be defined using uniform families of quantum circuits.

4. *A quantum algorithm for linear equations.* Let A be an $N \times N$ matrix and b a vector of length N. Suppose we want to solve the system of linear equations given by A and b, which means that we want to find a vector x such that $Ax = b$. A classical algorithm needs time at least N^2 because it has to read the input data.

   Now suppose that the matrix is huge, but we have an efficient algorithm to compute the entries of A and b. Moreover, suppose that we only need to know some properties of the solution. Then, in principle, we may be able to compute these properties in time essentially less than N. A.W. Harrow, A. Hassidim and S. Lloyd [113] found a quantum algorithm that for certain matrices and certain properties of solutions, can solve the task more efficiently than any known classical algorithm. Moreover, they proved that their algorithm is faster than any classical algorithm if **P** $\neq$ **BQP**.

   The essence of their algorithm is to compute the quantum state

$$\frac{1}{\sum |x_i|^2} \sum_{i=1}^{N} x_i |i\rangle,$$

   where $x_1, \dots, x_n$ is a solution. Note that we need only $\lceil \log_2 N \rceil$ bits to represent the indices i. For some matrices, the state can also be computed by quantum circuits of size polynomial in $\log N$. Although the complete information about the solution (except for the normalizing factor) is present in the state, we can get very little from this state. A simple measurement gives an index i with probability $|x_i|^2$. So we only learn that $x_i \neq 0$ (or that with high certainty, $|x_i|$ is not very small). More sophisticated measurements, which may require further quantum computations, can produce information that is hard to obtain by classical means.

5. *How much information is in one quantum bit?* A quantum bit is, by definition, a system from which we can get only two possible values (more precisely it is a system and a particular measurement). Thus the basic tenet is that one quantum bit carries the same amount of information as the classical bit. This is in spite of the infinitely many states in which the quantum bit can be. Yet, a schema has been devised by C.H. Bennett and S.J. Wiesner in which this rule is seemingly violated [20].

   Suppose Alice wants to send two bits to Bob. The natural way is simply to send two bits to Bob. Bennett and Wiesner showed that instead Bob can send one quantum bit to Alice and then Alice will need to send only one quantum bit to Bob. Thus it looks as if the quantum bit sent by Alice carries information of two classical bits. But notice that if we interpret the law about quantum bits corresponding to classical bits more liberally, namely, that *they have to exchange two quantum bits in order to exchange two bits of information*, then the law is not violated.

Let us try to deduce what Alice has to do. Suppose Alice succeeds in sending two bits using a single quantum bit. Then the above law could be violated in a different way. Remember that Alice received one bit from Bob, so the total exchange of bits could be three bits. Should Alice not break the law, she must dispose of the bit that she received. The only way she can do it is to send it back to Bob. Alas, now her task seems even harder—she has to send three bits using one quantum bit!

The clue is actually in sending Bob's quantum bit back. The trick is to modify Bob's bit without looking at it and send it back. There are four possible ways to modify it because Alice can flip the bit and switch the sign of the amplitude. Since Bob keeps his own copy of the quantum bit, he can determine in which way the bit was modified. Four possibilities means two bits.

In quantum physics we are never sure that the idea is correct, unless we check it by a mathematical argument. The schema works as follows. Bob prepares the superposition of two bits

$$\frac{1}{\sqrt{2}}|00\rangle + \frac{1}{\sqrt{2}}|11\rangle.$$

This means that the two bits are completely entangled. Then he sends the first bit to Alice and keeps the second. Alice applies one of the following four unitary mappings to the received quantum bit

$$\begin{pmatrix} 1 & 0 \\ 0 & 1 \end{pmatrix} \begin{pmatrix} 1 & 0 \\ 0 & -1 \end{pmatrix} \begin{pmatrix} 0 & 1 \\ 1 & 0 \end{pmatrix} \begin{pmatrix} 0 & -1 \\ 1 & 0 \end{pmatrix}.$$

Then she sends the bit back to Bob. Thus Bob has one of the following four linear superpositions

$$\frac{1}{\sqrt{2}}|00\rangle + \frac{1}{\sqrt{2}}|11\rangle,\quad \frac{1}{\sqrt{2}}|00\rangle - \frac{1}{\sqrt{2}}|11\rangle,\quad \frac{1}{\sqrt{2}}|10\rangle + \frac{1}{\sqrt{2}}|01\rangle,\quad \frac{1}{\sqrt{2}}|10\rangle - \frac{1}{\sqrt{2}}|01\rangle.$$

As these vectors are orthogonal, Bob can determine them by a measurement. Equivalently, there exists a unitary transformation that maps these vectors to basis states of two bits $|00\rangle, |01\rangle, |10\rangle, |11\rangle$.

It is always instructive to represent such schemas by quantum circuits, see the circuit in Fig. 5.8. The ternary gate applies one of the four unitary transformations to the third quantum bit of its input bits. The choice of the unitary transformation is controlled by the first two input bits, in a similar way as in the control not gate. This gate can be replaced by two binary gates. I leave to the reader to figure out the unitary transformations computed by the gates. Once you know what they should do, it is easy.

6. *More details about computing the period of a function.* Let f be a periodic function defined on integers whose period is less than some number N. The length of N (which is $\approx \log_2 N$) will be our input size. We assume that f is

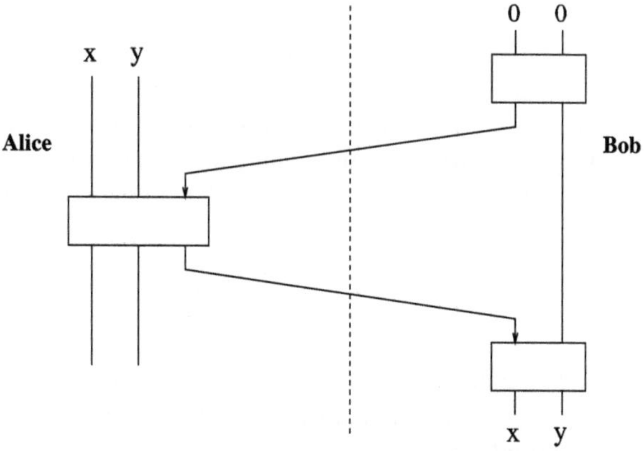

Fig. 5.8 Alice sends two bits using one qubit

efficiently computable, namely, it is computable by a classical deterministic algorithm that runs in polynomial time.

Let M be a number which is a power of 2, is sufficiently larger than N, and its length is polynomial in the length of N.

The first step is to compute the linear superposition of all numbers $0, 1, \ldots, M-1$, all with the same amplitudes:

$$\frac{1}{\sqrt{M}}\sum_{a=0}^{M-1}|a, 0\rangle.$$

The 0 in $|a, 0\rangle$ indicates that we use more registers for quantum bits than just those for a, and they are initially set to 0. (I have already explained how to produce such a superposition, see page 458.)

Next we compute $f(a)$ and put it into the free memory registers. So the next state is given by the following expression:

$$\frac{1}{\sqrt{M}}\sum_{a=0}^{M-1}\big|a, f(a)\big\rangle. \tag{5.12}$$

Again, we already know how to do it using a reversible, hence also quantum circuit (see page 461).

The last step is the quantum Fourier transform. This is the unitary transformation defined by

$$|x\rangle \mapsto \frac{1}{\sqrt{M}}\sum_{\ell=0}^{M-1} \mathrm{e}^{\frac{2\pi \mathrm{i}}{M}\ell x}\,|\ell\rangle.$$

Thus applying it to the first number of the state in which we left our quantum computer (5.12), we get

$$\frac{1}{M}\sum_{a=0}^{M-1}\sum_{\ell=0}^{M-1} \mathrm{e}^{\frac{2\pi \mathrm{i}}{M}\ell a}\,\big|\ell, f(a)\big\rangle. \tag{5.13}$$

Here it is important that no information about the computation of $f(a)$ was present before applying the Fourier transform, except for a and $f(a)$.

When observing this state (in terms of quantum physics, measuring this state) we see a pair (ℓ, d), where $d = f(a)$ for some a, with probability equal to the square of absolute value of the amplitude of this state. In order to compute the amplitude at $|\ell, d\rangle$, we have to add all terms with $|\ell, f(a)\rangle$ in which $f(a) = d$. Thus we get

$$\frac{1}{M} \sum_{a,\ f(a)=d} \mathrm{e}^{\frac{2\pi\mathrm{i}}{M}\ell a}\, |\ell, d\rangle .$$

Let r be the period of f; then $f(a) = d$ is equivalent to $a = kr + s$ for some s, $0 \leq s < r$, determined by d. Further, let $p = M/r$ and $h = 2\pi s/M$. Then the amplitude at $|\ell, d\rangle$ is

$$\frac{1}{M} \sum_k \mathrm{e}^{\frac{2\pi\mathrm{i}}{M}\ell(kr+s)} = \frac{1}{M} \sum_k \mathrm{e}^{\mathrm{i}(2\pi \frac{\ell}{p} k + h)} .$$

If r divides M, then p is an integer, and the analysis of this expression is easy (we have already done it). In such a case we see those ℓ that are random multiples of p. Having sufficiently many such numbers we can determine p by taking the greatest common divisor of them, whence we also get r.

If r does not divide M, which is typically true, the analysis is slightly more complicated. We take p to be the smallest integer larger than M/r. Since $p \neq M/r$, we get with non-negligible probability also some ℓ which is not a multiple of p; so taking the greatest common divisor does not work. Therefore, we need an argument that would determine r with a sufficiently large probability from a single ℓ. Suppose we obtain an ℓ which is a multiple of p, say $\ell = dp$. Then $\ell/M = dp/M$ is very close to d/r. Also we have a good probability that d and r are mutually prime. In such a case, applying the theory of rational approximations to ℓ/M will produce the fraction d/r, in particular it will give us r. This is the way to analyze the algorithm; to do it formally, however, requires some computations and the use of some results from number theory. I omit these details.

It is, however, important to realize that I skipped an essential part which is to show that the quantum Fourier transform can be computed by a polynomial size circuit. I will confine myself to defining the quantum gates that are used in the circuit in general, and drawing the circuit for $M = 2^4$, see Fig. 5.9.

One gate that we need is the unary gate defined by the matrix H, see (5.7), page 455. The others are binary gates defined by the matrices

$$S_k = \begin{pmatrix} 1 & 0 & 0 & 0 \\ 0 & 1 & 0 & 0 \\ 0 & 0 & 1 & 0 \\ 0 & 0 & 0 & \mathrm{e}^{\mathrm{i}\pi/2^k} \end{pmatrix},$$

for $0 < k < m$. The transformation S_k does nothing if at least one bit is 0; if both bits are 1, it rotates the phase by the angle $\pi/2^k$. The key for understand-

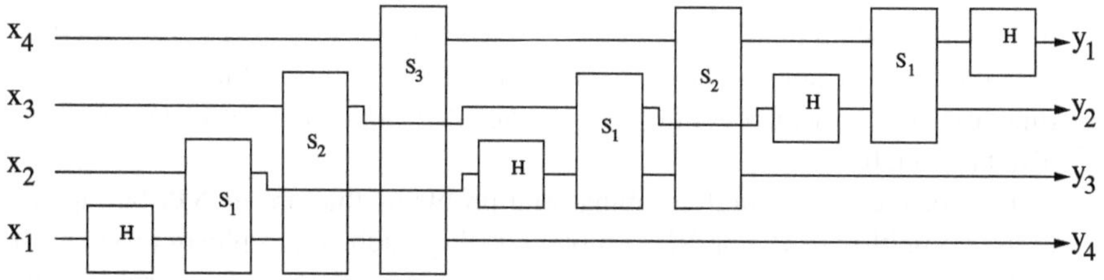

Fig. 5.9 The circuit for the quantum Fourier transform with $M = 2^4$

ing the circuit is the following formula of the quantum Fourier transform. Let $x_1, \dots, x_m$ denote the bits of the number x, with x_m being the most significant bit. Then the transform can be defined by

$$|x_1, \dots, x_m\rangle \mapsto \frac{1}{\sqrt{M}} \bigoplus_{j=1}^{m} \left(|0\rangle + e^{\pi i \sum_{t=0}^{j-1} x_{m-t}/2^t} |1\rangle\right).$$

I leave the verification to the reader.

7. *The hidden subgroup problem.* Having a set of quantum algorithms only for some particular problems is unsatisfactory from the point of view of theory. We would rather like to know what is the essential property of problems that makes them solvable by quantum computers in polynomial time. The so far best candidate for such a characterization is the *hidden subgroup problem*. This is the class of problems defined as follows. Given a finitely presented group G and an unknown subgroup H, one should find a set of generators of H. The group G is given quite explicitly, which means that we can decide in polynomial time if a string of bits is an element of the group, and we can compute the products and inverse elements in polynomial time. The hidden subgroup is determined by a function f which is constant on the cosets of H and takes on different values on different cosets.

   The most important polynomial time quantum algorithms fall into this class. Let us consider the period finding problem as an example. The group G is the additive group of integers $\mathbb{Z}$. The function f is the function whose period we want to compute. Suppose r is the period and f takes on r different values. Then the hidden subgroup is $r\mathbb{Z}$, the set of multiples of r. This subgroup is generated by r (or by $-r$), thus finding a set that generates it is equivalent to finding r.

   In the polynomial time quantum algorithms that we have, the group G is, moreover, commutative. Furthermore, one can solve the hidden subgroup problem in quantum polynomial time for every commutative group. This is very interesting, but not surprising, considering the fact that all commutative groups are products of cyclic groups, and the hidden subgroup problem for cyclic groups is essentially the period finding problem. The quantum Fourier transform can be defined for every finite group and for some noncommutative groups polynomial size quantum circuits have been found. Yet we do not have any nontrivial polynomial time quantum algorithm for the hidden subgroup problem for a class of noncommutative groups. More recently G. Kuperberg [174] found a *subexponential* (more precisely, running in time $2^{c\sqrt{n}}$) quantum algorithm for dihedral groups, which confirms the belief that noncommutative groups may also be tractable.

One of the few natural combinatorial problems that are in **NP**, but apparently are neither in **P**, nor **NP**-complete, is the graph isomorphism problem. It is the problem, for two given graphs, to determine if they are isomorphic. This problem can be presented as a hidden subgroup problem for a noncommutative group, hence we would get a polynomial time quantum algorithm for this problem if we showed that the general hidden subgroup problem is solvable in polynomial time on quantum machines.

8. *Quantum cryptography.* When quantum computers are built, we will need to develop new crypto-systems. It is possible that these systems will use coding and decoding functions that are efficiently computable only by quantum computers.

   Current quantum cryptography focuses on a different approach. Suppose we want to send a message x consisting of n bits. Let r be a random sequence of zeros and ones. Then we can encode x by the pair of strings (r, s), where $s = x \oplus r$, the bitwise sum of x and r modulo 2. If we send only one of the two strings, then a potential eavesdropper cannot learn any information because it is a random string. Now, it is possible to send quantum bits in such a way that the receiving party always detects any attempt to tamper with the message. This can be used to design protocols for secure communication. The basic idea of such a protocol is as follows. Suppose Alice wants to send x to Bob. Alice first sends r. If Bob determines that the message has not been tampered with, he confirms it to Alice. If Alice determines that Bob's confirmation is authentic, then she sends the second part s. The details are more complicated and would take us too far afield.

   The advantage of this approach is that one can prove that it is secure. Thus unlike the classical cryptographic protocols, the security of quantum cryptography is not based on any unproved assumptions. Such communication has been experimentally demonstrated.

9. *The many-worlds interpretation of quantum physics.* The classical (also called *Copenhagen*, or *Bohr*) interpretation of quantum physics is based on distinguishing between two phenomena: (1) a freely evolving system and (2) a measurement. A freely evolving system is governed by the Schrödinger equation. The essence of this equation is that the infinitesimal changes of the system are linear. The discrete approximation that we have considered here when discussing quantum computations is based on applying unitary transformations. Such a system is, in general, in a linear superposition of several states. A measurement is a process in which the system abruptly and randomly collapses to one of the states from the superposition. The probability of collapsing to a particular state is given by the square of the absolute value of the phase. I was using this interpretation when explaining the basics of quantum physics needed for quantum computations.

While nobody has any problem with the first part, the other part, the measurement, raises a lot of questions: what is a measurement? what is a measuring apparatus? is human presence necessary? and others. Among these the main one is: why can we not apply the same rules as we use for freely evolving systems also to measuring instruments and observers?

A solution in which measurement and observers obey the same laws as freely evolving systems was proposed by Hugh Everett III in his PhD thesis in 1956. The main idea is that a measurement is the process in which a measuring apparatus becomes *entangled* with the measured state.

Example Suppose an apparatus $\mathcal{A}$ has three states $A_?$, A_0 and A_1. The state $A_?$ the initial state, the two states A_0 and A_1 are the states in which $\mathcal{A}$ shows 0, respectively, 1. Suppose that we want to measure the quantum bit $2^{-1/2}|0\rangle + 2^{-1/2}|1\rangle$. The initial state of the system consisting of the apparatus and the quantum bit is

$$|A_?\rangle \otimes \left(2^{-1/2}|0\rangle + 2^{-1/2}|1\rangle\right) = 2^{-1/2}|A_?\rangle|0\rangle + 2^{-1/2}|A_?\rangle|1\rangle. \qquad (5.14)$$

After the measurement we get

$$2^{-1/2}|A_0\rangle|0\rangle + 2^{-1/2}|A_1\rangle|1\rangle. \qquad (5.15)$$

Observers, however, do not see the apparatus in a superposition; they only see one of the possible states. The reason is that observers also get entangled with the states of apparatuses. This presupposes that we allow superpositions of people in different states, an unacceptable idea for some philosophers. The typical argument against this interpretation is: why don't we see the superposition of people in different states? The answer is again: because of the entanglement. By watching a person we get immediately entangled with his or her state.

The complete picture of a measurement is more complex. First, the apparatus gets entangled with the system and then very quickly the world around gets entangled with it too because the macroscopic apparatus interacts very strongly with the matter around it. In particular, human observers also get entangled, most likely, already before they read the data. The entanglement then spreads through the universe. This is the explanation of the "collapse of the wave function". This, however, requires postulating the existence of alternative universes because now we have to talk about superpositions of whole chunks of the world, which is an even more controversial idea.

If we interpret $\mathcal{A}$ in the example above as the rest of the world when the quantum bit is taken away, we can use the same formulas (5.14) and (5.15). This is popularly explained as the world $A_?$ splitting into two worlds A_0 and A_1. This is not quite precise because what actually splits is only the part without the quantum bit. (Furthermore, whether some part of the universe splits or joins depends on the basis that we use to describe it.)

Most of the arguments against this interpretation point out that we do not observe superpositions of macroscopic objects and, therefore, it is unlikely that observers, not to say the whole universe, can be a superposition of several different states. The absence of such phenomena can, however, be easily explained by the concept of *environmental decoherence*. Decoherence means that a system that starts in a superposition of states gets spontaneously entangled with the environment and thus loses its quantum nature. This is because it is impossible to completely isolate a system in a superposition of states from the environment. Even individual elementary particles kept separately in vacuum interact with the matter outside. The bigger the object is, the stronger the interaction is and the stronger the interaction is, the shorter the time is before the superposition collapses to one of the states. The decoherence time is extremely short already for systems with a small number of atoms. Therefore, we do not observe quantum phenomena when large objects are involved.

In this interpretation it is also possible to explain the probabilities of collapsing to a particular state in the measurement process. The alternative universes in the superposition have amplitudes and the square of the absolute value of the amplitude of a universe U determines the probability that we are in U. Since we always are only in one of the alternative worlds, we cannot test these probabilities. What we can only do is to assume that we are in a world that is not "very unlikely". This sounds rather suspicious, but as a matter of fact, *this is the standard assumption in all probabilistic reasonings.* I will show in an example how one can use it to justify the probabilities of measurements.

Example Suppose we study the quantum bit $2^{-1/2}|0\rangle + 2^{-1/2}|1\rangle$. The probability $1/2$ that we get 0 (or 1) manifests itself only if we perform many measurements. So suppose that we measure n non-entangled copies of this quantum bit. When measuring these quantum bits we obtain very likely $n/2 \pm \sqrt{n}$ ones, according to the law of large numbers. The many-worlds explanation of this fact is that the total amplitude of the worlds in which the string has $n/2 \pm \sqrt{n}$ ones is in absolute value close to 1. So it is likely that we will end up in one of these worlds.

10. *Quantum proofs.* 'Quantum proofs' usually refers to interactive quantum protocols. There are a number of results showing that very likely one can define larger complexity classes by allowing quantum states and measurements instead of mere randomness. Here I will briefly describe a concept of non-interactive quantum proofs introduced in [228]. This concept was defined for propositional logic, but it can be generalized to first order logic. The basic idea is to allow the linear superposition of strings of formulas in the proof.

Our aim is to define a kind of proof systems that corresponds to the usual proof systems in propositional logic that are based on axiom schemas and deduction rules. Such systems are called Frege systems and we will learn more about them in the next chapter.

A *quantum deduction rule* is a unitary transformation U on a (small) finite set of strings of propositions S with the following property. If $\Gamma, \Delta \in S$ and Δ occurs in $U\Delta$ with nonzero amplitude, then the propositions of Δ logically follow from those in Γ. One can show that a quantum deduction rule is invertible—not only as a linear operator, but also logically. This means that U^{-1} is also a deduction rule.

In order to define a quantum proof, we have to view proofs as a rewriting process. The initial state is always the same, say, a string of $\top$s, where $\top$ is a constant for truth. In each step we rewrite the string by a deduction rule. A proposition is proved if it appears in the string obtained in this way.

In the quantum setting we apply quantum deduction rules and thus at each step we have a linear superposition of strings of propositions. To prove a proposition ϕ we only need to have ϕ occurring with a sufficiently large amplitude. To formalize the rewriting process we use quantum circuits. So, formally, a quantum proof is a quantum circuit with certain properties.

It is not difficult to show that such a proof encodes classical proofs of essentially the same size as the circuit (of the same propositions). However, there are arguments that support the conjecture that in general it is impossible to extract a classical proof from a quantum proof in polynomial time. It is conceivable that this is impossible even using quantum circuits. So it is theoretically possible that we will be able to prove a proposition with a quantum circuit, but we will not be able to produce a classical proof, although we will know that there is one that is not too long.

11. *Communication with extraterrestrials—continued from page 80*. Suppose we already could construct quantum computers. Then a reasonable message to send out would be to say "*We have quantum computers*". It will tell the recipients that our science is quite advanced, at least in physics and mathematics. One of the messages sent out, the *Arecibo Message*, contained a similar thing: a message that we know what life is based on—a picture of a double helix.

Whether or not it is a reasonable thing to do aside, it is an interesting problem how to formulate such a message. We would like to send a message that would show the solution of a problem that can only be solved using a quantum computer. Such a problem, as we believe, is factoring a large random number. But if we sent a large number with its factors, how would they know that we did not make it up? We can generate randomly two large primes and compute their product without a quantum computer. If we communicated bilaterally, it would be simple, we would ask *them* to send us a number to be factored, but the assumption is that they are too far away.

My proposal is to send the complete factorizations of all composite numbers in an interval $[n, n+a]$. The number n must be large enough, a should be small, so that we can do all these factorizations, but not too small. It seems that it is not possible to make up such an interval with all factorizations without being able to factor efficiently.

5.5 Descriptional Complexity

There is a type of complexity that is essentially different from what we have considered so far. Consider finite strings of symbols, finite graphs, finite algebras or in general any finite mathematical structures. Can one define the complexity of such entities? When defining the complexity of computations, we study a dynamical process, which quite naturally needs time and space to be performed. We clearly need something different when we consider static objects. For such objects, the most natural thing is to define the complexity to be the length of the shortest description. One can show in many examples that this is a good concept. For instance, we tend to think of symmetric objects as simpler than nonsymmetric and, indeed, one can use symmetries to give a more concise description. To determine an equilateral triangle we need only one number, whereas for a general triangle we need three. We often associate beauty with symmetry, and thus also with the possibility of short description. It is not only flowers whose symmetry we admire, but also theories. A theory whose equations are short and manifest symmetries seems to reflect the reality much better than a long list of apparently unrelated and complicated axioms.

The concept of descriptional complexity seems intuitively clear, but one has to be a bit careful and define it precisely. As we know, speaking vaguely about descriptions leads to Berry's paradox (the paradox of *the least number that cannot be described by a sentence with at most 100 letters*, see page 38). Therefore it is necessary to say clearly what the descriptions are. In general, we may use any formal language which is sufficiently universal to describe all entities that we are interested in. Each of these choices may define a different concept and there are almost no principles that would guide us which we should choose. In such a situation the natural thing is to take the simplest formal system which is universal. A sufficiently universal system should, among other things, be able to describe computations. So why not just focus on computations? Let the formal system used for descriptions be Turing machines (or anything that is equivalent to them).

Thus we are naturally led to the concept of *algorithmic complexity*. The algorithmic complexity of an object x is the size of the simplest algorithm that produces the object. This concept possesses several properties that make it interesting for further investigations. But one should bear in mind that many of the results proved about algorithmic complexity hold true for other versions of the descriptional complexity based on different formal systems.

The concept of algorithmic complexity was conceived in the 1960s independently by three mathematicians: G.J. Chaitin [39], A.N. Kolmogorov [160] and R.J. Solomonoff [277]. Most researchers refer to algorithmic complexity as the *Kolmogorov complexity* and I will follow this tradition.

The Algorithmic Complexity of Strings

As I said it is necessary to give a precise definition of the Kolmogorov complexity in order to avoid paradoxes, but not only because of that. If we want to state theorems, we need precise mathematical concepts. But as in other parts of this book, I will only try to convey the most important ideas and avoid inessential technicalities.

Our first convention will be restricting the class of studied structures to finite binary strings (sequences of zeros and ones). Finite binary strings are universal structures in the sense that they can encode any finite structures. Unlike in some other situations, for Kolmogorov complexity the particular way we code other structures is irrelevant; the only requirement is that the coding must be computable. Thus we only need to develop the theory for binary strings, keeping in mind that we can always apply it to arbitrary structures.

The next thing to choose is a computational model. We can take any model that defines all computable functions. I will use Turing machines, which is the most common approach, but this is not essential.

A key result used in algorithmic complexity is the existence of universal Turing machines. Recall that a universal Turing machine is a machine that can simulate all Turing machines (see page 130). This is a key concept in Kolmogorov complexity and therefore we have to state precisely what it means.

Definition 15 U is a universal Turing machine, if for every Turing machine M, there exists a binary string p such that for every input string x, U will halt on the input px if and only if M will halt, and if they halt then both output the same string.

This needs some explanation. We use here the convention that Turing machines compute only with binary strings, thus the 'code' of M must also be a binary string. The expression px is the *concatenation* of strings p and x, the string obtained by writing x after p. We cannot use additional symbols to separate p and q, but we can assume that the coding of Turing machines $M \mapsto p$ is chosen so that one can always determine where p ends and x starts.

Notice that we consider all Turing machines, not only those that halt on every input. This is an annoying complication, but it cannot be avoided—the class of Turing machines that always halt does not contain a universal one.

Let me also recall that a real computer is a good approximation of this theoretical concept. An input to a computer also can be split into two parts: a program p and data x to be processed by the program. The difference is that a real computer, of course, cannot work with arbitrarily large data.

Now we are ready to define the Kolmogorov complexity with respect to a universal Turing machine U.

Definition 16 The Kolmogorov complexity of a binary string y is the length of the shortest string x such that U eventually halts on the input x and outputs the string y.

Thus Kolmogorov complexity is a function that associates natural numbers to binary strings; it will be denoted by $C_U(y)$.

The first simple observation is that $C_U(y)$ is defined for all finite strings y and it is at most the length of y plus a constant. This is because the string y can always be a part of the program which simply prints the string. In the programming language C, for example, we can use:

```
printf(''11010001010111'')
```

if we need to print the string 11010001010111. In a computer this program is represented by a string of bits.

In this way we have defined infinitely many different measures of complexity one for each universal Turing machine U. Can we show that one of them is the right one? That would require one universal Turing machine to be distinguished from others by a special property. One possibility would be to posit that it is the smallest universal Turing machine. But how should we measure the size? The size depends on the particular formalization of Turing machines and then we have a problem again: what is the right formalization. What is even worse is that to find the smallest universal Turing machine seems to be a hopeless task.

Fortunately, the problem of having infinitely many different measures is not as bad as it looks at first glance. Although we have infinitely many different measures of Kolmogorov complexity, they do not differ much. It is not difficult to prove that every two universal Turing machines U_1 and U_2 give rise to measures that differ at most by an additive constant. Thus for long strings the difference is negligible. This is called the *Invariance Theorem* and it is one of the main results that justify the naturalness of the concept of the Kolmogorov complexity.

The proof of this basic result is very easy. Given U_1 and U_2, let p be the string by which we can simulate U_2 on U_1. Thus if $U_2(x)$ is defined (U_2 halts on the input x) and equals y, then also $U_1(px) = y$. Hence, the Kolmogorov complexity of y with respect to U_1 cannot be larger than the Kolmogorov complexity of y with respect to U_2 plus the length of p. This is expressed by the inequality

$$C_{U_1}(y) \leq C_{U_2}(y) + c,$$

where c is the length of p.

Incompressibility and Randomness

Let us assume that one reference universal Turing machine U is fixed and let us only use this machine from now on. Thus we can also suppress the subscript at $C(x)$.

The next basic result concerns the existence of strings with large Kolmogorov complexity. It asserts that, for every given length n, there is a string x of length n whose Kolmogorov complexity is at least n. Such strings are, quite naturally, called *incompressible*. If x is incompressible, it means that there is no way to encode it by a shorter string. We know that we can always program U to print x, which gives an upper bound $n + c$. Thus incompressible strings achieve this bound up to the constant c.

Again the proof of this theorem is very easy. It uses a counting argument that is not dissimilar to the proof of the existence of Boolean functions with exponential circuit complexity. We count the number of strings that have Kolmogorov complexity less than n and show that this number is less than the number of all strings of length n. The calculation is trivial. The number of all binary strings of length less than n is

$$1 + 2 + 4 + \cdots + 2^{n-1} = 2^n - 1 < 2^n.$$

Since every string with Kolmogorov complexity less than n is coded by a string whose length is less than n, we conclude that there must be at least one string of length n whose Kolmogorov complexity is at least n.[24]

Incompressible strings are very interesting objects. In order for a string x to be incompressible, there must be no regularity, no discernible pattern in the string. A simple example of a compressible string is a string of the form yy, a string whose first and the second halfs are the same. This can be defined by a program that instructs the machine to print y twice. Hence the Kolmogorov complexity of this string is at most half of its length plus a constant; the constant is the length of the part of the program that says '*print the following string twice*'. Recall that we used this idea for constructing a self-referential sentence, see page 273.

A remarkable property of incompressible strings is that they pass statistical tests of randomness. I will explain the idea on the simplest possible test. Consider binary strings of length n. The most basic property of random binary sequences is that they have approximately the same number of ones and zeros. (The difference between these two quantities is typically within $\pm\sqrt{n}$.) Suppose x is a string that has only $n/4$ ones. We may produce such a sequence by a random process, thus x may also look completely patternless, yet it can be compressed. The reason for that is that x belongs to a relatively small set that can be efficiently enumerated. The number of strings of length n with $n/4$ ones is approximately $2^{0.8113n}$. We can order them, say, lexicographically and enumerate them. Then, to specify one of them, we only need to give its number, which has at most $0.8113n$ binary digits. Hence the Kolmogorov complexity of every such string is at most $0.8113n + c$, for some constant c.

Thus the concept of incompressible strings enables us to specify particular strings that look completely random. This is not possible using only concepts from classical probability theory, where all strings of a given length have the same status. In probability theory we have to talk about properties of random strings (or other random structures) indirectly. For instance, when we need to express formally that '*a random string has approximately the same number of ones and zeros*', we say that '*the probability that a randomly chosen string has approximately the same number of ones and zeros tends to 1 as n* $\rightarrow \infty$'. So in probability theory we can talk about *properties* of random strings, but not about *a concrete random string.*

This reminds one of the possibility to talk about infinitely small/large numbers in nonstandard analysis. Whereas the concepts of infinitely small/large numbers can be treated only indirectly in classical analysis, in nonstandard analysis we can define such numbers. But Kolmogorov complexity gives us more than just the possibility to speak about random strings. It enables us to develop a new type of probability theory, *algorithmic probability*. In this theory it is possible, for example, to solve a problem which cannot be treated using classical probability theory—the problem of

[24] An attentive reader may have noticed that incompressibility is not invariant with respect to different choices of the universal Turing machine. An incompressible string of length n for one machine may have Kolmogorov complexity $n - c$ for a different machine, where c is a nonnegative constant (depending only on the pair of machines). But if n is very large with respect to c, the difference between having Kolmogorov complexity n or $n - c$ is not essential.

prediction from given data. The setting is very much like the questions in IQ tests: we are given a finite sequence of symbols that are assumed to be an initial part of an infinite sequence and we should predict the next symbol.

For example, given a string

$$0101010101010101010101010101010101010$$

the natural answer is that the next symbol is 1 because we conjecture that the infinite string is the string of alternating zeros and ones. According to classical probability theory, all strings are equally probable, thus we can only say that the next symbol is 0 or 1, each with probability $1/2$. The algorithmic approach is not to assume that all strings have equal probability, but to look for regularities that can be described by an algorithm. But again there are many algorithms that are consistent with given data, so we need a principle by which we choose one. The principle is to choose the algorithm with the shortest program. In terms of universal Turing machines, we look for the shortest binary string p such that, given p as the input, the universal machine U will print an infinite sequence starting with the given finite string. In the example above it seems clear that every program, say in the programming language C, that will extend the sequence by printing 0 must be longer than the simple program that prints a sequence of alternating zeros and ones.

Although it only looks like a rule of thumb, it is possible to justify this principle formally. It would take us too far afield to explain the necessary parts of the theory, therefore I will consider only a very special situation and sketch a simplified argument. Suppose X is an infinite sequence of zeros and ones defined by a program q. Suppose that we get the elements of the sequence X successively one after another and our task is to find a program that prints X (it does not have to be q itself). In order to make this example simpler, let us also assume that we have a computable upper bound $t(n)$ on the time that q needs to compute the nth digit of the infinite sequence.

To solve this problem we take an enumeration $p_1, p_2, p_3, \ldots$ of all programs. Then we try p_1 on longer and longer initial segments of X. We keep p_1 as long as it runs in the time limit $t(n)$ and prints the same bits as there are in X. When we discover that it runs longer, or prints a different bit, we discard it and test p_2 in the same way. In this way we may try many programs, but after a finite number of steps we will arrive at q or at a program that behaves like q. From that point on we will keep this program, although we will never be sure about it.

In other words, we make conjectures that X is defined by p_1, then by p_2 etc. We disprove some conjectures, but eventually we arrive at the correct conjecture that will be confirmed by all finite segments of X. Replace the word *program* by *theory* and X by *experimental data* and you get a description of how should science develop. In reality it is, of course, a much more complex process.

In this solution of the problem the lengths of programs are not mentioned at all. The only thing we need is that the enumeration is complete—it contains all programs so that we do not miss q. Notice, however, that the most natural way to enumerate programs is to order them by their lengths (and those of equal lengths arbitrarily). So the strategy to try the shortest program that has not been disproved as the current conjecture is just one of the possible strategies that ensure the completeness of the enumeration. This is the essence of why the rule of the shortest program works.

Noncomputability of the Kolmogorov Complexity

After good news there is also bad news: *The Kolmogorov complexity of finite binary strings, the function $x \mapsto C(x)$, is not algorithmically computable.* Thus it is a useful theoretical instrument, but in practice we cannot use it. One may still hope to at least approximate the function, but the matter is even much worse. Not only we cannot compute $C(x)$, but in fact *there is only a finite number of strings whose Kolmogorov complexity we can determine!*[25]

Let me first give an intuitive explanation of why the Kolmogorov complexity is not computable. Essentially, it is a consequence of the noncomputability of the halting problem. Recall that the halting problem is to decide if for a given Turing machine T and a finite string x, the machine T will eventually stop when started on x. A consequence of the undecidability of the halting problem is that if we take one *universal* Turing machine U, then the halting problem is still undecidable. Suppose we want to compute $C(x)$ for a string of length n. We know that $C(x) \leq n + c$, for some constant c (that we can compute), thus we only need to run the universal Turing machine U on all strings of length $\leq n + c$, which is a finite number of strings. If n is small, we may be lucky and find p such that $U(p) = x$ and on all shorter strings U halts and prints strings different from x. Then we know that $C(x)$ is the length of p. But there are strings on which U does not stop. Suppose q is such a string and suppose that $C(x)$ is larger than the length of q. If we are sufficiently patient, we will find the shortest p such that $U(p) = x$, but in order to know that p is the shortest one, we must be sure that U does not halt on q and that is undecidable.

In mathematics many quantities cannot be computed by an algorithm, but often we can determine them using mathematics; we can *prove* that they are equal to particular numbers. However even that is a problem if we are to determine the Kolmogorov complexity of a string.

Theorem 42 *Let T be a sound theory axiomatized by a finite list of axioms. Then there exists a natural number k_T such that for no concrete string x, T is able to prove that x has Kolmogorov complexity larger than k_T.*

The theorem is also true for recursively axiomatized theories (theories axiomatized by an infinite list of axioms for which it is algorithmically decidable whether a given sentence belongs to the list), but for the sake of avoiding distracting technicalities, I have only stated the weaker version.

The first reaction to this theorem is that there must be something wrong. Just a moment ago I proved that for every n, there exists at least one incompressible string.

[25] Using a fixed set of axioms.

The proof was very easy and certainly there is a true theory T in which it can be done. So if we take $n > k_T$, we get a contradiction. Or do we really get it? The gist is that the theorem talks about *concrete* strings. Indeed, a fairly weak theory T is able to formally prove that there exist incompressible strings. Then, if we take a specific number n, we can list all 2^n strings of length n and T proves that one of them is incompressible. We can reduce the list by proving in T that some strings are compressible. But for a sufficiently large n ($n > k_T$), we cannot reduce the list to a single item—this is what the theorem says.

The proof of this theorem is very simple. Let $Unpr_T$ be the unpredictable program for theory T defined on page 287. Recall that for every concrete string x, it is consistent with T that $Unpr_T$ prints x as the output. Let p_T be a binary encoding of $Unpr_T$ for the universal Turing machine U. We take some natural encoding so that the above property of $Unpr_T$ is preserved for $U(p_T)$. Thus, in particular, for every concrete string x, it is consistent with T that its Kolmogorov complexity $C(x)$ is at most the length of p_T. So we can take k_T equal to the length of p_T and the theorem is proved.

I will sketch a second proof, which is a simple application of Berry's paradox. Let ℓ_T be the length of a natural binary encoding of the axioms of T. We need to take k_T a little larger than ℓ_T; it will be clear from the proof that $k_T = \ell_T + c$ for a sufficiently large constant c (independent of T) will do. Suppose that there is at least one string x for which T proves that $C(x) > k_T$. Then we can write a program p that finds such a string. The program will systematically generate all proofs of T and, for each proof, it will check if it is a proof that a specific string x has $C(x) > k_T$. As soon as it finds such a string, p prints it and halts. It is clear that the essential part of the program will be the description of the axioms of T. The rest will be the same for all theories, hence it can be bounded by a constant. Since T proves only true sentences, the string that p finds has Kolmogorov complexity larger than k_T, but the string is also defined by a program that has the length at most k_T. This is a contradiction (the same as in Berry's paradox).

A striking property of the second proof, which is the reason why I gave the second proof, is that *it does not use self-reference.* (In the first proof self-reference was used to define the unpredictable program.) This is in contradiction with the commonly accepted presumption that every proof of the Incompleteness Theorem has to use self-reference. In this proof the only part that could be regarded as being related to self-reference is the fact that the program p should look for proofs that some string has the Kolmogorov complexity larger than p. Thus the program in a way refers to its length. Should this be called self-reference?

Let us rather look for similarities between Gödel's proof and this proof. Recall that the main trick in writing self-referential sentences was doubling the text—writing essentially the same text twice. As we already observed, this is a simple way to generate compressible strings, strings whose Kolmogorov complexities are less than their lengths. And this is, indeed, the essence of the trick: since the sentence should refer to itself, it should be possible to describe the sentence by a text that is shorter than the whole sentence. Thus if the two proofs of the incompleteness theorem do not share self-reference, they at least share something from Kolmogorov complexity.

Notice that Theorem 42 also implies that Kolmogorov complexity is not computable. Indeed, suppose that there were a Turing machine M that would compute $C(x)$, given x as the input. Let T be the theory axiomatized by some basic axioms of set theory plus the axiom:

M computes C.

Then to prove that a specific string x has the Kolmogorov complexity k we would only need to formalize the computation of M on the input x. Thus T would prove $C(x)$ to be equal to the numeral expressing its value for every given string. This would be in contradiction with Theorem 42 and the fact that there are strings of arbitrarily high Kolmogorov complexity.

For a true finitely axiomatized theory T let k_T denote the *least* number that satisfies the condition in Theorem 42. Chaitin proposed to use this number to measure the strength of theories. It is certainly an interesting parameter. Contrary to what the proofs of Theorem 42 may suggest, k_T is not the Kolmogorov complexity of the set of axioms of T. For example, if the set of axioms of T contains some axioms that do not influence what T proves about the Kolmogorov complexity, we can omit them and the new theory will have the same parameter. If S is stronger than T, we clearly have $k_T \leq k_S$. It may also happen that there is a stronger theory with much more concise presentation than T.

It would be very useful to scale the theories that we are working with (Peano Arithmetic, Zermelo-Fraenkel set theory and its extensions by large cardinal axioms etc.) using a numerical parameter such as k_T. Unfortunately, there is no algorithm for computing k_T for specific theories; we can only prove some bounds. Classical proof theory gives a scale based on constructive ordinals (see page 510). These ordinals can be determined for Peano Arithmetic and some extension of it, but unsurmountable complications arise when we want to do it for set theories.

Theorem 42 has an interesting corollary. Let us first observe there is no infinite algorithmic process that would enable us to extend a given sufficiently strong consistent theory S indefinitely. By '*extending indefinitely*' I mean that the theory resulting after adding all infinitely many axioms is *not* contained in a recursively axiomatized consistent theory T. This observation that such a process does not exist is easy—the process itself produces a recursively enumerable axiomatization and every such axiomatization can easily be transformed into a recursive one. (If we do not insist that the theory T should use the same language as S, we can even find a finitely axiomatized theory T with this property.) However, this is not true if we allow the process to be random and allow an error with small probability. Here is the precise statement, where again I state it only for finitely axiomatized theories.

Theorem 43 *For every sufficiently strong, sound and finitely axiomatized theory S and every $\varepsilon > 0$, there exists an algorithm (formally, a Turing machine M) with the following property. Given access to a source of random bits, M produces an infinite set of axioms such that with probability at least $1 - \varepsilon$, the resulting extension T of*

S is a sound theory, but it is not contained in any consistent recursively axiomatized theory.[26]

The idea of the proof is simple. Given S, although it is not possible to compute k_S precisely, it is possible to compute an upper bound $k \geq k_S$. If we now take n sufficiently larger than k and take a random sequence r of this length, we can make the probability that the sequence has Kolmogorov complexity larger than k arbitrarily small. Whence the axiom $C(r) > k$ will be true with high probability. Since $k \geq k_S$, it will be a new axiom. (For more details of the proof see Notes.)

What conclusion should we draw for the foundations of mathematics? The axioms $C(r) > k$ do not seem to be useful. Further, the way mathematicians discover new axioms is not completely random—they always have some reasons for adding a particular axiom. Yet randomness is involved in almost every discovery. The theorem above should be viewed as a theoretical possibility that *the presence of randomness in the process of discovering new axioms may enable us to extend set theory beyond any limits.*

Kolmogorov complexity can also be used to classify empirical theories. Suppose we have experimental data D and some theories T_i that explain the data. Which theory should we choose as the best? Now we cannot focus only on the Kolmogorov complexity of theories because the "agnostic" theory saying that "*all data are just random numbers*" is consistent with any data and is likely to have the smallest complexity. We have to take into account also the Kolmogorov complexity of data. Therefore J. Rissanen [245] proposed the following principle:

The Minimum Description Length Principle *The best theory T is the one that minimizes the* sum *of the length of the description of T* plus *the length of the description of data D using theory T.*

The idea behind it is that typical data have both some randomness and some regularity. We would like to separate these two by having a theory that would describe the regularities and a string of random bits that would encode the entropy contained in data. The principle forces us to optimize the choice of the theory: if we pick a too simple theory, then the description of data will be complicated; if we try to make the description of data too simple, the entropy contained in them will have to be moved to the theory.

Needless to say, the success of applying this concept in practice depends very much on how one interprets the notion of description.

Notes

1. *Proof of Theorem 43, more details.* If we do not insist that each time we add an axiom it be not provable from the theory constructed so far, we can define a

[26] Here I exceptionally deviate from the convention that theories in this book are always recursively axiomatizable.

sequence of axioms independently of the theory S. For a given $\varepsilon > 0$, pick a c sufficiently large and for every natural number $n \geq 1$, generate randomly a bit string r_n of length $2n + c$. We can choose c so that the probability that $C(r_n) \leq n$ is at most 2^{-n}. Hence the probability that $C(r_n) > n$ for all $n \geq 1$ is at least $1 - \varepsilon$.

Theorem 42 was stated for sound theories, but in fact one only needs consistent theories. Such a stronger version of this theorem implies that the set of all sentences $C(r_n) > n$ is not contained in any consistent recursively axiomatized theory.

A related question is whether it is possible to obtain by such a process a complete extension of a given theory. It has been proved that this is impossible for Peano Arithmetic [142], hence also for set theories in which Peano Arithmetic is interpretable. Let me stress that here I am talking about all consistent extensions, not only about the one that consists of all true arithmetical sentences.

2. *The prefix-free Kolmogorov complexity.* The concept of Kolmogorov complexity that we defined above seems very natural, but it still has some undesirable properties. Let us see what their source is. When we define computations of Turing machine on finite strings, we assume that they are delimited in some way, so that the machine knows where the word starts and where it ends. The usual convention is that there is an additional symbol, viewed as a blank, and the squares of the tape outside of the string contain this symbol (this is just an awkward way of saying that they are blank). Thus the machine can recognize the ends of the input string by reading the blank symbols. This apparently innocent technical detail has quite an important effect on the Kolmogorov complexity. The point is that if we mark the ends of the input string, then we give the machine additional information. One way to see this is to imagine that on the Turing machine tape we can only use the two symbols 0 and 1, no blank symbols, and we still need to delimit the input string. Then we must use some coding that necessarily increases the length.

   Thus the usual way of presenting input to a Turing machine gives more information than it should. The amount of additional information bits is small, but if we want to compute precisely, this is an essential error. Can we filter out the additional information about the length of the input? It seems that a quite satisfactory answer to this question has been found. The basic idea is that we should restrict the class of Turing machines to those that only read symbols of the input string and never look outside. As a Turing machine needs typically much more space than is the input length, we must use several tapes. We obviously need an input tape and a separate work tape, but it is also natural to have an output tape. Then we can assume that the head on the input tape moves only in one direction; say, starting on the first symbol it moves right. Thus we ensure that the machine never goes left from the input word.

   It is trickier to ensure that the machine does not go right from the input word. In fact, this is possible only because we can ignore inputs on which the machine does not print anything because it never stops. If the machine does not print any output on a given string, we can ignore it; it does not make any difference how much information it got. Thus, given a Turing machine T, we change it to a machine T' that is the same as T, except that if T' leaves the input string (which means it reads a blank symbol), then it goes into an infinite loop and hence does not print any output. Further, we stipulate that if T stops before reading the entire input, we will treat it as if it did not produce any output.

Let T' be such a Turing machine, let L be the set of inputs on which it stops and produces an output. Observe that if $x, y \in L$, then x is never a proper initial segment of y; we say x is not a *prefix* of y. Such sets are called prefix codes. Their advantage is that we can transmit the words from the code without any separating messages. Such Turing machines are, therefore, called *prefix Turing machines.* Kolmogorov complexity based on prefix Turing machines renders much better the intuitive notion of descriptive complexity. One can show that the basic results remain true in the new context. In particular, there exist universal prefix Turing machines and thus we also have the Invariance Theorem.

3. *Inductive reasoning.* Inductive reasoning was a problem that puzzled philosophers for a long time. On the one hand, it is clear that humans use it, on the other hand, there seemed to be no way to show the possibility of such reasoning formally. For *deductive* reasoning, there was logic, but there was no "inductive logic". What was missing was the concept of computability. If we allow observable phenomena to be governed by completely arbitrary rules, then we surely have no chance to discover them. If, however, we restrict the rules to computable ones, we may eventually discover every such rule using a sufficient number of examples.

I have already considered a simple example in which the data is given by a computable process. In that example the process of producing zeros and ones was completely deterministic. This is called *learning* because once we learn a concept (definable by an algorithm), prediction becomes easy. In general, we would like to make predictions also in the case of processes involving randomness.

An extreme case is the uniform probability distribution. (This means that zeros and ones are equally likely and their appearance on different positions in the sequence are independent.) In this case we would also like to be able to discover what is going on, namely, that the sequence is completely random. An example that combines randomness with a strong dependence are sequences generated by repeating the following action: toss a coin, if you get heads write zero, if you get tails write two ones. The problem of predicting the next element in the sequence has the following solution:

a. if the current finite sequence ends with an odd number of ones, the next element is 1;
b. otherwise the next element is zero or one, both with probability $1/2$.

In the 1960s, Solomonoff pioneered an approach to the problem of prediction based on algorithmic probability [277]. His results and subsequent work of other researchers clearly demonstrate that prediction is possible. I will only state a simple version of his main result.

First we need to formalize the concept of an infinite process that produces a binary infinite sequence. Let us denote the set of infinite binary sequences by

$\{0,1\}^\omega$. A probability measure on $\{0,1\}^\omega$ is a mapping μ defined on certain subsets of $\{0,1\}^\omega$ and taking on values in the real interval $[0,1]$. Let us leave aside the question for which subsets of $\{0,1\}^\omega$ the measure μ is defined, as it plays little role in what we are going to discuss. The conditions that μ has to satisfy in order to be called a probability measure is that the measure of the entire space is 1 and that it is σ-additive, which means that the measure of the union of a countable family of disjoint sets is the sum of their measures.

We have already used the uniform measure on $\{0,1\}^\omega$, which is a formalization of the process of randomly tossing a coin infinitely many times, when we discussed forcing (page 363). This measure is determined by the condition that, for every n and every finite sequence x of length n, the measure of the set of infinite sequences extending x is $1/2^n$. Let $\mathbf{u}$ denote the uniform probability measure and Γ_x the set of infinite sequences extending x. Then the defining condition for $\mathbf{u}$ is

$$\mathbf{u}(\Gamma_x) = 2^{-n},$$

for x of length n.

In general, every measure on $\{0,1\}^\omega$ is determined in such a way. Thus we only need to know the values $\mu(\Gamma_x)$ for all finite sequences x. The conditions defining measures reduce to

$$\mu(\Gamma_\Lambda) = 1,$$

where Λ denotes the empty sequence, and

$$\mu(\Gamma_x) = \mu(\Gamma_{x0}) + \mu(\Gamma_{x1}).$$

Hence we can identify measures with functions defined on finite binary sequences satisfying the two conditions above. Let us simplify notation by writing $\mu(x)$ for $\mu(\Gamma_x)$. In terms of probability, $\mu(x)$ is the probability that the random infinite sequence starts with x, where randomness is with respect to the probability measure μ.

The prediction problem that we are interested in is: *for a given finite sequence x, to determine (or at least to approximate) the probabilities that the sequence will continue with 0, respectively with 1.* Formally, this means that we want to determine the conditional probabilities $\mu(x0|x)$ and $\mu(x1|x)$ defined by

$$\mu(x0|x) = \frac{\mu(x0)}{\mu(x)} \quad \text{and} \quad \mu(x1|x) = \frac{\mu(x1)}{\mu(x)}.$$

Since $\mu(x0|x) + \mu(x1|x) = 1$, it suffices to have one of the two probabilities.

The *assumption* is that we get longer and longer segments of a random infinite sequence and the *goal* is to use this knowledge to make better and better predictions.

Note that we cannot repeat the experiment by starting over with another random infinite sequence. Therefore we cannot learn μ; if we ever can do anything, we can only learn a part of it.

The key idea is to restrict the class of measures to computable ones, but since μ is a real valued function we must define what computability means. It means that there exists an algorithm (a Turing machine) that for a given string x and a natural number k computes a pair of integers p, q such that $|\mu(x) - p/q| \leq 1/k$. Now we are ready to state Solomonoff's result.

Theorem 44 *There exists a measure* **M** *such that for every computable measure* μ*, the following is true. Let* X *be a random infinite sequence and let* X_n *denote the initial segment of length* n*. Then with probability* 1,

$$\lim_{n\to\infty} \frac{\mathbf{M}(X_n 0|X_n)}{\mu(X_n 0|X_n)} = 1.$$

The probability of this event is with respect to measure μ*.*

This theorem does not say anything about computability of **M** or the rate of convergence, still it is a striking result. The meaning of the theorem is that we do not have to try various hypothesis—there is one universal measure that suffices. How can a single measure approximate all computable measures? The point is in the last sentence of the theorem. It means that the sequences X are chosen according to the probability distribution μ. In order to get intuition, recall the problem of predicting a sequence defined by an algorithm on page 484. One can view it as a simplified version of the theorem above, where each measure μ has full weight on a single infinite sequence. The algorithm that solves this problem is the measure **M** that satisfies the theorem for such special measures μ.

Concerning computability, one can prove that **M** cannot be computable. It is, however, possible to generalize the concept of a computable measure so that we can compute universal measures at least approximately. We say that ν is a *semimeasure*, if instead of the equalities above it satisfies

$$\nu(\Lambda) \leq 1 \quad \text{and} \quad \nu(x0) + \nu(x1) \leq \nu(x).$$

We say that ν is *semicomputable from below*, if there exists an algorithm which for every x computes a nondecreasing sequence of rational numbers converging to $\nu(x)$. Then one can show that there exists a semimeasure **m** that is universal, in the sense of Theorem 44, for semimeasures semicomputable from below, and **m** *is also semicomputable from below.* It follows that we can, at least theoretically, approximate $\mathbf{m}(x0|x)$ with arbitrary precision. Thus Theorem 44, generalized in this way, gives us at least a theoretical possibility to compute predictions.

One can show several connections between these concepts and the Kolmogorov complexity of which I will mention only one. One can prove that there exists a constant c such that

$$2^{K(x)-c} \leq \mathbf{m}(x) \leq 2^{K(x)+c}.$$

4. *Using incompressibility instead of randomness.* The existence of incompressible strings is proved by a probabilistic proof (and moreover we know that we cannot do it by a constructive proof for sufficiently long length of the strings). Once we have such strings, we can use them to avoid probabilistic arguments in proofs such as Erdős' bound on the Ramsey number. Instead of estimating probabilities, we estimate Kolmogorov complexities. Such proofs are often conceptually simpler because they use formulas with fewer quantifier alternations.

Main Points of the Chapter

- Complexity is an inherent property of computational problems. We distinguish time complexity, space complexity, and other types.
- A number of fundamental problems in computational complexity is still open; the main one is whether $\mathbf{P} = \mathbf{NP}$.
- The use of random bits helps us compute solutions of some problems faster.
- Pseudorandomness can replace true randomness in some computations, and it is an essential concept in cryptography.
- To prove that secure cryptography is possible we would need to know that some conjectures stronger than $\mathbf{P} \neq \mathbf{NP}$ are true; namely, that one-way functions exist, or equivalently that pseudorandom generators exist.
- Some functions can be computed faster by splitting the task into many subtasks that can be computed in parallel.
- There are strong indications (but, again, we are not able to prove it formally) that quantum computers can solve some problems, e.g., factoring integers, much faster than classical ones. When quantum computers are built, we will be able to compute tasks that are not feasible on classical computers using algorithms that we know.
- Kolmogorov complexity of a string—the minimal length of a description of the string—is a useful theoretical concept. In particular, it enables us to define incompressible strings.

Chapter 6
Proof Complexity

"You're claiming that...mathematics might be strewn with primordial defects in consistency? Like space might be strewn with cosmic strings?"
"Exactly." She stared back at me, feigning nonchalance. "If space does not join up with itself smoothly, everywhere...why should mathematical logic?"

Greg Egan, *Luminous*

LOGIC became important when mathematicians realized that mathematics cannot be founded only on our intuition about mathematical entities and that it was necessary to introduce axiomatic systems and formalize reasoning. Moreover, it turned out that we will never have a complete set of axioms for number theory and set theory. Therefore logic has an important role to play, both in studying what is provable from particular sets of axioms and, even more, in studying what is not provable. To this end, two fields of logic have been developed: proof theory and model theory. The aim of proof theory is to study provability by syntactical means, namely, to actually study proofs. In model theory, on the other hand, we use models to show that a sentence is not provable. (Recall that in order to show that a sentence is not provable from a set of axioms, it suffices to find a model that satisfies the axioms, but not the sentence.) Both fields have, of course, a much wider range of applications.

When we are not able to prove a theorem from a set of axioms, the reason may be not that it is not provable, but that the proof is too difficult to find. This is the problem that mathematicians have to cope with in their everyday work. To prove a theorem with a nontrivial proof requires ingenuity, experience, patience, and often also a little bit of luck. Yet sometimes all this may not be enough. It may happen that every proof of the theorem is so complex that it cannot even be written down. In some rare cases, such as the Four Color Theorem, a computer can help us, but what if the lengths of proofs exceed even the capacity of computers? This can indeed happen. For every sufficiently strong theory, one can construct short sentences whose proofs are so long that they cannot be physically represented in our universe. Although known examples of such theorems are rather contrived, there is no reason why it cannot also happen for interesting open problems.

Thus there is yet another role for logic: to study the complexity of proofs. The basic measure of the complexity of a proof is its length. We may be unable to find

P. Pudlák, *Logical Foundations of Mathematics and Computational Complexity*, Springer Monographs in Mathematics, DOI 10.1007/978-3-319-00119-7_6,

a proof for various reasons, but large length is an absolute obstacle because an extremely long proof cannot ever be written down. Naturally, most of the research in proof complexity theory focuses on studying the lengths of proofs. However, there are other reasons for studying the lengths of proofs, the main one being the relations of proof complexity to computational complexity. The starting point of both theories is essentially the same. In computational complexity theory the classical concept of computability is replaced by feasible computability, usually represented by computability in polynomial time. Similarly, in proof complexity the concept of provability is replaced by feasible provability, where proofs of polynomial lengths are the standard representation of the imprecise concept of short proofs. There are several other analogies between these two theories, which strongly suggests that computational complexity and proof complexity are just two facets of the same concept.

Therefore the term *proof complexity* is now interpreted in a much broader sense than a few decades ago. It can be briefly described as the study of the phenomenon of complexity using the means of mathematical logic. In this field we study not only the lengths of proofs, but also calculi for propositional logic and theories that formalize reasoning with concepts of a given complexity.

6.1 Proof Theory

We have already encountered proof theory in this book several times. In previous parts I wrote mainly about the formalization of first order logic and Gödel's incompleteness theorem, which is the most important result in proof theory. In this section I will describe some other classical results of proof theory and some results about the lengths of proofs in first order logic.

How to Speed-Up Proofs

The lengths of proofs may depend very much on the formal systems used. It may happen that for two formal systems some theorems have much shorter proofs in one than in the other. In proof theory this phenomenon is called *speed-up*.[1] I will mention two important situations in which speed-up occurs.

The first one concerns an axiomatic theory its extension obtained by adding a new axiom. Let S be a theory and T an extension of S by an axiom α not provable in S. Such a theory T is denoted by $S + \alpha$. We will, of course, assume that both theories are consistent. If we add an unprovable sentence to S, there will be an infinite number of other sentences in the extension that are unprovable in S. But we cannot talk about speeding-up unprovable sentences, as they have no proofs in S. We are concerned with the sentences that are provable in both theories. If a sentence

[1]The choice of the word is not quite justified, as we may actually find a long proof more quickly than a short one.

ϕ is provable in both theories S and T, then it may have a shorter proof in T because we may also use the new axiom α.

Indeed, it is not very difficult to prove that under very mild assumptions about S, there is an extremely large speed-up in T for any α unprovable in S (it suffices to assume that one can formalize a relatively weak fragment of arithmetic in S). The speed-up is so large that it cannot be bounded by any computable function. This means that if we take a sentence ϕ provable in S and we only have a proof P of ϕ in T, we are not able to estimate, by any algorithm, the length of a proof P' of ϕ in S. The reason is that with increasing lengths of proofs P the lengths of proofs P' grow at a rate so fast that it is beyond all computable functions. This result is due to Mostowski [200].

For all practical purposes, it does not matter whether the speed-up is exponential, superexponential, or non-computable because it just means that some proofs in the weaker system are infeasible. What is remarkable about the result above is that such a speed-up appears when T is extended with any unprovable sentence α, however mild this extension is.

The second result, due to R. Parikh [211], concerns a single theory and two different ways of using the theory. Suppose a sentence ϕ is provable in a theory T, but all the proofs are extremely long. Thus we are not able to write any of them down. It may, however, happen that we can still learn that a proof of ϕ in T exists and that we can prove this fact using T. This is the case when the sentence

"*sentence ϕ is provable in T*"

has a short proof in T.

Let us denote the above sentence by $Pr_T(\phi)$. It is important to realize that the two sentences ϕ and $Pr_T(\phi)$ are essentially different. The first one can be an arbitrary mathematical sentence, whereas the second one is a sentence speaking about proofs in a formal system for the theory T. Furthermore, it is not possible to derive one from the other in T, in spite of the fact that they convey essentially the same information.

To be quite precise, we have to assume something about the theory T. First, it has to be able to talk about finite structures, and it has to be sufficiently strong that it can prove basic facts about syntax. Second, it has to be a sound theory, which means that all sentences provable in T are true sentences. (For example, we know, by the Second Gödel Incompleteness Theorem, that there are *consistent* theories T that prove that *they are inconsistent*; such a theory T is not sound, since it proves a sentence that is false.)

Let us get back to ϕ and $Pr_T(\phi)$ and consider the lengths of proofs of these sentences. One can show that if ϕ has a short proof, then also $Pr_T(\phi)$ has a short proof. More exactly, given a proof of ϕ we can construct a proof of $Pr_T(\phi)$ in polynomial time. The converse is not true. There are sentences ϕ such that $Pr_T(\phi)$ has a short proof, but ϕ has only very long proofs. Hence we can substantially shorten proofs by proving sentences $Pr_T(\phi)$ instead of proving ϕ directly. The speed-up is extremely large, but not as large as in the previous case. (In particular it can be bounded by a computable function, but this makes little difference from the practical point of view.)

What happens if we repeat this construction, namely, if *we prove that it is provable that ϕ is provable?* It is better to use a bit of formalism in such cases. So the question is: do we get a speed-up if we use $Pr_T(Pr_T(\phi))$ instead of $Pr_T(\phi)$? Yes, we do, and again the speed-up is extremely large. We can continue in this manner and each time we add another provability predicate, we get such a speed up.

In more intuitive terms, this phenomenon means that sometimes when we are stuck with a problem, we can solve it by considering it '*from a higher perspective*'. The '*higher perspective*' here means considering not the sentence itself, but its provability. The ability of humans to switch to a higher level is often contrasted with the limited abilities of computer programs in automated theorem proving. Furthermore, the fact that we have an infinite hierarchy of levels creates the illusion that in principle we can solve any problem—we need only to go to a sufficiently high level. I will explain in detail why this is wrong in Chap. 7. Here I will only show that iterating provability, although it does give a very large speed-up, is not as powerful as one would expect.

My argument will be based on comparing the two methods of speeding-up proofs considered in this subsection. The question is: given a theory T, do we get shorter proofs by extending by a new axiom, or by proving $Pr_T(\phi)$ instead of ϕ? To answer this question, we need an important concept, the *reflection principle for T*. Roughly speaking, it is the principle that allows us to derive the truth of a formula from the fact that it is provable. Formally, the reflection principle for T is the set of implications of the form

$$Pr_T(\phi) \rightarrow \phi,$$

for every sentence ϕ.

This principle is not derivable in T because a special instance of it implies the consistency of T (see page 615). Thus there are sentences ϕ for which the implications above are not provable in T. However, for some T' stronger than T, it is possible that T' proves (all implications of) the reflection principle for T. Furthermore, in typical cases the lengths of these proofs are short, namely, they can be bounded by a polynomial depending on the size of ϕ. Let T' be an extension of T that satisfies this property and let us now compare

- L_1: the length of the shortest proof of $Pr_T(\phi)$ in T, and
- L_2: the length of the shortest proof of ϕ in T'.

Since T' is an extension of T, a T-proof of $Pr_T(\phi)$ is also a T'-proof of $Pr_T(\phi)$. So the shortest proof of $Pr_T(\phi)$ in T' has length at most L_1. Also in T' we can prove $Pr_T(\phi) \rightarrow \phi$ by a short proof, say, of length m. Then, using *modus ponens,* we get a proof of ϕ of length at most $L_1 + m$. Thus we have obtained a bound

$$L_2 \leq L_1 + m,$$

where m is only polynomial in the length of ϕ.

In other words, in T' we can prove ϕ almost as quickly as we can prove $Pr_T(\phi)$ in T. There is essentially no speed-up between proofs of ϕ and proofs of $Pr_T(\phi)$ if we work in the stronger theory T'. But not only that: we can just as easily obtain

$Pr_T(\phi)$ from $Pr_T(Pr_T(\phi))$ and then derive ϕ as above. Clearly, we can continue in this way for any finite number of applications of the provability predicate. Thus T' eliminates speed-up for any number of iterations of Pr_T!

It is not difficult to come up with such extensions. Let us take as an example Zermelo-Fraenkel Set Theory, *ZF*. In order to obtain an extension of *ZF* in which the sentences expressing the reflection principle for *ZF* have short proofs, we need only to add the axiom postulating the existence of an inaccessible cardinal number, which is the weakest of all large-cardinal axioms. Essentially the same can be done with any strengthening of *ZF*. For example, let T be *ZF* augmented with an axiom expressing the existence of a measurable cardinal. Then it suffices to add an axiom saying that there exists an inaccessible cardinal above a measurable cardinal.

The conclusion is that *it is much more effective to use new mathematical axioms than "shifting to a higher perspective"*. However, we must be aware of the danger of introducing inconsistencies, if we use new axioms.

As for concrete examples of speed-up, the situation is similar as it is with concrete independent sentences. In fact, all concrete examples of speed-up have been constructed from concrete independent sentences by instantiation. One such example is due to H. Friedman.[2] He proved that the special case of Kruskal's Theorem for *labeled trees on a six-element set* has a small proof (one that can be easily written down) in the theory $\Pi^1_1 - CA$, but any proof in ATR_0 has length at least[3]

$$\left.2^{2^{\cdot^{\cdot^{\cdot^{2}}}}}\right\}\text{ 1000 times.}$$

Direct and Indirect Proofs

Imagine a proof as a path going from the assumptions of a theorem to the conclusion. The path can be straight or circuitous. In logic we call a proof that goes straight from the assumptions to the conclusion a *direct proof*; a roundabout proof is called *indirect*. Thinking of proofs in such a geometric way can, however, be rather misleading. It can happen that there is a short indirect proof, while all direct proofs are long.

So, what precisely does it mean when we say that a proof is direct? A direct proof is characterized by the property that it only uses concepts which occur in the theorem. We may also allow some simple combinations of these concepts, the precise definition of which depends on the particular logical calculus used. A typical direct proof is based on splitting the problem into several cases. An indirect proof, on the other hand, contains auxiliary theorems, which are usually called *lemmas*. Lemmas may be statements that formally have little to do with the theorem to be proved.

Direct proofs are as a rule easier to find. We need only to try combinations of concepts already contained in the statement of the theorem. Direct proofs are also

---

[2]Unpublished; the result was presented in [272], but the proof therein is incomplete.

[3]See page 643 for the definitions of these theories.

easier to understand and thus they are more useful when we want to understand why a theorem is true, which is in most cases the reason why we study a proof. This leads us to an important question: *can every theorem be proved using a direct proof?*

The short answer to the above question is: *yes*. The more detailed answer is: *in principle yes, but in practice not always.* We will see that direct proofs may be much longer than indirect ones and that this may prevent us from writing down such proofs.

I will describe two results that show that any proof can be transformed into a direct proof. The first one is the important *Herbrand Theorem* [121]. This theorem is due to Jacques Herbrand (1908–1931), a talented French mathematician who died very young in a mountaineering accident.

To avoid technicalities, I will explain only a special case of Herbrand's Theorem (a more general form is in Notes). We will consider statements of the form '*there exists an x which satisfies ϕ*', which we represent formally by

$$\exists x\ \phi(x),$$

where we are assuming that this is a purely existential formula, namely, that there are no quantifiers inside of ϕ. Herbrand's Theorem concerns provability of such sentences in pure first order logic.[4]

Theorem 45 *The sentence $\exists x\ \phi(x)$ is provable if and only if for some terms $t_1, t_2, \dots, t_n$, the disjunction*

$$\phi(t_1) \vee \phi(t_2) \vee \dots \vee \phi(t_n) \tag{6.1}$$

is provable.

The latter formula says that there are n objects, specified by terms $t_1, t_2, \dots, t_n$, such that ϕ holds for at least one of them. Notice that this can be interpreted as a proof by case distinction. There are n possible cases such that in case 1 we prove $\exists x\ \phi(x)$ by showing $\phi(t_1)$, in case 2 we prove $\exists x\ \phi(x)$ by showing $\phi(t_2)$, and so on.

Herbrand's Theorem moreover says that disjunction (6.1) is provable in propositional logic. This shows that we only need to use subformulas of ϕ and nothing else (more precisely, subformulas with various terms substituted in them). Thus we can construct a direct proof of $\exists x\ \phi(x)$. Furthermore, once we have such a disjunction, we can determine its validity by an algorithm, say, by trying all combinations of truth values.

Herbrand's theorem has other important applications. In particular the fact that (in a sense) it reduces provability in first order logic to provability in propositional logic makes it attractive for automated theorem proving. It should be noted that Herbrand's Theorem does not give us any bound on the number of disjuncts, if we only know the formula. If we could bound the number of disjuncts and the complexity of the terms, we would get a decision procedure for first order logic, which

---

[4] By the completeness theorem, logical validity and provability are the same things. As we are now interested in proofs, not just mere logical validity, I will use the latter term.

is impossible, as we know. Moreover, such a Herbrand proof may be much longer than the usual one, as we will see in a moment. Yet it seems easier to base theorem proving programs on searching for terms and testing propositional validity, than on generating sequences of formulas representing proofs.

Herbrand's theorem has also been used for proof mining—to analyze proofs in main-stream mathematics in order to obtain explicit bounds from purely existential proofs and to improve known bounds. A nice example is H. Luckhardt's "Herbrand analysis" of proofs of Roth's Theorem [187]. Roth's Theorem is a result in number theory that states that for every irrational algebraic number a and $\varepsilon > 0$, the inequality

$$\left| a - \frac{p}{q} \right| < \frac{1}{q^{2+\varepsilon}}$$

has only a finite number of solutions with p and q coprime integers. The essence of this result is that irrational algebraic numbers cannot be approximated by fractions with error asymptotically less than one over the square of the denominator. Luckhardt studied upper bounds on the number of solutions as functions of the degree of a (the degree of the irreducible polynomial whose solution is a) and ε. He improved the best bounds that were known at the time.

Sequent Calculus and Cut-Elimination

Another way of transforming general proofs into direct ones was invented by Gerhard Gentzen (1909–1945). Incidentally, he also died young and tragically; he was a victim of Czech maltreatment of Germans after Germany was defeated in 1945.

Originally Gentzen studied the natural deduction calculus. Recall that this is a calculus which formalizes in the best way what mathematicians actually do (see pages 97, 113). Obviously, for logical investigations of proofs, a calculus with fewer and simpler rules is better. Therefore Gentzen gradually simplified his system eventually arriving at the sequent calculus. The crucial idea on which this calculus is based is that one does not need elimination rules; it suffices to use introduction rules. So we can view the sequent calculus as a formal system in which complex formulas are gradually derived from simple ones. For each connective and quantifier, we have a few natural rules that enable us to derive compound formulas from their components.

Let us consider a couple of examples. Natural rules for conjunction and disjunction are:

- *from A and B derive* $A \wedge B$;
- *from A derive* $A \vee B$;
- *from B derive* $A \vee B$.

Unfortunately, this idea does not work without a technical modification of the system. Recall that a proof is a sequence of formulas such that each formula is an instance of an axiom schema or is derived from previous formulas by a logical rule available in the given system. So one step of a proof is one formula. The modification that Gentzen introduced is that one step may consist of several formulas. Such a string of formulas is called a *sequent*. The meaning of a sequent is the disjunction of the formulas in it. The reason why we do not write it in the form of a disjunction is that we want to define operations on *formulas* rather than on *subformulas*. In this context it is better to view a proof as a text whose lines are sequents. Therefore we also say *proof lines* instead of *proof steps*.

In a sequent calculus we apply rules to a fixed small number of formulas (usually just one) in one or more sequents and the rest of the sequents is simply copied to the next proof line. Thus an application of the rule for conjunction looks like:

$$\begin{array}{l} \vdots \\ \alpha, \gamma_1, \dots, \gamma_k \\ \vdots \\ \beta, \delta_1, \dots, \delta_l \\ \vdots \\ \alpha \wedge \beta, \gamma_1, \dots, \gamma_k, \delta_1, \dots, \delta_l \end{array}$$

The rules for other connectives and for quantifiers are treated in a similar fashion.

We also need some *initial sequents*, lines that can always be added. These are all sequents of the form $\alpha, \neg\alpha$, where α stands for an arbitrary formula. These sequents represent the law of excluded middle. Further, we need some *structural rules*, namely, *permutation, weakening* and *contraction*. The first one enables us to permute the formulas in a sequent. Using weakening we can add an arbitrary formula to a sequent (which makes the disjunction of the formulas weaker; thus this rule also preserves logical validity). To reduce the length of a sequent we are allowed to "contract" pairs of identical formulas, by replacing them with a single formula.

There is another structural rule that is treated separately because it plays a special role. This rule is called *cut* and enables us to eliminate a pair of complementary formulas. Here is how it looks:

$$\begin{array}{l} \vdots \\ \alpha, \gamma_1, \dots, \gamma_k \\ \vdots \\ \neg\alpha, \delta_1, \dots, \delta_l \\ \vdots \\ \gamma_1, \dots, \gamma_k, \delta_1, \dots, \delta_l \end{array}$$

Notice that contraction and cut are the only rules that reduce the complexity of sequents. Whereas contraction only discards repetitions of the same formula, using cut we can completely eliminate a formula from the rest of the proof. Thus it is only the presence of a cut that makes a proof indirect. We can view the first two sequents participating in an application of the cut rule as lemmas needed to derive the third sequent. The role of these sequents as lemmas is more apparent if, for example, the side formulas $\gamma_1, \dots, \gamma_k$ in the first sequent are missing. Then α itself is a lemma. The second sequent, $\neg\alpha, \delta_1, \dots, \delta_l$, can be interpreted as the lemma 'α *implies the disjunction of* $\delta_1, \dots, \delta_l$', formally expressed by

$$\alpha \to \delta_1 \vee \cdots \vee \delta_l.$$

Thus, from lemma α and this implication, we get $\delta_1 \vee \cdots \vee \delta_l$.

Gentzen proved that one can derive all true sentences in the system without the cut rule. The reason for having the cut rule in the system is that it makes the system more efficient. The cut rule is needed to simulate the rule of *modus ponens*, which is the main rule used in natural reasoning. We will see that without the cut rule some proofs must be extremely long. Moreover, Gentzen described an algorithm to transform a proof with cuts into a cut-free proof [89, 90]. This is called the *cut-elimination procedure.*

Recall that all propositional rules, except for cut, construct larger formulas from smaller ones. The quantifier rules do not change the propositional structure of formulas; they only change terms and add quantifiers. Hence the only parts that can possibly occur in a cut-free proof without occurring in the proved sequent are terms. Let us say that a formula α is a *quasi-subformula* of β if it is a subformula when we disregard terms and variables. For example, $\exists z(\neg P(z, f(z)))$ is a quasi-subformula of $\forall x(\exists y(\neg P(x, y)) \vee Q(x))$. Thus cut-free proofs have the following *subformula property*:

Every formula in a proof is a quasi-subformula of some formula in the last sequent.

The subformula property allows us to view cut-free proofs as direct proofs, and Gentzen's cut-elimination procedure as a way of transforming general proofs into direct ones.

Herbrand's Theorem can be easily derived from Gentzen's Cut-elimination Theorem. Consider a proof of a logically valid sentence $\exists x\, \phi(x)$. By Gentzen's Theorem, we can transform it into a cut-free proof. Then one can easily see that it is possible to permute the proof lines so that all instances of propositional rules precede instances of quantifier rules. Thus the resulting proof has the following form.

$$\vdots \qquad \text{[propositional rules]}$$

$$\phi(t_1), \phi(t_2), \dots, \phi(t_n)$$

$$\vdots \qquad \text{[}\exists\text{-introduction rule applied to formulas } \phi(t_i)\text{, and contraction applied to pairs } \exists x\, \phi(x), \exists x\, \phi(x)\text{]}$$

$$\exists x\, \phi(x)$$

Recall that the interpretation of a sequent is the disjunction of the formulas in it. So the sequent in the middle of the proof is a Herbrand disjunction and is derived in the propositional calculus.

Although the cut-elimination procedure is not extremely complicated, it may result in a tremendous increase in the length of the proof. The estimates of the length of the cut-free proof have the form of the superexponential function. If we denote by n the length of a proof with cuts, then one can only bound the length of the cut-free proof by

$$\left.2^{2^{2^{\cdots^{2}}}}\right\} n \text{ times.}$$

It would already be bad if it were a single exponential, 2^n, but this is much worse. It is not just because we do not know a better cut-elimination procedure, but because there is no better one. It has been proved that the superexponential cannot be replaced by a function that grows more slowly. Examples of sentences for which such an increase of length is unavoidable were found by R. Statman [284]. These bounds are quite general: they not only hold for cut-elimination in the sequent calculus, but also show that if we want to get any proof with formulas of low complexity, we have to increase the size of the proof this much. In particular, the bound also applies to Herbrand's Theorem.

In mathematics we distinguish elementary and nonelementary proofs. This is not a precisely defined concept; we usually say that a proof is nonelementary if it uses results from a branch of mathematics other than the one from which the theorem comes. In many cases, a nonelementary proof was discovered first and an elementary one was only found later. But in some cases we still only have nonelementary proofs. This concept is very much related to the formal concept of direct and indirect proofs and the fact that direct proofs may be much longer than indirect ones suggests that the reason why we do not have elementary proofs of some theorems is that the length of such proofs is too large.

We can draw a similar conclusion for automated theorem proving. Almost all automated theorem provers are based on Herbrand's Theorem. It is very likely that in this way we will never be able to prove some theorems that have already been proved by mathematicians. Therefore it is necessary to extend theorem provers so that they will also be able to generate lemmas. Some research into this problem has already been done, but finding useful lemmas is still more an art than a science.

Useful Inconsistent Theories

A logical theory classifies sentences into two categories: true and false. We know that we cannot fully formalize the true sentences, for example, in arithmetic. Thus we also have a third category—the independent sentences. However, what really matters is that a consistent theory gives us *some* classification. When a theory is inconsistent, everything is provable, so there is no classification. This is because in logic we can derive any sentence from a contradiction (ex falso sequitur quodlibet). Therefore the problem of consistency is the main problem in the foundations of mathematics.

Therefore inconsistent theories seem to be completely useless. But if we replace the qualitative concept of provability by the quantitative concept of the lengths of proofs, the situation looks different. If T is inconsistent, then we may still distinguish between sentences that have short proofs and those that have long proofs. This is only possible when the shortest proof of contradiction is long, since otherwise every short sentence also has a short proof.

Another reason for considering only consistent theories is that we believe that reality is consistent. But quite surprisingly, if we take into account the lengths of proofs, we can also use inconsistent theories to prove true facts. This is because it can happen that although the theory is inconsistent, all sentences that have short proofs are true.

The idea that inconsistent theories can be useful appeared in the seminal paper *Existence and Feasibility in Arithmetic* of R. Parikh [211] (the same paper where he showed the speed-up using provability predicates). To construct an inconsistent theory with the required properties, Parikh started with an arbitrary consistent extension T of Peano Arithmetic. Then he added a new predicate $F(x)$ (F standing for '*feasible number*') and several axioms. For the sake of simplicity, I will only use the most important ones:

1. $F(0)$ (*0 is feasible*);
2. *if* $F(x)$ *then* $F(x+1)$ (*if* x *is feasible, so is* $x+1$);
3. *there exists a number* x *such that* $\neg F(x)$ (*there exists an unfeasible number*).

It is not difficult to show that this theory is consistent. To this end it suffices to take a nonstandard model of arithmetic and interpret the predicate F as standard numbers. For the consistency, it is important that we do not specify the unfeasible number in axiom 3. If we replace the variable x by a concrete number, the theory becomes inconsistent. The key idea of Parikh's theory is to change the third axiom so that it does talk about a concrete number, but a very large one.

3'. $\neg F(t)$ (t *is unfeasible*).

Here one picks for t a suitable closed arithmetical term that represents a concrete large number m.

This theory, let us call it FPA_t, is clearly inconsistent. If we start with axiom 1 and apply axiom 2 m times we will prove that m is feasible. More precisely we prove that the number represented by

$$0+\underbrace{1+1+\cdots+1}_{m\text{-times}}$$

is feasible. Then we need only to prove that this term represents the same number as the term t. The length of such a proof can be estimated by some small multiple of m. However, if the term t is a concise representation of m, there can be much shorter proofs. In general, to ensure that no proof of contradiction in FPA_t is shorter than n, the value of the term t must be much larger than n. Parikh showed that if we take an approximately n times iterated exponential (a stack of twos of length several times n), then one can guarantee that there is no proof of contradiction of length less than n.

In effect he showed more: if one considers sentences that do not mention the predicate F, then such short proofs prove only true sentences, although the proof may use F. Thus if we pick n to be a safe upper bound on the lengths of all proofs that can ever be produced, and m to be the n times iterated exponential, then we can only prove true sentences about numbers.

Let us observe that for this theorem to be nontrivial it is necessary to have a short term t representing the large number m. Such a term cannot be constructed in the basic language of arithmetic that uses only $0, S, +$ and $\times$. We need to enrich the language by exponentiation *exp* (the function 2^x) and superexponentiation *supexp*. These functions can be described using axioms that correspond to their recursive definitions.

$$\begin{aligned} exp(0) &= 1 \\ exp(x+1) &= 2 \cdot exp(x) \\ supexp(0) &= 1 \\ supexp(x+1) &= exp\big(supexp(x)\big). \end{aligned}$$

Using the function symbol *supexp* we can write a very short term

$$supexp\big(supexp\big(supexp(1+1+1)\big)\big)$$

whose value is large enough to ensure that no contradiction will ever be proved in FPA_t and, moreover, all arithmetical sentences that one will ever be able to prove will be true statements.

This theory is used to justify the strict finitistic view called *ultrafinitism*. According to this approach to the foundations of mathematics we can only say that a number n exists if it is small enough to be physically represented. Large numbers are considered to be only abstractions that do not represent existing entities. In the theory above, the predicate F is used to distinguish between small and large, or as the ultrafinitists say, between feasible and unfeasible. But this theory does not render correctly our intuition about small (feasible) on the one hand, and large (unfeasible) on the other. We consider $2^{2^{100}}$ already to be a large number, as we certainly cannot represent it in decimal notation in the visible universe. But if we take $t = 2^{2^{100}}$, we will get a very short proof of contradiction in FPA_t. (I estimate it to be half a page, even if written completely formally.)

I think that the significance of this result is rather in showing that there are theories that have very short axiomatic systems and are inconsistent, but where every proof of contradiction is extremely long. Therefore we cannot exclude the possibility that the set theories that we are using as the foundations of mathematics are inconsistent, but we will never find a contradiction. The positive message is that if this is so, it may still be all right: if the shortest proof of contradiction is really very large, then what we are proving are true sentences anyway.

This is, of course, not the only possible scenario of what can happen if we admit the possibility that we are using inconsistent theories. The reason why we are not able to find a contradiction (and perhaps never will) may not be the large length of the proofs of contradiction. It could happen that there is a proof of contradiction of *medium length*, but one that is, so to speak, well-hidden. Recall that deciding whether a sentence has a proof of polynomial length, for some fixed polynomial, is an **NP**-complete problem. Thus if $\mathbf{P} \neq \mathbf{NP}$ (and if this manifests itself on the level of medium length inputs), there is no efficient algorithm to find a proof of contradiction, even if we know that it has medium length. Mathematicians are not algorithms, so this is only an indication that it may be hard. But if one believes that cryptography is possible (specifically, that there are secure encryption schemas), then one must also admit that it is possible to hide a contradiction so that nobody will find it.

The prevailing feeling is that Nature is not evil. She does not reveal secrets easily, but she also does not try hard to hide them from us. Therefore we believe that if there is a contradiction in set theory, or if some large cardinal is inconsistent, we will eventually discover this. Furthermore, we are not confined to simply searching the proofs in a given theory. We may look at the theory from higher perspectives and thus even be able to find a proof of contradiction whose length is beyond what can be written down.

My own view is that, most likely, Zermelo-Fraenkel Set Theory is consistent and is also consistent with most of the currently studied large-cardinal axioms. The reason I am talking about the possibility of inconsistency is that we need to have a complete picture of what can happen. We need it in order to understand the main problem of the foundations of mathematics, which is the consistency of the foundations.

Interlude—Life in an Inconsistent World

According to Gödel's Second Incompleteness Theorem, Peano Arithmetic augmented with the formal inconsistency of Peano arithmetic, the theory that we denote by $PA + \neg Con(PA)$, is consistent. By the Completeness Theorem, this theory has a model M. Imagine a world $\mathcal{W}$ in which the natural numbers are M. Since M is necessarily a nonstandard model, we should think of people living in $\mathcal{W}$ as being able to compute with nonstandard numbers like we are able to compute with standard natural numbers. I will leave to the reader's imagination how they can do it; for example, these people may have nonstandard size, or they have standard size, but they have computers of nonstandard size and nonstandard speed, etc. This is not important for our discussion.

Suppose that these people discover one of the proofs of contradiction in *PA*. Then they would be very frustrated when trying to determine the theory of the natural numbers. On the one hand they would see that the axioms of Peano Arithmetic are satisfied in their natural numbers, on the other hand they would know that the axioms are inconsistent. Since the axioms are true, but one can still derive a contradiction from them, they would conclude that *logic fails*. Their life would be miserable: they would only be able to watch what is going on, but not to make any predictions because predictions need deduction. (This is certainly only a fictitious situation and I am also exaggerating the problems that people in $\mathcal{W}$ would face.)

Why are we so sure that *we* do not live in an inconsistent world? The reason is that logic in our world seems to work perfectly; even the extremely long and complicated proofs in mathematics never fail. It is not only logic that is consistent, mathematics works perfectly as well—once a theorem is proven, it holds true in all circumstances without any exceptions. This confirms our belief in the soundness of mathematical theories. As a result we are always ready to extend our theories by statements expressing their soundness.

But let us have a closer look at the inconsistent world $\mathcal{W}$ before we condemn it as a very unlikely alternative to ours. From our perspective, the natural numbers M of $\mathcal{W}$ are a nonstandard model. Such a model contains nonstandard numbers, which are integers larger than all ours. Peano Arithmetic has an infinite number of axioms, hence in $\mathcal{W}$ there are also axioms of Peano Arithmetic that have nonstandard lengths ('nonstandard' now means 'longer than standard'). It is not true that all axioms of Peano Arithmetic are satisfied in M; we only know that the axioms of standard length are true in M. It may even be impossible to define satisfiability for axioms of nonstandard length. Furthermore, the proof of contradiction that is present in $\mathcal{W}$ is a proof from axioms of nonstandard lengths[5] and there is no proof of contradiction from the axioms of standard lengths.

We would like to advise people in $\mathcal{W}$ to only use axioms of standard length, but unfortunately they would not understand; they are not able to distinguish nonstandard numbers from standard ones. In principle, they might somehow determine the smallest axiom such that the contradiction is derivable from it and from the preceding axioms. Then they could use the initial segment of axioms below this axiom to avoid inconsistencies. But there is no effective procedure to find this axiom.

Here is what they always *can* do. Take the longest axiom in the proof of contradiction from the Peano Axioms and discard this axiom, along with all axioms of the same and larger lengths. When in the future you find a contradiction from the reduced set of axioms, reduce the set of admissible axioms in the same way again. After a finite number[6] of such reductions, they will not prove contradictions anymore, either just because they will not find them, or because they will reach a consistent subset of the axioms of Peano Arithmetic. In any case, the reduced set will contain all axioms of standard lengths.

Although this looks a little weird, it is not very dissimilar from what we are doing in set theory. We are studying various axioms postulating the existence of large cardinal numbers. When we discover that some large cardinal is inconsistent, we restrict our set of axioms to large cardinals below it.[7]

---

[5]More precisely, at least one of them must have nonstandard length.

[6]Finite from the point of view of $\mathcal{W}$, which can be a nonstandard number.

[7]Strictly speaking, if we say that a large cardinal is inconsistent, then it does not exist and we cannot talk about cardinals below it. So this is only an intuitive explanation. We can simulate such a situation by a model of $ZFC + A + \neg Con(ZFC + A)$ for some large cardinal axiom A that is consistent. This model will play the role of the inconsistent world $\mathcal{W}$.

We think we are lucky and do not have to do anything like that in our world. But if we view foundations from the point of view of logicism, we have, in fact, already imposed restrictions on the use of logic. It was Russell's Paradox, discovered in the logical system of Frege, that forced us to restrict the use of the Comprehension Axiom. As explained in Chap. 3, there were essentially two proposals about how to restrict the Comprehension Axiom. First, to use types and apply the axioms only when the variables are properly typed. Second, to apply it only when the defined set is in a certain sense small. The first way is used in various theories of types inspired by Russell's prototype. The second is the set theory developed from Zermelo's axioms. Today most people view set theory as a branch of mathematics, so they do not like the idea that it is "logic with restrictions". But whether or not set theory is logic with restrictions is only a matter of philosophy. If Frege's formal system introduced in *Grundgesetze der Arithmetik* were consistent, we would certainly call it logic.

It may seem that finding a contradiction in formal foundations are rare events in the history of mathematics, but the opposite is true: the quest for the foundations of mathematics has always been a struggle with contradictions. Here are some examples.

- As we know, in 1901 Russell found a contradiction in Frege's *Grundgesetze der Arithmetik*, the first formal system aiming to build foundations for mathematics.
- In 1932 Church published his formal system in the paper *A Set of Postulates for the Foundation of Logic* [42]. A contradiction in this system was found shortly after it appeared. In 1933 he published a revised version, but that also turned out to be inconsistent. The consistent part of these systems developed into the λ-calculus.
- In 1940 Quine published a book *Mathematical Logic* [233] in which he introduced an extension of his previous system from his 1937 article *New Foundations for Mathematical Logic*. Again a contradiction was found shortly after it was published and he had to revise his system. However, even the consistency of New Foundations is not clear—there is no proof of the relative consistency of New Foundations relative to Zermelo-Fraenkel Set Theory or its extensions by large cardinals, and some researchers suspect that the theory is in fact inconsistent.
- The first papers about large cardinals appeared already in the 1910s. The concept of a measurable cardinal was introduced around 1930. For a fairly long time many logicians suspected that the measurable cardinal was inconsistent, but new, much larger cardinals were studied and no inconsistency was found. This lasted until 1967, when W.N. Reinhardt proposed a large cardinal axiom based on elementary embeddings. The axiom was soon shown to be inconsistent, but the proof is not trivial. The largest cardinals that have been proposed and not found inconsistent are based on weakening the condition used by Reinhardt.
- In 1971 Martin-Löf introduced the first version of his intuitionistic type theory. Again the same story: it was inconsistent and had to be revised.

Ordinal Analysis of Theories

In mathematics we often use parameters which provide us with some basic information about studied structures. Usually, the parameter is a natural number and is called *order, rank, dimension, degree* etc. Such parameters enable us to compare the structures we are studying or help prove that they are different. For theories in first order logic, we do not have any useful numerical parameter (the number k_T proposed by Chaitin, which I mentioned in the last chapter, is practically impossible to determine, thus is only of theoretical interest). It is, however, possible to assign countable *infinite ordinals* to theories and in many cases the ordinal has also been computed. These ordinals are *constructive,* which means that they are explicitly described (see page 186). Thus, although infinite, they can be represented by finite objects. Since ordinals are linearly ordered, we can use them, for example, to compare the strength of theories.

The study of ordinals assigned to theories is an important branch of proof theory, called *ordinal analysis of theories*. This branch of proof theory started with the work of Gentzen, who proved the consistency of Peano Arithmetic using transfinite induction over ε_0. The ordinal ε_0 has a more intimate relation to Peano Arithmetic: it is the least ordinal for which Peano Arithmetic is not able to prove that it is well-ordered. The fact that Peano Arithmetic is unable to prove that ε_0 is well ordered follows from a combination of Gentzen's result with Gödel's. Indeed, if we had a proof that ε_0 were well ordered in Peano Arithmetic, then we could use Gentzen's proof to prove that Peano Arithmetic was consistent, which would violate Gödel's Theorem.

In this way the ordinal ε_0 is uniquely determined by Peano Arithmetic. In general, the least ordinal α for which a theory T does not prove that it is well-ordered is called *the proof-theoretic ordinal of the theory* T. This is not quite precise because rather than talking about ordinals, we should talk about their *representations.* In a typical set theory T we can prove that there are uncountably many countable ordinals, there are uncountable ordinals and more. This is not in contradiction with the existence of a bound on provably well-ordered orderings. What the theory is unable to prove is that *specific definitions* of orderings of natural numbers are well-orderings. Namely, it is unable to prove this for definitions of very large constructive ordinals. As the concept is rather subtle, we need a formal definition.

Definition 17 The proof-theoretic ordinal of T is the least ordinal α such that for no representation of α, T proves that it is well-ordered.

Equivalently, α is the limit of all ordinals that can be represented so that T proves that they are well-ordered.

In logic we often need to talk about "natural" representations of concepts in theories. So it is remarkable that the proof-theoretic ordinal of a theory is simply an ordinal, not a suitable representation of an ordinal. The trick that enables one to avoid talking about natural representations is to consider *all* possible representations.

In order to avoid problems with formalizing the concept of well-ordering, one only considers theories in which it is possible to talk about sets. Then we can use the usual set-theoretical definitions. This excludes Peano Arithmetic as it is usually defined, but one can define an extension in which sets are present and which is essentially of the same strength.

Naturally, we would like to find proof-theoretic ordinals for as many theories as possible. The motivation is not only to determine a parameter that reflects the strength of theories, but also to get consistency proofs for theories stronger than Peano Arithmetic. This is because having the proof-theoretic ordinal we are usually able to produce a constructive proof of the consistency of the theory. Of course, such consistency proofs do not solve the consistency problem, they only replace the assumption of the consistency of the theory with the assumption that the particular ordinal notation defines a well-ordered set, but it is additional evidence for believing that the theory is consistent. In any case, the ordinal analysis usually provides us with a better understanding of the theory and its relation to other theories.

Let us have a closer look at the case of Peano Arithmetic. This is the theory that postulates the principle of induction for all arithmetical formulas. It is an easy fact that the principle of induction is equivalent to the least number principle (that *every nonempty set has the least element*). This is nothing else but the fact that the set of natural numbers, which is the least infinite ordinal ω, is well-ordered. Now, this may easily lead to the confusion: *why did Gentzen reduce the consistency of the theory that says that ω is well-ordered to the fact that ε_0, which is a larger ordinal, is well-ordered?* The reason is that he wanted to find a *constructive* proof of the consistency of Peano Arithmetic, a proof that would only use simple concepts, which means algorithmically decidable relations. Therefore he could not use the general least number principle; he had to use the principle restricted to computable subsets. And this is the reason why he had to take a larger ordinal. Thus Gentzen's result can be presented as the reduction of

(a) the least number principle for *arithmetical sets* and the ordinal ω, to
(b) the least number principle for *computable sets* (a smaller class) and ε_0 (a larger ordinal).

How much did Gentzen fulfill Hilbert's vision of the proofs of consistency? One may be worried by the fact that the proof uses a transfinite ordinal much larger than ω, the ordinal representing natural numbers. Indeed, in the standard set-theoretical representation, ε_0 is the set of all ordinals less than ε_0, thus it is a set that contains infinite sets (which, in turn, also contain infinite sets etc.). But recall that Cantor normal form (Chap. 3, page 193) enables us to represent all ordinals below ε_0 by finite terms (or even natural numbers, if we prefer) so that the ordering relation is algorithmically decidable. As a matter of fact, any constructive ordinal has a representation of this kind *by definition*, but the representation of ordinals below ε_0 is especially simple.

So the complexity of (b) is not caused by the representation of the ordinal, but by its properties. It is good to recall the application of ε_0 to proving that Goodstein sequences always end at zero (page 321). For each Goodstein sequence, we constructed a strictly decreasing sequence of ordinals below ε_0. We also noticed that, for Goodstein sequences to get to zero, it takes an extremely large number of steps. Therefore, also the decreasing sequences of ordinals are extremely long. They are so long that it is quite nontrivial to prove that they actually reach zero. For a linear ordering, the condition that every decreasing sequence of elements is finite is equivalent to being well-ordered. Thus it is also quite nontrivial to prove using limited means that the representation of ε_0 by the Cantor normal form defines a well-ordering.

According to the Second Incompleteness Theorem one cannot reduce the consistency of Peano Arithmetic to a very simple statement; it must be a statement unprovable in Peano Arithmetic. Therefore, Gentzen's proof is not a *reduction* of the strength of the statement, it is rather a *transformation* into a different kind of statement. No matter whether we accept it as a solution of Hilbert's problem for Peano Arithmetic or not, it certainly shows the consistency from a different perspective and that is always useful.

Let us now look at how one can determine the proof-theoretic ordinal. To prove that α is the proof-theoretic ordinal of T, we have to show two things:

1. for every $\beta < \alpha$, T proves that β is well-ordered (a lower bound);
2. T does not prove that α is well-ordered (an upper bound).

The second part is certainly more interesting. All proofs of 2. are based on the Second Incompleteness Theorem, namely, one proves that the sentence "*α is constructively well-ordered*" implies the consistency of T. To give you some idea of these proofs, I will briefly sketch the main steps of a proof of the consistency of Peano Arithmetic using ε_0, which gives the upper bound on the proof-theoretic ordinal of Peano Arithmetic. This proof is due to the German proof-theorist Kurt Schütte (1909–1998) and is conceptually simpler than the original Gentzen's proof.

Having in mind that we should use only constructive means, the idea of the proof of consistency looks quite natural. This proof, like essentially all proofs of consistency, uses induction on the length of a proof to show that every formula in the proof is true, which in particular implies that a contradiction cannot appear in the proof. To this end, we need to be able to express that a formula is true. According to the Gödel-Tarski theorem on the undefinability of truth (see page 283), we cannot define the truth of arithmetical formulas in arithmetic, since this relation is of higher complexity. But we do not need to consider all proofs as we are interested only in proofs of contradiction. The important fact is that contradiction can be expressed by a quantifier free formula, say $0 = 1$. However, in the rest of the proof we can have arbitrarily complex arithmetical formulas, which is the main problem that we have to cope with.

Now recall what we learned in the first subsection: there are means to reduce the complexity of formulas in proofs. By Gentzen's Cut-Elimination Theorem, we can transform every proof into a cut-free one. If we have a cut-free proof, we know that all formulas in the proof are quasi-subformulas of the proved sentence. Then in order to prove that all formulas in the proof are true, we only need a truth definition for quasi-subformulas of the proved sentence. In particular, if the sentence has small quantifier complexity, we only need the truth definition for formulas of that quantifier complexity.

This looks promising, but what is the formula that we want to show to be unprovable? It is not $0 = 1$; we are considering provability from axioms, hence it is an implication of the form

$$(\alpha_1 \wedge \alpha_2 \wedge \cdots \wedge \alpha_n) \to 0 = 1,$$

where $\alpha_1, \ldots, \alpha_n$ are some axioms of Peano Arithmetic. So we need another idea. (Of course, we did not expect that the consistency proof would be so easy—if it were, we could do it in Peano Arithmetic, which is impossible by the Second Incompleteness Theorem.)

The next idea is, basically, to eliminate also the use of induction axioms. Suppose we have proved the assumption of an induction axiom

$$\phi(0) \wedge \forall x\big(\phi(x) \to \phi(x+1)\big), \tag{6.2}$$

and we want to derive the conclusion

$$\forall x\ \phi(x). \tag{6.3}$$

The crucial observation is that if we only need a numerical instance $\phi(n)$ of the sentence $\forall x\ \phi(x)$, we can derive it from the assumption (6.2) only using *logic*, namely, by $n+1$ applications of *modus ponens*.[8]

This is good because we want to eliminate the use of axioms so that we can then apply cut-elimination to reduce the complexity of formulas in the proof. What is not good is that we only get numerical instances, instead of the sentence with the universal quantifier (6.3). At this point another idea steps in, which is to allow a rule with an infinite number of premises (this is Schütte's idea). The rule is:

From $\phi(0), \phi(1), \phi(2), \ldots,$ *derive* $\forall x\ \phi(x)$.

Combining this rule with the proofs of numerical instances makes it possible to derive the conclusion of the induction axiom for ϕ. Thus we can eliminate induction axioms. What remains are the basic axioms that determine the properties of the successor, addition and multiplication. This is a finite set of axioms and, moreover, their quantifier complexity is low. In fact, if we introduce the predecessor function, we can present them using purely universal sentences.

The elimination of the induction axioms is obviously not for free. The price we have to pay is that we must use infinite proofs. It may look strange that we wanted to find a constructive proof and yet ended up using infinite proofs that use a very nonconstructive infinite rule. The point is that we are not going to study such proofs in general. We only need to describe the infinite proofs that result from finite proofs by eliminating the induction axioms. We can represent each such proof by the program that produces its parts. A program is a piece of finite text, so these infinite proofs are represented by finite entities.

The proof that we obtain by eliminating the induction axioms is infinite, but it shares an important property with finite proofs: the length from any initial sequent to

[8]Assume we represent n by the sum of n ones, $1+1+\cdots+1$. By the way, the reason why we believe in the induction principle is that it is a repetition of *modus ponens* infinitely many times.

the last sequent is finite. Since the proof tree is infinitely branching, due to the new rule, the latter property does not imply that there is a general bound on the length of the branches of the tree. Typically there will be arbitrarily long (finite) paths from initial sequents to the last sequent. The nice feature of such trees is that, although we cannot use numbers to bound their depths, we can measure the depth using ordinals.

The next step is to apply cut-elimination. This is done very much like for ordinary finite proofs, so I will not go into the details. Thus we finally obtain a proof all of whose formulas are of low complexity, namely, they define computable predicates. For such formulas, we do have a definition of truth in Peano Arithmetic. Now we may apply the standard way of proving consistency—proving by induction that all formulas in the proof are true. Since we are not in the standard situation, where proofs are finite, we have to use a stronger means. We use *transfinite induction* on the ordinals that bound the depths of such proofs to prove that all formulas in such a proof are true.

These are the basic ideas of the proof. The technical part is to estimate the ordinal depths of the infinite proofs obtained by the elimination of induction and cuts. Once it is established that the depth is always some ordinal below ε_0, all the pieces fit together and the consistency of Peano Arithmetic is proved.

One can prove that the proof-theoretic ordinal exists for every theory T that extends Peano Arithmetic. But this general theorem gives us little information about specific proof-theoretic ordinals. What we need is a representation of this ordinal based on a combinatorial construction, such as the Cantor normal form of ordinals less than ε_0, rather than a construction based on syntactical concepts related to the theory. This is already very hard for relatively weak set theories, and ingenious tricks are needed to make even slow progress in this field. There seems to be no hope to do the ordinal analysis of theories such as Zermelo-Fraenkel Set Theory with the means presently available. Ordinal analysis fails also at the opposite end of the spectrum of theories—for weak theories. One can analyze subtheories of Peano Arithmetic where induction is restricted to Σ_n formulas, but for theories weaker than arithmetic with induction for Σ_1 formulas, we are not able to use ordinals to distinguish different theories. This is quite unfortunate because such weak theories play a key role in proof complexity, as we will see in the next section, and we lack any means for proving that some theories are stronger than others.

Another way by which we can arrive at the same classification is based on fast growing functions. In Chap. 4 we saw that the unprovability of some sentences may be caused by the fact that they implicitly define very fast growing functions (see page 320). As n grows, the values of such a function $f(n)$ are so large that the theory is unable to prove that computations of the value of f always terminate. This suggests that we could compare theories according to how fast-growing functions they are able to handle.

We will only consider computable arithmetical functions and, accordingly, only definitions by Σ_1 arithmetical formulas. We say that *a function f is provably total in a theory T*, if there exists a Σ_1-definition of f in T such that it is provable in T that for every natural number n, the value of $f(n)$ is defined.

If T is finitely axiomatized or given by a computable set of axioms, we can effectively enumerate all definitions of computable functions in T and all proofs of their totality. Then we can take the "diagonal" over all provably total functions and obtain a computable function that is not provably total. So for every theory, there are functions that are not provably total. Hence, if we classify theories according to how fast growing functions are provably total in them, the resulting hierarchy is infinite.

In order to get a reasonable classification of theories, we must select a suitable subset of all computable functions. Furthermore, the set should be linearly ordered by the relation of growing faster. A natural way of defining such sets, called *hierarchies of functions,* is to define functions by transfinite recursion. To this end, we need a constructive ordinal β and for every limit ordinal $\alpha < \beta$, we need a sequence $\alpha_0 < \alpha_1 < \alpha_2 < \cdots$ that converges to α. I sketched the basic idea of these constructions in Chap. 4, page 336.

If we want to prove that the functions f_α are provably total in T for all ordinals α below some constructive ordinal β, we need to construct a representation of every such α in T and prove in T that it defines a well-ordered set. Then we can use transfinite induction over α to prove that all functions f_α, for $\alpha < \beta$, are provably total. If T does not prove that α is well-ordered for some $\alpha < \beta$, then it also does not prove that f_α is total. Thus using function hierarchies we get the same parameter as before, the proof-theoretic ordinal of the theory. So the possibility of formalizing large well-orderings is closely tied with the possibility of formalizing fast growing functions.

The proof-theoretic ordinal of T gives, certainly, only partial information about T. It is interesting, however, that in almost all of the cases in which it was determined, the ordinal completely determines the arithmetical sentences provable in T. One can show that, for these theories, provable arithmetical sentences are precisely those that are provable in Peano Arithmetic augmented with the schema of transfinite induction for the proof theoretical ordinal of T.

Notes

1. *Speed-up.* The speed-up phenomenon can be nicely demonstrated using sentences expressing finite consistencies. Let $Con_T(x)$ be a suitable formalization of the assertion that there is no proof of contradiction in T of length at most x. For fast growing functions that have simple formalizations in T, we will consider the sentences $Con_T(f(\bar{n}))$, where n denotes the nth numeral in binary representation. More precisely, if f is not provably total in T, this sentence should express: *either $f(n)$ is not defined, or $Con_T(f(n))$.*

   Let $S = T + Con_T$. Since Con_T is $\forall x\ Con_T(x)$, the sentences $Con_T(f(\bar{n}))$ have short proofs in S. On the other hand, by Theorem 58, page 565, proofs of these sentences in T must be almost as long as $f(n)$. This demonstrates speed-up by adding new axioms.

   To demonstrate speed-up by provability, one can use the same sentences except that the functions f must be provably total in T.

2. *Sequent calculi.* An efficient way of presenting proofs in the natural deduction calculus is to use sequents of the form

$$\phi_1, \ldots, \phi_n \Rightarrow \psi,$$

where ψ denotes the currently proved formula and $\phi_1, \ldots, \phi_n$ denote the currently active assumptions. In order to get a more symmetric calculus, one can allow strings of formulas also on the right hand side. The logical meaning of such a sequent

$$\phi_1, \ldots, \phi_n \Rightarrow \psi_1, \ldots, \psi_m$$

is

$$\phi_1 \wedge \cdots \wedge \phi_n \rightarrow \psi_1 \vee \cdots \vee \psi_m,$$

(the conjunction of the first sequence of formulas implies the disjunction of the second one). What I described above is a modification of Gentzen's sequent calculus proposed by Schütte [258], called *one sided sequent calculus.* I will define a version of this calculus in more detail.

Sequents are finite multisets of formulas using the connectives $\neg$, $\wedge$, $\vee$ and the two quantifiers $\forall$, $\exists$. A multiset is represented by a string of formulas separated by commas; viewing it as a multiset means that we ignore the order of formulas in the string. To save using the rules for negation, we only allow negations in atomic formulas. For a compound formula ϕ, we interpret $\neg\phi$ as a formula obtained by pushing the negation to the atomic formulas while implicitly using the rules

$$\neg(A \wedge B) \equiv \neg A \vee \neg B, \qquad \neg(A \vee B) \equiv \neg A \wedge \neg B, \qquad \neg\forall x A \equiv \exists x \neg A,$$

$$\neg\exists x A \equiv \forall x \neg A, \qquad \neg\neg A \equiv A.$$

The explicit rules are:

- *axiom* $A, \neg A$; (axiom)
- *from* Γ, *derive* Γ, A; (weakening)
- *from* Γ, A, A, *derive* Γ, A; (contraction)
- *from* Γ, A, B, *derive* $\Gamma, A \vee B$; ($\vee$-introduction)
- *from* Γ, A *and* Δ, B, *derive* $\Gamma, \Delta, A \wedge B$; ($\wedge$-introduction)
- *from* $\Gamma, A(y)$, *where* y *is not free in* Γ, *derive* $\Gamma, \forall x A$; ($\forall$-introduction)
- *from* $\Gamma, A(t)$, *derive* $\Gamma, \exists x A(x)$; ($\exists$-introduction)
- *from* Γ, A *and* $\Delta, \neg A$, *derive* Γ, Δ. (cut)

Here I am using A and B to represent formulas, Γ and Δ to represent strings of formulas, x and y to represent variables and t to represent a term. Proofs in sequent calculi are usually presented in a tree form with applications of rules shown by horizontal lines.

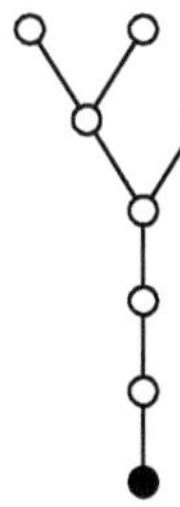

Fig. 6.1 The proof-tree of the example

Example Consider a proof of the propositional tautology $\neg\alpha \vee (\alpha \wedge \neg\beta) \vee (\beta \wedge \neg\gamma) \vee \gamma$.

$$\dfrac{\dfrac{\dfrac{\dfrac{\dfrac{\alpha, \neg\alpha \qquad \beta, \neg\beta}{\neg\alpha, (\alpha \wedge \neg\beta), \beta} \qquad \gamma, \neg\gamma}{\neg\alpha, (\alpha \wedge \neg\beta), (\beta \wedge \neg\gamma), \gamma}}{\neg\alpha \vee (\alpha \wedge \neg\beta), (\beta \wedge \neg\gamma), \gamma}}{\neg\alpha \vee (\alpha \wedge \neg\beta) \vee (\beta \wedge \neg\gamma), \gamma}}{\neg\alpha \vee (\alpha \wedge \neg\beta) \vee (\beta \wedge \neg\gamma) \vee \gamma}$$

The graph representing the tree structure of this proof is in Fig. 6.1.

Proofs in sequent calculi are quite cumbersome. These calculi are not practical for formalizing mathematical proofs, but are very convenient for proof theory. In particular, the transformation of indirect proofs into direct ones is quite natural—this is the cut-elimination procedure.

3. *Cut-elimination.* The main idea of cut-elimination is to replace cuts with complex formulas by cuts with less complex formulas. By eliminating one cut with a complex formula we may introduce a large number of cuts with simpler cut formulas, but we still make progress because the number of complex cuts is reduced.

   As an example, consider the elimination of a cut in which the cut formula is a disjunction of two other formulas. Let the cut formula be $\phi \vee \psi$. In the calculus introduced above, the negation of it is identified with $\neg\phi \wedge \neg\psi$. One has to consider several cases of how these formulas were introduced in the proof. The most typical case is that $\phi \vee \psi$ was introduced by $\vee$-introduction and $\neg\phi \wedge \neg\psi$ was introduced by $\wedge$-introduction. Then instead of the cut applied to $\phi \vee \psi$ and $\neg\phi \wedge \neg\psi$, we apply two cuts to the pairs $\phi, \neg\phi$ and $\psi, \neg\psi$. This is shown schematically as the transformation of the proof

$$\dfrac{\dfrac{\dfrac{\dfrac{\dfrac{\vdots}{\Gamma', \phi, \psi}}{\Gamma', \phi \vee \psi}}{\vdots}}{\Gamma, \phi \vee \psi} \qquad \dfrac{\dfrac{\dfrac{\dfrac{\vdots}{\Delta', \neg\phi} \quad \dfrac{\vdots}{\Delta'', \neg\psi}}{\Delta', \Delta'', \neg\phi \wedge \neg\psi}}{\vdots}}{\Delta, \neg\phi \wedge \neg\psi}}{\dfrac{\Gamma, \Delta}{\vdots}}$$

into the proof

$$\cfrac{\cfrac{\cfrac{\vdots}{\Gamma', \phi, \psi} \qquad \cfrac{\vdots}{\Delta', \neg\phi}}{\Gamma', \Delta', \psi} \qquad \cfrac{\vdots}{\Delta''\neg\psi}}{\cfrac{\Gamma', \Delta', \Delta''}{\cfrac{\vdots}{\cfrac{\Gamma, \Delta}{\vdots}}}}$$

In this particular type of proof the formula $\phi \vee \psi$ was introduced only once, therefore the proof after the elimination of the cut is not larger. In general, $\phi \vee \psi$ can be introduced several times. Then we have to introduce more cuts on the pairs $\phi, \neg\phi$ and $\psi, \neg\psi$ and the resulting proof is larger. Thus we have to choose a suitable strategy, i.e., an order for eliminating cuts, otherwise the procedure may not terminate.

One strategy that leads to a terminating procedure is to always eliminate the uppermost cuts. If the cut formula in such a cut is not atomic, the elimination will produce new cuts, but the complexity of the formulas in the new cuts will be smaller. Then we eliminate those, which will produce even more cuts, etc. But since elimination of cuts on atomic formulas does not produce new cuts, we will eventually eliminate all cuts.

This resembles the problem of Hercules and Hydra (see page 324), but here Hydra behaves differently: if we cut a head, the new heads will grow on upper, not lower, nodes. Also for cut elimination, one only needs a superexponential number of steps, whereas in the game of Hercules and Hydra the number is much larger. I invite the reader to design the corresponding game and to study strategies for it.

4. *Herbrand's Theorem.* There are constructive proofs that give an explicit construction of terms in the disjunction and nonconstructive proofs that only prove the existence of such terms. I will explain the idea of a nonconstructive proof. The proof is very similar to the proof of the Completeness Theorem that I sketched in Chap. 2 (page 107).

Suppose, as before, we have a sentence with only one quantifier and it is an existential quantifier, $\exists x\, \phi(x)$. Arguing by contradiction, suppose that none of the disjunctions of the form (6.1) is a propositional tautology. This is equivalent to the fact that any finite set of term instances

$$\{\neg\phi(t_1), \neg\phi(t_2), \dots, \neg\phi(t_n)\}$$

is consistent. This means that one can assign truth values to the atomic sentences so that all sentences $\neg\phi(t_1), \neg\phi(t_2), \dots, \neg\phi(t_n)$ are true. By the compactness of propositional calculus (page 115), the set of *all* sentences $\neg\phi(t)$ is consistent. Hence there exists a truth assignment A to atomic sentences such that *all* sentences $\neg\phi(t)$ are true. Using A, we construct a term model of the sentence $\forall x\neg\phi(x)$ in the same way as in the proof of the completeness theorem. Namely, the elements of the model are terms, a function symbol $f(x_1,\dots,x_k)$ is interpreted as the mapping $t_1,\dots,t_k \mapsto f(t_1,\dots,t_k)$ and, for a relation symbol $R(x_1,\dots,x_l)$, the corresponding relation is true for terms $s_1,\dots,s_l$, if and only if the atomic formula $R(s_1,\dots,s_l)$ is true in the assignment A.

Thus from the assumption that no disjunction of the form (6.1) is a propositional tautology, we have derived the existence of a model in which $\exists x\ \phi(x)$ is false, hence cannot be provable. This proves the theorem.

Herbrand's theorem for existential formulas with more existential quantifiers is essentially the same. The general form of Herbrand's theorem characterizes provability (which is the same as logical validity) of arbitrary sentences in prenex form (this means that all quantifiers precede the other parts of the formula). The general case is reduced to the existential case by eliminating universal quantifiers by means of new function symbols. This is the dual transformation to skolemization that I mentioned in Chap. 2. Consider, as an example, a sentence of the form

$$\exists x\forall y\exists z\forall u\ \phi(x,y,z,u).$$

The *herbrandization* of this formula is

$$\exists x\exists z\ \phi\big(x,F(x),z,G(x,z)\big).$$

A sentence in the prenex form is provable if and only if its herbrandization is. This is an immediate consequence of the fact that a sentence is consistent (has a model) if and only if its skolemization is consistent. Thus the same proof as above gives us that the sentence $\exists x\forall y\exists z\forall u\ \phi(x,y,z,u)$ is provable if and only if a disjunction of the form

$$\phi\big(t_1,F(t_1),s_1,G(t_1,s_1)\big)\vee\phi\big(t_2,F(t_2),s_2,G(t_2,s_2)\big)\vee\cdots$$
$$\vee\,\phi\big(t_n,F(t_n),s_n,G(t_n,s_n)\big)$$

is a propositional tautology for some terms $t_1,\dots,t_n,s_1,\dots,s_n$. Notice that the function symbols F and G may occur also inside of terms $t_1,\dots,t_n,s_1,\dots,s_n$.

One can also state a general form of Herbrand's theorem without additional function symbols. In general one has to state a rather complicated condition on terms, but for the prefix $\forall\exists\forall$ it is fairly simple: a sentence of the form

$$\forall x\exists y\forall z\ \phi(x,y,z)$$

is provable if and only if a disjunction of the form

$$\phi\big(x,t_1(x),y_1\big)\vee\phi\big(x,t_2(x,y_1),y_2\big)\vee\cdots\vee\phi\big(x,t_n(x,y_1,y_2,\dots,y_{n-1}),y_n\big)$$

is a propositional tautology for some terms $t_1,\dots,t_n$ that may only contain the variables explicitly shown.

5. *Parikh's proof.* The proof that there is no short proof of contradiction in Parikh's theory FPA_t has two main steps. Suppose a proof d is a short proof of contradiction.

In the first step we replace d by a direct proof d', which can be, for example, a Herbrand disjunction. The direct proof d' can be fairly large; we can only bound it by a superexponential function in the length of d. This is the reason why one has to take the term t so that it defines a large number.

Let m be the number that t defines (the value of t). The second step is a proof that any direct proof of contradiction must have size at least m. This is proved by showing that in a direct proof of contradiction there must be a term with value n for every n, $0 \leq n < m$. The basic idea is that one can obtain a contradiction only using term instances of the axiom

$$F(x) \rightarrow F(x+1)$$

and propositional logic. If there is no term s in the proof whose value is n, for $0 \leq n < m$, then we can interpret $F(n)$ as the formula $x \leq n$. With this interpretation (and the natural interpretation of the arithmetical operations and the inequality relation) all term instances of the axioms occurring in the proof become true. Hence the proof cannot contain a contradiction.

Since there are at least m different terms in the proof d', its length is at least m. This implies that if k is the length of d, then $supexp(k) \geq m$. This gives the required lower bound on the length of d.

6. *A well-hidden contradiction.* Given a one-way function, or a pseudorandom generator, one can construct a *bit commitment schema.* This is a function $\beta(x, y)$ computable in polynomial time whose values are 0, 1 and $*$ such that

   a. for every x, there exists y such that $\beta(x, y) = 0$ or $\beta(x, y) = 1$;
   b. for no x, y and y', $\beta(x, y) = 0$ and $\beta(x, y') = 1$;
   c. no polynomial algorithm is able, from a given x, to predict the bit 0 or 1 with probability substantially different from 1/2. (Here we assume that an n is fixed and x is a randomly chosen bit string of length n.)

   This function enables one to commit to a bit $b \in \{0, 1\}$ without having to reveal information about it. To this end one takes x and y so that $\beta(x, y) = b$ and publishes only x. When he has to prove that he was committed to b, he reveals y. (For the schema to be practical, one also needs to be able to generate instances x, y such that $\beta(x, y) \in \{0, 1\}$, which is not important for us now.)

   Suppose that we have such a β. Let s be a string of length n. Consider two theories

$$ZFC + \forall y\, \beta(s, y) \neq 0 \quad \text{and} \quad ZFC + \forall y\, \beta(s, y) \neq 1.$$

   According to the first condition one of these theories is inconsistent and there is a short contradiction—we only need to verify the computation of $\beta(s, t)$ for a suitable t. But if we could find a contradiction efficiently, any contradiction, not just the one given by a witness t, we would be able to decide which bit is encoded by s. So if the bit commitment schema is secure, it is difficult to find a contradiction for a randomly chosen s.

7. *Proofs, trees and ordinals.* When rules with infinite numbers of premises are used, the structures of the proofs are infinite well-founded trees. A tree is called

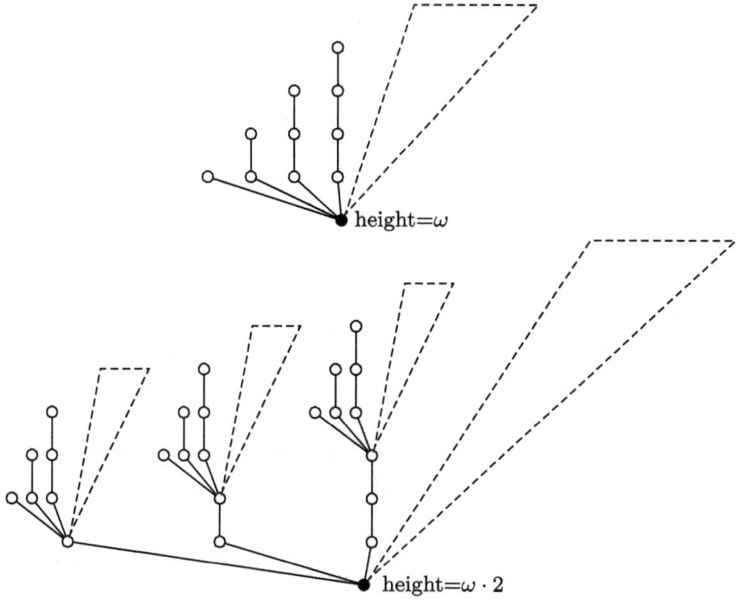

Fig. 6.2 Schematic figures of trees of heights ω and $\omega \cdot 2$

well-founded if all its branches are finite. This makes perfect sense because we want to derive sentences from axioms that are only at the leaves of the tree. If there were infinite branches, there would be deductions that would not based on any axiom.

A well-founded tree has a *height*, which is a countable ordinal defined inductively as follows. A tree consisting of a single vertex, the root, has height 0. If α_i, for $i \in I$, are the heights of the subtrees attached to the root, then the height of the tree is $\sup_{i \in I}(\alpha_i + 1)$. See an example in Fig. 6.2.

8. *Proof-theoretic ordinals.* Many types of proof-theoretic ordinals have been defined. The one defined in this section is the most basic one and is usually denoted by $|T|_{\text{sup}}$. It is tempting to define the proof-theoretic ordinal of a theory T to be *the least order type of a recursive definition of an ordering using which one can prove the consistency of* T. This definition fails very badly—every theory would have the ordinal equal to ω. The trick is simple; define the ordering to be the natural ordering of natural numbers up to the Gödel number of the first proof of contradiction in T and after that reverse it. However, if we fix a "natural" definition of a big constructive ordinal β, we can use this definition to analyze theories with proof-theoretical ordinals less than β.

   Ordinal analysis has been successfully applied to several theories stronger than *PA*. In particular, the Feferman-Schütte ordinal Γ_0 is associated with the subsystem of Second-Order Arithmetic ATR_0 and the Bachmann-Howard ordinal $\psi(\varepsilon_{\Omega+1})$ is associated with Kripke-Platek Set Theory (with the axiom of infinity).

9. *Theorems with short nonelementary proofs.* There are examples of theorems whose nonelementary proofs are shorter than all known elementary proofs. In many cases first a nonelementary proof is found and later an elementary, usually longer, proof is also discovered. An example of this is the Prime Number Theorem (page 61).

   Another area where problems are stated using elementary concepts, but are sometimes solved using nonelementary means, is finite combinatorics. *Kneser's Conjecture* (which is now a theorem of L. Lovász) is the following statement:

   Theorem 46 *Let n and k be natural numbers such that $n \geq 2k$. Let $C_1 \cup C_2 \cup \cdots \cup C_{n-2k+1}$ be a partition of k-element subsets of an n-element set. Then in one class C_i there are two disjoint sets.*

   Lovász solved the problem using topological methods. Soon after I. Baranyi gave a simple proof based on the Borsuk-Ulam Theorem:

   Theorem 47 *An antipodal continuous mapping from an n-dimensional sphere to an m-dimensional sphere exists only if $m \geq n$.*

   An antipodal mapping is a mapping that maps pairs of opposite points to pairs of opposite points. Kneser's Conjecture was stated in 1955 and was proved in 1978, but only in 2003 was an elementary proof found by J. Matoušek [194].

   In logic there are several theorems that can be proved either using proof theory, where the proofs are longer, or model theory, where the proofs are shorter. An example is Herbrand's theorem.

   Strictly speaking, it is not clear that the nonelementary proofs are shorter in the formal sense considered in this section. We would have to say explicitly what the theory is in which the nonelementary proof is formalized. Consider, for example, Kneser's Conjecture. The nonelementary proof is short, but refers to the Borsuk-Ulam Theorem. Should we consider the Borsuk-Ulam Theorem to be an axiom? This would not be fair, but if we do not, then we have to take into account also the length of the proof of the Borsuk-Ulam Theorem.

   Ideally, we would like to have an example of a theorem that has a long proof in Finite Set Theory (Zermelo-Fraenkel Set Theory without the axiom of infinity) and a short proof in Zermelo-Fraenkel Set Theory. I am not aware of any example of a mathematical result of this type. One problem is that a lot of methods that use infinite sets can be simulated in Finite Set Theory. Another is that deriving basic algebraic, geometric and topological results requires quite lengthy proofs.

   But even if we had such a theorem, the most difficult problem would still be there—to prove that an elementary proof cannot be short. We are only able to prove that for contrived theorems.

6.2 Theories and Complexity Classes

The relationship between proofs and computations has been studied since the time the fields of proof theory and computability theory started. A number of interesting results connecting proofs with computations have been proved. Among the logical calculi that can be used for formalizing computation the most prominent is the λ-calculus of Church.

After the advent of computational complexity theory it was natural to look for logical calculi that could be related to *polynomial time computations.* At that time Parikh (in the same paper I mentioned some pages back [211]) proposed "*an anthropomorphic system*" *PB* based on Peano Arithmetic with induction restricted to a subclass of formulas whose validity is efficiently testable.[9] In Parikh's theory *PB* induction is restricted to bounded arithmetical formulas, formulas in which the range of quantification is bounded to finite intervals. For such formulas, one can algorithmically decide their satisfiability in the domain of natural numbers (the algorithm runs in linear space). An important property of *PB*, as well as of many theories studied subsequently, is that it is not provable that for every x, there exists 2^x, in other words, the exponential function is not provably total. These two properties led Parikh to the conclusion that such a system is closer to what humans can actually do, if we view it from the philosophical point of view, and closer to the research in complexity theory, if view it from the mathematical point of view.

Only a little later, in 1975, Cook introduced his system *PV* of *polynomially verifiable* identities [49]. His aim was to design a system that was somehow connected with polynomial time computations. Building on these two formal systems, more theories were introduced in the work of J. Paris and A. Wilkie [215], S. Buss [34] and others. Although different, the essence of these systems is the same.

Another line of research that started at about the same time concerned proving lower bounds on the lengths of proofs in propositional calculus. This apparently unrelated subject has a very close connection to the theories mentioned above. The connection was discovered by Cook and described in the same paper in which he introduced his theory *PV*. It is the fact that proofs of sentences of a certain type can be translated into at most polynomially long proofs of sequences of tautologies in the propositional calculus. This in principle enables one to prove the independence of some sentences by proving lower bounds on tautologies in the propositional calculus. Thus the logical problem of proving independence is reduced to a combinatorial problem of studying the lengths of propositional proofs. It has turned out that proving lower bounds on propositional proofs is a very hard problem, so this connection has not produced many independence results. Nevertheless, it helped us more fully understand the essence of the problems concerning these theories and the propositional calculus.

Originally the term '*Proof Complexity*' was only used for the subfield that studies *lengths of proofs*. As the connections between theories and the propositional

---

[9] The notation *PB* has never been used afterwards, instead the theory was denoted by $I\Delta_0$, where I stands for *induction* and Δ_0 for bounded arithmetical formulas.

calculus became more apparent, the term acquired much wider meaning and it is now used for the whole field that includes not only the complexity of propositional proofs, but also the study of theories related to feasible computations.

Theories Corresponding to Complexity Classes

One of the key ideas of Proof Complexity is that it is possible to define theories that in some sense correspond to complexity classes. The link is not mere formal similarity; it is based on certain relations between a complexity class and the corresponding theory. We are not able to construct such a theory for every complexity class, but there are a number of important complexity classes for which theories have been defined which relate closely to their corresponding classes. Here I will only outline the basic idea of this correspondence.

As we are mostly interested in complexity classes that are near the bottom of the hierarchy, we may view the corresponding theories as subtheories of Peano Arithmetic. Recall that Peano Arithmetic is axiomatized by two sets of axioms:

1. *Basic Axioms.* These axioms state a few basic properties of the constant 0 and the arithmetical operations. Specifically these are the following axioms (where, for the sake of simplicity, I am omitting the axioms for the less-than-or-equal relation $\leq$):

   $$\begin{array}{lll} S(x) \neq 0 & x+0=x & x \cdot 0 = 0 \\ S(x) = S(y) \rightarrow x = y & x + S(x) = S(x+y) & x \cdot S(y) = S(x+y) \\ x \neq 0 \rightarrow \exists y\big(x = S(y)\big) & & \end{array}$$

2. *The Schema of Induction.* This is an infinite set of axioms, one axiom for every arithmetical formula with free variable x, that states the principle of induction for this formula. Formally, it is expressed by

   $$\big(\phi(0) \wedge \forall x\big(\phi(x) \rightarrow \phi\big(S(x)\big)\big)\big) \rightarrow \forall x\, \phi(x),$$

   where $\phi(x)$ is the arithmetical formula. In other words, we state the principle of induction in the maximal possible strength that we can achieve using only the language of arithmetical operations.

Let us also recall that Peano Arithmetic is in a well-defined sense equivalent to Finite Set Theory. The theory is fairly strong, in particular, essentially all mathematical results concerning numbers and finite structures can be formalized and proved within the framework of Peano Arithmetic and Finite Set Theory.

Now suppose we want to prove theorems about sets and functions from a certain class. Clearly, we will need the induction principle at least for sets from this class. Specifically, if we are interested in a complexity class C and the sets in this class are exactly those that can be defined by formulas from a class Φ, then we should postulate the induction axiom for every formula from the class Φ. The key idea is to study theories in which we state induction *only* for these formulas. In other words, we take something like the minimal theory that is able to describe the objects we are interested in and to prove general theorems about them.

Given a complexity class $\mathcal{C}$, I will use the following symbol

$$\Theta_{\mathcal{C}}$$

to denote the theory in which induction is restricted to sets of complexity $\mathcal{C}$. This is a rather informal notation for several reasons. In order to define this theory precisely, we need to specify a class of formulas Φ that represents the sets in $\mathcal{C}$. The language of basic arithmetical operations is usually not rich enough to get a natural class of such formulas, thus we must use more function and predicate symbols. Then we also have to extend the set of basic axioms in order to fix the interpretations of the new symbols. All these things are, however, just technicalities that we can ignore in our informal description of the theories. So the structure of the set of axioms of $\Theta_{\mathcal{C}}$ is essentially the same as for Peano Arithmetic:

1. *Basic Axioms.* A finite set of axioms concerning the primitive notions. In particular we can always include the basic axioms of Peano Arithmetic.
2. *The Schema of Induction for the class of formulas* Φ. An infinite set of axioms—one axiom for every formula of Φ with free variable x that states the principle of induction for this formula.

The theory *PB* introduced by Parikh was Peano Arithmetic with induction restricted to formulas with bounded quantifiers. We call a quantifier bounded if the quantified variable x occurs in a context of the form $x \leq t$ for some term t not containing x. Informally, the range of quantification is restricted to the finite domain of numbers between 0 and t. The theories introduced by Buss were also based on induction for bounded formulas, except that he used a richer language. Since complexity classes are always defined by bounding resources, the classes of formulas defining them consist of bounded formulas in some language. Therefore we use the term *Bounded Arithmetic* to denote all theories constructed in this way.

The schema described above enables us, in principle, to define the theory $\Theta_{\mathcal{C}}$ for every complexity class $\mathcal{C}$, but in order to view this theory as associated with the complexity class, we want to have closer relations between these two objects. For a number of complexity classes, such relations have been shown; I will say more about these relations shortly.

Let us observe one particular similarity between complexity classes and these theories. A complexity class is defined by *restricting computational resources*. A Bounded Arithmetic theory is defined by *restricting provability resources*. We will see other similarities of this type in the following subsections.

Several Reasons why Studying Weak Theories Is Important

According to the Incompleteness Theorem, there are unprovable true sentences in Peano Arithmetic. We can make the theory stronger by postulating induction for a larger class of formulas, but such a theory will still be incomplete. By stating induction for a subclass of arithmetical formulas we are making things worse—the theories are even more incomplete. In fact, the theories that we will consider will be much weaker than Peano Arithmetic. Although we can never fully determine a structure such as the natural numbers with arithmetical operations, we can at least try to make the theory as strong as possible. Thus it may seem strange to do the opposite, to work with a restricted set of axioms, but there are good reasons for this. One should realize that by defining theories of Bounded Arithmetic we are not trying to propose theories that should serve as "foundations for arithmetic". We are defining new objects of study—theories of a particular kind. The rule of thumb is that the weaker the theory is, the more one can prove about it. Thus considering weak theories is, in fact, an advantage.[10]

Now I will briefly review some reasons why researchers are interested in these theories. Then, in the following subsections, I will describe the theories in more detail.

1. *It is possible to study problems that concern theories and that are analogous to the problems about complexity classes.*

Let us consider an example. As sketched above, one can define theories for some complexity classes, in particular this is possible for **P** and **NP**. So let $\Theta_{\mathbf{P}}$ and $\Theta_{\mathbf{NP}}$ be the theories corresponding to the complexity classes **P** and **NP**. Then we can ask the question that corresponds to the fundamental problem **P** vs. **NP** in the realm of theories:

Do the theories $\Theta_{\mathbf{P}}$ and $\Theta_{\mathbf{NP}}$ prove the same theorems?

I will state it more succinctly as the question

$$\Theta_{\mathbf{P}} \equiv \Theta_{\mathbf{NP}}?$$

Most problems in complexity theory are stated as a question about whether or not two complexity classes are the same. Similarly, the main open problems in proof complexity ask for proving or disproving that two theories are equivalent. Although they concern different concepts, the stumbling blocks preventing us from solving these problems seem to be very similar. Thus we are led to the conclusion that logic must play an important role not only in proof complexity, but also in computational complexity. Therefore it seems conceivable that our inability to make any progress in solving problems in computational complexity may be caused not by their mathematical complexity, but by some obstacles of a logical nature. If this feeling is right, the study of theories of Bounded Arithmetic should help us solve both types of problems.

2. *It is possible to characterize functions from some complexity classes as functions that can be formalized in the corresponding theories.*

This is an important example of a *formally provable* relation between theories and complexity classes. Let us consider an example again. $\Theta_{\mathbf{P}}$, the theory associated

[10] Yet the theories are still too strong if one wants to apply the methods of main-stream model theory. Thus one has to develop methods specific for this field.

with the complexity class **P**, has the following constructive property: if we prove the existence of a number satisfying some polynomial time condition, then we can construct this number in polynomial time. More precisely, if for some polynomial time relation R, we prove that for every x, there exists y such that x is in relation R with y, then one can find an algorithm that for every x, constructs such a y in polynomial time.

Note that it is unlikely that in general such algorithms should exists for all polynomial time computable relations of this type. Thus the provability in $\Theta_{\mathbf{P}}$ guarantees us more than if we only knew that for every x such a y existed.

3. *It is natural to argue about complexity classes using concepts of the same nature.*

When Parikh defined his theory *PB*, he was motivated by such considerations. He also observed that in his theory it is not possible to prove that for every number n, there exists the number 2^n, that is, the exponential function is not provably total. Thus in proving a sentence about n one cannot use numbers much larger than n. We know that this may prevent us from using nonelementary arguments, such as arguments using probability and algebra, which are quite common in combinatorics and number theory. But our aim is not to use these theories to prove new theorems, rather to use them to understand what methods are needed for particular theorems.

Another example is Cook's theory *PV*. Cook's idea is that as **P** renders very well the intuitive concept of efficiently computable functions, we should also have a similar concept concerning proofs. I will explain his theory in more detail shortly.

4. *We want to determine the axioms needed for proving that a given algorithm does what it is expected to do.*

When we are studying a computational problem from the point of view of complexity, we ask what is the least time, space etc., that an algorithm needs for solving the problem. We thus consider various computational resources needed for computation. But an algorithm is not just a piece of code in a programing language; there is also an *idea* (or several ones) on which it is based. If we are to design a correct algorithm for a problem, we have to use the definition of the problem and we have to apply some *ideas, methods, tricks, principles etc.* The latter words can be formalized as the *axioms needed for proving the correctness of the algorithm.* These are not computational resources, but they can also be viewed as a kind of complexity of the algorithm. It is natural to call them the *proof complexity* of the problem.

It is possible, for example, that for some problem P, we cannot solve P efficiently using algorithms based on simple arguments, while there exists an algorithm based on deeper mathematics that solves it efficiently. Thus we may have a trade-off between the time complexity and the complexity of axioms needed for proving the correctness of an algorithm for P.[11] This shows that there exists an intrinsic connection between computational and proof complexities. In order to formalize this connection, we need a scale of formal systems. These formal systems need not only

[11] We do not have a concrete example for which we can prove this tradeoff, but we conjecture that testing primality is a problem of this kind.

be theories formalized in first order logic, but we will see that one can also use proof systems for the propositional calculus.

*Universal-***P** *Sentences and Formalization of Complexity Classes*

Although we are interested in all sentences provable in the theories that we study, some sentences are more important than others. I have already mentioned the set of sentences Π_1, which I also call *universal-finite sentences.* These are sentences of the form $\forall x\ \phi(x)$, where in ϕ all quantifiers are bounded. Hence for a given x we can algorithmically decide whether $\phi(x)$ is true or false. Such sentences are *empirically testable:* we can test the validity of $\phi(x)$ for various values x, but we will never be sure that $\forall x\ \phi(x)$ is true (unless we prove the sentence). On the other hand, a Σ_1 sentence, which is of the form $\exists x\ \phi(x)$, with ϕ a bounded formula, is *empirically verifiable:* given an element a that satisfies $\phi(a)$, we can verify the truth of $\exists x\ \phi(x)$.

Algorithmic decidability is a very rough approximation of what actually can be computed. Having the concept of polynomial time computations, it is natural to consider the sentences $\forall x\ \phi(x)$ with the predicate $\phi(x)$ being *computable in polynomial time.* Thus we get a much better approximation of the intuitive concept of empirically testable sentences. Not surprising, these sentences play an important role in proof complexity. I will call them *universal-***P** *sentences* and use the notation

$$\Pi_{\mathbf{P}}$$

for the set of these sentences.

This concept needs a more precise definition, otherwise it may lead to nonsensical conclusions. For instance, if a sentence of the form $\forall x\ \phi(x)$, where ϕ is *arbitrary*, is true, then the satisfiability of $\phi(x)$ is trivial, in particular, it is decidable in polynomial time. Applying the above definition literally, we would conclude that every true sentence starting with a universal quantifier is universal-**P**.

The point is that we must know *a priori* that ϕ is decidable in polynomial time. More precisely, every theory T to which we apply this concept should be "aware" of the fact that ϕ is decidable in polynomial time. This means that if we formalize Turing machine computations, then for some Turing machine M, T is able to prove that M accepts exactly those strings or numbers that satisfy ϕ and M runs in polynomial time.

So the main question is how one should pick the class $\Phi_{\mathbf{P}}$ of formulas that define sets in **P**. A simple solution is to take the formulas that define Turing machines with built-in clocks that ensure polynomial bounds on their running time. This suffices for formalizing $\Pi_{\mathbf{P}}$ and some other concepts that we will need. However, if we used it to define the theory $\Theta_{\mathbf{P}}$, we would run into problems. In order to prove properties of Turing machines, we would need to use induction, but the induction axioms would already be stated in terms of Turing machines. For defining $\Theta_{\mathbf{P}}$, it is better to use a class defined by simple syntactical conditions. Such a class can be defined if we are willing to use a language with infinitely many primitive symbols. This is the approach taken in the theory PV. For languages with finite sets of primitives, there is another solution. One can take a class of formulas Γ that defines only a subclass of **P**, but produces induction axioms that are strong enough to prove induction for all predicates in **P** (see Notes).

All this also concerns sets of formulas $\Phi_{\mathcal{C}}$ for other complexity classes $\mathcal{C}$. For some complexity classes, for example for **NP**, it is easy to define a class of formulas that is syntactically simple, but for others it may be even more difficult than for **P**.

It is clear that I am using the symbols $\Phi_{\mathcal{C}}$, $\Theta_{\mathcal{C}}$ and $\Pi_{\mathbf{P}}$ as generic names for sets that can be specified in various ways. But in spite of the ambiguity of the definition of $\Phi_{\mathcal{C}}$, the concept of the associated theory $\Theta_{\mathcal{C}}$ is quite robust: the theorems that one can derive from various axiomatizations are the same. To stress the fact that I am considering theories T and S as the set of theorems provable therein, I will use the notation $T \equiv S$ to express that T and S are equivalent in the sense that they prove the same theorems.

Some Relations Between Theories and Complexity Classes

It is interesting that we can formulate problems in proof complexity that are like problems in computational complexity, but one may wonder whether this similarity has deeper roots or is just a superficial analogy. I will explain, using the example of the problems **P** vs. **NP** and $\Theta_{\mathbf{P}}$ vs. $\Theta_{\mathbf{NP}}$, what we know about the relation between such pairs of problems.

Recall that the theories $\Theta_{\mathbf{P}}$ and $\Theta_{\mathbf{NP}}$ are defined by restricting the induction axiom schema to a set of formulas that define sets in **P** and to a set of formulas that define sets in **NP** respectively. Let these two sets of formulas be denoted by $\Phi_{\mathbf{P}}$ and $\Phi_{\mathbf{NP}}$. Naturally, we pick the classes so that $\Phi_{\mathbf{P}}$ is a subset of $\Phi_{\mathbf{NP}}$ and the basic axioms are the same. This ensures that (the set of sentences provable in) $\Theta_{\mathbf{NP}}$ is an extension of, or is equal to $\Theta_{\mathbf{P}}$, just as the class **NP** is an extension of, or is equal to the class **P**.

Suppose that $\mathbf{P} = \mathbf{NP}$. This means that formulas of $\Phi_{\mathbf{P}}$ define all sets that are definable by formulas of $\Phi_{\mathbf{NP}}$. So for every formula $\phi(x)$ from $\Phi_{\mathbf{NP}}$, there exists an equivalent formula $\psi(x)$ in $\Phi_{\mathbf{P}}$. However, the equivalence $\phi(x) \equiv \psi(x)$ does not have to be provable in $\Theta_{\mathbf{P}}$. It may happen that to prove this equivalence one would need stronger axioms than are available in $\Theta_{\mathbf{P}}$. Therefore it is possible that $\mathbf{P} = \mathbf{NP}$ and $\Theta_{\mathbf{P}}$ is not equivalent to $\Theta_{\mathbf{NP}}$.

Let us now consider the opposite relation. Suppose that $\Theta_{\mathbf{P}} \equiv \Theta_{\mathbf{NP}}$. Formally, this means that for every formula $\phi(x)$ from $\Phi_{\mathbf{NP}}$, there exists a finite set of formulas $\psi_1(x), \dots, \psi_n(x)$ from $\Phi_{\mathbf{P}}$ such that the induction axiom for $\phi(x)$ follows from the induction axioms for $\psi_1(x), \dots, \psi_n(x)$ and the basic axioms. Notice that this does not say anything about a formula from the class $\Phi_{\mathbf{NP}}$ being equivalent to a formula from $\Phi_{\mathbf{P}}$. So again, we cannot exclude the possibility that $\Theta_{\mathbf{P}} \equiv \Theta_{\mathbf{NP}}$ and $\mathbf{P} \neq \mathbf{NP}$ hold true simultaneously. However, although we are not able to prove such a direct connection, the assumption $\Theta_{\mathbf{P}} \equiv \Theta_{\mathbf{NP}}$ does imply an inclusion relation between two complexity classes.

Theorem 48 *If* $\Theta_{\mathbf{P}} \equiv \Theta_{\mathbf{NP}}$, *then* **NP** $\subseteq$ **nonuniform-P**.

The statement **P** = **NP** is equivalent to **NP** $\subseteq$ **P** because we know that the other inclusion is true. Since also **P** $\subseteq$ **nonuniform-P**, the conclusion of the theorem is a weakening of **P** = **NP**. This result is due to J. Krajíček, G. Takeuti and the author.

This theorem shows that there are provable relations between proof complexity and computational complexity. Further, it shows that it is reasonable to conjecture that $\Theta_{\mathbf{P}} \not\equiv \Theta_{\mathbf{NP}}$ since it follows from the generally accepted conjecture from computational complexity that **NP** $\not\subseteq$ **nonuniform-P**.

Several more implications of this kind have been proved, but no implication in the opposite direction is known. In spite of not having consequences in complexity theory, proving $\Theta_{\mathbf{P}} \not\equiv \Theta_{\mathbf{NP}}$ would be an extremely important result in proof complexity, since it is one of the central problems in this field. But it would also be interesting from the point of view of computational complexity, as it would show that **P** = **NP** cannot be proved using limited means.

Search Problems

A search problem is given by a binary relation $R(x, y)$. The task is, for a given input x, to find y such that x is in relation R with y. We will be interested in the special kind of search problems where

1. *the relation* R *is computable in polynomial time,*
2. *the length of* y *is polynomially bounded by the length of* x.

This is a familiar situation—I explained the **P** vs. **NP** problem using search problems of this kind. But if we are only interested in complexity classes, as we were when defining **P** and **NP**, we need only to decide whether or not the search problem has a solution for a given input x; we do not need to find a solution. So the **P** vs. **NP** problem concerns *decision problems*, whereas *search problems* concern actually finding y. The class of all problems satisfying 1. and 2. is called *Polynomial Search Problems* and I will abbreviate it by **PS**.

From the point of view of complexity, the class **PS** is not particularly interesting. One can easily prove that all **PS** problems are solvable in polynomial time if and only if **P** = **NP**. What is much more interesting is the subclass of *Total Polynomial Search Problems*, which will be abbreviated by **TPS**.[12] These are the **PS** problems that satisfy another condition:

3. *for every* x, *there exists* y *such that* $R(x, y)$ *holds true.*

They are called *total* because an algorithm solving the problem determines a function f such that for all x, $R(x, f(x))$ holds true. Note that there are no natural de-

[12] I deviate from the more common abbreviations **FNP** for **PS** and **TFNP** for **TPS**, since I find the usual notation rather confusing. In **FNP** and **TFNP** the letter 'F' refers to 'function', but these classes are not classes of functions.

cision problems associated with problems in **TPS**—since a solution always exists, there is nothing to decide.

Example An example of a total polynomial search problem is the problem of finding a proper factor of a composite number. As stated it is not total because prime numbers do not have proper factors. But since primality is decidable in polynomial time, we can make it total by stipulating that 0 is a solution for inputs that are primes.

Condition 3. is fundamentally different from the previous two. If we want to guarantee that an algorithm A for R runs in polynomial time, we can simply equip A with a step-counter and arrange it so that it always stops after the number of steps reaches a particular polynomial value. Similarly, we can guarantee that the length of y does not exceed some polynomial value. But if we are to produce a certificate that condition 3. is always true, we have to give a *proof* of the condition. For some **TPS** problems, the proof may be easy, for some it may be hard. The particular measure of hardness that we will be interested in is the minimal theory in which condition 3. is provable.

This brings us back to the theories that are studied in proof complexity. But before talking about theories in Bounded Arithmetic, I need to recall a general phenomenon concerning incompleteness. In the section about concrete independent combinatorial sentences I mentioned that these sentences are connected with fast growing functions. The sentences were of the form

$$\forall x \exists y \; R(x, y),$$

where $R(x, y)$ was some computable relation; these are the Π_2 sentences. We can also view such sentences as total search problems, but, of course, not polynomial because we do not bound the size of y. The particular sentences that I presented were not provable in Peano Arithmetic because, for increasing x, the least y satisfying the relation R was increasing so fast that Peano Arithmetic was not able to capture such a rate of growth. In general, every theory T (consistent and having a recursive set of axioms) can handle only a limited rate of growth. Thus we can always define a function that grows so fast that it is not possible to prove in T that the function is total. This enables us to use fast growing functions as a scale for measuring the strength of theories.

If measured by provably total functions, most of the theories of Bounded Arithmetic would be at the bottom of the scale. In a Bounded Arithmetic theory one can typically formalize only polynomial growth. More precisely, one can prove that a function is total only when the *length* of the function value grows at most as a polynomial in the length of the argument. But this does not mean that they are extremely weak if we consider sentences of low complexity. In fact, this is an advantage of studying these theories—we are naturally led to problems that concern low complexity classes. Namely, if a Π_2 sentence is provable in Bounded Arithmetic, then the length of y is polynomially bounded by the length of x, so condition 2. is automatically satisfied. It suffices then to assume condition 1. and we get a **TPS** problem. So instead of fast growing functions, we study problems in **TPS** and instead of their growth rate, we study their computational complexity.

Let us see what some particular theories can prove about **TPS** problems. Specifically, we will consider the theories $\Theta_{\mathbf{P}}$ and $\Theta_{\mathbf{NP}}$. In what follows we will always assume that R is a binary relation computable in polynomial time (condition 1). The next theorem, due to S.R. Buss [34], expresses formally the relation between $\Theta_{\mathbf{P}}$ and **P** that I explained in paragraph 2. of the previous subsection.

Theorem 49 *If a sentence $\forall x \exists y\ R(x, y)$, with $R(x, y)$ representing a polynomial time computable relation, is provable in $\Theta_{\mathbf{P}}$, then the associated search problem is solvable in polynomial time, that is, there exists a polynomial time algorithm that for every x, constructs y such that $R(x, y)$.*

Theorems of this type are called *witnessing theorems* because one can efficiently find a witness for the existential quantifier $\exists y$, given x an as input.

The class of search problems solvable in polynomial time is denoted by **FP**. Thus the theorem says that the problems for which $\Theta_{\mathbf{P}}$ can verify condition 3. are from the class **FP**. One can also show the opposite: every problem in **FP** can be formalized in such a way that $\Theta_{\mathbf{P}}$ proves condition 3.

Since **FP** is a sort of a functional version of the class **P**, we get another reason for associating the theory $\Theta_{\mathbf{P}}$ with the complexity class **P**.

In order to state the corresponding result for $\Theta_{\mathbf{NP}}$, we need to define another class of search problems, *Polynomial Local Search*, abbreviated by **PLS**. This class was introduced by D.S. Johnson, C.H. Papadimitriou and M. Yannakakis [144]. A **PLS** problem S is defined as follows. Given an input string a we have

1. a set U_a;
2. a subset $F_a \subseteq U_a$ whose members are called *feasible solutions* (but they are not necessarily solutions of S); we assume that a fixed element, say 0, is always in F_a;
3. a *neighborhood function* h_a that maps U_a into itself;
4. and a *cost function* c that assigns a rational number to every $x \in U_a$ (we may assume that c is the same function for all inputs a).

To motivate the definition of solutions of S, I will first give an interpretation of these notions. We think of feasible solutions as potential solutions and we are interested in their cost. We do not insist on finding a solution with a maximal cost. Instead, we are satisfied with a solution that cannot be improved using the neighborhood function. Thus the neighborhood function should be thought of as a heuristic that helps improve the cost of a feasible solution. We always have some feasible solution, namely 0, and we can try to improve the cost of a feasible solution that we have by applying the function h to it. Since the set of feasible solutions is finite, we cannot go on improving the cost forever; the process has to terminate. There may be two reasons for stopping at some feasible solution x: (1) $h_a(x)$ is not a feasible solution, or (2) the cost does not increase, that is, $c(h_a(x)) \leq c(x)$. Such an x is defined to be a solution of the search problem S.

Formally, an x is a *solution* of S for input value a if

1. *x is a feasible solution, that is, $x \in F_a$, and*
2. *$h_a(x)$ is not a feasible solution, or $c(h_a(x)) \leq c(x)$.*

It should be stressed that we do not require x to be obtained by the process described above. I used the process only to motivate the definition. Also note that a solution x does not have to be a feasible solution with maximal cost; the definition only states that it cannot be improved using the heuristic h—it is *locally maximal.*

As we are defining a class of **TPS** problems, we require that the primitives by which a **PLS** problem is defined are computable in polynomial time. In particular, the set U_a is a set of all strings of length n, where n is polynomial in the length of the input a, one can decide in polynomial time if $x \in F_a$ and one can compute in polynomial time the functions h_a and c.

Example One can view *Linear Programming* as the following problem: for a given convex polyhedron P in n-dimensional Euclidean space and a linear function ℓ, find a vertex v of P on which ℓ attains the maximum value. *Dantzig's simplex algorithm* for solving Linear Programming is very efficient in practice, but on some inputs requires exponential time. The idea of the algorithm is to start with an arbitrary vertex v of P and consider all neighbors of v, which are vertices connected by edges to v. If there is among them u such that $\ell(u) > \ell(v)$, replace v by u and continue. If there is no such vertex, then $\ell(v)$ is the maximum.

If properly formalized, Dantzig's algorithm can be presented as an instance of a **PLS** problem. However, it is not a typical member, because first, the solution is always a global maximum, not a local one, and second, Linear Programming is solvable in polynomial time (by different algorithms), whereas we believe that there are **PLS** problems unsolvable in polynomial time.

Furthermore, we want **PLS** to be a true complexity class. As defined so far, it is an interesting class of search problems, but it is not closed under polynomial reductions. Let us, therefore, stipulate that **PLS** is the class of all total search problems that can be polynomially reduced to a search problem defined above.

The concept of polynomial reducibility of search problems is similar to the polynomial reducibility of sets mentioned in Chap. 5. Suppose we have two polynomial search problems P and Q, defined by relations $r(x, y)$ and $s(x, y)$ respectively. We say that *P is polynomially reducible to Q,* if there exists a polynomial time algorithm that for a given a computes b such that $r(a, b)$ *using queries to the problem Q.* This means that during the computation the algorithm may compute some c and request d such that $s(c, d)$. It may ask several such queries, and if they are answered correctly, it must produce a correct answer: some b such that $r(a, b)$.

The natural algorithm to compute a solution of a **PLS** problem which I used to motivate the definition does not help us to compute solutions efficiently. The size of U_a can be exponentially large and the number of values of the cost function can be exponential. Thus the natural algorithm typically runs in exponential time. It is a generally accepted conjecture that there are **PLS** problems that cannot be solved in polynomial time. On the other hand we also believe that **PLS** does not exhaust all total polynomial search problems, in fact, it seems to be just a small subclass.

After a rather lengthy description, we can return to our theory $\Theta_{\mathbf{NP}}$ and at last state a characterization of the total search problems that can be formalized in it. We will say that a search problem P defined by a relation $r(x, y)$ is *properly formalized* in a theory T if

1. the relation $r(x, y)$ is formalized by a formula from $\Phi_{\mathbf{P}}$ and
2. T proves that the search problem is total (condition 3. of the definition of **TPS**).

The following theorem is due to Buss and Krajíček [35].

Theorem 50 *If a sentence $\forall x \exists y\ R(x, y)$, with $R(x, y) \in \Phi_{\mathbf{P}}$, is provable in $\Theta_{\mathbf{NP}}$, then the associated search problem is in* **PLS**.

Less formally:

Every **TPS** *problem that can be properly formalized in $\Theta_{\mathbf{NP}}$ is in* **PLS**.

Again, one can also prove that every **PLS** problem can be properly formalized in $\Theta_{\mathbf{NP}}$. Thus **PLS** characterizes the search problems properly formalizable in $\Theta_{\mathbf{NP}}$.

The two characterizations give us another relation between the theories and the complexity classes.

Corollary 1 *If $\Theta_{\mathbf{P}} \equiv \Theta_{\mathbf{NP}}$, then* **FP** = **PLS**.

While the assumption is the same, the conclusion of the theorem is different from the conclusion of Theorem 48 stated in the previous subsection. We do not know if one conclusion follows from the other.

Bounded Arithmetic is a natural set up for studying search problems, but we can ask the same question for arbitrary theories, in particular for theories that are not based on postulating induction for a certain class of formulas. We only need the theory to be strong enough to formalize polynomial time computations. We can even take strong theories such as Zermelo-Fraenkel Set Theory. Each such theory defines a subclass of search problems of **TPS** thus it seems that we could measure the strength of theories by these search-problem-complexity classes. But there are two obstacles: first, we are not able to prove that the hierarchy of search problems increases; second, we do not have explicit descriptions of these classes, except for the theories at the very bottom. Still, the mere fact that there is such a possibility seems very interesting. Let me, therefore, state this fundamental fact explicitly.

It is possible that one could measure the strength of theories by complexity classes contained in **TPS**.

I will say more about it in the last section of this chapter.

Notes

1. *Bounded arithmetical formulas.* Bounded quantifiers are of the form $\forall x \leq t$ and $\exists x \leq t$, where x is the quantified variable and t is a term. The variables of t must

be different from x and must either be free or quantified before x in the formula in which the bounded quantifier is used. Bounded quantifiers may be treated as primitive notions or as abbreviations:

$$\forall x \leq t\, \phi \quad \text{as an abbreviation of } \forall x(x \leq t \rightarrow \phi),$$
$$\exists x \leq t\, \phi \quad \text{as an abbreviation of } \exists x(x \leq t \wedge \phi).$$

Formulas that contain no quantifiers or only bounded quantifiers are called *bounded formulas*.

Example The following bounded formula defines that y is the least proper divisor of x:

$$1 < y \wedge \exists z \leq x(y \cdot z = x) \wedge \forall w \leq y\big(\exists z \leq x(w \cdot z = x) \rightarrow (w = 1 \vee w = y)\big).$$

Since the ranges of quantified variables in a bounded formula are finite intervals, satisfiability of bounded formulas is algorithmically decidable.

2. *Definition of $\Theta_{\mathbf{P}}$.* According to the general paradigm we should define a set of formulas $\Phi_{\mathbf{P}}$ that define the sets in **P** and define $\Theta_{\mathbf{P}}$ as some set of basic axioms plus induction axioms for all formulas in $\Phi_{\mathbf{P}}$. We have met such a class in the subsection about the equational theory PV. So one possibility is to define $\Theta_{\mathbf{P}}$ as an extension of PV, or better of $PV1$, to first order logic. A rather unpleasant feature of such a theory is that the set of function symbols is infinite. Because of this, such a theory has an infinite number of axioms on top of the induction axioms.

Apparently, there is no natural set of formulas based on a finite set of predicates and functions that defines **P**. Nevertheless, one can define $\Theta_{\mathbf{P}}$ using a language with only a finite number of primitives. The trick is that one does not have to postulate induction for formulas defining all sets in **P**. It suffices to postulate induction for a set of formulas that define only some sets in **P** and derive induction for the others from these axioms. An elegant theory of this kind, which I am going to describe below, was proposed by E. Jeřábek [140].

The primitives of the language of the theory consist of the usual constant 0, the successor function $S(x)$, the arithmetical operations $+$ and $\cdot$, the binary relation $\leq$ and three more functions $|x|$, $x\#y$ and $\lfloor \frac{x}{2^y} \rfloor$. The intended interpretation of $|x|$ is the length of x in binary representation, which is in mathematical symbols $\lceil \log(x+1) \rceil$. The intended interpretation of $x\#y$ is $2^{|x|\cdot|y|}$. The intended interpretation of the third function is clear from the symbols representing it—it is x with y least significant bits truncated; it can also be defined as the y-times iterated truncation function $tr(x)$. (Recall that $\lceil \cdots \rceil$ and $\lfloor \cdots \rfloor$ denote rounding up to an integer and respectively rounding down to an integer.)

The binary operation # plays a special role. Let us observe that for a natural number n, the length of $n\#n$ is approximately quadratic in the length of n. Similarly, the length of $n\#(n\#n)$ is approximately cubic in the length of n, etc. Viewing numbers as binary strings, the presence of this operation (with the axioms that determine its properties) ensures the possibility of constructing strings of polynomial length for every polynomial. This is a necessary condition if we want to be able to formalize polynomial time computations, but it is also sufficient if it is part of a suitable axiomatization. One can also show that the terms of this language do not extend the length more than polynomially. Due to this fact and the special form of the axioms, one can show that the theory can prove that the length of a string can be extended polynomially, but not more than that.

Here are the basic axioms that determine the meaning of the primitive symbols.

$$x+0=x, \qquad x+S(y)=S(x+y),$$

$$x\cdot 0=0, \qquad x\cdot S(y)=x\cdot y+x,$$

$$\lfloor \tfrac{x}{2^0}\rfloor = x, \qquad \lfloor \tfrac{x}{2^y}\rfloor = 2\cdot\lfloor \tfrac{x}{2^{S(y)}}\rfloor \vee \lfloor \tfrac{x}{2^y}\rfloor = S(2\cdot\lfloor \tfrac{x}{2^{S(y)}}\rfloor),$$

$$|0|=0, \qquad x\neq 0\rightarrow |x| = S(|\lfloor \tfrac{x}{2^1}\rfloor|),$$

$$0\#1=1, \qquad x\neq 0\rightarrow x\#1 = 2\cdot(\lfloor \tfrac{x}{2^1}\rfloor\#1),$$

$$x\neq 0\rightarrow (y\#x) = (y\#1)\cdot(y\#\lfloor \tfrac{x}{2^1}\rfloor).$$

As usual, 1 and 2 are abbreviations for $S(0)$ and $SS(0)$.

A *sharply bounded quantifier* is a bounded quantifier in which the bounding term t has the form $|s|$, i.e., the outermost function is the length function. A *sharply bounded formula* is a formula with all quantifiers (if there are any) sharply bounded. The class of sharply bounded formulas is denoted by Σ^b_0. One can easily see that the range of the quantified variables in a sharply bounded formulas is bounded by a polynomial in the lengths of the free variables. Therefore, satisfiability of sharply bounded formulas can be computed in polynomial time. To obtain $\Theta_{\mathbf{P}}$, we now add an infinite set of axioms, the instances of *induction*

$$\big(\phi(0)\wedge\forall x\big(\phi(x)\rightarrow\phi\big(S(x)\big)\big)\big)\rightarrow\forall x\,\phi(x)$$

for sharply bounded formulas ϕ.

The class of sets definable by sharply bounded formulas is most likely smaller than **P**, yet this schema of induction suffices to prove induction for all sets in **P**. In $\Theta_{\mathbf{P}}$ we define the sets of **P** using bounded formulas that are not sharply bounded. The set of these formulas can be defined syntactically, hence it would be possible to have the induction schema stated for all such formulas. It is only for reasons of simplicity and elegance that induction is postulated only for sharply bounded formulas.

3. *Definition of* $\Theta_{\mathbf{NP}}$. To define $\Theta_{\mathbf{NP}}$ is a simpler task because there is a natural class of formulas that define all sets in **NP**. Also the theory is stronger, hence formalizing the necessary concepts in it is easier.

   The class of formulas that we need is denoted by Σ^b_1. It consists of formulas of the following form: a prefix of bounded existential quantifiers followed by a sharply bounded formula.[13] The prefix can be empty, so $\Sigma_0^b \subseteq \Sigma_1^b$. We define $\Theta_{\mathbf{NP}}$ as an extension of $\Theta_{\mathbf{P}}$ in which the induction schema is postulated for all Σ_1^b-formulas.

4. *Alternative definitions of theories associated with* **P**. Some authors associate different theories with **P**. In this book I associate a theory T with a complexity class $\mathcal{C}$ if T has induction axioms for formulas defining the sets of the complexity class $\mathcal{C}$. In other words, my criterion is based on induction axioms. Others consider witnessing theorems (such as Theorem 49) to be the criterion. This means that they associate with a theory the smallest complexity class of functions for which the witnessing theorem holds, and then they take the corresponding complexity class of sets. If we use witnessing functions to determine theories corresponding to complexity classes, we may postulate a somewhat stronger form of induction. In particular, in the case of the class **P**, we may define the theory using Polynomial Induction
$$\phi(0, y) \wedge \forall x\big(\phi\big(tr(x), y\big) \rightarrow \phi(x, y)\big) \rightarrow \forall x\, \phi(x, y)$$
for the set of Σ_1^b-formulas. This theory, denoted by S_2^1, is stronger than $\Theta_{\mathbf{P}}$, assuming a plausible conjecture about complexity classes. Yet the witnessing functions for S_2^1 are still only **FP**.

   Nevertheless, this discrepancy is not significant. The most important theorems are those that are $\Pi_{\mathbf{P}}$ sentences and these theorems are the same in all proposed theories for **P**.
5. *On the proof of Theorem* 48. The idea of the proof is first to derive nontrivial information about computations from the provability of Σ_1^b-induction from $\Theta_{\mathbf{P}}$ and then reduce this information to a relation between two complexity classes.

   Let $\gamma(x, y)$ be a formula defining a relation computable in polynomial time; think of the relation as a kind of search problem, in which x is a parameter and y is a solution to x. Suppose that 0 is always a solution and no solution is greater than x, i.e.,
$$\forall x\big(\gamma(x, 0) \wedge \forall y\big(\gamma(x, y) \rightarrow y \leq x\big)\big)$$
is true. We will, moreover, assume that this is provable in $\Theta_{\mathbf{P}}$. Then it is provable in $\Theta_{\mathbf{NP}}$ that for every x, there exists a largest solution. Expressed formally, the following sentence is provable in $\Theta_{\mathbf{NP}}$:
$$\forall x \exists y \forall z\big(\gamma(x, y) \wedge \big(\gamma(x, z) \rightarrow z \leq y\big)\big). \tag{6.4}$$
Hence, if $\Theta_{\mathbf{P}} \equiv \Theta_{\mathbf{NP}}$ were true, then this sentence would also be provable in $\Theta_{\mathbf{P}}$. To derive a computational consequence of such an assumption, we will use the following general result.

Lemma 15 *Let $\phi(x, y, z)$ be a formula defining a polynomial time relation. Suppose that*
$$\forall x \exists y \forall z\, \phi(x, y, z)$$

[13] Σ_1^b is often defined as a more general class where in the prefix sharply bounded universal quantifiers are also allowed.

is provable in $\Theta_{\mathbf{P}}$. Then there exists a number k and polynomial time computable functions $f_0, \ldots, f_{k-1}$ such that

$$\phi\big(n, f_0(n), m_1\big) \vee \phi\big(n, f_1(n, m_1), m_2\big) \vee \phi\big(n, f_2(n, m_1, m_2), m_3\big) \vee \cdots \\ \vee \phi\big(n, f_{k-1}(n, m_1, \ldots, m_{k-1}), m_k\big)$$

is true for all numbers $n, m_1, m_2, \ldots, m_k$.

This lemma was proved using Herbrand's theorem (see the version without additional function symbols on page 519). The best way to understand the complicated formula is to interpret it as interactive computation with two players. The two players are called Teacher and Student. Student is limited to polynomial time computations and his goal is, for a given n, to compute a solution s, which is a number satisfying $\forall z\, \phi(n, s, z)$. Student may make several attempts and each time Teacher helps him by providing a counterexample to his s. This means that Teacher gives Student some t such that $\phi(n, s, t)$ is false, unless Student has already found a solution.

The meaning of the formula is that Student, represented by the functions $f_0, \ldots, f_{k-1}$, succeeds in at most k steps. Given n, Student first computes $f_0(n)$. If it is not a solution, he gets m_1 from Teacher and computes $f_2(n, m_1)$, and so on. Teacher cannot give counterexamples in all k rounds because that would make the formula false. Hence Student must succeed.

Now, let us look at what would happen if sentence (6.4) were provable in $\Theta_{\mathbf{P}}$. Under this assumption we would be able to compute greatest solutions for every search problem defined by a polynomial time relation, using only polynomial time and a constant number of counterexamples. Without counterexamples, this would imply $\mathbf{P} = \mathbf{NP}$. It seems likely that a constant number of counterexamples cannot be of much help because the number of possible solutions can be exponential. This intuition is confirmed by proving that in such a case we would have $\mathbf{NP} \subseteq$ **nonuniform-P**. Roughly speaking, Teacher's advice can be simulated by nonuniformity.

6. $\mathbf{P} = \mathbf{NP}$ *and collapsing theories.* We are not able to use the assumption $\mathbf{P} = \mathbf{NP}$ to prove that two theories are equal. But if we moreover know that $\mathbf{P} = \mathbf{NP}$ is provable in a specific theory, then we can prove such a consequence.

   Let us consider an example. We know that if $\mathbf{P} = \mathbf{NP}$, then $\mathbf{EXP} = \mathbf{NEXP}$. The proof of this implication can be formalized in $\Theta_{\mathbf{EXP}}$, hence if $\Theta_{\mathbf{EXP}}$ proves $\mathbf{P} = \mathbf{NP}$, then it also proves $\mathbf{EXP} = \mathbf{NEXP}$. But if it proves that $\mathbf{EXP} = \mathbf{NEXP}$, then we can derive induction for **NEXP** formulas from induction for **EXP** formulas. Putting things together we get:

   If $\Theta_{\mathbf{EXP}}$ proves $\mathbf{P} = \mathbf{NP}$, then $\Theta_{\mathbf{EXP}} = \Theta_{\mathbf{NEXP}}$.

   Thus the provability of $\mathbf{P} = \mathbf{NP}$ in some theory implies a collapse of two theories.

   But, if $\mathbf{P} = \mathbf{NP}$ is true, how would we find it out? Only by proving it in some theory. The theory could be so strong that it does not correspond to any natural complexity class; however, it is more likely that such a theory would not be too strong (in particular, it is conceivable that it would be $\Theta_{\mathbf{EXP}}$). Hence it is likely that we would get a collapse of theories.

7. *More on search problems.* In the standard usage **FP** stands for the class of *functions* computable in polynomial time. I am abusing the notation by using it for a class of search problems.

   If we have the correspondence **P** $\leftrightarrow$ $\Theta_{\mathbf{P}}$ $\leftrightarrow$ **FP**, and **NP** $\leftrightarrow$ $\Theta_{\mathbf{NP}}$ $\leftrightarrow$ **PLS**, the curious reader may wonder what we can get from it about the relations between the complexity classes. Namely, we have the following correspondences **P** $\leftrightarrow$ **FP**, and **NP** $\leftrightarrow$ **PLS**. The first one is easily explicable: one is a class of decision problem, the other is a class of search problems with the same bounds on the computational resources. In the case of **NP** $\leftrightarrow$ **PLS**, however, I do not have any explanation that does not mention $\Theta_{\mathbf{NP}}$.

   Corollary 1 gives us some provable connection between the theory $\Theta_{\mathbf{NP}}$ and a complexity class of search problems, but does not relate the theory $\Theta_{\mathbf{NP}}$ to the complexity class **NP**. In order to obtain such a connection, we have to generalize search problems. We will consider computations that run in polynomial time and have free access to an **NP** set (use **NP** sets as oracles). Using such computations we define the class $\mathbf{P}^{\mathbf{NP}}$ as the class of decision problems computable in polynomial time with free access to **NP**. The following is a version of Theorem 49 for $\Theta_{\mathbf{NP}}$, due to Buss [34].

   Theorem 51 *If a sentence* $\forall x \exists y\, R(x, y)$, *with* $R(x, y)$ *representing a* $\mathbf{P}^{\mathbf{NP}}$ *relation, is provable in* $\Theta_{\mathbf{NP}}$, *then there exists a polynomial time algorithm that uses a set in* **NP** *as an oracle such that, for every* x, *it constructs* y *such that* $R(x, y)$.

   In this way we get an almost direct connection between the theory $\Theta_{\mathbf{NP}}$ and the class **NP**. This relation does not determine $\Theta_{\mathbf{NP}}$ uniquely from **NP**, but it is a good indication that our choice of theory is good.

   Theorems 49 and 51 suggest a paradigm that one can use for finding more pairs of complexity classes and associated theories. Indeed, associated theories have been constructed for many important complexity classes.

8. *The Bounded Arithmetic Hierarchy.* One can define a hierarchy of theories that corresponds to the Polynomial Hierarchy. This was done by Buss in [34]. For each class of sets Σ_n^p, he defined the class of bounded formulas Σ_n^b that define the sets in Σ_n^p. Then he defined theories T_2^n using a finite set of basic axioms plus induction for Σ_n^b formulas. In my notation these theories are representatives of $\Theta_{\Sigma_n^p}$. We call the series of theories *the Bounded Arithmetic Hierarchy*. The union of all theories T_2^n is denoted T_2.

   On page 402 we noted that if $\mathbf{NP} \subseteq$ **nonuniform-P**, then $\Sigma_2^p = \Pi_2^p$. Hence, by Theorem 48, $\Theta_{\mathbf{NP}}$ is stronger than $\Theta_{\mathbf{P}}$ if $\Sigma_2^p \neq \Pi_2^p$. One can extend this result to higher levels of the Bounded Arithmetic Hierarchy and show that it does not collapse if the Polynomial Hierarchy does not collapse.

9. *Theories with oracles.* A natural way to extend theories in Bounded Arithmetic is to enrich the language so that one can talk about sets. The simplest such extensions are obtained by adding a new predicate R to the language and extending the induction axioms to the corresponding class of formulas in the extended language. Given a theory T, we denote such an extension by $T[R]$.

For example, to obtain $\Theta_{\mathbf{P}}[R]$, consider the formalization of $\Theta_{\mathbf{P}}$ described above. First we define sharply bounded formulas in the extended language in the same way as in the original one, the only difference being that we can now also use R. Then we extend the induction scheme to these formulas.

The interpretation of a sentence involving R is that the sentence is true *for all* subsets R. The computational interpretation of R is as an "oracle", a means to obtain additional information for free. Indeed, theories augmented with R behave very much like complexity classes with oracles. The difference is that whereas in complexity theory an oracle is an explicit set, in logic it is just a variable. We follow the complexity-theoretic terminology by calling the extended theories *relativized.*

As in complexity theory, we can prove relativized separations. In particular, we can prove *without any unproved assumptions* that

$$\Theta_{\mathbf{P}}[R] \neq \Theta_{\mathbf{NP}}[R] \neq \Theta_{\Sigma_2^p}[R] \neq \Theta_{\Sigma_3^p}[R] \ldots,$$

see [167]. This is proved by reducing the collapse of the relativized Bounded Arithmetic Hierarchy to the collapse of the Polynomial Hierarchy relativized to an oracle. J. Håstad proved that there exists an oracle with respect to which the Polynomial Hierarchy does not collapse [115].

6.3 Propositional Proofs

Let us now return to the beginnings of Proof Complexity in the 1970's. After defining the classes **P** and **NP**, Cook looked for a concept in logic that would correspond to the class **P**. Once we accept polynomial time computations as a natural formalization of feasible computations, it seems that there should be some concept representing feasible proofs in a similar way. Polynomial time computations are computations that perform a polynomial number of elementary steps in order to produce the result. The corresponding concept in logic should be proofs consisting of a polynomial number of elementary steps, so we need only to define elementary proof steps. The question, however, is not as easy as it looks at first glance.

Feasibly Constructive Proofs

In 1975 Cook introduced a system intended to formalize feasible proofs. It was in the seminal paper *Feasibly constructive proofs and the propositional calculus* [49], which I consider to be the founding paper of the field of proof complexity, for all the fundamental ideas of this field are more or less explicitly present in it. Cook called his formal system *PV*, for *Polynomial Verifiability*. I will not describe all details of the system *PV*; I will rather focus on explaining the motivation behind some particular concepts used in it.

Let us put ourselves in Cook's position back in the 1970s and try to devise such a formal system. The first thing we have to do is decide what formulas such a system should use. It is natural to require the following properties from the system:

1. *the sentences should have low logical complexity;*
2. *the system should be able to formalize basic statements about feasible computations.*

With hindsight, we guess that these sentences should be universal-**P** sentences, but we have to be more specific about how we represent polynomial time computable sets. Since we want to formalize polynomial time algorithms, a natural thing is to have a name for every such algorithm. Therefore, in *PV* there is a term $t(x_1, \dots, x_k)$ for every polynomial time algorithm that has k numbers as inputs. The converse is also true: every term defines a function computable in polynomial time. Although *PV* formally talks about natural numbers, it treats them also as binary strings. In particular, there is a term for every polynomial time algorithm whose input and output values are binary strings.

In *PV* all formulas are equations of the form

$$s(x_1, \dots, x_k) = t(x_1, \dots, x_k)$$

with s and t terms. The meaning is that the equation holds for all natural numbers $x_1, \dots, x_k$, which is written in first order logic as $\forall x_1 \dots \forall x_k\ s(x_1, \dots, x_k) = t(x_1, \dots, x_k)$, but as usual in equational theories, we omit the universal quantifiers in *PV*. Using such formulas, we are able to express that two algorithms define the same function. This gives us a particular formalization of $\Pi_{\mathbf{P}}$ sentences.

Proceeding further to the axioms and rules of the system, our next goal is to satisfy the following requirement:

3. *the nature of proofs should also be feasible.*

It is much harder to give a more precise meaning to this vague statement, although intuitively it seems reasonable. Here comes another useful idea—*polynomial verifiability.* The idea behind this concept is:

> *Given an equality with a feasible proof, one should be able to verify the equality in a polynomial number of steps.*

Using this concept we can specify the system that we are trying to design as the strongest system in which all provable equalities are polynomially verifiable.

So what is needed now is a definition of the polynomial verifiability of *PV* equations. A naive definition of the polynomial verifiability of $s(x) = t(x)$ would be to say that we can verify in polynomial time that $s(N)$ represents the same number as $t(N)$ for every given number N. But this does not say anything about provability, since every true equality of this form is verifiable in this sense (we can simply run the polynomial time algorithms corresponding to s and t on the input N and test if they produce the same number). In order to obtain a nontrivial concept, one has to add a condition about the uniformity of such verifications. Specifically, for a given input length n the verification of $s(N) = t(N)$ for *all* numbers N of length n should have the same form. When formalized, it means that for every input length n, there exists *one* verification that works for all N of length n. In such a verification the number is represented by a *variable* x, so technically it is a proof of a general statement, except that the domain of the variable is finite (the set of numbers of length n). What we gain by restricting x to numbers of length n is that we do not need essentially any general axioms in such proofs.

These ideas cannot be fully understood without seeing at least some details of the formalism. So let us start with the language of *PV*. It contains the basic arithmetical operations $+$ and $\times$, and the constant 0, but on top of these also several less common primitives. What will only be important for us are the two successor functions $s_0(x)$ and $s_1(x)$ and the trimming function $tr(x)$. The intended interpretations of them are:

$$s_0(x) = 2x, \qquad s_1(x) = 2x + 1, \qquad tr(x) = \lfloor x/2 \rfloor,$$

where $\lfloor y \rfloor$ is the largest integer less than or equal to y. Notice that these are bit operations on the binary representations of the number x: s_0 adds 0 to the binary representation of x, s_0 adds 1 to the binary representation of x, and $tr(x)$ trims off the last bit (except for 0, where $tr(0) = 0$). As usual, one can form compound terms from the primitive function symbols and constants.

Furthermore, one can introduce new function symbols that represent functions defined by *recursion on notation*. Given functions $f(y)$, $g_0(x, y)$ and $g_1(x, y)$, we say that $h(x, y)$ is defined by recursion on notation, if it satisfies:

$$\begin{aligned} h(0, y) &= f(y) \\ h\big(s_0(x), y\big) &= g_0\big(h(x, y), y\big) \\ h\big(s_1(x), y\big) &= g_1\big(h(x, y), y\big). \end{aligned} \tag{6.5}$$

(The variable y plays the role of a parameter; for the sake of simplicity I am using only one parameter.) The advantage of this kind of recursion is that for a number N of length n and any number M, we need only $n+1$ applications of the functions f, g_0, g_1 in order to compute the value $h(N, M)$. Indeed, if $a_n a_{n-1} \dots a_1$ is the binary representation of N, then

$$N = s_{a_n}\big(s_{a_{n-1}}\big(\dots s_{a_1}(0)\dots\big)\big),$$

hence

$$h(N, M) = g_{a_n}\big(g_{a_{n-1}}\big(\dots g_{a_1}\big(f(M), M\big)\dots, M\big), M\big). \tag{6.6}$$

This suggests that if the initial functions f, g_0, g_1 are computable in polynomial time, then so should h be. In general this is not true, because it can happen that the numbers grow exponentially with the length of the first parameter. Therefore it is necessary to add a bound on the size of the values of h. This is just an inessential complication and I skip the technical details of how the growth of h is controlled in *PV*.

Let us now proceed to the axioms and rules. *PV* contains basic axioms about the primitive constants and functions. For example, we have the following axioms connecting the two successors with the trimming function:

$$tr\big(s_0(x)\big) = x, \qquad tr\big(s_1(x)\big) = x.$$

When we introduce a new symbol h for a function defined by recursion on notation, we also add the corresponding equalities (6.5). The key axioms are the axioms of induction. In PV we are only allowed to use equalities, thus one cannot state induction in the usual form. There is a way to simulate induction using only equations of PV terms, but since I want to avoid technicalities, I will switch to an extension $PV1$ of PV in which one can use propositions made of equations.

The schema of induction in $PV1$ is motivated by recursion on notation, therefore it is called *Induction on Notation* (sometimes '*Polynomial Induction*' is also used). Translating recursion on notation literally, we obtain the following form of induction on notation:[14]

$$\phi(0, y) \wedge \forall x \big(\phi(x, y) \to \phi\big(s_0(x), y\big) \wedge \phi\big(s_1(x), y\big)\big) \to \forall x\, \phi(x, y),$$

where the formula $\phi(x, y)$ is an equality $s(x, y) = t(x, y)$ for some PV terms s and t. (Again I am simplifying it by only using one parameter y.) Instead of this, people prefer the more concise equivalent version:

$$\phi(0, y) \wedge \forall x \big(\phi\big(tr(x), y\big) \to \phi(x, y)\big) \to \forall x\, \phi(x, y).$$

If we state induction as an axiom (in either way) we cannot avoid the use of the universal quantifier in the antecedent. Therefore, in $PV1$ we represent the induction principle by a deduction rule:

from $\phi(0, y)$ and $\phi(tr(z), y) \to \phi(z, y)$, it is possible to derive $\phi(x, y)$.

Notice that the induction rule is the only general principle used in $PV1$; otherwise we only have axioms of equality, a finite set of axioms about the primitive concepts and the equations that determine the functions introduced by recursion on notation.

Now we can return to the concept of polynomial verifiability and explain it in more detail. Suppose we prove an equality $s(x) = t(x)$ and we want to verify it for numbers of length at most n. To this end we have to formally express that x has length at most n. This can be done by the following equation:

$$tr\big(tr\big(\dots tr(x)\dots\big)\big) = 0, \tag{6.7}$$

where the function symbol tr is applied n-times. I will abbreviate the long term at the left hand side by $tr^n(x)$. As I said, verifying the equality $s(x) = t(x)$ for numbers of length at most n is nothing else but proving it only using elementary means. This means that we should use an assumption saying that the length of x is at most n and then only elementary rules. The assumption about the length will be represented by equation (6.7) above. Since $PV1$ is based on elementary axioms and rules and one general principle, we can define:

Polynomial verification is a proof that uses the axioms of $PV1$ *without the induction rule.*

[14]It turns out that one can also use the standard form of induction without increasing the power of PV. Nevertheless, for showing the connection with polynomial verifiability induction on notation is more convenient.

Such a definition may not seem quite satisfactory. Why do we say that some axioms are elementary and others not? A more detailed analysis of the axioms is clearly needed. But I am afraid that this would not help very much if one just wants to understand the essence of the problem. It is more instructive to watch how the induction rule can be eliminated. If we accept the definition of the polynomial verifiability as proposed above, elimination of the induction rule is all we need in order to prove that the equalities provable in $PV1$ are polynomially verifiable.

So let us look at how induction can be eliminated. Let us consider an equality $s(x) = t(x)$ provable in $PV1$; I will abbreviate it by $\phi(x)$. Let a proof P of the equality $\phi(x)$ be given and assume we need a polynomial verification for numbers of length n. The idea is to use our assumption that x has length at most n expressed by the equation (6.7) to transform P into a proof that does not use the rule of Induction on Notation. Suppose P ends with an application of the induction rule, namely, that $\phi(x)$ was obtained from assumptions $\phi(0)$ and $\phi(tr(z)) \rightarrow \phi(z)$. Since we have $tr^n(x) = 0$, we get

$$\phi\big(tr^n(x)\big)$$

by substitution from $\phi(0)$. Now we will use the second assumption of the induction rule. We will successively substitute terms $tr^{n-1}(x), tr^{n-2}(x), \dots, tr(x)$ and x for the variable z.[15] Thus we obtain

$$\begin{gathered}\phi\big(tr^n(x)\big) \rightarrow \phi\big(tr^{n-1}(x)\big),\\ \phi\big(tr^{n-1}(x)\big) \rightarrow \phi\big(tr^{n-2}(x)\big),\\ \dots\\ \phi\big(tr(x)\big) \rightarrow \phi(x).\end{gathered}$$

Now we apply *modus ponens* n times and obtain $\phi(x)$. Thus we have eliminated one application of the induction rule and the length of the resulting proof is bounded by a polynomial in n. Repeating essentially the same procedure for all occurrences of the induction rule we get a proof that does not use this rule and has polynomial size. This proof is a polynomial verification of the equation $s(x) = t(x)$.

The fact that universal sentences provable in PV (or $PV1$) can be polynomially verified is one good reason for associating PV with the complexity class **P**. But, of course, any theory that is weaker than PV also has this property. Therefore we need another property to impose a lower bound on the strength a theory that we want to associate with **P**. The ability of PV to prove that polynomial verifiability is sound is such a property. More precisely, PV proves the formalization of the following sentence:

If x has length n and y (a code of a proof) is a polynomial verification of $\phi(z)$ for numbers of length n, then $\phi(x)$ is true.[16]

---

[15]Strictly speaking, we cannot use substitution to obtain the following formulas. Instead we have to reproduce the proof of $\phi(tr(z)) \rightarrow \phi(z)$ for z replaced by each of the terms $tr^{n-1}(x), tr^{n-2}(x), \dots, tr(x)$.

[16]The general reflection principle is stronger than consistency, hence not provable in the corresponding theory, but in this case the class of sentences is very restricted, so the sentence is provable PV.

This property pretty much determines PV: essentially, PV is the weakest system that proves this reflection principle.

The concept of polynomial verifiability is important and deserves more detailed study. There is, however, an alternative approach which is, in some sense, much cleaner and therefore is the choice that is usually preferred over polynomial verifiability. It is based on showing a similar relation of proofs in first order logic to proofs in the propositional calculus. This means that we replace equational proofs of equalities $s(N) = t(N)$ by propositional proofs of tautologies that express the same fact as $s(N) = t(N)$.

Down to Propositional Logic

When I was a student I wondered why some authors of logic books pay so much attention to the propositional calculus. Back then I thought that there was nothing interesting about the propositional calculus. The two main reasons for me were: first, I thought one could express only elementary facts using propositional logic; second, the calculus is decidable. Both are true, but on a closer look, one can still find very interesting problems. As a matter of fact, my current perception of the propositional calculus is quite the opposite—I believe that some of the most fundamental problems are in this area.

I will start with some examples of what can be formalized in propositional logic. The prevailing wrong impression that only trivialities can be expressed in propositional logic is caused by the fact that introductory logic texts mention only very simple formulas. Once we are willing to work with long formulas, we can express fairly complex statements.

The first set of examples are arithmetical statements. With n propositional variables we can encode n-bit numbers. Let number x be represented by propositional variables $p_1, p_2, \ldots, p_n$ and number y by $q_1, q_2, \ldots, q_n$. Then we can express the equality $x = y$ by

$$(p_1 \equiv q_1) \wedge (p_2 \equiv q_2) \wedge \cdots \wedge (p_n \equiv q_n).$$

This is not a tautology, it is a formula talking about two unknown n-bit numbers, which is true for some assignments and false for others.

A more interesting example is a formula that expresses $x \le y$ for two n-bit numbers. Let the two numbers be represented as above with p_1 and q_1 being the most significant bits, p_2 and q_2 the second most significant etc. Then the formula reads:

$$\begin{aligned}
&\big((p_1 \equiv q_1) \wedge \cdots \wedge (p_{n-2} \equiv q_{n-2}) \wedge (p_{n-1} \equiv q_{n-1}) \wedge (p_n \equiv q_n)\big)\vee \\
&\big((p_1 \equiv q_1) \wedge \cdots \wedge (p_{n-2} \equiv q_{n-2}) \wedge (p_{n-1} \equiv q_{n-1}) \wedge \neg p_n \wedge q_n\big)\vee \\
&\big((p_1 \equiv q_1) \wedge \cdots \wedge (p_{n-2} \equiv q_{n-2}) \wedge \neg p_{n-1} \wedge q_{n-1}\big)\vee \\
&\ldots \\
&(\neg p_1 \wedge q_1).
\end{aligned}$$

Using this formula (and the one obtained by interchanging p_i's with q_i's), we can express the true arithmetical fact

$$(x \leq y) \vee (y \leq x),$$

for n-bit numbers, by a propositional *tautology*. We cannot express the above dichotomy by a single tautology for all numbers. If we want to talk about all numbers, we have to use a sequence of tautologies parameterized by the lengths of the numbers.

Let us now consider the formula $x + y = z$ where x and y are (at most) n-bit numbers and z is an (at most) $n+1$-bit number. Similarly as we did for $x = y$, we want to say that each bit of $x + y$ is equal to the corresponding bit of z. Therefore, it suffices, for every i, to construct a formula which is true if the ith bit of $x + y$ is 1 and false if it is 0. To construct such a formula, the standard algorithm based on computing the carry bits is used. Thus one obtains a formula whose size is polynomial in n.

For $x \cdot y = z$, we can also construct a polynomial size formula, although a it is little more complicated. Then we can combine these formulas to construct tautologies that express various true arithmetical sentences. For example, we can construct a tautology that expresses the distributivity law for n-bit numbers. All these tautologies have polynomial sizes, when measured by the length of the numbers about which they talk.

A more interesting example of an arithmetical tautology is the following. Let P be a fixed prime number. The fact that P is prime can be expressed in first order logic by

$$x \cdot y = P \rightarrow (x = 1 \vee y = 1).$$

As we have formulas for equality and multiplication, we can write a formula, in fact a tautology, expressing the primality of P. In this tautology the bits of x and y will be represented by propositional variables, whereas the bits of P will be the corresponding truth constants. If P has n bits, we need only to consider numbers x and y with at most n bits, so we can write it as a single tautology.

Arithmetical tautologies are an ample source of interesting tautologies, but there are others. Perhaps the most heavily researched set of tautologies are the tautologies expressing the *Pigeonhole Principle*. The principle says that if there are $n + 1$ pigeons in n holes, then there is at least one hole in which there are at least two pigeons. Let us consider propositional variables p_{ij} indexed by two indices $i = 1, 2, \ldots, n+1$ and $j = 1, 2, \ldots, n$. One should think of the variable p_{ij} as being true if and only if pigeon i sits in hole j. I will use $\bigwedge$ and $\bigvee$ to denote conjunctions and disjunctions of several terms. The pigeonhole tautology PHP_n then reads:

$$\bigwedge_{i=1}^{n+1} \bigvee_{j=1}^{n} p_{ij} \rightarrow \bigvee_{j=1}^{n} \bigvee_{i=1}^{n+1} \bigvee_{1 \leq i' \leq n+1;\ i' \neq i} (p_{ij} \wedge p_{i'j}).$$

Notice that this formula closely follows the English description of the principle, except for the last part, where instead of saying that *there are two* pigeons, we say that *there exist a pigeon i and there exists a pigeon i′ different from i*.

Let us now express the Pigeonhole Principle in first order logic. Let $P(x, y)$ be a symbol for a binary relation, where x is a number from 1 to $n+1$ and y is a number from 1 to n. The first order sentence reads:

$$\forall x \exists y P(x, y) \rightarrow \exists y \exists x \exists x' \big(x' \neq x \wedge P(x, y) \wedge P\big(x', y\big)\big).$$

The similarity of the two expressions above is quite conspicuous. The quantifiers correspond to the big conjunctions and disjunctions and the binary relation is represented by propositional variables indexed by pairs. This suggests that we can translate any first order sentence true in some finite relational structure into a propositional tautology; this is indeed possible.

So what can actually be expressed by propositional formulas? First recall that *every* Boolean function of n variables can be defined using a propositional formula (we could also say *Boolean formula,* meaning the same thing) with n variables. Such representations are, for example, the well-known disjunctive and conjunctive normal forms. These forms have exponential size, except for some very special Boolean functions. We cannot use them when we want to express, say, that a medium size number P is a prime because the formula would be too large.

Thus the right question is: which functions can be defined by medium size Boolean formulas? In precise mathematical terms the question is: *which sets of 0–1 strings can be defined by sequences of formulas of polynomial size?* We know that if we use Boolean circuits, instead of formulas, we can define every set in **P** by a sequence of polynomial size circuits. Formulas can be viewed as a special kind of circuits, circuits whose underlying graph is a tree. It seems that this is a weaker computational model. Using sequences of polynomial size Boolean formulas, or equivalently using sequences of polynomial size tree-like Boolean circuits, one can define the sets in a class called $\mathbf{NC}^1$, a subclass of **P**. Whether the two classes are equal or not, is one of the wide-open problems. The commonly accepted conjecture is that $\mathbf{NC}^1$ is a proper class of **P**.

In this way the propositional calculus is closely connected with computational complexity, but it is important to realize that there are also fundamental differences between the two fields. In complexity theory we study the circuit complexity of Boolean functions and we also study the formula complexity of Boolean functions. In the propositional calculus we are primarily interested in tautologies. A tautology is also represented by a Boolean formula, but the formula is true for every assignment of the propositional variables. Hence the formula defines a very simple Boolean function—the function identically equal to 1 (where 1 represents the truth). The relation to the complexity of Boolean functions manifests itself only when we look at the components from which tautologies are constructed.

Example Let $\phi(p_1, \ldots, p_n)$ and $\psi(p_1, \ldots, p_n)$ be two propositional formulas. Suppose that

$$\phi(p_1, \ldots, p_n) \equiv \psi(p_1, \ldots, p_n)$$

is a tautology. Since it is a tautology, it defines the Boolean function constantly equal to 1. But the function defined by ϕ, which is the same as the function defined by ψ, may be an arbitrary one.

In particular, for proof complexity it is more important to know which *sentences* can be represented by sequences of Boolean formulas. Since we can define $\mathbf{NC}^1$ sets, we can also represent *universal-NC^1 sentences*. The reason is that in logic we are interested in tautologies. A tautology, although represented by Boolean formula, is interpreted as if there were universal quantifiers in front of it. Saying that the formula is satisfied by all truth assignments is the same as if we had these universal quantifiers and said that the quantified propositional formula is true. Since these implicit quantifiers range only over finite domains, we need infinite sequences of Boolean formulas to represent a sentence in first order logic. The relation between the sentence and the sequence of propositional formulas is that the sentence is true if and only if all propositional formulas are tautologies.

Example Above I have sketched how to represent the true arithmetical sentence $\forall x \forall y (x \leq y \vee y \leq x)$.

Given a sentence ϕ of first order logic, we call the sequence of Boolean formulas representing ϕ a *translation of ϕ into propositional logic*. The word 'translation' is justified by the fact that we often want to preserve at least part of the syntactical structure of ϕ in the translation, which enables us to prove interesting relations between proofs in predicate logic and proofs in the propositional calculus.

It is not difficult to show that in fact one can translate universal-**P** sentences into sequences of propositional formulas. The universal quantifier enables us to bootstrap the $\mathbf{NC}^1$ predicate into a **P** predicate. Thus we can represent our fundamental type of sentences in propositional logic.

Propositional Proof Systems

Let us turn to the most important part of the propositional calculus, the proofs. We have a variety of interesting tautologies and we wonder how difficult it is to prove them. There are various axiomatizations of propositional logic. Essentially in all these axiomatizations one can prove every tautology of size n by a proof of at most exponential size in n, but exponentially large proofs are not very useful. The crucial question is which tautologies have polynomial size proofs. Of course, this makes sense only for infinite sequences of tautologies.

Example We may ask if the pigeonhole tautologies PHP_n have polynomial size proofs. The bound should be polynomial in the lengths of the tautologies, but since PHP_n has polynomial size in n, this is the same as being polynomial in n. Although these are fairly easy tautologies, we know that in some weak proof systems they have only exponential proofs.

In order to be able to answer such questions we must first specify what a propositional proof is. The presence of different axiomatizations suggests that the answer may depend on how we define proofs. Therefore it is useful to introduce the concept of a *propositional proof system*. A typical proof system is based on axiom schemas and derivation rules. A proof in such a system is a sequence of formulas that follow either from the schemas or from the previous formulas by applying a rule. The choice of the schemas and rules influences the minimal length of the proof of a tautology. In order to get more efficient proof systems we can extend these basic systems in various ways. We may add infinite families of tautologies, such as the pigeonhole tautologies. Or we can add some special kinds of rules. One of the strong rules is the *substitution rule.* According to this rule, if we derive a tautology ϕ, we may pick any variable p in ϕ and any proposition ψ and substitute ψ for p in ϕ. Another important rule is the *extension rule.* This rule allows us to abbreviate derived tautologies by symbols in order to avoid the use of extremely large formulas.

One can also consider proof systems that differ more radically from the standard concept. To give you a glimpse on how such proof systems might look, I will describe a classical result in algebra called *Hilbert's Nullstellensatz.* Suppose a system of m algebraic equations with n complex variables is given by

$$\begin{aligned} p_1(x_1,\dots,x_n) &= 0 \\ p_2(x_1,\dots,x_n) &= 0 \\ &\dots \\ p_m(x_1,\dots,x_n) &= 0 \end{aligned} \tag{6.8}$$

where $p_1, p_2, \dots, p_m$ are polynomials. Suppose that we can find polynomials

$$q_1(x_1,\dots,x_n), \quad q_2(x_1,\dots,x_n) \quad ,\dots, \quad q_m(x_1,\dots,x_n)$$

such that

$$\sum_{i=1}^{m} q_i(x_1,\dots,x_n)p_i(x_1,\dots,x_n) \tag{6.9}$$

is the constant 1 polynomial. Then we know that the system of equations (6.8) does not have a solution. Indeed, if a string of numbers $a_1,\dots,a_n$ were a solution, then after substituting $a_1,\dots,a_n$ into the sum (6.9) we would get a sum of zeros, which is zero. But the sum is identically equal to 1, hence a solution cannot exist.

Thus using such strings of polynomials we can prove that a system of equations does not have a solution. (We can check that the polynomial (6.9) is 1 by expanding it into a sum of monomials.) Hilbert's Nullstellensatz says that this method always works, namely, *if the system does not have a solution, then such a string of polynomials $q_1, q_2, \dots, q_m$ exists.* So we can view it as a proof system for unsatisfiable systems of equations. The strings of polynomials $p_1, p_2, \dots, p_m$ play the role of a tautology and strings of polynomials $q_1, q_2, \dots, q_m$ that satisfy the condition above play the role of a proof. In this system we are assuming that the polynomials are represented as sums of monomials. The system is formally not a propositional proof system because it concerns complex numbers, but a modification of this system also works for the two element field.[17] Interpreting 1 as truth and 0 as falsehood, as usual, we get a natural propositional proof system.

---

[17]For the two-element field, we must always include the equation $x_j^2 - x_j = 0$ for all variables, which expresses that variables represent zeros and ones.

What is remarkable about this system is that it is not based on sequential derivations of tautologies. A string of polynomials $q_1, q_2, \ldots, q_m$, which is a proof in this system, in general does not have parts that are proofs of the unsatisfiability of other systems of equations. The proof makes sense only as a whole.

Another noteworthy fact is the use of a mathematical theorem for designing a proof system. Hilbert's Nullstellensatz is a typical characterization theorem in which a *universal* property is characterized by an *existential* property. The universal property that is characterized is: '*no* string of number is a solution'; the existential property is: 'there *exists* a string of polynomials such that the sum (6.9) reduces to 1'. In general, this is exactly what proofs should do. If we want to prove a general statement, we need to characterize the fact that something holds in all structures (of some given type), by the existence of a string of symbols (with some properties).

With these examples in mind we can now isolate the basic properties of propositional proof systems:

1. *Soundness and Completeness.*
2. *Verifiability.*

Soundness means that only tautologies are provable; completeness means that all tautologies are provable. Thus the first condition says that the set of provable formulas is the set of tautologies. This seems to be in contrast to the situation in first order logic, where different theories may prove different sets of theorems. However, what we are actually interested in are the sets of tautologies that have short proofs; these sets are in general different for different systems.

The second condition talks about the nature of proofs. We would like a proof to be undeniable evidence. Whether or not we accept it as such depends also on whether or not we believe in the soundness of the system, but if we believe in it, then we need only to know for sure that a given text is a proof in the system. Put otherwise, we should be able to verify the correctness of a proof. The verification should be a mechanical process and it should not require any ingenuity of the verifier. We formalize it by the requirement that

2′. *the correctness of proofs can be verified by an algorithm running in polynomial time.*

The general definition of a propositional proof system, due to S.A. Cook and R.A. Reckhow [51], is based on these two conditions. The definition below is different, but equivalent to theirs.

Definition 18 A proof system is given by a binary relation R defined on strings in a finite alphabet such that

1″. for any x there exists y such that $R(x, y)$ if and only if x represents a propositional tautology;
2″. $R(x, y)$ is decidable in polynomial time.

Since we are interested in proofs that have at most polynomial length, the central problem about propositional proof systems is:

Problem 3 Does there exist a propositional proof system P such that for some polynomial $p(x)$, every tautology ϕ has a proof in the proof system P of length at most $p(|\phi|)$, where $|\phi|$ denotes the length of ϕ.

It is not difficult to see that this problem is equivalent to the following problem about complexity classes:

Problem 4 Is $\mathbf{NP} = \mathbf{coNP}$?

Most researchers, though not all, believe that the answer to these problems is negative: there is no such proof system, which is equivalent to $\mathbf{NP} \neq \mathbf{coNP}$. The inequality $\mathbf{NP} \neq \mathbf{coNP}$ implies $\mathbf{P} \neq \mathbf{NP}$, hence proving this conjecture could be even more difficult than proving $\mathbf{P} \neq \mathbf{NP}$.

Let us call a proof system *polynomially bounded* when it satisfies the condition from Problem 3. So the conjecture is that no propositional proof system is polynomially bounded. Can we prove at least that some specific proof systems are not polynomially bounded? Such results have been obtained and researchers have been working on extending them to stronger proof systems with the aim of developing new lower-bound methods that eventually may lead to the proof of the conjecture. There are other reasons for proving lower bounds on the lengths of proofs in concrete proof systems, the most important of which is the close connection of propositional proof systems with first order theories.

Theories and Proof Systems

On our journey through formal systems we arrived at the bottom level—propositional proof systems. However, we will see that the bottom and the top are not as far apart as it may seem on the first sight. In plain words, propositional proof systems are closely connected with first-order theories.

In order to demonstrate these connections, we have to identify certain types of proof system. The most common proof systems are based on axioms, or axiom schemas, and derivation rules. One such system was described on page 112. The proof systems of this type are called *Frege systems*. Frege was the first to introduce a formal system for first-order logic. The propositional part of his proof system was essentially what we call a Frege system today, except that we have to exclude his substitution rule. All Frege systems are equivalent in the sense that I will explain shortly.

Let us say that a propositional proof system P *polynomially simulates* a propositional proof system Q if, given a tautology ϕ and a proof d of ϕ in Q, one can construct in polynomial time a proof d' of ϕ in P. When P polynomially simulates Q we think of P as *being at least as strong as* Q.

One can prove that every two Frege systems P and Q polynomially simulate each other. The proof is very easy when the two systems use the same set of connectives.

Then in order to simulate a proof d in Q by a proof in P, we need only to simulate each step in the proof d. To simulate one step in d we need a constant number of steps in P. Thus the proof d' in P is larger at most by a constant factor. Since any two Frege systems simulate each other, Frege systems form a natural class of proof systems.

The most important class of proof systems is called *Extended Frege* proof systems. An Extended Frege proof system is a Frege System augmented with the Extension Rule. As mentioned before, the rule is used to abbreviate long propositions and is formalized as follows. For every proposition ψ, we may introduce the proposition

$$q \equiv \psi,$$

where q is a variable that does not occur in the previous part of the proof. Furthermore, q must not occur in the proposition that we are proving. The reason for using this rule is that we may have proofs that have a small (polynomial) number of steps, but some of the formulas occurring in the proof may be (exponentially) large. In particular, there are proofs that are exponentially large, but have only polynomial number of steps. Repeated applications of the extension rule enable us to transform such proofs into polynomial size proofs.

However, we are not able to prove that it is necessary to use the Extension Rule in order to avoid the exponential blowup of formulas, although it seems very likely to be the case. On the positive side, we know that every two Extended Frege systems polynomially simulate each other. Thus Extended Frege proof systems form another natural class of proof systems, which we denote by **EF**.

The reason why we consider Extended Frege systems important is their relation to the theory $\Theta_{\mathbf{P}}$. The main relation is the content of the following theorem of Cook [49].

Theorem 52

1. *In* $\Theta_{\mathbf{P}}$ *it is provable that Extended Frege systems are sound.*
2. *If it is provable in* $\Theta_{\mathbf{P}}$ *that a propositional proof system* P *is sound, then Extended Frege proof systems polynomially simulate* P.

Thus the Extended Frege proof systems are determined by the condition of *being the strongest systems whose soundness is provable in* $\Theta_{\mathbf{P}}$.

We have obtained the triad of closely related concepts

P	$\Theta_{\mathbf{P}}$	**EF**

one in computational complexity, one in first order logic and one in propositional logic. There are more such triads, though the propositional proof systems of them are not as natural as **EF**.

There is another important connection between $\Theta_{\mathbf{P}}$ and **EF**. Let $\forall x\ \phi(x)$ be a true $\Pi_{\mathbf{P}}$ sentence. Let a sequence of tautologies τ_n, $n = 1, 2, \ldots$, be the propositional translations of $\forall x\ \phi(x)$. Thus τ_n expresses in propositional logic that $\phi(x)$ holds true for all numbers x of length n. Now we can ask whether the tautologies have feasible proofs. The feasibility can be interpreted in a weaker sense as the existence of proofs of polynomial lengths, and in the stronger sense, as the existence of an algorithm that for a given n constructs in polynomial time a proof of τ_n. According to our intuition, we expect that if $\forall x\, \phi(x)$ is an "easy" theorem, then the tautologies τ_n should have short proofs (that are also easy to find). The intuition is correct; we just have to define easiness as provability in a weak theory and to consider a suitable propositional proof system. The prime example is the pair $\Theta_{\mathbf{P}}$ and **EF** (also from [49]).

Theorem 53 1. *Suppose a $\Pi_{\mathbf{P}}$ sentence $\forall x\, \phi(x)$ is a theorem of $\Theta_{\mathbf{P}}$. Then there exists a polynomial time algorithm that constructs proofs of the propositional translations τ_n of this sentence in an Extended Frege proof system. In particular, the propositional translations have polynomial size proofs.*

2. *If a proof system P has the property stated in* 1. *for Extended Frege systems, then P polynomially simulates Extended Frege systems. In plain words, Extended Frege systems are the weakest systems for which* 1. *holds true.*

The opposite of 1. is not true, there are sentences $\forall x\, \phi(x)$ that are not provable in $\Theta_{\mathbf{P}}$, but whose propositional translations have polynomial size proofs in **EF** proof systems (moreover the proofs are constructible in polynomial time).

In order to prove 1., one has to transform a proof in first order logic into a proof in propositional logic. This is similar to what we have already seen: starting with a general proof in first order logic, we want to reduce the complexity of formulas in the proof. In the case of Theorem 53, we want to reduce the complexity all the way down to propositional logic. To this end we can use polynomially verifiable proofs. Recall that such proofs, although they are in first order logic, use only elementary axioms. Thus one only needs to simulate the elementary axioms and rules $PV1$, which is only a technical problem. In this way one can translate proofs of theorems of $PV1$ into proofs in an Extended Frege system. Since $PV1$ and $\Theta_{\mathbf{P}}$ prove the same $\Pi_{\mathbf{P}}$ sentences, we obtain the same for $\Theta_{\mathbf{P}}$.

The relation of $\Theta_{\mathbf{P}}$ to **EF** is similar to the relation of **P** to **nonuniform-P**. Recall that $\mathbf{P} \subseteq$ **nonuniform-P** and a language A in the alphabet $\{0, 1\}$ is in **nonuniform-P** is if and only if there exists a sequence of polynomial size Boolean circuits C_n such that for every n, the section of A consisting of inputs of length n is the set of inputs accepted by the circuit C_n. So we have a polynomial time Turing machine on the one hand and a sequence of circuits on the other. Similarly, we have a proof of a sentence ψ on the one hand, and a sequence of propositional proofs of the translations of ψ on the other. The similarity of these relations suggests that we should view **EF**, the class of proof systems associated with theory $\Theta_{\mathbf{P}}$, as the nonuniform version of the theory. It is therefore natural to extend the triad to the quadruple:

P	**nonuniform-P**
$\Theta_{\mathbf{P}}$	**EF**

The two rows represent complexity classes and formal systems; the two columns represent uniform and nonuniform concepts.

I have mentioned that there are other triads of complexity classes, theories and propositional proof systems. In particular, there exists a propositional proof system G_1 associated with $\Theta_{\mathbf{NP}}$, which means that G_1 is the strongest propositional proof system whose soundness is provable in $\Theta_{\mathbf{NP}}$. The associated proof systems for $\Theta_{\mathbf{P}}$ and $\Theta_{\mathbf{NP}}$ are a means that potentially can be used to show that these two theories are different. Here is the idea of such a proof. Suppose that $\Theta_{\mathbf{P}} \equiv \Theta_{\mathbf{NP}}$. Since $\Theta_{\mathbf{NP}}$ proves the soundness of G_1 and we are assuming $\Theta_{\mathbf{P}} \equiv \Theta_{\mathbf{NP}}$, $\Theta_{\mathbf{P}}$ also proves the soundness of G_1. Since Extended Frege systems are the strongest systems whose soundness $\Theta_{\mathbf{P}}$ proves, they must polynomially simulate G_1. In particular, if P is an Extended Frege system, then every tautology has at most a polynomially longer proof in P than in G_1. Hence a possible strategy for proving $\Theta_{\mathbf{P}} \not\equiv \Theta_{\mathbf{NP}}$ is to find a sequence of tautologies τ_n, $n = 1, 2, \ldots$ such that for some polynomial $p(n)$ the tautologies have proofs of lengths at most $p(n)$ in G_1, but do not have proofs of polynomial length in P.

In this way the problem of separating $\Theta_{\mathbf{P}}$ from $\Theta_{\mathbf{NP}}$ has been reduced to a combinatorial problem about proving lower bounds on the lengths of proofs in **EF**. Unfortunately, we are not able to use this reduction because we are not able to prove superpolynomial lower bounds on **EF** proofs for any sequence of tautologies. In other words, we are not able to disprove that **EF** proof systems are polynomially bounded. If, contrary to our expectation, **EF** proof systems were polynomially bounded, then in particular we would have **NP** = **coNP**. Thus not only are we not able to prove that **NP** = **coNP**, which means that no propositional proof system is polynomially bounded, but we are also not able to prove the (probably) weaker statement that Extended Frege proof systems are not polynomially bounded.

As it is one of the central problems in the area, I will state this explicitly.

Problem 5 Are **EF** proof systems polynomially bounded?

We are not able to prove superpolynomial lower bounds even for the (probably) weaker class of Frege proof systems. These are the most natural propositional proof systems, yet we do not know how to prove such bounds. Although this problem does not seem to have any direct connection with complexity, it seems as difficult as the open problems about complexity classes. In particular, it looks very much like the problem of proving superpolynomial lower bounds on Boolean circuits computing Boolean functions defined by **NP** sets. A Boolean circuit can be represented by a sequence of elementary Boolean operations; likewise a Frege proof is a sequence of applications of elementary logical rules; and an Extended Frege proof is only slightly more complex. We do not have any methods for proving more than linear lower bounds for any concrete Boolean function; likewise we lack methods for proving more than linear lower bounds on Extended Frege proofs of any tautology.

Proving lower bounds can not only help us prove separation of theories, it can also be used to prove independence results. According to Theorem 53, to prove that a $\Pi_{\mathbf{P}}$ sentence ϕ is unprovable in $\Theta_{\mathbf{P}}$, it suffices to prove superpolynomial lower bounds on the lengths of proofs in **EF** of the propositional translations of ϕ. The same is true for any theory that has an associated proof system (in fact, as we will see shortly, for all theories). Such independence results have actually been proved for some weaker systems, but it would take us too far afield to explain these results.

Proving that some sentence is unprovable in $\Theta_{\mathbf{P}}$ seems to be a very weak result. Indeed, $\Theta_{\mathbf{P}}$ is much weaker than Peano Arithmetic, Peano Arithmetic is much weaker than Zermelo-Fraenkel set theory and Zermelo-Fraenkel set theory is much weaker than the strongest set theories that have been considered (obtained by adding large cardinal axioms to Zermelo-Fraenkel set theory). But our goal is not to show independence result for the strongest possible theories. We want to know what is provable in some particular theories. We have a good reason for asking what is provable in $\Theta_{\mathbf{P}}$, namely, we want to prove that it is weaker than $\Theta_{\mathbf{NP}}$. We also hope that once we are able to prove independence results for $\Theta_{\mathbf{P}}$, we will have a method that will enable us to prove such results for other theories.

There is another important reason for trying to prove independence results in this way. Unless a theory T is very weak, we know only one general method of proving the independence of $\Pi_{\mathbf{P}}$ sentences with respect to T—using Gödel's theorem. Unfortunately, the Gödel sentences are not useful for proving independence of concrete mathematical theorems. Therefore we are looking for other methods; proving lower bounds on propositional proofs is an alternative.

Proof Complexity and Computational Complexity

When comparing proof complexity with computational complexity the most striking fact is that many concepts in one field have a counterpart in the other. Here are some pairs of such concepts.

computational complexity	*proof complexity*
complexity classes	theories
elementary operations	basic axioms
computations with bounded resources	induction restricted to a class of formulas
the **P** vs. **NP** problem	the $\Theta_{\mathbf{P}}$ vs. $\Theta_{\mathbf{NP}}$ problem
circuits	propositional proofs

Although in most cases the correspondence is not buttressed by a formal definition, the similarities are very conspicuous. This suggests that the two fields study the same general phenomenon from different perspectives. Yet the two fields are essentially different. They have different goals and, in spite of some relations, the open problems in proof complexity are not equivalent to open problems in computational complexity.

In particular, researchers working in proof complexity are not trying to solve **P** vs. **NP** or some other problem about separating complexity classes. Proof complexity studies how difficult it is to prove certain theorems. Thus in the case of **P** vs. **NP**, the questions relevant for proof complexity are: *how strong a theory do we need to prove* $\mathbf{P} = \mathbf{NP}$ (if it is true), and *how strong a theory do we need to prove* $\mathbf{P} \neq \mathbf{NP}$ (if it is true)? When a proof of $\mathbf{P} = \mathbf{NP}$ or of $\mathbf{P} \neq \mathbf{NP}$ is found, we will be able to give an upper estimate on the theory in which it is provable. Meanwhile we can only try to show the unprovability of such statements in weak theories.

Concerning the unprovability of such statements, very little has been achieved so far. In fact, it seems that to prove such results is as hard as to solve the corresponding problems. For example, proving that $\Theta_{\mathbf{P}} \not\equiv \Theta_{\mathbf{NP}}$, which would show that $\mathbf{P} = \mathbf{NP}$ is not provable in $\Theta_{\mathbf{P}}$, seems as difficult as actually proving $\mathbf{P} \neq \mathbf{NP}$. But there are some results in this direction. One of the early results, due to Buss [34], is the following.

Theorem 54 $\mathbf{P} \neq \mathbf{NP} \cap \mathbf{coNP}$ *is not provable in* $\Theta_{\mathbf{P}}$.

This is an easy corollary of Theorem 49 (see Notes for the proof). The conjecture $\mathbf{P} \neq \mathbf{NP} \cap \mathbf{coNP}$ is important in cryptography because if it were true that $\mathbf{P} = \mathbf{NP} \cap \mathbf{coNP}$, then the protocols currently used would probably be insecure.[18] Specifically, the existence of one-way length preserving bijections implies $\mathbf{P} \neq \mathbf{NP} \cap \mathbf{coNP}$.

In principle, it should be possible to use proof complexity to show that the currently known methods are not adequate to solve the **P** vs. **NP** problem by proving superpolynomial lower bounds on the circuit complexity of functions in **NP**. Unfortunately, we do not have such a result yet. Partial progress in this direction has been made by Razborov [241, 242]. He considered circuits of a special form: two disjoint circuits C and D connected by the *XOR* output gate. He defined a theory in which the two parts of such circuits are represented by two predicates R and S. The restrictions on induction axioms depend on whether or not the formulas contain both predicates R and S, or only one. For formulas containing both predicates, the restriction is much stronger. For this kind of circuit and this theory, he was able to prove that the theory cannot prove superpolynomial lower bounds on the size of the circuit in general, but if the circuit has constant depth, then one can use the random restriction technique to prove exponential lower bounds.[19] The result about the unprovability of lower bounds is only conditional; it requires the PRG-conjecture.

Although the setting is rather unnatural, this result confirms the conjecture that there is a qualitative difference between proving lower bounds for restricted types of circuit and for general ones.

Once I was asked why we study proof complexity when it seems that the problems in proof complexity are as difficult as in those computational complexity. If,

---

[18]The conjecture only talks about polynomial time algorithms, but for practical purposes it is important what the polynomial bound is in these algorithms. The protocols would be insecure in practice if the polynomial bounds were small.

[19]In fact, in order to formalize the lower-bound proof, he had to find a new proof of that result.

for example, we are interested in the **P** vs. **NP** problem, we should address the problem directly. My answer is, to put it shortly, as follows. Firstly, as I explained above, our aim in proof complexity is not to solve problems from computational complexity (but, of course, if we discover a possible way to do this using proof complexity we will go after it). Secondly, I am a logician and am interested in the foundations of mathematics. Proof complexity is directly connected with the foundations of mathematics, whereas the connection with computational complexity is not so tight. Furthermore, I believe that there may be hurdles of a fundamental nature that prevent us from solving the basic problems of computational complexity. If this is the case, we are more likely to find them using logic.

Nevertheless, there are a few cases in which proof complexity has indeed helped solve a problem in computational complexity. In the rest of this subsection I will describe such a result. It concerns an important problem called *Integer Linear Programming*. An instance of Integer Linear Programming is given by a finite set of linear inequalities with rational coefficients:

$$\begin{aligned} a_{11}x_1 + a_{12}x_2 + \cdots + a_{1n}x_n &\geq b_1 \\ a_{21}x_1 + a_{22}x_2 + \cdots + a_{2n}x_n &\geq b_2 \\ &\cdots \\ a_{m1}x_1 + a_{m2}x_2 + \cdots + a_{mn}x_n &\geq b_m, \end{aligned} \tag{6.10}$$

where a_{ij} are rational numbers and x_j are unknowns. The aim is to find an integral solution, which means to find integers $c_1, c_2, \ldots, c_n$ that satisfy the inequalities.

This problem is closely related to the *Linear Programming* problem, which is defined in the same way, with the difference that what we are looking for now are any *rational* solutions. Furthermore, if a solution exists, then we want to find a solution that optimizes some linear function. For Linear Programming, polynomial time algorithms are known, whereas Integer Linear Programming is not solvable in polynomial time unless **P** = **NP**. Deciding whether the set of inequalities (6.10) has an integral solution is an **NP**-complete problem. Linear Programming is a potent tool in designing efficient algorithms. Integer Linear Programming could play an even more important role, if it were solvable in polynomial time. Still, it makes sense to design algorithms for Integer Linear Programming and study how they perform on particular types of data, even if we believe that **P** ≠ **NP**. On the other hand, if somebody believes that **P** = **NP** and wants to prove it, then it would be a good strategy to try to find a polynomial time algorithm for Integer Linear Programming.

Let us now focus on decision problems, since they determine the complexity of these two problems. From high school, we know two operations that preserve solutions of linear inequalities:

1. *adding two equations;*
2. *multiplying an equation by a positive constant.*

According to a well-known result, the system of inequalities (6.10) does not have a rational solution if and only if we can derive from it the contradictory inequality

$$0 \geq 1$$

using these two deduction rules. The proof is always short, since it suffices to multiply the equations (6.10) by nonnegative constants and then add them. The tricky thing is how to find the constants.

It can happen that a system of linear inequalities has a rational solution, but does not have any integral solutions. Thus we need something else to characterize the solvability of (6.10) in the domain of integers. In 1975, building on work of R.E. Gomory [104], V. Chvátal [45] proved that it suffices to add another simple rule in order to characterize unsolvability of (6.10) over the integers.

3. *If the constants $c_1, c_2, \ldots, c_n$ are integers, then from*

$$c_1x_1 + c_2x_2 + \cdots + c_nx_n \geq d$$

we can derive

$$c_1x_1 + c_2x_2 + \cdots + c_nx_n \geq \lceil d \rceil,$$

where $\lceil d \rceil$ denotes the least integer such that $\lceil d \rceil \geq d$.

In other words, if the coefficients on the left hand side are integral, we can round up the number on the right hand side. This rule is clearly true in the domain of integers because if the coefficients on the left hand side are integral, then the sum on the left hand side is an integer. The method based on 1., 2. and 3. is called the method of *cutting-planes*.[20]

Knowing that 1. and 2. suffice for designing a polynomial time algorithm for Linear Programming, it is natural to ask whether 1., 2. and 3. suffice for designing a polynomial time algorithm for Integer Linear Programming. Since we believe that **P** $\neq$ **NP**, we expect the answer to be negative. So the question is: *can we prove that there is no polynomial time algorithm for Integer Linear Programming based on the method of cutting-planes?* Proving **P** $\neq$ **NP** may be extremely difficult, but what we are asking for is much weaker: we only want to prove that a certain type of algorithms cannot run in polynomial time.

Such a result was proved in 1996 [225] and the proof uses methods developed in proof complexity. It is quite natural to view the three rules of the method of cutting-planes as the derivation rules of a proof system. Since the variables $x_1, x_2, \ldots, x_n$ range over integers, adding the inequalities $0 \leq x_1 \leq 1, 0 \leq x_2 \leq 1, \ldots, 0 \leq x_n \leq 1$ further restricts the solutions to zeros and ones. Thus we can think of inequalities as a kind of Boolean formulas. In particular, we can express disjunctions of variables and negated variables by inequalities (for example, the disjunction $x \vee y \vee \neg z$ is equivalent to $x + y + (1 - z) \geq 1$). This enables us to represent various interesting tautologies by unsatisfiable sets of inequalities.

Let τ be a propositional tautology and let S be the system of unsatisfiable inequalities representing τ. If d is a derivation of the contradictory inequality $0 \geq 1$ from S, we say that d is a *cutting-planes* proof of τ. This defines the *cutting-planes*

[20] I am only talking about characterizing unsolvability, for the sake of simplicity, but Gomory and Chvátal proved the more general result that the method of cutting-planes suffices to derive all inequalities that are consequences of the initial ones.

proof system. It is one of the few proof systems for which we are able to prove exponential lower bounds on the lengths of proofs. Translated back to Integer Linear Programming, it means that there are systems of inequalities that are not satisfied by integers, but for which any derivation of a contradiction based on the rules 1., 2. and 3. is of exponential length. In particular, whatever strategy we use in an algorithm that is based only on these three rules, the algorithm must run in exponential time.

Similar results have been obtained for some other methods for solving Integer Linear Programming.

Notes

1. *The method of feasible interpolation.* Feasible interpolation, invented by J. Krajíček [162], is a useful way of proving lower bounds on the lengths of proofs in propositional calculus. It is based on the classical result called *Craig's Interpolation Theorem,* due William Craig [53]. That theorem is normally stated for first order logic, but also has a version for propositional logic:

   Theorem 55 *Let $\phi(\bar{p},\bar{q})$ denote a proposition with propositional variables $p_1, p_2, \ldots, p_n$, $q_1, q_2, \ldots, q_m$ and $\psi(\bar{p},\bar{r})$ denote a proposition with propositional variables $p_1, p_2, \ldots, p_n$, $r_1, r_2, \ldots, r_k$. Suppose that*

   $$\phi(\bar{p},\bar{q}) \rightarrow \psi(\bar{p},\bar{r})$$

   is a tautology. Then there exists a proposition $\iota(\bar{p})$ that contains only the common variables $p_1, p_2, \ldots, p_n$ such that the two implications

   $$\phi(\bar{p},\bar{q}) \rightarrow \iota(\bar{p}), \quad \iota(\bar{p}) \rightarrow \psi(\bar{p},\bar{r})$$

   are tautologies.

   Proposition $\iota(\bar{p})$ is called an *interpolant* of $\phi(\bar{p},\bar{q}) \rightarrow \psi(\bar{p},\bar{r})$. The intuition behind this theorem is that the logical connection between two formulas must be mediated by common primitive parts of the formulas. The proof of this theorem is not difficult if one uses semantic considerations.

   In general the size of the interpolant can be exponentially larger than the two propositions. Krajíček's crucial idea was that one may be able to bound the size of the interpolant by the size of the *proof* of $\phi(\bar{p},\bar{q}) \rightarrow \psi(\bar{p},\bar{r})$. This is indeed true for some propositional proof systems with one proviso: one has to consider the size of the interpolant not as a formula, but as a *circuit*. Using circuits we can define Boolean functions in a more concise way. Let us define this property of proof systems precisely.

   Definition 19 A proof system P has the *feasible interpolation property*, if there is a polynomial time algorithm such that given an implication of the form

$\phi(\bar{p},\bar{q}) \rightarrow \psi(\bar{p},\bar{r})$ and its proof d in the proof system P, the algorithm constructs a circuit $C(\bar{p})$ that interpolates the implication. In particular, the size of the circuit C is bounded by a polynomial in the size of the proof d.

If a proof system P has the feasible interpolation property, then we can reduce the problem of proving lower bounds on proofs in P to proving lower bounds on Boolean circuits. To this end we need to find a suitable implication which is a tautology and prove a lower bound on the size of any circuit that interpolates it. This reduces one difficult problem to another, probably even more difficult one. Yet this method is useful. There are plausible conjectures about the sizes of circuits from which one can prove that some concrete tautologies must be hard for P. The most remarkable fact is, however, that in many cases we can prove unconditional results (results that are not based on any unproved conjectures). This is because quite often we can prove that the interpolating circuits must have special properties.

In particular, for some proof systems one can prove that if the common variables $\bar{p}$ occur only positively in ϕ, then one can construct a *monotone* circuit. For monotone circuits, exponential lower bounds have been proved. The fact that one can prove exponential lower bounds on the size of monotone circuits that compute certain function does not automatically give us lower bounds on the proofs; we need more. We need to show that small monotone circuits cannot interpolate certain implications. It is quite remarkable that the lower bounds on monotone circuits that had been proved ten years before feasible interpolation was discovered give us exactly what is needed. We do not know whether this is only a fluke or if there is a deeper reason behind it.

Lower bounds based on monotone feasible interpolation have been proved for the resolution calculus [162], the cutting-planes proof system [225], the propositional intuitionistic calculus [134] and several other systems. Unfortunately, the application of this method is limited to relatively weak proof systems. It has been shown that Frege systems do not have the feasible interpolation property assuming that factoring is hard, a conjecture widely accepted in computational complexity, [29, 166].

2. *Sentences unprovable in* $\Theta_\mathbf{P}$. Although we lack methods for proving the independence of arithmetical sentences from $\Theta_\mathbf{P}$, we can at least prove some partial results.

The first kind of result concerns proofs of unprovability based on conjectures from computational complexity. An example of such an independence is the following theorem [166] (independence of a sentence expressing a consequence of the Euler-Fermat theorem):

Theorem 56 *If the cryptographic protocol RSA (see page 430) is secure, then* $\Theta_\mathbf{P}$ *does not prove the following sentence*

$$\forall x \forall y \big(x > 1 \wedge (x,y) = 1 \rightarrow \exists z (y^z \equiv 1 \bmod x)\big).$$

(The meaning of the expression $(x,y) = 1$ is that x and y are coprime.)

The second kind of result involves sentences that are likely to be false. An example of such a sentence was given in Theorem 54. It was stated rather informally as '**P** $\neq$ **NP** $\cap$ **coNP** *is not provable in* $\Theta_{\mathbf{P}}$'. The precise statement is as follows.

If $\Theta_{\mathbf{P}}$ *proves that* $\forall x\ \phi(x) \equiv \psi(x)$, *where* ϕ *and* ψ *are* Σ_1^b *and* Π_1^b *sentences respectively, then the sentences define a set in* **P**.

Proof Let $\phi(x)$ have the form $\exists y\ \alpha(x, y)$ where α is sharply bounded. Let $\psi(x)$ have the form $\forall z\ \beta(x, z)$ where β is sharply bounded. We are assuming that the bounds on the quantifiers $\exists x$ and $\forall y$ are implicit in the formulas. The assumption about the provability of $\forall x\ \phi(x) \equiv \psi(x)$ in $\Theta_{\mathbf{P}}$ implies that

$$\forall x \exists y \exists z \big(\alpha(x, y) \vee \neg\beta(x, z)\big)$$

is provable in $\Theta_{\mathbf{P}}$. According to Theorem 49, there exist polynomial time computable functions f and g such that the following is true:

$$\forall x \big(\alpha\big(x, f(x)\big) \vee \neg\beta\big(x, g(x)\big)\big).$$

Whence,

$$\phi(x) \equiv \alpha\big(x, f(x)\big).$$

Since $\alpha(x, y)$ is sharply bounded and f computable in polynomial time, $\alpha(x, f(x))$ defines a set in **P**. This finishes the proof. □

3. *Unprovability of circuit lower bounds.* In [162], Krajíček proved the following result, which captures the essence of Razborov's result about the unprovability of circuit lower bounds.

Theorem 57 *Assuming the PRG-Conjecture* (*see page* 435) *the following is true. If P is a propositional proof system with the feasible interpolation property, then for every Boolean function* f, *it is hard to prove in* P *any exponential lower bound on the circuit complexity of* f.

This theorem is stated very informally and requires specification. Let n and m be numbers. We represent Boolean functions of n variables by 0–1 strings of length 2^n that list the values of the function, the truth tables. Then we construct a formula $\alpha(\bar{x}, \bar{y})$ where $\bar{x}$ is a string of 2^n variables and $\bar{y}$ is some string of auxiliary variables whose length depends on the parameter m. In α the variables $\bar{x}$ encode a truth table of a Boolean function of n variables and the variables y encode a circuit of size m. The formula expresses that $\bar{x}$ is a Boolean function that does not have a circuit of size m or less.

Let P be a propositional proof system with the feasible interpolation property. Let f be an n-variable Boolean function whose complexity is strictly larger than m and let $\bar{a}$ be the truth table of the function. Then $\alpha(\bar{a}, \bar{y})$ is a tautology. The content of the theorem is that, assuming the PRG-Conjecture and $m \geq 2^{n^{\varepsilon}}$, for some $\varepsilon > 0$, then $\alpha(\bar{a}, \bar{y})$ does not have a polynomial size proof in P.

(Here we are comparing the sizes of proofs with the sizes of formulas. The size of the formula $\alpha(\bar{a}, \bar{y})$ is 2^{cn} for some constant c. Thus a proof of *polynomial size* means one of size 2^{dn}, for some constant d.)

6.4 Feasible Incompleteness

In the last section, we saw some interesting connections between complexity theory and logic. However, the fact that complexity theory is connected with logic does not imply that there must also be such a connection to the foundations of mathematics. Therefore in this section I will present more concrete connections between complexity theory and the foundations. I will be mostly talking about conjectures, so the arguments are rather speculative. Whether or not these argument do give such connections depends very much on whether or not such conjectures are true. I believe that they are true, but they seem even more elusive than the main problems in complexity theory. In any case the conjectures are concrete mathematical statements, hence they can be proved or refuted (at least in principle), and those who do not like speculations can view the conjectures simply as difficult open problems.

Before considering specific conjectures, I will state a general thesis. The thesis, which is not a mathematical statement, is intended to encompass various related conjectures that researchers working in proof complexity have contemplated, although most of them have never been stated explicitly. It is possible that we will eventually find a single conjecture that will imply all the others, which will allow us to proclaim this conjecture to be the formalization of the thesis.

The Feasible Incompleteness Thesis *The phenomenon of incompleteness manifests itself at the level of polynomial time computations.*

There are two types of consideration that lead to the thesis. The first one concerns the problem of consistency. Because of Gödel's Incompleteness Theorem the program of proving the consistency of the foundations of mathematics by finitistic means, as proposed by Hilbert, is not realizable. The foundations of all contemporary mathematics are based on Zermelo-Fraenkel Set Theory. To prove the consistency of Zermelo-Fraenkel Set Theory, not only do finitistic means not suffice, but we need an even stronger theory than the theory itself. Being unable to prove consistency, we may ask if we can prove at least that there are no *short* proofs of contradiction. If, in particular, we could show that the smallest proof of contradiction has to be very large, say of length at least 10^9, we would be sure that we will never encounter such a proof in practice. We could view this as a partial realization of Hilbert's program. Also recall that, as we have seen in Sect. 6.1, inconsistent theories may be useful in the sense that their theorems with short proofs are true. So for practical purposes we do not need absolute consistency.

Given a bound n we can always find a finitistic proof of at most exponential length of the statement expressing that a given consistent theory T does not prove contradiction by proofs of size n. This is because we can list all strings of length n and for each of them prove that it is not a proof of contradiction "by inspection". The question is whether there are shorter proofs than these exponential size "brute force" proofs. So we again encounter the problem of exponential expansion. It should not then come as a surprise that the problem is connected with open problems in complexity theory.

The second source of conjectures is proof complexity, namely, the theories that we call Bounded Arithmetic. The main open problems concern the separation of pairs of theories associated with the pairs of complexity classes that are conjectured to be different. We not only want to show that the sets of theorems provable in these theories are different, but moreover we would like to show that the theories differ already for low complexity theorems. Again, we are not able to accomplish this goal, thus we can only make conjectures.

It seems that there is nothing special about theories corresponding to low complexity classes, thus it is natural to state conjectures about general theories. Some conjectures are quite specific, with little relation to classical incompleteness results, but some have the distinctive flavor of the incompleteness phenomenon. They say that every theory has only a limited ability to formalize complex functions and complex sets, which means that if the complexity of sets and functions is too high, the theory is unable to prove their properties. In other words, the conjectures say that *a possible cause of incompleteness is computational complexity.*

When we do not have proofs we argue by analogy. The analogy here is that we know that the unprovability of certain sentences can be caused by the fact that they describe a function that grows very fast, so fast that the theory is unable to prove that such large numbers exist. If the extremely fast growth of a function can cause unprovability, then why not the extreme complexity of a function?

These two kinds of reasoning that lead to the Feasible Incompleteness Thesis also produce two different types of conjectures. The first type concerns the constructibility of proofs and the length of proofs of certain sentences. The second type concerns the provability of sentences connected with problems in computational complexity. However, I will show that there is a conjecture that can be interpreted in both ways, which supports my feeling that there should only be one thesis.

All but one of the conjectures that I am going to state have equivalent versions stated using only concepts from computational complexity. So we can also view them as conjectures from complexity theory that have important consequences in proof complexity. This fact again confirms the close ties between the two fields. The form of these versions is typically a statement saying that some class of problems does not have a complete member, that is, one to which all others can be efficiently reduced. The equivalences and the implications between conjectures are fairly easy and I will sketch the proofs of most of them.

The reason I state the Feasible Incompleteness conjectures is that it is unlikely that the related problems are going to be solved soon. It may even happen that the set-theoretical axioms that we use, however strong they are, are not suited for problems in the theory of computational and proof complexity. If that is the case, then we need two things:

1. Some arguments (which do not have to be completely formal) showing that this is indeed the case.
2. New axioms that will enable us to decide the problems.

I imagine that this could be achieved by a system of conjectures that would do both—explain why the problems are not provable in the current formal systems and also decide them. Therefore we should look for general principles that could be accepted as such new postulates. Whether or not the Feasible Incompleteness Thesis points in the right direction is yet to be seen.

The Feasible Consistency Problem

I will now consider the first kind of problem in more detail. As we have seen before, the standard way of formalizing the intuitive concept of feasibility is by considering arbitrarily large instances and using polynomial bounds. So again, instead of bounds for concrete numbers we will rather study how the lengths of proofs of consistency grow as we increase the bounds on the lengths of proofs. Further, it is natural not to limit ourselves to specific theories, but rather to consider a general situation. So I will assume that we have two theories S and T, both consistent and both formalized by a finite number of axioms. The latter condition excludes some important theories, such as Peano Arithmetic and Zermelo-Fraenkel Set Theory, but all that I am going to say will also hold true for them. It is simply a convenient assumption to avoid the complications caused by infinite sets of axioms. T is the theory whose consistency we are studying and S is the theory in which we are proving facts about T. You can think of T as being a strong theory that we might need for the foundation of mathematics (say, a finite part of Zermelo-Fraenkel Set Theory) and S as being a fairly weak theory about whose soundness we do not have any doubts (a theory that formalizes "*finite means*", such as a finite part of Peano Arithmetic). I will need to distinguish between proofs in S and T using short expressions, so I will call the former *S-proofs* and the latter *T-proofs*.

Given such theories S and T, we want to know how difficult is to prove the sentences:

> "*there is no proof of contradiction derivable from the axioms of T whose length is at most n,*"

for specific numbers $n = 1, 2, \dots$. I will abbreviate these sentences by saying "*T is consistent up to length n*" and I will use the formulas

$$Con_T(n)$$

to denote them. I will call '*the feasible consistency problem*' for these theories the following question:

> *If one is allowed to argue only in the theory S, how difficult is it to prove the sentence expressing that T is consistent up to length n, where n is a concrete number?*

The most interesting values of n are those that estimate proofs that can actually be written down, but we prefer to study the more mathematical question of how the length of such proofs increases with increasing n. This problem is stated informally, so let us consider how it can be formalized. One possibility is:

Problem 6 Do the sentences $Con_T(n)$ have S-proofs of length at most $p(n)$ for some polynomial p?

We can measure the difficulty of proving the sentences $Con_T(n)$ not only by the length of S-proofs, but also by the difficulty of finding them. Thus we have another formalization of the problem above:

Problem 7 Is there a polynomial time algorithm for finding S-proofs of $Con_T(n)$?

Clearly, a positive answer to Problem 7 implies a positive answer to Problem 6, since an S-proof constructed by a polynomial time algorithm can only be of polynomial length. I conjecture that if S is essentially weaker than T, then the answer to Problem 6 is negative, which implies a negative answer to Problem 7.

These problems are most interesting when T is strong and S is weak, but let us also see what happens in general.

1. **Suppose that T is weak and S is strong** (which is the opposite of what we are actually interested in). The consistency of T is clearly equivalent to the statement that $Con_T(n)$ is true for all n. Formally, this is the sentence $\forall x\ Con_T(x)$. Suppose that S is so strong that it proves the consistency of T, thus it proves $\forall x\ Con_T(x)$. Then, to get $Con_T(n)$, an instance of the general statement $\forall x\ Con_T(x)$, we need only to apply a single logical rule, the rule of deriving a specific instance from a general statement. Thus in such a case the proofs are always short and easy to construct. In fact, if n is represented in binary, the lengths of such proofs will only be logarithmic in n, as opposed to polynomial in n.

2. **Suppose that** $T = S$. This is an important special case, since it corresponds to Gödel's Second Incompleteness Theorem. This similarity suggests that Problems 7 and 6 should have negative answers, but the opposite is true. There are polynomial size proofs of $Con_T(n)$ in T and they can be constructed in polynomial time [223]. Although the analogy with the Second Incompleteness Theorem fails, there is still a difference between this and the previous case. In the previous case (when S proved the consistency of T) we had logarithmic size proofs, assuming n was represented in binary. In the present case ($T = S$) this is not true anymore. No matter how concisely we represent n, the lengths of the proofs of $Con_T(n)$ are still close to n. So, in a sense, we do get an exponential difference between these two cases. This lower bound was proved by H. Friedman in 1979 [81].

Theorem 58 *For every consistent and finitely axiomatized theory T, there exists $\varepsilon > 0$ such that, for every n, every T-proof of $Con_T(n)$ has size at least n^{ε}.*

In other words, the length of the smallest T-proof of $Con_T(n)$ grows asymptotically at least as fast n^{ε}. (There is a polynomial gap between the lower and upper bounds; the true value seems to be close to n.)

3. **Suppose that T is essentially stronger than** S. To ensure this relation we may suppose, for example, that T proves the consistency of S. The feasible consistency problem in this case is wide open. Since T is at least as strong as S, the sentence $Con_T(n)$ implies $Con_S(n')$, for n' at most polynomially smaller than n. Thus Friedman's Theorem (applied to S) tells us that the shortest S-proofs of $Con_T(n)$ have lengths at least $n^{\varepsilon'}$, for some $\varepsilon' > 0$. But for all we know, the shortest S-proofs of $Con_T(n)$ may be of exponential length. It would indeed be interesting to prove this, but again it is the kind of statement which we do not have the slightest idea how to prove.

The Relation to the P *Versus* NP *Problem*

In fact, it may be as difficult as proving $\mathbf{P} \neq \mathbf{NP}$. To see the relation between the feasible incompleteness problem and problems in complexity theory, I will present an argument that in several steps naturally leads to a conjecture about the feasible incompleteness problem.

(i) *Assume* $\mathbf{P} \neq \mathbf{NP}$. *Then there is no polynomial time algorithm for deciding if a sentence has a proof of length*[21] *at most n in first order logic.*

(Polynomial time means that the algorithm runs in time polynomial in n and the length of the sentence.) This is because this problem (properly formalized) is **NP**-complete. Thus, in fact, the latter statement is equivalent to $\mathbf{P} \neq \mathbf{NP}$.

(ii) Hence *if* $\mathbf{P} \neq \mathbf{NP}$, *there do not exist a consistent finitely axiomatized theory S and an algorithm A with the following property: given a sentence ϕ and a number n such that ϕ does not have a proof of length at most n in first order logic, A constructs an S-proof of this fact in polynomial time.*

This is because constructing a proof of some sentence is at least as hard as deciding if the sentence is true. Naturally, the reason why there is no such algorithm for a theory S may be that there are no such proofs. But since the sentences in question talk about finite things, every sufficiently strong S can prove them. So the reason may be rather that the proofs are more than polynomially long. If we assume a little bit more, then this is indeed what happens. I will state this explicitly.

(iii) *Assume* $\mathbf{coNP} \neq \mathbf{NP}$. *Then there do not exist a consistent finitely axiomatized theory S and a polynomial p with the following property: for every number n and*

[21] The length of a proof is the length of a string of symbols that represents the proof.

every sentence ϕ, if ϕ does not have a proof of length n in first order logic, then there is an S-proof of length at most $p(n)$ of (the formalization of) the statement that ϕ does not have a proof of size n.

Again, the assumption and the conclusion are in fact equivalent. Let us reformulate the nonexistence of S and p as follows:

For every consistent finitely axiomatized theory S, there exists a sequence of sentences $\{\phi_1, \phi_2, \phi_3, \ldots\}$ such that

1. *no ϕ_n has a proof of length n in first-order logic, and*
2. *the sentences 'ϕ_n* does not have a proof of length at most n' *do not have polynomially bounded S-proofs.*

We would certainly like to know what the sentences ϕ_n are, but the mere assumption that **coNP** $\neq$ **NP** does not give us any hint. We may also wonder whether the sequence is in some sense uniform. The most uniform sequence is a sequence in which all elements are the same. If all sentences ϕ_n are one fixed sentence ϕ, then the condition about the nonexistence of proofs of length n is simply the condition that ϕ is unprovable in first order logic. Let T be the theory axiomatized by $\neg\phi$; then the condition that ϕ is unprovable is equivalent to T being consistent. Similarly, the conclusion of the statement above is that the sentences expressing the consistency of T up to proofs of length n do not have polynomially bounded S-proofs. Thus we have arrived at the main conjecture associated with the Feasible Incompleteness Thesis.

Conjecture 4 *For every consistent finitely axiomatized theory S, there exists a consistent finitely axiomatized theory T such that the sentences $Con_T(n)$ do not have polynomially bounded S-proofs.*

Let us analyze further why the conjecture seems probable. To this end I will state a simple consequence of Gödel's Second Incompleteness Theorem.

For every consistent finitely axiomatized theory S, there exists a consistent finitely axiomatized theory T such that S does not prove the consistency of T.

We know that one can simply take $T = S$, which is Gödel's theorem, but suppose we did not know it. Then we could explain the theorem as follows. If one randomly produces a finite set of axioms (which is the same as producing a single sentence), then it is difficult to decide if the set is consistent or not. If there were a theory S that, contrary to the statement above, could prove the consistency of every consistent theory, then it would help us to decide this difficult question; in fact, it would give us an algorithm. But since there is no algorithm for deciding consistency, there cannot be such a theory.

The Feasible Incompleteness Thesis says that if we limit ourselves to polynomial time computations and polynomial size proofs, we should see essentially the same picture as in the unlimited case. Thus it seems unlikely that we can prove the consistency of T up to n by a short S-proof when T is much stronger than S.

Let us return to the connections with problems in complexity theory. I started with a statement equivalent to $\mathbf{P} \neq \mathbf{NP}$, then I made it stronger. The stronger version was equivalent to $\mathbf{coNP} \neq \mathbf{NP}$. Then I made it more specific, thus again stronger. So the resulting Conjecture 4 is at least as strong as $\mathbf{coNP} \neq \mathbf{NP}$, which in turn implies $\mathbf{P} \neq \mathbf{NP}$. Thus Conjecture 4 not only talks about an important problem in proof complexity and the foundations of mathematics, it also decides the most important problem in complexity theory.

In general, the fact that some conjecture implies $\mathbf{P} \neq \mathbf{NP}$ is not surprising; one can easily produce lots of such conjectures. What is remarkable, however, is that Conjecture 4 talks about consistency and provability. Thus we can support our belief in $\mathbf{P} \neq \mathbf{NP}$ by an essentially different type of reasoning.

Conjecture 4 can be equivalently expressed using concepts from computational complexity theory; I will show this in the Notes. Here I will mention another equivalent formulation that is based on the concept of the propositional proof system.

Call a propositional proof system P *length-optimal* if for every propositional proof system Q, there exists a polynomial p such that for every tautology τ, if there is a proof of length n of τ in Q, then there is a proof of length at most $p(n)$ of τ in P. In plain words, if there exists a short proof of a tautology in any proof system, then there also exists a short proof of the tautology in P. Conjecture 4 is equivalent to the following statement:

Conjecture 4′ *There exists no length-optimal proof system.*

The proof of the equivalence of Conjectures 4 and 4′ is based on two constructions:

1. a construction of a propositional proof system for an arbitrary (sufficiently strong) consistent theory;
2. a construction of a theory for an arbitrary propositional proof system.

We already know that some theories can be naturally associated with propositional proof systems; but this can be done only for some special theories. Fortunately, for proving the equivalence, we do not need theories and propositional proof system to be associated so tightly. The nature of the statements that we want to prove to be equivalent is, so to speak, asymptotic. Thus we only need

1. for every theory a propositional proof system that is strong enough, to simulate all provable $\Pi_{\mathbf{P}}$-sentences;
2. for every propositional proof system a theory that is strong enough, to prove its soundness.

Such calculi and theories can be constructed quite easily. Given a theory T we define a propositional proof system P_T as follows. Let *taut*(x) be a formula expressing that x is a tautology. Then a proof of a tautology ϕ in P_T is simply a proof of *taut*(ϕ) in T. As concerns the second construction, given a propositional proof system P, we simply take a sufficiently strong base theory and extend it by the axiom saying that P is a sound proof system.

This enables us to switch between theories and proof systems. If S were a theory that could prove $Con_T(n)$ shortly for all finite consistent theories, then the proof system P_S would be length-optimal. Vice versa, if a proof system P were length-optimal, then the theory S with the axiom that P is a sound proof system would prove the sentences $Con_T(n)$ shortly. (For more details see the proof of the equivalence of related Conjectures 7 and 7′ in Notes.)

Conjecture 4 is not specific about how strong a theory T must be, given the theory S, and how big the lower bound is. The most common way of making a theory stronger is to add its consistency (which is unprovable in it according to the Second Incompleteness Theorem). Concerning the lower bound, we can conjecture that it is close to the best upper bound that we have, which is exponential. Thus we arrive at the following stronger conjecture (proposed by J. Mycielski when we were discussing Conjecture 4).

Conjecture 5 *Let S and T be finitely axiomatized consistent theories. If T proves the consistency of S, then the shortest S-proofs of $Con_T(n)$ have exponential lengths.*

Can Computational Complexity Cause Incompleteness?

The second kind of conjecture is about the provability of sentences of the form

$$\forall x \exists y\, \phi(x, y), \tag{6.11}$$

where ϕ represents a binary relation computable in polynomial time. In Chap. 4 (page 320) we saw concrete examples of independent sentences of this form. These sentences had the property that they described very fast growing functions and the independence of these sentences was caused by the fact that the theories were not able to prove the existence of such rapidly growing functions. In this chapter we saw (page 514) that one can define hierarchies of functions indexed by constructive ordinals, by which one can measure the strength of such sentences.

We met sentences of this form again in Sect. 6.2 of this chapter. There they were used to define **TPS**, the class of total polynomial search problems. Recall that *total* means that the sentence (6.11) is true (thus the sentence implicitly defines a function defined for all x) and *polynomial* means that the relation ϕ is decidable in polynomial time and, for a given x, the length of every y satisfying $\phi(x, y)$ is polynomially bounded by the length of x. I briefly mentioned the possibility that with increasing strength, theories can formalize more and more search problems. This is essentially the conjecture I am going to describe in more detail now.

The basic idea is as follows. We know that the unprovability of a sentence of the form $\forall x \exists y\, \phi(x, y)$ can be caused by the fact that the function it implicitly defines grows very fast. If the length of y is polynomially bounded by the length of x, then the unprovability cannot be caused by the growth of the implicitly defined function, but it is possible that it is caused by the *computational complexity* of the function.

Let x be a fixed number of length n, and suppose that the bound on the length of y is $p(n)$ for some polynomial p. Although the length of y is polynomially bounded, we only know that it is somewhere in an exponentially large set (namely, a set of size $2^{p(n)}$, if we consider binary strings). If one can only use polynomial time algorithms, it does not help that this set is finite; it is like exploring an infinite set. Therefore it is conceivable that there is an infinite hierarchy of total polynomial search problems, like there is an infinite hierarchy of fast growing functions, and that every theory can formalize search problems only up to certain level. If this is indeed the case, we can say that the incompleteness concerning such sentences is caused by computational complexity.

Recall that a theory T properly formalizes a search problem defined by a binary relation r if $r(x, y)$ is defined by a formula $\phi(x, y)$ from the class $\Phi_{\mathbf{P}}$ and T proves $\forall x \exists y\, \phi(x, y)$. We will say that a search problem R_1 is *strictly stronger* than a search problem R_2, if R_2 is polynomially reducible to R_1, but R_1 is not polynomially reducible to R_2. Now we are ready to state the conjecture.

Conjecture 6 *For every finitely axiomatized consistent theory T, there exists a total polynomial search problem Q which is strictly stronger than all search problems that T properly formalizes.*

To see that the conjecture is about incompleteness, consider T and Q with the properties stated in the conjecture. Then for any formula $\psi(x, y)$ that defines Q, the theory T does not prove $\forall x \exists y\, \psi(x, y)$.

But there is a caveat here: the fact that there exist unprovable sentences of this form is a trivial consequence of Gödel's Incompleteness Theorem. In fact, for every total polynomial search problem R, we can find a formula $\phi(x, y)$ which defines R such that $\forall x \exists y\, \phi(x, y)$ is not provable in T by incorporating the consistency of T into $\phi(x, y)$ (see Notes). The essence of the conjecture is that *the unprovability of $\forall x \exists y\, \psi(x, y)$ is caused by the complexity of Q*, which is expressed by saying that the sentence is unprovable for representations of Q by *any $\Phi_{\mathbf{P}}$-formula.*

Again, this conjecture has several interesting consequences in computational complexity theory. An immediate consequence of the conjecture is that the hierarchy of total polynomial search problems is infinite in the following sense: there is no total polynomial search problem to which one could reduce every total polynomial search problem. The latter statement implies $\mathbf{P} \neq \mathbf{NP}$. Furthermore, it has an equivalent stated only using the concepts of search problems and reducibilities.

Conjecture 6′ *There exists no complete problem among total polynomial search problems, that is, no problem to which total polynomial search problems are reducible.*

The equivalence of Conjectures 6 and 6′ is a consequence of the existence of a complete problem among those for which the condition of totality is provable in the theory T (see Notes).

The above conjecture concerns sentences with two quantifiers and a polynomial time computable relation. The natural question is whether one can state a conjecture of this kind also for sentences with only one quantifier, namely for $\Pi_{\mathbf{P}}$-sentences. I will present such a conjecture in the next subsection and give another example in Notes.

A Connection Between the Two Parts of the Thesis

We have seen two types of conjectures that are instances of the Feasible Incompleteness Thesis. At first they may seem very different. The first type of conjectures concern the lengths of proofs, whereas the second type is about unprovability. Yet they are not so different. I will give an example of a conjecture that is slightly weaker than Conjecture 4, but can also be stated in the form very similar to Conjecture 6.

Conjecture 7 *For every consistent finitely axiomatized theory S, there exists a consistent finitely axiomatized theory T such that S-proofs of sentences $Con_T(n)$ cannot be produced by an algorithm in time $p(n)$, for p a polynomial.*

Notice that Conjecture 4 gives a stronger reason why sentences $Con_T(n)$ are hard for S, namely, that their proofs are more than polynomially long. The above conjecture only says that it is difficult to find these S-proofs. Like Conjecture 4, it can also be stated in terms of propositional proof systems. To this end, define that a *propositional proof system P is optimal* if it polynomially simulates every propositional proof system. This property is, clearly, stronger than being length-optimal.

Conjecture 7′ *Optimal propositional proof systems do not exist.*

The fact that Conjectures 4 and 7 can be stated in terms of propositional proof systems (as Conjectures 4′ and 7′) shows that, in a sense, the hardness of sentences $Con_T(n)$ is caused by complexity. More precisely, it is caused by the existence of infinite hierarchies of propositional proof systems. But we can get a statement whose form is much closer to Conjecture 6 by blending the two versions of Conjecture 7 into one. Loosely speaking, the conjecture says that the complexity of propositional proof systems causes the unprovability of their soundness.

Conjecture 7″ *For every finitely axiomatized consistent theory T, there exists a propositional proof system P such that T does not prove the soundness of any formalization of P by a $\Phi_{\mathbf{P}}$-formula.*

Recall that soundness means that all propositional formulas provable in the proof system are tautologies. The soundness of P can be expressed by the $\Pi_{\mathbf{P}}$-sentence:

> *For every P-proof d, every propositional formula ϕ and every truth assignment a to the variables of ϕ, if d is a proof of ϕ, then ϕ is satisfied by a.*

Thus we have a natural conjecture about the unprovability of $\Pi_{\mathbf{P}}$-sentences different from Gödel's sentences. (For the proof of the equivalence of these conjectures, see Notes.)

The diagram below shows the known dependencies between the conjectures.

$$\begin{array}{ccccc} 2 \longrightarrow & 1 \Leftrightarrow 1' & \longrightarrow & \mathbf{NP} \neq \mathbf{coNP} & \\ & \downarrow & & \downarrow & \\ 5 \Leftrightarrow 5' \longrightarrow & 4 \Leftrightarrow 4' \Leftrightarrow 4'' & \longrightarrow & \mathbf{P} \neq \mathbf{NP} & \longleftarrow 3 \Leftrightarrow 3'' \end{array}$$

The subject of feasible incompleteness has not received the attention it deserves. Most researchers are convinced that solving the big open problems in complexity theory, such as proving $\mathbf{P} \neq \mathbf{NP}$, can be achieved by developing better combinatorial methods or using other parts of mathematics which have deeper theories, such as algebraic geometry, group representation theory etc. But we have seen in this chapter that computational complexity has many connection with logic and in particular with proof complexity. The connections with feasible incompleteness show a link to the foundations of mathematics. In light of this evidence it is conceivable that the open problems in complexity theory may require a fundamentally different approach. If one wants to embark on this project, the natural starting point is the study of feasible incompleteness conjectures.

Notes

1. *Lower bounds on the proof complexity of $Con_T(n)$—proof of Theorem* 58. We will assume that T is finitely axiomatized, consistent and sufficiently strong for formalizing syntax. The following notation comes in handy for this proof. For a sentence ϕ, we will denote by $\|\phi\|_T$ the length of the shortest proof of ϕ in T, if there is one, and otherwise $\|\phi\|_T = \infty$. Further, we will denote by $\bar{n}$ the binary numeral representing a natural number n. The binary numeral is the term of length approximately $\log n$ defined as follows. If we have the successor functions s_0 and s_1, then we use the representation given by the formula (6.6), page 542. Otherwise we replace the successor functions by the terms that define them and use the same formula, or use a formula that defines this expression, in case there are no such terms. We will also assume that the rule of *modus ponens* is in the proof system for first order logic considered here.

   Now the theorem we want to prove is stated simply as the following inequality for all n:

$$\left\| Con_T(\bar{n}) \right\|_T > n^{\varepsilon}, \tag{6.12}$$

   where $\varepsilon > 0$ is a constant. However, instead of this, we will only prove a closely related inequality

$$\left\| Con_T\left(\bar{n}^c\right) \right\|_T > n - c' \log n, \tag{6.13}$$

for some constants c and c'. Intuitively this is the same as (6.12) with ε smaller than $\frac{1}{c}$, but formally it needs a little argument which I am omitting.

The proof is based on the Diagonal Lemma 2 (page 291). According to that lemma, one can construct a formula $\delta(x)$ such that the sentence

$$\forall x \left(\delta(x) \equiv \left\|\delta(\bar{x})\right\|_T > x\right) \tag{6.14}$$

is provable in T.

The first step is to prove that $\delta(\bar{n})$ is true for every n which is, by (6.14),

$$\left\|\delta(\bar{n})\right\|_T > n. \tag{6.15}$$

Indeed, suppose the contrary $\|\delta(\bar{n})\|_T \leq n$. Then in particular, $\delta(\bar{n})$ is provable in T. By Σ completeness, $\|\delta(\bar{n})\|_T \leq n$ is provable in T. On the other hand, since T proves $\delta(\bar{n})$, it also proves $\|\delta(\bar{n})\|_T > n$ according to (6.14). Thus T is inconsistent. Since we are assuming that T is consistent, we have obtained a contradiction, hence (6.15) must be true.

The part of the proof that we have just done corresponds to the proof of the First Incompleteness Theorem. As in the proof of the Second Incompleteness Theorem, we will now formalize the above proof in T. However we need more—we need also to estimate the lengths of proofs of the formulas occurring in the proof.

When proving (6.15), we have shown that the assumption $\|\delta(\bar{n})\|_T \leq n$ implies that T is inconsistent. One can estimate the lengths of proofs and show that if $\delta(\bar{n})$ has a proof of length at most n in T, then a contradiction has length at most n^c for some constant c; written formally

$$\left\|\delta(\bar{n})\right\|_T \leq n \;\rightarrow\; \|0=1\|_T \leq n^c.$$

Let us look at the contrapositive implication where the negation of the consequent is replaced by consistency up to n^c

$$Con_T\left(\bar{n}^c\right) \;\rightarrow\; \left\|\delta(\bar{n})\right\|_T > n.$$

Using (6.14) we can replace the consequent by $\delta(\bar{n})$. Thus we get

$$Con_T\left(\bar{n}^c\right) \;\rightarrow\; \delta(\bar{n}). \tag{6.16}$$

Since $\delta(\bar{n})$ follows from $Con_T(\bar{n}^c)$ and the sentence above by *modus ponens*, we have

$$\left\|\delta(\bar{n})\right\|_T \leq \left\|Con_T\left(\bar{n}^c\right)\right\|_T + \left\|Con_T\left(\bar{n}^c\right) \rightarrow \delta(\bar{n})\right\|_T.$$

We can also write it as

$$\left\|\delta(\bar{n})\right\|_T - \left\|Con_T\left(\bar{n}^c\right) \rightarrow \delta(\bar{n})\right\|_T \leq \left\|Con_T\left(\bar{n}^c\right)\right\|_T.$$

We already have a lower bound on $\|\delta(\bar{n})\|_T$, (6.15), so in order to prove a lower bound on $\|Con_T(\bar{n}^c)\|_T$, it remains to show an upper bound on $\|Con_T(\bar{n}^c) \rightarrow \delta(\bar{n})\|_T$. The reason why it has a proof of logarithmic length is that (6.16) is provable in T as the universal sentence

$$\forall x\left(Con_T\left(\bar{x}^c\right) \;\rightarrow\; \delta(\bar{x})\right).$$

Thus to prove a numeric instance of it, we only need to substitute the numeral representing n, which has logarithmic length. This finishes the proof of (6.13).

2. *Upper bounds on the proof complexity of $Con_T(n)$.* For stating the upper bound on the lengths of proofs of finite consistencies we need the concept of a *sequential theory*. Essentially, a theory T is sequential if there are formulas for coding and decoding finite sequences of arbitrary objects of the theory. The basic axioms these formulas must satisfy are, stated informally:

   a. an empty sequence has a code,
   b. given a sequence x and an object y, one can code a sequence that is the prolongation of x by object y.

   More important than the details of the definition is the fact that it is a natural property satisfied by many theories.

   Theorem 59 [223] *For every finitely axiomatized sequential theory T, there exists a constant c such that for all n,*

$$\left\| Con_T(\bar{n}) \right\|_T \leq n^c.$$

   The proof of the theorem uses the fact that in sequential theories it is possible to define *partial truth predicates.* A partial truth predicate is a formula that defines truth for a subset of all formulas. (Recall that by the Gödel-Tarski Theorem 4.2 it is not possible to define truth for all formulas inside a theory.) Given a number n, it is possible to construct a formula $Tr_n(x)$ that defines truth for formulas of length at most n. The size of $Tr_n(x)$ and the proofs of its basic properties are polynomial in n. Specifically, there are polynomial size proofs of the sentences expressing:

   - if ϕ is an axiom of length at most n, then it is true;
   - if ϕ and $\psi_1, \ldots, \psi_k$ are sentences of length at most n and ϕ follows from $\psi_1, \ldots, \psi_k$ by a single application of a logical rule, then if $\psi_1, \ldots, \psi_k$ are true then also ϕ is true;
   - $\phi \wedge \neg\phi$ is not true.

   From this, we get polynomial size proofs of $Con_T(m) \to Con_T(m+1)$ for all $m < n$. Noting that $Con_T(0)$ is trivially provable, we get a polynomial size proof of $Con_T(n)$.

   The gap between the lower and upper bounds has been reduced to the factor $c(\log n)^2$, where c is a constant, for a particular proof system for first-order logic [224].

3. *Theories with infinite sets of axioms.* The lower bound on $Con_T(n)$ can be generalized to theories with an infinite number of axioms. The only reason for using finitely axiomatized theories was that we needed the proof system to be defined by a polynomial time decidable relation. In order for the relation '*d is a proof of sentence ϕ*' to be decidable in polynomial time, we do not need a finite set of axioms; it suffices to have a set of axioms such that there exists a polynomial time algorithm for deciding if a sentence is an axiom. If this condition is

satisfied and the proof relation is formalized by $\Phi_{\mathbf{P}}$-formulas, then the proof goes through. In particular *PA* and *ZFC* have sets of axioms defined by simple syntactical conditions, hence decidable in polynomial time.

The upper bound on $Con_T(n)$ can be generalized to theories that are axiomatized by a finite number of axiom schemas. Thus we get polynomial upper bounds for *PA* and *ZFC*.

4. *Conjecture 4 in purely computational terms.* I have mentioned that Conjecture 4 implies some separations of pairs of complexity classes. In particular we know the following implications

$$\text{Conjecture 4} \rightarrow \mathbf{NEXP} \neq \mathbf{coNEXP} \rightarrow \mathbf{NP} \neq \mathbf{coNP} \rightarrow \mathbf{P} \neq \mathbf{NP}.$$

Thus **NEXP** $\neq$ **coNEXP** is the strongest separation among the three; yet it does not seem to be equivalent to the conjecture. To state the computational version of the conjecture we need the following concept. We say that a set X of strings in a finite alphabet is *polynomially sparse* if there exists a polynomial $p(x)$ such that for every number n, the number of strings of length n is at most $p(n)$.

The following statement is equivalent to Conjecture 4:

Let A be a **coNP**-*complete set. Then there is no nondeterministic Turing machine M with the following properties*:

a. *M accepts A* (*i.e., A is the set of strings accepted by M*);
b. *for every polynomially sparse set* $X \subseteq A$, $X \in \mathbf{P}$, *there exists a polynomial* $q(x)$ *such that M accepts every string* $w \in X$ *in time* $q(|w|)$.

The inequality **NP** $\neq$ **coNP** is equivalent to the statement that a **coNP**-complete set cannot be accepted by a nondeterministic Turing machine in polynomial time. The above statement strengthens it by saying that even some sparse subsets of A that are in **P** cannot be accepted by such a machine in polynomial time.

Although rather complicated, this version of the conjecture has an advantage—since it is stated purely in terms of Turing machines and computations, it can be relativized. An oracle has been constructed with respect to which the conjecture is true [164]. Thus we have some evidence supporting Conjecture 4.

5. *Making formulas artificially stronger.* Suppose a total polynomial search problem P is defined by a formula $\forall x \exists y\, \phi(x, y)$ and let T be a consistent finitely axiomatized theory. We can make a formula that defines the same search problem and that is unprovable in T by forming the conjunction of $\forall x \exists y\, \phi(x, y)$ with Con_T. If we write this as $\forall x \exists y\, \phi(x, y) \wedge Con_T$, it does not look like a definition of a search problem, but we can express it equivalently by:

$$\forall x \exists y \left(x \text{ is not a proof of contradiction in } T \wedge \phi(x, y)\right).$$

This shows that it is very important which formula we use to represent a search problem in a theory T.

6. *"Syntactical" and "semantical" complexity classes.* Complexity theorists distinguish complexity classes that are defined syntactically and those that are defined semantically. A class $\mathcal{C}$ is syntactically defined if, for a Turing machine M, we

can decide whether or not it defines an element of $\mathcal{C}$ "simply by looking at it", whereas if $\mathcal{C}$ is only defined semantically, we have to "test M for what it actually does". I propose the following definition instead of this vague description.

Definition 20 A class $\mathcal{C}$ is *syntactically* defined if there exists an algorithmically decidable set of Turing machines $\mathcal{M}$, such that for every X, $X \in \mathcal{C}$ if and only if there exists an $M \in \mathcal{M}$ such that M defines X (i.e., X is the set of inputs accepted by M). Otherwise the class is defined *semantically*.

Examples of syntactical classes are **P** and **NP**. The standard definition of **P** is not syntactical, since it is algorithmically undecidable if a Turing machine runs in polynomial time. Therefore, for the class **P**, we take $\mathcal{M}$ to be Turing machines *equipped with a clock running in polynomial time*. This means that the program for the Turing machine contains a simple routine that counts the number of steps and stops the machine after polynomial time, disregarding whether or not the other computations are finished. The set of Turing machines for **NP** is defined in the same way except that we use nondeterministic machines. A class that is not known to be syntactical is **BPP**. (But if the conjecture **BPP** = **P** is true, then it is syntactical.)

The rule of thumb is that syntactical classes possess complete sets, while semantical classes do not. Thus there are **NP**-complete sets but no **BPP**-complete sets are known.

One can easily show that syntactically defined classes have a property that links this concept with provability in theories.

Proposition 12 *If $\mathcal{C}$ is syntactically defined, then there exists a true theory T such that for every $X \in \mathcal{C}$, there exists a Turing machine M such that M defines X and T proves the sentence* 'the set of inputs accepted by M is in $\mathcal{C}$'.

If $\mathcal{C}$ is not syntactically definable, then it is possible that there is no such theory. If that is the case, it still does not imply that the sets in $\mathcal{C}$ form an infinite hierarchy with respect to the quasiordering by polynomial reductions; there may exist a complete problem in $\mathcal{C}$. The conjecture that they do form an infinite hierarchy can be viewed as an instance of the Feasible Incompleteness Thesis. Conjectures 4′ and 6′ are statements of this form, except that they talk about search problems (objects determined by binary relations) rather than sets. Next is another such example.

7. *A conjecture about $\Pi_{\mathbf{P}}$-sentences.* An interesting concept related both to computational complexity and proof complexity is the concept of pairs of disjoint **NP** sets. The words define this concept quite precisely—we consider pairs (A, B), where $A \cap B = \emptyset$ and $A \in \mathbf{NP}$, $B \in \mathbf{NP}$. We say that a pair (A, B) is *polynomially reducible* to a pair (A', B') if there exists a polynomial time computable function f which maps A into A' and B into B'. We say that a pair (A, B) is *polynomially separable* if it is polynomially reducible to the pair $(\{0\}, \{1\})$.

In complexity theory such problems are called '*promise problems*'. The "promise" is that we only get inputs from A or from B. The problem is to decide from which set the input comes. The promise is what makes the definition of the class of problems "semantical".

Clearly, if (A, B) is polynomially reducible to (A', B') and (A', B') is polynomially separable, then so is (A, B). We conjecture that there are pairs that are not polynomially separable. This conjecture is a consequence of the conjecture $\mathbf{NP} \cap \mathbf{coNP} \neq \mathbf{P}$. The latter conjecture, in turn, follows from some cryptographic conjectures, such as the conjecture that there exist one-way permutations.

The following is a stronger conjecture than the conjecture about the existence of polynomially inseparable pairs.

Conjecture 8 *There is no complete disjoint* **NP** *pair. In more detail, there is no pair of disjoint* **NP** *sets* (C, D) *such that every pair of disjoint* **NP** *sets is polynomially reducible to* (C, D).

This conjecture is supported by a result of C. Glaßer, A. Selman, S. Sengupta and L. Zhang [94] that there exists an oracle with respect to which the conjecture is true. An equivalent conjecture concerning provability in theories is the following:

Conjecture 8′ *For every finitely axiomatized consistent theory* T*, there exists a pair of disjoint* **NP** *sets* (A, B) *such that for no formal definitions of the sets* A *and* B *is it provable in* T *that* A *is disjoint from* B.

This is similar to Conjecture 6, but the sentences it concerns are $\Pi_{\mathbf{P}}$. Indeed, let $\exists y\, \phi(x, y)$ be a definition of A and $\exists y\, \psi(x, y)$ be a definition of B. Then the disjointness of A and B can be expressed by:

$$\forall x \forall y \forall z\, \big(\neg\phi(x, y) \vee \neg\psi(x, y)\big).$$

Since $\phi(x, y)$ and $\psi(x, y)$ define polynomial time computable relations, this sentence is $\Pi_{\mathbf{P}}$.

One can show that Conjecture 8 implies Conjecture 7.

8. *The equivalence of Conjectures* 6 *and* 6′.

(6) ⇒ (6′). This direction is easy. Suppose Conjecture 6′ is false. Let S be a complete **TPS**. Let T be a sufficiently strong finitely axiomatized fragment of arithmetic or set theory augmented with the axiom expressing that R is total. Then such a T shows that Conjecture 6 is false.

(6′) ⇒ (6). This implication requires us to be more specific about how we formalize **TPS** problems. Suppose a search problem S is defined using a binary relation $R(x, y)$, where x's are inputs and y's are potential solutions. Further we have a polynomial bound p on the lengths of solutions y, namely, we require $|y| \leq p(|x|)$ for some polynomial p. There is one more implicit polynomial bound, which is the bound on the running time of the Turing machine computing the relation R. When S is formalized in some theory, we also need that the theory is able to prove the inequality $|y| \leq p(|x|)$ for the solutions and the polynomial upper bound on the running time of the Turing machine.

For **TPS** problems S_1 and S_2, we will use the inequality $S_1 \leq S_2$ to denote that S_1 is polynomially reducible to S_2. An easy fact is that for every S_1 and S_2, there exists an S such that $S_1 \leq S$ and $S_2 \leq S$. Simply take the product of the relations defining S_1 and S_2. Slightly less trivial is the following lemma.

Lemma 16 *For every finitely axiomatized consistent theory T, there exists a* **TPS** *problem S_T, such that every* **TPS** *problem properly formalized in T is polynomially reducible to it.*

Proof. We will define the binary relation $R(x, y)$ for S_T. Let us view x as consisting of three parts. The first part is an input z to a search problem; the second part is a description of a search problem S and a T-proof that it is total. The third part is an arbitrary string whose purpose is to pad the input to a sufficient size so that its length bounds the length of the solutions of S and the time needed to compute the binary relation of S. If x has such a form, then we define $R(x, y)$ to be true if y is a solution for z in S. If x is not of this form, then we define $R(x, y)$ to be true for all y. I leave to the reader to verify all the conditions that R has to satisfy. □

With this lemma in hand, the rest of the proof is easy. Suppose Conjecture 6′ is true. Let T be a finitely axiomatized consistent theory. According to our assumption, there exists a **TPS** problem Q' which is not reducible to S_T. Let Q be a **TPS** problem such that both $Q' \leq Q$ and $S_T \leq Q$. Then Q is also not polynomially reducible to S_T, due to the transitivity of $\leq$. Again, by transitivity, if P is properly formalized in T, then $Q \not\leq P$. So Q has all the properties required by Conjecture 6.

9. *The equivalence of Conjectures* 7, 7′ *and* 7″. We will consider finitely axiomatized consistent theories. Furthermore, we will assume that they are sufficiently strong. Consistency and the latter condition imply that such theories prove only true $\Pi_{\mathbf{P}}$ sentences.

Recall that, given a theory T satisfying the conditions above, one can construct a propositional proof system P_T which simulates T for $\Pi_{\mathbf{P}}$ sentences. This means that for a $\Pi_{\mathbf{P}}$ sentence ϕ, one can define a sequence of tautologies τ_n, $n = 1, 2, \ldots$ expressing the truth of the finite instances of ϕ such that these sentences have polynomial size proofs in P_T.

Given a propositional proof system P, one can easily construct a theory T_P that proves the soundness of the system. This means that for a formula $pr(x, y)$ formalizing '*y is a P-proof of x*' and a formula $taut(x)$ formalizing '*x is a tautology*', T proves the sentence $\forall x \forall y\, (pr(x, y) \rightarrow taut(x))$.

I will now outline the proofs of implications between these conjectures.

(7′) ⇒ (7). Suppose (7) is false and let S be the theory that witnesses this fact. We will show that the propositional proof system P_S is optimal.

Let P be an arbitrary propositional proof system. Let d be a P-proof of a tautology ϕ. According to the definition of P_S, it suffices to construct in polynomial time an S-proof of the sentence $taut(\bar{\phi})$.

We will now argue in T. "*Suppose* ϕ *were not a tautology; let a be a falsifying assignment of* ϕ. *Then* T_P, *the theory associated with* P, *would prove that* ϕ *is falsified by a (since* T_P *is sufficiently strong, it satisfies* Σ*-completeness). The length of this proof is polynomial in the length of* ϕ, *which is at most the length of d. Also by* Σ*-completeness,* T_P *proves that d is a* P*-proof of* ϕ. *This proof of polynomial length in the length of d. But* T_P *proves the soundness of* P, *thus we get that* T_P *proves a contradiction by a proof whose length is polynomial in the length of d.*"

The above proof can actually be constructed in polynomial time. Since we also assume that we can construct in polynomial time S-proofs of $Con_{T_P}(\bar{n})$, we can construct an S-proof of *taut*$(\bar{\phi})$, hence a P_S-proof of ϕ. This algorithm runs in time that is polynomial in the length of d. Thus it is a polynomial simulation of the proof system P by the proof system P_S. This finishes the proof that P_S is optimal.

(7″) ⇒ (7′). Suppose (7′) is false. Let Q be an optimal propositional proof system. We will show that T_Q, the theory that contains an axiom that Q is sound, falsifies (7). The proof system Q is defined by a polynomial time relation $R(x, y)$ that defines that y is a proof of x. We are assuming that R is $\Phi_{\mathbf{P}}$-formalized in T_Q.

Let P be an arbitrary propositional proof system. Since Q polynomially simulates all propositional proof systems, there exists a polynomial time computable function f such that d is a P-proof of ϕ if and only if $R(\phi, f(d))$. Hence we can define the proof system P using the polynomial time computable relation $R(x, f(y))$. To obtain a $\Phi_{\mathbf{P}}$-formalization of P in T_Q, use the formalization of $R(x, y)$ and a $\Phi_{\mathbf{P}}$-formalization of the function f. Since T_Q proves the soundness for the formalization of $R(x, y)$, it also proves soundness for the formalization of $R(x, f(y))$.

(7) ⇒ (7″). Suppose (7″) is false. Let S be a theory that falsifies (7″). We will show that it also falsifies (7).

Let T be a consistent finitely axiomatized theory. Let τ_n be the propositional translations of $Con_T(\bar{n})$. We can define these translations so that S proves that $Con_T(\bar{n})$ is equivalent to *taut*(τ_n) by proofs that can be constructed in polynomial time. Let P be an arbitrary propositional proof system (say, a Frege system). Extend P by postulating that τ_n is a proof of itself. Let P' be the extended propositional proof system. According to our assumption, S proves the soundness of P'. Hence to prove that τ_n is a tautology in S, we only need to verify that it has the right form. This only requires polynomial size proofs that can easily be constructed. Hence proofs of $Con_T(\bar{n})$ can also be constructed in polynomial time.

10. *Impagliazzo's worlds and other scenarios.* Since we do not know answers to basic problems in complexity theory, we have to be prepared for possibilities of how the true relations between the complexity classes can be. In 1995 Russell Impagliazzo attempted to organize the possible scenarios into five "worlds" for which he also invented suitable names: *Algorithmica, Heuristica, Pessiland, Minicrypt, Cryptomania,* [135].

Algorithmica is the world in which $\mathbf{P} = \mathbf{NP}$ (or at least $\mathbf{NP} \subseteq \mathbf{BPP}$). This is the best world for somebody who is concerned with efficient computations. In this world not only can one solve all **NP** problems, but also polynomial search problems, find optimal solutions, and more.

The other extreme, *Cryptomania,* is a world in which there are hard problems, but which offers a different advantage: one can do all the tricks in cryptography that one needs. This is the world that most people assume is ours. To prove it, it would suffice to show that factoring is hard; however, some experts doubt that the latter is true.

But there is also a bad alternative, *Pessiland* (the worst land), where you can do neither: you cannot compute efficiently problems in **NP**, and you cannot do cryptography either. The hard problems in this land cannot be used to conceal messages; in particular, one-way functions do not exist.

The other two are intermediate worlds. In *Heuristica* $\mathbf{P} \neq \mathbf{NP}$, but one can still solve random instances of **NP** problems efficiently with high probability. In *Minicrypt* one can do some basic cryptography, because there are one-way functions and pseudorandom generators, but there are no secure public-key systems.

Impagliazzo's classification focuses on cryptography and comparison of worst-case complexity with average-case complexity. Different point of views would, clearly, lead to a different sets of worlds. The following is a natural classification from the point of view of proof complexity:

a. $\mathbf{P} = \mathbf{NP}$; in this world it is decidable in polynomial time if a proposition is a tautology. Consequently, tautologies have polynomial size proofs that can be constructed in polynomial time. Also for first order logic there is an efficient algorithm to find a proof of a logically valid sentence, if a short proof exists.
b. $\mathbf{P} \neq \mathbf{NP}$ *and* $\mathbf{NP} = \mathbf{coNP}$; here tautologies have polynomial size proofs, but we do not have a polynomial time algorithm to find these proofs.
c. $\mathbf{NP} \neq \mathbf{coNP}$ *and there exists a propositional proof system that polynomially simulates all propositional proof systems.* For all we know, such a system can be a Frege system. If we know the optimal proof system P, then it makes no sense to use other systems.
d. $\mathbf{NP} \neq \mathbf{coNP}$ *and there exists a length-optimal proof system, but not one that polynomially simulates all proof systems.* Here it is reasonable to try different proof systems when proving a tautology.
e. *There does not exist a length-optimal proof system.* (This implies $\mathbf{NP} \neq \mathbf{coNP}$.)

As before, also here most researchers believe that our world is the most complex one, which is the last world in the list.

Main Points of the Chapter

- Every proof in first order logic can be transformed into a direct proof, but it may expand superexponentially.

- Theorems may have much shorter proofs if we

  1. either add a new axiom to the axiomatic system that we use,
  2. or instead of proving the theorem, we only prove that the theorem is provable.

  The second way is an instance of what is called "considering from a higher perspective", but the first one is more effective.
- A theory can be useful even if it is inconsistent because it can happen that contradiction has a very long proof and all sentences provable by short proofs are true.
- Countable ordinals can be used to gauge the strength of theories.
- There are many analogies between computational complexity and proof complexity. In particular, there are theories that we view as counterparts of particular complexity classes, such as **P** and **NP**.
- Proof complexity, although closely related to computational complexity, is nevertheless a different theory. In particular, the open problems of proof complexity are not mere reformulations of problems from computational complexity.
- The main problems in proof complexity are whether or not certain pairs of theories prove the same theorems. In some cases it is possible to derive separations of pairs of theories using conjectures about separations of pairs of complexity classes.
- The problem of separating theories can often be reduced to proving lower bounds on the lengths of proofs in the propositional calculus. Therefore, problems about the lengths of proofs in proof systems for the propositional calculus also belong among the central problems in proof complexity.
- According to the Feasible Incompleteness Thesis, the phenomenon of incompleteness should already manifest itself at the level of polynomial length proofs and polynomial time computations.
- Several conjectures that can be viewed as instances of the Feasible Incompleteness Thesis have been put forward. They have various implications in computational complexity; in particular $\mathbf{P} \neq \mathbf{NP}$ would be a consequence of some of these conjectures.
- This connection hints that the essence of the big open problems in complexity theory could be logical, rather than combinatorial.

Chapter 7
Consistency, Truth and Existence

And God created every living creature that now moveth, and one was man. Mud as man alone could speak. God leaned close as mud as man sat up, looked around and spoke. Man blinked.
"What is the purpose *of all this?" he asked politely.*
"Everything must have a purpose?" asked God.
"Certainly," said man.
"Then I leave it to you to think of one for all this," said God.
Kurt Vonnegut, Jr. *Cat's Cradle*

IN previous chapters I surveyed mathematical results that were often only loosely connected with the foundations of mathematics. I will now turn to questions more intimately connected with foundations. These questions have already appeared at several places in the book, but they were discussed only briefly.

At the end of the 19th century, when the formalization of mathematics started, it seemed quite possible that a complete axiomatization of the whole of mathematics could be achieved. If this were accomplished, we would have one set of axioms for everything that we would ever need in mathematics. It also seemed possible to prove the consistency of such a formal system. Hilbert was among the first who realized the importance of these problems and he promoted the program whose aim was to find such an axiomatization and to prove its consistency. As we know, these hopes were dashed by Gödel's Incompleteness Theorem. So we may need new axioms to solve open problems now, or in the future, and we always have to worry about the consistency of our formal systems. But what should we do about it? Stated more explicitly, we need answers to the following two fundamental questions:

1. How can we find new mathematical principles (axioms) and recognize that they are true?
2. Is contemporary mathematics safe from contradictions?

If Hilbert's program did achieve its goals, these two questions not only would have been answered, but we would also have *mathematical proofs* justifying the answers. Hilbert envisioned that we would *prove* that certain axiomatization of numbers and sets of numbers were complete and thus we would not need to look for new axioms. Furthermore, he expected that we would also *prove* that the axiomatization was consistent, answering the second question once for ever.

P. Pudlák, *Logical Foundations of Mathematics and Computational Complexity*, Springer Monographs in Mathematics, DOI 10.1007/978-3-319-00119-7_7,

As the questions stand now, it seems that we cannot solve them as mathematical problems. Not having a mathematical answer to these questions, we can only build systems of beliefs that support particular answers to these questions. In other words, we have to turn to philosophy. Once we allow metaphysical considerations we immediately arrive at another important question, closely related to the above two:

3. Do mathematical entities exist in reality?

How is this related to the questions above? Suppose we are studying some mathematical structure M, such as the natural numbers, and we propose a theory T for it. If we believe that M satisfies the axioms of T and we believe that M is real, then we also know that T is consistent because reality is consistent; consistency is a basic attribute of reality. Assuming, moreover, that M is a uniquely determined structure, there is a unique way how to extend T, if T is not complete. Hence if the Incompleteness Theorem applies to theories describing M, as is the case with natural numbers, we cannot fully formalize the truth about M, but we know that for each sentence ϕ, one of the sentences ϕ and $\neg\phi$ is an axiom that we can add. This does not tell us *how to find* extensions of T that are true in M, but at least we know that there is an answer to the question above for every sentence ϕ.

In fact, question 1. is also metaphysical because it talks about truth. One has to assume reality if one claims that an axiom is true. The need for new methods, principles, or axioms is felt also by those who reject the reality of abstract mathematical concepts. Instead of studying reality, it is possible to assume a pragmatic view and just look for useful axioms. But it is clear that the way one approaches this problem influences the way one looks for new axioms.

While mathematical reality is elusive, most people believe in physical reality. But mathematics is used to describe physical reality with great success, so the two realities must be connected. Therefore I consider the following question also to be one of the most fundamental problems.

4. How is physical reality connected with mathematics?

I will start this chapter with a brief survey of some schools in the philosophy of mathematics and explain how they approach the first three problems. In the second section I will discuss some mathematical results that are relevant to these questions. I will mention some relatively simple, but important observations about logical theories, and several deep results in set theory, that have bearings on these questions. In the last section I will address question 4. I will not attempt to give a comprehensive answer; I will only state some basic questions and offer a view that is radically different from the standard one. Its appeal is in its connection with computational complexity.

7.1 Consistency and Existence

Geometrical Foundations

During the history of mathematics the role of the most fundamental concept alternated between the natural numbers and the geometry of space. For Pythagoreans, numbers were not only the most basic concept of mathematics, but they were also attributed with the mystical role of being the essence of everything. In contrast, Euclid's approach to the foundations was geometrical. This is clear from the fact that the first four books of *Elements* are completely devoted to geometry. Euclid used three notions that correspond to numbers as we view them today: *magnitudes, ratios* and *numbers*. Magnitudes are lengths of lines, angles, areas and volumes. For us, they are just positive real numbers. Ratios are ratios of two magnitudes. A contemporary mathematician may be puzzled why ratios are entities different from magnitudes. A physicist is, however, more likely to guess the true reason: Euclid connected mathematical concepts with physical reality. Magnitudes have physical *dimension*, which means that they are measured in physical units, such as meters, square meters, kilograms etc., while ratios are dimensionless, so their nature is different. Numbers are defined in *Elements* by choosing a *unit* and saying that a number is what is composed of units. So they are positive integers—natural numbers without zero. We can think of numbers as certain lines and use geometry to derive arithmetical theorems.

One would expect that having made the connection between natural numbers and geometry, Euclid derived all arithmetic from geometry. Unfortunately, the physical interpretation of magnitudes prevented him from doing so. He did note that addition of numbers corresponds to connecting two lines, but he did not interpret multiplication as an operation producing a line from two lines. The reason is that the product of two lines has a different physical dimension, thus it should be associated with *area*, rather than length. The mutual interpretations of arithmetic and geometry, though based on easy constructions that Euclid knew, were explicitly stated by Descartes much later.

What is important for us here is that Euclid's contemporaries did not have a problem with the foundations of mathematics. Although the interpretation of arithmetic in geometry does not explicitly appear in *Elements*, it is clear that they believed that they could fully rely on geometry. Since geometry was viewed as a science about physical space, mathematical entities were also part of the physical world. Assuming such geometrical foundations, all questions about meaning, consistency, existence etc., can be easily answered by pointing to physical reality.

Due to the authority of Euclid, the geometric view of numbers persisted for a long time. For example, when explaining basic arithmetical concepts at the beginning of *Arithmetica Universalis*, Isaac Newton mentions a geometric construction of the product of two quantities (see also Fig. 7.1):

> "And so if you were to multiply any two Lines, *AC* and *AD*, by one another, take *AB* for Unity, and draw *BC*, and parallel to it *DE*, and *AE* will be the Product of this Multiplication; because it is to *AD* as *AC*, to Unity *AB*, ..."[1]

In the 19th century, when the first systematic treatments of the foundations of mathematics appeared, mathematics differed from ancient Greeks' mathematics in an essential way. New kinds of geometries had been discovered and mathematicians

[1] *Universal Arithmetick* (English translation of *Arithmetica Universalis*), (1720), [208], page 4.

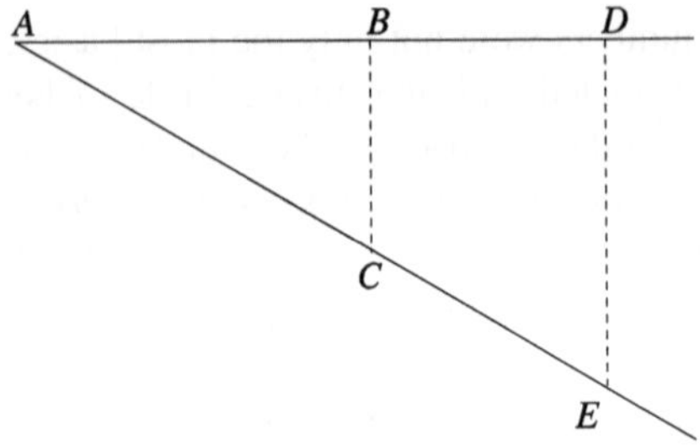

Fig. 7.1 Multiplying two real numbers using geometry

realized that physical space did not have to be Euclidean. Furthermore, a lot of new concepts had been defined, many of which had little to do with geometry. In particular, algebraic structures were more related to the natural and real numbers than to geometry. Therefore it seemed more natural to base the foundations on arithmetic. When geometry is replaced by arithmetic, one can still keep the connection with reality; after all, the natural numbers are precisely the concept invented for counting physical objects. However, what most philosophers of mathematics wanted was not to explain numbers by physical reality, but conversely, *to explain physical reality by mathematics.* Thus the natural numbers had to be treated as a primitive concept.[2]

Research in the logical foundations at the beginning of the 20th century showed that the natural numbers and pure logic (more precisely, first order logic) do not suffice—we also need sets. But natural numbers can be defined using sets, so sets alone suffice for the foundations. Unfortunately, if we base the foundations on sets, we completely lose any connection with the physical world and we have to turn to metaphysics.

A Platonist's View of Mathematics

In philosophy *realism* is the belief that there is a reality that is independent of the human mind. When talking about physical entities that we can observe using our senses, most people would agree with realism. The problems arise when one asks about the existence of abstract concepts. For example, we may all agree that a particular chair in the room exists, but does the abstract idea of a chair exist? Mathematics only studies abstract entities which makes accepting realism more difficult.

Since the publication of *On Platonism in Mathematics* [22] by Paul Bernays many authors have used the term '*platonism*' for '*mathematical realism*', although others consider platonism to be only a branch of it. Here I will use the term '*platonism*' in the more general sense.

Platonism is attractive for several reasons. Assuming that the axioms describe a real set-theoretical universe, the problem of their consistency is eliminated—in

[2]Ironically, Frege, who contributed so much to the foundations based on numbers, seeing his project fail, also contemplated geometrical foundations (see [30], page 147).

reality there cannot be an inconsistency because either a sentence is true or it is false, but not both. Another very important consequence of this assumption is that the main mathematical structures are uniquely determined. In particular, the natural numbers, as a mathematical structure, are unique. Therefore the object of the study of number theory is absolutely clear. Real numbers are also uniquely determined. Problems start when one asks about the cardinality of the set of real numbers. The axioms of Zermelo-Fraenkel Set Theory are not able to decide what this cardinality is. (This is the problem of the validity of the Continuum Hypothesis mentioned in Chap. 3.) Similar problems arise in the study of the properties of subsets of real numbers.

Logicians have another reason for accepting platonism: the *inner consistency* of this belief. Recall that in order to avoid semantic paradoxes, one has to distinguish between the object language, the language that we study, on the one hand, and the metalanguage, the language in which we reason about the object language, on the other. So when we are talking about sentences in set theory, the object language is the language of set theory and the metalanguage is usually English. Metalanguage can also be a formal system. For instance, we may reason about Zermelo-Fraenkel Set Theory in itself. The two languages may be the same, but we should distinguish the two levels: the object level and the level of discourse. When we prove theorems about formal systems, we apply mathematics. Thus we have to use some set theory (or arithmetic in which syntax can also be formalized). From the point of view of set theory, formal systems describe real objects. We can prove that natural numbers are uniquely determined and that formal systems, such as Peano Arithmetic, prove true theorems about them. Similarly, axiom systems for set theory prove true theorems about sets, cardinal numbers etc.

To sum up, if we assume platonism, then we can prove that it is right. This is, obviously, a circular argument. As such, it is not acceptable as a proof, but it can be used as a test of consistency: it shows that an assumption applied to itself does not lead to contradiction. That is why I call the above argument inner consistency.

Platonistic belief also entails the fact that every mathematical statement is either true or false.[3] This does not imply that we should always be able to learn which of the two possibilities is true for every particular mathematical sentence. Nevertheless, a typical platonist believes that such a decision is always possible at least in principle.

In some fields of mathematics the source of this belief is the presence of a variety of proof methods. This is especially apparent in number theory. The ancient Greeks used elementary methods for proving theorems about numbers, but one can also trace back to them some methods that use higher order concepts. The method of sieves, first used by Erastothenes, is based on using sets of numbers, which is a higher order concept. After the discovery of calculus, analytical methods became popular in number theory. Contemporary number theorists often use deep theorems

[3]This view is not shared by all philosophers, see Notes. Other philosophers of mathematics argue that one *should* focus on questions about the truth value of mathematical statements. I will come to this shortly.

from algebra and algebraic geometry. This shows that, although the object of the study, the set of natural numbers, remains the same, number theorists are able to use more and more tools. For some problems, tools are not available, but number theorists hope they will eventually be developed.

Logicians generally agree on the fact that essentially all known results in number theory can be proved in Peano Arithmetic because the higher order concepts needed in nonelementary proofs, such as infinite sets and functions defined on integers (which are not expressible in Peano Arithmetic) can be simulated by suitable concepts. In Chap. 4, I discussed some mathematical theorems not provable in Peano Arithmetic, but these were not typical mathematical theorems. They were based on fast growing functions, whereas a typical problem is not of this kind. Moreover, the unprovable arithmetical sentences that use fast growing functions can be proved using set theory. It is possible to construct arithmetical sentences that can only be proved using much stronger axioms, even large-cardinal axioms, but we do not know of any arithmetical sentences that are not likely to be proved in set theory extended with large-cardinal axioms. There are, of course, many notorious open problems that we do not know how to solve, but mathematicians believe that they only need new *methods*, not new *axioms*. Apparently there is no immediate need of new axioms in number theory.

An interesting case is set theory because many statements here have been shown to be independent of the axioms of Zermelo-Fraenkel Set Theory (assuming this theory is consistent, which is not a problem for a platonist). Platonism in set theory does not necessarily mean believing in the truth of Zermelo-Fraenkel axioms, but since other set theories are not used in mathematics, a platonist believing in some other axiom system of set theory is rather an exception. The test case, and the most popular among the independent statements, is the Continuum Hypothesis. Since it is independent of the axioms of Zermelo-Fraenkel Set Theory, the problem is not caused just by the lack of a method—one has to come up with new axioms in order to decide it. We can resolve it formally by adding the Continuum Hypothesis, or its negation as an axiom, but this is not acceptable because it would be an arbitrary choice. If the universe of sets is a reality, we should find out what is true by studying it, not by postulating axioms in an arbitrary fashion. The new axioms should be general principles that we recognize as true.

Set theorist have proposed a number of new axioms that seem to be true general principles. The most important ones are axioms about the existence of large cardinals. I presented the concept of a large cardinal in Chap. 3. Roughly speaking, such an axiom postulates the existence of an essentially higher type of infinity. A platonist views the Incompleteness Theorem as the impossibility of capturing all types of infinity by one axiom system. Therefore, extending a system of axioms by postulating the existence of higher infinities is what one necessarily has to do in order to get closer to the complete truth. (We know, however, that large cardinal axioms alone do not suffice to prove all true sentences.)

How does one find large-cardinal axioms? An interesting observation is that when platonists are to decide the truth of a large-cardinal axiom, they use consistency as the criterion. This looks like a rule: a cardinal number κ with some properties *exists*, if it is *consistent* to assume that it exists. Of course, there is no such formal rule, but in concrete cases this is indeed what is used. There cannot be such a rule for several reasons, one of which is the impossibility of deciding the consistency of an axiom. Therefore in practice a large-cardinal axiom is accepted if it survives serious attempts to be disproved.

I mentioned in Chap. 3 that all large-cardinal axioms that have been proposed so far are compatible, except for a few inessential cases. Thus we do not have problems with choosing the right ones from them—we can accept them all. This is very important because, besides consistency, the most difficult problem when choosing axioms is which axioms should be chosen from a set of axioms that are incompatible. Moreover, the large-cardinal axioms form a linear scale. As we do not have an explanation for the linearity other than that these axioms describe real entities, this fact seems to support platonism.

Probably only few mathematicians are pure platonists in the sense that they believe in the reality of all mathematical entities ever considered. A moderate platonist believes in the reality of the natural numbers and other countable structures, but not more. A typical mathematician is a less moderate platonist and also believes in the existence of the real numbers, real and complex functions etc. This view is supported by physical reality—space around us is a continuous structure, hence real numbers have a physical model. It is harder to imagine cardinalities higher than the continuum (the cardinality of the real numbers) because there are no physical entities that we could associate with such sets. Therefore the extreme form of platonism, the belief in the reality of all infinite cardinalities studied in set theory, is professed only by set theorists.

In the 1970s, Michael Dummett and Haim Gaifman independently proposed an approach to ontological questions in mathematics based on *factually meaningful questions* instead of the elusive notion of *existence* [64, 84, 85]. The basic idea is as follows. While it is difficult to determine what a person means when he says that a certain abstract entity exists, one can learn his views by asking whether or not he considers questions about the entity meaningful. The person should say whether or not the questions can, in principle, be answered yes or no. If he says that the question has a definite truth value, we can infer that he believes that the entity exists. For mathematicians, this approach is certainly attractive because it looks more scientific and avoids metaphysical discussions.

In a recent article Gaifman presented a list of seven questions that can be used to determine to what degree a mathematician is a platonist [86]. One can test mathematicians' views by asking them which questions in the list have definite truth values. For example, the first question in the list is

"*What is the largest prime number dividing* $2^{2^{43112609}} - 1$*?*"

A mathematician who says that this question is meaningless is classified as *ultrafinitist.* Similarly one can characterize finitists. Note that this approach can give us more detailed information about mathematicians' views than if we simply talked about existence. This is because not only we ask whether questions about a particular mathematical structure are meaningful, but also *which questions* are. For instance, it is conceivable that somebody may claim that all Π_1 (universal-finite) sentences about natural numbers are either true or false, but deny it for more complex arithmetical sentences. This difference cannot be determined by asking whether the natural numbers (as a mathematical structure) exist.

The most famous platonist was Gödel, whose platonistic conception included the universe of sets. He did not consider platonism as the more acceptable alternative, but as "the only tenable view". Here is what he wrote about the Continuum Hypothesis in 1947 (when its independence had not been proved yet):

> "It is to be noted, however, that even if one should succeed in proving its undemonstrability as well, this would (in contradistinction, for example, to the proof for the transcendency of π) by no means settle the question definitively. Only someone who (like the intuitionist) denies that the concepts and axioms of classical set theory have any meaning (or any well-defined meaning) could be satisfied with such a solution, not *someone who believes them to describe some well-determined reality*. For *in this reality Cantor's conjecture*[4] *must be either true or false*, and its undecidability from the axioms as known today can only mean that these axioms do not contain a complete description of this reality;"[5]

In the same paper Gödel proposes to use large-cardinal axioms to settle the truth of statements that are independent of the axioms of Zermelo-Fraenkel Set Theory, which is called the *Gödel Program* nowadays. The theory of large cardinals is one of the most active areas of set theory. The list of the studied large cardinals is fairly long and the cases where a proposed large-cardinal axiom turned out to be inconsistent are rare. Each stronger cardinal enables us to prove more arithmetical sentences. It is therefore conceivable that, in principle, one may be able to derive every true arithmetical sentence from some large-cardinal axiom. However, there is a limit to what large-cardinal axioms can help us prove—they cannot decide sentences that concern sets of higher cardinality, in particular, they cannot decide the Continuum Hypothesis. I will say more about this in the next section.

Gödel expected that large-cardinal axioms alone might not suffice, but he still considered it likely that for any independent sentence, axioms would be found that would decide it. But how are we going to find them? Here is what he wrote about it.

> "There might exist axioms so abundant in their verifiable consequences, shedding so much light upon a whole discipline, and furnishing such powerful methods for solving given problems (and even solving them, as far as that is possible, in a constructive way) that quite irrespective of their intrinsic necessity they would have to be assumed at least in the same sense as any well established physical theory."[6]

His point is that we will recognize new true axioms fairly easily, because they will distinguish themselves by their power and usefulness. There is historical evidence that supports this idea: an example is the Replacement Axiom Schema, added to Zermelo's original axiom system. We know that adding just one new infinity makes set

---

[4]Here he refers to the Continuum Hypothesis.

[5]Kurt Gödel, *What is Cantor's Continuum Problem?* (1947) [98], pages 519–520. The italics are mine.

[6]Ibidem, page 521; both quotations reprinted with the permission of The American Mathematical Monthly.

theory stronger. But what happens when we add Replacement is an explosion of infinities; the resulting increase in the strength of the theory is tremendous. Moreover, the schema is also useful; for example, it enables us to generalize certain constructions to arbitrary cardinalities. The '*verifiable consequences*' probably refer to the parts of mathematics that we assume to be clear. These may be theorems that have difficult proofs, but are easily provable using the new axioms, and theorems about natural numbers that we can test experimentally.

Since this is an important idea, let me give it a name.

The Principle of Power and Usefulness *The universe of sets is determined by the axioms that furnish powerful methods and are abundant in their verifiable consequences.*

This idea reminds me of the anthropic principle from cosmology. This principle is used to explain why fundamental constants of physics seem to be tuned extremely favorably for us: even a small change in the values would have prevented life from forming because there would be no suns and no planets—the universe would be simple and dull. With a grain of salt, the principle above can also be justified by the anthropic principle: we live in the world not only with the most interesting physics, but also with the most interesting mathematics.

Intuitionistic Mathematics

I briefly mentioned intuitionistic mathematics when I was describing constructive mathematics in Chap. 2. Now I will say more about the philosophical aspects of intuitionism. Intuitionism is a radically different approach to the foundations of mathematics. Initially intuitionism did not only concern the philosophy of mathematics; intuitionists proposed an essential revision of mathematics and rebuilding it on constructive foundations. Research in intuitionistic mathematics is still going on, but it does not attract the attention of working mathematicians as much as before. Nevertheless, most of the current attempts to revise the foundations are still connected with intuitionism.

The main proponent of intuitionism was the Dutch mathematician L.E.J. Brouwer. After a short and successful period of activity in algebraic topology (he proved, for example, the well-known Brouwer's Fixed Point Theorem) he turned his mathematical activity to the foundations of mathematics and set to work revising mathematics on intuitionistic grounds. Other mathematicians, notably Hermann Weyl, who had similar views embraced this program as a realization of their ideas about the foundations of mathematics. This stream of thought went in the opposite direction than formalism and to some extent was a reaction to it. Hilbert became worried by the influence that intuitionism was gaining and the possible consequences in case it prevailed. He considered it his duty to defend traditional mathematics against it, which resulted in the well-known controversy between the two great mathematicians.

Below is my attempt to define intuitionism by five main principles, accompanied with quotations from Arend Heyting and Hermann Weyl.

1. Mathematics studies *mental mathematical constructions.*

> "[Brouwer's program] consisted in the investigation of mental mathematical constructions as such, without reference to questions regarding the nature of the constructed objects, such as whether these objects exist independently of our knowledge of them."[7]

2. Classical logic is only adequate in the domain of finite structures; infinite structures require a different kind of logic, *intuitionistic logic.*

> "...in mathematics from the very beginning we deal with the infinite, whereas ordinary logic is made for reasoning about finite collections."[8]

3. Basic mathematical concepts do not need definitions or explanations because people have natural *intuition* about them. (This is sometimes called *primordial intuition* and from that the term *intuitionism* is derived.) The most important one is the concept of natural numbers.

> "A mathematical construction ought to be so immediate to the mind and its result so clear that it needs no foundations whatsoever."[9]

4. Language and formalism is not able to describe mathematics precisely.

> "As the meaning of a word can never be fixed precisely enough to exclude every possibility of misunderstanding, we can never be mathematically sure that the formal system expresses correctly our mathematical thoughts."[10]

5. The natural numbers should be viewed only as a *potential infinity.* (Recall that potential infinity is the possibility to extend a mathematical object without limits, although at each stage the object is finite. In particular, for every natural number n, we can construct the number $n + 1$, but we should never view such a process as completed, which would be viewing it as an *actual infinity*.)

> "Brouwer made it clear, as I think beyond any doubt, that there is no evidence supporting the belief in the existential character of the totality of all natural numbers.... The sequence of numbers which grows beyond any stage already reached by passing to the next number, is a manifold of possibilities open towards infinity; it remains forever in the status of creation, but is not a closed realm of things existing in themselves."[11]

According to the first principle, mathematics is subjective. Intuitionists admit that the individual sensations must correspond to some abstract entities, otherwise there would be no correspondence between the mathematics of different individuals, but they say that mathematics should not study such metaphysical concepts. According to intuitionists, banning metaphysical consideration is a way to avoid paradoxes.

---

[7] A. Heyting, *Intuitionism, An introduction*, (1971) [122], page 1.

[8] *ibidem*, page 1.

[9] *ibidem*, page 6.

[10] *ibidem*, page 4.

[11] H. Weyl, *Mathematics and logic*, (1947) [308], page 9.

In intuitionism existence is replaced with constructibility. In this intuitionists revert to the common approach before the advent of nonconstructive proofs of existential sentences. As Bernays noted

> "Euclid postulates: *One can join* two points by a straight line; Hilbert states the axioms: Given any two points, *there exists* a straight line on which both are situated."[12]

Contemporary mathematicians also often say '*one can construct* x' or '*one can find* x' instead of '*there exists* x', but they do not mean that there exists an algorithm for finding x. It is just easier to imagine an existential statement as an idealized construction. As we noted in Chap. 5, nonconstructive proofs are ubiquitous, but mathematicians are still interested in explicit constructions or descriptions of the objects claimed to exist. Intuitionistic logic has a means to distinguish between constructive and nonconstructive proofs of an existential formula. In the first case the existential formula is true, in the second case only the doubly negated existential formula is true. Such doubly negated sentences are expressed in words by 'it is not possible that x does not exist'. For most mathematicians, it seems more effective to use 'there exists' in general and, if they want to stress that the proof is constructive, to specify how the element can be constructed (for example, by an algorithm running in polynomial time).

For intuitionists the algorithm or the process of construction is more essential than what the construction actually produces, whereas platonists and formalist are only interested in the object and do not care what means are used to justify its existence. There is an aspect of practical importance in the intuitionistic approach: if we want to use the fact that an object with certain properties exists, we need to have an explicit example of such an object, which means we have to construct one. But what is a construction? Aren't constructions just another kind of mathematical structures? Indeed, in mainstream non-intuitionistic mathematics algorithms are treated in the same way as any other kind of structures.

Intuitionists do not see any reason for reducing mathematics to logic. They claim that we have better intuition about basic mathematical concepts, in particular, about the natural numbers, than about logic. They argue that the principle of mathematical induction is as obvious as any logical rule. According to Brouwer, "*mathematics is the exact part of our thinking*" (which is very close to Immanuel Kant's view of mathematics). Probably everybody would agree that the principle of mathematical induction is intuitively as clear as logical axioms and rules, but this fact should not be used against formalizing mathematics. When we formalize natural numbers in first order logic using the axioms of Peano Arithmetic, we are not saying that logic is more important; we are only making the rules precise about what constitutes a proof of an arithmetical sentence and what does not.

If we identify mathematics with our informal ideas, then the fourth principle, the impossibility of formalizing mathematics, is to some extent true. Whenever we formally define a mathematical concept, we have many possibilities of how to do it

[12]Paul Bernays, *Sur le platonism dans les mathématiques* [22], (1935) page 53. The italics are mine.

and each of these gives a different flavor to the defined entity. Consider, for instance, the definition of a group. One can define groups using the unit, the operation of multiplication and the operation of inversion. One can also define groups using only the operation of multiplication, and there are other possibilities. We may prefer one of these possibilities for various reasons (historical, practical, etc.), but none has the right to be identified with the concept of a group. So if we interpret the fourth principle as *the impossibility to uniquely determine one formalization of a given concept*, then it is true. But this is not the intended meaning of 4. What it intended to say was that *one can never completely determine a mathematical concept by a formal definition.* But formalization of mathematics has reached such a high level that we can safely claim that *every mathematical concept is fully formalizable.* I dwelt on this subject in Chap. 2 at length, so let me only continue with the example of groups. The fact that we have many formalizations of groups is unimportant. We know that *each* of the formalizations captures what we want to express and they are equivalent in a sense that can be precisely defined.

Intuitionists maintained that one reason why paradoxes appeared in mathematics was not respecting the fourth principle. Indeed, if we speak informally about concepts such as *true, definable*, etc., we may produce paradoxes. Such paradoxes are called *semantical* because of their connection with the semantics of the language (an example is Berry's paradox mentioned in Chap. 1, page 38). As I have explained in previous chapters, these paradoxes have two causes: (1) not distinguishing between the syntactical concepts that are objects of the study and the syntactical concepts used in our arguments, and (2) the use of certain concepts without precise definitions. Once we define the concepts properly and distinguish the levels of discourse, they disappear. Thus, in fact, formalism helps to get rid of semantical paradoxes.

Intuitionists circumvent the problem of consistency by pointing out that intuitionistic mathematics is based on our intuition about the basic concepts which is *a priori* consistent. It would be too easy to dismiss this argument by simply pointing to numerous errors in mathematical writings. Instead, let us compare it with the view of platonists. Platonists say that mathematics is consistent because it describes reality. They refer to a basic tenet of every realistic philosophy that reality is consistent. Intuitionism is an idealistic philosophy, so it naturally seeks the justification for consistency in the mind. Thus whether or not one accepts the intuitionistic justification for the consistency of mathematics depends only on whether one accepts idealistic or realistic philosophy. However, mathematical practice rather shows that intuition is not a good criterion for consistency. In general, it is true that a complicated proof that is not based on an intuitively clear idea is more likely to be wrong. But there are cases when a wrong proof was accepted as correct because it looked precisely like the right kind of proof for the given theorem. In some cases the theorem was eventually proved using different means, but in some cases the theorem was disproved. One could argue that the wrong proof was based on wrong intuition, but how do we tell apart a wrong intuition from a correct one?

Rejecting intuitionistic philosophy does not necessarily mean declaring its mathematics to be unworthy. Intuitionistic mathematics, stripped of philosophy, is a respectable branch of mathematics with interesting results and problems that surpass the realm of intuitionism. For instance, some formal systems developed for constructive mathematics and based on intuitionistic logic turned out to be very useful in computer science.

I conclude this part with an observation concerning a connection between the intuitionistic conception of mathematics and quantum physics. Consider a sentence ϕ such that neither a proof of ϕ, nor a proof of $\neg\phi$ is known. For a classical mathematician, either ϕ or $\neg\phi$ is true anyway; intuitionistic mathematicians reject the statement that either ϕ or $\neg\phi$ is true unless they know a proof of one of these two sentences.

Now consider a physicist measuring a quantum bit b (which is represented by a suitable physical system, say, the state of an atom). Measuring a quantum bit means that the result will either be $b = 0$ or $b = 1$. Assume also that the physical interaction of the measuring apparatus with the quantum bit is negligible. A classical physicist, one that would not use quantum theory, would conclude that before measuring the bit was either 0 or 1. However, if the bit was a nontrivial superposition of 0 and 1, then this is not true.

If one wants to argue about quantum systems without using the mathematical description of the states, it is possible to use *quantum logic*. This is a logic that is different from classical logic, and is also different from intuitionistic logic, so the connection between intuitionistic mathematics and quantum physics is not the use of the same logic. What is in common is that what a subject learns depends on his actions. In the first case, by proving one of the two possibilities, the intuitionistic mathematician makes the sentence $\phi \vee \neg\phi$ true. In the second case, by measuring the bit b, the physicist makes $b = 0 \vee b = 1$ true.

By this example, I want to prepare the reader for the idea, which I will discuss later in this chapter, that an observer could play an active role not only in physics, but also in mathematics.

Logicism

After Dedekind had shown that it is possible to reduce all mathematics to the natural numbers, it seemed that the ultimate foundations had been established. But soon after that mathematicians and philosophers working in the foundations of mathematics realized that the reduction was not to the natural numbers alone, but to the natural numbers *and logic*. Logic cannot be eliminated for obvious reasons, but if we use logic, do we need arithmetic at all? If we could do without arithmetic, then we would indeed have the best foundations we could hope for. The basic thesis of *logicism* is that this is possible, that we only need logic. If we accept this thesis, then mathematics is a part of logic.

For further discussion it will help to state the basic thesis of logicism explicitly.

The Logicist Thesis *Mathematics can be founded purely on logic.*

The founder of logicism was Frege. When he started his research, mathematical logic did not exist, so he had to invent the logical calculus himself. As I mentioned in Chap. 2, he did it in his *Begriffsschrift* and developed further in his later works. As we know, his system was inconsistent because of Russell's paradox. He considered this to be a serious problem and did not try to fix the system to avoid the paradox. Although he did not present a consistent system, his work has had a tremendous impact. In particular it contained the first formalization of what we now call first order logic. His ideas are still used by modern logicists.

Russell was the next most important figure in logicism. The following quotation from his work expresses the Logicist Thesis in different words.

> "The fundamental thesis of the following pages, that *mathematics and logic are identical*, is one which I have never since seen any reason to modify."[13]

He proposed to avoid the paradox that he had found by introducing types. I presented his Theory of Types in the chapter about set theory, but in fact, Russell did not view the Theory of Types as an axiomatic system for set theory. As his aim was to prove the Logicist Thesis, he viewed his system as a higher order logic rather than as set theory. Also the systems proposed by logicists later were mostly versions of the Theory of Types.

Another important logicist was Church. In his paper *A Set of Postulates for the Foundations of Logic*, published in 1932 [42], he talks about the foundations of logic, rather than mathematics, but he compares his system with Russell's and Zermelo's (saying that those "*appear somewhat artificial*"). So he, clearly, assumed the Logicist Thesis and intended his system to give foundations for mathematics. His idea was to avoid the paradoxes by disallowing the law of excluded middle. Nevertheless, the system was found to be inconsistent, and so also was its modification published in 1933. It turned out that the important feature of the system was not disallowing the law of excluded middle, but the particular way of using terms. This consistent part has been studied under the name λ-calculus.

The Logicist Thesis is not a mathematical statement that one could prove or disprove formally, but in contrast to the principles of other philosophical views of mathematics it is a fairly concrete statement. In principle, we could prove it by defining a logical calculus and showing that all mathematics can be formalized in it. It will be more difficult to disprove it conclusively because we will need to make more precise what logic in general is and what mathematics is, but I think that the arguments against the Logicist Thesis are quite convincing. I will now present a couple of such arguments.

Russell's paradox certainly undermined the position of logicism, but Russell's solution, the Theory of Types that appeared shortly afterwards, gave new impetus to this approach. The detrimental blow to logicism was actually Gödel's Incompleteness Theorem. The consequence of the Incompleteness Theorem is that starting from second order logic, higher order logics are not axiomatizable (by decidable

---

[13]Bertrand Russell, *The Principles of Mathematics*, Introduction to the second edition, (1903), [251]; the italics are mine.

sets of axioms). In particular, it is not possible to formalize them by a finite set of axiom schemas and rules. In plain words it means that we will never be able to state all principles that are true in these logics. As I argued in Chap. 2, this is the reason why we should not consider these logics to be genuine logics. What then remains is first order logic. But first order logic is too weak—we cannot formalize the structure of natural numbers in first order logic without assuming axioms specific for arithmetic.

One may play down the problem of incompleteness by saying that we actually do not need all true sentences of higher order logics, or that we can define completely different logics. Incompleteness is, however, a deeper problem. It concerns all formal systems and the independent sentences are of very low logical complexity (they are Π_1). These sentences can be equivalently stated as expressing that certain Diophantine equations do not have solutions. Therefore incompleteness concerns very concrete and important mathematical problems. Consequently, we have to view mathematics as an open ended system. We have to discover new axioms not only in order to decide what is true about infinite sets, but also because we need to decide such simple arithmetical sentences.

Example As we noted in Chap. 4 (see page 308), one can construct a Diophantine equation that is unsolvable if and only if Zermelo-Fraenkel set theory is consistent. An example is the equation on page 306 with suitable parameters **A**, **B**, **C**, **D**. If the Logicist Thesis is true, every true mathematical sentence should be derivable from logic. Hence there should be a *logical principle* that implies the unsolvability of this equation, a principle so strong that it implies the consistency of *ZFC*.

So I only see two ways in which one can still defend the Logicist Thesis.

1. To propose a formal system having a logical character and say that what it does not cover are only artificial problems that are not genuinely mathematical.
2. To say that set theory is a part of logic.

There are rationales behind both approaches. Indeed, practically all results in mainstream mathematics can be proved in Finite Set Theory (the axiomatic system equivalent in its strength to Peano Arithmetic) provided that they are suitably formalized. But 1 violates one of the basic principles of science: the freedom to investigate everything. According to this principle we should not prohibit studying any mathematical problem, however artificial it may look. Accepting 1 would mean renouncing any mathematics that cannot be done in the proposed formal system. In contrast, the standard foundations based on Zermelo-Fraenkel Set Theory are viewed as an open system to which we can add more axioms when needed.

Concerning 2, one can argue that the formal system for set theory that we use today, Zermelo-Fraenkel Set Theory, which developed from Frege's system by restricting the Comprehension Schema to certain formulas, is part of logic. But mathematicians working in set theory do not view it as logic. They rather view it as the study of particular mathematical structures. Recall that Cantor started to study sets because of a problem in mathematical analysis. He needed to understand what kinds of sets of reals there are. This is still a central problem in set-theoretical topology and descriptive set theory. The latter two areas can hardly be viewed as logic. Another area of set theory that cannot be considered logic is the theory of large cardinal numbers. While we could say that the axioms of Zermelo-Fraenkel Set Theory are of a logical nature, I do not think we can say the same about the large-cardinal axioms. Those axioms are often statements about combinatorial structures, properties of measures and other purely mathematical sentences. Admittedly, one cannot exclude that an extremely strong logical principle will be found that would be stronger than the strongest large-cardinal axioms that we assume to be consistent. But I do not think we have one now.[14]

In modern logicism, formal systems are based on typed λ-calculi. These calculi originated from Russell's Theory of Types and from systems proposed by Church. Typed λ-calculi are studied mainly because of their key role in functional programming, but they also have several features that are very attractive for logicists. Before stating these properties, let us briefly recall that in typed λ-calculi there are two sorts of objects: types and terms. Types define kinds of functions and they are also interpreted as formulas. Terms represent functions and can also be viewed as proofs. The duality of the interpretations of types and terms is the *Curry-Howard isomorphism.*

Here are the properties that are useful for the foundations.

1. Logical formulas and proofs are objects of the system and do not have to be introduced via coding or numbering, unlike in arithmetical theories or set theories. Therefore, one can view such a system as *logic.*
2. Since terms represent computable functions, the semantics of the system is very concrete. Unlike in set theories, one does not need to consider infinite sets and uncountable cardinalities.
3. In order to prove the consistency of the system, one only needs to prove the convergence of an associated procedure.
4. Structures definable by recursion have typically very simple definitions; this also concerns natural numbers. This property is useful in programming and computer assisted proof search and proof verification.

So why bother with set-theoretical foundations? Why don't we switch to typed λ-calculi? The reason is that the above properties, however attractive they look, do not present big gains. Let us look at them more closely.

1. Formalization of syntax in arithmetic and set theory is a one time task. Once we do it, we do not have to refer to the way it was done. All we need are some basic properties that the formalization satisfies. Hence property 1. cannot be a reason for abandoning set theory.
2. I exaggerated when presenting property 2. In fact, when complicated types are introduced, one often refers to their set-theoretical interpretation in order to justify them. On the other hand, when studying a classical system based on set theory,

[14]The axioms about elementary embeddings (see page 204) are not of purely logical nature because they need the concept of ordinal.

we may focus on computable functions that are provably total in the system. Then we can say that these functions are the true semantics, whereas the other objects of the system are only ideal objects. (This is roughly the way Hilbert interpreted mathematical concepts.)

3. For weak systems, it is easy to prove their consistency by proving the convergence of normalization procedures, but for strong ones, one faces the same kind of problems as when proving the consistency of classical systems. Note that consistency is expressed by a Π_1 sentence, whereas the convergence of a normalization procedure is a Π_2 sentence. This is not surprising because we know that it is easier to find a combinatorial interpretation of the Σ_1-reflection principle than an interpretation of consistency. The principles independent of Peano Arithmetic that I mentioned in Chap. 4 are equivalent to the Σ_1-reflection principle of Peano Arithmetic. When using ordinal analysis to prove that a theory T is consistent, we need to show, for a particular definition of the associated proof-theoretic ordinal α, that every *algorithm* that produces a decreasing sequence of ordinals below α terminates after a finite number of steps. So ordinal analysis is a "strategy" similar to proving termination of a normalization procedure.
4. While some concepts do have simple definitions in typed λ-calculi, I am not sure that this is true of the majority of mathematical concepts. I am afraid that complicated definitions in typed λ-calculi may often obfuscate the true nature of many mathematical problems.

Thus I view these systems as a very interesting and useful research area, but I do not consider them to be a solution of the fundamental problems stated at the beginning of this chapter.

Formalism and Hilbert's Philosophy of Mathematics

It is difficult to define schools in the philosophy of mathematics precisely. Philosophers of mathematics often disagree in classifying the views of particular mathematicians. This also concerns formalism, but one obvious feature on which all authors agree is that formalism stresses the importance of formalizing mathematics. Specifically, a formalist maintains that every field of mathematics can be and should be formalized because this is the only way to achieve absolute precision in mathematics. But this view is now accepted by almost all mathematicians, at least to the extent that one should always explicitly state the axioms used in the field under study. So a positive attitude towards the formalization of mathematics is not sufficient for classifying somebody as a formalist. What distinguishes formalists from platonists is *ontology*. For a pure platonist, such as Gödel, all sets, whatever their cardinality is, are as real as finite sets. A formalist distinguishes between real entities and ideal ones. The real entities are typically finite structures, such as natural numbers, but often also real numbers. Ideal mathematical entities are those that do not have natural representations in physical reality.

When we classify ontological views, we cannot use only two categories; instead we have to consider a whole spectrum of views. Some mathematicians claim that only finite structures are real and the rest is only a fiction. This is called *finitism*. Others may accept real numbers and subsets of real numbers as objective reality, while rejecting sets of larger cardinality. And so on up to pure platonism. So there does not seem to be a clear line between platonism and formalism and I will not attempt to make the distinction more precise.

The most extreme form of formalism is *game formalism*. According to this conception, mathematical concepts have no meaning; they only serve as symbols in formal games. Proving theorems in a particular field of mathematics is just playing the game determined by the rules of that field. It is hard to believe that any mathematician ever really viewed mathematics in this way. It is more likely that game formalism was just an attempt to solve, or rather to avoid, the philosophical question about what the subject of mathematics is. Let us see what J. Thomae, a proponent of this approach, wrote about it.

> "For the formalist conception, arithmetic is a game with signs that are well referred to as empty, by which one means to say that they (in the calculating game) have no content given to them beside that which comes from their role in the rules of combination (rules of the game).
>
> ...
>
> The formal standpoint allows us to throw off all metaphysical difficulties. That's the gain it offers us."[15]

But it is not true that this is a safe way to avoid all metaphysical problems because then the questions arise: What is the nature of games? Are they real entities or only mental constructions? And why is mathematics so useful in applications?

A prominent formalist was H.B. Curry, the co-inventor of combinatory logic. His philosophy, roughly speaking, was that the subject of mathematics is the study of formal systems.

> "The definition I advocate is briefly this: Mathematics is the science of formal systems.
>
> ...
>
> The essence of mathematics is that we make definitions by recursion, and then draw particular consequences by applying the definition and general consequences by mathematical induction."[16]

But he rejects the idea that mathematics is merely a formal game where the symbols have empty meaning. According to Curry, propositions do have meaning whose truth is "*determined by the fundamental definitions*".

When it was shown in the works of Frege, Russell and Whitehead that logic can be formalized, Hilbert realized that formalization is not only a way to make mathematical reasoning absolutely precise, but can also be a way to avoid paradoxes. Because, once a theory is fully formalized, the question about its consistency becomes a *mathematical problem*, and, in fact, a purely combinatorial one. This idea must have been really new and unusual because Frege asked Hilbert repeatedly

[15]J. Thomae, *Elementare Theorie der analytischen Functionen einer complexen Veränderlichen*, Halle, 1898; as cited by M. Detlefsen in [63], page 301.

[16]H.B. Curry, *Outlines of a Formalist Philosophy of Mathematics*, [54], pages 56 and 57.

about it in his correspondence with him. According to Frege, the only way to show the consistency of a system is to interpret it.[17]

Hilbert was fascinated by the possibility of proving consistency using mathematical means and spoke about it on many occasions. It was necessary to limit the means by which the consistency was to be proved; proving the consistency of a theory T using T itself would be a circular argument (but now we know that even this is not possible). The natural choice for limited means was *'finite means'*, which means *reasoning only about finite mathematical objects.* Finite mathematics was considered to be safe from contradictions by all philosophers of mathematics. Hence a proof of consistency by finite means could not be questioned.

Hilbert was not a game formalist in the sense that he would deny that mathematical concepts had any meaning. When he talked about a *"formula game"* (*"Formelspiel"*), he rather used this term to explain that one can formalize mathematical reasoning so that it can be mechanically checked without knowing the meaning of the symbols. His views about the existence of mathematical entities were more complex and, in fact, very liberal: he was willing to identify existence with consistency.

> "If contradictory attributes be assigned to a concept, I say that mathematically the concept does not exist. So, for example, a real number whose square is -1 does not exist mathematically. But if it can be proved that the attributes assigned to the concept can never lead to a contradiction by the application of a finite number of logical processes, I say that the mathematical existence of the concept (for example, of a number or a function which satisfies certain conditions) is thereby proved. In the case before us, where we are concerned with the axioms of real numbers in arithmetic, the proof of the compatibility of the axioms is at the same time the proof of the mathematical existence of the complete system of real numbers or of the continuum."[18]

In this he agreed with the French mathematician Henri Poincaré (1854–1912). (Incidentally Poincaré is considered to be an intuitionist by some authors,[19] which demonstrates how difficult is to classify the schools in the philosophy of mathematics.)

> "Mathematics is independent of the existence of material objects; in mathematics the word exist can have only one meaning, it means free from contradiction."[20]

It should be noted that at that time it was still conceivable that a complete axiomatization of real numbers would be found. For further reference I will state this view as a principle.

---

[17]Ironically, Frege was right, at least in the case of strong systems, as it turned out when Gödel proved the Second Incompleteness Theorem.

[18]D. Hilbert, *Mathematical Problems*, (1902) page 446 [125]. This is from his list of problems presented in 1900, the section about his second problem, *The Compatibility of the Arithmetical Axioms.* In this English translation the German word '*Widerspruchslosigkeit*' was translated as '*compatibility*'. The literal translation of the German word is '*contradiction-freeness*', which we today translate as '*consistency*'. Hilbert proposed to prove the consistency of a theory in which one can formalize both the natural numbers and the real numbers. A natural theory for this purpose is Second-Order Arithmetic, see page 295.

[19]E.g., by A.A. Fraenkel, Y. Bar-Hillel and A. Levy, in *Foundations of Set Theory*, [75].

[20]H. Poincaré, *The Foundations of Science*, page 454, [220].

The Existence from Consistency Principle *If a mathematical concept is consistent, then there exist entities representing this concept.*

Later Hilbert moderated his claim and presented it only as an open problem, in fact, as one of the main problems of foundational studies.

> "In this [*that they are to a large extent behind the times*] I see the reason, too, why these most recent investigations in fact stop short of the great problems of the theory of foundations, for example, the question of the construction of functions, the proof or refutation of Cantor's continuum hypothesis, *the question whether all mathematical problems are solvable*, and *the question whether consistency and existence are equivalent for mathematical objects*."[21]

In this quotation there is another problem of the foundations of mathematics that he often mentioned: the problem whether all mathematical problems are solvable. Above it is stated as an open problem, but earlier he called it '*the axiom of solvability*'. It was no coincidence that Hilbert mentioned the Existence from Consistency Principle and his Axiom of Solvability in the same sentence. These two principles are closely related. If there is no proof that no object satisfies property P, then it is consistent to assume that there is such an object. If we accept the Existence from Consistency Principle, then an object satisfying P indeed exists. Thus, in principle, the problem to find an x satisfying P is always solvable—either there is a proof that there is no such x, or there exists an x such that $P(x)$ holds true. What Hilbert did not realize was that proving consistency could be an unsolvable problem. The idea of the axiom of solvability can be traced back to his lecture at the International Congress of Mathematics in Paris in 1900, where the well-known '*no ignorabimus*' quotation appeared:

> "We hear within us the perpetual call: There is the problem. Seek its solution. You can find it by pure reason, for in mathematics there is no ignorabimus."[22]

In his lecture *On the infinite* presented in 1925 he was more explicit about his views about the existence of mathematical entities ([126], page 376). He distinguished between concrete mathematical objects and ideal ones which '*are not to be found in reality*'. Ideal objects are introduced by what he called '*the method of ideal elements*'. He mentioned the points at infinity in projective geometry, complex numbers and 'number ideals' (ideals in rings) as examples. These ideal elements are used to extend structures to make some laws universally valid. (For example, the points at infinity are added because we want to guarantee that every pair of lines intersects; we need to add points at which parallel lines intersect.) His explanation of infinite sets was that they are such convenient ideal elements.

---

[21] D. Hilbert, *The foundations of mathematics* (*Die Grundlagen der Mathematik, 1927*), [127], page 437, (the italics are mine). As this lecture was given two years before Gödel proved the Completeness Theorem, Hilbert may have had in mind the problem of the completeness of the calculus for first order logic. At the time when Gödel worked on the Completeness Theorem, it was known as an open problem posed by Hilbert.

[22] D. Hilbert, *Mathematical Problems*, (1902), page 446 [125]. '*Ignorabimus*' means '*we shall never know*'. By this word, he referred to the doctrine advocated by some scientists in the late 19th century which declared that there are a priori limitations to the knowledge we can acquire.

Hilbert also talked about '*ideal propositions*', propositions that talk about *ideal elements.* He, more or less, accepted the view of intuitionists and others who maintained that classical logic is only adequate for reasoning about finite objects. But instead of using a different logic he proposed the concepts of ideal elements and ideal propositions. Ideal elements are introduced in order to facilitate reasoning about real elements. We are only interested in propositions about real elements, but we may use ideal propositions, meaning propositions about ideal elements, to prove real propositions.

Example The concept of an *ideal*, which is very important in many branches of mathematics, comes from the "*ideal numbers*" introduced by Ernst Kummer. Since the theorem about the unique factorization of numbers is false in some rings, Kummer introduced the ideal numbers to have a substitute for this theorem in such rings. This enabled him to prove Fermat's Last Theorem for many exponents. His method cannot be used for all exponents because in some rings the unique factorization theorem fails even if one uses prime ideals instead of prime numbers.

Hilbert maintained that ideal elements and ideal propositions can solve the problem that we use the logic of *finite* structures to argue about *infinite* ones. I am not aware of him giving a particular example of this, so I made up one myself. Let us assume the viewpoint that the natural numbers are only potentially infinite, which means that their totality does not form a real object. Then it also is reasonable to assume, as intuitionists do, that the law of excluded middle does not hold true for sentences such as '*either all numbers have property P, or there exists a number that does not have property P*'. Indeed, how can we talk about all numbers before they are all created? But there is another solution, one that is used by the majority of mathematicians: let us pretend that the set of all natural numbers exists, or in Hilbert's words, let us add it as an ideal element. Then we can retain classical logic.

According to Hilbert, the part of mathematics that deals with real structures is consistent, but when we introduce ideal propositions we may inadvertently produce contradictions, as happened to Frege and Cantor. This problem, he thought, could be solved by providing proofs of consistency.

Hilbert always stressed that one should in no way limit the concepts and the methods that mathematicians use. The only reason for not allowing a method should be the fact that it produces contradictions. This was fully in accord with his views on the existence of mathematical entities. It was also this liberal approach that led him to promote the problem of consistency as the main goal of research in the foundations. He described the reaction to the paradoxes in set theory (clearly, referring to intuitionism) as follows:

> "The reaction was so violent that the commonest and most fruitful notions and the very simplest and most important modes of inference in mathematics were threatened and their use was to be prohibited."[23]

[23] D. Hilbert, *On the infinite*, (1925), [126] page 375.

Then he went on to say that on the contrary, we should look for new ways of proving theorems and just make them secure by proving their consistency.

> "But there is a completely satisfactory way of escaping the paradoxes without committing treason against our science. . . .
> (1) We shall carefully investigate those ways of forming notions and those modes of inference that are fruitful; we shall nurse them, support them, and make them usable, wherever there is the slightest promise of success. No one shall be able to drive us from the paradise that Cantor created for us.
> (2) It is necessary to make inferences everywhere as reliable as they are in ordinary elementary number theory, which no one questions and in which contradictions and paradoxes arise only through our carelessness."[24]

The main contribution of Hilbert's formalist philosophy is the identification of the consistency problem as a mathematical problem. The proposed solution, proving consistencies by finite means, turned out to be impossible, but had Hilbert not asked this problem, it is possible that Gödel would not have worked on it and proved the Incompleteness Theorems.

Quine's Web of Belief and Consistency

Quine was one of the most influential philosophers of science of the 20th century. He is also famous for his work in logic, in particular in set theory, which I mentioned in Chap. 3. His philosophy is a modern form of naturalism that puts science and scientific methodology above traditional philosophy. In particular, according to Quine, one should apply the scientific method to epistemology, which means that we should study, using psychology, neurology etc., how various beliefs are formed in human brains. (The subsection below is an example of such an approach.) This kind of naturalism is attractive for most natural scientists because it resonates with their belief in the superiority of science over philosophy.

The main aspect of Quine's philosophy is *holism.* It is the view that one cannot consider a thing separately because it only makes sense in the context of all other things. This concerns both results in science and science as a whole. Whenever we perform an experiment we are using some theory to design it and to interpret the result. There is no experiment without a theory. But not only this—each theory only makes sense in the context of other theories. Everything is just a part of a "*web of belief*". He thought of knowledge as a field whose periphery consists of empirical statements and in which, as one progresses towards the center, the empirical statements are gradually replaced by concepts and theories whose abstractness increases. In particular, mathematics occupies a central region.

> "The totality of our so-called knowledge or beliefs, from the most casual matters of geography and history to the profoundest laws of atomic physics or even pure mathematics and logic, is a man made fabric which impinges on experience only along the edges. Or

[24] *Ibidem*, page 375–376.

to change the figure, total science is like a field of force whose boundary conditions are experience."[25]

But unlike the solutions of typical partial differential equations, the closer to the center we go, the more ambiguous the web is.

"The edge of the system must be kept squared with experience; the rest, with all its elaborate myths or fictions, has its objective the simplicity of laws."[26]

According to Quine, mathematics is as empirical as other sciences. It is empirically tested whenever we test any theory that uses mathematics. Abstract mathematical concepts are posited; they are chosen from alternatives according to their usefulness. He was rather skeptical about the usefulness of the most abstract concepts in mathematics. The reason for accepting them is only the necessity to keep the whole system of mathematical theorems coherent. Let me suggest my own example. Suppose we do not believe in (the usefulness of) uncountable cardinalities. We may still accept them in our set theory, because prohibiting Cantor's proof would mutilate set theory too much, but we should not expect them to be very useful.

One interesting idea that concerns the discovery of new concepts and axioms is his *Principle of Minimal Changes.* According to this principle one should always choose the modification of the current system that results in minimal disturbances. This is similar to the Principle of Minimum Description Length that I mentioned in Chap. 5 (page 488): a theory with a simple description is more likely to be in accord with the large systems of other theories than a complicated and contrived one.

The most interesting aspect of holism, coherentism and similar approaches is that they suggest a way to justify the belief that the *whole* of mathematics is consistent. The idea is that if there were a contradiction in current mathematical knowledge, it would have soon been discovered because it would have had a big impact on the consistency of the whole web of belief. Or perhaps it would not be discovered, but we would see that things somehow do not fit together. In other words, holism suggests that it is unlikely that a contradiction would be hidden in a distant corner without giving us any sign of its presence.

I think this is actually the way most mathematicians perceive the consistency problem. The justification by coherence is also used in individual theories. Recall, for example, the case of non-Euclidean geometries. Although Lobachevsky and Bolyai did not construct models of these theories, the fact that they were able to develop coherent theories persuaded them, and other mathematicians as well, that the theories were consistent. In set theory, where consistency is the most pressing problem, such an argument is often used too. Set theorists believe that certain axioms are consistent, or even true because the resulting theory "behaves well".

Although this seems very plausible, we do not have any mathematical result supporting this idea. On the contrary, as we noted in the previous chapter, one can construct a theory with a concealed contradiction, provided that some conjectures in cryptography are true (see page 520).

---

[25] W.V.O. Quine, *From a Logical Point of View*, [232], page 42.

[26] *Ibidem*, page 45.

Psychological Reasons for Accepting Certain Beliefs

I now digress to psychology because I think little attention is paid to it when ontological commitments in mathematics are discussed. Most mathematicians are not philosophers, therefore their beliefs are not based on well-developed philosophical systems; rather they are influenced by education, by what they hear from their colleagues, and by their daily experience with mathematical work. Hence their choice of a particular approach to the existence of mathematical objects is guided more by unconscious psychological processes than conscious logical deductions. I will only consider platonism and intuitionism.

There are several psychological factors that make platonism an attractive belief. I mentioned one at the very beginning of the book. When people are thinking about an object for a long time, the object becomes more and more real in their minds, no matter whether it is real or imaginary. The natural tendency of the human mind is to believe in the existence of the objects perceived by sight. The reason is that sight gives considerably more information about them than any other sense, and also more than one can learn if the object is only reported. However, if a lot of information is provided to the brain by different means, it can compensate for not seeing the object. The information about the object can also be provided by the thinking subject itself. One can add more features to the mental image of the object by logical deduction, or by fantasy. An example of the former is a mathematician thinking about a mathematical concept; the latter happens in mystical ecstasies. The job of a mathematician is to read, think, talk and write about mathematical objects. The more a mathematician learns and proves about an object, the more real it seems to him.

Another factor is that platonism is to an extent indoctrinated into students during their studies. This is not because teachers would like to influence students' philosophical views. The reason is simply the fact that when teaching science it is best to assume the viewpoint of a realist. The human mind naturally prefers learning about reality to learning about ideas that are not represented by real objects. Knowledge concerning real things is useful in practical life. Furthermore, it is much easier to explain things from the perspective of a realist. No alternative to platonism has such a clarity and inner consistency. For this reason, I have also explained mathematics from this standpoint in previous chapters.

The concept of intuition clearly belongs to psychology. Intuition is the ability to reach a correct conclusion from given or observed data without apparent use of deductive reasoning. This concerns all intellectual activities, but let us focus on intuition in mathematics. What is referred to as can be classified into three types:

1. inborn mathematical abilities,
2. mental pictures of mathematical concepts,
3. subconscious reasoning.

The first kind includes logical reasoning and the ability to generalize from examples. Further, it includes the ability to count with small numbers and to understand elementary concepts of three-dimensional geometry. Here we can also include the important ability to learn algorithms. By this I mean performing a series of actions to reach a desired goal. When somebody learns enough mathematics, these actions can be mathematical operations. Another inborn mathematical ability is probabilistic reasoning.

The mental picture of a concept is the total knowledge about it, a sort of database of everything related to it. If the database is well organized, the subject is able to answer questions about the concept very quickly. A mental picture is formed by experience, education and contemplation. Inborn mathematical abilities are also based on experience, but not individual experience; they are the result of the experience of ancestors which has been encoded in the genomes by means of natural selection.

Routine activities are often performed subconsciously. Mathematics is no exception—a good mathematician is able to solve a simple problem very quickly without consciously performing all the necessary deductions. As most of these deductions are done subconsciously, they are carried through much faster. Since it happens so fast and since even the mathematician himself does not realize it, it looks like a little miracle, but instead we call it *intuition*. When an actual proof is required, the mathematician can slow down and report all the steps in the deduction he is doing.

The *primordial intuition about numbers*, the term used by intuitionists, should clearly refer to an *inborn* intellectual ability. My skeptical view is that the inborn abilities concerning arithmetic do not go beyond comparing the cardinalities of very small sets. The intuition about geometry is certainly much more developed, because it is needed for image recognition, planning movement, etc. (It may be developed even better in birds because they move in three dimensions.) I think that most intuition of adults about numbers comes from education, and the inborn ability of logical reasoning is more developed than the intuition of numbers. Although logic is rarely taught outside of universities, most people understand logical deductions and are able to use them. On the other hand, children are taught counting at a very early age, so it is difficult to determine what is an inborn ability and what is not. We would have to test people with no education.

Having said that, if you explain mathematical induction to people who never heard about it, they will immediately recognize it as a valid principle. I do not have an explanation for this, but I think it just needs deeper psychological study. Accepting the fact that humans have intuition about mathematical induction simply as a dogma is an unscientific approach.

Notes

1. *Forms of platonism.* Platonism in mathematics has been a matter of intensive research in philosophy. This is quite natural, as it bears on fundamental problems of philosophy. To give you a glimpse of this research I will list several forms that have been identified (see [41, 184]).

   a. *Working realism* is a methodological approach (not a philosophical view) to present mathematics *as if platonism were true.* I assumed this position when explaining mathematical structures and the semantics of first-order logic.

b. *Epistemological and ontological platonisms.* Epistemological platonism is the view that mathematical entities exist and humans can acquire knowledge about their properties, whereas ontological platonism is only the claim that mathematical entities exist. It is disputed whether purely ontological platonism is a tenable view.
c. *Truth-value realism* is defined as the view that every mathematical statement has a definite truth value: either it is true, or it is false. This definition certainly needs an explanation. In mathematics we study various theories and various structures, hence the truth of mathematical statements depends on a particular theory or a particular structure. Therefore the definition should only be applied to statements about concrete structures, such as the natural numbers, the real numbers, the universe of sets etc. Hence this belief presupposes that one can uniquely determine these structures. Most mathematicians agree that the natural numbers are unique, but not everybody accepts the uniqueness of the universe of sets.
d. *Full-blooded, or plenitudinous platonism* is an attempt to reconcile platonism with the incompleteness of axiomatic set theories. According to this view there exist a multitude of set-theoretical universes that have different properties. For example, in some universes the Continuum Hypothesis is true, in others it is false. If applied to all mathematical structures, this view essentially boils down to the Existence from Consistency Principle, discussed in connection with Hilbert's philosophy of mathematics.

2. *A definition of the natural numbers in the polymorphic λ-calculus.* The natural numbers are objects of type

$$(X \to X) \to (X \to X)$$

where X is a type variable.

This is an example of a very concise definition of a fundamental mathematical structure. To explain why this simple formula defines the natural numbers, we first observe that objects of this type are functions that map functions of some unknown type to functions of the same type. (The type is unknown because X is a type variable.) If we do not know anything about the nature of functions, then there is very little we can do with them. Given a function f, we can do precisely the following: we can replace it by the identity function (which is present in every type), or leave it as it is, or iterate it several times. If we express it in symbols,

$$f \mapsto id, \quad f \mapsto f, \quad f \mapsto ff, \quad f \mapsto fff, \quad \dots$$

we see that this naturally corresponds to $0, 1, 2, 3, \dots$. This is not only a convenient *representation* of numbers, but one can also argue that it is an *explanation* of this concept: natural numbers are quantities expressing the "amount of iteration" of a process. One such process is counting objects.

Note that in first order logic the terms that we can form from one unary function symbol,

$$x, \quad f(x), \quad f(f(x)), \quad f(f(f(x))), \quad \dots$$

correspond to the numbers defined in the λ-calculus, but they are only syntactical concepts and cannot be defined inside of logic.

7.2 The Attributes of Reality

The concept of truth in mathematics is elusive and some philosophical schools even reject it as meaningless. Yet looking at concrete examples we feel that we have intuition of what is true and what is not. But there is no mystical faculty of the brain that is above logical reasoning. We accept certain theories as true and others as false because in the back of our mind we have some idea of how reality should behave. The main attribute of reality, we believe, is that it is consistent. Therefore we not only assume that our theories are consistent, but also that they do not imply their own inconsistency. Consistency is the most important attribute of true theories, but there are also reflection principles, which are stronger than consistency. We can add these principles to the theory to make our assumptions explicit. Then we can add reflection principles for these extended theories and so on. We will see how far one can get in this way.

Most set-theorists believe that any axiomatic system is able to cover only a small part of all infinities. Therefore one has to study axioms that postulate higher infinities—the large cardinals. This is a very natural way to extend the axioms of set theory, but it is not sufficiently universal; for example, one cannot decide the Continuum Hypothesis by adding only large cardinal axioms. Still, the study of the consequences of large cardinal axioms is useful and it also gives us hints how to choose other axioms, as we will see shortly.

The Importance of Universal-Finite Sentences

Recall that a *universal-finite sentence* is a sentence asserting that some computable property P holds for all numbers. Thus the form of this sentences is:

$$\forall x\ \psi(x)$$

where the formula $\psi(x)$ expresses that the property P holds for x. The finiteness in the name means that, for a given x, we can test in finite time whether it has the property P. This is not a completely formal definition because it does not say how we express the property. The class of sentences of this kind that has a natural syntactical definition is the class in which the formulas ψ contain only bounded quantifiers. Since the ranges of quantification in these formulas are bounded, we can test $\psi(x)$ efficiently. The class of these sentences is denoted by Π_1 and the sentences are called *Π_1 sentences.*

I also introduced the class $\Pi_{\mathbf{P}}$, the *universal-***P** *sentences*, the subclass of universal-finite sentences in which the property is computable in polynomial time, and called them *empirically testable sentences.* Here the condition of computability is strengthened to polynomial time computability, which captures more accurately what can be computed in practice. The importance of these sentences stems from the fact that these sentences are the output of mathematical research that can be applied in practice.

More specifically, applicable results of mathematics are of two kinds:

1. predictions and
2. algorithms.

In general, scientific theories should give us predictions of the form: if we apply a function f (describing an action) to x (data about an object), we obtain $y = f(x)$ (data about the result of the measurement) that satisfies some relation $R(x, y)$. If the scientific theory is from a field that uses mathematics, such as physics, chemistry etc., the predicted relation between the initial situation and the result, which I denoted by $R(x, y)$, is derived using mathematics. Thus the sentence $\forall x\ R(x, f(x))$ is a mathematical theorem. Should the prediction be practically testable, R and f must be efficiently computable, which implies that the sentence $\forall x\ R(x, f(x))$ is universal-**P**. So predictions are, from the point of view of logic, universal-**P** sentences.

Algorithms are also associated with empirically testable sentences. The reason is that when designing an algorithm it does not suffice only to write down the program; we must also prove that the proposed algorithm does what it is supposed to do. The situation can be formally described in a similar way as above: if f is the algorithm and R is the relation that we want to satisfy, we need to prove $\forall x\ R(x, f(y))$.

In the rest of this section the distinction between universal finite and universal-**P** will not be significant, thus I will only talk about Π_1 sentences. These sentences play an important role in the foundations of mathematics apart from their place in applications of mathematics. In spite of their special form, they occur very often in mathematics. Many open problems, including the Riemann Hypothesis, can be equivalently stated as Π_1 sentences. The statement $\mathbf{P} \neq \mathbf{NP}$ is not a Π_1 sentence and we do not know an equivalent Π_1 form of this conjecture, but if we state a specific lower bound on the complexity of a specific **NP**-complete problem, instead of $\mathbf{P} \neq \mathbf{NP}$, we obtain a Π_1 sentence. Some other conjectures can also be strengthened in such a way in order to obtain a Π_1 sentence. What is especially important for foundations is that the sentences that express consistencies of theories are Π_1.

A simple, but important observation is that it is not possible to prove the independence of a Π_1 sentence without actually proving it. For further reference, I will state it formally and give it a name.

Proposition 13 (The Unambiguity of Π_1 Sentences) *If a Π_1 sentence is independent of a sufficiently strong theory, then it is true.*

The condition of being sufficiently strong is precisely the property called Σ-completeness. Recall that Σ_1 sentences are existential sentences. Formally, they are defined in the same way as Π_1 sentences with the universal quantifier replaced by an existential quantifier. A theory is Σ-complete if it proves every true Σ_1 sentence. Peano Arithmetic is Σ-complete, but even much weaker theories have this property. The simplest Σ-complete theory is *Robinson Arithmetic* (which is essentially Peano Arithmetic without the axioms of induction, see page 116). The point is that in order to prove a Σ_1 sentence, we only have to find the witness for the existentially quantified variable and check that it satisfies the computable property ψ. Essentially, the computation is the proof that the witness satisfies the property.

The principle above is just a reformulation of Σ-completeness. Suppose that a Π_1 sentence ϕ is independent of a Σ-complete theory T. Then, in particular, $\neg\phi$ is not provable in T (and this is the only property needed). But $\neg\phi$ is equivalent to a Σ_1 sentence and T is Σ-complete. So $\neg\phi$ cannot be true, hence ϕ is true. (It is also instructive to see a model-theoretical proof, see Notes.)

The independence of a sentence ϕ from a theory T means that neither ϕ nor $\neg\phi$ is provable in T. We cannot prove the independence of a Π_1 sentence ϕ from a Σ-complete theory without deciding the truth of ϕ because such a proof would entail proving that $\neg\phi$ is not provable and hence ϕ is true. But if we only prove that ϕ is unprovable in T, we do not learn anything about the truth of ϕ. If, for instance, we showed that the Riemann Hypothesis were unprovable in Peano Arithmetic, we would only learn that it must be difficult to prove it if it is true, but it can still go either way. The bottom line is that, for Π_1 sentences, we can only prove that they are difficult to prove, in other words, that they have high proof complexity.

Let us now turn to more philosophical matters. We can restate the Unambiguity of Π_1 Sentences as follows:

If it is consistent to assume that there exists no counterexample to a Π_1 sentence ϕ, then there is none, hence ϕ is true.

If we compare this with the Existence from Consistency Principle (page 602), then the two principles seem to be in conflict. According to the Existence from Consistency Principle, if it is consistent that a number with certain properties exists, then it exists. But for some Π_1 sentences ϕ it can be consistent (with respect to some theory) both that a counterexample to ϕ exists and that it does not. Then the two principles are apparently in conflict because, according to the Unambiguity the counterexample should not exist and according to the Existence from Consistency Principle it should.

However, it is possible to avoid this conflict by interpreting the Existence from Consistency Principle correctly. If the existence of a number with some computable property is consistent with, say, Peano Arithmetic, then there exists a model M of Peano Arithmetic in which there exists such a number n. If a number with this property does not exist in the standard natural numbers, the model M has to be nonstandard. Then we should interpret the Existence from Consistency Principle as saying that *the number exists in some mathematical structure*, the nonstandard model M. Furthermore, we should assign the same status of existence to the nonstandard model M as to the standard model.

The importance of Π_1 sentences should also be reflected in the formation of theories. Indeed, it seems that when we choose between the alternatives of whether to accept a sentence α or $\neg\alpha$ as an axiom, we often prefer the choice that has more Π_1 sentences among its consequences. The simplest example is adding the sentence Con_T (expressing the consistency of T) to a consistent theory T. Should we accept Con_T or $\neg Con_T$? We accept the former because it is true. But this is also the choice that produces more Π_1 consequences: T extended by Con_T trivially proves the Π_1 sentence Con_T, whereas T extended by $\neg Con_T$ does not prove any new Π_1 sentences (see Notes).

Similarly, by adding a large-cardinal axiom we obtain a new Π_1 sentence, while by accepting the negation of this axiom we do not get any. When neither α nor $\neg\alpha$ implies any Π_1 sentence, we do not know what to do. Incidentally, this is the case with the Continuum Hypothesis—both it and its negation have no Π_1 consequences and we still do not have an argument that would conclusively decide which is true.

The Consistency Strength

The purpose of theories proposed as the foundations of mathematics is to provide a framework in which one can formalize the concepts studied in mathematics and represent proofs of theorems about these concepts. Logicians also use these theories to prove that other theories are consistent. It may seem that proving consistencies is only a very special use of these theories, but the opposite is true. If a theory T proves the Completeness Theorem for first order logic, then, working in T, the consistency of a theory S is equivalent to the existence of a model M of S. The theory S may be a set of axioms that specify a certain kind of structures. In mathematics we only study concepts that are "nonempty", which means that they are represented by at least one structure; empty concepts are useless. When we are proving that there exists a structure of the type S that satisfies some additional property ϕ, we are, in fact, proving the consistency of the theory S extended by ϕ.

A theory that should serve as foundations for mathematics should, clearly, be sufficiently strong to prove all basic theorems. In particular, it should be able to prove the Completeness Theorem. A mathematician may object that this theorem belongs to logic and is not important for mainstream mathematics. But the proof of the completeness theorem only needs very weak set-theoretical axioms, hence a theory not able to formalize the proof would be very weak. Therefore it is desirable that theories intended as foundations should prove as many consistencies as possible. We say that they should have big *consistency strength.*

For the rest of this section, I need to change the notation for consistency statements. I will use $Con(T)$ instead of Con_T to improve readability since I will need to substitute complicated expressions for T.

Formally, we define the *consistency strength* of a theory T as the set of all sentences of the form $Con(S)$ provable in T, where $Con(S)$ is a formalization of the consistency of a theory S. If T is Σ-consistent, then it proves only the consistency of consistent theories. We can also define consistency strength using the concept of Π_1 sentences. This is due the following fact.

Proposition 14 *The sentences of the form $Con(S)$ are Π_1, and for every Π_1 sentence ϕ, Peano Arithmetic proves $\phi \equiv Con(Q+\phi)$.*[27]

[27] Recall that in general the expression $T+\phi$ denotes the theory obtained from a theory T by adding a sentence ϕ as an additional axiom and Q is Robinson's Arithmetic—Peano Arithmetic without induction.

Thus the consistency strength of a theory T is simply the set of all Π_1 sentences provable in T. Since theorems that can be presented in the form of a Π_1 sentence play a key role in mathematics, the consistency strength of the foundations matters, whether one is interested in logic or not.

I believe that the crucial reason for Zermelo-Fraenkel Set Theory becoming the number one choice for axioms of set theory was its high consistency strength. This is supported by the ease of formalizations of mathematical concepts, a property of Zermelo-Fraenkel Set Theory inherited from Cantor's presentation of set theory. Furthermore, there is a way to make the consistency strength of Zermelo-Fraenkel Set Theory still bigger, apparently without any limit. Every axiom that introduces a large cardinal bigger than the previous ones has bigger consistency strength because it proves the consistency of the axioms introducing smaller large cardinals. When Gödel talked about *axioms furnishing powerful methods* and Hilbert talked about *modes of inference that are fruitful*, none of them explicitly referred to consistency strength, but in fact this was the essence of what they said.

Big consistency strength of a theory T is a positive property, but we pay for it by taking a bigger risk that T may turn out to be inconsistent. Most theorems only need relatively weak fragments of Zermelo-Fraenkel Set Theory, but there are some that need a substantial part of it and there are a few that even need large-cardinal axioms. So sometimes we have to take the risk of using an inconsistent theory. We have seen how some philosophical approaches cope with the problem of consistency, but there is no satisfactory solution. Apparently, we have to live with the risk.

We can also measure the complexity of a theory T by the difficulty of proving its consistency. We say that T_1 *is more likely consistent than* T_2, if $Con(T_2) \to Con(T_1)$ is provable in some weak base theory, say *PA*. We also say that T_1 *is equiconsistent to* T_2, if $Con(T_1) \equiv Con(T_2)$ is provable in *PA*. Equiconsistency and equal consistency strength are related, but different concepts.

Sound Theories

When we decide to use some theory T as the foundations of mathematics we want T to be consistent, but this is, certainly, not the only requirement. We want much more: T should only prove *true sentences*. But what does this mean? If T is the only formal system that provides us with the information about mathematical structures, how can we then test that T only proves true sentences? Does this requirement make sense if we are not platonists?

Let us call a theory T *true*, if it proves only true sentences. This is just a vague notion that will serve us to start the discourse. (A platonist may accept it as a clear definition, but it will still remain a metaphysical concept, not a mathematical one.) I will show that one can partially formalize this concept by presenting mathematical properties that a true theory must satisfy. The formalized concept will be called *soundness*. The fact that soundness can be defined is very important for the foundations because it gives us hints how to make theories stronger. It is also important for philosophers to know that such a concept can be formalized. The most important thing is to realize that there is a big difference between assuming the axioms of T and assuming that T is sound.

To define soundness we have to model a situation in which there is a theory T and there is a structure M that is intended to be described by T. Due to Gödel's Incompleteness Theorem, the theories we are interested in are always incomplete, so T cannot fully determine the sentences that are true in M. To describe such a situation, we need a metatheory S in which we can talk about the theory T and in which it is also possible to define M. Furthermore, we also need to be able to define satisfiability of formulas of T in M. The metatheory must be sufficiently strong to be able to formalize these things, but otherwise it may be relatively weak. In particular, we do not need the consistency of T to be provable in S.

Suppose now that T is intended to describe the natural numbers using the language of arithmetic with the non-logical symbols $0, 1, +, \times$ and $\leq$. Then we need to have a model M of the natural numbers in S and we should be able to define satisfiability of arithmetical formulas in M. To this end we can take S to be a set theory and we only need fairly week axioms. In such a set theory we have many models that satisfy the axioms of Peano Arithmetic, but there is one that is uniquely determined up to isomorphism, which we call *the standard model of arithmetic.* It is, in a sense, the shortest one, one that is an initial segment of any other. This is the structure that we use to formalize natural numbers in set theory and which we denote by $\mathbb{N}$. It is this structure to which we refer when defining soundness.

This setting enables us to define the soundness of arithmetical theories. So we define in S that an arithmetical theory T is sound, if all axioms of T, hence also all theorems of T, are true in $\mathbb{N}$. A little more generally, we can define *arithmetical soundness* of any theory in which the natural numbers are formalized, which, in particular, concerns set theories.

Definition 21 (Arithmetically sound theories) A theory T in which the natural numbers are formalized is *arithmetically sound* if all arithmetical theorems of T are true in the structure $\mathbb{N}$. (This is a definition in a theory S in which the structure $\mathbb{N}$ is defined.)

In this and the following subsection we will restrict ourselves to the study of arithmetical soundness. One can define more general concepts of soundness, but they are not so natural. Arithmetical soundness is natural because the concept of natural numbers is very robust. For one thing, when considering models of arithmetic, we have one that clearly stands out—the standard model. For another, the only axioms which we use for natural numbers, except for the basic ones, are induction axioms for various classes of formulas. In contrast to this, if we assume in S that there exists a model M of set theory, there is no natural choice of a canonical one. Also it is not clear which axioms of set theory we should assume to hold true in M.

The following is a basic fact about arithmetically sound theories.

Proposition 15 *If T is an arithmetically sound theory, then $T + Con(T)$ is also arithmetically sound, in particular, $T + Con(T)$ is consistent.*

Let me stress that this is a genuine mathematical theorem—here is the proof. Arguing in the metatheory S we observe that if T is arithmetically sound, then T is consistent because it has $\mathbb{N}$ as a model. Hence the arithmetical sentence $Con(T)$ holds in $\mathbb{N}$. Therefore, all axioms of $T + Con(T)$ are true in $\mathbb{N}$, which means that $T + Con(T)$ is arithmetically sound.

This proposition shows, in particular, that the sentence 'T *is arithmetically sound*' is stronger than the sentence 'T *is consistent*'. Indeed, from the mere consistency of T we cannot deduce that $T + Con(T)$ is consistent. For example, if $T = T_0 + \neg Con(T_0)$, where T_0 is consistent, then T is consistent, but $T + Con(T)$ is inconsistent because $Con(T)$ implies $Con(T_0)$.

If T is an arithmetical theory, then arithmetical soundness formalizes the concept of the soundness of T completely. It is therefore interesting to study this concept for arithmetical theories. The assumption that T is arithmetically sound is not expressible by an arithmetical sentence, but there are a lot of interesting arithmetical sentences that are consequences of it. They are based on consistency statements and reflection principles. Starting with a sound arithmetical theory T, we can repeatedly apply Proposition 15 to obtain stronger and stronger theories. We will study this and similar processes shortly.

Reflection Principles

We have seen that soundness entails consistency. While soundness is not expressible in the language of arithmetic, consistency is. It will certainly be interesting to learn other arithmetical consequences of soundness. These sentences may depend on the metatheory, but there are facts that hold true essentially for any metatheory that is able to formalize soundness. The arithmetical consequences of soundness serve us as a paradigm of how we can make theories stronger in general. For the sake of simplicity, I will only consider arithmetical theories in the following two subsections.

Reflection principles are natural generalizations of consistency statements. Recall from the previous chapter (page 498) that the reflection principle for a theory T is the schema (one axiom for every sentence ϕ in the language of T)

Rfn(T): $Pr_T(\ulcorner\phi\urcorner) \to \phi.$

The sentence $Pr_T(\ulcorner\phi\urcorner)$ expresses that ϕ is provable in T. Hence $Pr_T(\ulcorner\phi\urcorner) \to \phi$ expresses that an arithmetical sentence ϕ follows from the provability of ϕ in the theory T. So the schema Rfn(T) is a way to formalize the soundness of T. Although it looks like the right way of expressing soundness, it captures only some consequences of it. The soundness of T cannot be stated using only the language of T, but there are consequences of soundness stronger than Rfn(T) that can be stated in this language, as we will see shortly.

The sentence $Con(T)$ formalizing the consistency of T is a special instance of the reflection principle. Namely, if ϕ is any sentence provably false in T, then $Pr_T(\ulcorner\phi\urcorner) \to \phi$ is equivalent to consistency. For example, the sentence

$$Pr_T(\ulcorner 0 = 1\urcorner) \to 0 = 1,$$

is equivalent to $Con(T)$ because it is, clearly, equivalent to $\neg Pr_T(\ulcorner 0 = 1\urcorner)$. This shows that some instances of the schema are not provable in T; in fact, $Pr_T(\ulcorner\phi\urcorner) \to \phi$ is provable in T only if ϕ is provable in T. (This is *Löb's Theorem.*)

If T is arithmetically sound, then the sentences Rfn(T) are also arithmetically sound. The proof of this fact is essentially the same as for $Con(T)$: If T is arithmetically sound, then all arithmetical sentences that it proves are also true. The set of sentences Rfn(T) is a formalization of the latter fact, thus these sentences are true.

The sentences of Rfn(T) are not consequences of any finite part of the set Rfn(T). Therefore this schema cannot be axiomatized by a single axiom. To state a reflection principle as a single axiom one has to restrict the complexity of sentences used in the schema. For the sake of simplicity, from now until the end of the next subsection, I will consider only arithmetical theories.

Recall that the classes of arithmetical formulas Π_n and Σ_n, for $n = 1, 2, \ldots$, are defined by restricting the number of alternations of unbounded quantifiers to n.[28] By the Gödel-Tarski Theorem 4.2 (page 283), it is not possible to define truth for all arithmetical formulas using an arithmetical formula, but it is possible to define truth for Π_n and Σ_n formulas for every fixed number n. This means that it is possible to define formulas

$$Sat_{\Pi_n}(x, y) \quad \text{and} \quad Sat_{\Sigma_n}(x, y),$$

for which one can prove in T that they satisfy the Tarski inductive conditions[29] for Π_n and Σ_n formulas. The meaning of the variables x and y is that the formula with Gödel number x is satisfied by the string of numbers coded by the number y.

A sentence ϕ is true if it is satisfied by every string of numbers, which is the same as being satisfied by the empty string, because ϕ does not contain free variables. Thus from $Sat_{\Pi_n}(x, y)$ and $Sat_{\Sigma_n}(x, y)$, we can define the truth predicates

$$Tr_{\Pi_n}(x) \quad \text{and} \quad Tr_{\Sigma_n}(x).$$

A consequence of the validity of Tarski's conditions is that, for every sentence $\phi \in \Pi_n$ (and similarly for Σ_n),

$$Tr_{\Pi_n}(\ulcorner\phi\urcorner) \equiv \phi$$

is provable in T.

Now we can state the Π_n- and Σ_n-reflection principles as single sentences:

$$\text{RFN}_{\Pi_n}(T): \quad \forall x\,(P_n(x) \wedge Pr_T(x) \to Tr_{\Pi_n}(x)),$$
$$\text{RFN}_{\Sigma_n}(T): \quad \forall x\,(S_n(x) \wedge Pr_T(x) \to Tr_{\Sigma_n}(x)),$$

---

[28] See page 139.

[29] See page 88.

where $P_n(x)$ and $S_n(x)$ are formalizations of '*x is a* Π_n *sentence*' and '*x is a* Σ_n *sentence*'. Notice that these sentences are different from the corresponding schemata. In a schema, such as $\mathrm{Rfn}(T)$, we state the reflection principle for each concrete sentence. But in $\mathrm{RFN}_{\Pi_n}(T)$ and $\mathrm{RFN}_{\Sigma_n}(T)$ we quantify sentences in the formulas. This makes these principles stronger than the schema $\mathrm{Rfn}(T)$ restricted to Π_n and Σ_n formulas, except for the class of formulas Π_1.

The weakest of these reflection principles is $\mathrm{RFN}_{\Pi_1}(T)$. One can prove that it is equivalent to $Con(T)$. The next one, $\mathrm{RFN}_{\Sigma_1}(T)$ is a very interesting case. As stated, it is an artificial sentence based on syntactical concepts. But often it can be equivalently stated as a truly mathematical sentence. In particular, if T is Peano Arithmetic, then all known "mathematical" independence results are equivalent to $\mathrm{RFN}_{\Sigma_1}(PA)$ (see Chap. 4, Concrete independence). The concrete independent sentences found for stronger theories also have this property: they are equivalent to the Σ_1-reflection principles of these theories.

To see that $\mathrm{RFN}_{\Sigma_1}(T)$ is stronger than $Con(T)$, we observe that

$$\mathrm{RFN}_{\Sigma_1}(T) \to \mathrm{RFN}_{\Sigma_1}\big(T + Con(T)\big),$$

provably in a weak metatheory. Compare this proposition with Proposition 15. The only difference is that the property that T and $T + Con(T)$ are arithmetically sound is replaced by the Σ_1-reflection principle. So we can view Σ_1-reflection principle as an approximation of the arithmetical soundness, one that suffices to prove that the consistency statement can be added to T. Thus we can prove in $T + \mathrm{RFN}_{\Sigma_1}(T)$ that all theories in the progression

$$T_0^{Con} = T, \quad T_1^{Con} = T_0^{Con} + Con\big(T_0^{Con}\big), \quad T_2^{Con} = T_1^{Con} + Con\big(T_1^{Con}\big), \quad \dots \tag{7.1}$$

are consistent.

The set of all axioms $\mathrm{RFN}_{\Sigma_n}(T)$, $n = 1, 2, \dots$ is called *the uniform reflection principle* and is denoted by $\mathrm{RFN}(T)$. This principle is stronger than $\mathrm{Rfn}(T)$ (as the notation suggests).

Assuming T is arithmetically sound, we can add reflection principles to T while preserving truth. Thus we get strengthenings of Proposition 15 such as:

Proposition 16 *If T is an arithmetically sound theory, then $T + \mathrm{RFN}(T)$ is also arithmetically sound.*

To see how strong the uniform reflection principle is, consider, as an example, the following classical result of G. Kreisel and A. Levy [169].

Theorem 60 *Elementary Arithmetic augmented by the uniform reflection principle proves the same sentences as Peano Arithmetic. Formally,*

$$EA + \mathrm{RFN}(EA) \equiv PA.$$

Elementary Arithmetic is a convenient basic subtheory of Peano Arithmetic obtained by restricting the induction schema to Δ_0 formulas and adding an axiom saying that exponentiation is a total function. (This theory is often denoted more conspicuously by $I\Delta_0 + Exp$.) So *EA* is very weak when compared with *PA*. The theorem shows that one can replace essentially all induction axioms by the uniform reflection principle.

Iterating the extension by the uniform reflection principle, we obtain the progression of theories

$$T_0^{\mathrm{RFN}} = T, \quad T_1^{\mathrm{RFN}} = T_0^{\mathrm{RFN}} + \mathrm{RFN}\big(T_0^{\mathrm{RFN}}\big), \quad T_2^{\mathrm{RFN}} = T_1^{\mathrm{RFN}} + \mathrm{RFN}\big(T_1^{\mathrm{RFN}}\big), \quad \dots \tag{7.2}$$

all of which are arithmetically sound provided that T is an arithmetically sound theory. It is a strictly increasing progression, since T_{n+1}^{RFN} proves the consistency of T_n^{RFN}. This progression is increasing faster than (7.1) because adding the uniform reflection principle is more than adding the consistency. In fact, we can insert an infinite progression of the type (7.1) between every two consecutive elements of (7.2).

Transfinite Iterations of Consistency and Reflection Principles

If we want to obtain stronger consequences of the assumption that the theory T is arithmetically sound, the natural next step is to prolong the progressions of theories transfinitely. This idea first appeared in Turing's paper (based on his PhD thesis) in 1939 [294].

Let us first consider the operation of adding consistency. To prolong the progression (7.1), define T_ω^{Con} to be the union of all theories T_n^{Con}, $n \in \omega$. I am now using ω to denote the natural numbers because this is the symbol that one uses in the context of ordinal numbers. We need ordinals to define transfinite iterations. Informally, we can express T_ω^{Con} by

$$T_\omega^{Con} \equiv T + Con(T) + Con\big(T + Con(T)\big) + Con\big(T + Con\big(T + Con(T)\big)\big) + \cdots.$$

The next theory is

$$T_{\omega+1}^{Con} \equiv T_\omega^{Con} + Con\big(T_\omega^{Con}\big).$$

In general, we want to define, for every constructive ordinal α, a theory T_α^{Con} so that the progression satisfies the natural conditions of the definition by transfinite recursion:

1. $T_0^{Con} := T$;
2. $T_{\alpha+1}^{Con} \equiv T_\alpha^{Con} + Con(T_\alpha^{Con})$;
3. if λ is a limit ordinal, then T_λ^{Con} is equivalent to the union of theories T_α^{Con} for $\alpha \prec \lambda$.

I will use the symbol $\prec$ for a strict ordering on ordinals. Since ordinals will be represented by natural numbers, we need to distinguish between the ordering of numbers and ordinals.

Note that one can only construct such progressions for *constructive* ordinals. The reason is that in order to formally define T_{α}^{Con} we have to formalize theories T_{β}^{Con} for all $\beta \prec \alpha$, in particular, we have to formalize these ordinals. In fact, one cannot formalize such progressions uniformly for all constructive ordinals. It is only possible, for every given constructive ordinal α, to define a progression of theories indexed by ordinals less than α. I will denote such a progression by $\{T_{\beta}^{Con}\}_{\beta \prec \alpha}$, but this notation is rather misleading. These progressions are not uniquely determined by the theory T and the ordinal α. They are sensitive to the particular choice of the formalization of ordinals. For a constructive ordinal α, a well-ordering of natural numbers of type α can be defined by various arithmetical formulas $\nu(x, y)$. Different choices of ν, for the same ordinal α, may define theories that prove different theorems (even if we only considered Π_1 theorems). In concrete instances one uses ordinals that possess certain standard notations. When such notations are formalized in a natural way, the ambiguity of the definition of transfinite progressions disappears. Therefore, throughout this subsection I will assume that the ordinals and the transfinite progressions are formalized in a natural way. In particular I will assume that a natural ordering of a sufficiently large type is fixed and denote it by $\prec$.

Transfinite progressions based on iterating consistency, as defined above, can be used to measure the strength of true Π_1 sentences. Namely, given such a progression $\{T_{\beta}^{Con}\}_{\beta \prec \alpha}$ and a true Π_1 sentence ϕ, we can ask what is the least ordinal β such that ϕ is provable in the theory T_{β}^{Con}, provided that there is any such theory in the progression. In some cases one can even characterize *all* Π_1 sentences provable in the theories T_{β}^{Con}. The prototype of such results is the theorem of U.R. Schmerl [256].

Theorem 61 *The Π_1 theorems of $EA_{\varepsilon_0}^{Con}$ are exactly the Π_1 theorems of Peano Arithmetic.*

In this theorem $EA_{\varepsilon_0}^{Con}$ is the last theory in the progression $\{EA_{\alpha}^{Con}\}_{\alpha \preceq \varepsilon_0}$ that starts with *EA*.[30]

Interesting results have also been proved for transfinite iterations of reflection principles. I have mentioned that the Σ_1-reflection principle is linked with fast growing functions. This connection was used by L. Beklemishev to show that transfinite hierarchies of fast growing functions can be characterized by transfinite progressions based on iterating the Σ_1-reflection principle [18]. Namely, for a particular hierarchy of recursive functions f_α, the axiom saying that f_α is a total function is equivalent to the αth theory in the progression.

Transfinite progressions are a very useful tool for analyzing theories, but let us now look at this concept from the point of view of the foundations of mathematics. The first impression is that they may be an ideal means for discovering truth in arithmetic. It looks as if we could start with a very weak theory, such as *EA*, and only assuming that it is arithmetically sound we could get to much stronger theories. It is true that we can get beyond *PA* but not very far.

---

[30]Recall that ε_0 is the limit of $\omega, \omega^\omega, \omega^{\omega^\omega}, \ldots$.

In order to prove that EA_α^{Con} is arithmetically sound, we need two assumptions:

1. that EA is arithmetically sound, and
2. transfinite induction up to α.

If α is sufficiently large, then 2. is not a consequence of 1.; in fact, it can be much stronger than 1. So in such a case we are using an assumption about ordinals, not the soundness of EA. One possible way to avoid the additional assumption about transfinite induction was proposed by G. Kreisel and S. Feferman [70, 168]. The idea is based on the concept of *autonomous progression.* These are, roughly speaking, progressions $\{T_\beta^{Con}\}_{\beta \prec \alpha}$ such that on stage β the theories defined so far can prove transfinite induction up to β. Thus starting with a theory that proves transfinite induction only for small ordinals, we may still get fairly long progressions because we may prove transfinite induction for larger ordinals on the way.

When we are adding only consistency statements, this does not work. For example, although we can get a little beyond EA by assuming that it is arithmetically sound and adding consistency statements, we cannot reach PA. The reason is that when adding consistency statements we are only adding Π_1 sentences. These sentences do not change the ordinals for which one can prove transfinite induction. In our example the autonomous progression $\{EA_\alpha^{Con}\}$ stops at ω^2 because EA proves transfinite induction only for ordinals less than ω^2.[31]

The situation changes dramatically when we use reflection principles (at least Σ_1-reflection principle) instead of consistency. Then the self-enforcing effect takes place: the further the progression goes, the more ordinals we can use; then we can go even further and prove well-foundedness for even more ordinals etc. But however strong it may seem, it still has an upper bound; it stops at some constructive ordinal. The general reason for the existence of an upper bound is that this is an algorithmic construction and therefore it cannot exhaust all constructive ordinals.

As the theory of autonomous progression is rather complicated, it is better to look at a concrete example. Consider the transfinite progression that starts at PA and is based on iterating the uniform reflection principle. Thus the progression satisfies the clauses:

1. $T_0^{\mathrm{RFN}} := PA$;
2. $T_{\alpha+1}^{\mathrm{RFN}} := T_\alpha^{\mathrm{RFN}} + \mathrm{RFN}(T_\alpha^{\mathrm{RFN}})$;
3. if λ is a limit ordinal, then T_λ^{RFN} is the union of theories T_α^{RFN} for $\alpha \prec \lambda$.

Define that this *progression is autonomous at an ordinal* β if transfinite induction up to $\beta + 1$ is provable from T_γ^{RFN} for some $\gamma \prec \beta + 1$. Say that it is *autonomous below* α if it is autonomous at every $\beta < \alpha$.

Transfinite induction up to β is the schema: *for all arithmetical formulas* $\phi(x)$,

$$\forall y \prec \beta\big((\forall z \prec y\, \phi(z)) \rightarrow \phi(y)\big) \rightarrow \forall x \prec \beta\, \phi(x).$$

This is the set of sentences that is often used to express that $\prec$ is a well-ordering up to β. The particular form is not important; what is, however, important is that

[31] Here I am talking about transfinite induction for Δ_0 formulas.

we state it as a *schema for all arithmetical formulas*. Thus it is a generalization of the usual induction that is postulated in Peano Arithmetic. We can view the usual induction as a special case of transfinite induction, namely, as transfinite induction up to ω.

Although in *PA* the axioms postulate induction only up to ω, one can derive in this theory transfinite induction for every $\alpha < \varepsilon_0$, and, as we noted in Chap. 6, ε_0 is the precise bound. For the other terms in the progression $\{T_\alpha^{\mathrm{RFN}}\}$, the bounds were computed by Schmerl [256].

Theorem 62 *For every α, the ordinal ε_α is the largest ordinal ξ such that T_α^{RFN} proves transfinite induction up to β for every $\beta < \xi$. Moreover, T_α^{RFN} is equivalent to PA augmented with the schema of transfinite induction up to β for every arithmetical formula $\phi(x)$ and every $\beta < \varepsilon_\alpha$.*

Let us recall the definition of the ordinals ε_α. These are ordinals ε satisfying $\omega^\varepsilon = \varepsilon$, indexed in the increasing order. Thus ε_0 is the least such ordinal, ε_1 is the second least and so on.

One can nicely demonstrate the self-enforcing effect of autonomous progressions by the following example. Since *PA* (which is the first term T_0^{RFN}) proves transfinite induction for every $\alpha < \varepsilon_0$, we know that the progression is autonomous below every ε_α for $\alpha < \varepsilon_0$. Since $\lim\{\varepsilon_\alpha;\ \alpha < \varepsilon_0\} = \varepsilon_{\varepsilon_0}$, the progression is autonomous below $\varepsilon_{\varepsilon_0}$. The same argument shows that it is autonomous below $\lim\{\varepsilon_\alpha;\ \alpha < \varepsilon_{\varepsilon_0}\} = \varepsilon_{\varepsilon_{\varepsilon_0}}$ and so on. Hence the progression is autonomous below

$$\lim\{\varepsilon_0, \varepsilon_{\varepsilon_0}, \varepsilon_{\varepsilon_{\varepsilon_0}}, \ldots\}.$$

The name of this ordinal is $\phi_2(0)$. Since for every $\alpha < \phi_2(0)$, we have $\varepsilon_\alpha < \phi_2(0)$, this is as far as we can go; the progression is not autonomous at $\phi_2(0)$.

The theory $T_{\phi_2(0)}^{\mathrm{RFN}}$ is the union of T_α^{RFN} for $\alpha < \phi_2(0)$, so it is the theory that characterizes what is "autonomously provable". How strong is this theory? If compared with Peano Arithmetic, it is much stronger; if compared with typical set theories, it is much weaker. One can show that very weak set-theoretical axioms give much stronger consequences. The natural framework for studying the strength of autonomous progressions is in subsystems of Second-Order Arithmetic. These are theories in which the language of arithmetic is extended so that one can speak about sets of numbers. Introducing these theories would take us too far afield, so I confine myself to a few remarks about it in Notes.

Interlude—Incompleteness and the Human Mind

In 1961 the British philosopher John Lucas published a paper in which he argued that the Incompleteness Theorem implies that humans are qualitatively different from machines, in particular, human thinking cannot be simulated by computer. Although logicians immediately rejected his argument, Lucas's argument was acclaimed by some philosophers as a proof that artificial intelligence is impossible, that is, that human thinking is superior to what can be achieved by any man-made device. Several variations of the argument appeared afterwards. The typical form of the argument is this:

> *Let C be a computer that supposedly simulates human thinking. Represent the set of sentences that C is able to prove by a formal system T and apply the Incompleteness Theorem. Thus we get the sentence γ_T that is not provable in T, hence not provable by the computer C. But we, humans, know that γ_T is true, hence we know more.*

Here γ_T is Gödel's self-referential sentence expressing that γ_T is unprovable in T, which is equivalent to the consistency of T, assuming very mild conditions about T. Hence if T is consistent, then γ_T is true.

It is difficult to point to mistakes in these attempted proofs because the arguments are always very vague. Moreover, they are wrapped into lengthy discussions which makes it difficult to distinguish between comments and what should be the steps in the alleged proof. When explaining their "proofs" the authors not only use different wordings, but also change the essence of the argument. But if someone is patient enough to extract the logical structure of the argument, in spite of all these obstacles, a logical gap becomes immediately apparent.

But why am I so sure that the proof is wrong? It is very simple. If you want to prove that X proves (knows, ascertains, ...) more than Y, then you must assume more about X. Hence if you want to prove that humans are different from machines, you must assume that they have certain properties that machines do not. But this is exactly what you want to prove. So these alleged proofs either contain a logical error or tacitly use assumptions that, more or less, trivially imply the conclusion.

Therefore, if one wants to show that machines cannot do what human brains can, one has to use some *empirical facts* about human brains. The authors should be fair and state these facts explicitly as the assumptions of the proof. It is certainly wrong when the empirical fact used in the proof only appears in the form such as "*we can see to be true*" or "*a rational being can, standing outside the system, see that it is true*". Once the empirical assumptions are stated explicitly, the discourse can focus on the only relevant question (which has nothing to do with mathematics): whether or not these assumptions are true facts.

Let us have look at the errors that occur in these attempted proofs anyway. I was able to find three kinds of errors which I explain below. Note, however, that some arguments can be classified in more than one way because we always have to guess what the intended meaning of the argument was.

1. *Failing to distinguish an object in the proof from the subject presenting the proof.*

In my presentation of the argument above the gap is clear: the claim that we know that γ_T is true is not justified. We know that γ_T is true, if we know that T is consistent, but how do we know that T is consistent? Suppose we believe that T is consistent. Then C also believes that T is consistent, since it simulates our thinking.

The problem is that the authors of the argument confuse the subject that is making this argument with the objects about which they are talking. So let us sort it out and distinguish who is who. There is the computer C, there is a human H that C is supposed to simulate, and there is another person presenting the argument. I will call the human H 'she', the person presenting the argument 'we', and the computer C is, naturally, 'it'.

The first thing to note is that the argument requires the assumption that T, the set of sentences that C can prove, is consistent. Otherwise C can prove any sentence and we cannot deduce that H proves more than C. This assumption was not stated in the paragraph presenting the alleged proof above, but it is assumed either implicitly or explicitly in these arguments.

So we want to prove that if T is consistent, then H proves more sentences than C. The consistency of T is an assumption in our proof, not a fact that H knows. We cannot use this assumption to deduce that H knows that T is consistent. Note that we also do not know if T is consistent, we are only assuming it in order to prove the implication. Thus the argument breaks down.

In order to expose the error more clearly I distinguished H and we, but it does not matter if they are the same person. One only needs to realize that if we make an assumption about H, or about ourselves in an argument, it is only an *assumption*, it is not a fact that we know.

Example Lucas's original argument is an example of this kind of error. His paper starts with an explanation why Gödel's self-referential sentence for a formal system is not provable in the system—he just repeats the basic argument of Gödel's proof of the Incompleteness Theorem. A few paragraphs below he states the essence of his argument:

> "Gödel's theorem must apply to cybernetical machines because it is of the essence of being a machine, that it should be a concrete instantiation of a formal system. It follows that *given any machine which is consistent* and capable of doing simple arithmetic, there is a formula which it is incapable of producing as being true—i.e., the formula is unprovable-in-the-system-but *which we can see to be true*. It follows that no machine can be a complete or adequate model of the mind, that minds are essentially different from machines."[32]

Look at the pieces of text that I italicized. In terms of C, H and 'we', the first italicized piece of text is the assumption that C is consistent. So in the proof we know that C is consistent (the second italicized piece). But this is only *our assumption* in the proof—it does not imply that H knows that C is consistent.

Another example is in Penrose's book *Shadows of the Mind.* Penrose's argument concludes with the following paragraph:

> "Moreover, if we *know* that A is sound, then we *know* that $C_k(k)$ does not stop. Thus we know something that A is unable to ascertain. It follows that A *cannot* encapsulate our understanding."[33]

In that argument A stands for the sentences available to the computer (which I denoted by T above) and '$C_k(k)$ *does not stop*' is the Gödel sentence for the formal

[32] J.R. Lucas, *Minds, machines and Gödel.* Philosophy, 36 (1961), pages 120–124; the italics are mine.

[33] R. Penrose, *Shadows of the Mind*, Oxford Univ. Press (1994), page 73.

system A (which I denoted by γ_T above). Penrose deduces that "*we know something that A is unable to ascertain*" from the assumption that "*if we know that A is sound*" and then he goes on, as if the latter one was not only an assumption, but a proven fact.

Some people may still doubt that it is an error not to distinguish the object from the subject. So I made up an example where it is absolutely clear. I will use the fact that no mathematician knows that the Riemann Hypothesis is true. In my argument I replace 'we' by 'I'. The argument is about me, so 'H' (the human) also represents me. The computer 'C' is replaced by a mathematician. I will "prove" that I know more than any other mathematician.

I will suppose that RH is true. Thus I know that RH is true. Since I am H, H knows that RH is true. Hence H knows more than any other mathematician.

It is an absolute nonsense, but the reasoning is exactly the same as in the arguments above.

2. *Using the assumption that humans know that they are consistent.*

Here we have to be more careful and distinguish between '*know*' and '*believe*'. A person knows ϕ if he believes that ϕ is true and, moreover, ϕ is true. Gödel's formula concerns belief, not knowledge, because truth cannot be defined in the language about which it talks.

It seems perfectly all right to assert:

What I believe, including this sentence, is consistent.

But if you carefully read Chap. 4 (see page 293), you already know that although such sentences can be written formally, they are always inconsistent (provided that the belief includes sufficiently strong axioms of arithmetic). This is an immediate consequence of the Second Incompleteness Theorem. Thus if H believes that she is consistent, she is inconsistent, and then C is also inconsistent because we assume that C simulates H.

One can also use a weaker assumption: for every consistent sentence ϕ, H can recognize that ϕ is consistent. This assumption does not imply that H is inconsistent, but it implies that H can decide an algorithmically unsolvable problem. To conclude from this assumption that humans can do more than computers, we do not need the Incompleteness Theorem.

Variations of this use soundness or truth instead of consistency. But the point is not so much what the assumption is, but the fact that there is an additional assumption that does not appear in the claim. So instead of proving:

humans can do more than machines,

the proof only gives:

if humans can recognize consistent (sound, true, ...) sentences, then they can do more than machines.

Such a statement is not interesting because we know that recognizing consistent (sound, true, ...) sentences is an algorithmically unsolvable problem.

Example An example of this kind of error is in Penrose's Reply to Commentaries on Shadows of the Mind:

> "We try to suppose that the totality of methods of (unassailable) mathematical reasoning that are in principle humanly accessible can be encapsulated in some (not necessarily computational) sound formal system F. A human mathematician, if presented with F, could argue as follows (bearing in mind that the phrase "I am F" is merely a shorthand for "F encapsulates all the humanly accessible methods of mathematical proof"):
> (A) "Though I don't know that I necessarily am F, I conclude that *if I were, then the system F would have to be sound* and, more to the point, F' would have to be sound, where F' is F supplemented by the further assertion "I am F". ..."[34]

Note the italicized part where he uses the assumption that human beings believe that they are sound.

3. *The failure to distinguish between consistency and soundness.*

A typical spurious argument based on this misunderstanding is as follows:

> *According to Gödel's Theorem a formal system T (of the sentences available to a computer, etc.) is unable to prove: 'if T is consistent, then also $T + Con(T)$ is consistent'. But we, humans, immediately see that it is true.*

> The author goes on and explains why it is true:

> *Since T is sound, and $Con(T)$ is a true sentence, $T + Con(T)$ is also sound, hence consistent.*

The last line is perfectly correct—it is exactly Proposition 15 from the previous subsection. What is wrong is an unjustified shift: the assumption above that T is consistent has been replaced by the assumption below that T is sound. While C is only allowed to use $Con(T)$, we can use the soundness of T. As we know from the previous two subsections, the second assumption is much stronger.

Example It seems that this kind of confusion was a source of a mistake in the following argument in Lucas's paper:

> "Even if we adjoin to a formal system the infinite set of axioms consisting of Gödelian formulae, the resulting system is still incomplete, and contains a formula which cannot be proved-in-the-system, although a rational being can, standing outside the system, see that it is true."[35]

There is nothing mysterious in the conclusion that we can add consistency to the resulting system if we are assuming that we start with a sound system as we have seen in the previous subsection. The soundness of T can be formalized, and thus it

[34] R. Penrose, *Beyond the Doubting of a Shadow*, PSYCHE, 2(23), (1996); the italics are mine.

[35] J.R. Lucas, *Minds, machines and Gödel*, Philosophy 36, 120–124, (1961).

can be one of the sentences given to the computer. If this is the case, the computer can make the same deduction, namely, it can also derive that $T + Con(T)$ is sound, hence consistent. So a computer knowing the soundness would also recognize all these axioms as true.

Incidentally, Gödel did believe that the human mind has abilities superior to artificial devices. In a lecture given in 1951 he said:

> "...if the human mind were equivalent to a finite machine, then objective mathematics not only would be incompletable in the sense of not being contained in any well-defined axiomatic system, but moreover there would exist absolutely unsolvable problems..., where the epithet 'absolutely' means that they would be undecidable, not just within some particular axiomatic system, but by any mathematical proof the mind can conceive."[36]

But notice that unlike Lucas and Penrose, Gödel did not claim to prove the superiority of humans. Gödel only argued that the assumption that human thinking could be simulated by machines had an unlikely consequence—the existence of problems that we would never solve. But the question whether the consequence is unlikely is a matter of the philosophy that one accepts.

Another often quoted part of the same paper is referred to as *Gödel's dichotomy:*

> "Either mathematics is incompletable in this sense, that its evident axioms can never be comprised in a finite rule, that is to say, the human mind (even within the realm of pure mathematics) infinitely surpasses the powers of any finite machine, or else there exist absolutely unsolvable Diophantine problems of the type specified.[37] "

The essence of both statements is the same. In the second one he talks about formal systems, instead of machines, and is more specific about the nature of the unsolvable sentences. In both statements the empirical evidence that should enable us to show the difference between the human mind and machines is stated explicitly. This evidence, or rather belief, is the non-existence of absolutely unsolvable problems.

The human mind is certainly not equivalent to a formal system, but in spite of its complexity, changing content, etc., we can estimate the *logical strength* of the mathematical assumptions that a particular person uses. Logical strength is not a function of the complexity of knowledge; it only depends on the mathematical axioms the person believes. Then, using this estimate, we can indeed produce a sentence that the person is unable to prove. But in order to deduce that the sentence is true, we have to use stronger axioms than those that the person uses.

---

[36] K. Gödel, *Some basic theorems on the foundations of mathematics and their implications*, in [99], page 310.

[37] Because at that time Hilbert's 10th problem was still open, Gödel refers to the representation of Π_1 sentences in the form $\forall x_1 \dots \forall x_n \exists y_1 \dots \exists y_m\ p(x_1, \dots, x_n, y_1, \dots, y_m) = 0$, where p is a polynomial.

What Do We Actually Mean when We Accept Some Axioms?

Most of the discussion concerning the applicability of Gödel's Theorem to the human mind revolves around the problem that the collection of facts that a person accepts as true is too complex to be formalized. What we have in our brains is, indeed, a very complex structure composed of pieces of information. Most likely, if we applied logic strictly, we would also find out that it is inconsistent. But all this complexity, and possibly inconsistency, is due to facts that have nothing to do with the logical strength of our mathematical assumptions. Furthermore, to determine the logical strength of the mathematical facts we do not need the totality of all the theorems that a person knows, but only the axioms he uses. So if we only ask about the axioms, we can learn the strength of somebody's mathematical assumption fairly easily—we can simply ask the person which axioms of set theory he believes. If we only use these axioms, and not the full description of his knowledge, the construction of the Gödel sentence can also be fairly simple.

The problem is, however, that the axioms which a person believes are not precisely delimited. Mathematicians would typically say that their assumptions are exactly the axioms of Zermelo-Fraenkel Set Theory, but when we ask them if they also accept $Con(ZFC)$ as true, they would say yes. And we can go on and ask about $Con(ZFC + Con(ZFC))$, get the same answer and so on.

Some philosophers of mathematics proposed the explanation that humans have some higher faculty that enables them to recognize truth. But this is an unscientific approach because there is a very simple explanation that does not refer to any as yet unknown phenomena: if mathematicians accept a set theory (or another kind of theory) T as foundations for mathematics, then they always assume that T is sound. We know that this entails that extensions such as $T + Con(T)$ are also sound. It would certainly be very unreasonable to accept a theory that is not sound. In particular, if we assume arithmetical soundness, we get all reflection principles, and even some iterations of them.

Let us continue our interview and ask the mathematician if his assumptions can be characterized by $ZFC + ArithSound(ZFC)$, where $ArithSound(ZFC)$ is a sentence expressing the arithmetical soundness of ZFC. At this point he would probably say that he does not understand, or does not care. But suppose he would say that this is what he means. Then we can continue and ask: what about the true sentence expressing the consistency of his theory—the sentence that we denote by $Con(ZFC + ArithSound(ZFC))$?

But how do *we* know that this sentence is true? What we have in the back of our mind is that we can again use the arithmetical soundness assumption. But now it is not enough to assume that ZFC is arithmetically sound because the sentence $ArithSound(ZFC)$ is not arithmetical. We need a higher order soundness that talks about relations on natural numbers. Thus we are leaving the firm ground of arithmetic and have to use concepts that are not as clear as the numbers. At this stage we can still find a reasonable formalization of soundness because the set-theoretical formulas that we need are of very low complexity. But the situation will recur and we will need still higher and higher concepts of soundness. We are being dragged more deeply into set theory and things are becoming less and less clear. The consequence is that the arguments saying that we can go on making the theory stronger by adding assumptions about soundness will be less and less convincing. In any case, eventually we will need to define soundness with respect to all set-theoretical formulas. This requires us to go beyond *ZFC*. One can certainly do this, but if our mathematician were to say that this is what he believes, he would be contradicting himself because originally he said *ZFC* and now he would be talking about a completely different theory.

This is obviously just an idealized example. In reality, set theorists know that extending set theory in such a way is very inefficient and unnecessarily complicated. Instead of trying to formalize the concept of soundness and extending the theory by axioms based on this concept, we can add an axiom saying that there exists an inaccessible cardinal. This is a very simple axiom and is at least as strong as those one would get by using soundness. Its simplicity is an advantage not only when using the axiomatic system, but also in the theoretical study of it. It help us to assess how likely it is that this assumption is consistent (and if we are platonists, how likely it is that it is true).

Therefore, research into the set-theoretical foundations of mathematics focuses on large-cardinal axioms. The ideas of truth, soundness and reflection principles are, nonetheless, used in this area and some principles studied in large-cardinal theory are called reflection principles. In this area one can also study "attributes of reality", that is, what we implicitly expect to be true when we postulate some large-cardinal axiom. One such attribute is the belief that if a large cardinal with property P exists, then there exist many such cardinals.

Let us get back to our mathematician. What can we conclude about his mathematical assumptions? I think we have to conclude that the strength of his belief is not the same for all assumptions that he accepts. As we go beyond *ZFC* (in our example situation) the strength gradually decreases. If we imagine that the strength can be measured by real numbers, then these numbers would gradually approach zero. But there would be no particular place where it would fall from a positive value to zero. If we, however, go far enough it could be zero. This is not in contradiction with the fact that he has finite memory. The scale concerns all assumptions that he can *in principle* consider.

Since the structure of the assumptions that one can consider is very complicated, I will give a simplified example.

Example Let a theory T be fixed and consider the progression of theories T_α^{Con} obtained by adding consistency for ordinals $\alpha \leq \omega^2$. Recall that every ordinal $\alpha < \omega^2$ can be written as $\alpha = \omega \cdot m + n$, where m and n are natural numbers. Suppose that the strength of our mathematician's belief in $T_{\omega \cdot m+n}^{Con}$ is expressed by the number $1/m$ and it is 0 for $T_{\omega^2}^{Con}$. Then the degree of belief in a particular theory S in this progression is the same as the degree for $S + Con(S)$. If, however, we go from S to S_ω^{Con}, then the degree decreases but is not zero, assuming it is not zero for S. If we make the leap bigger by going from S to $S_{\omega^2}^{Con}$, it will drop from nonzero to zero.

This is not a realistic example because for most mathematicians, their belief will not decrease with the jump from S to S_{ω}^{Con}. Our mathematician believing in *ZFC* would probably accept all reflection principles with the same confidence, but he may have less belief in $ZFC + ArithSound(ZFC)$. The strength of his belief in *ArithSound*(*ZFC*) must be positive, otherwise he would not use it to support his beliefs in reflection principles. There is nothing wrong in believing in the general principle *ArithSound*(*ZFC*) less than in its consequences, although it may seem a little strange at the first sight. Such situations are not uncommon in science, or even in daily life. In science we often extrapolate and generalize our partial knowledge and we are aware of the fact that the generalizations are only conjectures that can be refuted. Note that I am not claiming that the more general a principle is, the less we believe in it; there are some very general principles, such as formal logic, that we trust the most. I am only saying that in some cases we are not quite sure that a general principle to which special cases seem to point is correct. In theoretical physics even inconsistent theories are used to calculate with very high precision what elementary particles do. In set theory, logicians propose very big leaps by postulating large-cardinal axioms with increasing strength. That we have more confidence in *ZFC* than in its extension by a very strong large-cardinal axiom is quite natural.

The Large-Cardinal Program

There are some *ad hoc* definitions of large-cardinal axioms that are used to prove some general theorems about them, but it seems that it is impossible to define them in such a way that the definition would capture completely our idea of what a large cardinal is. Here is my attempt to give at least an approximate description.

Large-Cardinal Axioms *A large-cardinal axiom postulates the existence of an infinite cardinal number κ that cannot be obtained from smaller large cardinals by using methods by which the smaller ones were defined.*

Let us consider some "non-examples". The axiom saying that κ is an inaccessible cardinal below which there is a smaller inaccessible cardinal, in other words, that there exist at least two inaccessible cardinals, is not provable from the axiom saying that there exists at least one. Yet we do not accept it as a large cardinal axiom because the step from one inaccessible cardinal to two is "too small"—it is not based on a new idea. A much stronger axiom is this:

> *There exists an inaccessible κ such that there are κ inaccessible cardinals below κ.*

This is still not considered a new large-cardinal axiom, since the idea of this definition is essentially the same as in the definition of an inaccessible cardinal, except that it is applied to inaccessible cardinals instead of all cardinals.

What one would like to put in the definition is that '*the condition defining κ speaks inherently about largeness*'. But this is impossible because each new large-cardinal axiom introduces *a new kind of largeness.*

Axioms postulating the existence of very large cardinalities were the first axioms that were used to make *ZFC* stronger. Unlike some other axioms extending *ZFC*, large cardinals axioms are not controversial. Set theorists do not dispute whether to accept a large-cardinal axiom, or rather its negation. The only issue on which they may disagree is whether it is likely that a particular large-cardinal axiom is consistent or not. It seems that large-cardinal axioms are the kind of sentences to which the Existence from Consistency Principle is applicable: if it is consistent that a particular large cardinal exists, then it exists.

We have also noted that large-cardinal axioms enable us to prove more Π_1 sentences, thus they may have even practical consequences. But they also decide some sentences about sets of real numbers (see page 214).

The usefulness and the noncontroversial nature of these axioms is the reason for trying to use large-cardinal axioms as a sort of universal means to decide sentences that are not decidable using only the axioms of *ZFC*. An early proponent of this approach was Gödel; therefore this program is also called *the Gödel Program.* Here is what Gödel wrote about it in 1946.

> "It is certainly impossible to give a combinational and decidable characterization of what an axiom of infinity is; but there might exist, e.g., a characterization of the following sort: An axiom of infinity is a proposition which has a certain (decidable) formal structure and which in addition is true. Such a concept of demonstrability might have the required closure property, i.e., the following could be true: Any proof for a set-theoretic theorem in the next higher system above set theory (i.e., any proof involving the concept of truth which I just used) is replaceable by a proof from such an axiom of infinity. It is not impossible that for such a concept of demonstrability some completeness theorem would hold which would say that every proposition expressible in set theory is decidable from the present axioms plus some true assertion about the largeness of the universe of all sets."[38]

Before turning to results in this program, which often require rather technical concepts, I should mention another application of large cardinals, which can be explained very easily. This is the empirical evidence that large-cardinal axioms can be used to calibrate all proposed set-theoretical axioms. More precisely, for almost all axioms α that have not been shown inconsistent with *ZFC*, either the consistency of $ZFC + \alpha$ is provable already in *ZFC* or $ZFC + \alpha$ is equiconsistent to $ZFC + \beta$, where β is some large-cardinal axiom. Recall that the large cardinals so far considered form an almost linear scale; specifically, except for a few rare exceptions, the consistency strengths of any two large cardinals are comparable. Thus large-cardinal axioms serve as a linear scale for measuring consistency strength. This scale could be in principle used for other theories, not only for extensions of *ZFC*, but the consistency of essentially all other theories is provable already in *ZFC*.

Although large cardinal axioms are statements about very large sets, they are more likely to decide the truth of low complexity sentences than those that are

[38] *K. Gödel, Collected Works, Vol II, Oxford Univ. Press, p.* 151.

higher in the hierarchy. Therefore I will start with sentences talking about finite sets, more precisely about *hereditarily finite sets.* These are finite sets whose elements are finite and the elements of whose elements are finite, and so on. These sets can be easily enumerated by natural numbers; hence one can equivalently study arithmetical sentences instead of these sentences.

We know that the set of provable arithmetical sentences increases with the increasing strength of large-cardinal axioms.[39] It is, therefore, conceivable that for every true arithmetical sentence ϕ, a large cardinal axiom eventually will be found which will imply ϕ. Unfortunately there is no evidence for this in mathematical practice—none of the well-known open problems in number theory has been solved by a large cardinal axiom.

Problem 8 Can every arithmetical sentence be decided by a large-cardinal axiom?

It should be stressed that this is not a mathematical statement that can be proved or disproved. In order to answer this question one will have to use some formal definition of large-cardinal axioms, or at least some obvious property that must be satisfied by all such axioms.

It may be easier to solve some special instances of this problem. In particular it seems likely that the problem has a positive answer if restricted to Π_1 sentences. We know that every Π_1 sentence follows from its consistency (and the axioms of Robinson Arithmetic). Hence Problem 1 restricted to Π_1 sentences can be equivalently stated as follows:

Problem 9 Can the consistency of any consistent theory be proved in *ZFC* plus some large-cardinal axiom?

Until now all proofs of consistency have been done in *ZFC*, possibly augmented with large-cardinal axioms. Is this evidence for the claim that large cardinal axioms suffice, or is it just because we do not have better ideas about how to prove consistencies?

Large Cardinals and Forcing

There are sentences about infinite sets that are decided by large cardinal axioms, but they are scarce. Therefore a more pragmatic approach has been proposed, based on the following idea: it is certainly not possible to avoid incompleteness, but maybe we can at least avoid the known methods for proving independence. As we are now considering set theory, namely extensions of *ZFC*, we have only two methods: Gödel's

[39]For this, we do not need the strong condition on large cardinal axioms stated above. For example, the axiom saying that there are at least two inaccessible cardinals also has more arithmetical consequences than the axiom postulating the existence of at least one.

Second Incompleteness Theorem (the unprovability of consistency) and Cohen's forcing. Gödel's method produces independent Π_1 sentences, but we know that if such sentences are independent, then they are true. By this method combinatorial Π_2 sentences were also shown to be independent from Peano Arithmetic and from some stronger systems, including *ZFC* plus some large cardinals. But again, it is clear that these sentences are true. Gödel's Second Incompleteness Theorem is also used to prove that a large-cardinal axiom is not provable from large-cardinal axioms postulating smaller cardinals. Also in this case it is clear which sentence (if consistent) we should consider true.

Hence from the practical point of view, Gödel's incompleteness is harmless. But we cannot say this about Cohen's forcing. On the positive side, it is conceivable that one can avoid the incompleteness results produced by the forcing method by extending *ZFC* with some axioms. The best would be to add axioms for which we have reasons to assume that they are true. What we have at hand are large-cardinal axioms. As we will see, they alone cannot do the job, but they still perform very well at the bottom of the hierarchy of set-theoretical sentences.

The aim of this approach is to obtain an extension of *ZFC* that is invariant under forcing in the following sense. Let T be an extension of *ZFC* and ϕ a sentence. Say that *in* T *the sentence* ϕ *is invariant under forcing, if the independence of* ϕ *from* T *is not provable using forcing.* To be more specific, I will also define this using generic extensions of models, on which the forcing method is based.

Definition 22 In an extension T of *ZFC* a sentence ϕ is invariant under forcing, if for every model M of T and every generic extension $M[G]$ that is also a model of T, the following is true: if ϕ is true in M, then ϕ is also true in $M[G]$.

If a sentence ϕ is provable in T, then it is, clearly, invariant under forcing in T. As we will see shortly, there are also sentences that are not provable in T and are invariant under forcing in T.

It seems that there is no consistent extension T of *ZFC* in which all sentences are invariant under forcing.[40] Therefore this approach focuses on sentences of low complexity. What has been achieved is best explained using the following three sets of sets of low complexity. Let $H(\omega_0)$ denote the hereditarily finite sets, $H(\omega_1)$ denote the hereditarily countable sets and $H(\omega_2)$ denote the sets that are hereditarily of cardinality at most $\aleph_1$. Let $\Phi(\omega_0)$, $\Phi(\omega_1)$ and $\Phi(\omega_2)$ denote the sets of sentences that speak about sets in $H(\omega_0)$, $H(\omega_1)$ and $H(\omega_2)$ respectively. Note that in $\Phi(\omega_i)$ the range of quantification is also restricted to $H(\omega_i)$. We can equivalently define $\Phi(\omega_i)$, for $i = 0, 1, 2$, by saying that $\Phi(\omega_i)$ is the set of all first order sentences of the structure $(H(\omega_i); \in)$.[41]

[40] Recall that in this book I am always assuming that theories are axiomatized by algorithmically decidable sets of axioms.

[41] The explanation of the notation $H(\omega_i)$ is that these are sets of sets of hereditary cardinality *less than* ω_i. The notation $\Phi(\omega_i)$ is introduced only for this section.

Sentences of $\Phi(\omega_0)$ are studied in finite combinatorics and in the theory of natural numbers. As we have already noted, we do not have to take the whole of $\Phi(\omega_0)$; it suffices to consider only arithmetical sentences. These sentences are already invariant under forcing in *ZFC*. The reason is very simple: a generic extension $M[G]$ has the same ordinals as M, hence, in particular, $M[G]$ has the same natural numbers as M. Thus $M[G]$ satisfies the same $\Phi(\omega_0)$ sentences as M.

Sentences in $\Phi(\omega_1)$ are a very interesting case—there exists a large-cardinal axiom α such that $ZFC + \alpha$ makes all sentences in $\Phi(\omega_1)$ invariant under forcing. To state the axiom α we need the concept of a Woodin cardinal, mentioned in Chap. 3 (see page 229). This invariance result is the theorem of Woodin [309, 310].

Theorem 63 *In the theory*

$$ZFC + \text{'for every cardinal } \kappa\text{, there exists a larger Woodin cardinal'}$$

every sentence of $\Phi(\omega_1)$ *is invariant under forcing.*

If we extend *ZFC* by classes, we can express the axiom by saying that there are so many Woodin cardinals that they do not form a set; in other words, *there exists a proper class of Woodin cardinals.* Although it is formally not correct to use it in the context of *ZFC*, logicians prefer this version.

The axiom used in this theorem is not strictly speaking a large-cardinal axiom because it does not postulate the existence of a single large cardinal. Most authors, however, call also such axioms large-cardinal axioms.[42]

This theorem is probably the most remarkable result in the large cardinal program. It shows that we can indeed sharpen our blurred image of the universe of sets using large cardinal assumptions.

That said, in practice this large cardinal axiom is not very convenient. It speaks about very large sets, while we are only interested in countable sets. It would be better to have a low complexity axiom that would do the same job. There is no single axiom that could be used instead of the large cardinal assumption, but one axiom is very close to it. It is the *Axiom of Projective Determinacy*, abbreviated by *PD*. This axiom suffices for all practical purposes, which means that the $\Phi(\omega_1)$ sentences that occur in mathematical practice are invariant in $ZFC + PD$. Projective determinacy is provable in *ZFC* extended with the axiom postulating the existence infinitely many Woodin cardinals (one does not need a proper class of Woodin cardinals).

To explain this axiom, recall the concept of an infinite game and the Axiom of Determinacy (see page 219). A game is given by a set X of countably infinite sequences of zeros and ones; the sequences in X are the winning positions of the first Player 1; the other sequences are winning position for Player 2. The Axiom of Determinacy says that every game is determined, which means that for every X, one of the players has a winning strategy. This axiom is inconsistent with *ZFC*,

---

[42]It is very easy to define a large-cardinal axiom that is only slightly stronger. Such is the axiom postulating the existence of an inaccessible cardinal λ such that there are λ Woodin cardinals below λ. If one wants to use an established large cardinal, then the supercompact cardinal would suffice.

that is, it is provable in *ZFC* that there is a nondetermined game. The construction of such a game uses the Axiom of Choice in an essential way, hence the game, or more precisely the set X that defines the game, is not explicitly definable. For some classes of definable sets, it has been proved that the sets in the classes always define determined games. The Axiom of Projective Determinacy says that for a very large class of definable sets the games are determined. Namely, it postulates that all games definable in the structure $(\mathbb{N}, \mathcal{P}(\mathbb{N}), +, \times, \in)$ are determined. (The name 'projective' comes from the name for sets definable in this structure—these sets are called projective.)

The case of the structure $H(\omega_2)$ and the sentences $\Phi(\omega_2)$ is more complicated. In 1961 Solovay proved that the existence of a measurable cardinal implies a property of the universe of sets that points to the negation of *CH* (see page 214). But soon after that Levy and Solovay proved that a measurable cardinal cannot decide *CH* and some other statements that were proved independent by forcing [183]. Their result, stated more precisely, is as follows.

Theorem 64 *Measurable cardinals are preserved by generic extensions that use small sets of forcing conditions.*[43]

What this means is that given a model with a measurable cardinal, we can do essentially all forcing constructions and we will still have this cardinal in the models. In particular, we can change the validity of the Continuum Hypothesis in both directions while keeping the measurable cardinal in the model. (The technical condition about small sets of forcing conditions is satisfied in these constructions.) A generalization of this result states that, for a particular definition of large cardinals, no large cardinal can make all $\Phi(\omega_2)$ sentences invariant under forcing. In particular, none of the studied large-cardinal axioms makes all $\Phi(\omega_2)$ sentences invariant under forcing.

So an axiom of a different kind is needed. Such an axiom was proposed by Woodin and denoted by $(\star)$.

Theorem 65 *In the theory*

$$ZFC + \text{'there exists a proper class of Woodin cardinals'} + (\star)$$

every sentence of $\Phi(\omega_2)$ *is invariant under forcing.*

The innocent-looking $(\star)$ is a rather complicated sentence that I am not going to explain. Let me just say that it belongs to the so called *forcing axioms*, one of which is the Martin Axiom (see page 363).

Although it seems to settle the case of $\Phi(\omega_2)$ sentences, it is not exactly what platonists would like to have. Woodin proved that if the theory in Theorem 65 without $(\star)$ is consistent then it is consistent also with it, but what is not known is whether

[43]The latter condition on generic extensions means that the cardinality of the partial order defining the forcing conditions is less than the measurable cardinals that should be preserved.

this theory is consistent with *larger cardinals*. Why is this important when we do not know if large cardinals are consistent anyway? The point is that large cardinals seem to go in the right direction. As we have seen, postulating their negation would be rather unnatural. So once we believe that large cardinal axioms are true, all sentences that are incompatible with them must be false.

There is more to say about this case, but things get really complicated. Without going into details, I will state some results concerning the Continuum Hypothesis. As originally stated, *CH* is not a sentence about $H(\omega_2)$, but there is a sentence in $\Phi(\omega_2)$ that is equivalent to it in *ZFC*. Thus the study of the invariance of sentences of $\Phi(\omega_2)$ could give us some clue whether one should accept *CH* or reject it. The axiom $(\star)$ implies the negation of *CH*; in fact, it implies that $2^{\aleph_0} = \aleph_2$. But this fact is a rather weak argument for the failure of *CH* because there may be other axioms that make $\Phi(\omega_2)$ sentences invariant under forcing and imply *CH*.

Woodin developed a very complicated theory aiming to prove that this cannot happen. He proved (see [312]):

Theorem 66 *Assuming the Strong Ω-conjecture and the existence of a proper class of Woodin cardinals, if $ZFC + \alpha$ is an extension of ZFC in which $\Phi(\omega_2)$ sentences are invariant under forcing and which is compatible with large-cardinal axioms, then $ZFC + \alpha$ refutes CH.*

(The Strong Ω-conjecture will be explained in Notes.) This theorem would be a good evidence for refuting *CH*, but unfortunately there is still something missing—the proof of the Ω-conjecture. There are a number of results that hint that the Ω-conjecture should be true, but even if the Ω-conjecture were proved, the theorem above could not be accepted as a very strong evidence for the failure of *CH* because there are other results that point to the truth of *CH*. For example, an earlier result of Woodin shows that if one wants to make another set of sentences invariant under forcing, the Σ_1^2 sentences, then one has to make *CH* true (see Theorem 67 in Notes).

Although there are results that help us understand what is going on at these small levels of the set-theoretical universe, it is still not possible to say conclusively whether *CH* or its negation is true. My personal view is that one may eventually be able to decide *CH* by presenting a collection of axioms and a persuasive justification of this extension of *ZFC*. If this happens, it will be a great achievement, but we will still have to cope with incompleteness in set theory. I am afraid that the justifications for sentences that are higher in the hierarchy will become less and less compelling as we climb the hierarchy upward. This trend is quite apparent already in the three levels discussed in this subsection.

Do We Need Axioms of an Essentially New Kind?

Set theory is the field where most of the search for new axioms is going on. Most of this research concerns principles about infinite sets. Here I want to discuss axioms that could help us prove theorems about finite sets. I will focus on the most basic type of sentences, the Π_1 sentences. I view the problem of finding axioms that decide Π_1 sentences as one of the most fundamental problems in the foundations of mathematics, for the reasons I have already explained.

It may seem that this problem is not urgent because there are no examples of truly mathematical Π_1 theorems that needed special axioms. It is possible that all open problems that are of this form are provable in the axiomatic systems that we are already using, such as Zermelo-Fraenkel Set theory, or even Finite Set Theory. Indeed, it seems that even such difficult proofs as Wiles' proof of the Fermat's Last Theorem can be done in Finite Set Theory. But it is also possible that we do not have such examples of independent sentences simply because we do not have suitable methods to prove independence, and that, in fact, some open problems are independent of the theories that we are using. In any case, we know that there are true Π_1 sentences that are not provable in our theories and it is likely that among them there are some that we need now or will need in the future.

Let us recall that true Π_1 sentences are the simplest sentences to which the Incompleteness Theorem applies; thus, in particular, there is no formal system in which they are all provable. But we also know that they have a very special property, the unambiguity (see page 610), which suggests that deciding their truth may be simpler than for sentences of higher complexity. Another special property is that every Π_1 sentence follows from the statement expressing its consistency. So we only need to find ways to prove arithmetical sentences that express the consistency of theories.

Recall also that ordinal analysis, one of the most important technique in proof theory, reduces the proof of the consistency of a theory T to the statement that, roughly speaking, a particular constructive ordinal α is well-ordered. More precisely, the consistency of T is reduced to a sentence expressing the fact that a particular representation of α defines a well-ordering. If we disregard the "small complication" that one has to find a suitable representation of the ordinal, it seems that we do have a method for systematically generating theories that will exhaust all true Π_1 sentences. We will construct larger and larger ordinals α and, from the sentences expressing transfinite induction up to α, we will derive more and more true Π_1 sentences. There is no systematic way of constructing large ordinals, but it seems that it is easier to come up with a large constructive ordinal than to invent a completely new axiom. Furthermore, it seems easier to recognize a well-ordering than a true Π_1 sentence.

Unfortunately, this does not work in practice. Pictures of ordinals, such as those on page 186, may give the impression that we can easily recognize what is an ordinal and what isn't. But these pictures are deceiving since these ordinals are very small. Imagine rather that you are given a very complicated definition of a binary relation. How would you then find out that it defines a well-ordering? Finding (representations of) large constructive ordinals is also very difficult. Already for moderately strong theories one needs very large constructive ordinals with highly nontrivial proofs of the well-ordering property of their notations. For Second-Order Arithmetic, which is just a tiny fragment of set theory, these ordinals are beyond the scope of current techniques.

However, we are now discussing not practice, but philosophy, so we should rather ask: can this work at least *in principle?* Namely:

Problem 10 Is it the case that for every true Π_1 sentence ϕ, there exists a constructive ordinal α such that ϕ follows from the sentence expressing transfinite induction up to α?

This question is still stated too vaguely. In order to state it as a mathematical problem we would need to say precisely for which representations of the ordinal α we state transfinite induction. If we fix a representation A of a large constructive ordinal α and watch what is going on for ordinals $\beta < \alpha$ represented by initial segments of A, then usually everything works perfectly: not only do we get an increasing hierarchy of Π_1 sentences as the ordinals grow, but we also get that the arithmetical part of the theories corresponding to these ordinals can be fully axiomatized just using transfinite induction, suitably stated for these ordinals. All of this only works for fixed α and A. If we want to consider all constructive ordinals, we have to talk about arbitrary representations of ordinals, and then we run into trouble. We obtain the trivial answer that every Π_1 sentence follows from transfinite induction stated for a suitable notation for ω because one can encode the truth of the sentence in the definition of the notation. We, clearly, need to restrict the set of possible definitions of well-orderings, but we do not know how. As a matter of fact, it is even not clear whether one can formalize the question above in a nontrivial and meaningful way.

Transfinite induction stated for a representation of a constructive ordinal is a Π_2 sentence. We met concrete Π_2 sentences independent of Peano Arithmetic in Chap. 4 and such sentences are also known for fairly strong theories. These sentence are very special in that they define fast growing computable numerical functions. Constructive ordinals are also used to define hierarchies of fast growing functions and the fact that these functions are well-defined is proved using transfinite induction over these ordinals. But fast growing functions are interesting independently of constructive ordinals. In particular, they can be used to derive more Π_1 sentences. One can show by concrete examples that adding an axiom that a fast growing function is total (defined for all numbers) results in more Π_1 sentences being provable.

Why should we use more complex Π_2 sentences to derive simple Π_1 sentences? Again, as in the case of ordinals, fast growth is a concrete combinatorial property and it apparently enables us to state stronger and stronger principles without any upper limit. Unfortunately this idea suffers the same problem as the one based on constructive ordinals: we do not know how to formalize it. We would like to describe functions so that the only information that we get from the definition is their asymptotic growth. We need to avoid trivial examples in which information about the truth of Π_1 sentences is encoded in the definitions. It seems that there are only two ways to define fast growing functions: (1) using ordinals, or (2) using specific combinatorial principles, such as those that we saw in Chap. 4. If we use ordinals then we only transform this problem into another one. If we use combinatorial principles to define fast growing functions, then we cannot say that the Π_1 sentences that we derive from these functions are derived only using their fast rate of growth. So again we do not know if this idea is sound and can be formalized.

There is another type of axioms that produces new true Π_1 sentences; these are axioms postulating the existence of large cardinals, which I discussed in the previous subsection. If α and β are two large-cardinal axioms and it is provable in *ZFC* that β defines a cardinal larger than α, then also $ZFC+\beta$ proves the consistency of $ZFC+\alpha$. Thus the consistency of $ZFC+\alpha$ is a Π_1 sentence unprovable in $ZFC+\alpha$ and provable in $ZFC+\beta$.

All three types of axioms mentioned above talk about infinity: transfinite ordinals, functions whose values go to infinity and infinite cardinals. The higher the infinity is, the more Π_1 sentences we get. This is not only empirically verified, but one can also prove theorems confirming this fact. What is not clear is whether we can get all Π_1 true sentences in this way. I have stated this problem for large cardinals and constructive ordinals, but it is a more general problem.

Problem 11 Does infinity alone (such as large constructive ordinals, fast growing functions and large cardinals) suffice for proving all true universal-finite sentences, or do we also have to look for axioms of an essentially different nature?

Gödel believed that in fact one can decide *every sentence of set theory* using a "true assertion about the largeness of the universe of all sets". We have noted that this is not true unless one comes up with a completely new type of large-cardinal axioms, but the answer may still be positive for Π_1 sentences.

The statement of Problem 11 is only an informal question, a question from the philosophy of mathematics rather than a mathematical problem. We do not have a formal mathematical statement that expresses its meaning because we do not have a formal definition of axioms of infinity. Finding a formalization of this problem, even if it were only of a special case, would already be big progress.

Although the history of mathematics points rather to the first possibility, that infinity is all that we need, I think we should also seriously consider the second possibility—that we may need axioms that do not postulate infinities in any form. One piece of supporting evidence for the second possibility comes from set theory. We know that the large cardinal program has certain limitations, namely, large-cardinal axioms cannot decide some relatively low complexity sentences about infinite sets. This is still a long way from arithmetical sentences, not to say from Π_1 sentences, but it does show a certain weakness of the axioms of infinity.

When looking for Π_1 sentences that do not follow from axioms of infinity what comes to mind are open mathematical problems of this logical complexity (such as the Riemann Hypothesis). I think, however, that we should rather look for such sentences in logic and complexity theory. These fields are closely connected with the foundations of mathematics and therefore it is more likely that in these fields there are some problems that are not decidable using standard means.

Consider Conjecture 6 (page 570) that says that for every finitely axiomatized consistent theory T, there exists a total polynomial search problem Q which is strictly stronger than all search problems that T properly formalizes. This conjecture itself could be such an axiom, but it is more interesting to see what it would imply. If the conjecture were true, we would have an essentially different way of extending our theories by new Π_1 sentences: rather than postulating that some fast-growing functions are total, these sentences would say that certain polynomial search problems are total. Admittedly, it is neither clear whether such sentences would be useful, nor how we would find them, but this is not important for our theoretical discussion. What is important is that in this way we would be able to state axioms of a different nature.

Assuming that we do need axioms of an essentially new nature, how will we ascertain this fact? How will we prove that we need a new kind of axioms? This is a big problem and it may not have a mathematical solution. All I can suggest is to scrutinize the arithmetical consequences of axioms of infinity and look for a property that all consequences of axioms of infinity have. When we find one, then we should look for true sentences that do not have this property.

Notes

1. *A model-theoretical proof of the Unambiguity of Π_1 Sentences.* This proof rests on the fact that every model of Peano Arithmetic contains the standard model as an initial segment. Another fact needed in the proof is that if a Π_1 sentence is true in a model M, then it is also true in its standard submodel. The proof is now easy: If a Π_1 sentence ϕ is consistent with Peano Arithmetic, then there exists a model in which it is true, but then it is also true in its standard part which is isomorphic to the natural numbers. Hence ϕ is true.
2. *Are independence proofs of* $\mathbf{P} \neq \mathbf{NP}$ *possible?* The sentence $\mathbf{P} \neq \mathbf{NP}$ can be equivalently stated as a Π_2 sentence by formalizing the following sentence:

   For every polynomial time algorithm A, there exists a CNF formula ϕ for which A decides the satisfiability of ϕ incorrectly.

   Here we are using the fact that satisfiability of Boolean CNF formulas is **NP**-complete. It seems unlikely that we can push this down to Π_1, so the Principle of Unambiguity of Π_1 Sentences does not apply here. But note two things.

   a. When proving lower bounds, we usually do not talk about all polynomials, but rather about one function that grows faster than all polynomials. Suppose f is such a function and suppose moreover that it is efficiently computable. For example, $f(x) = x^{\log x}$. Then the formalization of the following sentence is a Π_1 sentence.

      For every n and every circuit C of size at most $f(n)$, there exists a formula ϕ of size n for which C does not decide the satisfiability of ϕ correctly.

      (There is an existential quantifier, but it is a bounded one, since it only quantifies formulas of size n.) This sentence is stronger than $\mathbf{P} \neq \mathbf{NP}$, but if f grows very slowly, the difference is not essential.

b. If we strongly believe that $\mathbf{P} \neq \mathbf{NP}$, then showing that $\mathbf{P} = \mathbf{NP}$ is unprovable in a theory T is less interesting than showing that $\mathbf{P} \neq \mathbf{NP}$ is unprovable in T. If $\mathbf{P} \neq \mathbf{NP}$ is indeed true, the former fact would only confirm our belief, but would not say anything about the theory T. The latter fact would, however, show that it is hard to prove $\mathbf{P} \neq \mathbf{NP}$—one would need a theory stronger than T to prove it. Now, if we proved that $\mathbf{P} \neq \mathbf{NP}$ were unprovable in T, then also all the Π_1 statements of the form above would be unprovable. But this still would have no consequences concerning the truth or falsehood of $\mathbf{P} \neq \mathbf{NP}$.

Let me recap in terms of models. It is unlikely that we will be able to construct a model in which $\mathbf{P} \neq \mathbf{NP}$ without actually knowing that $\mathbf{P} \neq \mathbf{NP}$ is true. But it is conceivable that we will be able to construct a model in which $\mathbf{P} = \mathbf{NP}$ and that would show that it is difficult to prove that $\mathbf{P} \neq \mathbf{NP}$.

3. *The role of the metatheory.* Let S be our metatheory. To define the arithmetical soundness of a theory T in S, we must have a definition of arithmetical truth. For example, *PA* does not suffice because, by the Gödel-Tarski Theorem, we cannot define arithmetical truth by an arithmetical formula.

   Suppose we can define satisfiability of arithmetical formulas in S. Then we can define the concept of arithmetical soundness, but this may still not be enough. We also need to prove some properties about it. In particular, we would like to be able to show that if an arithmetical sentence ϕ is true, then every sentence derivable from ϕ, hence also the theory axiomatized by ϕ, is sound. To this end we have to be able to apply induction to the definition of satisfiability of arithmetical formulas.

   In theories such as *ZFC*, or Z_2 (see below), we have induction for all formulas. Hence, whenever we have a definition of satisfaction for a class of formulas, we can apply induction. This is not true in general.

   In *PA*, where we do not have a definition of satisfaction for arithmetical formulas, the closest approximation to arithmetical soundness is the uniform reflection principle $\mathrm{RFN}(T)$, which can only be expressed by an infinite set of sentences.

   The following proposition is an illustration of what is going on in *PA*.

Proposition 17 *For every arithmetical sentence ϕ, $PA \vdash \phi \to \mathrm{RFN}(\phi)$, hence in particular $PA \vdash \phi \to Con(\phi)$.*

Note that in $\mathrm{RFN}(\phi)$ and $Con(\phi)$ we consider the theory axiomatized by the single axiom ϕ; the proposition would be false if we used $PA + \phi$.

Proof Suppose that n is given such that ϕ is a Σ_n sentence, or a sentence of lower complexity. We are to prove $PA \vdash \phi \to \mathrm{RFN}_{\Sigma_n}(\phi)$. In *PA* we can prove the cut-elimination theorem, which implies that if a Σ_n sentence ψ is provable from ϕ, then there exists a proof d that uses only Σ_n sentences. Using the definition of satisfiability Sat_{Σ_n} we can now prove by induction that all formulas in d are satisfied by all assignment of numbers to their free variables.

In particular, ψ is true. (See also Lemma 2.34 and Theorem 2.35 in Part III of [111].) □

To sum up, we need two conditions for formalizing the concept of soundness:

- the language of the metatheory should be able to define truth of the sentences under consideration;
- induction should hold for a class of formulas that includes the formula defining truth.

In informal reasoning about foundations people often make assumptions that are not justified; one such mistake is taking the above two conditions for granted. It is important to keep in mind that every theory, including metatheories, has limitations on what the language can express.

4. *Proof of Proposition* 14.

   a. The provability of $\phi \to Con(Q+\phi)$ in *PA* is an immediate consequence of Proposition 17.
   b. Let ϕ be $\forall x\ \psi(x)$ where $\psi(x)$ is a bounded formula. We will argue in *PA*. Suppose $Con(Q+\phi)$ is true. If ϕ is false, we have $\neg\psi(n)$ for some n. By Σ-completeness, Q proves $\neg\psi(\bar{n})$. But $\forall x\ \psi(x)$ is an axiom of $Q+\phi$, hence $Q+\phi$ is inconsistent contrary to our assumption. Hence ϕ is true. Thus we have proved $Con(Q+\phi) \to \phi$ in *PA*.

5. *Equal consistency strength vs. equiconsistency.* If T is an arithmetical theory axiomatized by a *finite* number of axioms, then *PA* proves $T \equiv T + Con(T)$ by Proposition 17. So T and $T + Con(T)$ is an example of a pair of equiconsistent theories that do not have the same consistency strength.
6. *Extending T by $\neg Con(T)$ does not add new Π_1 sentences.* The precise statement of this fact is:

Proposition 18 *Let T be a theory in which the Completeness Theorem for predicate calculus and the Second Incompleteness Theorem are provable. Then $T + \neg Con(T)$ proves the same Π_1 sentences as T.*

We will derive the proposition from the following lemma.

Lemma 17 *For every model M of T, there exists a model M' of $T + \neg Con(T)$ such that $\mathbb{N}^M$ (the natural numbers of M) is an initial segment of $\mathbb{N}^{M'}$ (the natural numbers of M').*

Let M be a model of T. If $M \models \neg Con(T)$, we can take $M' = M$. So suppose that $M \models Con(T)$. Using the Incompleteness Theorem we get $M \models Con(T + \neg Con(T))$. Using the Completeness Theorem we can define in M a model M' of $T + \neg Con(T)$. From the point of view of M, $\mathbb{N}^M$ is the standard model, so it is an initial segment of every model of natural numbers, in particular $\mathbb{N}^M$ is an initial segment of $\mathbb{N}^{M'}$.

Let us now prove the proposition. Suppose that $T + \neg Con(T)$ proves a Π_1 sentence ϕ. To prove that ϕ is provable already in T, we need to show that ϕ

holds true in every model of T. So let M be an arbitrary model of T and let M' be the model from the lemma; then $\mathbb{N}^{M'} \models \phi$. Since $\mathbb{N}^M$ is an initial segment of $\mathbb{N}^{M'}$, ϕ is true also in N^M, which means that it is true in M.

The Completeness and the Incompleteness theorems are provable already in $I\Sigma_1$, the subtheory of *PA* in which the induction schema is restricted to Σ_1 formulas.

7. *More reflection principles.* The uniform reflection principle RFN(T) is usually stated as the schema

$$\forall x \left(Pr_T\left(\ulcorner \phi(\bar{x}) \urcorner\right) \rightarrow \phi(x)\right)$$

for all formulas $\phi(x)$. This is equivalent to the definition I used, which is the union of $\mathrm{RFN}_{\Sigma_n}(T)$, $n = 1, 2, \dots$. Further, it is equivalent to the *formalized ω-rule*, which is

$$\forall x\ Pr_T\left(\ulcorner \phi(\bar{x}) \urcorner\right) \rightarrow \forall x\ \phi(x).$$

The formalized ω-rule follows from the uniform reflection principle using only logic. The opposite implication requires a proof. Recall that the ω-rule is used to eliminate induction axioms. This is a hint for the proof of Theorem 60.

ω-consistency, introduced by Gödel, is the schema

$$\forall x\ Pr_T\left(\ulcorner \phi(\bar{x}) \urcorner\right) \rightarrow \neg Pr_T\left(\ulcorner \exists x\ \neg\phi(x) \urcorner\right).$$

It can also be viewed as a kind of reflection principle.

8. *The definition of transfinite progressions of theories.* Consider progressions starting with some arithmetical theory T and obtained by adding consistencies. First observe that we can replace the three conditions 1., 2. and 3. on page 618 by the following informally stated condition:

$$T_\alpha^{Con} := T + \left\{Con(T_\beta^{Con}); \beta \prec \alpha\right\}.$$

In order to formalize this condition in arithmetic, we need an arithmetical formula $\sigma(x, y)$ whose meaning is '*x is an axiom of T_y^{Con}*'. Let $\tau(x)$ be a formula that defines the axioms of T. Then the condition above is formalized by

$$\forall x \forall y \left(\sigma(x, y) \equiv \tau(x) \vee \exists z \left(z \prec y \wedge x = \ulcorner Con(\sigma_{\bar{z}}) \urcorner\right)\right), \qquad (7.3)$$

where $\ulcorner Con(\sigma_{\bar{z}}) \urcorner$ is a formalization of the consistency of the theory axiomatized by sentences whose Gödel numbers x satisfy $\sigma(x, z)$. This is a self-referential formula to which one can apply the Diagonal Lemma 2 (page 291). If one applies this lemma to the theory T, which ensures that (7.3) is provable in T, then the progression is uniquely defined. This means that for any choice of $\sigma(x, y)$ such that T proves (7.3), we get equivalent theories in the progressions.

Ambiguity may be caused by the choice of the formula for the axioms of T and the choice of the formula for the ordering of ordinals $\prec$. If one chooses natural formulas representing these concepts, the progressions behave as we expect. One can specify some properties that natural formalizations should satisfy, but there is no general definition. "Pathological" formalizations were found by Turing [294] and Feferman [70]. Turing showed that for every true Π_1 sentence

ϕ, one can formalize $PA_{\omega+1}$ so that it proves ϕ. This has, however, no practical meaning because in order to see that such a formalization is correct, one must know *a priori* that ϕ is true.

9. *Subsystems of Second-Order Arithmetic.* Second-Order Arithmetic, Z_2, is a fairly strong theory. Five subsystems of Z_2 have been identified as important benchmarks: RCA_0, WKL_0, ACA_0, ATR_0 and $\Pi^1_1 - CA_0$ (listed in order of increasing strength). For us, only ACA_0 and ATR_0 are important, but I will also mention another theory, ACA.

   ACA_0 is the subsystem of Z_2 in which the Schema of Comprehension is restricted to arithmetical formulas with sets only as parameters. In other words, ACA_0 has a comprehension axiom for every formula in which sets are not quantified. This theory is a conservative extension of *PA*, which means that it proves the same arithmetical sentences as *PA*. The acronym stands for *Arithmetical Comprehension Axiom* and the zero indicates that we only have an induction axiom of the form stated above instead of a schema of induction for all formulas.

   ATR_0 is an extension of *PA* in which an axiom schema postulates that it is possible to define sets by transfinite recursion using arithmetical formulas. Thus *ATR* stands for *Arithmetical Transfinite Recursion*. The schema can be informally described as follows.

   Let $\phi(x, Y)$ be an arithmetical formula with a set parameter Y, let $\prec$ be a well-ordering of natural numbers. Then there exist sets Y_a for every number a such that

$$x \in Y_a \equiv \phi\big(x, \{Y_b\}_{b \prec a}\big).$$

   To state the schema formally, first represent the indexed system Y_a by the binary relation $Z = \{(a, x); x \in Y_a\}$, and then represent the binary relations $\prec$ and Z as sets using the pairing function on natural numbers.

   The schema of transfinite recursion of ATR_0 works in a similar manner as autonomous progressions. If we want to prove the existence of a set by transfinite recursion, we first need to prove that the relation representing the ordering is a well-ordering. This connection is not superficial. Feferman defined a theory *IR* that proves the same arithmetical sentences as ATR_0 and used *IR* to characterize the sentences provable in a certain autonomous progression [71].

   In the example of a transfinite progression that I described on page 621 the resulting theory $T^{\mathrm{RFN}}_{\phi_2(0)}$ is only a small subset of arithmetical theorems of ATR_0. This is quite apparent from the comparison of the ordinals for which the theories prove transfinite induction (for arithmetical formulas). While $T^{\mathrm{RFN}}_{\phi_2(0)}$ proves transfinite induction only for ordinals less than $\phi_2(0)$, ATR_0 proves transfinite induction for all ordinals less than Γ_0. Recall that Γ_0 is the least ordinal γ such that $\gamma = \phi_\gamma(0)$, so it is much bigger than $\phi_2(0)$. This shows that simple set-theoretical axioms are much stronger than transfinitely iterated arithmetical principles.

   ACA_0 is said to be a *predicative extension* of *PA*, because the formulas in the comprehension schema of ACA_0 use sets only as parameters. Hence when defining a new set we do not refer to the totality of all sets, but only to parameters which are interpreted as sets that we have already constructed at previous stages. Repeated applications of such a comprehension axiom can be viewed as defining sets by recursion. Kreisel's idea from the late 1950's is that there is no reason to restrict the recursion to ω. If the theory is able to see that α is well-ordered, we should allow recursion up to α. This was the seminal idea that led to introducing various predicative systems. The ordinal Γ_0 and the theory ATR_0 are often presented as upper bounds on predicative ordinals and theories.

ACA is the strengthening of ACA_0 in which induction is postulated for all formulas in L_2. While in the theories with the induction axiom (all five theories mentioned above) induction holds for the same class of formulas as comprehension, here this is not the case.

10. *Predicativism.* This is a stream in the philosophy of mathematics whose aim is to develop mathematics on the basis of predicative theories. It is based on the idea, due to Poincaré and Russell, that one can secure the consistency of formal systems by using only predicative definitions of sets. Restricting to predicative definitions presents rather severe limitations since non-predicative definitions are used very often in mathematics. Hermann Weyl developed methods to overcome these difficulties at least in some cases [307]. Currently the main proponent of predicativism is Solomon Feferman.
11. *Well-orderings and autonomous progressions.* An informal argument suggests that an upper bound on autonomous progressions $\{T_\beta^{\mathrm{RFN}}\}_{\beta<\alpha}$ is any theory S in which one can prove that T is arithmetically sound. In other words, the arithmetical soundness of T should suffice to prove that all theories in the progression are sound. The idea is to use transfinite induction. In order to get to T_β^{RFN} we need transfinite induction up to β. By the induction assumption S knows that all theories T_γ^{RFN}, for $\gamma < \beta$, are arithmetically sound. Since these theories prove transfinite induction up to β, and because of the autonomy of the progression, S knows that transfinite induction up to β is true. So apparently we have enough transfinite induction in S to prove that all theories in the progression are sound. However attempting to formalize this argument one immediately finds the hidden flaw. When we formalize transfinite induction up to β in T, we can state it only for formulas in the language of T. In S we must be able to express the truth for all formulas of T and this is not possible in the same language. Since the theory S must have a richer language, we also need transfinite induction for formulas in the richer language. The theory S does prove transfinite induction up to β for formulas in the language of T, but this is not enough. (This mistake appears in otherwise excellent book of T. Franzén [76].) Additional assumption about well-orderings are needed.

If the metatheory is strong, we may be able to prove that the theories in an autonomous progression are true by other means. For example, we may know that S proves transfinite induction up to $\phi_2(0)$ (for appropriate formulas) and thus deduce that the theories PA_α^{RFN}, for $\alpha < \phi_2(0)$, are true. In some theories it is also possible to amplify the transfinite induction stated for sets to transfinite induction for a class of more general formulas. This principle is called the *Bar Rule*. Using this rule one can define a more precise upper bound on $\{PA_\alpha^{\mathrm{RFN}}\}_{\alpha<\phi_2(0)}$: all sentences provable in the latter arithmetic theory are provable in the subsystem of Second Order Arithmetic $ACA+BR$, where BR stands for the Bar Rule stated for all formulas in L_2. (See [219], where Δ_0^1-CA is used instead of ACA.)

12. *Reflection principles in set theory.* Reflection principles have a different meaning in set theory, although there are similarities between the uses in arithmetic and set theory. In set theory reflection principles are formalizations of the following idea.

    The class of all sets V is so big that it is impossible to describe it. Therefore any property of V must already be satisfied by some V_α of the cumulative hierarchy.

    We can also say that *the class of all ordinal numbers ON is so big that every property of ON is already satisfied by some cardinal κ.* Indeed, ON looks very much like a large cardinal, except that it is not a set. For example, it satisfies the conditions that define inaccessible cardinals (see page 198).

    Formalizations of this rule of thumb were used to define large cardinals. In particular, for a class Γ of set-theoretical sentences with one class variable, we define that κ is *Γ-indescribable* if for every $R\subseteq V_\kappa$ and every $\phi\in\Gamma$,

$$(V_\kappa;\in,R)\models\phi \quad\Rightarrow\quad \exists\alpha<\kappa\ (V_\alpha;\in,R\cap V_\alpha)\models\phi.$$

    By choosing suitable sets of sentences Γ, one can give equivalent definitions of inaccessible, weakly compact and other cardinals. These cardinals are, however, relatively low in the hierarchy of large cardinals.

13. *Ω-logic and the Ω-conjecture.* Woodin introduced Ω-logic in order to describe sentences that are invariant under forcing. The definition of semantics is not difficult. Given a sentence of set theory ϕ, we say that it is Ω-valid, or write $\models_\Omega \phi$, if for every ordinal α, ϕ is true in every generic extension of V_α which is a model of ZFC. I will not define when a sentence is Ω-provable, which is written as $\vdash_\Omega \phi$. Let me just say that an Ω-proof is represented by an infinite object, a subset of ω^ω with certain properties. Although Ω-proofs do not have the structure of usual proofs (a sequence or a tree of formulas), all theorems of ZFC are Ω-provable and modus ponens is an admissible rule. Further, Ω-provability is sound with respect to Ω-validity, that is, if $\vdash_\Omega \phi$ then $\models_\Omega \phi$. What is missing is the completeness—this is the Ω-conjecture.

    The Ω-conjecture *Every Ω-valid sentence is Ω-provable.*

    Unfortunately, one often needs a stronger assumption, the *Strong Ω-conjecture*. The definition of this conjecture (see [157]) is too technical to be included here.

    The result about Σ_1^2 mentioned after Theorem 66 is the following theorem of Woodin (see [157]).

Theorem 67 *Assuming the Strong Ω-conjecture and the existence of a proper class of Woodin cardinals, the following is true.*

a. *$ZFC + CH$ makes all Σ_1^2 sentences invariant under forcing.*
b. *For every Σ_1^2 sentences ϕ that makes Σ_1^2 sentences invariant under forcing, ϕ is Ω-equivalent to CH.*

Σ_1^2 sentences are the sentences in the 3rd-order arithmetic that start with an existential 3rd-order quantifier and the rest is a 2nd-order formula.

It is well-known that *CH* can be stated as a Σ_1^2 sentence. Hence Theorems 66 and 67 imply that, assuming the strong Ω-conjecture, there is no extension of *ZFC* compatible with large-cardinal axioms that makes all sentences invariant under forcing.

7.3 Finitism and Physical Reality

The geometry of three dimensions developed in antiquity, as presented in Euclid's *Elements*, was intended to describe physical space. In the 19th century, when non-Euclidean geometries were discovered, some scientists began to suspect that Euclidean geometry was not the geometry of physical space. But only after Einstein's Theory of Relativity was tested did it become clear that this is the case. The reason is that bodies with mass cause curvature of space around them. Even more dramatic things happen with space in black holes. Furthermore, the geometry of space changes over time, which 19th century scientists certainly did not expect to be possible.

Today the fact that flat Euclidean geometry is not the geometry of physical space is accepted as a natural phenomenon. Maybe, there are other mathematical structures that we identify with physical reality, but one day we will discover that they are different. In this section I will discuss the possibility that the discrepancy between physical reality and mathematical models may concern even so basic concept as the natural numbers.

Mathematical and Physical Natural Numbers

The natural numbers can be represented in various ways in physical reality. When we use numbers to count objects, they are represented by sets of physical objects. But instead of using objects, let us consider the representation of numbers in Euclid's style. Let ℓ be a line going from a point A to infinity. Let a unit length be fixed, say 1 meter. Imagine that we make marks every 1 meter on the line ℓ. This will be our representation of the numbers. As we know, addition and multiplication can be defined geometrically, so we have a representation of the structure with these operations. I will call this structure *the small physical natural numbers*, or *lengths*, and denote it by $\mathbb{L}_{phys}$.

We can use another physical structure to represent the natural numbers. In this structure a number is represented by a string of zeros and ones that we place on the marks on a finite initial segment of the line ℓ. In this case addition and multiplication cannot not be defined geometrically, but, in principle, these operations can be done mechanically for arbitrary long strings. Thus we get the second structure, which I will call *the physical natural numbers* and denote by $\mathbb{N}_{phys}$.

One can think of many other representations of numbers by physical objects. We may use physical quantities other than length, such as time, or the strengths of fields. Thus we may obtain more than two different structures, but it seems that all representations should reduce in some natural way to the two structures defined above, and maybe even down to one, if the two are isomorphic.

The Natural Numbers in a Finite Universe

If the universe is finite, then the line ℓ has a finite length and both ways represent only numbers of finite segments of the mathematical natural numbers. We have $\mathbb{L}_{phys} = \{0, 1, 2, 3, \dots, n\}$ for some number n, and $\mathbb{N}_{phys} = \{0, 1, 2, 3, \dots, 2^n - 1\}$. So the physical and mathematical natural numbers are different and, moreover, $\mathbb{L}_{phys} \neq \mathbb{N}_{phys}$. The largest numbers in $\mathbb{L}_{phys}$ and $\mathbb{N}_{phys}$ depend on the unit that we use. By taking smaller units we can represent larger numbers. Hence the largest number in $\mathbb{L}_{phys}$ is defined by convention.

Recall the informal classification of natural numbers to small, medium and large that I used to motivate concepts in computational complexity in Chap. 5. A small number is a number n such that we can let a computer run for n steps. In particular, we can do exhaustive search of numbers less than or equal to n. Medium numbers are those n which we can write down and compute with, but for which it is impossible to do exhaustive search up to n. Large numbers are the numbers that are larger than medium numbers—we cannot even write them down in binary notation. This classification was based on what humans can do with numbers. In a finite universe the three categories can be given a physical meaning: the small numbers are $\mathbb{L}_{phys}$, the medium numbers are $\mathbb{N}_{phys}$ and large numbers are all the rest. Although we cannot do it systematically, we can define also some large numbers, represent them by symbols and compute with them. Therefore the fact that large numbers do not have natural representations should not be considered a reason to prohibit them. They are just a different kind of entity. I will call the set of all three kinds of numbers *the mathematical natural numbers*, and I will use the standard notation $\mathbb{N}$ for them.

In a finite universe, complexity theory can be based on this classification of numbers, instead of the concept of polynomial growth. What I used before only as a motivation for defining the basic concepts of computational complexity now becomes real. Therefore in a finite universe computational complexity should play an important role. According to current theories, physical processes are efficiently computable, which means that one can compute the results of experiments in polynomial time (except that we may need to use quantum Turing machines, not just the classical ones). If this is the case, then complexity theory should tell us not only what humans can efficiently compute, but also what is physically possible.

Clearly, all this is a very simplified picture of a possible reality. First, infinity can manifest itself in various ways as we noted above. Second, in computational complexity there are two basic computational resources—time and space. It is reasonable to conjecture that if space is finite, then so is time, but it is not clear what would be the relation between these two finite entities. To get a theory similar to standard complexity theory, we would need to have the structure of time be essentially the same as $\mathbb{L}_{phys}$ so that we could say that what is physically computable is computable in time t bounded by the largest element of $\mathbb{L}_{phys}$, or its small multiple.

I do not want to go into detail about how this could be formalized because it would require a model of a discrete universe and I do not know any such model. Instead, let us see if it is possible at all that the universe could be finite. Some people still think that this is the kind of question that will never be resolved, but this is a misconception. The slight variations of the cosmic microwave background can be used not only for studying the evolution of the universe from the big bang, but also to test hypotheses about the shape and the finiteness of space. The more precise recent measurements were compared with predictions of various models of the universe. It turned out that some finite models fit the data better than models assuming infinite space. To test the conjectures about the finiteness of space it is necessary to come up with a specific model that also determines the shape. The model that fitted the data best was the Poincaré dodecahedral space, which always has finite size [188]. It would be very interesting if space had such an exotic structure, instead of the familiar kinds of structure, such as the sphere or the torus.

The problem of the finiteness of space has not been conclusively resolved yet, but there is, however, a good chance that this or some other model will be confirmed with high reliability when more data are available and more computations are performed. That said, one cannot rule out that sceptics are right after all. For example, the universe may be finite, but too big for us to be able to ascertain this fact.

If the universe is finite, its size is at least the size of the *observable universe*, whose diameter is estimated to be 8×10^{26} meters, which is approximately 10^{62} Planck lengths.[44]

The Natural Numbers in an Infinite Universe

The prevailing opinion is that the universe is infinite. For the sake of simplicity, I will assume that both space and time are infinite. This will suffice for explaining in what sense the physical and the mathematical natural numbers could be different.

---

[44] Physicists believe that it is impossible to distinguish two events occurring in a distance shorter than the Planck length.

The standard view of philosophers of mathematics and physics is that if the universe is infinite, then the mathematical natural numbers are exactly what one can physically represent; in particular, all representations are isomorphic. It is difficult to state this as a thesis because of the vagueness of the concepts involved, but I will try it anyway.

The Natural Number Thesis *The concept of natural numbers studied in mathematics is unique and any reasonable physical representation of natural numbers is a model of this concept.*

We can express a special case of the thesis for the two particular representations mentioned above by the equations:

$$\mathbb{N} = \mathbb{L}_{phys} = \mathbb{N}_{phys}.$$

Recall that the discovery that the geometry of space is not Euclidean was possible only after non-Euclidean geometries had been found. Hence to challenge the Natural Number Thesis one must first propose a different kind of natural numbers. Therefore my main aim in this section is to present an alternative view of the physical natural numbers. This is certainly not the first attempt to offer a different theory of natural numbers. I have explained one such theory in Chap. 3, Vopěnka's Alternative Set Theory. Vopěnka and others, however, did not distinguish between the mathematical natural numbers and the physical natural numbers. Their aim was to propose different kinds of foundations of mathematics.

The theory that I am going to present is not supported by observations of physical reality. I mean it only as an example of a different approach. It is definitely a controversial proposal, but I will also mention even bolder extensions of this theory. I believe that it is important to look for alternatives to the classical view of the natural numbers. Once we have more theories, we can compare them and decide which is more likely to be the right one.

An important concept that we have met several times in this book is the concept of fast growing functions. We have seen explicit examples of extremely fast growing functions in Chap. 4. We also know that there are functions that grow much faster and that every theory is only able to formalize functions up to some limited growth. In the "ideal" natural numbers, which these theories try to describe, we have all these wildly growing functions. In more technical terms, we believe that if it is consistent to assume that a fast growing function exists, then it does exist.

How is it in the physical world? Let us consider the exponential function $y = 2^x$. From the point of view of the theory of fast growing functions the exponential function represents only a very mild growth. On the other hand, in complexity theory it is the paradigm of a function growing too fast. The question is whether $\mathbb{L}_{phys}$ is closed under this function (which would mean that for every number n in $\mathbb{L}_{phys}$, there is a number representing 2^n in $\mathbb{L}_{phys}$). Our intuition tells us that this must be so. The argument behind this feeling is that if some n is in $\mathbb{L}_{phys}$ and 2^n is not, then we would find a number m less than n such that 2^m is in $\mathbb{L}_{phys}$ but 2^{m+1} is not. But if we can represent 2^m, we should also be able to represent 2^{m+1}, because it is only twice as large as 2^m and we assume that space is infinite. Hence $\mathbb{L}_{phys}$ must be closed under exponentiation.

In this argument it is reasonable to assume that if length x exists then also length $2x$ does. So let us assume that $\mathbb{L}_{phys}$ is closed under multiplying by 2. What is, however, wrong is the assumption that one can test the existence of a representation of 2^x for all numbers x up to m. When we test numbers $0, 1, 2, \ldots$ in order to find an m with the above properties, the numbers $2^0, 2^1, 2^2, \ldots$ go beyond any number in $\mathbb{L}_{phys}$ because 2^n is not in $\mathbb{L}_{phys}$ and $\mathbb{L}_{phys}$ is closed under multiplying by 2. Thus we will have to search an infinite part of the line ℓ in order to find m. It seems unrealistic that one could do this in finite time.

A Theory of the Physical Natural Numbers

The new theory *TPhN* that I am going to propose is based on postulating that the exponential function is not defined on all physical natural numbers because it grows too fast. This approach is a particular form of finitism. When speaking about finitism, one should distinguish *subjective* from *objective finitisms.* Intuitionism is an example of a subjective finitism. Intuitionists reject actual infinity because, for them, mathematics is a mental activity. Since humans are finite beings, they can only imagine finite structures. Another form of finitism is based on the idea that, although mathematics is real, our limited means enable us to handle only finite structures. This is also a subjective approach. Objective finitism, on the other hand, rejects actual infinity on the basis that finiteness is an essential property of the universe. An extreme form of finitism, called ultrafinitism, considers only small numbers as meaningful. The system I am going to present can be classified as objective ultrafinitism. But I should stress that this is only an attempt to present an alternative view of physical reality. I do not mean it as a basis for the foundations of mathematics.

I will present the theory *TPhN* using four axioms stated informally and mention possible ways to make the system a formal theory later. In the informal presentation I will tacitly assume that we already have some basic axioms that enable us to formalize basic concepts mentioned in the axioms. All objects of this theory are numbers; their intended interpretation is $\mathbb{N}_{phys}$. The predicate L denotes logarithms, or lengths of numbers, and its intended interpretation is $\mathbb{L}_{phys}$. It will also be convenient to use the symbol N for the universe of all numbers, although there are no other elements in the theory.

Axiom 1 (Infinity) *L contains 0 and 1 and is closed under $+$ and $\times$.*

I am stating this simple axiom explicitly because the question which functions are total plays an important role in this theory. This axiom, in particular, implies that the numbers are an infinite structure. Note that this axiom can be stated without referring to L; we only need to require stronger closure properties of the universe of numbers.

Axiom 2 (Feasible Computations) *For every algorithm A (presented as a Turing machine), every number m in L, and every number n, there exist m steps of the computation of A on the input n.*

This axiom is intended to make the theory strong enough to formalize computations. We have to ensure that computations can be formalized so that one can prove that computations of length m exist for every m in L. Since L is known only to be closed under $+$ and $\times$, such computations are very much like computations in polynomial time.

Axiom 3 (Induction) *For every predicate $P(x)$, every number m in L, and every number n, if there exists an algorithm A that can test the predicate P for numbers $x = 0, 1, 2, \ldots, n$, and such that A needs time at most m for every such x, then the following holds true. If $P(0)$ and $P(x) \to P(x+1)$ for all $x = 0, 1, 2, \ldots, n-1$, then $P(n)$.*

Axiom 3 is a special case of the general principle of induction. In this axiom I only postulate induction for efficiently testable predicates. The reason is that I assume that physical processes correspond to efficient computations. In other words, efficiently testable means *experimentally testable.* It is natural not to postulate induction for predicates that we cannot test.

So far the axioms are satisfied by the mathematical natural numbers $\mathbb{N}$. The last axiom makes *TPhN* incompatible with $\mathbb{N}$.

Axiom 4 (Limited Universe) *There exists a number ν such that 2^ν does not exist.*

The axiom can also be stated in physical terms as

$$\mathbb{L}_{phys} \neq \mathbb{N}_{phys}.$$

To present the theory *TPhN* formally, one has to specify the language of the theory and add several more axioms. So one should view Axioms 1–4 not as a formal theory, but as a specification of a class of theories. This is similar to the generic name Θ_C for theories corresponding to a complexity class C. In fact, the theories that conform to Axioms 1, 2 and 3 in the best way are the theories of $\Theta_\mathbf{P}$, the theories associated with the complexity class **P**. Hence *TPhN* can be specified by:

$$\Theta_\mathbf{P} + \text{Axiom 4}.$$

In other words, if we want a formal theory, we only need to take one of the theories formalizing $\Theta_\mathbf{P}$ and add Axiom 4.

I will now give a physical justification of Axiom 3. Induction, as stated in the axiom, is equivalent to the statement that if $P(0)$ and $\neg P(n)$, then there exists an $x < n$ such that $P(x)$ is true and $P(x+1)$ is false. In the theory *TPhN* we identify physical processes with algorithms, so we should show that such an x can be produced by an algorithm. Note, however, that, in general, n does not have to be in $\mathbb{L}_{phys}$, so we do not have enough time and space to check all numbers between 0 and

n for the property P, although we can do it for each of them individually because $m \in \mathbb{L}_{phys}$. The algorithm is based on the *binary search*, a well-known trick used in many algorithms. Take $n/2$ if it is an integer, otherwise round it down to an integer. Let this number be n_1.[45] Test if $P(n_1)$ is true. If it is false, then we focus on the interval $[0, n_1]$ because we know that there is an $x < n_1$ such that $P(x)$ is true and $P(x+1)$ is false. Otherwise we focus on $[n_1, n]$ and search such an x between n_1 and n. Then we divide the interval again into two approximately equal parts, and so on, until the interval has length 2, in which case we are done. In each step the length of the interval decreases approximately by a factor of 2. Hence we need to do this approximately $\log_2 n$-times. Checking whether $P(x)$ is true or false takes time m. We know that m and $\log_2 n$ are in $\mathbb{L}_{phys}$. Since, by Axiom 1, $\mathbb{L}_{phys}$ is closed under multiplication, we have $m \cdot \log_2 n \in \mathbb{L}_{phys}$. Hence we have enough time and space to run this algorithm and find and x such that $P(x)$ and not $P(x+1)$.

What this argument actually shows is that induction for N follows from induction for L (a well-known fact in proof complexity). Induction for L is justified by the fact that we have enough time to test every number below n if n is in L.

Axiom 3 raises the natural question whether it is possible to use more efficient ways of computation to make the induction axiom a stronger. What comes immediately to mind are probabilistic and quantum algorithms. Theories that formalize probabilistic polynomial time computations have been studied; formalizing quantum computations is more difficult and to date we do not have theories associated with polynomial time quantum computations. After all, it is too early to think about generalizations—first we should test this simpler theory.

The theory *TPhN* certainly reminds one of Vopěnka's Alternative Set Theory. In that theory there are also two types of natural numbers: *N* and a proper initial segment *FN* called the 'finite numbers'. There is, however, a huge difference between the properties of our numbers and Vopěnka's numbers. For example, both *FN* and *N* are closed under very fast growing functions. Since, in particular, *FN* is closed under exponentiation, we cannot interpret *FN* as $\mathbb{L}_{phys}$ and *N* as $\mathbb{N}_{phys}$. It would be more in the spirit of Vopěnka's philosophy to interpret *FN* as the physical natural numbers and *N* as the mathematical natural numbers.

The Inconsistency of Ultrafinitistic Systems

Should the theory *TPhN* describe the physical natural numbers, the number ν in Axiom 4 must be a concrete number that we can physically represent. How big this number is should be established by observation. It could be only a little larger than the size of the observable universe (10^{62} in Planck lengths), it could be Archimedes's number $10^{8 \cdot 10^{16}}$, or it could be much larger.

[45] n_1 can also be defined as n with the last binary digit deleted.

If we view it from the higher perspective of pure mathematics, the theory *TPhN* is consistent, but as soon as we postulate Axiom 4 for a concrete number ν, it is inconsistent.[46]

Let ν be a fixed concrete number and let us first look at a simple proof of inconsistency (essentially, the argument already considered above). Axiom 4 says that ν is not in L, but according to Axiom 1, L is closed under the successor function. Thus we can gradually prove that $0, 1, 2, \ldots$ are in L, so eventually we prove that ν is in L, which is a contradiction. But notice that this proof is very long; it is has length at least ν. Since ν is not in $\mathbb{L}_{phys}$, this proof is not physically representable and we can ignore it. Unfortunately there are much shorter proofs of inconsistency—proofs whose lengths are in $\mathbb{L}_{phys}$. I will sketch two such inconsistency proofs.

1. For the sake of simplicity, assume that ν is a power of 2, say $\nu = 2^k$. Recall that $L(x)$ is the formula saying that 2^x exists. So we have $L(k)$. Let $\lambda(x)$ be defined by

$$\lambda(x) \equiv L(2^x).$$

We have $\lambda(0)$ and

$$\forall x\ (\lambda(x) \to \lambda(x+1)).$$

This is because if $L(2^x)$ then also $L(2^{x+1})$, since $2^{x+1} = 2^x + 2^x$ and L is closed under addition. We cannot use induction to prove that $\lambda(k)$. Instead we take all instances of the above formula up to $k-1$:

$$\lambda(0) \to \lambda(1), \quad \lambda(1) \to \lambda(2), \quad \lambda(2) \to \lambda(3), \quad \ldots \quad \lambda(k-2) \to \lambda(k-1),$$
$$\lambda(k-1) \to \lambda(k).$$

Using $\lambda(0)$ and these formulas, we derive $\lambda(k)$ by applying modus ponens k-times. From the definitions of formulas λ and L we get that $L(2^k)$, which is $L(\nu)$, and then $\exists y\ (y = 2^\nu)$. This is a contradiction because we assumed that 2^ν does not exist.

2. (Suggested by S. Buss.) Suppose that the language of the theory contains a symbol for the squaring function, say x^2. I will again assume that $\nu = 2^k$. Then the term

$$(\ldots(((2^2)^2)^2)^2\ldots)^2 \quad \text{(}k\text{ squarings)}$$

has value 2^{2^k}. In fact, one can prove it using a proof of length polynomial in k and formalize it in our *TPhN*. Hence 2^ν exists.

So our theory *TPhN* describes an inconsistent world. This is bad, but as we noted above, one can cope with inconsistencies. One can argue that although the proofs

[46]This can also be explained by pointing to the fact that the theory *TPhN* is not ω-consistent.

are physically representable, they are not valid because they use concepts that have no physical meaning.

The first proof uses the formula $\lambda(x)$. As we noted above, we cannot test whether a number n satisfies $\lambda(x)$ because for a number n that does not satisfy $\lambda(x)$, we would need an infinite number of steps. Therefore, it is natural to disallow the use of $\lambda(x)$ and similar formulas.

The second proof uses the assumption that *if we can write down a closed arithmetical term, then the number equal to the value of the term must exist.* This is a rather dubious assumption. I can write '*unicorn*', but this does not imply that unicorns exist. So, again, it is natural to restrict the use of terms and disallow terms that would define very large numbers.

A natural way to avoid such proofs of inconsistency is to consider a proof system for first order logic in which

1. only direct proofs are allowed (for example, cut-free sequent calculus), and
2. the language does not allow constructions of terms that define large numbers (for example, we can use $+$ and $\times$, but not a symbol for the unary function x^2).

This will ensure that no proofs of inconsistency in such a proof system will be physically representable.

Since I have been rather vague, one may get the impression that these are only intuitive arguments. But there is also mathematical content behind them. Let us consider *TPhN* where ν is not a concrete number. This theory is consistent, but proves its own inconsistency. Specifically, it proves that in the usual proof systems for first order logic there exists a proof of contradiction whose length is in L. If we, however, restrict the proof system as explained above, then *TPhN* does not prove the existence of a proof of contradiction whose length is in L.

From the point of view of a finitist, the interesting sentences are only those that speak about finite structures. In number theory we can represent these statements as sentences about numbers less than some given number n and a polynomial time computable predicate $P(x)$. Natural calculi for these sentences are polynomially verifiable proofs and Extended Frege proofs. The soundness of such proofs is provable in $\Theta_{\mathbf{P}}$, hence also in *TPhN*. Therefore, we can safely use polynomially verifiable proofs and Extended Frege proofs in *TPhN*.

The Axiom of Solvability and Alternative Arithmetics

The big advantage of using $\Theta_{\mathbf{P}}$ instead of stronger theories is that it enables us to model interesting situations, which would be impossible if we used stronger theories. But to do this, we still have to assume some unproven hypotheses even for $\Theta_{\mathbf{P}}$. These constructions are possible due to the flexibility of models of $\Theta_{\mathbf{P}}$ with respect to removing and adding elements, which can be used to change **NP** properties in the models. However, changing **NP** properties is only possible if the model is not closed under exponentiation (and certain hypotheses are true). So the interesting models of $\Theta_{\mathbf{P}}$ are those that satisfy Axiom 4.

Let M be a nonstandard model of $\Theta_{\mathbf{P}}$. Suppose a nonstandard number $f \in M$ encodes a propositional formula $\phi(p_1, \ldots, p_k)$ that is unsatisfiable in M. To make ϕ satisfiable in an extension N, we must add a satisfying assignment $(a_1, \ldots, a_k)$ to N. Such an assignment is a string shorter than ϕ, hence in natural encoding it is coded by a number $m < f$. Since satisfiability of Boolean formulas is **NP**-complete, we can represent any **NP** property as satisfiability of a Boolean formula.

Since $\Theta_{\mathbf{P}}$ proves that Extended Frege systems are sound, we can only make such a change if there is no Extended Frege proof of $\neg\phi(p_1, \ldots, p_k)$. Surprisingly, this is also a sufficient condition for the existence of such an extension. We believe that Extended Frege systems are not polynomially bounded, which means that there are tautologies that do not have Extended Frege proofs of polynomial length, hence there are models which can be extended in this way. Note that for this to be possible, it is essential that the model is not closed under exponentiation (Axiom 4) because every tautology has an Extended Frege proof of at most exponential length.

Using this characterization we can construct a chain of extension in which we satisfy more and more **NP** properties. The limit of this process is a model in which all tautologies without an Extended Frege are turned to non-tautologies. Put otherwise, every tautology in the model has an Extended Frege proof in the model. Thus we may consistently extend *TPhN* by the following axiom.

Axiom 5 (Solvability) *Every* **NP** *predicate $P(x)$ is either satisfiable, or there exists a feasibly constructive proof that it is not satisfiable.*

Here 'feasibly constructive' is just a fancy name for 'Extended Frege'. I have several reasons for using this rather vague term. First, I am presenting the theory *TPhN* in an informal way and do not want to specify a particular proof system. Second, the reason for using 'feasibly' is to stress that in the theory with Axiom 5 proofs of long tautologies have subexponential length. We know that Extended Frege system is complete, which means that every tautology has a proof. But, in general, we only know that they have exponentially long proofs. In *TPhN* the set of lengths is not closed under exponentiation. Hence in *TPhN* + Axiom 5, if m is in L and 2^m is not, then every tautology whose length is at least m has a proof of subexponential length. In plain words, I imagine L as "feasible numbers", in contrast to those that are not in L; so the lengths of proofs in the theory are feasible numbers.

I borrowed the name 'Axiom of Solvability' from Hilbert because Axiom 5 can be loosely rephrased as: *for every finite problem, either there is a solution or a proof that the solution does not exist.*

The most interesting fact is that, assuming a certain plausible conjecture in complexity theory, there are models of $\Theta_{\mathbf{P}}$, and also of *TPhN*, that can be extended in two mutually incompatible ways. Specifically, we can construct a model M of *TPhN* and two **NP** predicates $P_0(x)$ and $P_1(x)$ such that there are two extensions of M, one in which $P_0(x)$ is satisfied and another in which $P_1(x)$ is satisfied, but there is no extension in which both are satisfied. Furthermore, there seems to be no reason to prefer one extension to the other.

Knowing this fact and accepting Axiom 5, we can speculate how **NP** predicates are satisfied in the physical natural numbers. One can imagine a scenario in which the natural numbers originated as a very basic structure and then evolved, by a *random process*, to a structure satisfying Axiom 5. In this scenario our physical natural numbers would be just one randomly chosen structure from an infinite set of possible structures.

Such a scenario can be used to explain the phenomenon of pseudorandomness. In number theory there are various sets and sequences that are defined by simple conditions or algorithms, but nevertheless look like random objects. A prime example of such a function is the Möbius function. Since the Möbius function has three values, it is better to consider a closely related *Liouville function*, which has only values 1 and -1. It has been conjectured that this function behaves, in a certain sense, like a random sequence of 1 s and -1 s. Slightly more precisely, the conjecture says that there is no simple test that can distinguish the Liouville sequence from a random sequence.[47] If we believe that our natural numbers are just one randomly chosen structure from many possible ones, we should also expect that the randomness will manifest itself in our natural numbers. The apparently random behavior of the Liouville function could be a trace of a random process by which the natural numbers originated.

Accepting the possibility of alternative arithmetics, one can also speculate that there are intrinsically unsolvable problems. Here is the argument. We can only see a finite part of the natural numbers. If there are alternative structures, we cannot deduce all properties of the natural numbers because what we see may be part of many possible structures and different structures satisfy different properties. On the positive side, we might be able to apply the standard assumption of statistics that we live in a random world. For example, if the values of the Liouville function resulted from a random process, they must look random because our world is most likely not a special one.

It will surely be difficult to lift such speculations to the level of a scientific discourse. I do have a result that may be the beginning of such a research program, but it is too early to talk about it [229]. Instead, let us turn to the question of whether it is conceivable that there are alternative arithmetics.

Intuitively the existence of alternative arithmetics seems impossible. The structure of the natural numbers is determined by some basic arithmetical laws and logic. To allow different structures, we would have to reject either the basic arithmetical laws or logic. A world in which logic or the basic arithmetical laws are different would be so alien to us that we would not consider it as an alternative to ours.

This argument, however, ignores an important issue. There may be a world with the same arithmetical laws and the same logic, but with different *proofs*. Proofs are just finite structures that can be coded by numbers; so if we can alter the arithmetical

---

[47] In [108] B. Green mentions the *Möbius Randomness Principle* proposed by P. Sarnak. This conjecture was not stated exactly because it did not specify the complexity of the tests. Natural formalizations of this conjecture are equivalent to the corresponding conjectures for the Liouville function.

structure we can also alter the structure of proofs. This is exactly what happens in nonstandard models of arithmetic. There is a model of Peano Arithmetic in which Peano Arithmetic is consistent and there is another model in which Peano Arithmetic is inconsistent. Both models satisfy the axioms of Peano Arithmetic and formalize logic in the same way. But in the second model there is a proof (a proof of a contradiction from the axioms of Peano Arithmetic) that is not present in the first model.

Can we imagine an arithmetic different from ours? It is certainly difficult to imagine a world in which 91 is a prime, but a world in which RSA-129 is a prime does not look so impossible.

Undetermined Mathematics

As I was about to start writing this chapter I received as a present a book of science-fiction stories by Greg Egan. In two loosely connected stories in the book, *Luminous* and *Dark Integers*, the plot is based on mathematics.[48] The idea is that there are inconsistencies derivable from the basic principles governing natural numbers. The inconsistencies manifest themselves only for very large numbers; so they have not been noticed by mathematicians, but can be produced by computers. It turns out that these inconsistencies form a border between our universe and an alien universe. Furthermore, it is possible to move the border by doing calculations with numbers at the border. As typical in Egan's stories, the talk about fictitious science is interleaved with scenes in which the two worlds fight to extend their territories of consistency.

What I find most interesting in these two stories is the idea that mathematics changes as we are doing it, which Egan called "*malleable mathematics*". The idea that mathematics depends on mathematicians' activities is not new. Recall that one of the basic tenets of intuitionism is that we should not claim that '*A or B is true*' before we either establish A or establish B. Hence only at the moment when we prove one of the two propositions can we assert that '*A or B*' is true. Intuitionism views mathematics as a mental activity and this principle is in accord with this. So the truth of a disjunction is subjective—it depends on whether a particular mathematician has already got a proof of one of the disjuncts. Egan's view is different in these stories. Mathematical facts can be changed not only by people, but by anything, for example, by a computer that performs the computations, and the facts that are changed become objective reality.

I had secretly entertained a related idea already before I read the stories, but had not developed any theory of such mathematics. The fact that a similar idea has appeared in print, although only in the form of science fiction, encourages me to mention it in this book.

The first step is to accept the possibility of worlds with different arithmetics. Above I have presented a theory that allows such a possibility. This theory has a

[48] Greg Egan, *Dark integers and other stories*, Subterranean Press, 2008.

serious problem, inconsistency, which may be solved in a way I sketched. But there may be better ways to model alternative arithmetics.

The second step is to admit that arithmetic in such worlds does not have to be completely fixed from the very beginning, in other words, that the multiverse splits only gradually into universes with different arithmetics. Namely, the difference between worlds increases as time passes.

The third step is to consider what can cause the splittings. Since we are talking about *information*, observers should play a role in the process. For example, we can view testing a large number for primality as an experiment. It is conceivable that for some very large numbers, the property of primality is undetermined and the primality of such numbers will only be fixed after an experiment (primality testing) produces an answer.

The crucial question is whether such a theory can predict anything experimentally testable. I think that this is not totally excluded. But if there are such experiments, then very likely we will not be able to do them with the currently available means. The reason is that the properties of the numbers that we can handle with our limited means must already be fixed.

Towards Unified Foundations of All Science

All science can in principle be reduced to physics and mathematics. So it seems that when we find good foundations for each of these two sciences, we will have a foundation for science as a whole. But I think we can hardly have a good foundation for all science if we do not find *unified* foundations for mathematics and physics. Presently we are not happy with the current foundations of either of the two, so it seems premature to think about the unified foundations. But searching for unified foundations may be a cure for foundational problems in both fields.

Before considering such a project we should fully understand the difference between physics and mathematics. In simple examples it is clear: The quadratic function

$$y = ax^2$$

occurs frequently in both mathematics and physics. We can view it as a "universal tool" that is useful in many situations. In mathematics we only study how this tool works together with other tools. In physics, however, we associate it with very concrete situations. For example, a physicist, using different letters

$$v = gt^2,$$

may interpret it as the dependence of the velocity on time; say, as the velocity of a stone falling from the Leaning Tower of Pisa. Thus some aspect of reality is closely connected with this equation, although the real event that is described by it is not actually identified with the equation. The stone consists of atoms which have nothing to do with the equation.

As theoretical physics develops, reality recedes and mathematics advances. In our example, the stone was just a stone at the time of Galileo; but now we also have a theory that explains atoms. Thus, for example, we can write down equations for a free-falling hydrogen atom. We still believe that a free-falling hydrogen atom is a real thing because we have personal experience with free fall and we know that hydrogen is one of the two constituents of water. We imagine a hydrogen atom to be two particles, one orbiting the other, but what the equations actually describe is a rather sophisticated mathematics: they describe how the wave functions of an electron and proton develop in time; there are no particles looking like points or balls. We can go on and describe the particles themselves as special solutions of equations in a more general theory, and so on. The further we go, the more these models are like mathematics and the less like the reality we are used to.

This does not mean that theoretical physics and mathematics will eventually become one science. Mathematics will aways use structures that are not physically representable. But if space is infinite, then one can, in principle, represent any finite mathematical structure by a physical object. Hence a theorem about finite mathematical structures is also a physical law. So at least in the realm of finite structures these two sciences coincide.[49]

In a lecture given in 2002, Stephen Hawking argued that physical theories must be incomplete because mathematics is a part of physical reality and hence must also be described by physical theories. Since mathematical theories are incomplete because of Gödel's Theorem, some physical theories must be incomplete as well.

> "So if there are mathematical results that can not be proved, there are physical problems that can not be predicted. One example might be the Goldbach conjecture. Given an even number of wood blocks, can you always divide them into two piles, each of which can not be arranged in a rectangle? That is, it contains a prime number of blocks."[50]

Thus the foundations of physics have to cope with the same problem as mathematics—incompleteness.

One feature that distinguishes physics from mathematics is that in physics we need experiments to verify theories. But let us look at this putative difference more closely. What is in physics a theory and an experiment testing the theory is in mathematics a conjecture and a computation of specific instances of the conjecture. It seems that in a mathematical experiment it does not matter what physical device one uses to do the computation, whereas a physical experiment is always dependent on a particular form of matter or field. But consider the experiments that test quantum theory, such as the two slit experiment. In this experiment a particle is shot through a pair of narrow slits in a board onto a screen. After shooting many particles, in spite of sending at most one particle at a time, one can observe an interference pattern on the screen with alternating brighter and darker bands. This experiment *does not depend* on the type of particle and the materials of which the board and the

---

[49] Here I am assuming the Natural Number Thesis that identifies the mathematical natural numbers with the physical ones.

[50] S. Hawking, *Gödel and the end of physics*, a public lecture held in 2002, [116].

screen are made. All that matters are the widths of slits, the distance between them and the energy of the particle. In other words, the experiment does not test *matter*, it only tests a *principle* of quantum theory.

In the two slit experiment we still feel that it is physics, not mathematics, that is being tested. But now suppose that we had a quantum computer and used it to factor a large number whose factors we do not know. Would this be a mathematical, or a physical experiment?

The *Hilbert-Pólya approach to the Riemann Hypothesis* is based on the conjecture that the imaginary parts of the nontrivial zeros of the zeta function are the eigenvalues of a Hermitian operator. In quantum mechanics, measurable quantities are defined as the eigenvalues of Hermitian operators. Thus, in a sense, the Pólya-Hilbert approach rests upon trying to establish a physical interpretation of the Riemann Hypothesis. The conjecture that there is such an interpretation is supported by a number of results, most notably the theorem called *Selberg's trace formula.* According to A. Odlyzko, the Hilbert-Pólya conjecture "*is often regarded as the most promising way to prove the Riemann Hypothesis*".[51]

Further results suggest a connection with the eigenvalues of random Hermitian matrices, which are used to study the distributions of the energy levels of systems of atoms. It would be a great achievement if the suggested connections were confirmed. But it would be even more interesting if, for example, one could *explain why* the distribution of the nontrivial zeros of the zeta function is similar to the distributions of the energy levels of systems of atoms. To this end one would not have to design common foundations for all science; it would only suffice to find deeper connections between the foundations of physics and mathematics.

One could go on and give more examples of connections between physics and mathematics, but this would not be very interesting; we know that there are many. In contrast, it is surprising how few connections we know between the foundations of these two sciences. One reason for the lack of such connections is the amount and the complexity of results in these fields. Because of this, for somebody interested in foundations, it is very difficult *just to learn* the facts that could be relevant for foundational studies in mathematics and physics, let alone to find connections. I hope this book will help those who want to start thinking about foundations.

Concluding Remark

In spite of the progress that has been made, the most fundamental problems in the foundations of mathematics and complexity theory are still open. Obviously, we would like to live to see these problems solved, but it seems that only a few, if any, will be solved in the near future. But rather than being frustrated, we should view this positively. The greatness of science lies in the fact that, however much we achieve in the quest to understand the world around us, there will still remain mysteries that will occupy future generations of scientists.

---

[51] See http://www.dtc.umn.edu/~odlyzko/polya/index.html.

Notes

1. *The strength of the theory of the physical natural numbers.* One problem is that the theory *TPhN*, formalized by $\Theta_{\mathbf{P}}$ + Axiom 4, is rather weak. But it is not as weak as it appears. The reason is that physical quantities are bounded by numbers in $\mathbb{L}_{phys}$ and the arithmetic of $\mathbb{L}_{phys}$ is in some sense stronger than the arithmetic of $\mathbb{N}_{phys}$. Most proofs in finite combinatorics and number theory only need to assume that for a given number n, 2^n also exists, which condition is satisfied by numbers in $\mathbb{L}_{phys}$. Furthermore, we can add more axioms to make the theory stronger.

   In any case, some standard mathematical facts will have to be revised. Consider, for example, the harmonic series. In *TPhN* we have

$$\sum_{1\leq m\in L}\frac{1}{m}\leq\sum_{1\leq m\leq n}\frac{1}{m}\leq\ln n+C,$$

   where n is an arbitrary number $n\notin L$. Thus the sum over all numbers in L is bounded by a number in L. Unfortunately, we cannot consistently assume that the sum is a real number. Still, it is conceivable that in *TPhN* there may be new ways to cope with divergent series, which is a problem in some physical theories.

   I introduced the theory *TPhN* only as a way to describe the physical natural numbers. In order to use this theory to formalize physical theories, one would have to extend it so that higher order concepts (real numbers, functions, etc.) could be formalized in it.
2. *Second order theories for* $\Theta_{\mathbf{P}}$. In this book I only considered theories for complexity classes whose elements are numbers. In the Notes to the previous section I mentioned some subsystems of Second-Order Arithmetic. Similar, but much weaker subsystems can also be used to define theories corresponding to complexity classes. Sometimes this is very useful, in particular for classes below **P**. Such second order systems for **P** were introduced by S. Buss (the theory V_1^1) and S. Cook and P. Nguyen (the theory V^1), [34, 50]. In the context of *TPhN* it would also be natural to use such theories because the two physical representations of numbers would then correspond to two different kinds of objects: L would be the numbers in such a theory, and N would be the finite sets of numbers in the theory. (Finite sets of numbers can be identified with 0–1 strings.)
3. *Changing* **NP** *properties in extensions of nonstandard models of* $\Theta_{\mathbf{P}}$. Let $P(x)$ be an **NP** property. Such a property is defined by a binary relation $R(x, y)$ decidable in polynomial time and a polynomial p: $P(x)\equiv\exists y\,(|y|\leq p(|x|)\wedge R(x,y))$. For the sake of simplicity, let us assume that the bound on the length of y is implicit in R. Given a model M and a nonstandard element $a\in M$, we would like to change the property P of a from being true to being false and vice versa.

   The first idea that comes to mind is that if $M\models\neg P(a)$, then we can change it to $P(a)$ by adding a b to M such that $R(a,b)$. This task looks very similar to what is done in algebra when one needs to add roots of polynomials to a field. Similarly in set theory, given a model of *ZFC*, one can extend it to a model of

ZFC by adding a generic set G and all sets definable from G. In this way one can change properties of models of set theory, in particular, one can change the arithmetic of infinite cardinals, but forcing, as used in set theory, cannot change the arithmetic of natural numbers. There are some techniques for constructing models of arithmetic based on the idea of forcing, but so far they have rather limited reach [163].

Adding b to M means, more precisely, the following. Given a model M of a theory T and an element $a \in M$, we want to find an extension $N \supseteq M$ and $b \in N$ such that $N \models R(a, b)$ and N is still a model of T. An obstacle to this may be an element $d \in M$ that is a proof of the sentence $\forall y \neg R(a, y)$. This is because d is an element of any extension of M and, assuming T proves the soundness of the proof d, no extension can tolerate a b that would make $R(a, b)$ false. As we noted above, this is the only obstacle. I will state this result only for the theory $\Theta_{\mathbf{P}}$.

Theorem 68 *Given a model $M \models \Theta_{\mathbf{P}}$, an element $a \in M$ and an* **NP** *property $P(x)$, it is possible to extend M to a model $N \models \Theta_{\mathbf{P}}$ such that $N \models P(a)$ if and only if in M there is no Extended Frege proof of the propositional translation of the sentence $\forall y \neg R(a, y)$.*

Recall that Extended Frege proof systems are naturally associated with $\Theta_{\mathbf{P}}$ by being the strongest proof systems whose soundness $\Theta_{\mathbf{P}}$ proves (Theorem 52, page 552).

The first thing we observe is that if $\Theta_{\mathbf{P}}$ proves $\mathbf{NP} = \mathbf{coNP}$, then we cannot make $P(a)$ true in an extension of M, unless it already holds true in M. This is because the condition that $\Theta_{\mathbf{P}}$ proves $\mathbf{NP} = \mathbf{coNP}$ implies that $\Theta_{\mathbf{P}}$ proves that Extended Frege proof systems are complete. If the latter is true, then the propositional translation of the sentence $\forall y \neg R(a, y)$ has an Extended Frege proof in the model whenever it is true in the model.

This observation suggests a way to prove that $\mathbf{NP} = \mathbf{coNP}$ is not provable in $\Theta_{\mathbf{P}}$: it suffices to find $M \subseteq N$ models of $\Theta_{\mathbf{P}}$ such that $M \models \neg P(a)$ and $N \models P(a)$, for some **NP** property. Therefore researchers in proof complexity believe that when they succeed in finding a technique for constructing models of arithmetic as powerful as forcing is in set theory, they will obtain the independence results they are looking for.

4. **NP**-*full models.* Using Theorem 68 we can construct an interesting model of $\Theta_{\mathbf{P}}$.

Theorem 69 *There exists a nonstandard model N of $\Theta_{\mathbf{P}}$ with a nonstandard element a such that*

1. *in N every tautology has an Extended Frege proof;*
2. *for every element $c \in N$, there exists a standard number k such that $|c| \leq |a|^k$.*

To construct this model, one starts with a countable nonstandard model satisfying condition 2. and applies repeatedly Theorem 68 to make every unprovable tautology false.

We call models satisfying condition 1. above **NP**-*full*, because such models cannot be extended to models in which an **NP** false property of a number is changed to a true property. One can obtain an **NP**-full model from any model by extending it cofinally. In the informal presentation above I used Axiom 5 to define this property.

5. *Is* $\Theta_{\mathbf{P}}$ *consistent with* $\mathbf{NP} = \mathbf{coNP}$*?* The model in the theorem above shows the consistency of $\Theta_{\mathbf{P}}$ with $\mathbf{NP} = \mathbf{coNP}$ in a rather weak sense: every tautology has an Extended Frege proof whose length is $|a|^k$, for some standard k, which, unfortunately, depends on the tautology. The problem of showing the consistency of $\Theta_{\mathbf{P}}$ with $\mathbf{NP} = \mathbf{coNP}$ in the natural formalization is still open.
6. *Mutually inconsistent extensions of* $\Theta_{\mathbf{P}}$. Assuming that factoring integers is not computable in polynomial time, we can construct a model of $\Theta_{\mathbf{P}}$ that has two incompatible extensions.

Theorem 70 *Suppose that factoring integers is not computable in polynomial time. Then there exists a model* $M \models \Theta_{\mathbf{P}}$*, an element* $a \in M$*, two* **NP** *predicates* P_1 *and* P_2 *and two cofinal extensions* N_1 *and* N_2 *of* M *to models of* $\Theta_{\mathbf{P}}$ *such that*

a. $M \models \neg P_1(a) \wedge \neg P_2(a)$;
b. $N_1 \models P_1(a)$ and $N_2 \models P_2(a)$;
c. *there exists no extension* $N \models \Theta_{\mathbf{P}}$ *such that* $N \models P_1(a) \wedge P_2(a)$.

The proof of this theorem uses a formalization of a bit commitment schema (see page 520).

7. *The Möbius Randomness Principle.* This principle, proposed by Peter Sarnak, asserts that:

the Möbius function μ *is asymptotically orthogonal to any low-complexity function* $F : \mathbb{N} \rightarrow [-1, 1]$.

Asymptotic orthogonality means that $\sum_{i=1}^{n} \mu(i)F(i)/n$ tends to 0 as n goes to infinity. The '*low-complexity*' property is not quite specified; according to what we know, the principle could be valid for all polynomial time computable functions F. The principle for the special case of the constant function $F(x) = 1$ follows from the Prime Number Theorem (and, in a certain sense, is equivalent to it). Some special cases of the principle were recently proved by B. Green [108] and J. Bourgain [33].

Suppose we interpret the low-complexity property as meaning computable in polynomial time. Then we obtain a property of μ which is a kind of pseudorandomness property, similar to those used in cryptography. We can use the assumption that functions with this property exist to construct incompatible extensions of a model of $\Theta_{\mathbf{P}}$. In fact, one can prove more. One can construct a set of extensions with a probability distribution that reflects the pseudorandomness of these functions [229].

8. *Injecting inconsistencies into models of arithmetic.* As we have observed, for stronger theories (theories that prove that exponentiation is total) we cannot use

extensions to change an **NP** property of an element. So instead a different question has been studied: how large an initial segment can we preserve when changing an **NP** property? In the paper [281] whose title I have used for this note R. Solovay studied the special case where the property is the existence of a contradiction in the given theory. His bounds were improved by several authors; I will only state the best known bound [165]. Recall that $Con_T(x)$ denotes a formula expressing that there is no proof of contradiction of length $\leq x$ in T. ($Con_T(x)$ is an **NEXP** property; if one wants to talk about an **NP** property, one should replace it by $Con_T(|x|)$.)

Theorem 71 *Let T be a consistent theory axiomatized by an* **NP** *set of axioms and suppose T is sufficiently strong. Let M be a nonstandard countable model of T and let $a, c \in M$ be nonstandard elements. Then there exists a model K of T such that*

a. $[0, a] \subseteq M \cap K$ *(the models M and K agree at least on the initial segment $[0, a]$);*
b. $K \models \neg Con_T(a^c)$.

The theorem talks about all models of T, but the interesting case is when in M there is no proof of contradiction in T. The intended application is that for every nonstandard a, we can take arbitrarily small nonstandard c and construct a model K with the properties above.

This theorem was proved using bounds on the lengths of the proofs of $Con_T(n)$ in T, (Theorem 58, page 565).

Main Points of the Chapter

- Mathematical realism, often called platonism, assumes that mathematical entities are as real as physical entities.
- Intuitionists view mathematics as a science studying mental constructions. They think that no formalism is able to present mathematics precisely.
- Logicism is the view that mathematics and logic are the same: every mathematical theorem follows from principles of logic.
- According to formalism, only formal systems based on axioms and logic enable us to achieve mathematical rigour.
- The arguments claiming that Gödel's incompleteness theorem implies that mind is superior to machines are based on logical errors.
- One can strengthen a theory by iteratively adding consistency statements, but other simple axioms make theories substantially stronger.
- Axioms postulating the existence of large cardinal numbers (higher infinities) help us to cope with the incompleteness of axiomatic set theory, at least at low levels of the hierarchy of sets.

- Higher infinities, whether stated as large-cardinal axioms, or fast growing functions, or large constructive ordinals, enable us to prove more arithmetical sentences. It is an open problem whether there are arithmetical sentences that need a different kind of axiom.
- The mathematical and the physical natural numbers could have different structures.

Bibliographical Remarks

The literature relevant to the topics treated in this book is vast and it would be an immense task to give a survey of it. Therefore here I only list sources that I used more extensively, the articles explicitly mentioned in the text and a few books that I either consider classical or find useful as further reading. In most cases I refer to the original publications, but the interested reader may find many old papers reprinted (and translated to English) in anthologies. The literature is commented in the main text and the notes after sections, so the purpose of the notes below is only to provide some additional information and suggest further reading.

Chapter 1

The approach to the foundations of mathematics based on mathematical structures is explained in the first volume of the Bourbaki series of monographs [32]. The standard reference for Ramsey theory is [107]. Cantor's most important paper about set theory is [36]. Paradoxes are treated at length in Fraenkel, Bar-Hillel and Levy [75], which is a classics in the literature about the foundations of set theory. Bertrand Russell wrote about the discovery of his paradox in [253]; Zermelo's discovery of this paradox is described in his biography written by Ebbinghaus [65].

To mention a few more classics in the foundations of mathematics, the two volumes of Hilbert and Bernays [128, 129] present Hilbert's proof-theoretical approach to the foundations, Beth [24] treats the foundations from a historical and philosophical standpoint, Kleene [154] focuses on computability and Feferman's book [72] is a more recent presentation of foundations from the position of a predicativist.

Chapter 2

There are many books devoted to formal language theory. The concept of a context-free grammar can be found in most textbooks about theoretical computer science.

P. Pudlák, *Logical Foundations of Mathematics and Computational Complexity*, Springer Monographs in Mathematics, DOI 10.1007/978-3-319-00119-7,

Essentially the same concerns other concepts treated in this chapter: proofs, models and computations. Thus I only mention a few that are among the most popular ones. Shoenfield [267] and Manin [190] are general introductions to logic and set theory. Grzegorczyk [110] is a readable introduction to logic. For further reading about proof theory, I recommend Takeuti [288], and the book by Troelstra and Schwichtenberg [291].

The history of first-order logic is treated in Moore's article [199]. The main part of Gödel's paper [96] is a proof of the First Incompleteness Theorem. The last section contains a sketch of a proof of the Second Incompleteness Theorem. Gödel intended to publish another paper on the subject with a detailed proof (with number II). That paper has not been written, probably because the sketch of the proof in the published paper turned out to be sufficiently clear. Three Brouwer's papers about intuitionism are reprinted in the anthology [118], which also contains other classical papers in the foundations of mathematics. The Church-Turing Thesis appeared for the fist time as *Thesis I.* in Kleene's book [154, page 300]. A classical monograph on the λ-calculus is Barendregt [13]. For a brief presentation of Churches type theory, see Andrews [4]. Many mathematicians discovered parts of the Curry-Howard Isomorphism and often these ideas were published much later (see the history of the λ-calculus, including the Curry-Howard Isomorphism, described in [130]). The key publications are Curry and Feys [55] and Howard [133].

Chapter 3

The standard reference for set theory is Jech [138]. Kanamori [148] is a survey of the history of set theory. Part II of Boolos's book [30] contains 13 essays about Frege's work and some proposals how to fix his system in order to be consistent. For a brief presentation of Type Theory, see Coquand [52]. I am grateful to Ilan Vardi for drawing my attention to his paper *Archimedes, The Sand Reckoner* [296]. The theory of large cardinals is treated in Kanamori [147]. For a survey on large cardinals and algebra, see Dehornoy [62].

Chapter 4

My presentation of the Galois theory is inspired by the readable introduction to this theory by Stewart [285]. The sketch of the unsolubility of a concrete quintic equation is based on Stewart's elementary approach. An excellent exposition of Matyasevich's Theorem and related results in logical investigations of arithmetic is Smoryński [276]. In Table 4.1 the number $3 \cdot 2^{402653211} - 2$ is from [152]; I have just computed the sequence a few steps back to obtain some more values. The collection of articles about the Collatz problem, edited by Lagarias [177], is a useful overview of this area. It contains an article by Collatz describing the history of the problem; Conway's article [47] is reprinted there too. Kunen [173] is an outstanding introduction to the independence proofs in set theory.

Chapter 5

Two monographs on computational complexity have appeared recently: Arora and Barak [8] and Goldreich [101]. There are a number of books that cover special topics in computational complexity such as circuit complexity, algebraic complexity, randomized computations and cryptography, but somebody with little background in complexity theory should rather start with general introductions such as the two above. Some of the standard reference monographs are Bürgisser, Clausen, and Shokrollahi [31] in algebraic complexity, Goldreich [102, 103] in cryptography, Nielsen and Chuang [205] in quantum computing, and Li and Vitanyi [185] in Kolmogorov complexity.

Chapter 6

I have already mentioned some books about proof theory in the notes for Chap. 2. The standard reference for ordinal analysis is Pohlers's monograph [219] and I also recommend a more readable survey of Rathjen [237]. The first book about Bounded Arithmetic was Buss [34]. Krajíček's monograph [161] covers all main subjects in proof complexity. A more recent monograph of Cook and Nguyen [50] focuses on theories associated with complexity classes.

Conjecture 2 of Sect. 6.3 first appeared in print in [223]; Conjectures 1 and 4 and Conjectures 3 and 5 appeared in [227] and [164] respectively.

Chapter 7

The 26 chapters of *The Oxford Handbook of Philosophy of Mathematics and Logic* [265], edited by Shapiro, cover all main positions in the philosophy of mathematics and several other topics. An on-line source that I find very useful is *The Stanford Encyclopedia of Philosophy* [316], which contains many articles about the philosophy of mathematics and history of logic. For a survey on reflection principles, see Smoryński [273]. Readable surveys about the axioms providing invariance with respect to forcing and other results on the Continuum Hypothesis are Woodin [310, 311] and Dehornoy [61].

References

1. Ajtai, M.: Σ_1^1 formulae on finite structures. Ann. Pure Appl. Log. **24**, 1–48 (1983)
2. Agrawal, M., Kayal, N., Saxena, N.: PRIMES is in P. Ann. Math. **160**(2), 781–793 (2004)
3. Alekhnovich, M.: Mutilated chessboard problem is exponentially hard for resolution. Theor. Comput. Sci. **310**(1–3), 513–525 (2004)
4. Andrews, P.: Church's type theory. In: Zalta, E.N. (ed.) Stanford Encyclopedia of Philosophy, Spring 2009 edn. (2009). http://plato.stanford.edu/archives/spr2009/entries/type-theory-church
5. Appel, K., Haken, W.: Every planar map is four colorable. Part I. Discharging. Ill. J. Math. **21**, 429–490 (1977)
6. Appel, K., Haken, W., Koch, J.: Every planar map is four colorable. Part II. Reducibility. Ill. J. Math. **21**, 491–567 (1977)
7. Arora, S., Lund, C., Motwani, R., Sudan, M., Szegedy, M.: Proof verification and the hardness of approximation problems. J. ACM **45**(3), 501–555 (1998)
8. Arora, S., Barak, B.: Computational Complexity: A Modern Approach. Cambridge University Press, Cambridge (2009)
9. Arora, S., Safra, S.: Probabilistic checking of proofs: A new characterization of NP. J. ACM **45**(1), 70–122 (1998)
10. Avigad, J., Sommer, R.: A model-theoretical approach to ordinal analysis. Bull. Symb. Log. **3**, 17–59 (1997)
11. Babai, L.: Trading group theory for randomness. In: Proc. 17th ACM Symp. on Theory of Computing, pp. 421–429 (1985)
12. Banach, S., Tarski, A.: Sur la décomposition des ensembles de points en parties respectivement congruentes. Fundam. Math. **6**, 244–277 (1924)
13. Barendregt, H.P.: The Lambda Calculus: Its Syntax and Semantics. North-Holland, Amsterdam (1984)
14. Barwise, J.: Admissible Sets and Structures: An Approach to Definability Theory. Springer, Berlin (1975)
15. Bernstein, A., Robinson, A.: Solution of an invariant subspace problem of K.T. Smith and P.R. Halmos. Pac. J. Math. **16**(3), 421–431 (1966)
16. Bachmann, H.: Die Normalfunktionen und das Problem der ausgezeichneten Folgen von Ordnungszahlen. Vierteljschr. Naturforsch. Ges. Zürich **95**, 115–147 (1950)
17. Baker, T.P., Gill, J., Solovay, R.: Relativizations of the P = ? NP question. SIAM J. Comput. **4**(4), 431–442 (1975)
18. Beklemishev, L.D.: A proof-theoretic analysis of collection. Arch. Math. Log. **34**(4–5), 216–238 (1998)
19. Bennett, C.H.: Logical reversibility of computation. IBM J. Res. Dev. **17**(6), 525–532 (1973)

P. Pudlák, *Logical Foundations of Mathematics and Computational Complexity*, Springer Monographs in Mathematics, DOI 10.1007/978-3-319-00119-7,

20. Bennett, C.H., Wiesner, S.J.: Communication via one- and two-particle operators on Einstein-Podolsky-Rosen states. Phys. Rev. Lett. **69**, 2881–2884 (1992)
21. Berger, R.: The undecidability of the domino problem. Mem. Am. Math. Soc. **66** (1966)
22. Bernays, P.: Sur le platonism dans les mathematiques. Enseign. Math. **34**, 52–69 (1935)
23. Bernays, P.: A system of axiomatic set theory I. J. Symb. Log. **2**, 65–77 (1937)
24. Beth, E.W.: The Foundations of Mathematics. A Study in the Philosophy of Science. North-Holland, Amsterdam (1959)
25. Blum, L., Shub, M., Smale, S.: On a theory of computation and complexity over the real numbers: NP-completeness, recursive functions and universal machines. Bull. Am. Math. Soc. **21**, 1–46 (1989)
26. Blum, L., Cucker, F., Shub, M., Smale, S.: Complexity and Real Computation. Springer, Berlin (1997)
27. Blum, N.: A boolean function requiring 3n network size. Theor. Comput. Sci. **28**, 337–345 (1984)
28. Bolzano, B.: Paradoxien des Unendlichen. C.H. Reclam, Leipzig (1951)
29. Bonet, M.L., Pitassi, T., Raz, R.: On interpolation and automatization for Frege systems. SIAM J. Comput. **29**(6), 1939–1967 (2000)
30. Boolos, G.: Logic, Logic, and Logic. Harvard University Press, Cambridge (1998)
31. Bürgisser, P., Clausen, M., Shokrollahi, M.A.: Algebraic Complexity Theory. Springer, Berlin (1997)
32. Bourbaki, N.: Theory of Sets. Elements of Mathematics, vol. 1. Addison-Wesley, Reading (1974)
33. Bourgain, J.: On the Fourier-Walsh spectreum of the Moebius function (2011). http://arxiv.org/pdf/1112.1423.pdf, arXiv:1112.1423
34. Buss, S.R.: Bounded Arithmetic. Bibliopolis, Naples (1986)
35. Buss, S.R., Krajíček, J.: An application of boolean complexity to separation problems in bounded arithmetic. Proc. Lond. Math. Soc. **69**(3), 1–21 (1994)
36. Cantor, G.: Grundlagen einer allgemeinen Mannichfaltigkeitslehre. Ein mathematisch-philosophischer Versuch in der Lehre des Unendlichen. Teubner, Leipzig (1882)
37. Cantor, G.: Über eine elementare Frage der Mannigfaltigkeitslehre. Jahresbericht der Deutsch. Math. Vereing. I, 75–78 (1890/91)
38. Carnap, R.: Logische Syntax der Sprache. Springer, Berlin (1934)
39. Chaitin, G.J.: On the simplicity and speed of programs for computing infinite sets of natural numbers. J. ACM **16**(3), 407–422 (1969)
40. Cheng, Q.: Straight-line programs and torsion points on elliptic curves. Comput. Complex. **12**(1), 150–161 (2003)
41. Cheyne, C.: Knowledge, Cause, and Abstract Objects: Causal Objections to Platonism. Kluwer Academic, Dordrecht (2001)
42. Church, A.: A set of postulates for the foundation of logic (1). Ann. Math. **33**, 346–366 (1932)
43. Church, A.: An unsolvable problem of elementary number theory. Am. J. Math. **58**, 345–363 (1936)
44. Church, A.: A formulation of the simple theory of types. J. Symb. Log. **5**, 56–68 (1940)
45. Chvátal, V.: Edmonds polytops and a hierarchy of combinatorial problems. Discrete Math. **4**, 305–337 (1973)
46. Cohen, P.: Set Theory and the Continuum Hypothesis, Benjamin, New York (1963)
47. Conway J.H: Unpredictable iterations. In: Proc. 1972 Number Th. Conf, pp. 49–52. University Press of Colorado, Boulder (1972)
48. Cook, S.A.: The complexity of theorem proving procedures. In: Proc. 3rd Annual ACM Symposium on Theory of Computing, pp. 151–158 (1971)
49. Cook, S.A.: Feasibly constructive proofs and the propositional calculus. In: Proc. Seventh Annual ACM Symposium on Theory of Computing, pp. 83–97. ACM, New York (1975)
50. Cook, S., Nguyen, P.: Logical Foundations of Proof Complexity. ASL Perspectives in Logic. Cambridge University Press, Cambridge (2010)

51. Cook, S.A., Reckhow, R.A.: The relative efficiency of propositional proof systems. J. Symb. Log. **44**(1), 36–50 (1979)
52. Coquand, T.: Type theory. In: Zalta, E.N. (ed.) Stanford Encyclopedia of Philosophy, Spring 2010 edn. (2010). http://plato.stanford.edu/archives/spr2010/entries/type-theory/
53. Craig, W.: Three uses of the Herbrand-Gentzen theorem in relating model theory and proof theory. J. Symb. Log. **22**(3), 269–285 (1957)
54. Curry, H.B.: Outlines of a Formalist Philosophy of Mathematics. Studies in Logic and Foundations of Mathematics. North-Holland, Amsterdam (1951)
55. Curry, H.B., Feys, R.: Combinatory Logic. Vol. I. North-Holland, Amsterdam (1958)
56. Davis, M., Putnam, H.: A computing procedure for quantification theory. J. ACM **7**(3), 201–215 (1960)
57. Davis, M., Putnam, H., Robinson, J.: The decision problem for exponential Diophantine equations. Ann. Math. (2) **74**(3), 425–436 (1961)
58. Dawson, J.W. Jr.: The reception of Gödel's incompleteness theorems. In: Drucker, T. (ed.) Perspectives on the History of Mathematical Logic, pp. 84–100. Birkhäuser, Boston (1991)
59. Dedekind, R.: Was sind und was sollen die Zahlen? 1. Auflage. Vieweg, Braunschweig (1888)
60. Dehornoy, P.: Braid groups and left distributive operations. Trans. Am. Math. Soc. **345**(1), 115–150 (1994)
61. Dehornoy, P.: Recent progress about the Continuum Hypothesis (after Woodin). Séminaire Bourbaki, exposé 915, mars 2003 (2003)
62. Dehornoy, P.: Elementary embeddings and algebra. In: Foreman, M., Kanamori, A. (eds.) Handbook of Set Theory. Springer, Berlin (2010)
63. Detlefsen, M.: Formalism. In: Shapiro, S. (ed.) The Oxford Handbook of Philosophy of Mathematics and Logic, pp. 236–317. Oxford University Press, London (2005)
64. Dummett, M.: Realism. In: Truth and Other Enigmas, pp. 145–165. Harvard University Press, Cambridge (1978)
65. Ebbinghaus, H.-D., Peckhaus, V.: Ernst Zermelo: An Approach to His Life and Work. Springer, Berlin (2007)
66. Elitzur, A.C., Vaidman, L.: Quantum mechanical interaction-free measurements. Found. Phys. **23**, 987–997 (1993)
67. Erdős, P.: Some remarks on the theory of graphs. Bull. Am. Math. Soc. **53**, 292–294 (1947)
68. Erdős, P.: On a new method in elementary number theory which leads to an elementary proof of the prime number theorem. Proc. Natl. Acad. Sci. USA **35**, 374–384 (1949)
69. Erdős, P., Szekerés, G.: A combinatorial problem in geometry. Compos. Math. **2**, 463–470 (1935)
70. Feferman, S.: Transfinite recursive progressions of axiomatic theories. J. Symb. Log. **27**(3), 259–315 (1962)
71. Feferman, S.: Systems of predicative analysis. J. Symb. Log. **29**(1), 1–30 (1964)
72. Feferman, S.: In the Light of Logic. Oxford University Press, London (1998)
73. Feynman, R.: Simulating physics with computers. Int. J. Theor. Phys. **21**(6–7), 467 (1982)
74. Fraenkel, A.A.: Zu den Grundlagen der Cantor-Zermeloschen Mengenlehre. Math. Ann. **86**, 230–237 (1922)
75. Fraenkel, A.A., Bar-Hillel, Y., Levy, A.: Foundations of Set Theory. North-Holland, Amsterdam (1973)
76. Franzén, T.: Inexhaustibility: A Non-exhaustive Treatment. Lecture Notes in Logic, vol. 16. Association for Symbolic Logic, AK Peters, Wellesley (2004)
77. Frege, G.: Begriffsschrift: eine der arithmetischen nachgebildete Formelsprache des reinen Denkens. Halle (1879)
78. Frege, G.: Grundgesetze der Arithmetik I. Hermann Pohle, Jena (1893)
79. Frege, G.: Grundgesetze der Arithmetik II. Hermann Pohle, Jena (1903)
80. Friedberg, R.M.: Two recursively enumerable sets of incomparable degrees of unsolvability. Proc. Natl. Acad. Sci. USA **43**, 236–238 (1957)

81. Friedman, H.: On the consistency, completeness and correctness. Unpublished typescript (1979)
82. Friedman, H.: Finite functions and the necessary use of large cardinals. Ann. Math. **148**, 803–893 (1998)
83. Furst, M.L., Saxe, J.B., Sipser, M.: Parity, circuits, and the polynomial-time hierarchy. Math. Syst. Theory **17**(1), 13–27 (1984)
84. Gaifman, H.: Ontology and conceptual framework part I. Erkenntnis **9**, 329–353 (1975)
85. Gaifman, H.: Ontology and conceptual framework part II. Erkenntnis **10**, 21–85 (1976)
86. Gaifman, H.: Ontology and realism in mathematics. Rev. Symb. Log. **5**(3), 480–512 (2012)
87. Gál, A., Hansen, K.A., Koucký, M., Pudlák, P., Viola, E.: Tight bounds on computing error-correcting codes by bounded-depth circuits with arbitrary gates. In: Proc. STOC 2012, pp. 479–494 (2012)
88. Galilei, G.: Discorsi e dimostrazioni matematiche intorno a due nuove scienze attinenti la mecanica e i movimenti locali. Elzevir, Leiden (1638)
89. Gentzen, G.: Untersuchungen über das logische Schließen I. Math. Z. **39**(2), 176–210 (1934)
90. Gentzen, G.: Untersuchungen über das logische Schließen II. Math. Z. **39**(3), 405–431 (1935)
91. Gentzen, G.: Die Widerspruchsfreiheit der reinen Zahlentheorie. Math. Ann. **112**, 493–565 (1936)
92. Gentzen, G.: Neue fassung des widerspruchsfreiheitsbeweises für die reine Zahlentheorie. Forsch. Logik Grundlegung exakten Wiss. **4**, 19–44 (1938)
93. Girard, J.-Y.: Proof Theory and Logical Complexity. Bibliopolis, Naples (1987)
94. Glaßer, C., Selman A, A., Sengupta, S., Zhang, L.: Disjoint NP-pairs. SIAM J. Comput. **33**(6), 1369–1416 (2004)
95. Gödel, K.: Die Vollständigkeit der Axiome des logischen Funktionenkalküls. Monatshefte Math. Phys. **37**, 349–360 (1930)
96. Gödel, K.: Über formal unentscheidbare Sätze der Principia Mathematica und verwandter Systeme, I. Monatshefte Math. Phys. **38**, 173–198 (1931)
97. Gödel, K.: The Consistency of the Axiom of Choice and of the Generalized Continuum Hypothesis with the Axioms of Set Theory. Annals of Mathematical Studies, vol. 3. Princeton University Press, Princeton (1940)
98. Gödel, K.: What is Cantor's continuum problem? Am. Math. Mon. **54**(9), 515–525 (1947)
99. Gödel, K.: Collected Works: Volume III. Unpublished Essays and Lectures. Feferman, S., Dawson, J.W., Goldfarb, W., Parsons, C., Sieg, W. (eds.). Oxford University Press, London (1995)
100. Gödel, K. (ed.): Collected Works: Volume V. Correspondence, H.-Z. Feferman, S., Dawson, J.W., Goldfarb, W., Parsons, C., Sieg, W. (eds.). Oxford University Press, London (2003)
101. Goldreich, O.: Computational Complexity, a Conceptual Perspective. Cambridge University Press, London (2008)
102. Goldreich, O.: The Foundations of Cryptography, Volume 1. Cambridge University Press, London (2001)
103. Goldreich, O.: The Foundations of Cryptography, Volume 2. Cambridge University Press, London (2004)
104. Gomory, R.E.: Outline of an algorithm for integer solutions to linear programs. Bull. Am. Math. Soc. **64**, 275–278 (1958)
105. Goodstein, R.L.: On the restricted ordinal theorem. J. Symb. Log. **9**, 33–41 (1944)
106. Gonthier, G.: Formal proof–the four-color theorem. Not. Am. Math. Soc. **55**(11), 1382–1393 (2008)
107. Graham, R.L., Rothschild, B.L., Spencer, J.H: Ramsey Theory. Wiley, New York (1990)
108. Green, B.: On (not) computing the Möbius functions using bounded depth circuits. Comb. Probab. Comput. **21**(6), 942–951 (2012)
109. Grover, L.K.: A fast quantum mechanical algorithm for database search. In: Proc. 28th Annual ACM Symposium on the Theory of Computing, pp. 212–218 (1996)

110. Grzegorczyk, A.: An Outline of Mathematical Logic: Fundamental Results and Notions Explained with all Details. Reidel, Dordrecht (1974)
111. Hájek, P., Pudlák, P.: Metamathematics of First Order Arithmetic. ASL Perspectives in Logic. Springer, Berlin (1993)
112. Hales, T.C.: A proof of the Kepler conjecture. Ann. Math. (2) **162**(3), 1065–1185 (2005)
113. Harrow, A.W., Hassidim, A., Lloyd, S.: Quantum algorithm for linear systems of equations. Phys. Rev. Lett. **103**, 150502 (2009)
114. Hartmanis, J., Stearns, R.E.: On the computational complexity of algorithms. Trans. Am. Math. Soc. **117**, 285–306 (1965)
115. Håstad, J.: Almost optimal lower bounds for small depth circuits. In: Micali, S. (ed.) Randomness and Computation, Advances in Computing Research, vol. 5, pp. 143–170. JAI Press, London (1989)
116. Hawking, S.: Godel and the end of physics. A public lecture held in 2002. http://www.hawking.org.uk/godel-and-the-end-of-physics.html
117. Heath, T.L.: The Thirteen Books of Euclid's Elements. Cambridge University Press, Cambridge (1908)
118. van Heijenoort, J.: A Source Book in Mathematical Logic, 1879–1931. Harvard University Press, Cambridge (1976)
119. Henkin, L., Monk, J.D., Tarski, A.: Cylindric Algebras I. North-Holland, Amsterdam (1975)
120. Henkin, L., Monk, J.D., Tarski, A.: Cylindric Algebras II. North-Holland, Amsterdam (1985)
121. Herbrand, J.: Recherches sur la theorie de la demonstration. Travaux de la Societe des Sciences et des Lettres de Varsovie, Class III, Sciences Mathematiques et Physiques 33, 33–160 (1930)
122. Heyting, A.: Intuitionism: An introduction. Studies in Logic and the Foundations of Mathematics, vol. 16. North-Holland, Amsterdam (1971)
123. Hilbert, D.: Über die Endlichkeit des Invariantensystems für binären Grundformen. Math. Ann. **33**, 223–226 (1889)
124. Hilbert, D.: Grundlagen der Geometrie. Teubner, Berlin (1899)
125. Hilbert, D.: Mathematical problems. Bull. Am. Math. Soc. **8**, 437–479 (1902)
126. Hilbert, D.: On the infinite. In: van Heijenoort, J. (ed.) From Frege to Gödel: A Source Book in Mathematical Logic, 1879–1931, pp. 367–392. Harvard University Press, Cambridge (1967)
127. Hilbert, D.: Die Grundlagen der Mathematik (a lecture given in Hamburg in 1927). English translation "The foundations of mathematics". In: van Heijenoort, J. (ed.) From Frege to Gödel: A Source Book in Mathematical Logic, 1879–1931, pp. 464–479. Harvard University Press, Cambridge (1967)
128. Hilbert, D., Bernays, P.: Grundlagen der Mathematik. I. Die Grundlehren der mathematischen Wissenschaften, vol. 40. Springer, Berlin (1934)
129. Hilbert, D., Bernays, P.: Grundlagen der Mathematik. II. Die Grundlehren der mathematischen Wissenschaften, vol. 50. Springer, Berlin (1939)
130. Hindley, J.R., Cardone, F.: History of λ-calculus and combinatory logic. In: Gabbay, D.M., Woods, J. (eds.) Handbook of the History of Logic. Elsevier, Amsterdam (2006)
131. Hirschfeld, J.: The nonstandard treatment of Hilbert's fifth problem. Trans. Am. Math. Soc. **321**(1), 379–400 (1990)
132. Hopcroft, J., Paul, W.J., Valiant, L.G.: On time vs. space. J. ACM **24**(2), 332–337 (1977)
133. Howard, W.A.: The formulae-as-types notion of construction. In: Seldin, J.P., Hindley, J.R. (eds.) To H.B. Curry: Essays on Combinatory Logic, Lambda Calculus and Formalism, pp. 479–490. Academic Press, Boston (1980)
134. Hrubeš, P.: A lower bound for intuitionistic logic. Ann. Pure Appl. Log. **146**, 72–90 (2007)
135. Impagliazzo, R.: A personal view of average-case complexity. In: 10th Annual Structure in Complexity Theory Conference (SCT'95), pp. 134–147 (1995)
136. Impagliazzo, R., Wigderson, A.: P = BPP unless E has subexponential circuits: Derandomizing the XOR lemma. In: Proc. 29th STOC, pp. 220–229 (1997)

137. Jech, T.: OTTER experiments in a system of combinatory logic. J. Autom. Reason. **14**(3), 413–426 (1995)
138. Jech, T.: Set Theory, the Third Millennium Edition. Springer, Berlin (2003)
139. Jensen, R.B.: On the consistency of a slight(?) modification of Quine's NF. Synthese **19**, 250–263 (1969)
140. Jeřábek, E.: The strength of sharply bounded induction. Math. Log. Q. **52**(6), 613–624 (2006)
141. Jockusch, C.G., Jr.: Ramsey's theorem and recursion theory. J. Symb. Log. **37**, 268–280 (1972)
142. Jockusch, C.G., Soare, R.I.: Π_1^0 classes and degrees of theories. Trans. Am. Math. Soc. **173**, 33–56 (1972)
143. Jones, J.P.: Universal Diophantine equation. J. Symb. Log. **47**, 549–571 (1982)
144. Johnson, D., Papadimitriou, C., Yannakakis, M.: How easy is local search? J. Comput. Syst. Sci. **37**, 79–100 (1988)
145. Kabanets, V., Impagliazzo, R.: Derandomizing polynomial identity tests means proving circuit lower bounds. Comput. Complex. **13**(1–2), 1–46 (2004)
146. Kahr, A.S., Moore, E.F., Wang, H.: Entscheidungsproblem reduced to the AEA case. Proc. Natl. Acad. Sci. USA **48**(3), 365–377 (1962)
147. Kanamori, A.: The Higher Infinite, Large Cardinals in Set Theory from Their Beginnings. Springer, Berlin (1994)
148. Kanamori, A.: The mathematical development of set theory from Cantor to Cohen. Bull. Symb. Log. **2**, 1–71 (1996)
149. Kanamori, A., McAloon, K.: On Gödel incompleteness and finite combinatorics. Ann. Pure Appl. Log. **33**(1), 23–41 (1987)
150. Katz, V.J.: A History of Mathematics – an Introduction. Harper Collins, New York (1993)
151. Ketonen, J., Solovay, R.: Rapidly growing Ramsey functions. Ann. Math. **113**, 267–314 (1981)
152. Kirby, L., Paris, J.: Accessible independence results for Peano arithmetic. Bull. Lond. Math. Soc. **14**, 285–293 (1982)
153. Kleene, S.C.: General recursive functions of natural numbers. Math. Ann. **112**, 727–742 (1936)
154. Kleene, S.C.: Introduction to Metamathematics. Van Nostrand, New York (1952)
155. Kleene, S.C.: Extension of an effectively generated class of functions by enumeration. Colloq. Math. **6**(1), 67–78 (1958)
156. Koblitz, N.: A Course in Number Theory and Cryptography. Springer, New York (1987)
157. Koellner, P., Woodin, W.H.: Incompatible Omega-complete theories. J. Symb. Log. **74**(4), 1155–1170 (2009)
158. Kohlenbach, U.: Applied Proof Theory: Proof Interpretations and Their Use in Mathematics. Springer, Berlin (2008)
159. Kollár, J., Rónyai, L., Szabó, T.: Norm-graphs and bipartite Turán numbers. Combinatorica **16**(3), 399–406 (1996)
160. Kolmogorov, A.: On tables of random numbers. Sankhya, Ser. A **25**, 369–375 (1963)
161. Krajíček, J.: Bounded Arithmetic, Propositional Logic, and Complexity Theory. Encyclopedia of Mathematics and Its Applications, vol. 60. Cambridge University Press, Cambridge (1995)
162. Krajíček, J.: Interpolation theorems, lower bounds for proof systems, and independence results for bounded arithmetic. J. Symb. Log. **62**(2), 457–486 (1997)
163. Krajíček, J.: Forcing with Random Variables. London Math. Soc. Lecture Note Series, vol. 382. Cambridge University Press, Cambridge (2011)
164. Krajíček, J., Pudlák, P.: Propositional proof systems, the consistency of first order theories and the complexity of computations. J. Symb. Log. **54**(3), 1063–1079 (1989)
165. Krajíček, J., Pudlák, P.: On the structure of initial segments of models of arithmetic. Arch. Math. Log. **28**, 91–98 (1989)

166. Krajíček, J., Pudlák, P.: Some consequences of cryptographical conjectures for S_2^1 and EF. In: Leivant, D. (ed.) Logic and Computational Complexity, (Proceedings of Meeting Held in Indianapolis 1994). LNCS, vol. 960, pp. 210–220. Springer, Berlin (1995)
167. Krajíček, J., Pudlák, P., Takeuti, G.: Bounded arithmetic and polynomial hierarchy. Ann. Pure Appl. Log. **52**, 143–154 (1991)
168. Kreisel, G.: Ordinal logics and the characterization of informal concepts of proof. In: Proc. of the 8th International Congress of Mathematicians, pp. 289–299. Edinburgh University Press, Edinburgh (1958). 1960
169. Kreisel, G., Levy, A.: Reflection principles and their use for establishing the complexity of axiomatic systems. Z. Math. Log. Grundl. Math. **14**, 97–142 (1968)
170. Kripke, S.A.: A completeness theorem in modal logic. J. Symb. Log. **24**(1), 1–14 (1959)
171. Kruskal, J.B.: Well-quasi-ordering, the tree theorem, and Vázsonyi's conjecture. Trans. Am. Math. Soc. **95**, 210–225 (1960)
172. Kunen, K.: Combinatorics. In: Barwise, J. (ed.) Handbook of Mathematical Logic. North-Holland, Amsterdam (1977)
173. Kunen, K.: Set Theory: An Introduction to Independence Proofs. North-Holland, Amsterdam (1980)
174. Kuperberg, G.: A subexponential-time quantum algorithm for the dihedral hidden subgroup problem. SIAM J. Comput. **35**(1), 170–188 (2005)
175. Laczkovich, M.: Conjecture and Proof. TypoTeX, Budapest (1998)
176. Ladner, R.E.: On the structure of polynomial time reducibility. J. ACM **22**, 155–171 (1975)
177. Lagarias, G. (ed.): The Ultimate Challenge: The $3x+1$ Problem. Am. Math. Soc., Providence (2010)
178. Lakatos, I.: Proofs and Refutations. Cambridge University Press, Cambridge (1976)
179. Laver, R.: The left-distributive law and the freeness of an algebra of elementary embeddings. Adv. Math. **91**, 209–231 (1992)
180. Laver, R.: On the algebra of elementary embeddings of a rank into itself. Adv. Math. **110**, 334–346 (1995)
181. Lawvere, F.W.: An elementary theory of the category of sets. Proc. Natl. Acad. Sci. USA **52**, 1506–1511 (1964)
182. Levin, L.: Universal'nye perebornye zadachi. Probl. Inf. Transm. **9**(3), 265–266 (1973). (Russian)
183. Levy, A., Solovay, R.M.: Measurable cardinals and the continuum hypothesis. Isr. J. Math. **5**, 234–248 (1967)
184. Linnebo, Ø.: Platonism in the philosophy of mathematics. In: Zalta, E.N. (ed.) Stanford Encyclopedia of Philosophy (2011). http://plato.stanford.edu/entries/platonism-mathematics/
185. Li, M., Vitanyi, P.: An Introduction to Kolmogorov Complexity and Its Applications. Springer, Berlin (2008)
186. Löwenheim, L.: Über Möglichkeiten im Relativkalkül. Math. Ann. **76**(4), 447–470 (1915)
187. Luckhardt, H.: Herbrand-Analysen Zweier Beweise Des Satzes von Roth: Polynomiale Anzahlschranken. J. Symb. Log. **54**(1) (1989)
188. Luminet, J.-P., Weeks, J., Riazuelo, A., Lehoucq, R., Uzan, J.-P.: Dodecahedral space topology as an explanation for weak wide-angle temperature correlations in the cosmic microwave background. Nature **425**(9), 593–595 (2003)
189. Mac Lane, S., Moerdijk, I.: Sheaves in Geometry and Logic: A First Introduction to Topos Theory. Springer, New York (1992)
190. Manin, Yu.I.: A Course in Mathematical Logic for Mathematicians. Springer, New York (2010)
191. Margulis, G.A.: Explicit constructions of expanders. Probl. Pereda. Inf. **9**(4), 71–80 (1973)
192. Martin, D.A., Solovay, R.M.: Internal Cohen extensions. Ann. Math. Log. **2**(2), 143–178 (1970)
193. Matiyasevich Yu, V.: Enumerable sets are Diophantine. Dokl. Akad. Nauk SSSR **191**(2), 279–282 (1970). (In Russian; English translation: Sov. Math. Dokl. **11**(2), 354–358)

194. Matoušek, J.: A combinatorial proof of Kneser's conjecture. Combinatorica **2**(1), 163–170 (2004)
195. McCarthy, J.: A tough nut for proof procedures. Stanford Artificial Intelligence Project, Memo No. 16 (1964)
196. McCarthy, C.: What does it take to prove Fermat's Last Theorem? Grothendieck and the logic of number theory. Bull. Symb. Log. **16**(3), 359–377 (2010)
197. McCune, W.: Solution of the Robbins problem. J. Autom. Reason. **19**(3), 263–276 (1997)
198. Miller, G.L.: Riemann's hypothesis and tests for primality. J. Comput. Syst. Sci. **13**(3), 300–317 (1976)
199. Moore, G.H.: The emergence of first-order logic. In: Aspray, W., Kitcher, P. (eds.) Minnesota Studies of the Philosophy of Science, XI: History and Philosophy of Modern Mathematics, pp. 95–135. University of Minnesota Press, Minneapolis (1988)
200. Mostowski, A.: Sentences Undecidable in Formalized Arithmetic: An Exposition of the Theory of Kurt Gödel. Studies in Logic. North-Holland, Amsterdam (1952)
201. Mostowski, A.: A generalization of the incompleteness theorem. Fundam. Math. **49**, 205–232 (1961)
202. Muchnik, A.A.: On the unsolvability of the problem of reducibility in the theory of algorithms. Dokl. Akad. Nauk SSSR **108**, 194–197 (1956) (Russian)
203. Mulmuley, K., Sohoni, M.: Geometric complexity theory I: An approach to the P vs. NP and related problems. SIAM J. Comput. **31**(2), 496–526 (2001)
204. Mycielski, J., Steinhaus, H.: A mathematical axiom contradicting the axiom of choice. Bull. Acad. Pol. Sci., Sér. Sci. Math. Astron. Phys. **10**, 1–3 (1962)
205. Nielsen, M.A., Chunag, I.L.: Quantum Computation and Quantum Information: 10th Anniversary Edition. Cambridge University Press, Cambridge (2010)
206. Németi, I., Dávid, Gy.: Relativistic computers and the Turing barrier. Appl. Math. Comput. **178**, 118–142 (2006)
207. von Neumann, J.: Eine Axiomatisierung der Mengenlehre. J. Reine Angew. Math. **154**, 219–240 (1925)
208. Newton, I.: Universal Arithmetick: Or, a Treatise of Arithmetical Composition and Resolution. J. Senex, London (1720)
209. Nisan, N., Wigderson, A.: Hardness vs. randomness. J. Comput. Syst. Sci. **49**(2), 149–167 (1994)
210. Odlyzko, A.M., te Riele, H.J.J.: Disproof of the Mertens conjecture. J. Reine Angew. Math. **357**, 138–160 (1985)
211. Parikh, R.: Existence and feasibility in arithmetic. J. Symb. Log. **36**(3), 494–508 (1971)
212. Paris, J.B.: Some independence results for Peano arithmetic. J. Symb. Log. **43**(4), 725–731 (1978)
213. Paris, J.B.: A hierarchy of cuts in models of arithmetic. In: Model Theory of Algebra and Arithmetic, Karpacz, 1979. Springer Lecture Notes in Math., vol. 834, pp. 312–337 (1980)
214. Paris, J., Harrington, L.: A mathematical incompleteness in Peano arithmetic. In: Barwise, J. (ed.) Handbook of Mathematical Logic, pp. 1133–1142. North-Holland, Amsterdam (1977)
215. Paris, J., Wilkie, A.: Δ_0 sets and induction. In: Proc. Jadswin Logic Conference (Poland), pp. 237–248. Leeds University Press, Leeds (1981)
216. Peano, G.: Arithmetices Principia, Nova Methodo Exposita. Fratres Bocca, Torino (1889)
217. Peano, G., Cassina, U.: Formulario Matematico. Fratres Bocca, Torino (1908)
218. Planck, M.: Where Is Science Going? Norton, New York (1932)
219. Pohlers, W.: Proof Theory. The First Step to Impredicativity. Springer, Berlin (1989)
220. Poincaré, H.: The Foundations of Science. The Science Press, New York (1908)
221. Post, E.: Finite combinatory processes—formulation 1. J. Symb. Log. **1**(3), 103–105 (1936)
222. Pudlák, P.: Cuts, consistency statements and interpretations. J. Symb. Log. **50**(2), 423–441 (1985)
223. Pudlák, P.: On the length of proofs of finitistic consistency statements in first order theories. In: Logic Colloquium, vol. 84, pp. 165–196. North-Holland, Amsterdam (1986)

224. Pudlák, P.: Improved bounds to the length of proofs of finitistic consistency statements. In: Contemporary Mathematics, vol. 65, pp. 309–331. Am. Math. Soc., Providence (1987)
225. Pudlák, P.: Lower bounds for resolution and cutting planes proofs and monotone computations. J. Symb. Log. **62**(3), 981–998 (1997)
226. Pudlák, P.: Complexity theory and genetics: The computational power of crossing over. Inf. Comput. **171**, 201–223 (2001)
227. Pudlák, P.: Gödel and computations. SIGACT News **37**(4), 13–21 (2006)
228. Pudlák, P.: Quantum deduction rules. Ann. Pure Appl. Log. **157**, 16–29 (2009)
229. Pudlák, P.: Randomness, pseudorandomness and models of arithmetic. arXiv:1210.4692
230. Pudlák, P., Rödl, V., Sgall, J.: Boolean circuits, tensor ranks and communication complexity. SIAM J. Comput. **26**(3), 605–633 (1997)
231. Quine, W.V.: New foundations for mathematical logic. Am. Math. Mon. **44**, 70–80 (1937)
232. Quine, W.V.: From a Logical Point of View: Nine Logico-Philosophical Essays, 2nd edn. Harvard University Press, Cambridge (2003)
233. Quine, W.V.: Mathematical Logic. Norton, New York (1940)
234. Rabin, M.O.: Digital signatures and public-key functions as intractable as factorization. MIT Laboratory of Computer Science Technical Report 212, (1979).
235. Ramsey, F.P.: On a problem of formal logic. Proc. Lond. Math. Soc. **30**(1), 264–286 (1930)
236. Rathjen, M.: The higher infinite in proof theory. In: Makowsky, J., Ravve, E. (eds.) Logic Colloquium '95. Springer Lecture Notes in Logic, vol. 11, pp. 275–304 (1998)
237. Rathjen, M.: The realm of ordinal analysis. In: Cooper, S.B., Truss, J.K. (eds.) Sets and Proofs, pp. 219–279. Cambridge University Press, Cambridge (1999)
238. Razborov, A.: Lower bounds for the monotone complexity of some boolean functions. Dokl. Akad. Nauk SSSR **281**(4), 798–801 (1985). (In Russian; English translation in: Sov. Math. Dokl. **31**, 354–357 (1985))
239. Razborov, A.: Lower bounds on the size of bounded-depth networks over a complete basis with logical addition. Mat. Zametki **41**(4), 598–607 (1987). (In Russian; English translation in: Math. Notes Acad. Sci. USSR **41**(4), 333–338 (1987))
240. Razborov, A.: On the method of approximation. In: Proc. of the 21st ACM STOC, pp. 169–176 (1989)
241. Razborov, A.: Unprovability of lower bounds on the circuit size in certain fragments of bounded arithmetic. Izv. Math. **59**(1), 201–224 (1995)
242. Razborov, A.: Bounded arithmetic and lower bounds in Boolean complexity. In: Clote, P., Remmel, J. (eds.) Feasible Mathematics II, pp. 344–386. Birkhauser, Basel (1995)
243. Razborov, A., Rudich, S.: Natural proofs. J. Comput. Syst. Sci. **55**(1), 24–35 (1997)
244. Reisch, S.: Hex ist PSPACE-vollstñdig (Hex is PSPACE-complete). Acta Inform. **15**, 167–191 (1981)
245. Rissanen, J.: Modeling by the shortest data description. Automatica **14**, 465–471 (1978)
246. Rivest, R., Shamir, A., Adleman, L.: A method for obtaining digital signatures and public-key cryptosystems. Commun. ACM **21**(2), 120–126 (1978)
247. Robinson, A.: Non-standard Analysis. North-Holland, Amsterdam (1966)
248. Robinson, J.A.: A machine-oriented logic based on the resolution principle. J. ACM **12**(1), 23–41 (1965)
249. Rosser, J.B.: Extensions of some theorems of Gödel and Church. J. Symb. Log. **1**, 87–91 (1936)
250. Rosser, J.B.: Logic for Mathematicians. McGraw-Hill, New York (1953)
251. Russell, B.: The Principles of Mathematics. Cambridge University Press, Cambridge (1903)
252. Russell, B.: Mathematical logic as based on the theory of types. Am. J. Math. **30**, 222–262 (1908)
253. Russell, B.: My Philosophical Development. Allen & Unwin, London (1959)
254. Russell, B.: The Autobiography of Bertrand Russell, vol. 1. Allen & Unwin, London (1967)
255. Savitch, W.J., Stimson, M.J.: Time bounded random access machines with parallel processing. J. ACM **26**(1), 103–118 (1979)

256. Schmerl, U.R.: A fine structure generated by reflection formulas over primitive recursive arithmetic. In: Boffa, M., van Dalen, D., McAloon, K. (eds.) Proc. Logic Colloquium'78, pp. 335–350. North-Holland, Amsterdam (1979)
257. Schönhage, A., Strassen V, V.: Schnelle multiplikation grosser Zahlen. Computing **7**, 281–292 (1971)
258. Schütte, K.: Bewiestheorie. Springer, Berlin (1960)
259. Schwartz, J.: Fast probabilistic algorithms for verification of polynomial identities. J. ACM **27**, 701–717 (1980)
260. Scott, D.: Measurable cardinals and constructible sets. Bull. Acad. Pol. Sci. **9**, 521–524 (1961)
261. Scott, D.: Continuous Lattices. Oxford Univ. Computing Lab. Technical Monograph PRG-7 (1971)
262. Scott, D., Solovay, R.: Boolean-Valued Models for Set Theory. Proc. AMS Summer Institute on Set Theory, Los Angeles. University of California, Berkeley (1967)
263. Selberg, A.: An elementary proof of the prime-number theorem. Ann. Math. (2) **50**, 305–313 (1949)
264. Shannon, C.E.: The synthesis of two-terminal switching circuits. Bell Syst. Tech. J. **28**, 59–98 (1949)
265. Shapiro, S. (ed.): The Oxford Handbook of Philosophy of Mathematics and Logic. Oxford University Press, London (2005)
266. Shelah, S.: Can you take Solovay's inaccessible away? Isr. J. Math. **48**(1), 1–47 (1984)
267. Shoenfield, J.R.: Mathematical Logic. Addison-Wesley, Reading (1967)
268. Shor, P.W.: Polynomial-time algorithms for prime factorization and discrete logarithms on a quantum computer. SIAM J. Comput. **26**(5), 1484–1509 (1997)
269. Simpson, S.G.: Nonprovability of certain combinatorial properties of finite trees. In: Harrington, L.A., et al. (eds.) Harvey's Friedman Research on the Foundations of Mathematics, pp. 87–117. North-Holland, Amsterdam (1985)
270. Skolem, T.: Logisch-kombinatorische Untersuchungen über die Erfüllbarkeit oder Beweisbarkeit mathematischer Sätze nebst einem Theoreme über dichte Mengen. Videnskapsselskapet Skrifter, I. Mat.-Naturvidensk. Kl. **6**, 1–36 (1920)
271. Skolem, T.: Einige Bemerkungen zur axiomatischen Begründung der Mengenlehre. In: Fünften Kongress der Skandinavischen Mathematiker in Helsingfors 1922. Helsingfors, pp. 217–232 (1923)
272. Smith, R.L.: The consistency strengths of some finite forms of the Higman and Kruskal theorems. In: Harrington, L.A., et al. (eds.) Harvey's Friedman Research on the Foundations of Mathematics, pp. 119–136. North-Holland, Amsterdam (1985)
273. Smoryński, C.: The incompleteness theorems. In: Barwise, J. (ed.) Handbook of Mathematical Logic, pp. 821–865. North-Holland, Amsterdam (1977)
274. Smoryński, C.: The varieties of arboreal experience. Math. Intell. **4**, 182–188 (1982)
275. Smoryński, C.: Nonstandard models and related developments. In: Harrington, L.A., et al. (eds.) Harvey's Friedman Research on the Foundations of Mathematics, pp. 179–229. North-Holland, Amsterdam (1985)
276. Smoryński, C.: Logical Number Theory I, an Introduction. Springer, Berlin (1991)
277. Solomonoff, R.: A Preliminary Report on a General Theory of Inductive Inference. Report V-131, Cambridge, Ma., Zator Co. (1960)
278. Solovay, R.M.: A model of set-theory in which every set of reals is Lebesgue measurable. Ann. Math. **92**(1), 1–56 (1970)
279. Solovay, R.M.: Real valued measurable cardinals. In: Scott, D.S. (ed.) Axiomatic Set Theory. Proc. Sym. in Pure Math. XIII, vol. 1, pp. 387–428 (1971)
280. Solovay, R.M.: Provability interpretations of modal logic. Isr. J. Math. **25**, 287–304 (1976)
281. Solovay, R.: Injecting inconsistencies into models of PA. Ann. Pure Appl. Log. **44**(1–2), 101–132 (1989)
282. Solovay, R.M., Strassen, V.: A fast Monte-Carlo test for primality. SIAM J. Comput. **6**(1), 84–85 (1977)

283. Specker, E.P.: Dualität. Dialectica **12**, 451–465 (1958)
284. Statman, R.: Proof-search and speed-up in the predicate calculus. Ann. Math. Log. **15**, 225–287 (1978)
285. Stewart, I.: Galois Theory. Chapman and Hall, New York (1998)
286. Strassen, V.: Gaussian elimination is not optimal. Numer. Math. **13**, 354–356 (1969)
287. Struik, D.J.: A Concise History of Mathematics. Dover, New York (1948)
288. Takeuti, G.: Proof Theory, 2nd. edn. North-Holland, Amsterdam (1987)
289. Tao, T.: Every odd number greater than 1 is the sum of at most five primes. Math. Comput. (to appear)
290. Tarski, A.: Der Wahrheitsbegriff in den formalisierten Sprachen. Stud. Philos. **1**, 261–405 (1936)
291. Troelstra, A., Schwichtenberg, H.: Basic Proof Theory, 2nd edn. Cambridge University Press, Cambridge (2000)
292. Tsfasman, M.A., Vlăduţ, S.G., Zink T, T.: Modular curves, Shimura curves and Goppa codes, better than Varshamov-Gilbert bound. Math. Nachr. **104**, 13–28 (1982)
293. Turing, A.M.: On computable numbers, with an application to the Entscheidungsproblem. Proc. Lond. Math. Soc. (2) **42**, 230–265 2(1937)
294. Turing, A.M.: Systems of logic based on ordinals. Proc. Lond. Math. Soc. **s2-45**(1), 161–228 (1939)
295. Uhlig, D.: On the synthesis of self-correcting schemes from functional elements with a small number of reliable elements. Mat. Zametki **15**(6), 937–944 (1974)
296. Vardi, I.: Archimedes, the Sand Reckoner. http://www.lix.polytechnique.fr/Labo/Ilan.Vardi/sand_reckoner.ps
297. Valiant, L.: The complexity of computing permanent. Theor. Comput. Sci. **8**, 189–201 (1979)
298. Vaught, R.L.: Axiomatizability by a schema. J. Symb. Log. **32**(4), 473–479 (1967)
299. Vinogradov, I.M.: Representation of an odd number as a sum of three primes. C. R. Acad. Sci. USSR **15**, 191–249 (1937)
300. Vopěnka, P.: On ∇-model of set theory. Bull. Acad. Pol. Sci., Sér. Sci. Math. Astron. Phys. **13**, 267–272 (1965)
301. Vopěnka, P.: Mathematics in the Alternative Set Theory. Teubner, Leipzig (1979)
302. Wang, H.: The formalization of mathematics. J. Symb. Log. **19**, 241–266 (1954)
303. Wang, H.: Toward mechanical mathematics. IBM J. Res. Dev. **4**(1), 2–22 (1960)
304. Wang, H.: Proving theorems by pattern recognition–II. Bell Syst. Tech. J. **40**(1), 1–41 (1961)
305. Wang, H.: A Survey of Mathematical Logic. Science Press, Peking (1962)
306. Wang, H.: Some facts about Kurt Gödel. J. Symb. Log. **46**(3), 653–659 (1981)
307. Weyl, H.: Das Kontinuum. Kritische Untersuchungen über die Grundlagen der Analysis. Veit, Leipzig (1918)
308. Weyl, H.: Mathematics and logic. A brief survey serving as a preface to a review of "The philosophy of Bertrand Russell". Am. Math. Mon. **53**, 2–13 (1946)
309. Woodin, W.H.: The Axiom of Determinacy, Forcing Axioms, and the Nonstationary Ideal. de Gruyter, Berlin (1999)
310. Woodin, W.H.: The continuum hypothesis, I. Not. Am. Math. Soc. **48**(6), 567–576 (2001)
311. Woodin, W.H.: The continuum hypothesis, II. Not. Am. Math. Soc. **48**(7), 681–690 (2001)
312. Woodin, W.H.: The Continuum Hypothesis and the Ω-Conjecture. Coxeter Lectures, Fields Institute, Toronto (2002)
313. Whitehead, A.N., Russell, B.: Principia Mathematica, I. Cambridge University Press, Cambridge (1910)
314. Whitehead, A.N., Russell, B.: Principia Mathematica, II. Cambridge University Press, Cambridge (1912)
315. Whitehead, A.N., Russell, B.: Principia Mathematica, III. Cambridge University Press, Cambridge (1913)
316. Zalta, E.N., Principal (eds.): The Stanford Encyclopedia of Philosophy. The Metaphysics Research Lab Center for the Study of Language and Information, Stanford University, Stanford. http://plato.stanford.edu/

317. Zermelo, E.: Beweis, dass jede Menge wohlgeordnet werden kann. Math. Ann. **59**(4), 514–516 (1904)
318. Zermelo, E.: Untersuchungen über die Grundlagen der Mengenlehre. I. Math. Ann. **65**, 261–281 (1908)
319. Zermelo, E.: Über Grenzzahlen und Mengenbereiche. Fundam. Math. **16**, 29–47 (1930)
320. Zippel, R.E.: Probabilistic algorithms for sparse polynomials. In: Proc. EUROSAM'79. Springer Lecture Notes in Computer Science, vol. 72, pp. 216–226 (1979)

Name Index

P. Pudlák, *Logical Foundations of Mathematics and Computational Complexity*, Springer Monographs in Mathematics, DOI 10.1007/978-3-319-00119-7,

Subject Index

P. Pudlák, *Logical Foundations of Mathematics and Computational Complexity*, Springer Monographs in Mathematics, DOI 10.1007/978-3-319-00119-7,

Symbols and Abbreviations

P. Pudlák, *Logical Foundations of Mathematics and Computational Complexity*, Springer Monographs in Mathematics, DOI 10.1007/978-3-319-00119-7,

MIX
Papier aus verantwortungsvollen Quellen
Paper from responsible sources
FSC® C105338

If you have any concerns about our products,
you can contact us on
ProductSafety@springernature.com

In case Publisher is established outside the EU,
the EU authorized representative is:
Springer Nature Customer Service Center GmbH
Europaplatz 3, 69115 Heidelberg, Germany

Printed by Libri Plureos GmbH
in Hamburg, Germany